Peter Schneider

Extragalactic Astronomy

Peter Schneider

Extragalactic Astronomy and Cosmology

An Introduction

With 446 figures, including 266 color figures

Springer

Prof. Dr. Peter Schneider

Argelander-Institut für Astronomie
Universität Bonn
Auf dem Hügel 71
D-53121 Bonn, Germany

e-mail: peter@astro.uni-bonn.de

ISBN 978-3-642-06971-0

ISBN 978-3-540-33175-9 (eBook)

Cover: The cover shows an HST image of the cluster RXJ 1347−1145, the most X-ray luminous cluster of galaxies known. The large number of gravitationally lensed arcs, of which only two of them have been detected from ground-based imaging previously, clearly shows that this redshift $z = 0.45$ cluster is a very massive one, its mass being dominated by dark matter. The data have been processed by Tim Schrabback and Thomas Erben, Argelander-Institut für Astronomie of Bonn University.

Springer is a part of Springer Science+Business Media
springer.com

Softcover reprint of the hardcover 1st edition 2006

Cover design: Erich Kirchner, Heidelberg

Preface

This book began as a series of lecture notes for an introductory astronomy course I have been teaching at the University of Bonn since 2001. This annual lecture course is aimed at students in the first phase of their studies. Most are enrolled in physics degrees and choose astronomy as one of their subjects. This series of lectures forms the second part of the introductory course, and since the majority of students have previously attended the first part, I therefore assume that they have acquired a basic knowledge of astronomical nomenclature and conventions, as well as of the basic properties of stars. Thus, in this part of the course, I concentrate mainly on extragalactic astronomy and cosmology, beginning with a discussion of our Milky Way as a typical (spiral) galaxy. To extend the potential readership of this book to a larger audience, the basics of astronomy and relevant facts about radiation fields and stars are summarized in the appendix.

The goal of the lecture course, and thus also of this book, is to confront physics students with astronomy early in their studies. Since their knowledge of physics is limited in their first year, many aspects of the material covered here need to be explained with simplified arguments. However, it is surprising to what extent modern extragalactic astronomy can be treated with such arguments. All the material in this book is covered in the lecture course, though not all details are written up here. I believe that only by covering this wide range of topics can the students be guided to the forefront of our present astrophysical knowledge. Hence, they learn a lot about issues which are currently not settled and under intense discussion. It is also this aspect which I consider of great importance for the role of astronomy in the framework of a physics program, since in most other sub-disciplines of physics the limits of our current knowledge are approached only at a later stage in the student's education.

In particular, the topic of cosmology is usually met with interest by students. Despite the large amount of material, most of them are able to digest and understand what they are taught, as evidenced from the oral examinations following this course – and this is not small-number statistics: my colleague Klaas de Boer and I together grade about 100 oral examinations per year, covering both parts of the introductory course. Some critical comments coming from students concern the extent of the material as well as its level. However, I do not see a rational reason why the level of an astronomy lecture should be lower than that of one in physics or mathematics.

Why did I turn this into a book? When preparing the concept for my lecture course, I soon noticed that there is no book which I can (or want to) follow. In particular, there are only a few astronomy textbooks in German, and they do not treat extragalactic astronomy and cosmology nearly to the extent and depth as I wanted for this course. Also, the choice of books on these topics in English is fairly limited – whereas a number of excellent introductory textbooks exist, most shy away from technical treatments of issues. However, many aspects can be explained better if a technical argument is also given. Thus I hope that this text presents a field of modern astrophysics at a level suitable for the aforementioned group of people. A further goal is to cover extragalactic astronomy to a level such that the reader should feel comfortable turning to more professional literature.

When being introduced to astronomy, students face two different problems simultaneously. On the one hand, they should learn to understand astrophysical arguments – such as those leading to the conclusion that the central engine in AGNs is a black hole. On the other hand, they are confronted with a multitude of new terms, concepts, and classifications, many of which can only be considered as historical burdens. Examples here are the classification of supernovae which, although based on observational criteria, do not agree with our current understanding of the supernova phenomenon, and the classification of the various types of AGNs. In the lectures, I have tried to separate these two issues, clearly indicating when facts are presented where the students should "just take note", or when astrophysical connections are uncovered which help to understand the properties of cosmic objects. The latter aspects are discussed in considerably more detail. I hope this distinction can still be clearly seen in this written version.

The order of the material in the course and in this book accounts for the fact that students in their first year of physics studies have a steeply rising learning curve; hence, I have tried to order the material partly according to its difficulty. For example, homogeneous world models are described first, whereas only later are the processes of structure formation discussed, motivated in the meantime by the treatment of galaxy clusters.

The topic and size of this book imply the necessity of a selection of topics. I want to apologize here to all of those colleagues whose favorite subject is not covered at the depth that they feel it deserves. I also took the freedom to elaborate on my own research topic – gravitational lensing – somewhat disproportionately. If it requires a justification: the basic equations of gravitational lensing are sufficiently simple that they and their consequences can be explained at an early stage in astronomy education.

With a field developing as quickly as the subject of this book, it is unavoidable that parts of the text will become somewhat out-of-date quickly. I have attempted to include some of the most recent results of the respective topics, but there are obvious limits. For example, just three weeks before the first half of the manuscript was sent to the publisher the three-year results from WMAP were published. Since these results are compatible with the earlier one-year data, I decided not to include them in this text.

Many students are not only interested in the physical aspects of astronomy, they are also passionate observational astronomers. Many of them have been active in astronomy for years and are fascinated by phenomena occurring beyond the Earth. I have tried to provide a glimpse of this fascination at some points in the lecture course, for instance through some historical details, by discussing specific observations or instruments, or by highlighting some of the great achievements of modern cosmology. At such points, the text may deviate from the more traditional "scholarly" style.

Producing the lecture notes, and their extension to a textbook, would have been impossible without the active help of several students and colleagues, whom I want to thank here. Jan Hartlap, Elisabeth Krause and Anja von der Linden made numerous suggestions for improving the text, produced graphics or searched for figures, and TEXed tables – deep thanks go to them. Oliver Czoske, Thomas Erben and Patrick Simon read the whole German version of the text in detail and made numerous constructive comments which led to a clear improvement of the text. Klaas de Boer and Thomas Reiprich read and commented on parts of this text. Searching for the sources of the figures, Leonardo Castaneda, Martin Kilbinger, Jasmin Pierloz and Peter Watts provided valuable help. A first version of the English translation of the book was produced by Ole Markgraf, and I thank him for this heroic task. Furthermore, Kathleen Schrüfer, Catherine Vlahakis and Peter Watts read the English version and made zillions of suggestions and corrections – I am very grateful to their invaluable help. Thomas Erben, Mischa Schirmer and Tim Schrabback produced the cover image very quickly after our HST data of the cluster RXJ 1347−1145 were taken. Finally, I thank all my colleagues and students who provided encouragement and support for finishing this book.

The collaboration with Springer-Verlag was very fruitful. Thanks to Wolf Beiglböck and Ramon Khanna for their encouragement and constructive collaboration. Bea Laier offered to contact authors and publishers to get the copyrights for reproducing figures – without her invaluable help, the publication of the book would have been delayed substantially. The interaction with LE-TEX, where the book was produced, and in particular with Uwe Matrisch, was constructive as well.

Furthermore, I thank all those colleagues who granted permission to reproduce their figures here, as well as the public relations departments of astronomical organizations and institutes who, through their excellent work in communicating astronomical knowledge to the general public, play an invaluable role in our profession. In addition, they provide a rich source of pictorial material of which I made ample use for this book. Representative of those, I would like to mention the European Southern Observatory (ESO), the Space Telescope Science Institute (STScI), the NASA/SAO/CXC archive for Chandra data and the Legacy Archive for Microwave Background Data Analysis (LAMBDA).

List of Contents

5. Active Galactic Nuclei

6. Clusters and Groups of Galaxies

8. Cosmology III: The Cosmological Parameters

9. The Universe at High Redshift

1. Introduction and Overview

1.1 Introduction

The Milky Way, the galaxy in which we live, is but one of many galaxies. As a matter of fact, the Milky Way, also called the Galaxy, is a fairly average representative of the class of spiral galaxies. Two other examples of spiral galaxies are shown in Fig. 1.1 and Fig. 1.2, one of which we are viewing from above (face-on), the other from the side (edge-on). These are all stellar systems in which the majority of stars are confined to a relatively thin disk. In our own Galaxy, this disk can be seen as the band of stars stretched across the night sky, which led to it being named the Milky Way. Besides such disk galaxies, there is a second major class of luminous stellar systems, the elliptical galaxies. Their properties differ in many respects from those of the spirals.

It was less than a hundred years ago that astronomers first realized that objects exist outside our Milky Way and that our world is significantly larger than the size of the Milky Way. In fact, galaxies are mere islands in the Universe: the diameter of our Galaxy[1] (and other galaxies) is much smaller than the average separation between luminous galaxies. The discovery of the existence of other stellar systems and their variety of morphologies raised the question of the origin and evolution of these galaxies. Is there anything between the galaxies, or is it just empty space? Are there any other cosmic bodies besides galaxies? Questions like these motivated us to explore the Universe as a whole and its evolution. Is our

[1] We shall use the terms "Milky Way" and "Galaxy" synonymously throughout.

Fig. 1.1. The spiral galaxy NGC 1232 may resemble our Milky Way if it were to be observed from "above" (face-on). This image, observed with the VLT, has a size of $6\overset{\prime}{.}8 \times 6\overset{\prime}{.}8$, corresponding to a linear size of 60 kpc at its distance of 30 Mpc. If this was our Galaxy, our Sun would be located at a distance of 8.0 kpc from the center, orbiting around it at a speed of $\sim 220\,\mathrm{km/s}$. A full revolution would take us about 230×10^6 years. The bright knots seen along the spiral arms of this galaxy are clusters of newly-formed stars, similar to bright young star clusters in our Milky Way. The different, more reddish, color of the inner part of this galaxy indicates that the average age of the stars there is higher than in the outer parts. The small galaxy at the lower left edge of the image is a companion galaxy that is distorted by the gravitational tidal forces caused by the spiral galaxy

Peter Schneider, Introduction and Overview.
In: Peter Schneider, Extragalactic Astronomy and Cosmology. pp. 1–33 (2006)
DOI: 10.1007/11614371_1

Fig. 1.2. We see the spiral galaxy NGC 4013 from the side (edge-on); an observer looking at the Milky Way from a direction which lies in the plane of the stellar disk ("from the side") may have a view like this. The disk is clearly visible, with its central region obscured by a layer of dust. One also sees the central bulge of the galaxy. As will be discussed at length later on, spiral galaxies like this one are surrounded by a halo of matter which is observed only through its gravitational action, e.g., by affecting the velocity of stars and gas rotating around the center of the galaxy

Universe finite or infinite? Does it change over time? Does it have a beginning and an end? Mankind has long been fascinated by these questions about the origin and the history of our world. But for only a few decades have we been able to approach these questions in an empirical manner. As we shall discuss in this book, many of the questions have now been answered. However, each answer raises yet more questions, as we aim towards an ever increasing understanding of the physics of the Universe.

The stars in our Galaxy have very different ages. The oldest stars are about 12 billion years old, whereas in some regions stars are still being born today: for instance in the well-known Orion nebula. Obviously, the stellar content of our Galaxy has changed over time. To understand the formation and evolution of the Galaxy a view of its (and thus our own) past would be useful. Unfortunately, this is physically impossible. However, due to the finite speed of light, we see objects at large distances in an earlier state, as they were in the past. One can now try to identify and analyze such distant galaxies, which may have been the progenitors of galaxies like our own Galaxy, in this way reconstructing the main aspects of the history of the Milky Way. We will never know the exact initial conditions that led to the evolution of the Milky Way, but we may be able to find some characteristic conditions. Emerging from such initial states, cosmic evolution should produce galaxies similar to our own, which we would then be able to observe from the outside. On the other hand, only within our own Galaxy can we study the physics of galaxy evolution *in situ*.

We are currently witnessing an epoch of tremendous discoveries in astronomy. The technical capabilities in observation and data reduction are currently evolving at an enormous pace. Two examples taken from ground-based optical astronomy should serve to illustrate this.

In 1993 the first 10-m class telescope, the Keck telescope, was commissioned, the first increase in light-collecting power of optical telescopes since the completion of the 5-m mirror on Mt. Palomar in 1948. Now, just a decade later, about ten telescopes of the 10-m class are in use, and even more are soon to come. In recent years, our capabilities to find very distant, and thus very dim, objects and to examine them in detail have improved immensely thanks to the capability of these large optical telescopes.

A second example is the technical evolution and size of optical detectors. Since the introduction of CCDs in astronomical observations at the end of the 1970s, which replaced photographic plates as optical detectors, the sensitivity, accuracy, and data rate of optical observations have increased enormously. At the end of the 1980s, a camera with 1000×1000 pixels (*pic*ture *el*ements) was considered a wide-field instrument. In 2003 a camera called Megacam began operating; it has $(18\,000)^2$ pixels and images a square degree of the sky at a sampling rate of $0\farcs2$ in a single exposure. Such a camera produces roughly 100 GB of data every night, the reduction of which requires fast computers and vast storage capacities. But it is not only optical astronomy that is in a phase of major development; there has also been huge progress in instrumentation in other wavebands. Space-based observing platforms are playing a crucial role in this. We will consider this topic in Sect. 1.3.

These technical advances have led to a vast increase in knowledge and insight in astronomy, especially in extragalactic astronomy and cosmology. Large telescopes and sensitive instruments have opened up a window to the distant Universe. Since any observation of distant objects is inevitably also a view into the past, due to the finite speed of light, studying objects in the early Universe has become possible. Today, we can study galaxies which emitted the light we observe at a time when the Universe was less than 10% of its current age; these galaxies are therefore in a very early evolutionary stage. We are thus able to observe the evolution of galaxies throughout the past history of the Universe. We have the opportunity to study the history of galaxies and thus that of our own Milky Way. We can examine at which epoch most of the stars that we observe today in the local Universe have formed because the history of star formation can be traced back to early epochs. In fact, it has been found that star formation is largely hidden from our eyes and only observable with space-based telescopes operating in the far-infrared waveband.

One of the most fascinating discoveries of recent years is that most galaxies harbor a black hole in their center, with a characteristic mass of millions or even billions of solar masses – so-called supermassive black holes. Although as soon as the first quasars were found in 1963 it was proposed that only processes around a supermassive black hole would be able to produce the huge amount of energy emitted by these ultra-luminous objects, the idea that such black holes exist in normal galaxies is fairly recent. Even more surprising was the finding that the black hole mass is closely related to the other properties of its parent galaxy, thus providing a clear indication that the evolution of supermassive black holes is closely linked to that of their host galaxies.

Detailed studies of individual galaxies and of associations of galaxies, which are called galaxy groups or clusters of galaxies, led to the surprising result that these objects contain considerably more mass than is visible in the form of stars and gas. Analyses of the dynamics of galaxies and clusters show that only 10–20% of their mass consists of stars, gas and dust that we are able to observe in emission or absorption. The largest fraction of their mass, however, is invisible. Hence, this hidden mass is called *dark matter*. We know of its presence only through its gravitational effects. The dominance of dark matter in galaxies and galaxy clusters was established in recent years from observations with radio, optical and X-ray telescopes, and it was also confirmed and quantified by other methods. However, we do not know what this dark matter consists of; the unambiguous evidence for its existence is called the "dark matter problem".

The nature of dark matter is one of the central questions not only in astrophysics but also poses a challenge to fundamental physics, unless the "dark matter problem" has an astronomical solution. Does dark matter consist of non-luminous celestial bodies, for instance burned-out stars? Or is it a new kind of matter? Have astronomers indirectly proven the existence of a new elementary particle which has thus far escaped detection in terrestrial laboratories? If dark matter indeed consists of a new kind of elementary particle, which is the common presumption today, it should exist in the Milky Way as well, in our immediate vicinity. Therefore, experiments which try to directly detect the constituents of dark matter with highly sensitive and sophisticated detectors have been set up in underground laboratories. Physicists and astronomers are eagerly awaiting the commissioning of the Large Hadron Collider (LHC), a particle accelerator at the European CERN research center which, from 2007 on, will produce particles at significantly higher energies than accessible today. The hope is to find an elementary particle that could serve as a candidate constituent of dark matter.

Without doubt, the most important development in recent years is the establishment of a standard model of cosmology, i.e., the science of the Universe as a whole. The Universe is known to expand and it has a finite age; we now believe that we know its age with a precision of as little as a few percent – it is $t_0 = 13.7$ Gyr. The Universe has evolved from a very dense and very hot state, the Big Bang, expanding and cooling over time. Even today, echoes of the Big Bang can be observed, for example in the form of the cosmic microwave background radiation. Accurate observations of this background radiation, emitted some 380 000 years after the Big Bang, have made an important contribution to what we know today about the composition of the Universe. However, these results raise more questions than they answer: only $\sim 4\%$ of the energy content of the Universe can be accounted for by matter which is well-known from other fields of physics, the *baryonic matter* that consists mainly of atomic nuclei and electrons. About 25% of the Universe consists of dark matter, as we already discussed in the context of galaxies and galaxy clusters. Recent observational results have shown that the mean density of dark matter dominates over that of baryonic matter also on cosmic scales.

Even more surprising than the existence of dark matter is the discovery that about 70% of the Universe consists of something that today is called vacuum energy, or dark energy, and that is closely related to the cosmological constant introduced by Albert Einstein. The fact that various names do exist for it by no means implies that we have any idea what this dark energy is. It reveals its existence exclusively in its effect on cosmic expansion, and it even dominates the expansion dynamics at the current epoch. Any efforts to estimate the density of dark energy from fundamental physics have failed hopelessly. An estimate of the vacuum energy density using quantum mechanics results in a value that is roughly *120 orders of magnitude* larger than the value derived from cosmology. For the foreseeable future observational cosmology will be the only empirical probe for dark energy, and an understanding of its physical nature will probably take a substantial amount of time. The existence of dark energy may well pose the greatest challenge to fundamental physics today.

In this book we will present a discussion of the extragalactic objects found in astronomy, starting with the Milky Way which, being a typical spiral galaxy, is considered a prototype of this class of stellar systems. The other central topic in this book is a presentation of modern astrophysical cosmology, which has experienced tremendous advances in recent years. Methods and results will be discussed in parallel. Besides providing an impression of the fascination that arises from astronomical observations and cosmological insights, astronomical methods and physical considerations will be our prime focus. We will start in the next section with a concise overview of the fields of extragalactic astronomy and cosmology. This is, on the one hand, intended to whet the reader's appetite and curiosity, and on the other hand to introduce some facts and technical terms that will be needed in what follows but which are discussed in detail only later in the book. In Sect. 1.3 we will describe some of the most important telescopes used in extragalactic astronomy today.

1.2 Overview

1.2.1 Our Milky Way as a Galaxy

The Milky Way is the only galaxy which we are able to examine in detail. We can resolve individual stars and analyze them spectroscopically. We can perform detailed studies of the interstellar medium (ISM), such as the properties of molecular clouds and star-forming regions. We can quantitatively examine extinction and reddening by dust. Furthermore, we can observe the local dynamics of stars and gas clouds as well as the properties of satellite galaxies (such the Magellanic Clouds). Finally, the Galactic center at a distance of only 8 kpc[2] gives us the unique opportunity to examine the central region of a galaxy at very high resolution. Only through a detailed understanding of our own Galaxy can we hope to understand the properties of other galaxies. Of course, we implicitly assume that the physical processes taking place in other galaxies obey the same laws of physics that apply to us. If this were not the case, we would barely have a chance to understand the physics of other objects in the Universe, let alone the Universe as a whole. We will return to this point shortly.

We will first discuss the properties of our own Galaxy. One of the main problems here, and in astronomy in general, is the determination of the distance to an object. Thus we will start by considering this topic. From the analysis of the distribution of stars and gas in the Milky Way we will then derive its structure. It is found that the Galaxy consists of several distinct components:

- a thin disk of stars and gas with a radius of about 20 kpc and a scale-height of about 300 pc, which also hosts the Sun;
- a $\sim$ 1 kpc thick disk, which contains a different stellar population compared to the thin disk;
- a central bulge, as is also found in other spiral galaxies;
- and a nearly spherical halo which contains most of the globular clusters and some old stars.

Figure 1.3 shows a schematic view of our Milky Way and its various components. For a better visual impression, Figs. 1.1 and 1.2 show two spiral galaxies, the

[2]One parsec (1 pc) is the common unit of distance in astronomy, with 1 pc $= 3.086 \times 10^{18}$ cm. Also used are 1 kpc $= 10^3$ pc, 1 Mpc $= 10^6$ pc, 1 Gpc $= 10^9$ pc. Other commonly used units and constants are listed in Appendix C.

Fig. 1.3. Schematic structure of the Milky Way consisting of the disk, the central bulge with the Galactic center, and the spherical halo in which most of the globular clusters are located. The Sun orbits around the Galactic center at a distance of about 8 kpc

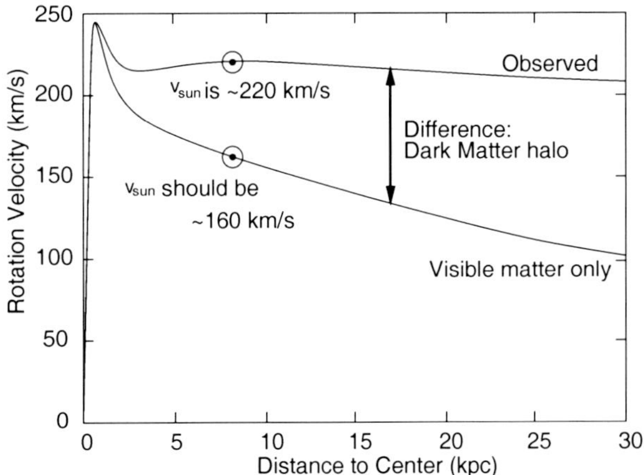

Fig. 1.4. The upper curve is the observed rotation curve $V(R)$ of our Galaxy, i.e., the rotational velocity of stars and gas around the Galactic center as a function of their galacto-centric distance. The lower curve is the rotation curve that we would predict based solely on the observed stellar mass of the Galaxy. The difference between these two curves is ascribed to the presence of dark matter, in which the Milky Way disk is embedded

former viewed from "above" (face-on) and the latter from the "side" (edge-on). In the former case, the spiral structure, from which this kind of galaxy derives its name, is clearly visible. The bright knots in the spiral arms are regions where young, luminous stars have recently formed. The image shows an obvious color gradient: the galaxy is redder in the center and bluest in the spiral arms – while star formation is currently taking place in the spiral arms, we find mainly old stars towards the center, especially in the bulge.

The Galactic disk rotates, with rotational velocity $V(R)$ depending on the distance R from the center. We can estimate the mass of the Galaxy from the distribution of the stellar light and the mean mass-to-light ratio of the stellar population, since gas and dust represent less than $\sim 10\%$ of the mass of the stars. From this mass estimate we can predict the rotational velocity as a function of radius simply from Newtonian mechanics. However, the observed rotational velocity of the Sun around the Galactic center is significantly higher than would be expected from the observed mass distribution. If $M(R_0)$ is the mass inside a sphere around the Galactic center with radius $R_0 \approx 8$ kpc, then the rotational velocity from Newtonian mechanics[3] is

$$V_0 = \sqrt{\frac{G\,M(R_0)}{R_0}}\,. \tag{1.1}$$

From the visible matter in stars we would expect a rotational velocity of ~ 160 km/s, but we observe $V_0 \sim 220$ km/s (see Fig. 1.4). This, and the shape of the rotation curve $V(R)$ for larger distances R from the Galactic center, indicates that our Galaxy contains significantly more mass than is visible in the form of stars.[4] This additional mass is called *dark matter*. Its physical nature is still unknown. The main candidates are weakly interacting elementary particles like those postulated by some elementary particle theories, but they have yet not been detected in the laboratory. Macroscopic objects (i.e., celestial bodies) are also in principle possible candidates if they emit very little light. We will discuss experiments which allow us to identify such macroscopic

[3] We use standard notation: G is the Newtonian gravitational constant, c the speed of light.
[4] Strictly speaking, (1.1) is valid only for a spherically symmetric mass distribution. However, the rotational velocity for an oblate density distribution does not differ much, so we can use this relation as an approximation.

objects and come to the conclusion that the solution of the dark matter problem probably can not be found in astronomy, but rather most likely in particle physics.

The stars in the various components of our Galaxy have different properties regarding their age and their chemical composition. By interpreting this fact one can infer some aspects of the evolution of the Galaxy. The relatively young age of the stars in the thin disk, compared to that of the older population in the bulge, suggests different phases in the formation and evolution of the Milky Way. Indeed, our Galaxy is a highly dynamic object that is still changing today. We see cold gas falling into the Galactic disk and hot gas outflowing. Currently the small neighboring Sagittarius dwarf galaxy is being torn apart in the tidal gravitational field of the Milky Way and will merge with it in the (cosmologically speaking) near future.

One cannot see far through the disk of the Galaxy at optical wavelengths due to extinction by dust. Therefore, the immediate vicinity of the Galactic center can be examined only in other wavebands, especially the infrared (IR) and the radio parts of the electromagnetic spectrum (see also Fig. 1.5). The Galactic center is a highly complex region but we have been able to study it in recent years thanks to various substantial improvements in IR observations regarding sensitivity and angular resolution. Proper motions, i.e., changes of the positions on the sky with time, of bright stars close to the center have been observed. They enable us to determine the mass M in a volume of radius ~ 0.1 pc to be $M(0.1\,\mathrm{pc}) \sim 3 \times 10^6\,M_\odot$. Although the data do not allow us to make a totally unambiguous interpretation of this mass concentration there is no plausible alternative to the conclusion that the center of the Milky Way harbors a supermassive black hole (SMBH) of roughly this mass. And yet this SMBH is far less massive than the ones that have been found in many other galaxies.

Unfortunately, we are unable to look at our Galaxy from the outside. This view from the inside renders it difficult to observe the global properties of the Milky Way. The structure and geometry of the Galaxy, e.g., its spiral arms, are hard to identify from our location. In addition, the extinction by dust hides large parts of the Galaxy from our view (see Fig. 1.6), so that the global

Fig. 1.5. The Galactic disk observed in nine different wavebands. Its appearance differs strongly in the various images; for example, the distribution of atomic hydrogen and of molecular gas is much more concentrated towards the Galactic plane than the distribution of stars observed in the near-infrared, the latter clearly showing the presence of a central bulge. The absorption by dust at optical wavelengths is also clearly visible and can be compared to that in Fig. 1.2

Fig. 1.6. The galaxy Dwingeloo 1 is only five times more distant than our closest large neighboring galaxy, Andromeda, yet it was not discovered until the 1990s because it hides behind the Galactic center. The absorption in this direction and numerous bright stars prevented it being discovered earlier. The figure shows an image observed with the Isaac Newton Telescope in the V-, R-, and I-bands

parameters of the Milky Way (like its total luminosity) are difficult to measure. These parameters are estimated much better from outside, i.e., in other similar spiral galaxies. In order to understand the large-scale properties of our Galaxy, a comparison with similar galaxies which we can examine in their entirety is extremely helpful. Only by combining the study of the Milky Way with that of other galaxies can we hope to fully understand the physical nature of galaxies and their evolution.

1.2.2 The World of Galaxies

Next we will discuss the properties of other galaxies. The two main types of galaxies are spirals (like the Milky Way, see also Fig. 1.7) and elliptical galaxies (Fig. 1.8). Besides these, there are additional classes such as irregular and dwarf galaxies, active galaxies, and starburst galaxies, where the latter have a very high star-formation rate in comparison to normal galaxies. These classes differ not only in their morphology, which forms the basis for their classification, but also in their physical properties such as color (indicating a different stellar content), internal reddening (depending on their dust

Fig. 1.7. NGC 2997 is a typical spiral galaxy, with its disk inclined by about 45° with respect to the line-of-sight. Like most spiral galaxies it has two spiral arms; they are significantly bluer than other parts of the galaxy. This is caused by ongoing star formation in these regions so that young, hot and thus blue stars are present in the arms, whereas the center of the galaxy, especially the bulge, consists mainly of old stars

Fig. 1.8. M87 is a very luminous elliptical galaxy in the center of the Virgo Cluster, at a distance of about 18 Mpc. The diameter of the visible part of this galaxy is about 40 kpc; it is significantly more massive than the Milky Way ($M > 3 \times 10^{12} M_{\odot}$). We will frequently refer to this galaxy: it is not only an excellent example of a central cluster galaxy but also a representative of the family of "active galaxies". It is a strong radio emitter (radio astronomers also know it as Virgo A), and it has an optical jet in its center

content), amount of interstellar gas, star-formation rate, etc. Galaxies of different morphologies have evolved in different ways.

Spiral galaxies are stellar systems in which active star formation is still taking place today, whereas elliptical galaxies consist mainly of old stars – their star formation was terminated a long time ago. The S0 galaxies, an intermediate type, show a disk similar to that of spiral galaxies but like ellipticals they consist mainly of old stars, i.e., stars of low mass and low temperature. Ellipticals and S0 galaxies together are often called *early-type galaxies*, whereas spirals are termed *late-type galaxies*. These names do not imply any interpretation but exist only for historical reasons.

The disks of spiral galaxies rotate differentially. As for the Milky Way, one can determine the mass from the rotational velocity using the Kepler law (1.1). One finds that, contrary to the expectation from the distribution of light, the rotation curve does not decline at larger distances from the center. *Like our own Galaxy, spiral galaxies contain a large amount of dark matter; the visible matter is embedded in a halo of dark matter.* We can only get rough estimates of the extent of this halo, but there are strong indications that it is substantially larger than the extent of the visual matter. For instance, the rotation curve is flat up to the largest radii where one still finds gas to measure the velocity. Studying dark matter in elliptical galaxies is more complicated, but the existence of dark halos has also been proven for ellipticals.

The Hertzsprung–Russell diagram of stars, or their color–magnitude diagram (see Appendix B), has turned out to be the most important diagram in stellar astrophysics. The fact that most stars are aligned along a one-dimensional sequence, the main sequence, led to the conclusion that, for main-sequence stars, the luminosity and the surface temperature are not independent parameters. Instead, the properties of such stars are in principle characterized by only a single parameter: the stellar mass. We will also see that the various properties of galaxies are not independent parameters. Rather, dynamical properties (such as the rotational velocity of spirals) are closely related to the luminosity. These scaling relations are of similar importance to the study of galaxies as the Hertzsprung–Russell diagram is for stars. In addition, they turn out to be very convenient tools for the determination of galaxy distances.

Like our Milky Way, other galaxies also seem to harbor a SMBH in their center. We obtained the astonishing result that the mass of such a SMBH is closely related to the velocity distribution of stars in elliptical galaxies or in the bulge of spirals. The physical reason for this close correlation is as yet unknown, but it strongly suggests a joint evolution of galaxies and their SMBHs.

1.2.3 The Hubble Expansion of the Universe

The radial velocity of galaxies, measured by means of the Doppler shift of spectral lines (Fig. 1.9), is positive for nearly all galaxies, i.e., they appear to be moving away from us. In 1928, Edwin Hubble discovered that

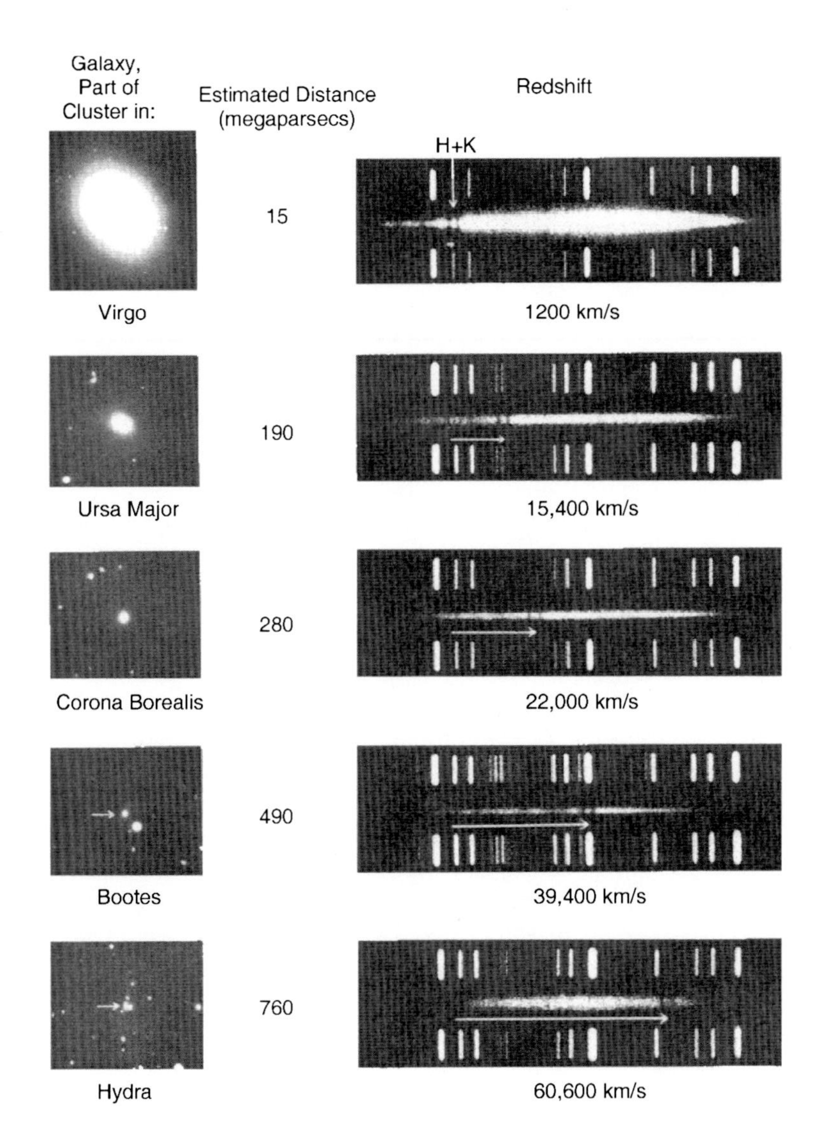

Fig. 1.9. The spectra of galaxies show characteristic spectral lines, e.g., the H + K lines of calcium. These lines, however, do not appear at the wavelengths measured in the laboratory but are in general shifted towards longer wavelengths. This is shown here for a set of sample galaxies, with distance increasing from top to bottom. The shift in the lines, interpreted as being due to the Doppler effect, allows us to determine the relative radial velocity – the larger it is, the more distant the galaxy is. The discrete lines above and below the spectra are for calibration purposes only

this escape velocity v increases with the distance of the galaxy. He identified a linear relation between the radial velocity v and the distance D of galaxies, called a Hubble law,

$$\boxed{v = H_0 D} \,, \tag{1.2}$$

where H_0 is a constant. If we plot the radial velocity of galaxies against their distance, as is done in the Hubble diagram of Fig. 1.10, the resulting points are approximated by a straight line, with the slope being determined by the constant of proportionality, H_0, which is called the *Hubble constant*. The fact that all galaxies seem to move away from us with a velocity which increases linearly with their distance is interpreted such that the Universe is expanding. We will see later that this *Hubble expansion* of the Universe is a natural property of cosmological world models.

The value of H_0 has been determined with appreciable precision only in recent years, yielding the conservative estimate

$$60\,\mathrm{km\,s^{-1}\,Mpc^{-1}} \lesssim H_0 \lesssim 80\,\mathrm{km\,s^{-1}\,Mpc^{-1}} \,, \tag{1.3}$$

obtained from several different methods which will be discussed later. The error margins vary for the differ-

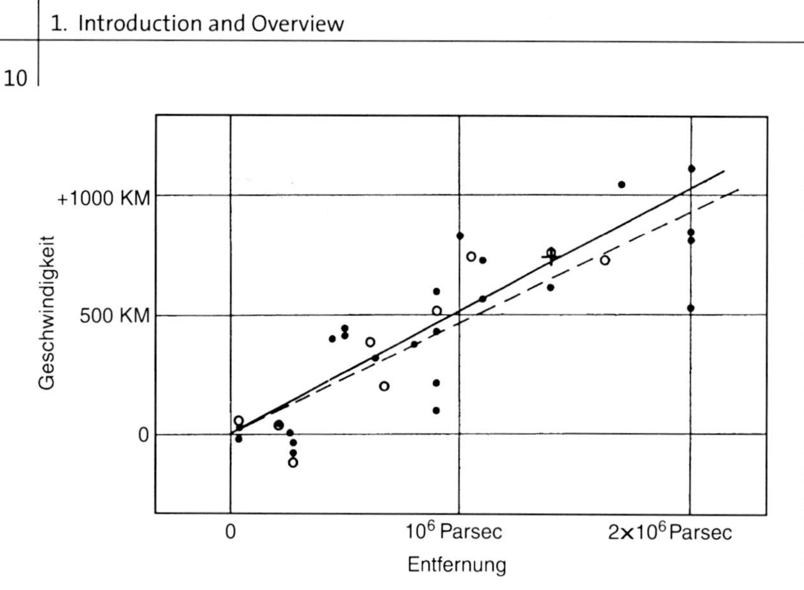

Fig. 1.10. The original 1929 version of the Hubble diagram shows the radial velocity of galaxies as a function of their distance. The reader may notice that the velocity axis is labeled with errornous units – of course they should read km/s. While the radial (escape) velocity is easily measured by means of the Doppler shift in spectral lines, an accurate determination of distances is much more difficult; we will discuss methods of distance determination for galaxies in Sect. 3.6. Hubble has underestimated the distances considerably, resulting in too high a value for the Hubble constant. Only very few and very close galaxies show a blueshift, i.e., they move towards us; one of these is Andromeda (= M31)

ent methods and also for different authors. The main problem in determining H_0 is in measuring the absolute distance of galaxies, whereas Doppler shifts are easily measurable. If one assumes (1.2) to be valid, the radial velocity of a galaxy is a measure of its distance. One defines the *redshift*, z, of an object from the wavelength shift in spectral lines,

$$z := \frac{\lambda_{\rm obs} - \lambda_0}{\lambda_0} , \qquad \lambda_{\rm obs} = (1+z)\lambda_0 , \tag{1.4}$$

with λ_0 denoting the wavelength of a spectral transition in the rest-frame of the emitter and $\lambda_{\rm obs}$ the observed wavelength. For instance, the Lyman-α transition, i.e., the transition from the first excited level to the ground state in the hydrogen atom is at $\lambda_0 = 1216$ Å. For small redshifts,

$$v \approx zc , \tag{1.5}$$

whereas this relation has to be modified for large redshifts, together with the interpretation of the redshift itself.[5] Combining (1.2) and (1.5), we obtain

$$D \approx \frac{zc}{H_0} \approx 3000\, z\, h^{-1}\,{\rm Mpc} , \tag{1.6}$$

where the uncertainty in determining H_0 is parametrized by the scaled Hubble constant h, defined as

$$\boxed{H_0 = h\, 100\,{\rm km\,s^{-1}\,Mpc^{-1}}} . \tag{1.7}$$

Distance determinations based on redshift therefore always contain a factor of h^{-1}, as seen in (1.6). It needs to be emphasized once more that (1.5) and (1.6) are valid only for $z \ll 1$; the generalization for larger redshifts will be discussed in Sect. 4.3. Nevertheless, z is also a measure of distance for large redshifts.

1.2.4 Active Galaxies and Starburst Galaxies

A special class of galaxies are the so-called active galaxies which have a very strong energy source in their center (active galactic nucleus, AGN). The best-known representatives of these AGNs are the quasars, objects typically at high redshift and with quite exotic properties. Their spectrum shows strong emission lines which can be extremely broad, with a relative width of $\Delta\lambda/\lambda \sim 0.03$. The line width is caused by very high

[5] What is observed is the wavelength shift of spectral lines. Depending on the context, it is interpreted either as a radial velocity of a source moving away from us – for instance, if we measure the radial velocity of stars in the Milky Way – or as a cosmological escape velocity, as is the case for the Hubble law. It is in principle impossible to distinguish between these two interpretations, because a galaxy not only takes part in the cosmic expansion but it can, in addition, have a so-called peculiar velocity. We will therefore use the words "Doppler shift" and "redshift", respectively, and "radial velocity" depending on the context, but always keeping in mind that both are measured by the shift of spectral lines. Only when observing the distant Universe where the Doppler shift is fully dominated by the cosmic expansion will we exclusively call it "redshift".

random velocities of the gas which emits these line: if we interpret the line width as due to Doppler broadening resulting from the superposition of lines of emitting gas with a very broad velocity distribution, we obtain velocities of typically $\Delta v \sim 10\,000\,\mathrm{km/s}$. The central source of these objects is much brighter than the other parts of the galaxy, making these sources appear nearly point-like on optical images. Only with the Hubble Space Telescope (HST) did astronomers succeed in detecting structure in the optical emission for a large sample of quasars (Fig. 1.11).

Many properties of quasars resemble those of Seyfert type I galaxies, which are galaxies with a very luminous nucleus and very broad emission lines. For this reason, quasars are often interpreted as extreme members of this class. The total luminosity of quasars is extremely large, with some of them emitting more than a thousand times the luminosity of our Galaxy. In addition, this radiation must originate from a very small spatial region whose size can be estimated, e.g., from the variability time-scale of the source. Due to these and other properties which will be discussed in Chap. 5, it is concluded that the nuclei of active galaxies must contain a supermassive black hole as the central powerhouse. The radiation is produced by matter falling towards this black hole, a process called accretion, thereby converting its gravitational potential energy into kinetic energy. If this kinetic energy is then transformed into internal energy (i.e., heat) as happens in the so-called accretion disk due to friction, it can get radiated away. This is in fact an extremely efficient process of energy production. For a given mass, the accretion onto a black hole is about 10 times more efficient than the nuclear fusion of hydrogen into helium. AGNs often emit radiation across a very large portion of the electromagnetic spectrum, from radio up to X-ray and gamma radiation.

Spiral galaxies still form stars today; indeed star formation is a common phenomenon in galaxies. In addition, there are galaxies with a considerably higher star-formation rate than "normal" spirals. These galaxies are undergoing a burst of star formation and are thus known as *starburst galaxies*. Their star-formation rates are typically between 10 and 300 $M_\odot/\mathrm{yr}$, whereas our Milky Way gives birth to about 2 $M_\odot/\mathrm{yr}$ of new stars. This vigorous star formation often takes place in localized regions, e.g., in the vicinity of the center of the respective galaxy. Starbursts are substantially affected, if not triggered, by disturbances in the gravitational field of the galaxy, such as those caused by galaxy interactions. Such starburst galaxies (see Fig. 1.12) are extremely luminous in the far-infrared (FIR); they emit up to 98% of their total luminosity in this part of the spectrum. This happens by dust emission: dust in these galaxies absorbs a large proportion of the energetic UV radiation

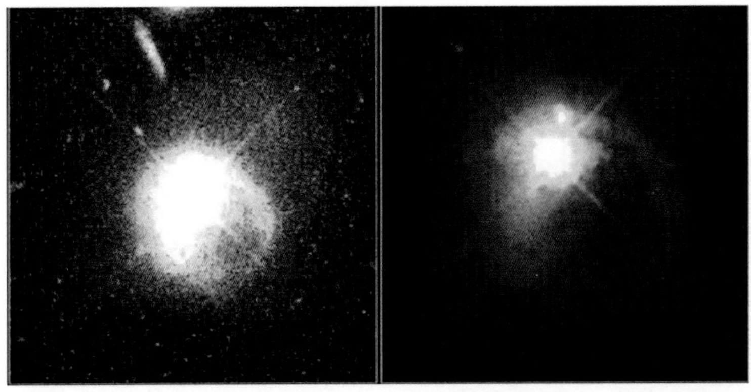

Fig. 1.11. The quasar PKS 2349 is located at the center of a galaxy, its host galaxy. The diffraction spikes (diffraction patterns caused by the suspension of the telescope's secondary mirror) in the middle of the object show that the center of the galaxy contains a point source, the actual quasar, which is significantly brighter than its host galaxy. The galaxy shows clear signs of distortion, visible as large and thin tidal tails. The tails are caused by a neighboring galaxy that is visible in the right-hand image, just above the quasar; it is about the size of the Large Magellanic Cloud. Quasar host galaxies are often distorted or in the process of merging with other galaxies. The two images shown here differ in their brightness contrast

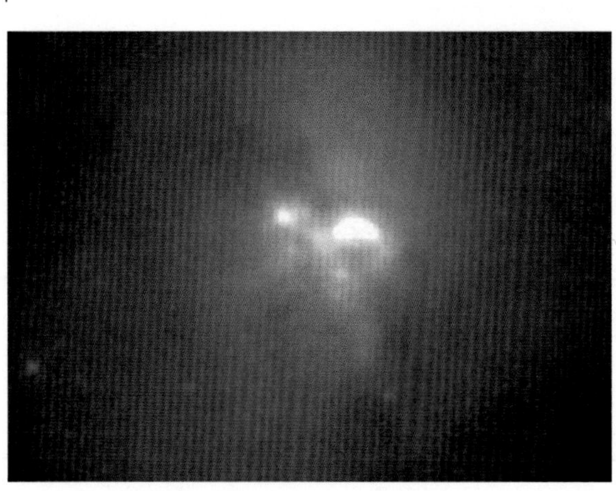

Fig. 1.12. Arp 220 is the most luminous object in the local Universe. Originally cataloged as a peculiar galaxy, the infrared satellite IRAS later discovered its enormous luminosity in the infrared (IR). Arp 220 is the prototype of ultra-luminous infrared galaxies (ULIRGs). This near-IR image taken with the Hubble Space Telescope (HST) unveils the structure of this object. With two colliding spiral galaxies in the center of Arp 220, the disturbances in the interstellar medium caused by this collision trigger a starburst. Dust in the galaxy absorbs most of the ultraviolet (UV) radiation from the young hot stars and re-emits it in the IR

produced in the star-formation region and then re-emits this energy in the form of thermal radiation in the FIR.

1.2.5 Voids, Clusters of Galaxies, and Dark Matter

The likelihood of galaxies interacting (Fig. 1.13) is enhanced by the fact that galaxies are not randomly distributed in space. The projection of galaxies on the celestrial sphere, for instance, shows a distinct structure. In addition, measuring the distances of galaxies allows a determination of their three-dimensional distribution. One finds a strong correlation of the galaxy positions. There are regions in space that have a very high galaxy density, but also regions where nearly no galaxies are seen at all. The latter are called *voids*. Such voids can have diameters of up to $30\,h^{-1}$ Mpc.

Clusters of galaxies are gravitationally bound systems of a hundred or more galaxies in a volume of diameter $\sim 2\,h^{-1}$ Mpc. Clusters predominantly contain early-type galaxies, so there is not much star formation taking place any more. Some clusters of galaxies seem to be circular in projection, others have a highly elliptical or irregular distribution of galaxies; some even have more than one center. The cluster of galaxies closest to us is the Virgo Cluster, at a distance of ~ 18 Mpc; it is a cluster with an irregular galaxy distribution. The closest regular cluster is Coma, at a distance of ~ 90 Mpc.[6] Coma (Fig. 1.14) contains about 1000 luminous galaxies, of which 85% are early-type galaxies.

[6] The distances of these two clusters are not determined from redshift measurements, but by direct methods that will be discussed in Sect. 3.6; such direct measurements are one of the most successful methods of determining the Hubble constant.

Fig. 1.13. Two spiral galaxies interacting with each other. NGC 2207 (on the left) and IC 2163 are not only close neighbors in projection: the strong gravitational tidal interaction they are exerting on each other is clearly visible in the pronounced tidal arms, particularly visible to the right of the right-hand galaxy. Furthermore, a bridge of stars is seen to connect these two galaxies, also due to tidal gravitational forces. This image was taken with the Hubble Space Telescope

Fig. 1.14. The Coma cluster of galaxies, at a distance of roughly 90 Mpc from us, is the closest massive regular cluster of galaxies. Almost all objects visible in this image are galaxies associated with the cluster – Coma contains more than a thousand luminous galaxies

In 1933, Fritz Zwicky measured the radial velocities of the galaxies in Coma and found that they have a dispersion of about 1000 km/s. From the total luminosity of all its galaxies the mass of the cluster can be estimated. If the stars in the cluster galaxies have an average mass-to-light ratio (M/L) similar to that of our Sun, we would conclude $M = (M_\odot/L_\odot)L$. However, stars in early-type galaxies are on average slightly less massive than the Sun and thus have a slightly higher M/L.[7] Thus, the above mass estimate needs to be increased by a factor of ~ 10.

Zwicky then estimated the mass of the cluster by multiplying the luminosity of its member galaxies with the mass-to-light ratio. From this mass and the size of the cluster, he could then estimate the velocity that a galaxy needs to have in order to escape from the gravitational field of the cluster – the escape velocity. He found that the characteristic peculiar velocity of cluster galaxies (i.e., the velocity relative to the mean velocity) is substantially larger than this escape velocity. In this case, the galaxies of the cluster would fly apart on a time-scale of about 10^9 years – the time it takes a galaxy to cross through the cluster once – and, consequently, the cluster would dissolve. However, since Coma seems to be a relaxed cluster, i.e., it is in equilibrium and thus its age is definitely larger than the dynamical time-scale of 10^9 years, Zwicky concluded that the Coma cluster contains significantly more mass than the sum of the masses of its galaxies. Using the virial theorem[8] he was able to estimate the mass of the cluster from the velocity distribution of the galaxies. This was the first clear indicator of the existence of dark matter.

X-ray satellites later revealed that clusters of galaxies are strong sources of X-ray radiation. They contain hot gas, with temperatures ranging from 10^7 K up to 10^8 K (Fig. 1.15). This gas temperature is another measure for the depth of the cluster's potential well, since the hotter the gas is, the deeper the potential well has to be to prevent the gas from escaping via evaporation. Mass estimates based on the X-ray temperature result in values that are comparable to those from the velocity dispersion of the cluster galaxies, clearly confirming the hypothesis of the existence of dark matter in clusters. A third method for determining cluster masses, the so-called gravitational lensing effect, utilizes the fact that light is deflected in a gravitational field. The angle through which light rays are bent due to the presence of a massive object depends on the mass of that object. From observation and analysis of the gravitational lensing effect in clusters of galaxies, cluster masses are derived that are in agreement with those from the two other methods. Therefore, clusters of galaxies are a second class of cosmic objects whose mass is dominated by dark matter.

Clusters of galaxies are cosmologically young structures. Their dynamical time-scale, i.e., the time in which the mass distribution in a cluster settles into an equilibrium state, is estimated as the time it takes a member galaxy to fully cross the cluster once. With a characteristic velocity of $v \sim 1000$ km/s and a diameter of $2R \sim 2$ Mpc one thus finds

$$t_{\rm dyn} \sim \frac{2R}{v} \sim 2 \times 10^9 \,\mathrm{yr}\,. \tag{1.9}$$

[7]In Chap. 3 we will see that for stars in spiral galaxies $M/L \sim 3 M_\odot/L_\odot$ on average, while for those in elliptical galaxies a larger value of $M/L \sim 10\,M_\odot/L_\odot$ applies. Here and throughout this book, mass-to-light ratios are quoted in Solar units.

[8]The virial theorem in its simplest form says that, for an isolated dynamical system in a stationary state of equilibrium, the kinetic energy is just half the potential energy,

$$E_{\rm kin} = \frac{1}{2}|E_{\rm pot}|\,. \tag{1.8}$$

In particular, the system's total energy is $E_{\rm tot} = E_{\rm kin} + E_{\rm pot} = E_{\rm pot}/2 = -E_{\rm kin}$.

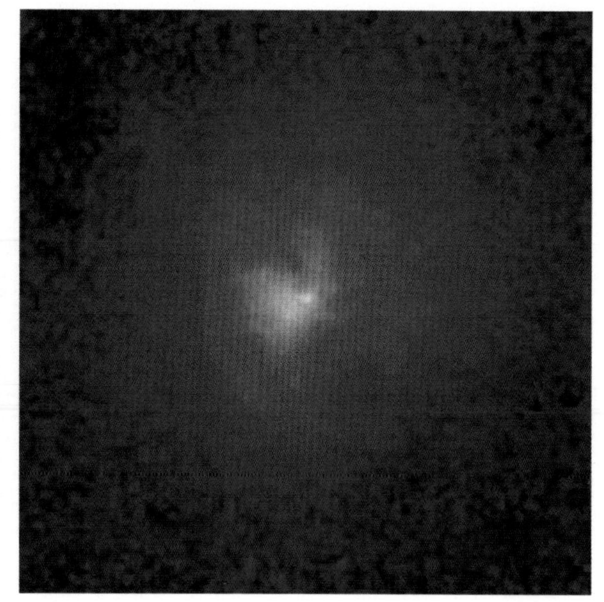

Fig. 1.15. The Hydra A cluster of galaxies. The left-hand figure shows an optical image, the one on the right an image taken with the X-ray satellite Chandra. The cluster has a redshift of $z \approx 0.054$ and is thus located at a distance of about 250 Mpc. The X-ray emission originates from gas at a temperature of 40×10^6 K which fills the space between the cluster galaxies. In the center of the cluster, the gas is cooler by about 15%

As we will later see, the Universe is about 14×10^9 years old. During this time galaxies have not had a chance to cross the cluster many times. Therefore, clusters still contain, at least in principle, information about their initial state. Most clusters have not had the time to fully relax and evolve into a state of equilibrium that would be largely independent of their initial conditions. Comparing this with the time taken for the Sun to rotate around the center of the Milky Way – about 2×10^8 years – galaxies thus have had plenty of time to reach their state of equilibrium.

Besides massive clusters of galaxies there are also galaxy groups, which sometimes contain only a few luminous galaxies. Our Milky Way is part of such a group, the Local Group, which also contains M31 (Andromeda) which is another dominant galaxy, as well as some far less luminous galaxies such as the Magellanic Clouds. Some groups of galaxies are very compact, i.e., their galaxies are confined within a very small volume (Fig. 1.16). Interactions between these galaxies cause the lifetimes of many such groups to be much smaller than the age of the Universe, and the galaxies in such groups will merge.

1.2.6 World Models and the Thermal History of the Universe

Quasars, clusters of galaxies, and nowadays even single galaxies are also found at very high redshifts where the simple Hubble law (1.2) is no longer valid. It is therefore necessary to generalize the distance–redshift relation. This requires considering world models as a whole, which are also called cosmological models. The dominant force in the Universe is gravitation. On the one hand, weak and strong interactions both have an extremely small (subatomic) range, and on the other hand, electromagnetic interactions do not play a role on large scales since the matter in the Universe is on average electrically neutral. Indeed, if it was not, currents would immediately flow to balance net charge densities. The accepted theory of gravitation is the theory of General Relativity (GR), formulated by Albert Einstein in 1915.

Based on the two postulates that (1) our place in the Universe is not distinguished from other locations and that (2) the distribution of matter around us is isotropic, at least on large scales, one can construct

Fig. 1.16. The galaxy group HCG87 belongs to the class of so-called compact groups. In this HST image we can see three massive galaxies belonging to this group: an edge-on spiral in the lower part of the image, an elliptical galaxy to the lower right, and another spiral in the upper part. The small spiral in the center is a background object and therefore does not belong to the group. The two lower galaxies have an active galactic nucleus, whereas the upper spiral seems to be undergoing a phase of star formation. The galaxies in this group are so close together that in projection they appear to touch. Between the galaxies, gas streams can be detected. The galaxies are disturbing each other, which could be the cause of the nuclear activity and star formation. The galaxies are bound in a common gravitational potential and will heavily interfere and presumably merge on a cosmologically small time-scale, which means in only a few orbits, with an orbit taking about 10^8 years. Such merging processes are of utmost importance for the evolution of the galaxy population

homogeneous and isotropic world models (so-called Friedmann–Lemaître models) that obey the laws of General Relativity. Expanding world models that contain the Hubble expansion result from this theory naturally. Essentially, these models are characterized by three parameters:

- the current expansion rate of the Universe, i.e., the Hubble constant H_0;
- the current mean matter density of the Universe ρ_m, often parametrized by the dimensionless *density parameter*

$$\Omega_\mathrm{m} = \frac{8\pi G}{3H_0^2}\rho_\mathrm{m}\,; \tag{1.10}$$

- and the density of the so-called vacuum energy, described by the cosmological constant Λ or by the corresponding density parameter of the vacuum

$$\Omega_\Lambda = \frac{\Lambda}{3H_0^2}\,. \tag{1.11}$$

The cosmological constant was originally introduced by Einstein to allow stationary world models within GR. After the discovery of the Hubble expansion he called the introduction of Λ into his equations his greatest blunder. In quantum mechanics Λ attains a different interpretation, that of an energy density of the vacuum.

The values of the cosmological parameters are known quite accurately today (see Chap. 8), with values of $\Omega_\mathrm{m} \approx 0.3$ and $\Omega_\Lambda \approx 0.7$. The discovery of a non-vanishing Ω_Λ came completely unexpectedly. To date, all attempts have failed to compute a reasonable value for Ω_Λ from quantum mechanics. By that we mean a value which has the same order-of-magnitude as the one we derive from cosmological observations. In fact, simple and plausible estimates lead to a value of Λ that is $\sim 10^{120}$ times larger than that obtained from observation, a tremendously bad estimate indeed. This huge discrepancy is probably one of the biggest challenges in fundamental physics today.

According to the Friedmann–Lemaître models, the Universe used to be smaller and hotter in the past, and it has continuously cooled down in the course of expansion. We are able to trace back the cosmic expansion under the assumption that the known laws of physics were also valid in the past. From that we get the Big Bang model of the Universe, according to which our Universe has evolved out of a very dense and very hot state, the so-called *Big Bang*. This world model makes a number of predictions that have been verified convincingly:

1. About 1/4 of the baryonic matter in the Universe should consist of helium which formed about 3 min after the Big Bang, while most of the rest consists

of hydrogen. This is indeed the case: the mass fraction of helium in metal-poor objects, whose chemical composition has not been significantly modified by processes of stellar evolution, is about 24%.

2. From the exact fraction of helium one can derive the number of neutrino families – the more neutrino species that exist, the larger the fraction of helium will be. From this, it was derived in 1981 that there are 3 kinds of neutrinos. This result was later confirmed by particle accelerator experiments.
3. Thermal radiation from the hot early phase of the Universe should still be measurable today. Predicted in 1946 by George Gamow, it was discovered by Arno Penzias and Robert Wilson in 1965. The corresponding photons have propagated freely after the Universe cooled down to about 3000 K and the plasma constituents combined to neutral atoms, an epoch called recombination. As a result of cosmic expansion, this radiation has cooled down to about $T_0 \approx 2.73$ K. This microwave radiation is nearly perfectly isotropic, once we subtract the radiation which is emitted locally by the Milky Way (see Fig. 1.17). Indeed, measurements from the COBE satellite showed that the cosmic microwave background (CMB) is the most accurate blackbody spectrum ever measured.
4. Today's structures in the Universe have evolved out of very small density fluctuations in the early cosmos. The seeds of structure formation must have already been present in the early phases of cosmic evolution. These density fluctuations should also be visible as small temperature fluctuations in the microwave background emitted about 380 000 years after the Big Bang at the epoch of recombination. In fact, COBE was the first to observe these predicted anisotropies (see Fig. 1.17). Later experiments, especially the WMAP satellite, observed the structure of the microwave background at much improved angular resolution and verified the theory of structure formation in the Universe in detail (see Sect. 8.6).

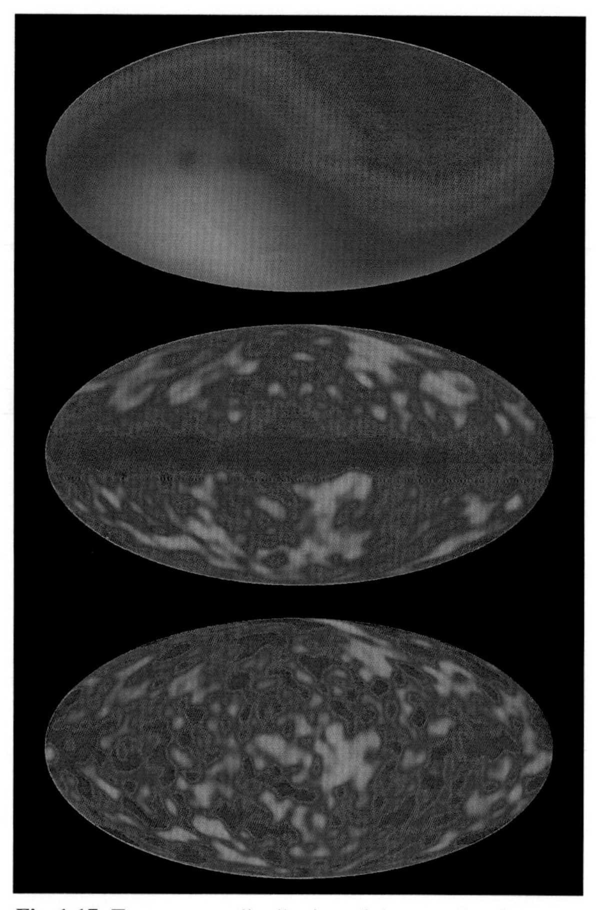

Fig. 1.17. Temperature distribution of the cosmic microwave background on the sky as measured by the COBE satellite. The uppermost image shows a dipole distribution; it originates from the Earth's motion relative to the rest-frame of the CMB. We move at a speed of ~ 600 km/s relative to that system, which leads to a dipole anisotropy with an amplitude of $\Delta T/T \sim v/c \sim 2 \times 10^{-3}$ due to the Doppler effect. If this dipole contribution is subtracted, we get the map in the middle which clearly shows the emission from the Galactic disk. Since this emission has a different spectral energy distribution (it is not a blackbody of $T \sim 3$ K), it can also be subtracted to get the temperature map at the bottom. These are the primordial fluctuations of the CMB, with an amplitude of about $\Delta T/T \sim 2 \times 10^{-5}$

With these predictions so impressively confirmed, in this book we will exclusively consider this cosmological model; currently there is no competing model of the Universe that could explain these very basic cosmological observations in such a natural way. In addition, this model does not seem to contradict any fundamental observation in cosmology. However, as the existence of a non-vanishing vacuum energy density shows, together with a matter density $\rho_{\rm m}$ that is about six times the mean baryon density in the Universe (which can be derived from the abundance of the chemical elements formed in the Big Bang), the physical nature of about 95% of the content of our Universe is not yet understood.

The CMB photons we receive today had their last physical interaction with matter when the Universe was about 3.8×10^5 years old. Also, the most distant galaxies and quasars known today (at $z \sim 6.5$) are strikingly young – we see them at a time when the Universe was less than a tenth of its current age. The exact relation between the age of the Universe at the time of the light emission and the redshift depends on the cosmological parameters H_0, Ω_m, and Ω_Λ. In the special case that $\Omega_m = 1$ and $\Omega_\Lambda = 0$, called the *Einstein–de Sitter model*, one obtains

$$t(z) = \frac{2}{3H_0} \frac{1}{(1+z)^{3/2}} . \tag{1.12}$$

In particular, the age of the Universe today (i.e., at $z = 0$) is, according to this model,

$$t_0 = \frac{2}{3H_0} \approx 6.5 \times 10^9 \, h^{-1} \, \text{yr} . \tag{1.13}$$

The Einstein–de Sitter (EdS) model is the simplest world model and we will sometimes use it as a reference, but recent observations suggest that $\Omega_m < 1$ and $\Omega_\Lambda > 0$. The mean density of the Universe in the EdS model is

$$\rho_0 = \rho_{cr} \equiv \frac{3H_0^2}{8\pi G} \approx 1.9 \times 10^{-29} \, h^2 \, \text{g cm}^{-3} , \tag{1.14}$$

hence it is really, really small.

1.2.7 Structure Formation and Galaxy Evolution

The low amplitude of the CMB anisotropies implies that the inhomogeneities must have been very small at the epoch of recombination, whereas today's Universe features very large density fluctuations, at least on scales of clusters of galaxies. Hence, the density field of the cosmic matter must have evolved. This structure evolution occurs because of gravitational instability, in that an overdense region will expand more slowly than the mean Universe due to its self-gravity. Therefore, any relative overdensity becomes amplified in time. The growth of density fluctuations in time will then cause the formation of large-scale structures, and the gravitational instability is also responsible for the formation of galaxies and clusters. Our world model sketched above predicts the abundance of galaxy clusters as a function of redshift, which can be compared with the observed cluster counts. This comparison can then be used to determine cosmological parameters.

Another essential conclusion from the smallness of the CMB anisotropies is the existence of dark matter on cosmic scales. The major fraction of cosmic matter is dark matter. The baryonic contribution to the matter density is $\lesssim 20\%$ and to the total energy density $\lesssim 5\%$. The energy density of the Universe is dominated by the vacuum energy.

Unfortunately, the spatial distribution of dark matter on large scales is not directly observable. We only observe galaxies or, more precisely, their stars and gas. One might expect that galaxies would be located preferentially where the dark matter density is high. However, it is by no means clear that local fluctuations of the galaxy number density are strictly proportional to the density of dark matter. The relation between the dark and luminous matter distributions is currently only approximately understood.

Eventually, this relation has to result from a detailed understanding of galaxy formation and evolution. Locations with a high density of dark matter can support the formation of galaxies. Thus we will have to examine how galaxies form and why there are different kinds of galaxies. In other words, what decides whether a forming galaxy will become an elliptical or a spiral? This question has not been definitively answered yet, but it is supposed that ellipticals can form only by the merging of galaxies. Indeed, the standard model of the Universe predicts that small galaxies will form first; larger galaxies will be formed later through the ongoing merger of smaller ones.

The evolution of galaxies can actually be observed directly. Galaxies at high redshift (i.e., cosmologically young galaxies) are in general smaller and bluer, and the star-formation rate was significantly higher in the earlier Universe than it is today. The change in the mean color of galaxies as a function of redshift can be understood as a combination of changes in the star formation processes and an aging of the stellar population.

1.2.8 Cosmology as a Triumph of the Human Mind

Cosmology, extragalactic astronomy, and astrophysics as a whole are a heroic undertaking of the human mind and a triumph of physics. To understand the Universe we apply physical laws that were found empirically under completely different circumstances. All the known laws of physics were derived "today" and, except for General Relativity, are based on experiments on a laboratory scale or, at most, on observations in the Solar System, such as Kepler's laws which formed the foundation for the Newtonian theory of gravitation. Is there any a priori reason to assume that these laws are also valid in other regions of the Universe or at completely different times? However, this is apparently indeed the case: nuclear reactions in the early Universe seem to obey the same laws of strong interaction that are measured today in our laboratories, since otherwise the prediction of a 25% mass fraction of helium would not be possible. Quantum mechanics, describing the wavelengths of atomic transitions, also seems to be valid at very large distances – since even the most distant objects show emission lines in their spectra with frequency ratios (which are described by the laws of quantum mechanics) identical to those in nearby objects.

By far the greatest achievement is General Relativity. It was originally formulated by Albert Einstein since his Special Theory of Relativity did not allow him to incorporate Newtonian gravitation. No empirical findings were known at that time (1915) which would not have been explained by the Newtonian theory of gravity. Nevertheless, Einstein developed a totally new theory of gravitation for purely theoretical reasons. The first success of this theory was the correct description of the gravitational deflection of light by the Sun, measured in 1919, and of the perihelion rotation of Mercury.[9] His theory permits a description of the expanding Universe, which became necessary after Hubble's discovery in 1928. Only with the help of this theory can we reconstruct the history of the Universe back into the past. Today this history seems to be well understood up to the time when the Universe was about 10^{-6} s old and had a temperature of about 10^{13} K. Particle physics models allow an extrapolation to even earlier epochs.

The cosmological predictions discussed above are based on General Relativity describing an expanding Universe, therefore providing a test of Einstein's theory. On the other hand, General Relativity also describes much smaller systems and with much stronger gravitational fields, such as neutron stars and black holes. With the discovery of a binary system consisting of two neutron stars, the binary pulsar PSR 1913+16, in the last ~ 25 years very accurate tests of General Relativity have become possible. For example, the observed perihelion rotation in this binary system and the shrinking of the binary orbit over time due to the radiation of energy by gravitational waves is very accurately described by General Relativity. Together, General Relativity has been successfully tested on length-scales from 10^{11} cm (the characteristic scale of the binary pulsar) to 10^{28} cm (the size of the visible Universe), that is over more than 10^{17} orders of magnitude – an impressive result indeed!

[9] This was already known in 1915, but it was not clear whether it might not have any other explanation, e.g., a quadrupole moment of the mass distribution of the Sun.

1.3 The Tools of Extragalactic Astronomy

Extragalactic sources – galaxies, quasars, clusters of galaxies – are at large distances. This means that in general they appear to be faint even if they are intrinsically luminous. They are also seen to have a very small angular size despite their possibly large linear extent. In fact, just three extragalactic sources are visible to the naked eye: the Andromeda galaxy (M31) and the Large and Small Magellanic Clouds. Thus for extragalactic astronomy, telescopes are needed that have large apertures (photon collecting area) and a high angular resolution. This applies to all wavebands, from radio astronomy to gamma ray astronomy.

The properties of astronomical telescopes and their instruments can be judged by different criteria, and we will briefly describe the most important ones. The *sensitivity* specifies how dim a source can be and still be observable in a given integration time. The sensitivity depends on the aperture of the telescope as well as on the efficiency of the instrument and the sensitivity of the detector. The sensitivity of optical telescopes, for instance, was increased by a large factor when CCDs replaced photographic plates as detectors in the early 1980s. The sensitivity also depends on the sky background, i.e., the brightness of the sky caused by non-astronomical sources. Artificial light in inhabited regions has forced optical telescopes to retreat into more and more remote areas of the world where *light pollution* is minimized. Radio astronomers have similar problems caused by radio emission from the telecommunication infrastructure of modern civilization. The *angular resolution* of a telescope specifies down to which angular separation two sources in the sky can still be separated by the detector. For diffraction-limited observations like those made with radio telescopes or space-born telescopes, the angular resolution $\Delta\theta$ is limited by the diameter D of the telescope. For a wavelength λ one has $\Delta\theta = \lambda/D$. For optical and near-infrared observations from the ground, the angular resolution is in general limited by turbulence in the atmosphere, which explains the choice of high mountain tops as sites for optical telescopes. These atmospheric turbulences cause, due to scintillation, the smearing of the images of astronomical sources, an effect that is called *seeing*. In interferometry, where one combines radiation detected by several telescopes, the angular resolution is limited by the spatial separation of the telescopes. The *spectral resolution* of an instrument specifies its capability to separate different wavelengths. The *throughput* of a telescope/instrument system is of particular importance in large sky surveys. For instance, the efficiency of photometric surveys depends on the number of spectra that can be observed simultaneously. Special multiplex spectrographs have been constructed for such tasks. Likewise, the efficiency of photometric surveys depends on the region of sky that can be observed simultaneously, i.e., the field-of-view of the camera. Finally, the efficiency of observations also depends on factors like the number of clear nights at an astronomical site, the fraction of an observing night in which actual science data is taken, the fraction of time an instrument cannot be used due to technical problems, the stability of the instrumental set-up (which determines the time required for calibration measurements), and many other such aspects.

In the rest of this section some telescopes will be presented that are of special relevance to extragalactic astronomy and to which we will frequently refer throughout the course of this book.

1.3.1 Radio Telescopes

With the exception of optical wavelengths, the Earth's atmosphere is transparent only for very large wavelengths – radio waves. The radio window of the atmosphere is cut off towards lower frequencies, at about $\nu \sim 10$ MHz, because radiation of a wavelength larger than $\lambda \sim 30$ m is reflected by the Earth's ionosphere and therefore cannot reach the ground. Below $\lambda \sim 5$ mm radiation is increasingly absorbed by oxygen and water vapor in the atmosphere. Therefore, below about $\lambda \sim 0.3$ mm ground-based observations are no longer possible.

Mankind became aware of cosmic radio radiation – in the early 1930s – only when noise in radio antennae was found that would not vanish, no matter how quiet the device was made. In order to identify the source of this noise the AT&T Bell Labs hired Karl Jansky, who constructed a movable antenna called "Jansky's Merry-Go-Round" (Fig. 1.18).

After some months Jansky had identified, besides thunderstorms, one source of interference that rose and set every day. However, it did not follow the course of the Sun which was originally suspected to be the source. Rather, it followed the stars. Jansky finally discovered that the signal originated from the direction of the center of the Milky Way. He published his result in 1933, but this publication also marked the end of his career as the world's first radio astronomer.

Inspired by Jansky's discovery, Grote Reber was the first to carry out real astronomy with radio waves. When AT&T refused to employ him, he built his own radio "dish" in his garden, with a diameter of nearly 10 m. Between 1938 and 1943, Reber compiled the first sky maps in the radio domain. Besides strong radiation from the center of the Milky Way he also identified sources in Cygnus and in Cassiopeia. Through Reber's research and publications radio astronomy became an accepted field of science after World War II.

The largest single-dish radio telescope is the Arecibo telescope, shown in Fig. 1.19. Due to its enormous area, and thus high sensitivity, this telescope, among other achievements, detected the first pulsar in a binary system, which is used as an important test laboratory for General Relativity (see Sect. 7.7). Also, the first extra-solar planet, in orbit around a pulsar, was discovered with the Arecibo telescope. For extragalactic astronomy Arecibo plays an important role in measuring the redshifts and line widths of spiral galaxies, both determined from the 21-cm emission line of neutral hydrogen (see Sect. 3.4).

The Effelsberg 100-m radio telescope of the Max-Planck-Institut für Radioastronomie was, for many years, the world's largest fully steerable radio telescope, but since 2000 this title has been claimed by the new

Fig. 1.18. "Jansky's Merry-Go-Round". By turning the structure in an azimuthal direction, a rough estimate of the position of radio sources could be obtained

Fig. 1.19. With a diameter of 305 m, the Arecibo telescope in Puerto Rico is the largest single-dish telescope in the world; it may also be known from the James Bond movie "Goldeneye". The disadvantage of its construction is its lack of steerability. Tracking of sources is only possible within narrow limits by moving the secondary mirror

Green Bank Telescope (see Fig. 1.20) after the old one collapsed in 1988. With Effelsberg, for example, star-formation regions can be investigated. Using molecular line spectroscopy, one can measure their densities and temperatures. Magnetic fields also play a role in star formation, though many details still need to be clarified. By measuring the polarized radio flux, Effelsberg has mapped the magnetic fields of numerous spiral galaxies. In addition, due to its huge collecting area Effelsberg plays an important role in interferometry at very long baselines (see below).

Because of the long wavelength, the angular resolution of even large radio telescopes is fairly low, compared to optical telescopes. For this reason, radio

Fig. 1.20. The world's two largest fully steerable radio telescopes. Left: The 100-m telescope in Effelsberg. It was commissioned in 1972 and is used in the wavelength range from 3.5 mm to 35 cm. Eighteen different detector systems are necessary for this. Right: The Green Bank Telescope. It does not have a rotationally symmetric mirror; one axis has a diameter of 100 m and the other 110 m

astronomers soon began utilizing interferometric methods, where the signals obtained by several telescopes are correlated to get an interference pattern. One can then reconstruct the structure of the source from this pattern using Fourier transformation. With this method one gets the same resolution as one would achieve with a single telescope of a diameter corresponding to the maximum pair separation of the individual telescopes used.

Following the first interferometric measurements in England (around 1960) and the construction of the large Westerbork Synthesis Radio Telescope in the Netherlands (around 1970), at the end of the 1970s the Very Large Array (VLA) in New Mexico (see Fig. 1.21) began operating. With the VLA one achieved an angular resolution in the radio domain comparable to that of optical telescopes at that time. For the first time, this allowed the combination of radio and optical images with the same resolution and thus the study of cosmic sources over a range of several clearly separated wavelength regimes. With the advent of the VLA radio astronomy experienced an enormous breakthrough, particularly in the study of AGNs. It became possible to examine the large extended jets of quasars and radio galaxies in detail (see Sect. 5.1.2). Other radio interferometers must also be mentioned here, such as the British MERLIN, where seven telescopes with a maximum separation of 230 km are combined.

In the radio domain it is also possible to interconnect completely independent and diverse antennae to form an interferometer. For example, in Very Long Baseline Interferometry (VLBI) radio telescopes on different continents are used simultaneously. These frequently also include Effelsberg and the VLA. In 1995 a system of ten identical 25-m antennae was set up in the USA, exclusively to be used in VLBI, the Very Long Baseline Array (VLBA). Angular resolutions of better than a milliarcsecond (mas) can be achieved with VLBI. Therefore, in extragalactic astronomy VLBI is

Fig. 1.21. The Very Large Array (VLA) in New Mexico consists of 27 antennae with a diameter of 25 m each that can be moved on rails. It is used in four different configurations that vary in the separation of the telescopes; switching configurations takes about two weeks

particularly used in the study of AGNs. With VLBI we have learned a great deal about the central regions of AGNs, such as the occurrence of apparent superluminal velocities in these sources.

Some of the radio telescopes described above are also capable of observing in the millimeter regime. For shorter wavelengths the surfaces of the antennae are typically too coarse, so that special telescopes are needed for wavelengths of 1 mm and below. The 30-m telescope on Pico Veleta (Fig. 1.22), with its exact surface shape, allows observations in the millimeter range. It is particularly used for molecular spectroscopy at these frequencies. Furthermore, important observations of high-redshift galaxies at 1.2 mm have been made with this telescope using the bolometer camera MAMBO. Similar observations are also conducted with the SCUBA (Submillimeter Common-User Bolometer Array) camera at the James Clerk Maxwell Telescope (JCMT; Fig. 1.23) on Mauna Kea, Hawaii. Due to its size and excellent location, the JCMT is arguably the most productive telescope in the submillimeter range; it is operated at wavelengths between 3 mm and 0.3 mm. With the SCUBA-camera, operating at 850 μm (0.85 mm), we can observe star-formation regions in distant galaxies for which the optical emission is nearly completely absorbed by dust in these sources. These dusty star-forming galaxies can be observed in the (sub-)millimeter regime of the electromagnetic spectrum even out to large redshifts, as will be discussed in Sect. 9.2.3.

To measure the tiny temperature fluctuations of the cosmic microwave background radiation one needs extremely stable observing conditions and low-noise detectors. In order to avoid the thermal radiation of the atmosphere as much as possible, balloons and satellites were constructed to operate instruments at very high altitude or in space. The American COBE (Cosmic Background Explorer) satellite measured the anisotropies of the CMB for the first time, at wavelengths of a few millimeters. In addition, the frequency spectrum of the CMB was precisely measured with instruments on COBE. The WMAP (Wilkinson Microwave Anisotropy Probe) satellite obtained, like COBE, a map of the full sky in the microwave regime, but at a significantly improved angular resolution and sensitivity. The first results from WMAP, published in February 2003, were an enormously important milestone for cosmology, as will be discussed in Sect. 8.6.5. Besides observing the CMB these missions are also of great importance for millimeter astronomy; these satellites not only measure the cosmic background radiation but of course also the microwave radiation of the Milky Way and of other galaxies.

1.3.2 Infrared Telescopes

In the wavelength range $1\,\mu\text{m} \lesssim \lambda \lesssim 300\,\mu\text{m}$, observations from the Earth's surface are always subject to very difficult conditions, if they are possible at all. The atmosphere has some windows in the near-infrared (NIR,

Fig. 1.22. The 30-m telescope on Pico Veleta was designed for observations in the millimeter range of the spectrum. This telescope, like all millimeter telescopes, is located on a mountain to minimize the column density of water in the atmosphere

Fig. 1.23. The JCMT has a 15-m dish. It is protected by the largest single piece of Gore-Tex, which has a transmissivity of 97% at submillimeter wavelengths

$1\,\mu\text{m} \lesssim \lambda \lesssim 2.4\,\mu\text{m}$) which render ground-based observations possible. In the mid-infrared (MIR, $2.4\,\mu\text{m} \lesssim \lambda \lesssim 20\,\mu\text{m}$) and far-infrared (FIR, $20\,\mu\text{m} \lesssim \lambda \lesssim 300\,\mu\text{m}$) regimes, observations need to be carried out from outside the atmosphere, i.e., using balloons, high-flying airplanes, or satellites. The instruments have to be cooled to very low temperatures, otherwise their own thermal radiation would outshine any signal.

The first noteworthy observations in the far-infrared were made by the Kuiper Airborne Observatory (KAO), an airplane equipped with a 91-cm mirror which operated at altitudes up to 15 km. However, the breakthrough for IR astronomy had to wait until the launch of IRAS, the InfraRed Astronomical Satellite (Fig. 1.24). In 1983, with its 60-cm telescope, IRAS compiled the first IR map of the sky at 12, 25, 60, and 100 μm, at an angular

Fig. 1.24. The left-hand picture shows an artist's impression of IRAS in orbit. The project was a cooperation of the Netherlands, the USA, and Great Britain. IRAS was launched in 1983 and operated for 10 months; after that the supply of liquid helium, needed to cool the detectors, was exhausted. During this time IRAS scanned 96% of the sky at four wavelengths. The ISO satellite, shown on the right, was an ESA project and observed between 1995 and 1998. Compared to IRAS it covered a larger wavelength range, had a better angular resolution and a thousand times higher sensitivity

resolution of $30''$ ($2'$) at 12 μm (100 μm). It discovered about a quarter of a million point sources as well as about 20 000 extended sources. The positional accuracy for point sources of better than $\sim 20''$ allowed an identification of these sources at optical wavelengths. Arguably the most important discovery by IRAS was the identification of galaxies which emit the major fraction of their energy in the FIR part of the spectrum. These sources, often called IRAS galaxies, have a very high star-formation rate where the UV light of the young stars is absorbed by dust and then re-emitted as thermal radiation in the FIR. IRAS discovered about 75 000 of these so-called ultra-luminous IR galaxies (ULIRGs).

In contrast to the IRAS mission with its prime task of mapping the full sky, the Infrared Space Observatory ISO (Fig. 1.24) was dedicated to observations of selected objects and sky regions in a wavelength range 2.5–240 μm. Although the telescope had the same diameter as IRAS its angular resolution at 12 μm was about a hundred times better than that of IRAS, since the latter was limited by the size of the detector elements. The sensitivity of ISO topped that of IRAS by a factor ~ 1000. ISO carried four instruments: two cameras and two spectrographs. Among the most important results from ISO in the extragalactic domain are the spatially-resolved observations of the dust-enshrouded star-formation regions of ULIRGs. Although the mission itself came to an end, the scientific analysis of the data continues on a large scale, since to date the ISO data are still unique in the infrared.

In 2003 a new infrared satellite was launched (the Spitzer Space Telescope) with capabilities that by far outperform those of ISO. With its 85-cm telescope, Spitzer observes at wavelengths between 3.6 and 160 μm. Its IRAC (Infrared Array Camera) camera, operating at wavelengths below ~ 9 μm, has a field-of-view of $5\rlap{.}'2 \times 5\rlap{.}'2$ and 256×256 pixels, significantly more than the 32×32 pixels of ISOCAM on ISO that had a comparable wavelength coverage. The spectral resolution of the IRS (Infrared Spectrograph) instrument in the MIR is about $R = \lambda/\Delta\lambda \sim 100$.

1.3.3 Optical Telescopes

The atmosphere is largely transparent in the optical part of the electromagnetic spectrum ($0.3\,\mu\mathrm{m} \lesssim \lambda \lesssim 1\,\mu\mathrm{m}$), and thus we are able to conduct observations from the ground. Since for the atmospheric windows in the NIR one normally uses the same telescopes as for optical astronomy, we will thus not distinguish between these two ranges here.

Although optical astronomy has been pursued for many decades, it has evolved very rapidly in recent years. This is linked to a large number of technical achievements. A good illustration of this is the 10-m Keck telescope which was put into operation in 1993; this was the first optical telescope with a mirror diameter of more than 6 m. Constructing telescopes of this size became possible by the development of adaptive optics, a method to control the surface of the mirror. A mirror of this size no longer has a stable shape but is affected, e.g., by gravitational deformation as the telescope is steered. It was also realized that part of the air turbulence that generates the seeing is caused by the telescope and its dome itself. By improving the thermal condition of telescopes and dome structures a reduction of the seeing could be achieved. The aforementioned replacement of photographic plates by CCDs, together with improvements to the latter, resulted in a vastly enhanced quantum efficiency of $\sim 70\%$ (at maximum even more than 90%), barely leaving room for further improvements.

The throughput of optical telescopes has been immensely increased by designing wide-field CCD cameras, the largest of which nowadays have a field-of-view of a square degree and $\sim 16\,000 \times 16\,000$ pixels, with a pixel scale of $\sim 0\overset{\prime\prime}{.}2$. Furthermore, multi-object spectrographs have been built which allow us to observe the spectra of a large number of objects simultaneously. The largest of them are able to get spectra for several hundred sources in one exposure. Finally, with the Hubble Space Telescope the angular resolution of optical observations was increased by a factor of ~ 10. Further developments that will revolutionize the field even more, such as interferometry in the near IR/optical and adaptive optics, will soon be added to these achievements.

Currently, about 13 optical telescopes of the 4-m class exist worldwide. They differ mainly in their location and their instrumentation. For example, the Canada–France–Hawaii Telescope (CFHT) on Mauna Kea (Fig. 1.25) has been a leader in wide-field photometry for many years, due to its extraordinarily good seeing. This is again emphasized by the installation of Megacam, a camera with $18\,000 \times 18\,000$ pixels. The Anglo-Australian Telescope (AAT) in Australia, in contrast, has distinctly worse seeing and has therefore specialized, among other things, in multi-object

Fig. 1.25. Telescopes at the summit of Mauna Kea, Hawaii, at an altitude of 4200 m. The cylindrical dome to the left and below the center of the image contains the Subaru 8-m telescope; just behind it are the two 10-m Keck telescopes. The two large domes at the back house the Canada–France–Hawaii telescope (CFHT, 3.6 m) and the 8-m Gemini North. The telescope at the lower right is the 15-m James Clerk Maxwell submillimeter telescope (JCMT)

spectroscopy, for which the 2dF (two-degree field) instrument was constructed. Most of these telescopes are also equipped with NIR instruments. The New Technology Telescope (NTT, see Fig. 1.26) is especially noteworthy due to its SOFI camera, a near-IR instrument that has a large field-of-view of $\sim 5' \times 5'$ and an excellent image quality.

Hubble Space Telescope. To avoid the greatest problem in ground-based optical astronomy, the rocket scientist Hermann Oberth had already speculated in the 1920s about telescopes in space which would not be affected by the influence of the Earth's atmosphere. In 1946 the astronomer Lyman Spitzer took up this issue again and discussed the possibilities for the realization of such a project.

Shortly after NASA was founded in 1958, the construction of a large telescope in space was declared a long-term goal. After several feasibility studies and ESA's agreement to join the project, the HST was finally built. However, the launch was delayed by the explosion of the space shuttle *Challenger* in 1986, so that it did not take place until April 24, 1990. An unpleasant surprise came as soon as the first images were taken: it was found that the 2.4-m main mirror was ground into the wrong shape. This problem was remedied in December 1993 during the first "servicing mission" (a series of Space Shuttle missions to the HST; see Fig. 1.27), when a correction lens was installed. After this, the HST became one of the most successful and best-known scientific instruments.

The refurbished HST has two optical cameras, the WFPC2 (Wide-Field and Planetary Camera) and, since 2002, the ACS (Advanced Camera for Surveys). The latter has a field-of-view of $3\overset{'}{.}4 \times 3\overset{'}{.}4$, about twice as large as WFPC2, and 4000×4000 pixels. Another instrument was STIS (Space Telescope Imaging Spectrograph), operating mainly in the UV and at short optical wavelengths. Due to a defect it was shut down in 2004. The HST also carries a NIR instrument, NICMOS (Near Infrared Camera and Multi-Object Spectrograph). The greatly reduced thermal radiation, compared to that on the surface of the Earth, led to progress in NIR astronomy, albeit with a very small field-of-view.

HST has provided important insights into our Solar System and the formation of stars, but it has achieved milestones in extragalactic astronomy. With HST observations of the nucleus of M87 (Fig. 1.8), one has derived from the Doppler shift of the gas emission that the center of this galaxy contains a black hole of two billion solar masses. HST has also proven that black holes exist in other galaxies and AGNs. The enormously improved angular resolution has allowed us to study galaxies to a hitherto unknown level of detail. In this book we will frequently report on results that were achieved with HST.

Arguably the most important contribution of the HST to extragalactic astronomy are the Hubble Deep Fields.

Fig. 1.26. The La Silla Observatory of ESO in Chile. On the peak in the middle, one can see the New Technology Telescope (NTT), a 3.5-m prototype of the VLT. The silvery shining dome to its left is the MPG/ESO 2.2-m telescope that is currently equipped with the Wide-Field Imager, a 8096^2 pixel camera with a $0.5°$ field-of-view. The picture was taken from the location of the 3.6-m telescope, the largest on La Silla

Fig. 1.27. Left: The HST mounted on the manipulator arm of the Space Shuttle during one of the repair missions. Right: The Hubble Deep Field (North) was taken in December 1995 and the data released one month later. To compile this multicolor image, which at that time was the deepest image of the sky, images from four different filters were combined

Scientists managed to convince Robert Williams, then director of the Space Telescope Science Institute, to use the HST to take a very deep image in an empty region of the sky, a field with (nearly) no foreground stars and without any known clusters of galaxies. At that time it was not clear whether anything interesting at all would come from these observations. Using the observing time that is allocated to the Director, the "director's discretionary time", in December 1995 HST was pointed at such a field in the Big Dipper, taking data for 10 days. The outcome was the Hubble Deep Field North (HDFN), one of the most important astronomical data sets, displayed in Fig. 1.27. From the HDFN and its southern counterpart, the HDFS, one obtains information about the early states of galaxies and their evolution. One of the first conclusions was that most of the early galaxies are classified as irregulars. In 2002, the Hubble Ultra-Deep Field (HUDF) was observed with the then newly installed ACS camera. Not only did it cover about twice the area of the HDFN but it was even deeper, by about one magnitude, owing to the higher sensitivity of ACS compared to WFPC2.

Large Telescopes. For more than 40 years the 5-m telescope on Mt. Palomar was the largest telescope in the western world – the Russian 6-m telescope suffered from major problems from the outset. The year 1993 saw the birth of a new class of telescope, of which the two Keck telescopes (see Fig. 1.28) were the first, each with a mirror diameter of 10 m.

The site of the two Kecks at the summit of Mauna Kea (at an altitude of 4200 m) provides ideal observing conditions for many nights per year. This summit is now home to several large telescopes. The new Japanese telescope Subaru, and Gemini North are also located here, as well as the aforementioned CFHT and JCMT. The significant increase in sensitivity obtained by Keck, especially in spectroscopy, permitted completely new insights, for instance through absorption line spectroscopy of quasars. Keck was also essential for the spectroscopic verification of innumer-

Fig. 1.28. The two Keck telescopes on Mauna Kea. With Keck I the era of large telescopes was heralded in 1993

Fig. 1.29. The left panel shows a map of the location of the VLT on Cerro Paranal. It can be reached via Antofagasta, about a two-hour flight north of Santiago de Chile. Then another three-hour trip by car through a desert (see Fig. 1.30) brings one to the site. Paranal is shown on the right during the construction phase; in the foreground we can see the construction camp. The top of the mountain was flattened to get a leveled space (of diameter ~ 300 m) large enough to accommodate the telescopes and the facilities used for optical interferometry (VLTI)

able galaxies of redshift $z \gtrsim 3$, which are normally so dim that they cannot be examined with smaller telescopes.

The largest ground-based telescope project to date was the construction of the Very Large Telescope (VLT) of the European Southern Observatory (ESO), consisting of four telescopes each with a diameter of 8.2 m. ESO already operates the La Silla Observatory in Chile (see Fig. 1.26), but a better location was found for the VLT, the Cerro Paranal (at an altitude of 2600 m). This mountain is located in the Atacama desert, one of the driest regions on Earth. To build the telescopes on the mountain a substantial part of the mountain top first had to be cut off (Fig. 1.29).

In contrast to the Keck telescopes, which have a primary mirror that is segmented into 36 hexagonal elements, the mirrors of the VLT are monolithic, i.e., they consist of a single piece. However, they are very thin compared to the 5-m mirror on Mt. Palomar, far too thin to be intrinsically stable against gravity and other effects such as thermal deformations. Therefore, as for the Kecks, the shape of the mirrors has to be controlled electronically (see Fig. 1.30, right). The monolithic structure of the VLT mirrors results in better image quality than that of the Keck telescopes, resulting in an appreciably simpler point-spread function.

Each of the four telescopes has three accessible foci; this way, 12 different instruments can be installed at the VLT at any time. Switching between the three instruments is done with a deflection mirror. The permanent installation of the instruments allows their stable operation.

The VLT (Fig. 1.31) also marks the beginning of a new form of ground-based observation with large optical telescopes. Whereas until recently an astronomer proposing an observation was assigned a certain number and dates of nights in which she could observe with the telescope, the VLT is mainly operated in the so-called service mode. The observations are performed by local astronomers according to detailed specifications that must be provided by the principal investigator of the observing program and the data are then transmitted to the astronomer at her home institution. A significant advantage of this procedure is that one can better account for special requirements for observing conditions. For example, observations that require very good seeing can be carried out during the appropriate atmospheric conditions. With service observing the chances of getting a useful data set are increased. At present about half of the observations with the VLT are performed in service mode. Another aspect of service observing is that the astronomer does not have to make the long journey (see Fig. 1.30), at the expense of also missing out on the adventure and experience of observing.

Fig. 1.30. Left: Transport of one of the VLT mirrors from Antofagasta to Paranal. The route passes through an extremely dry desert, and large parts of the road are not paved. The VLT thus clearly demonstrates that astronomers search for ever more remote locations to get the best possible observing conditions. Right: The active optics system at the VLT. Each mirror is supported at 150 points; at these points, the mirror is adjusted to correct for deformations. The primary mirror is always shaped such that the light is focused in an optimal way, with its form being corrected for the changing gravitational forces when the telescope changes the pointing direction. In adaptive optics, in contrast to active optics, the wavefront is controlled: the mirrors are deformed with high frequencies in such a way that the wavefront is as planar as possible after passing through the optical system. In this way one can correct for the permanently changing atmospheric conditions and achieve images at diffraction-limited resolution, though only across a fairly small region of the focal plane

Fig. 1.31. The Paranal Observatory after completion of the domes for the four VLT unit telescopes. The tracks seen in the foreground were installed for additional smaller telescopes that are now jointly used with the VLT unit telescopes for interferometric observations in the NIR

1.3.4 UV Telescopes

Radiation with a wavelength shorter than $\lambda \lesssim 0.3\,\mu\mathrm{m} = 3000\,\text{Å}$ cannot penetrate the Earth's atmosphere but is instead absorbed by the ozone layer, whereas radiation at wavelengths below 912 Å is absorbed by neutral hydrogen in the interstellar medium. The range between these two wavelengths is the UV part of the spectrum, in which observation is only possible from space.

The Copernicus satellite (also known as the Orbiting Astronomical Observatory 3, OAO-3) was the first long-term orbital mission designed to observe high-resolution spectra at ultraviolet wavelengths. In addition, the satellite contained an X-ray detector. Launched on 21 August, 1972, it obtained UV spectra of 551 sources until its decommissioning in 1981. Among the achievements of the Copernicus mission are the first detection of interstellar molecular hydrogen H_2 and of CO, and measurements of the composition of the interstellar medium as well as of the distribution of O VI, i.e., five-time ionized oxygen.

The IUE (International Ultraviolet Explorer) operated between 1978 and 1996 and proved to be a remarkably productive observatory. During its more than 18 years of observations more than 10^5 spectra of galactic and extragalactic sources were obtained. In particular, the IUE contributed substantially to our knowledge of AGN.

The HST, with its much larger aperture, marks the next substantial step in UV astronomy, although no UV instrument is operational onboard HST after the failure of STIS in 2004. Many new insights were gained with the HST, especially through spectroscopy of quasars in the UV, insights into both the the quasars themselves and, through the absorption lines in their spectra, into the intergalactic medium along the line-of-sight towards the sources. In 1999 the FUSE (Far Ultraviolet Spectroscopic Explorer) satellite was launched. From UV spectroscopy of absorption lines in luminous quasars this satellite provided us with a plethora of information on the state and chemical composition of the intergalactic medium.

While the majority of observations with UV satellites were dedicated to high-resolution spectroscopy of stars and AGNs, the prime purpose of the GALEX satellite mission, launched in 2003, is to compile an extended photometric survey. GALEX observes at wavelengths $1350\,\text{Å} \lesssim \lambda \lesssim 2830\,\text{Å}$ and will perform a complete sky survey as well as observe selected regions of the sky with a longer exposure time. In addition, it will perform several spectroscopic surveys. The results from GALEX will be of great importance, especially for the study of the star-formation rate in nearby and distant galaxies.

1.3.5 X-Ray Telescopes

As mentioned before, interstellar gas absorbs radiation at wavelengths shortward of 912 Å, the so-called Lyman edge. This corresponds to the ionization energy of hydrogen in its ground state, which is 13.6 eV. Only at energies about ten times this value does the ISM become transparent again and this denotes the low-energy limit of the domain of X-ray astronomy. Typically, X-ray astronomers do not measure the frequency of light in Hertz (or the wavelength in μm), but instead photons are characterized by their energy, measured in electron volts (eV).

The birth of X-ray astronomy was in the 1960s. Rocket and balloon-mounted telescopes which were originally only supposed to observe the Sun in X-rays also received signals from outside the Solar System. UHURU, the first satellite to observe exclusively the cosmic X-ray radiation, compiled the first X-ray map of the sky, discovering about 340 sources. This catalog of point sources was expanded in several follow-up missions, especially by NASA's High-Energy Astrophysical Observatory (HEAO-1) which also detected a diffuse X-ray background radiation. On HEAO-2, also known as the Einstein satellite, the first Wolter telescope was used for imaging, increasing the sensitivity by a factor of nearly a thousand compared to earlier missions. The Einstein observatory also marked a revolution in X-ray astronomy because of its high angular resolution, about $2''$ in the range of 0.1 to 4 keV. Among the great discoveries of the Einstein satellite is the X-ray emission of many clusters of galaxies that traces the presence of hot gas in the space between the cluster galaxies. The total mass of this gas significantly exceeds the mass of the stars in the cluster galaxies and therefore represents the main contribution to the baryonic content of clusters.

The next major step in X-ray astronomy was ROSAT (ROentgen SATellite; Fig. 1.32), launched in 1990. During the first six months of its nine-year mission ROSAT produced an all-sky map at far higher resolution than UHURU; this is called the ROSAT All Sky Survey. More than 10^5 individual sources were detected in this survey, the majority of them being AGNs. In the subsequent period of pointed observations ROSAT examined, among other types of sources, clusters of galaxies and AGNs. One of its instruments (PSPC) provided spectral information in the range between 0.1 and 2.4 keV at an angular resolution of $\sim 20''$, while the other (HRI) instrument had a much better angular resolution ($\sim 3''$) but did not provide any spectral information. The Japanese X-ray satellite ASCA (Advanced Satellite for Cosmology and Astrophysics), launched in 1993, was able to observe in a significantly higher energy range 0.5–12 keV and provided spectra of higher energy resolution, though at reduced angular resolution.

Since 1999 two new powerful satellites are in operation: NASA's Chandra observatory and ESA's XMM-Newton (X-ray Multi-Mirror Mission; see Fig. 1.32). Both have a large photon-collecting area and a high angular resolution, and they also set new standards in X-ray spectroscopy. Compared to ROSAT, the energy range accessible with these two satellites has been extended to 0.1–10 keV. The angular resolution of Chandra is about $0\farcs5$ and thus, for the first time, comparable to that of optical telescopes. This high angular resolution already led to major discoveries in the early years of operation. For instance, well-defined sharp structures in the X-ray emission from gas in clusters of galaxies were discovered, and X-ray radiation from the jets of AGNs which had been previously observed in the radio was detected. Furthermore, Chandra discovered a class of X-ray sources, termed ultra-luminous compact X-ray sources (ULXs), in which we may be observing the formation of black holes (Sect. 9.6). XMM-Newton has a larger sensitivity compared to Chandra, however at a somewhat smaller angular resolution. Among the most important observations of XMM-Newton at the beginning of its operation was the spectroscopy of AGNs and of clusters of galaxies.

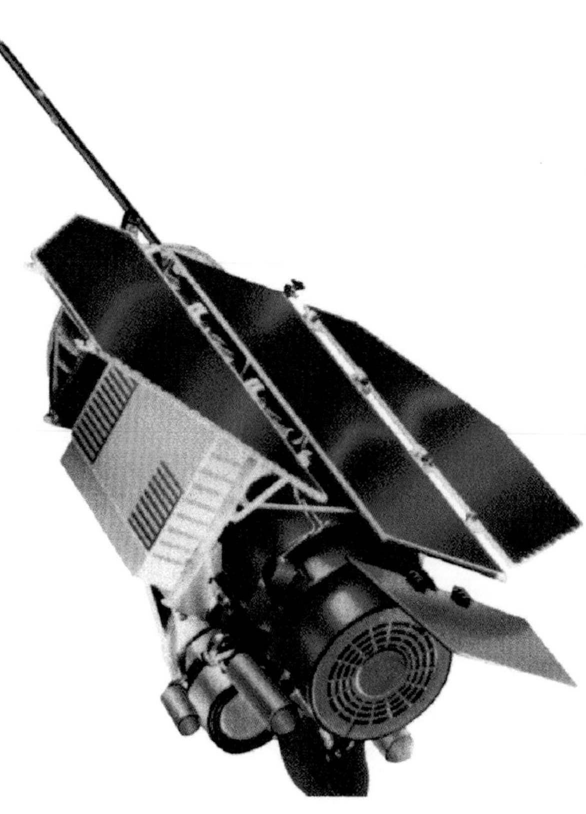

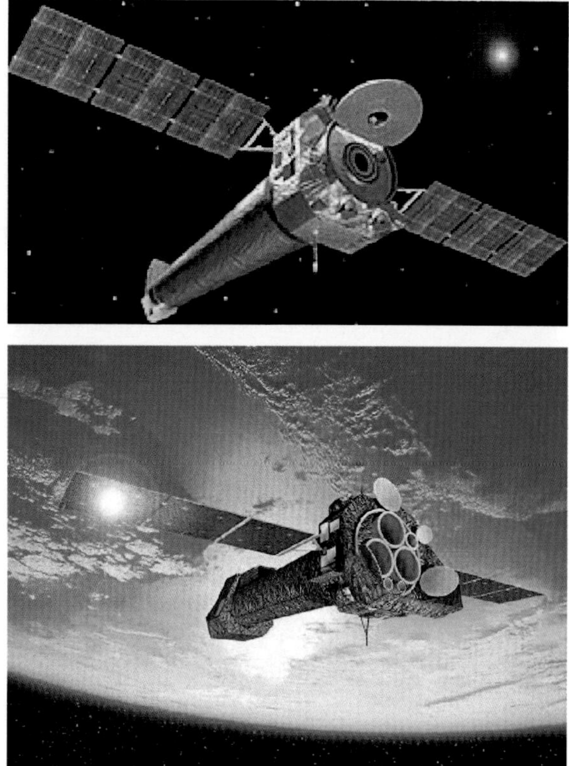

Fig. 1.32. Left: ROSAT, a German–US–British cooperation, was in orbit from 1990 to 1999 and observed in the energy range between 0.1 and 2.5 keV (soft X-ray). Upper right: Chandra was launched in July 1999. The energy range of its instruments lies between 0.1 and 10 keV. Its highly elliptical orbit permits long uninterrupted exposures. Lower right: XMM-Newton was launched in December 1999 and is planned to be used for 10 years. Observations are carried out with three telescopes at energies between 0.1 and 15 keV

1.3.6 Gamma-Ray Telescopes

The existence of gamma radiation was first postulated in the 1950s. This radiation is absorbed by the atmosphere, which is fortunate for the lifeforms on Earth. The first observations, carried out from balloons, rockets, and satellites, have yielded flux levels of less than 100 photons. Those gamma photons had energies in the GeV range and above.

Detailed observations became possible with the satellites SAS-2 and COS-B. They compiled a map of the galaxy, confirmed the existence of a gamma background radiation, and for the first time observed pulsars in the gamma range. The first Gamma Ray Bursts (GRB), extremely bright and short-duration flashes on the gamma-ray sky, were detected in the 1970s by military satellites. Only the Italian-Dutch satellite Beppo-SAX (1996 to 2002) managed to localize a GRB with sufficient accuracy to allow an identification of the source in other wavebands, and thus to reveal its physical nature; we will come back to this subject later, in Sect. 9.7.

An enormous advance in high-energy astronomy was made with the launch of the Compton Gamma Ray Observatory (CGRO; Fig. 1.33) in 1991; the observatory was operational for nine years. It carried four different instruments, among them the Burst And Transient Source Experiment (BATSE) and the Energetic Gamma Ray Experiment Telescope (EGRET). During its lifetime BATSE discovered more than 2000 GRBs and contributed substantially to the understanding of the nature of these mysterious gamma-ray flashes. EGRET

Fig. 1.33. The left image shows the Compton Gamma Ray Observatory (CGRO) mounted on the Space Shuttle manipulator arm. This NASA satellite carried out observations between 1991 and 2000. It was finally shut down after a gyroscope failed, and it burned up in the Earth's atmosphere in a controlled re-entry. ESA's Integral observatory, in operation since 2002, is shown on the right

discovered many AGNs at very high energies above 20 MeV, which hints at extreme processes taking place in these objects.

The successor of the CGRO, the Integral satellite, was put into orbit as an ESA mission by a Russian Proton rocket at the end of 2002. At a weight of two tons, it is the heaviest ESA satellite that has been launched thus far. It is primarily observing at energies of 15 keV to 10 MeV in the gamma range, but has additional instruments for observation in the optical and X-ray regimes.

2. The Milky Way as a Galaxy

The Earth is orbiting around the Sun, which itself is orbiting around the center of the Milky Way. Our Milky Way, the Galaxy, is the only galaxy in which we are able to study astrophysical processes in detail. Therefore, our journey through extragalactic astronomy will begin in our home Galaxy, with which we first need to become familiar before we are ready to take off into the depths of the Universe. Knowing the properties of the Milky Way is indispensable for understanding other galaxies.

2.1 Galactic Coordinates

On a clear night, and sufficiently far away from cities, one can see the magnificent band of the Milky Way on the sky (Fig. 2.1). This observation suggests that the distribution of light, i.e., that of the stars in the Galaxy, is predominantly that of a thin disk. A detailed analysis of the geometry of the distribution of stars and gas confirms this impression. This geometry of the Galaxy suggests the introduction of two specially adapted coordinate systems which are particularly convenient for quantitative descriptions.

Spherical Galactic Coordinates (ℓ, b). We consider a spherical coordinate system, with its center being "here", at the location of the Sun (see Fig. 2.2). The *Galactic plane* is the plane of the Galactic disk, i.e., it is parallel to the band of the Milky Way. The two *Galactic coordinates* ℓ and b are angular coordinates on the sphere. Here, b denotes the *Galactic latitude*, the angular distance of a source from the Galactic plane, with $b \in [-90°, +90°]$. The great circle $b = 0°$ is then located in the plane of the Galactic disk. The direction $b = 90°$ is perpendicular to the disk and denotes the North Galactic Pole (NGP), while $b = -90°$ marks the direction to the South Galactic Pole (SGP). The second angular coordinate is the *Galactic longitude* ℓ, with $\ell \in [0°, 360°]$. It measures the angular separation between the position of a source, projected perpendicularly onto the Galactic disk (see Fig. 2.2), and the Galactic center, which itself has angular coordinates $b = 0°$ and $\ell = 0°$. Given ℓ and b for a source, its location on the sky is fully specified. In order to specify its three-dimensional location, the distance of that source from us is also needed.

The conversion of the positions of sources given in Galactic coordinates (b, ℓ) to that in equatorial coordinates (α, δ) and vice versa is obtained from the rotation between these two coordinate systems, and is described by spherical trigonometry.[1] The necessary formulae can be found in numerous standard texts. We will not reproduce them here, since nowadays this transformation is done nearly exclusively using computer programs. Instead, we will give some examples. The following figures refer to the Epoch 2000: due to the precession

[1] The equatorial coordinates are defined by the direction of the Earth's rotation axis and by the rotation of the Earth. The intersections of the Earth's axis and the sphere define the northern and southern poles. The great circles on the sphere through these two poles, the meridians, are curves of constant *right ascension* α. Curves perpendicular to them and parallel to the projection of the Earth's equator onto the sky are curves of constant *declination* δ, with the poles located at $\delta = \pm 90°$.

Fig. 2.1. An unusual optical image of the Milky Way. This total view of the Galaxy is composed of a large number of individual images

Peter Schneider, The Milky Way as a Galaxy.
In: Peter Schneider, Extragalactic Astronomy and Cosmology. pp. 35–85 (2006)
DOI: 10.1007/11614371_2

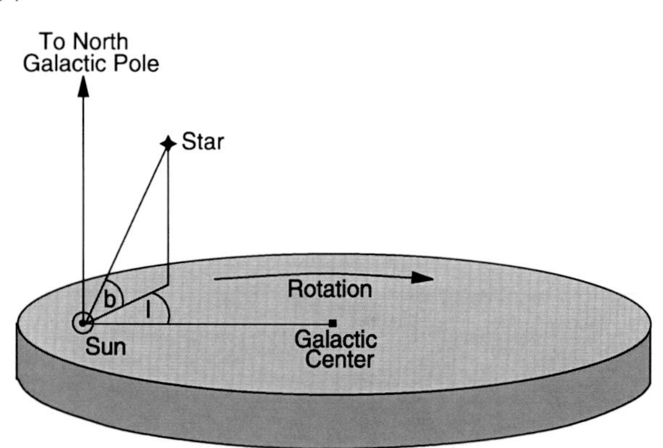

Fig. 2.2. The Sun is at the origin of the Galactic coordinate system. The directions to the Galactic center and to the North Galactic Pole (NGP) are indicated and are located at $\ell = 0°$ and $b = 0°$, and at $b = 90°$, respectively

of the rotation axis of the Earth, the equatorial coordinate system changes with time, and is updated from time to time. The position of the Galactic center (at $\ell = 0° = b$) is $\alpha = 17^{\rm h}45.6^{\rm m}$, $\delta = -28°56\overset{\prime}{.}2$ in equatorial coordinates. This immediately implies that at the La Silla Observatory, located at geographic latitude $-29°$, the Galactic center is found near the zenith at local midnight in May/June. Because of the high stellar density in the Galactic disk and the large extinction due to dust this is therefore not a good season for extragalactic observations from La Silla. The North Galactic Pole has coordinates $\alpha_{\rm NGP} = 192.859\,48° \approx 12^{\rm h}51^{\rm m}$, $\delta_{\rm NGP} = 27.128\,25° \approx 27°7\overset{\prime}{.}7$.

Zone of Avoidance. As already mentioned, the absorption by dust and the presence of numerous bright stars render optical observations of extragalactic sources in the direction of the disk difficult. The best observing conditions are found at large $|b|$, while it is very hard to do extragalactic astronomy in the optical regime at $|b| \lesssim 10°$; this region is therefore often called the "Zone of Avoidance". An illustrative example is the galaxy Dwingeloo 1, which was already mentioned in Sect. 1.1 (see Fig. 1.6). This galaxy was only discovered in the 1990s despite being in our immediate vicinity: it is located at low $|b|$, right in the Zone of Avoidance.

Cylindrical Galactic Coordinates (R, θ, z). The angular coordinates introduced above are well suited to describing the angular position of a source relative to the Galactic disk. However, we will now introduce another three-dimensional coordinate system for the description of the Milky Way geometry that will prove very convenient in the study of the kinematic and dynamic properties of the Milky Way. It is a cylindrical coordinate system, with the Galactic center at the origin (see also Fig. 2.13). The radial coordinate R measures the distance of an object from the Galactic center in the disk, and z specifies the height above the disk (objects with negative z are thus located below the Galactic disk, i.e., south of it). For instance, the Sun has a distance from the Galactic center of $R \approx 8$ kpc. The angle θ specifies the angular separation of an object in the disk relative to the position of the Sun, seen from the Galactic center. The distance of an object with coordinates R, θ, z from the Galactic center is then $\sqrt{R^2 + z^2}$, independent of θ. If the matter distribution in the Milky Way were axially symmetric, the density would then depend only on R and z, but not on θ. Since this assumption is a good approximation, this coordinate system is very well suited for the physical description of the Galaxy.

2.2 Determination of Distances Within Our Galaxy

A central problem in astronomy is the estimation of distances. The position of sources on the sphere gives us a two-dimensional picture. To obtain three-dimensional information, measurements of distances are required. Furthermore, we need to know the distance to a source if we want to draw conclusions about its physical parameters. For example, we can directly observe the angular diameter of an object, but to derive the physical size we need to know its distance. Another example is the determination of the *luminosity* L of a source, which can be derived from the observed *flux* S only by means of its distance D, using

$$\boxed{L = 4\pi S\, D^2} \,. \tag{2.1}$$

It is useful to consider the dimensions of the physical parameters in this equation. The unit of the luminosity is $[L] = {\rm erg\,s^{-1}}$, and that of the flux $[S] = {\rm erg\,s^{-1}\,cm^{-2}}$. The flux is the energy passing through a unit area per unit time (see Appendix A). Of course, the physical properties of a source are characterized by the luminosity L and not by the flux S, which depends on its distance from the Sun.

In the following section we will review various methods for the estimation of distances in our Milky Way, postponing the discussion of methods for estimating extragalactic distances to Sect. 3.6.

2.2.1 Trigonometric Parallax

The most important method of distance determination is the *trigonometric parallax*, and not only from a historical point-of-view. This method is based on a purely geometric effect and is therefore independent of any physical assumptions. Due to the motion of the Earth around the Sun the positions of nearby stars on the sphere change relative to those of very distant sources (e.g., extragalactic objects such as quasars). The latter therefore define a fixed reference frame on the sphere (see Fig. 2.3). In the course of a year the apparent position of a nearby star follows an ellipse on the sphere, the semimajor axis of which is called the *parallax p*. The axis ratio of this ellipse depends on the direction of the star relative to the ecliptic (the plane that is defined by the orbits of the planets) and is of no further interest. The parallax depends on the radius r of the Earth's orbit, hence on the Earth–Sun distance which is, by definition, one astronomical unit.[2] Furthermore, the parallax depends on the distance D of the star,

$$\frac{r}{D} = \tan p \approx p \,, \tag{2.2}$$

where we used $p \ll 1$ in the last step, and p is measured in radians as usual. The trigonometric parallax is also used to define the common unit of distance in astronomy: one *parsec* (pc) is the distance of a hypothetical source for which the parallax is exactly $p = 1''$. With the conversion of arcseconds to radians ($1'' \approx 4.848 \times 10^{-6}$ radians) one gets $p/1'' = 206\,265\,\mathrm{p}$, which for a parsec yields

$$\boxed{1\,\mathrm{pc} = 206\,265\,\mathrm{AU} = 3.086 \times 10^{18}\,\mathrm{cm}} \,. \tag{2.3}$$

[2]To be precise, the Earth's orbit is an ellipse and one astronomical unit is its semimajor axis, being $1\,\mathrm{AU} = 1.496 \times 10^{13}$ cm.

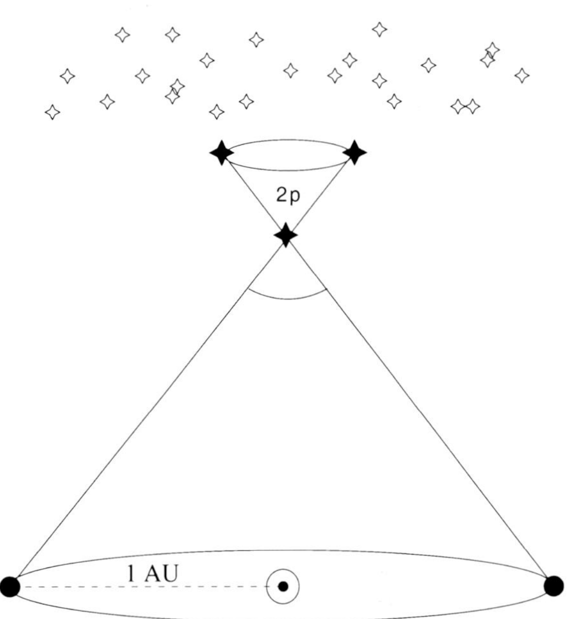

Fig. 2.3. Illustration of the parallax effect: in the course of the Earth's orbit around the Sun the apparent positions of nearby stars on the sky seem to change relative to those of very distant background sources

The distance corresponding to a measured parallax is then calculated as

$$\boxed{D = \left(\frac{p}{1''}\right)^{-1}\,\mathrm{pc}} \,. \tag{2.4}$$

To determine the parallax p, precise measurements of the position of an object at different times are needed, spread over a year, allowing us to measure the ellipse drawn on the sphere by the object's apparent position. For ground-based observation the accuracy of this method is limited by the atmosphere. The seeing causes a blurring of the images of astronomical sources and thus limits the accuracy of position measurements. From the ground this method is therefore limited to parallaxes larger than $\approx 0\farcs01$, implying that the trigonometric parallax yields distances to stars only within ~ 30 pc.

An extension of this method towards smaller p, and thus larger distances, became possible with the astrometric satellite HIPPARCOS. It operated between November 1989 and March 1993 and measured the positions and trigonometric parallaxes of about 120 000 bright stars, with a precision of $\sim 0\farcs001$ for the brighter targets. With HIPPARCOS the method of trigonometric parallax could be extended to stars up to distances of ~ 300 pc. The satellite GAIA, the successor mission to HIPPARCOS, is scheduled to be launched in 2012. GAIA will compile a catalog of $\sim 10^9$ stars up to $V \approx 20$ in four broad-band and eleven narrow-band filters. It will measure parallaxes for these stars with an accuracy of $\sim 2 \times 10^{-4}$ arcsec, with the accuracy for brighter stars even being considerably better. GAIA will thus determine the distances for $\sim 2 \times 10^8$ stars with a precision of 10%, and tangential velocities (see next section) with a precision of better than 3 km/s.

The trigonometric parallax method forms the basis of nearly all distance determinations owing to its purely geometrical nature. For example, using this method the distances to nearby stars have been determined, allowing the production of the Hertzsprung–Russell diagram (see Appendix B.2). Hence, all distance measures that are based on the properties of stars, such as will be described below, are calibrated by the trigonometric parallax.

2.2.2 Proper Motions

Stars are moving relative to us or, more precisely, relative to the Sun. To study the kinematics of the Milky Way we need to be able to measure the velocities of stars. The radial component v_r of the velocity is easily obtained from the Doppler shift of spectral lines,

$$\boxed{v_r = \frac{\Delta\lambda}{\lambda_0} c} , \tag{2.5}$$

where λ_0 is the rest-frame wavelength of an atomic transition and $\Delta\lambda = \lambda_{obs} - \lambda_0$ the Doppler shift of the wavelength due to the radial velocity of the source. The sign of the radial velocity is defined such that $v_r > 0$ corresponds to a motion away from us, i.e., to a redshift of spectral lines.

In contrast, the determination of the other two velocity components is much more difficult. The tangential component, v_t, of the velocity can be obtained from the *proper motion* of an object. In addition to the motion caused by the parallax, stars also change their positions on the sphere as a function of time because of the transverse component of their velocity relative to the Sun. The proper motion μ is thus an angular velocity, e.g., measured in milliarcseconds per year (mas/yr). This angular velocity is linked to the tangential velocity component via

$$\boxed{v_t = D\mu \quad \text{or} \quad \frac{v_t}{\text{km/s}} = 4.74 \left(\frac{D}{1\,\text{pc}}\right)\left(\frac{\mu}{1''/\text{yr}}\right)} . \tag{2.6}$$

Therefore, one can calculate the tangential velocity from the proper motion and the distance. If the latter is derived from the trigonometric parallax, (2.6) and (2.4) can be combined to yield

$$\frac{v_t}{\text{km/s}} = 4.74 \left(\frac{\mu}{1''/\text{yr}}\right)\left(\frac{p}{1''}\right)^{-1} . \tag{2.7}$$

HIPPARCOS measured proper motions for $\sim 10^5$ stars with an accuracy of up to a few mas/yr; however, they can be translated into physical velocities only if their distance is known.

Of course, the proper motion has two components, corresponding to the absolute value of the angular velocity and its direction on the sphere. Together with v_r this determines the three-dimensional velocity vector. Correcting for the known velocity of the Earth around the Sun, one can then compute the velocity vector $\boldsymbol{v}$ of the star relative to the Sun, called the *heliocentric velocity*.

2.2.3 Moving Cluster Parallax

The stars in an (open) star cluster all have a very similar spatial velocity. This implies that their proper motion vectors should be similar. To what extent the proper motions are aligned depends on the angular extent of the star cluster on the sphere. Like two railway tracks that run parallel but do not appear parallel to us, the vectors of proper motions in a star cluster also do not appear parallel. They are directed towards a convergence point, as depicted in Fig. 2.4. We shall demonstrate next how to use this effect to determine the distance to a star cluster.

We consider a star cluster and assume that all stars have the same spatial velocity $\boldsymbol{v}$. The position of the i-th star as a function of time is then described by

$$\boldsymbol{r}_i(t) = \boldsymbol{r}_i + \boldsymbol{v}t , \tag{2.8}$$

where $\boldsymbol{r}_i$ is the current position if we identify the origin of time, $t = 0$, with "today". The direction of a star

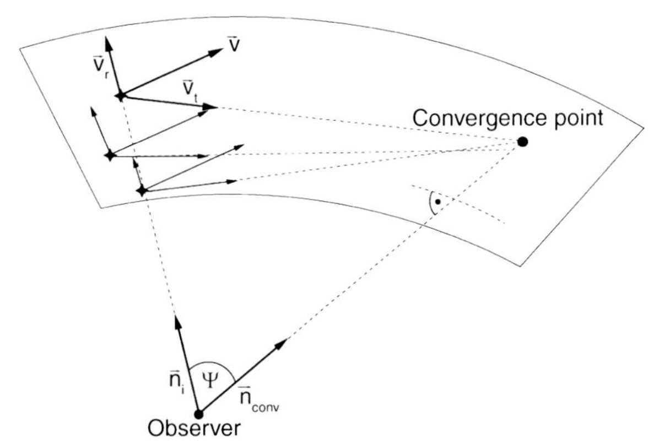

Fig. 2.4. The moving cluster parallax is a projection effect, similar to that known from viewing railway tracks. The directions of velocity vectors pointing away from us seem to converge and intersect at the convergence point. The connecting line from the observer to the convergence point is parallel to the velocity vector of the star cluster

relative to us is described by the unit vector

$$\boldsymbol{n}_i(t) := \frac{\boldsymbol{r}_i(t)}{|\boldsymbol{r}_i(t)|} . \tag{2.9}$$

From this, one infers that for large times, $t \to \infty$, the direction vectors are identical for all stars in the cluster,

$$\boldsymbol{n}_i(t) \to \frac{\boldsymbol{v}}{|\boldsymbol{v}|} =: \boldsymbol{n}_{\rm conv} . \tag{2.10}$$

Hence for large times all stars will appear at the same point $\boldsymbol{n}_{\rm conv}$: the convergence point. This only depends on the direction of the velocity vector of the star cluster. In other words, the direction vector of the stars is such that they are all moving towards the convergence point. Thus, $\boldsymbol{n}_{\rm conv}$ (and hence $\boldsymbol{v}/|\boldsymbol{v}|$) can be measured from the direction of the proper motions of the stars in the cluster, and so can $\boldsymbol{v}/|\boldsymbol{v}|$. On the other hand, one component of $\boldsymbol{v}$ can be determined from the (easily measured) radial velocity $v_{\rm r}$. With these two observables the three-dimensional velocity vector $\boldsymbol{v}$ is completely determined, as is easily demonstrated: let ψ be the angle between the line-of-sight $\boldsymbol{n}$ towards a star in the cluster and $\boldsymbol{v}$. The angle ψ is directly read off from the direction vector $\boldsymbol{n}$ and the convergence point, $\cos\psi = \boldsymbol{n} \cdot \boldsymbol{v}/|\boldsymbol{v}| = \boldsymbol{n}_{\rm conv} \cdot \boldsymbol{n}$. With $v \equiv |\boldsymbol{v}|$ one then obtains

$$v_{\rm r} = v \cos\psi , \quad v_{\rm t} = v \sin\psi ,$$

and so

$$v_{\rm t} = v_{\rm r} \tan\psi . \tag{2.11}$$

This means that the tangential velocity $v_{\rm t}$ can be measured without determining the distance to the stars in the cluster. On the other hand, (2.6) defines a relation between the proper motion, the distance, and $v_{\rm t}$. Hence, a distance determination for the star is now possible with

$$\mu = \frac{v_{\rm t}}{D} = \frac{v_{\rm r} \tan\psi}{D} \quad \to \quad D = \frac{v_{\rm r} \tan\psi}{\mu} . \tag{2.12}$$

This method yields accurate distance estimates of star clusters within ~ 200 pc. The accuracy depends on the measurability of the proper motions. Furthermore, the cluster should cover a sufficiently large area on the sky for the convergence point to be well defined. For the distance estimate, one can then take the average over a large number of stars in the cluster if one assumes that the spatial extent of the cluster is much smaller than its distance to us. Targets for applying this method are the Hyades, a cluster of about 200 stars at a mean distance of $D \approx 45$ pc, the Ursa-Major group of about 60 stars at $D \approx 24$ pc, and the Pleiades with about 600 stars at $D \approx 130$ pc.

Historically the distance determination to the Hyades, using the moving cluster parallax, was extremely important because it defined the scale to all other, larger distances. Its constituent stars of known distance are used to construct a calibrated Hertzsprung–Russell diagram which forms the basis for determining the distance to other star clusters, as will be discussed in Sect. 2.2.4. In other words, it is the lowest rung of the so-called distance ladder that we will discuss in Sect. 3.6. With HIPPARCOS, however, the distance to the Hyades stars could also be measured using the trigonometric parallax, yielding more accurate values. HIPPARCOS was even able to differentiate the "near" from the "far" side of the cluster – this star cluster is too close for the assumption of an approximately equal distance of all its stars to be still valid. A recent value for the mean distance of the Hyades is

$$\overline{D}_{\rm Hyades} = 46.3 \pm 0.3 \ {\rm pc} . \tag{2.13}$$

2.2.4 Photometric Distance; Extinction and Reddening

Most stars in the color–magnitude diagram are located along the main sequence. This enables us to compile a calibrated main sequence of those stars whose trigonometric parallaxes are measured, thus with known distances. Utilizing photometric methods, it is then possible to derive the distance to a star cluster, as we will demonstrate in the following.

The stars of a star cluster define their own main sequence (color–magnitude diagrams for some star clusters are displayed in Fig. 2.5); since they are all located at the same distance, their main sequence is already defined in a color–magnitude diagram in which only apparent magnitudes are plotted. This cluster main sequence can then be fitted to a calibrated main sequence[3] by a suitable choice of the distance, i.e., by adjusting the distance modulus $m - M$,

$$m - M = 5\log(D/\mathrm{pc}) - 5 ,$$

where m and M denote the apparent and absolute magnitude, respectively.

In reality this method cannot be applied so easily since the position of a star on the main sequence does not only depend on its mass but also on its age and metallicity. Furthermore, only stars of luminosity class V (i.e., dwarf stars) define the main sequence, but without spectroscopic data it is not possible to determine the luminosity class.

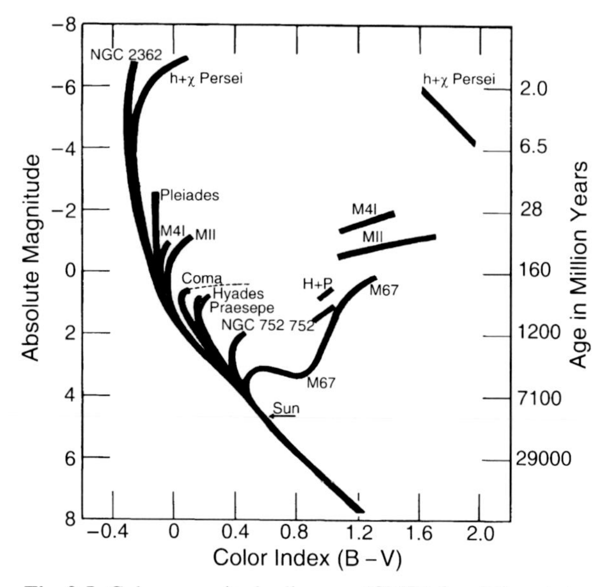

Fig. 2.5. Color–magnitude diagram (CMD) for different star clusters. Such a diagram can be used for the distance determination of star clusters because the absolute magnitudes of main-sequence stars are known (by calibration with nearby clusters, especially the Hyades). One can thus determine the distance modulus by vertically "shifting" the main sequence. Also, the age of a star cluster can be estimated from a CMD: luminous main-sequence stars have a shorter lifetime on the main sequence than less luminous ones. The turn-off point in the stellar sequence away from the main sequence therefore corresponds to that stellar mass for which the lifetime on the main sequence equals the age of the star cluster. Accordingly, the age is specified on the right axis as a function of the position of the turn-off point; the Sun will leave the main sequence after about 10×10^9 years

Extinction and Reddening. Another major problem is extinction. Absorption and scattering of light by dust affect the relation of absolute to apparent magnitude: for a given M, the apparent magnitude m becomes larger (fainter) in the case of absorption, making the source appear dimmer. Also, since extinction depends on wavelength, the spectrum of the source is modified and the observed color of the star changes. Because extinction by dust is always associated with such a change in color, one can estimate the absorption – provided one has sufficient information on the intrinsic color of a source or of an ensemble of sources. We will now demonstrate how this method can be used to estimate the distance to a star cluster.

We consider the equation of radiative transfer for pure absorption or scattering (see Appendix A),

$$\boxed{\frac{\mathrm{d}I_\nu}{\mathrm{d}s} = -\kappa_\nu I_\nu} , \tag{2.14}$$

where I_ν denotes the specific intensity at frequency ν, κ_ν the absorption coefficient, and s the distance coordinate along the light beam. The absorption coefficient has the dimension of an inverse length. Equation (2.14) says that the amount by which the intensity of a light beam is diminished on a path of length $\mathrm{d}s$ is proportional to the original intensity and to the path length $\mathrm{d}s$. The absorption coefficient is thus defined as the constant of proportionality. In other words, on the distance interval $\mathrm{d}s$, a fraction $\kappa_\nu\,\mathrm{d}s$ of all photons at frequency ν is absorbed or scattered out of the beam. The solution of the transport equation (2.14) is obtained by writing it in the form $\mathrm{d}\ln I_\nu = \mathrm{d}I_\nu/I_\nu = -\kappa_\nu\,\mathrm{d}s$ and integrating from 0 to s,

[3] i.e., to the main sequence in a color–magnitude diagram in which absolute magnitudes are plotted

$$\ln I_\nu(s) - \ln I_\nu(0) = -\int_0^s ds' \, \kappa_\nu(s') \equiv -\tau_\nu(s) ,$$

where in the last step we defined the *optical depth*, τ_ν, which depends on frequency. This yields

$$\boxed{I_\nu(s) = I_\nu(0)\, e^{-\tau_\nu(s)}} \,. \tag{2.15}$$

The specific intensity is thus reduced by a factor $e^{-\tau}$ compared to the case of no absorption taking place. Accordingly, for the flux we obtain

$$\boxed{S_\nu = S_\nu(0)\, e^{-\tau_\nu(s)}} \,, \tag{2.16}$$

where S_ν is the flux measured by the observer at a distance s from the source, and $S_\nu(0)$ is the flux of the source without absorption. Because of the relation between flux and magnitude $m = -2.5 \log S + \text{const}$, or $S \propto 10^{-0.4m}$, one has

$$\frac{S_\nu}{S_{\nu,0}} = 10^{-0.4(m-m_0)} = e^{-\tau_\nu} = 10^{-\log(e)\tau_\nu} ,$$

or

$$A_\nu := m - m_0 = -2.5 \log(S_\nu/S_{\nu,0}) = 2.5 \log(e)\, \tau_\nu = 1.086\tau_\nu \,. \tag{2.17}$$

Here, A_ν is the *extinction coefficient* describing the change of apparent magnitude m compared to that without absorption, m_0. Since the absorption coefficient κ_ν depends on frequency, absorption is always linked to a change in color. This is described by the *color excess* which is defined as follows:

$$E(X-Y) := A_X - A_Y = (X - X_0) - (Y - Y_0) = (X-Y) - (X-Y)_0 \,. \tag{2.18}$$

The color excess describes the change of the color index $(X-Y)$, measured in two filters X and Y that define the corresponding spectral windows by their transmission curves. The ratio $A_X/A_Y = \tau_{\nu(X)}/\tau_{\nu(Y)}$ depends only on the optical properties of the dust or, more specifically, on the ratio of the absorption coefficients in the two frequency bands X and Y considered here. Thus, the color excess is proportional to the extinction coefficient,

$$\boxed{E(X-Y) = A_X - A_Y = A_X\left(1 - \frac{A_Y}{A_X}\right) \equiv A_X\, R_X^{-1}} \,, \tag{2.19}$$

where in the last step we introduced the factor of proportionality R_X between the extinction coefficient and the color excess, which depends only on the properties of the dust and the choice of the filters. Usually, one uses a blue and a visual filter (see Appendix A.4.2 for a description of the filters commonly used) and writes

$$A_V = R_V\, E(B-V) \,. \tag{2.20}$$

For example, for dust in our Milky Way we have the characteristic relation

$$\boxed{A_V = (3.1 \pm 0.1) E(B-V)} \,. \tag{2.21}$$

This relation is not a universal law, but the factor of proportionality depends on the properties of the dust. They are determined, e.g., by the chemical composition and the size distribution of the dust grains. Fig. 2.6 shows the wavelength dependence of the extinction coefficient for different kinds of dust, corresponding to different values of R_V. In the optical part of the spectrum we have

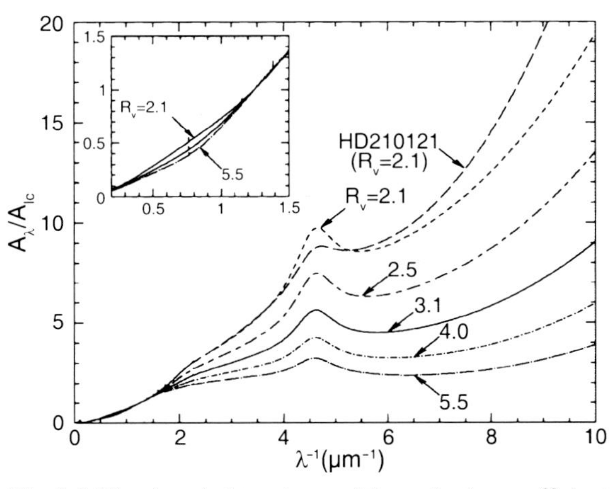

Fig. 2.6. Wavelength dependence of the extinction coefficient A_ν, normalized to the extinction coefficient A_I at $\lambda = 9000$ Å. Different kinds of clouds, characterized by the value of R_V, i.e., by the reddening law, are shown. On the x-axis we have plotted the inverse wavelength, so that the frequency increases to the right. The solid line specifies the mean Galactic extinction curve. The extinction coefficient, as determined from the observation of an individual star, is also shown; clearly the observed law deviates from the model in some details. The figure insert shows a detailed plot at relatively large wavelengths in the NIR range of the spectrum; at these wavelengths the extinction depends only weakly on the value of R_V

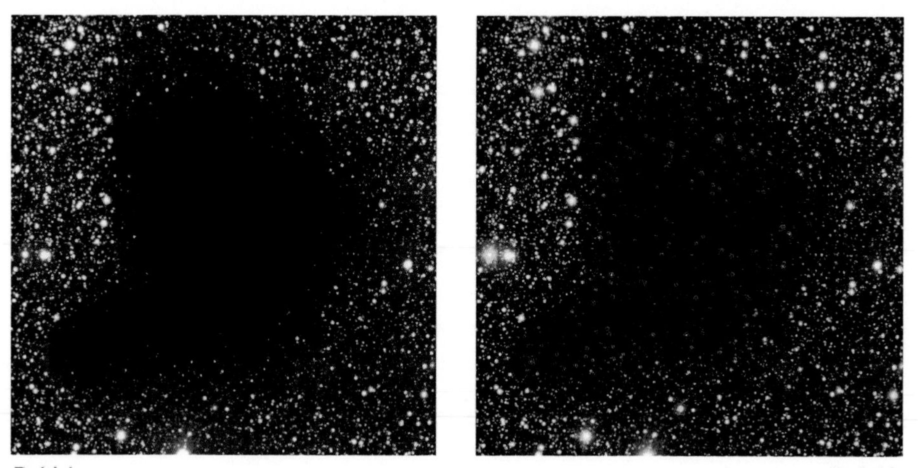

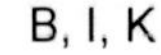

Fig. 2.7. These images of the molecular cloud Barnard 68 show the effects of extinction and reddening: the left image is a composite of exposures in the filters B, V, and I. At the center of the cloud essentially all the light from the background stars is absorbed. Near the edge it is dimmed and visibly shifted to the red. In the right-hand image observations in the filters B, I, and K have been combined (red is assigned here to the near-infrared K-band filter); we can clearly see that the cloud is more transparent at longer wavelengths

approximately $\tau_\nu \propto \nu$, i.e., blue light is absorbed (or scattered) more strongly than red light. The extinction therefore always causes a reddening.[4]

In the Solar neighborhood the extinction coefficient for sources in the disk is about

$$A_V \approx 1\,\mathrm{mag}\,\frac{D}{1\,\mathrm{kpc}}\,, \tag{2.22}$$

but this relation is at best a rough approximation, since the absorption coefficient can show strong local deviations from this law, for instance in the direction of molecular clouds (see, e.g., Fig. 2.7).

Color–color diagram. We now return to the distance determination for a star cluster. As a first step in this measurement, it is necessary to determine the degree of extinction, which can only be done by analyzing the reddening. The stars of the cluster are plotted in a *color–color diagram*, for example by plotting the colors $(U-B)$ and $(B-V)$ on the two axes (see Fig. 2.8). A color–color diagram also shows a main sequence along which the majority of the stars are aligned. The wavelength-dependent extinction causes a reddening *in both colors*. This shifts the positions of the stars in the diagram. The direction of the reddening vector depends only on the properties of the dust and is here assumed to be known, whereas the *amplitude* of the shift depends on the extinction coefficient. In a similar way to the CMD, this amplitude can now be determined if one has access to a calibrated, unreddened main sequence for the color–color diagram which can be obtained from the examination of nearby stars. From the relative shift of the main sequence in the two diagrams one can then derive the reddening and thus the extinction. The essential point here is the fact that the color–color diagram is independent of the distance.

This then defines the procedure for the distance determination of a star cluster using photometry: in the first step we determine the reddening $E(B-V)$, and thus with (2.21) also A_V, by shifting the main sequence in a color–color diagram along the reddening vector until it matches a calibrated main sequence. In the second step the distance modulus is then determined by vertically (i.e., in the direction of M) shifting the main sequence in the color–magnitude diagram until it matches a calibrated main sequence. From this, the distance is then obtained according to

$$\boxed{m - M = 5\log(D/1\,\mathrm{pc}) - 5 + A}\,. \tag{2.23}$$

[4] With what we have just learned we can readily answer the question of why the sky is blue and the setting Sun red.

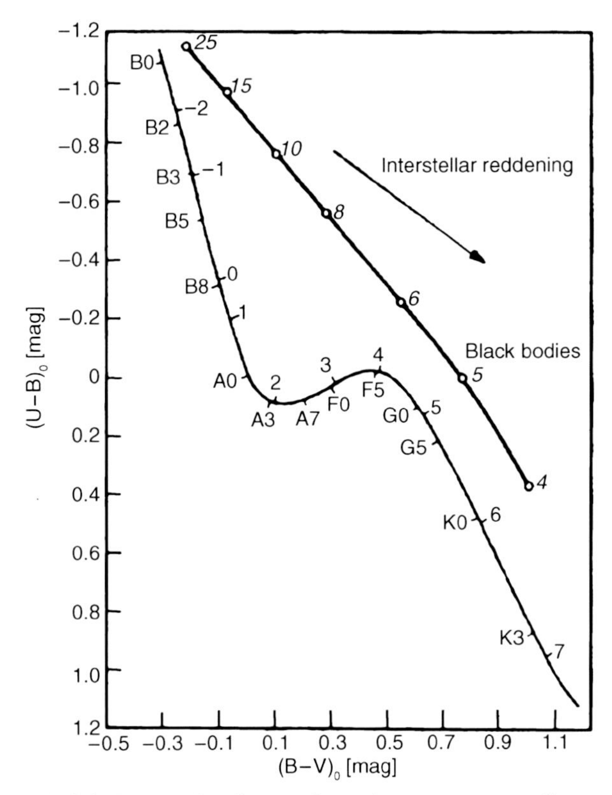

Fig. 2.8. Color–color diagram for main-sequence stars. Spectral types and absolute magnitudes are specified. Black bodies ($T/10^3$ K) would be located along the upper line. Interstellar reddening shifts the measured stellar locations parallel to the reddening vector indicated by the arrow

2.2.5 Spectroscopic Distance

From the spectrum of a star, the spectral type as well as the luminosity class can be determined. The former is determined from the strength of various absorption lines in the spectrum, while the latter is obtained from the width of the lines. From the line width the surface gravity of the star can be derived, and from that its radius (more precisely, M/R^2). From the spectral type and the luminosity class the position of the star in the HRD follows unambiguously. By means of stellar evolution models, the absolute magnitude M_V can then be determined. Furthermore, the comparison of the observed color with that expected from theory yields the color excess $E(B-V)$, and from that we obtain A_V. With this information we are then able to determine the distance using

$$\boxed{V - A_V - M_V = 5\log(D/\mathrm{pc}) - 5} \ . \tag{2.24}$$

2.2.6 Distances of Visual Binary Stars

Kepler's third law for a two-body problem,

$$\boxed{P^2 = \frac{4\pi^2}{G(m_1 + m_2)} a^3} \ , \tag{2.25}$$

specifies the relation between the orbital period P of a binary star, the masses m_i of the two components, and the semimajor axis a of the ellipse. The latter is defined by the distance vector between the two stars in the course of one period. This law can be used to determine the distance to a visual binary star. For such a system, the period P and the angular diameter 2θ of the orbit are direct observables. If one additionally knows the mass of the two stars, for instance from their spectral classification, a can be determined according to (2.25), and from this the distance follows with $D = a/\theta$.

2.2.7 Distances of Pulsating Stars

Several types of pulsating stars show periodic changes in their brightnesses, where the period of a star is related to its mass, and thus to its luminosity. This period–luminosity (PL) relation is ideally suited for distance measurements: since the determination of the period is independent of distance, one can obtain the luminosity directly from the period. The distance is thus directly derived from the measured magnitude using (2.24), if the extinction can be determined from color measurements.

The existence of a relation between the luminosity and the pulsation period can be expected from simple physical considerations. Pulsations are essentially radial density waves inside a star that propagate with the speed of sound, c_s. Thus, one can expect that the period is comparable to the sound crossing time through the star, $P \sim R/c_s$. The speed of sound c_s in a gas is of the same order of magnitude as the thermal velocity of the gas particles, so that $k_B T \sim m_p c_s^2$, where m_p is the proton mass (and thus a characteristic mass of particles in the stellar plasma) and k_B is Boltzmann's constant. According to the virial theorem, one expects that the gravitational binding energy of the star is about twice the kinetic (i.e., thermal) energy, so that for a proton

$$\frac{GMm_{\rm p}}{R} \sim k_{\rm B}T \; .$$

Combining these relations, for the pulsation period we obtain

$$P \sim \frac{R}{c_{\rm s}} \sim \frac{R\sqrt{m_{\rm p}}}{\sqrt{k_{\rm B}T}} \sim \frac{R^{3/2}}{\sqrt{GM}} \propto \overline{\rho}^{-1/2} \; , \qquad (2.26)$$

where $\overline{\rho}$ is the mean density of the star. This is a remarkable result – the pulsation period depends only on the mean density. Furthermore, the stellar luminosity is related to its mass by approximately $L \propto M^3$. If we now consider stars of equal effective temperature $T_{\rm eff}$ (where $L \propto R^2 T_{\rm eff}^4$), we find that

$$P \propto \frac{R^{3/2}}{\sqrt{M}} \propto L^{7/12} \; , \qquad (2.27)$$

which is the relation between period and luminosity that we were aiming for.

One finds that a well-defined period–luminosity relation exists for three types of pulsating stars:

- δ Cepheid stars (classical Cepheids). These are young stars found in the disk population (close to the Galactic plane) and in young star clusters. Owing to their position in or near the disk, extinction always plays a role in the determination of their luminosity. To minimize the effect of extinction it is particularly useful to look at the period–luminosity relation in the near-IR (e.g., in the K-band at $\lambda \sim 2.4\,\mu$m). Furthermore, the scatter around the period–luminosity relation is smaller for longer wavelengths of the applied filter, as is also shown in Fig. 2.9. The period–luminosity relation is also steeper for longer wavelengths, resulting in a more accurate determination of the absolute magnitude.
- W Virginis stars, also called Population II Cepheids (we will explain the term of stellar populations in Sect. 2.3.2). These are low-mass, metal-poor stars located in the halo of the Galaxy, in globular clusters, and near the Galactic center.
- RR Lyrae stars. These are likewise Population II stars and thus metal-poor. They are found in the halo, in globular clusters, and in the Galactic bulge. Their absolute magnitudes are confined to a narrow interval, $M_V \in [0.5, 1.0]$, with a mean value of about 0.6. This obviously makes them very good distance indicators. More precise predictions of their magnitudes are possible with the following dependence on metallicity and period:

$$\langle M_K \rangle = -(2.0 \pm 0.3)\log(P/1d) + (0.06 \pm 0.04)[{\rm Fe/H}] - 0.7 \pm 0.1 \; . \qquad (2.28)$$

Metallicity. In the last equation, the metallicity of a star was introduced, which needs to be defined. In astrophysics, all chemical elements heavier than helium are called *metals*. These elements, with the exception of some traces of lithium, were not produced in the early Universe but rather later in the interior of stars. The metallicity is thus also a measure of the chemical evolution and enrichment of matter in a star or gas cloud. For an element X, the metallicity index of a star is defined as

$$[{\rm X/H}] \equiv \log\left(\frac{n({\rm X})}{n({\rm H})}\right)_* - \log\left(\frac{n({\rm X})}{n({\rm H})}\right)_\odot \; , \qquad (2.29)$$

thus it is the logarithm of the ratio of the fraction of X relative to hydrogen in the star and in the Sun, where n is the number density of the species considered. For example, $[{\rm Fe/H}] = -1$ means that iron has only a tenth of its Solar abundance. The *metallicity* Z is the total mass fraction of all elements heavier than helium; the Sun has $Z \approx 0.02$, meaning that about 98% of the Solar mass are contributed by hydrogen and helium.

The period–luminosity relations are not only of significant importance for distance determination within our Galaxy. They also play an important role in extragalactic astronomy, since by far the most luminous of the three types of pulsating stars listed above, the Cepheids, are also found and observed in other galaxies; they therefore enable us to directly determine the distances of other galaxies, which is essential for measuring the Hubble constant. These aspects will be discussed in detail in Sect. 3.6.

2.3 The Structure of the Galaxy

Roughly speaking, the Galaxy consists of the disk, the central bulge, and the Galactic halo – a roughly spherical

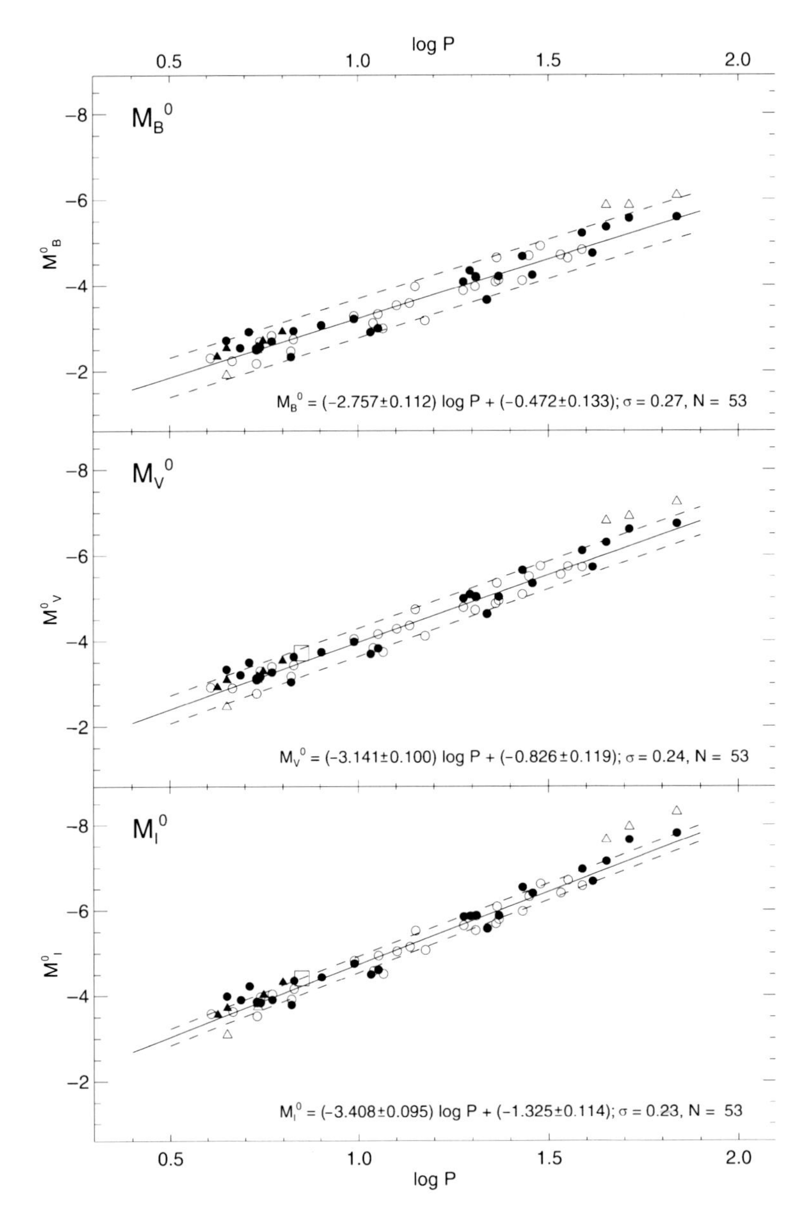

Fig. 2.9. Period–luminosity relation for Galactic Cepheids, measured in three different filters bands (B, V, and I, from top to bottom). The absolute magnitudes were corrected for extinction by using colors. The period is given in days. Open and solid circles denote data for those Cepheids for which distances were estimated using different methods; the three objects marked by triangles have a variable period and are discarded in the derivation of the period–luminosity relation. The latter is indicated by the solid line, with its parametrisation specified in the plots. The broken lines indicate the uncertainty range of the period–luminosity relation. The slope of the period–luminosity relation increases if one moves to redder filters

distribution of stars and globular clusters that surrounds the disk. The disk, whose stars form the visible band of the Milky Way, contains spiral arms similar to those observed in other galaxies. The Sun, together with its planets, orbits around the Galactic center on an approximately circular orbit. The distance R_0 to the Galactic center is not very well known, as we will discuss later. To have a reference value, the International Astronomical Union (IAU) officially defined the value of R_0 in 1985,

$$\boxed{R_0 = 8.5\,\text{kpc}} \quad \text{official value, IAU 1985} . \tag{2.30}$$

More recent examinations have, however, found that the real value is slightly smaller, $R_0 \approx 8.0\,\text{kpc}$. The diameter of the disk of stars, gas, and dust is $\sim 40\,\text{kpc}$.

A schematic depiction of our Galaxy is shown in Fig. 1.3. Its most important structural parameters are listed in Table 2.1.

2.3.1 The Galactic Disk: Distribution of Stars

By measuring the distances of stars in the Solar neighborhood one can determine the three-dimensional stellar distribution. From these investigations, one finds that there are different stellar components, as we will discuss below. For each of them, the number density in the direction perpendicular to the Galactic disk is approximately described by an exponential law,

$$n(z) \propto \exp\left(-\frac{|z|}{h}\right) , \qquad (2.31)$$

where the *scale-height h* specifies the thickness of the respective component. One finds that h varies between different populations of stars, motivating the definition of different components of the Galactic disk. In principle, three components need to be distinguished: (1) The *young thin disk* contains the largest fraction of gas and dust in the Galaxy, and in this region star formation is still taking place today. The youngest stars are found in the young thin disk, which has a scale-height of about $h_{\rm ytd} \sim 100$ pc. (2) The *old thin disk* is thicker and has a scale-height of about $h_{\rm otd} \sim 325$ pc. (3) The *thick disk* has a scale-height of $h_{\rm thick} \sim 1.5$ kpc. The thick disk contributes only about 2% to the total mass density in the Galactic plane at $z = 0$. This separation into three disk components is rather coarse and can be further refined if one uses a finer classification of stellar populations.

Molecular gas, out of which new stars are born, has the smallest scale-height, $h_{\rm mol} \sim 65$ pc, followed by the atomic gas. This can be clearly seen by comparing the distributions of atomic and molecular hydrogen in Fig. 1.5. The younger a stellar population is, the smaller its scale-height. Another characterization of the different stellar populations can be made with respect to the velocity dispersion of the stars, i.e., the amplitude of the components of their random motions. As a first approximation, the stars in the disk move around the Galactic center on circular orbits. However, these orbits are not perfectly circular: besides the orbital velocity (which is about 220 km/s in the Solar vicinity), they have additional random velocity components.

The formal definition of the components of the velocity dispersion is as follows: let $f(\boldsymbol{v})\mathrm{d}^3v$ be the number density of stars (of a given population) at a fixed location, with velocities in a volume element d^3v around $\boldsymbol{v}$ in the vector space of velocities. If we use Cartesian coordinates, for example $\boldsymbol{v} = (v_1, v_2, v_3)$, then $f(\boldsymbol{v})\mathrm{d}^3v$ is the number of stars with the i-th velocity component in the interval $[v_i, v_i + \mathrm{d}v_i]$, and $\mathrm{d}^3v = \mathrm{d}v_1\mathrm{d}v_2\mathrm{d}v_3$. The mean velocity $\langle\boldsymbol{v}\rangle$ of the population then follows from this distribution via

$$\langle\boldsymbol{v}\rangle = n^{-1}\int_{\mathbb{R}^3} \mathrm{d}^3v\, f(\boldsymbol{v})\, \boldsymbol{v} , \quad \text{where} \quad n = \int_{\mathbb{R}^3} \mathrm{d}^3v\, f(\boldsymbol{v}) \qquad (2.32)$$

denotes the total number density of stars in the population. The velocity dispersion σ then describes the mean squared deviations of the velocities from $\langle\boldsymbol{v}\rangle$. For

Table 2.1. Parameters and characteristic values for the components of the Milky Way. The scale-height denotes the distance from the Galactic plane at which the density has decreased to 1/e of its central value. σ_z is the velocity dispersion in the direction perpendicular to the disk

	Neutral gas	Thin disk	Thick disk	bulge	Stellar halo	Dm halo
M (10^{10}M$_\odot$)	0.5	6	0.2 to 0.4	1	0.1	55
$L_{\rm B}$ (10^{10}L$_\odot$)	–	1.8	0.02	0.3	0.1	0
$M/L_{\rm B}$ (M$_\odot$/L$_\odot$)	–	3	–	3	~ 1	–
diam. (kpc)	50	50	50	2	100	> 200
form	$e^{-h_z/z}$	$e^{-h_z/z}$	$e^{-h_z/z}$	bar?	$r^{-3.5}$	$(a^2+r^2)^{-1}$
scale-height (kpc)	0.13	0.325	1.5	0.4	3	2.8
σ_z (km s^{-1})	7	20	40	120	100	–
[Fe/H]	> 0.1	−0.5 to +0.3	−1.6 to −0.4	−1 to +1	−4.5 to −0.5	–

a component i of the velocity vector, the dispersion σ_i is defined as

$$\sigma_i^2 = \left\langle (v_i - \langle v_i \rangle)^2 \right\rangle = \left\langle v_i^2 - \langle v_i \rangle^2 \right\rangle$$
$$= n^{-1} \int_{\mathbb{R}^3} \mathrm{d}^3 v \, f(\boldsymbol{v}) \left(v_i^2 - \langle v_i \rangle^2 \right) . \quad (2.33)$$

The larger σ_i is, the broader the distribution of the stochastic motions. We note that the same concept applies to the velocity distribution of molecules in a gas. The mean velocity $\langle \boldsymbol{v} \rangle$ at each point defines the bulk velocity of the gas, e.g., the wind speed in the atmosphere, whereas the velocity dispersion is caused by thermal motion of the molecules and is determined by the temperature of the gas.

The random motion of the stars in the direction perpendicular to the disk is the reason for the finite thickness of the population; it is similar to a thermal distribution. Accordingly, it has the effect of a pressure, the so-called *dynamical pressure* of the distribution. This pressure determines the scale-height of the distribution, which corresponds to the law of atmospheres. The larger the dynamical pressure, i.e., the larger the velocity dispersion σ_z perpendicular to the disk, the larger the scale-height h will be. The analysis of stars in the Solar neighborhood yields $\sigma_z \sim 16$ km/s for stars younger than ~ 3 Gyr, corresponding to a scale-height of $h \sim 250$ pc, whereas stars older than ~ 6 Gyr have a scale-height of ~ 350 pc and a velocity dispersion of $\sigma_z \sim 25$ km/s.

The density distribution of the total star population, obtained from counts and distance determinations of stars, is to a good approximation described by

$$\boxed{n(R, z) = n_0 \left(\mathrm{e}^{-|z|/h_{\rm thin}} + 0.02 \mathrm{e}^{-|z|/h_{\rm thick}} \right) \mathrm{e}^{-R/h_R}} ; \quad (2.34)$$

here, R and z are the cylinder coordinates introduced above (see Sect. 2.1), with the origin at the Galactic center, and $h_{\rm thin} \approx h_{\rm otd} \approx 325$ pc is the scale-height of the thin disk. The distribution in the radial direction can also be well described by an exponential law, where $h_R \approx 3.5$ kpc denotes the *scale-length of the Galactic disk*. The normalization of the distribution is determined by the density $n \approx 0.2$ stars/pc^3 in the Solar neighborhood, for stars in the range of absolute magnitudes of $4.5 \le M_V \le 9.5$. The distribution described by (2.34) is not smooth at $z = 0$; it has a kink at this point and it is therefore unphysical. To get a smooth distribution which follows the exponential law for large z and is smooth in the plane of the disk, the distribution is slightly modified. As an example, for the luminosity density of the old thin disk (that is proportional to the number density of the stars), we can write:

$$\boxed{L(R, z) = \frac{L_0 \mathrm{e}^{-R/h_R}}{\cosh^2(z/h_z)}} , \quad (2.35)$$

with $h_z = 2h_{\rm thin}$ and $L_0 \approx 0.05 L_\odot/\mathrm{pc}^3$. The Sun is a member of the young thin disk and is located above the plane of the disk, at $z = 30$ pc.

2.3.2 The Galactic Disk: Chemical Composition and Age

Stellar Populations. The chemical composition of stars in the thin and the thick disks differs: we observe the clear tendency that stars in the thin disk have a higher metallicity than those in the thick disk. In contrast, the metallicity of stars in the Galactic halo and in the bulge is smaller. To paraphrase these trends, one distinguishes between stars of Population I (Pop I) which have a Solar-like metallicity ($Z \sim 0.02$) and are mainly located in the thin disk, and stars of Population II (Pop II) that are metal-poor ($Z \sim 0.001$) and predominantly found in the thick disk, in the halo, and in the bulge. In reality, stars cover a wide range in Z, and the figures above are only characteristic values. For stellar populations a somewhat finer separation was also introduced, such as "extreme Population I", "intermediate Population II", and so on. The populations also differ in age (stars of Pop I are younger than those of Pop II), in scale-height (as mentioned above), and in the velocity dispersion perpendicular to the disk (σ_z is larger for Pop II stars than for Pop I stars).

We shall now attempt to understand the origin of these different metallicities and their relation to the scale-height and to age. We start with a brief discussion of the phenomenon that is the main reason for the metal enrichment of the interstellar medium.

Metallicity and Supernovae. Supernovae (SNe) are explosive events. Within a few days, a SN can reach a luminosity of $10^9 L_\odot$, which is a considerable fraction of the total luminosity of a galaxy; after that the luminosity decreases again with a time-scale of weeks. In the explosion, a star is disrupted and (most of) the matter of the star is driven into the interstellar medium, enriching it with metals that were produced in the course of stellar evolution or in the process of the supernova explosion.

Classification of Supernovae. Depending on their spectral properties, SNe are divided into several classes. SNe of Type I do not show any Balmer lines of hydrogen in their spectrum, in contrast to those of Type II. A further subdivision of Type I SNe is based on spectral properties: SNe Ia show strong emission of SiII λ 6150 Å whereas no SiII at all is visible in spectra of Type Ib,c. Our current understanding of the supernova phenomenon differs from this spectral classification.[5] Following various observational results and also theoretical analyses, we are confident today that SNe Ia are a phenomenon which is intrinsically different from the other supernova types. For this interpretation, it is of particular importance that SNe Ia are found in all types of galaxies, whereas we observe SNe II and SNe Ib,c only in spiral and irregular galaxies, and here only in those regions in which blue stars predominate. As we will see in Chap. 3, the stellar population in elliptical galaxies consists almost exclusively of old stars, while spirals also contain young stars. From this observational fact it is concluded that the phenomenon of SNe II and SNe Ib,c is linked to a young stellar population, whereas SNe Ia occur in older stellar populations. We shall discuss the two classes of supernovae next.

Core-Collapse Supernovae. SNe II and SNe Ib,c are the final stages in the evolution of massive ($\gtrsim 8 M_\odot$) stars. Inside these stars, ever heavier elements are generated by fusion: once all the hydrogen is used up, helium will be burned, then carbon, oxygen, etc. This chain comes to an end when the iron nucleus is reached, the atomic nucleus with the highest binding energy per nucleon. After this no more energy can be gained from fusion to heavier elements so that the pressure, which is normally balancing the gravitational force in the star, can no longer be maintained. The star will thus collapse under its own gravity. This gravitational collapse will proceed until the innermost region reaches a density about three times the density of an atomic nucleus. At this point the so-called rebounce occurs: a shock wave runs towards the surface, thereby heating the infalling material, and the star explodes. In the center, a compact object probably remains – a neutron star or, possibly, depending on the mass of the iron core, a black hole. Such neutron stars are visible as pulsars[6] at the location of some historically observed SNe, the most famous of which is the Crab pulsar which has been identified with a supernovae explosion seen by Chinese astronomers in 1054. Presumably all neutron stars have been formed in such core-collapse supernovae.

The major fraction of the binding energy released in the formation of the compact object is emitted in the form of neutrinos: about 3×10^{53} erg. Underground neutrino detectors were able to trace about 10 neutrinos originating from SN 1987A in the Large Magellanic Cloud. Due to the high density inside the star after the collapse, even neutrinos, despite their very small cross-section, are absorbed and scattered, so that part of their outward-directed momentum contributes to the explosion of the stellar envelope. This shell expands at $v \sim 10\,000$ km/s, corresponding to a kinetic energy of $E_{\rm kin} \sim 10^{51}$ erg. Of this, only about 10^{49} erg is converted into photons in the hot envelope and then emitted – the energy of a SN that is visible in photons is thus only a small fraction of the total energy produced.

Owing to the various stages of nuclear fusion in the progenitor star, the chemical elements are arranged in shells: the light elements (H, He) in the outer shells, and the heavier elements (C, O, Ne, Mg, Si, Ar, Ca, Fe, Ni) in the inner ones – see Fig. 2.10. The explosion ejects them into the interstellar medium which is thus chemically enriched. It is important to note that mainly nuclei with an even number of protons and neutrons are formed. This is a consequence of the nuclear reaction chains

[5] This notation scheme (Type Ia, Type II, and so on) is characteristic for phenomena that one wishes to classify upon discovery, but for which no physical interpretation is available at that time. Other examples are the spectral classes of stars, which are not named in alphabetical order according to their mass on the main sequence, or the division of Seyfert galaxies into Type 1 and Type 2. Once such a notation is established, it often becomes permanent even if a later physical understanding of the phenomenon suggests a more meaningful classification.

[6] Pulsars are sources which show a *very* regular periodic radiation, most often seen at radio frequencies. Their periods lie in the range from $\sim 10^{-3}$ s (millisecond pulsars) to ~ 5 s. Their pulse period is identified as the rotational period of the neutron star – an object with about one Solar mass and a radius of ~ 10 km. The matter density in neutron stars is about the same as that in atomic nuclei.

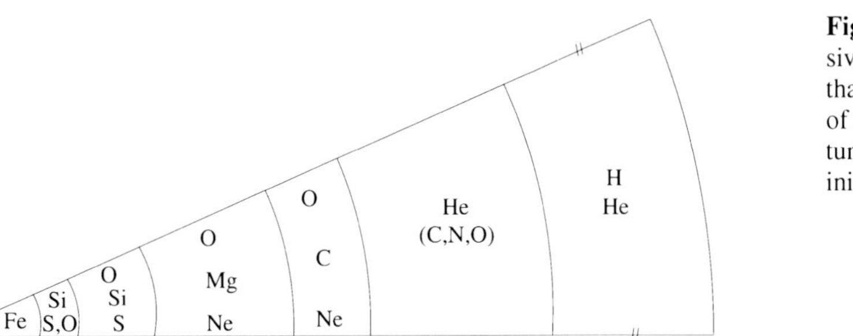

Fig. 2.10. Chemical shell structure of a massive star at the end of its life. The elements that have been formed in the various stages of the nuclear burning are ordered in a structure resembling that of an onion. This is the initial condition for a supernova explosion

involved, where successive nuclei in this chain are obtained by adding an α-particle (or ^{4}He-nucleus), i.e., two protons and two neutrons. Such elements are therefore called α-elements. The dominance of α-elements in the chemical abundance of the interstellar medium is thus a clear indication of nuclear fusion occurring in the He-rich zones of stars where the hydrogen has been burnt.

Supernovae Type Ia. SNe Ia are most likely the explosions of white dwarfs (WDs). These compact stars which form the final evolutionary stages of less massive stars no longer maintain their internal pressure by nuclear fusion. Rather, they are stabilized by the degeneracy pressure of the electrons – a quantum mechanical phenomenon. Such a white dwarf can only be stable if its mass does not exceed a limiting mass, the *Chandrasekhar mass*; it has a value of $M_{\rm Ch} \approx 1.44 M_\odot$. For $M > M_{\rm Ch}$, the degeneracy pressure can no longer balance the gravitational force. If matter falls onto a WD with mass below $M_{\rm Ch}$, as may happen by accretion in close binary systems, its mass will slowly increase and approach the limiting mass. At about $M \approx 1.3 M_\odot$, carbon burning will ignite in its interior, transforming about half of the star into iron-group elements, i.e., iron, cobalt, and nickel. The resulting explosion of the star will enrich the ISM with $\sim 0.6 M_\odot$ of Fe, while the WD itself will be torn apart completely, leaving no remnant star.

Since the initial conditions are probably very homogeneous for the class of SNe Ia (defined by the limiting mass prior to the trigger of the explosion), they are good candidates for *standard candles*: all SNe Ia have approximately the same luminosity. As we will discuss later (see Sect. 8.3.1), this is not really the case, but nevertheless SNe Ia play an important role in the cosmological distance determination, and thus in the determination of cosmological parameters.

This interpretation of the different types of SNe explains why one finds core-collapse SNe only in galaxies in which star formation occurs. They are the final stages of massive, i.e., young, stars which have a lifetime of not more than 2×10^7 yr. By contrast, SNe Ia can occur in all types of galaxies.

In addition to SNe, metal enrichment of the interstellar medium (ISM) also takes place in other stages of stellar evolution, by stellar winds or during phases in which stars eject part of their envelope which is then visible, e.g., as a planetary nebula. If the matter in the star has been mixed by convection prior to such a phase, so that the metals newly formed by nuclear fusion in the interior have been transported towards the surface of the star, these metals will then be released into the ISM.

Age–Metallicity Relation. Assuming that at the beginning of its evolution the Milky Way had a chemical composition with only low metal content, the metallicity should be strongly related to the age of a stellar population. With each new generation of stars, more metals are produced and ejected into the ISM, partially by stellar winds, but mainly by SN explosions. Stars that are formed later should therefore have a higher metal content than those that were formed in the early phase of the Galaxy. One would therefore expect that a relation should exists between the age of a star and its metallicity.

For instance, under this assumption [Fe/H] can be used as an age indicator for a stellar population, with the iron predominantly being produced and ejected in SNe of type Ia. Therefore, newly formed stars have a higher fraction of iron when they are born than their predecessors, and the youngest stars should have the highest iron abundance. Indeed one finds $[Fe/H] = -4.5$ for extremely old stars (i.e., 3×10^{-5} of the Solar iron abundance), whereas very young stars have $[Fe/H] = 1$, so their metallicity can significantly exceed that of the Sun.

However, this age–metallicity relation is not very tight. On the one hand, SNe Ia occur only $\gtrsim 10^9$ years after the formation of a stellar population. The exact time-span is not known because even if one accepts the scenario for SN Ia described above, it is unclear in what form and in what systems the accretion of material onto the white dwarf takes place and how long it typically takes until the limiting mass is reached. On the other hand, the mixing of the SN ejecta in the ISM occurs only locally, so that large inhomogeneities of the [Fe/H] ratio may be present in the ISM, and thus even for stars of the same age. An alternative measure for metallicity is [O/H], because oxygen, which is an α-element, is produced and ejected mainly in supernova explosions of massive stars. These begin only $\sim 10^7$ yr after the formation of a stellar population, which is virtually instantaneous.

Characteristic values for the metallicity are $-0.5 \lesssim [Fe/H] \lesssim 0.3$ in the thin disk, while for the thick disk $-1.0 \lesssim [Fe/H] \lesssim -0.4$ is typical. From this, one can deduce that stars in the thin disk must be significantly younger on average than those in the thick disk. This result can now be interpreted using the age–metallicity relation. Either star formation has started earlier, or ceased earlier, in the thick disk than in the thin disk, or stars that originally belonged to the thin disk have migrated into the thick disk. The second alternative is favored for various reasons. It would be hard to understand why molecular gas, out of which stars are formed, was much more broadly distributed in earlier times than it is today, where we find it well concentrated near the Galactic plane. In addition, the widening of an initially narrow stellar distribution in time is also expected. The matter distribution in the disk is not homogeneous and, along their orbits around the Galactic center, stars experience this inhomogeneous gravitational field caused by other stars, spiral arms, and massive molecular clouds. Stellar orbits are perturbed by such fluctuations, i.e., they gain a random velocity component perpendicular to the disk from local inhomogeneities of the gravitational field. In other words, the velocity dispersion σ_z of a stellar population grows in time, and the scale-height of a population increases. In contrast to stars, the gas keeps its narrow distribution around the Galactic plane due to internal friction.

This interpretation is, however, not unambiguous. Another scenario for the formation of the thick disk is also possible, where the stars of the thick disk were formed outside the Milky Way and only became constituents of the disk later, through accretion of satellite galaxies. This model is supported, among other reasons, by the fact that the rotational velocity of the thick disk around the Galactic center is smaller by ~ 50 km/s than that of the thin disk. In other spirals, in which a thick disk component was found and kinematically analyzed, the discrepancy between the rotation curves of the thick and thin disks is sometimes even stronger. In one case, the thick disk has been observed to rotate around the center of the galaxy in the opposite direction to the gas disk. In such a case, the aforementioned model of the evolution of the thick disk by kinematic heating of stars would definitely not apply.

Mass-to-Light Ratio. The total stellar mass of the thin disk is $\sim 6 \times 10^{10} M_\odot$, to which $\sim 0.5 \times 10^{10} M_\odot$ in the form of dust and gas has to be added. The luminosity of the stars in the thin disk is $L_B \approx 1.8 \times 10^{10} L_\odot$. Together, this yields a mass-to-light ratio of

$$\boxed{\frac{M}{L_B} \approx 3 \frac{M_\odot}{L_\odot} \quad \text{in thin disk}} \; . \tag{2.36}$$

The M/L ratio in the thick disk is higher. For this component, one has $M \sim 3 \times 10^9 M_\odot$ and $L_B \approx 2 \times 10^8 L_\odot$, so that $M/L_B \sim 15$ in Solar units. The thick disk thus does not play any significant role for the total mass budget of the Galactic disk, and even less for its total luminosity. On the other hand, the thick disk is invaluable for the diagnosis of the dynamical evolution of the disk. If the Milky Way were to be observed from the outside, one would find a M/L value for the disk of about 4 in Solar units; this is a characteristic value for spiral galaxies.

2.3.3 The Galactic Disk: Dust and Gas

The spiral structure of the Milky Way and other spiral galaxies is delineated by very young objects like O- and

B-stars and HII regions.[7] This is the reason why spiral arms appear blue. Obviously, star formation in our Milky Way takes place mainly in the spiral arms. Here, the molecular clouds – gas clouds which are sufficiently dense and cool for molecules to form in large abundance – contract under their own gravity and form new stars. The spiral arms are much less prominent in red light. Emission in the red is dominated by an older stellar population, and these old stars have had time to move away from the spiral arms. The Sun is located close to, but not in, a spiral arm – the so-called Orion arm.

Observing the gas in the Galaxy is made possible mainly by the 21-cm line emission of HI (neutral atomic hydrogen) and by the emission of CO, the second-most abundant molecule after H_2 (molecular hydrogen). H_2 is a symmetric molecule and thus has no electric dipole moment, which is the reason why it does not radiate strongly. In most cases it is assumed that the ratio of CO to H_2 is a universal constant (called the "X-factor"). Under this assumption, the distribution of CO can be converted into that of the molecular gas. The Milky Way is optically thin at 21 cm, i.e., 21-cm radiation is not absorbed along its path from the source to the observer. With radio-astronomical methods it is thus possible to observe atomic gas throughout the entire Galaxy.

To examine the distribution of dust, two options are available. First, dust is detected by the extinction it causes. This effect can be analyzed quantitatively, for instance by star counts or by investigating the reddening of stars (an example of this can be seen in Fig. 2.7). Second, dust emits thermal radiation observable in the FIR part of the spectrum, which was mapped by several satellites such as IRAS and COBE. By combining the sky maps of these two satellites at different frequencies the Galactic distribution of dust was determined. The dust temperature varies in a relatively narrow range between ~ 17 K and ~ 21 K, but even across this small range, the dust emission varies, for fixed column density, by a factor ~ 5 at a wavelength of 100 μm. Therefore, one needs to combine maps at different frequencies in order to determine column densities and temperatures. In addition, the zodiacal light caused by the reflection of solar radiation by dust inside our Solar system has to be subtracted before the Galactic FIR emission can be analyzed. This is possible with multifrequency data because of the different spectral shapes. The resulting distribution of dust is displayed in Fig. 2.11. It shows the concentration of dust around the Galactic plane, as well as large-scale anisotropies at high Galactic latitudes. The dust map shown here is routinely used for extinction correction when observing extragalactic sources.

Besides a strong concentration towards the Galactic plane, gas and dust are preferentially found in spiral arms where they serve as raw material for star formation. Molecular hydrogen (H_2) and dust are generally found at $3\,\mathrm{kpc} \lesssim R \lesssim 8\,\mathrm{kpc}$, within $|z| \lesssim 90$ pc of both sides of the Galactic plane. In contrast, the distribution of atomic hydrogen (HI) is observed out to much larger distances from the Galactic center ($R \lesssim 25$ kpc), with a scale-height of ~ 160 pc inside the Solar orbit, $R \lesssim R_0$. At larger distances from the Galactic center, $R \gtrsim 12$ kpc, the scale-height increases substantially to ~ 1 kpc. The gaseous disk is warped at these large radii though the origin of this warp is unclear. For example, it may be caused by the gravitational field of the Magellanic Clouds. The total mass in the two components of hydrogen is about $M(\mathrm{HI}) \approx 4 \times 10^9 M_\odot$ and $M(\mathrm{H}_2) \approx 10^9 M_\odot$, respectively, i.e., the gas mass in our Galaxy is less than $\sim 10\%$ of the stellar mass. The density of the gas in the Solar neighborhood is about $\rho(\mathrm{gas}) \sim 0.04 M_\odot/\mathrm{pc}^3$.

2.3.4 Cosmic Rays

The Magnetic Field of the Galaxy. Like many other cosmic objects, the Milky Way has a magnetic field. The properties of this field can be analyzed using a variety of methods and we list some of them in the following.

- Polarization of stellar light. The light of distant stars is partially polarized, with the degree of polarization being strongly related to the extinction, or reddening, of the star. This hints at the polarization being linked to the dust causing the extinction. The light scattered by dust particles is partially linearly polarized, with the direction of polarization depending on the alignment of the dust grains. If their orientation were random, the superposition of the scattered radiation from different dust particles would add up to a vanishing net polarization. However, a net polarization

[7] HII regions are nearly spherical regions of fully ionized hydrogen (thus the name HII region) surrounding a young hot star which photoionizes the gas. They emit strong emission lines of which the Balmer lines of hydrogen are strongest.

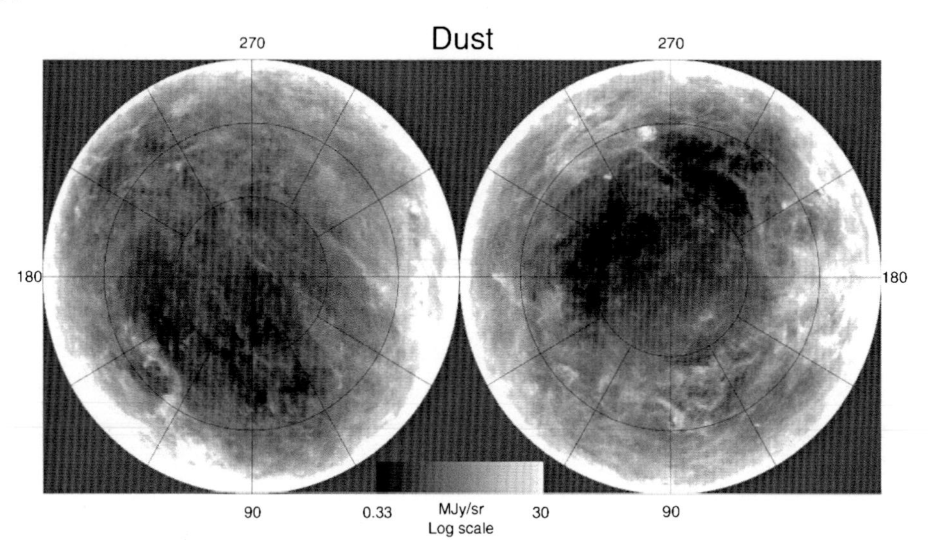

Fig. 2.11. Distribution of dust in the Galaxy, derived from a combination of IRAS and COBE sky maps. The northern Galactic sky in Galactic coordinates is displayed on the left, the southern on the right. We can clearly see the concentration of dust towards the Galactic plane, as well as regions with a very low column density of dust; these regions in the sky are particularly well suited for very deep extragalactic observations

is measured, so the orientation of dust particles cannot be random, rather it must be coherent on large scales. Such a coherent alignment is provided by a large-scale magnetic field, whereby the orientation of dust particles, measurable from the polarization direction, indicates the (projected) direction of the magnetic field.

- The Zeeman effect. The energy levels in an atom change if the atom is placed in a magnetic field. Of particular importance in the present context is the fact that the 21-cm transition line of neutral hydrogen is split in a magnetic field. Because the amplitude of the line split is proportional to the strength of the magnetic field, the field strength can be determined from observations of this Zeeman effect.
- Synchrotron radiation. When relativistic electrons move in a magnetic field they are subject to the Lorentz force. The corresponding acceleration is perpendicular both to the velocity vector of the particles and to the magnetic field vector. As a result, the electrons follow a helical (i.e., corkscrew) track, which is a superposition of circular orbits perpendicular to the field lines and a linear motion along the field. Since accelerated charges emit electromagnetic radiation, this helical movement is the source of the so-called synchrotron radiation (which will be discussed in more detail in Sect. 5.1.2). This radiation, which is observable at radio frequencies, is linearly polarized, with the direction of the polarization depending on the direction of the magnetic field.
- Faraday rotation. If polarized radiation passes through a magnetized plasma, the direction of the polarization rotates. The rotation angle depends quadratically on the wavelength of the radiation,

$$\Delta\theta = \mathrm{RM}\,\lambda^2 \,. \tag{2.37}$$

The *rotation measure* RM is the integral along the line-of-sight towards the source over the electron density and the component $B_\parallel$ of the magnetic field in direction of the line-of-sight,

$$\mathrm{RM} = 81\,\frac{\mathrm{rad}}{\mathrm{cm}^2}\int_0^D \frac{\mathrm{d}\ell}{\mathrm{pc}}\,\frac{n_\mathrm{e}}{\mathrm{cm}^{-3}}\,\frac{B_\parallel}{\mathrm{G}}\,. \tag{2.38}$$

The dependence of the rotation angle (2.37) on λ allows us to determine the rotation measure RM, and thus to estimate the product of electron density and magnetic field. If the former is known, one immediately gets information about B. By measuring the RM for sources in different directions and at different distances the magnetic field of the Galaxy can be mapped.

From applying the methods discussed above, we know that a magnetic field exists in the disk of our Milky Way. This field has a strength of about 4×10^{-6} G and mainly follows the spiral arms.

Cosmic Rays. We obtain most of the information about our Universe from the electromagnetic radiation that we observe. However, we receive an additional radiation component, the energetic cosmic rays. They consist primarily of electrically charged particles, mainly electrons and nuclei. In addition to the particle radiation that is produced in energetic processes at the Solar surface, a much more energetic cosmic ray component exists that can only originate in sources outside the Solar system.

The energy spectrum of the cosmic rays is, to a good approximation, a power law: the flux of particles with energy larger than E can be written as $S(>E) \propto E^{-q}$, with $q \approx 1.7$. However, the slope of the spectrum changes slightly, but significantly, at some energy scales: at $E \sim 10^{15}$ eV the spectrum becomes steeper, and at $E \gtrsim 10^{18}$ eV it flattens again.[8] Measurements of the spectrum at these high energies are rather uncertain, however, because of the strongly decreasing flux with increasing energy. This implies that only very few particles are detected.

Cosmic Ray Acceleration and Confinement. To accelerate particles to such high energies, highly energetic processes are necessary. For energies below 10^{15} eV, very convincing arguments suggest SN remnants as the sites of the acceleration. The SN explosion drives a shock front[9] into the ISM with an initial velocity of $\sim 10\,000$ km/s. Plasma processes in a shock front can accelerate some particles to very high energies. The theory of this diffuse shock acceleration predicts that the resulting energy spectrum of the particles follows a power law, the slope of which depends only on the strength of the shock (i.e., the ratio of the densities on both sides of the shock front). This power law agrees very well with the slope of the observed cosmic ray spectrum, if additional propagation processes in the Milky Way are taken into account. The presence of very energetic electrons in SN remnants is observed directly by their synchrotron emission, so that the slope of the produced spectrum is also directly observable.

Accelerated particles then propagate through the Galaxy where, due to the magnetic field, they move along complicated helical tracks. Therefore, the direction from which a particle arrives at Earth cannot be identified with the direction to its source of origin. The magnetic field is also the reason why particles do not leave the Milky Way along a straight path, but instead are stored for a long time ($\sim 10^7$ yr) before they eventually diffuse out, an effect also called confinement.

The sources of the particles with energy between $\sim 10^{15}$ eV and $\sim 10^{18}$ eV are likewise presumed to be located inside our Milky Way, because the magnetic field is sufficiently strong to confine them in the Galaxy. However, SN remnants are not likely sources for particles at these energies; in fact, the origin of these rays is largely unknown. Particles with energies larger than $\sim 10^{18}$ eV are probably of extragalactic origin. The radius of the helical tracks in the magnetic field of the Galaxy, i.e., their Larmor radius, is larger than the radius of the Milky Way itself, so they cannot be confined. Their origin is also unknown, but AGNs are the most probable source of these particles. Finally, one of the largest puzzles of high-energy astrophysics is the origin of cosmic rays with $E \gtrsim 10^{19}$ eV. The energy of these particles is so large that they are able to interact with the cosmic microwave background to produce pions and other particles, losing much of their energy in this process. These particles cannot propagate much further than ~ 10 Mpc through the Universe before they lose most of their energy. This implies that their acceleration sites should be located in the close vicinity of the Milky Way. Since the curvature of the orbits of such highly energetic particles is very small, it should, in principle, be possible to identify their origin: there are not many AGNs within 10 Mpc that are promising candidates for the origin of these ultra-high-energy cosmic rays. However, the observed number of these particles

[8] These energies should be compared with those reached in particle accelerators: LEP at CERN reached $\sim 100\,\text{GeV} = 10^{11}$ eV. Hence, cosmic accelerators are much more efficient than man-made machines.

[9] Shock fronts are surfaces in a gas flow where the parameters of state for the gas, such as pressure, density, and temperature, change discontinuously. The standard example for a shock front is the bang in an explosion, where a spherical shock wave propagates outwards from the point of explosion. Another example is the sonic boom caused, for example, by airplanes that move at a speed exceeding the velocity of sound. Such shock fronts are solutions of the hydrodynamic equations. They occur frequently in astrophysics, e.g., in explosion phenomena such as supernovae or in rapid (i.e., supersonic) flows such as those we will discuss in the context of AGNs.

is so small that no reliable information on these sources has thus far been obtained.

Energy Density. It is interesting to realize that the energy densities of cosmic rays, the magnetic field, the turbulent energy of the ISM, and the electromagnetic radiation of the stars are about the same – as if an equilibrium between these different components has been established. Since these components interact with each other – e.g., the turbulent motions of the ISM can amplify the magnetic field, and vice versa, the magnetic field affects the velocity of the ISM and of cosmic rays – it is not improbable that these interaction processes can establish an equipartition of the energy densities.

Gamma Radiation from the Milky Way. The Milky Way emits γ-radiation, as can be seen in Fig. 1.5. There is diffuse γ-ray emission which can be traced back to the cosmic rays in the Galaxy. When these energetic particles collide with nuclei in the interstellar medium, radiation is released. This gives rise to a continuum radiation which closely follows a power-law spectrum, such that the observed flux S_ν is $\propto \nu^{-\alpha}$, with $\alpha \sim 2$. The quantitative analysis of the distribution of this emission provides the most important information about the spatial distribution of cosmic rays in the Milky Way.

Gamma-Ray Lines. In addition to the continuum radiation, one also observes line radiation in γ-rays, at energies below ~ 10 MeV. The first detected and most prominent line has an energy of 1.809 MeV and corresponds to a radioactive decay of the Al^{26} nucleus. The spatial distribution of this emission is strongly concentrated towards the Galactic disk and thus follows the young stellar population in the Milky Way. Since the lifetime of the Al^{26} nucleus is short ($\sim 10^6$ yr), it must be produced near the emission site, which then implies that it is produced by the young stellar population. It is formed in hot stars and released to the interstellar medium either through stellar winds or core-collapse supernovae. Gamma lines from other radioactive nuclei have been detected as well.

Annihilation Radiation from the Galaxy. Furthermore, line radiation with an energy of 511 keV has been detected in the Galaxy. This line is produced when an electron and a positron annihilate into two photons, each with an energy corresponding to the rest-mass energy of an electron, i.e., 511 keV.[10] This annihilation radiation was identified first in the 1970s. With the Integral satellite, its emission morphology has been mapped with an angular resolution of $\sim 3°$. The 511 keV line emission is detected both from the Galactic disk and the bulge. The angular resolution is not sufficient to tell whether the annihilation line traces the young stellar population (i.e., the thin disk) or the older population in the thick disk. However, one can compare the distribution of the annihilation radiation with that of Al^{26} and other radioactive species. In about 85% of all decays Al^{26} emits a positron. If this positron annihilates close to its production site one can predict the expected annihilation radiation from the distribution of the 1.809 MeV line. In fact, the intensity and angular distribution of the 511 keV line from the disk is compatible with this scenario for the generation of positrons.

The origin of the annihilation radiation from the bulge, which has a luminosity larger than that from the disk by a factor ~ 5, is unknown. One needs to find a plausible source for the production of positrons in the bulge. There is no unique answer to this problem at present, but Type Ia supernovae and energetic processes near low-mass X-ray binaries are prime candidates for this source.

2.3.5 The Galactic Bulge

The Galactic bulge is the central thickening of our Galaxy. Figure 1.2 shows another spiral galaxy from its side, with its bulge clearly visible. The characteristic scale-length of the bulge is ~ 1 kpc. Owing to the strong extinction in the disk, the bulge is best observed in the IR, for instance with the IRAS and COBE satellites. The extinction to the Galactic Center in the visual is $A_V \sim 28$ mag. However, some lines-of-sight close to the Galactic center exist where A_V is significantly smaller, so that observations in optical and near IR light are possible, e.g., in Baade's window, located about 4° below the Galactic center at $\ell \sim 1°$, for which $A_V \sim 2$ mag (also see Sect. 2.6).

From the observations by COBE, and also from Galactic microlensing experiments (see Sect. 2.5), we know

[10] In addition to the two-photon annihilation, there is also an annihilation channel in which three photons are produced; the corresponding radiation forms a continuum spectrum, i.e., no spectral lines.

that our bulge has the shape of a bar, with the major axis pointing away from us by about 30°. The scale-height of the bulge is ~ 400 pc, with an axis ratio of ~ 0.6.

As is the case for the exponential profiles that describe the light distribution in the disk, the functional form of the brightness distribution in the bulge is also suggested from observations of other spiral galaxies. The profiles of their bulges, observed from the outside, are much better determined than in our Galaxy where we are located amid its stars.

The de Vaucouleurs Profile. The brightness profile of our bulge can be approximated by the de Vaucouleurs law which describes the surface brightness I as a function of the distance R from the center,

$$\boxed{\log\left(\frac{I(R)}{I_e}\right) = -3.3307\left[\left(\frac{R}{R_e}\right)^{1/4} - 1\right]}\,, \tag{2.39}$$

with $I(R)$ being the measured surface brightness, e.g., in $[I] = L_\odot/\mathrm{pc}^2$. R_e is the effective radius, defined such that half of the luminosity is emitted from within R_e,

$$\int_0^{R_e} \mathrm{d}R\, R\, I(R) = \frac{1}{2}\int_0^{\infty} \mathrm{d}R\, R\, I(R)\,. \tag{2.40}$$

This definition of R_e also leads to the numerical factor on the right-hand side of (2.39). As one can easily see from (2.39), $I_e = I(R_e)$ is the surface brightness at the effective radius. An alternative form of the de Vaucouleurs law is

$$\boxed{I(R) = I_e \exp\left(-7.669\left[(R/R_e)^{1/4} - 1\right]\right)}\,. \tag{2.41}$$

Because of its mathematical form, it is also called an $r^{1/4}$ law. The $r^{1/4}$ law falls off significantly more slowly than an exponential law for large R. For the Galactic bulge, one finds an effective radius of $R_e \approx 0.7$ kpc. With the de Vaucouleurs profile, a relation between luminosity, effective radius, and surface brightness is obtained by integrating over the surface brightness,

$$\boxed{L = \int_0^{\infty} \mathrm{d}R\, 2\pi R\, I(R) = 7.215\pi I_e\, R_e^2}\,. \tag{2.42}$$

Stellar Age Distribution in the Bulge. The stars in the bulge cover a large range in metallicity, $-1 \lesssim [\mathrm{Fe/H}] \lesssim +1$, with a mean of about 0.3, i.e., the mean metallicity is about twice that of the Sun. This high metallicity hints at a contribution by a rather young population, whereas the color of the bulge stars points towards a predominantly old stellar population. The bulge also contains about $10^8 M_\odot$ in molecular gas. On the other hand, one finds very metal-poor RR Lyrae stars, i.e., old stars. However, the distinction in membership between bulge and disk stars is not easy, so it is possible that the young component may actually be part of the inner disk.

The mass of the bulge is about $M_{\mathrm{bulge}} \sim 10^{10} M_\odot$ and its luminosity is $L_{B,\mathrm{bulge}} \sim 3 \times 10^9 L_\odot$, which results in a mass-to-light ratio of

$$\boxed{\frac{M}{L} \approx 3\frac{M_\odot}{L_\odot} \quad \text{in the bulge}}\,, \tag{2.43}$$

very similar to that of the thin disk.

2.3.6 The Visible Halo

The visible halo of our Galaxy consists of about 150 *globular clusters* and field stars with a high velocity component perpendicular to the Galactic plane. A globular cluster is a collection of typically several hundred thousand stars, contained within a spherical region of radius ~ 20 pc. The stars in the cluster are gravitationally bound and orbit in the common gravitational field. The old globular clusters with $[\mathrm{Fe/H}] < -0.8$ have an approximately spherical distribution around the Galactic center. A second population of globular clusters exists that contains younger stars with a higher metallicity, $[\mathrm{Fe/H}] > -0.8$. They have a more oblate geometrical distribution and are possibly part of the thick disk because they show roughly the same scale-height.

Most globular clusters are at a distance of $r \lesssim 35$ kpc (with $r = \sqrt{R^2 + z^2}$) from the Galactic center, but some are also found at $r > 60$ kpc. At these distances it is hard to judge whether these objects are part of the Galaxy or whether they have been captured from a neighboring galaxy, such as the Magellanic Clouds. Also, field stars have been found at distances out to $r \sim 50$ kpc, which is the reason why one assumes a characteristic value of $r_{\mathrm{halo}} \sim 50$ kpc for the extent of the visible halo.

The *density distribution* of metal-poor globular clusters and field stars in the halo is described by

$$n(r) \propto r^{-3.5} \; . \tag{2.44}$$

Alternatively, one can fit a de Vaucouleurs profile to the density distribution, which results in an effective radius of $r_e \sim 2.7$ kpc.

At large distances from the disk, neutral hydrogen is also found, in the form of clouds. Most of these clouds, visible in 21-cm line emission, have a negative radial velocity, i.e., they are moving towards us, with velocities of up to $v_r \sim -400$ km/s. These *high-velocity clouds* (HVCs) cannot be following the general Galactic rotation. We have virtually no means of determining the distances of these clouds, and thus their origin and nature are still subject to discussion. There is one exception, however: the Magellanic Stream is a narrow band of HI emission which follows the Magellanic Clouds along their orbit around the Galaxy (also see Fig. 6.6). This gas stream may be the result of a close encounter of the Magellanic Clouds with the Milky Way in the past. The (tidal) gravitational force that the Milky Way had imposed on our neighboring galaxies in such an encounter could strip away part of the interstellar gas from them.

2.3.7 The Distance to the Galactic Center

As already mentioned, our distance from the Galactic center is rather difficult to measure and thus not very precisely known. The general problem with such a measurement is the high extinction in the disk, prohibiting measurements of the distance of individual stars close to the Galactic center. Thus, one has to rely on more indirect methods, and the most important ones will be outlined here.

The visible halo of our Milky Way is populated by globular clusters and also by field stars. They have a spherical, or, more generally, a spheroidal distribution. The center of this distribution is obviously the center of gravity of the Milky Way, around which the halo objects are moving. If one measures the three-dimensional distribution of the halo population, the geometrical center of this distribution should correspond to the Galactic center.

This method can indeed be applied because, due to their extended distribution, halo objects can be observed at relatively large Galactic latitudes where they are not too strongly affected by extinction. As was discussed in Sect. 2.2, the distance determination of globular clusters is possible using photometric methods. On the other hand, one also finds RR Lyrae stars in globular clusters to which the period–luminosity relation can be applied. Therefore, the spatial distribution of the globular clusters can be determined. However, at about 150, the number of known globular clusters is relatively small, resulting in a fairly large statistical error for the determination of the common center. Much more numerous are the RR Lyrae field stars in the halo, for which distances can be measured using the period–luminosity relation. The statistical error in determining the center of their distribution is therefore much smaller. On the other hand, this distance to the Galactic center is based only on the calibration of the period–luminosity relation, and any uncertainty in this will propagate into a systematic error on R_0. Effects of the extinction add to this. However, such effects can be minimized by observing the RR Lyrae stars in the NIR, which in addition benefits from the narrower luminosity distribution of RR Lyrae stars in this wavelength regime. These analyses yield a value of $R_0 \approx 8.0$ kpc (see Fig. 2.12).

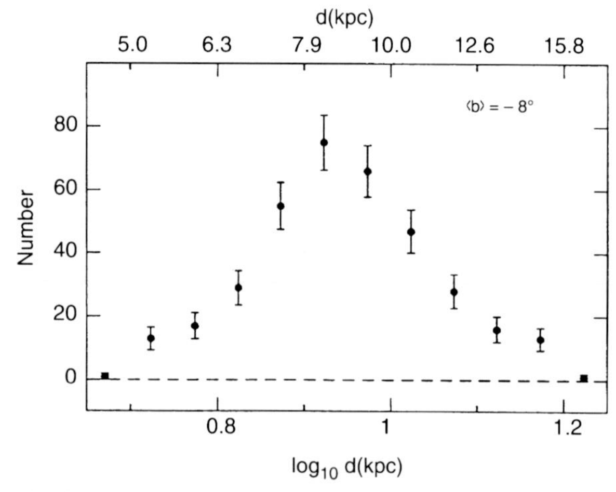

Fig. 2.12. The number of RR Lyrae stars as a function of distance, measured in a direction that closely passes the Galactic center, at $\ell = 0°$ and $b = -8°$. If we assume a spherically symmetric distribution of the RR Lyrae stars, concentrated towards the center, the distance to the Galactic center can be identified with the maximum of this distribution

2.4 Kinematics of the Galaxy

Unlike a solid body, the Galaxy rotates differentially. This means that the angular velocity is a function of the distance R from the Galactic center. Seen from above, i.e., from the NGP, the rotation is clockwise. To describe the velocity field quantitatively we will in the following introduce velocity components in the coordinate system (R, θ, z), as shown in Fig. 2.13. An object following a track $[R(t), \theta(t), z(t)]$ then has the velocity components

$$U := \frac{\mathrm{d}R}{\mathrm{d}t}, \quad V := R\frac{\mathrm{d}\theta}{\mathrm{d}t}, \quad W := \frac{\mathrm{d}z}{\mathrm{d}t}. \tag{2.45}$$

For example, the Sun is not moving on a simple circular orbit around the Galactic center, but currently inwards, $U < 0$, and with $W > 0$, so that it is moving away from the Galactic plane.

In this section we will examine the rotation of the Milky Way. We start with the determination of the velocity components of the Sun. Then we will consider the rotation curve of the Galaxy, which describes the rotational velocity $V(R)$ as a function of the distance R from the Galactic center. We will find the intriguing result that the velocity V does not decline towards large distances, but that it virtually remains constant. Because this result is of extraordinary importance, we will discuss the methods needed to derive it in some detail.

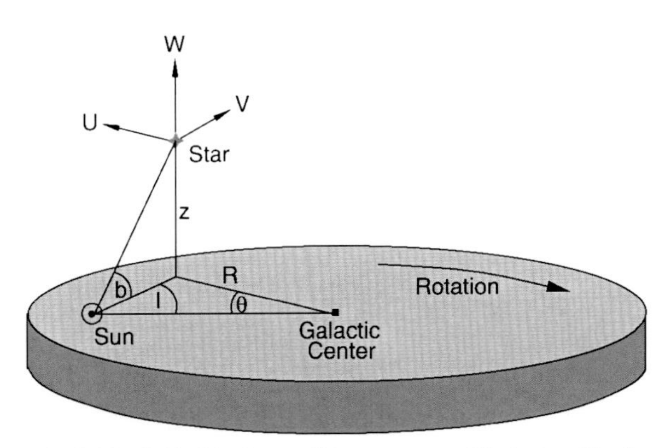

Fig. 2.13. Cylindrical coordinate system (R, θ, z) with the Galactic center at its origin. Note that θ increases in the clockwise direction if the disk is viewed from above. The corresponding velocity components (U, V, W) of a star are indicated

2.4.1 Determination of the Velocity of the Sun

Local Standard of Rest. To link local measurements to the Galactic coordinate system (R, θ, z), the *local standard of rest* is defined. It is a fictitious rest-frame in which velocities are measured. For this purpose, we consider a point that is located today at the position of the Sun and that moves along a perfectly circular orbit in the plane of the Galactic disk. The velocity components in the LSR are then by definition,

$$\boxed{U_{\mathrm{LSR}} \equiv 0\,, \quad V_{\mathrm{LSR}} \equiv V_0\,, \quad W_{\mathrm{LSR}} \equiv 0}\,, \tag{2.46}$$

with $V_0 \equiv V(R_0)$ being the orbital velocity at the location of the Sun. Although the LSR changes over time, the time-scale of this change is so large (the orbital period is $\sim 230 \times 10^6$ yr) that this effect is negligible.

Peculiar Velocity. The velocity of an object relative to the LSR is called its peculiar velocity. It is denoted by $\boldsymbol{v}$, and its components are given as

$$\boxed{\begin{aligned} \boldsymbol{v} \equiv (u, v, w) &= (U - U_{\mathrm{LSR}}, V - V_{\mathrm{LSR}}, W - W_{\mathrm{LSR}}) \\ &= (U, V - V_0, W) \end{aligned}}\,. \tag{2.47}$$

The velocity of the Sun relative to the LSR is denoted by $\boldsymbol{v}_\odot$. If $\boldsymbol{v}_\odot$ is known, any velocity measured relative to the Sun can be converted into a velocity relative to the LSR: let $\Delta\boldsymbol{v}$ be the velocity of a star relative to the Sun, which is directly measurable using the methods discussed in Sect. 2.2, then the peculiar velocity of this star is

$$\boldsymbol{v} = \boldsymbol{v}_\odot + \Delta\boldsymbol{v}\,. \tag{2.48}$$

Peculiar Velocity of the Sun. We consider now an ensemble of stars in the immediate vicinity of the Sun, and assume the Galaxy to be axially symmetric and stationary. Under these assumptions, the number of stars that move outwards to larger radii R equals the number of stars moving inwards. Likewise, as many stars move upwards through the Galactic plane as downwards. If these conditions are not satisfied, the assumption of a stationary distribution would be violated. The mean values of the corresponding peculiar velocity components must therefore vanish,

$$\langle u \rangle = 0\,, \quad \langle w \rangle = 0\,, \tag{2.49}$$

where the brackets denote an average over the ensemble considered. The analog argument is not valid for the v component because the mean value of v depends on the distribution of the orbits: if only circular orbits in the disk existed, we would also have $\langle v \rangle = 0$ (this is trivial, since then all stars would have $v = 0$), but this is not the case. From a statistical consideration of the orbits in the framework of stellar dynamics, one deduces that $\langle v \rangle$ is closely linked to the radial velocity dispersion of the stars: the larger it is, the more $\langle v \rangle$ deviates from zero. One finds that

$$\langle v \rangle = -C \left\langle u^2 \right\rangle , \tag{2.50}$$

where C is a positive constant that depends on the density distribution and on the local velocity distribution of the stars. The sign in (2.50) follows from noting that a circular orbit has a higher tangential velocity than elliptical orbits, which in addition have a non-zero radial component. Equation (2.50) expresses the fact that the mean rotational velocity of a stellar population around the Galactic center deviates from the corresponding circular orbit velocity, and that the deviation is stronger for a larger radial velocity dispersion. This phenomenon is also known as asymmetric drift. From the mean of (2.48) over the ensemble considered and by using (2.49) and (2.50), one obtains

$$\boxed{\boldsymbol{v}_\odot = \left(- \langle \Delta u \rangle \, , -C \left\langle u^2 \right\rangle - \langle \Delta v \rangle \, , - \langle \Delta w \rangle \right)} \; . \tag{2.51}$$

One still needs to determine the constant C in order to make use of this relation. This is done by considering different stellar populations and measuring $\left\langle u^2 \right\rangle$ and $\langle \Delta v \rangle$ separately for each of them. If these two quantities are then plotted in a diagram (see Fig. 2.14), a linear relation is obtained, as expected from (2.50). The slope C can be determined directly from this diagram. Furthermore, from the intersection with the $\langle \Delta v \rangle$-axis, $v_\odot$ is readily read off. The other velocity components in (2.51) follow by simply averaging, yielding the result:

$$\boxed{\boldsymbol{v}_\odot = (-10, 5, 7)\,\text{km/s}} \; . \tag{2.52}$$

Hence, the Sun is currently moving inwards, upwards, and faster than it would on a circular orbit at its location. We have therefore determined $\boldsymbol{v}_\odot$, so we are now able to analyze any measured stellar velocities relative to the LSR. However, we have not yet discussed how V_0, the rotational velocity of the LSR itself, is determined.

Velocity Dispersion of Stars. The dispersion of the stellar velocities relative to the LSR can now be determined, i.e., the mean square deviation of their velocities from the velocity of the LSR. For young stars (A stars, for example), this dispersion happens to be small. For older K giants it is larger, and is larger still for old, metal-poor red dwarf stars. We observe a very well-defined velocity-metallicity relation. When this is combined with the age–metallicity relation it appears that the oldest stars have the highest peculiar velocities. This effect is observed in all three coordinates. This result is in agreement with the relation between the age of a stellar population and its scale-height (discussed in Sect. 2.3.1), the latter being linked to the velocity dispersion via σ_z.

Asymmetric Drift. If one considers high-velocity stars, only a few are found that have $v > 65$ km/s and which are thus moving much faster around the Galactic center than the LSR. However, quite a few stars are found that have $v < -250$ km/s, so their orbital velocity is opposite to the direction of rotation of the LSR. Plotted in a $(u - v)$-diagram, a distribution is found which is narrowly concentrated around $u = 0\,\text{km/s} = v$ for young stars, as already mentioned above, and which gets increasingly wider for older stars. For the oldest stars,

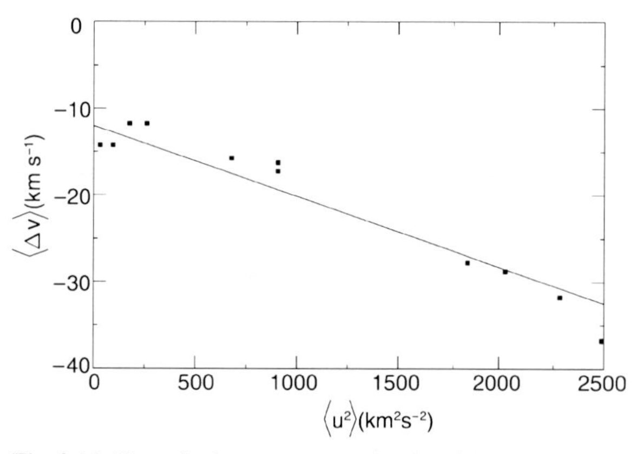

Fig. 2.14. The velocity components $\langle \Delta v \rangle = \langle v \rangle - v_\odot$ are plotted against $\left\langle u^2 \right\rangle$ for stars in the Solar neighborhood. Because of the linear relation, $v_\odot$ can be read off from the intersection with the x-axis, and C from the slope

which belong to the halo population, one obtains a circular envelope with its center located at $u = 0\,\mathrm{km/s}$ and $v \approx -220\,\mathrm{km/s}$ (see Fig. 2.15). If we assume that the Galactic halo, to which these high-velocity stars belong, does not rotate (or only very slowly), this asymmetry in the v-distribution can only be caused by the rotation of the LSR. The center of the envelope then has to be at $-V_0$. This yields the orbital velocity of the LSR

$$\boxed{V_0 \equiv V(R_0) = 220\,\mathrm{km/s}}\ . \tag{2.53}$$

Knowing this velocity, we can then compute the mass of the Galaxy inside the Solar orbit. A circular orbit is characterized by an equilibrium between centrifugal and gravitational acceleration, $V^2/R = GM(< R)/R^2$, so that

$$\boxed{M(< R_0) = \frac{V_0^2 R_0}{G} = 8.8 \times 10^{10} M_\odot}\ . \tag{2.54}$$

Furthermore, for the orbital period of the LSR, which is similar to that of the Sun, one obtains

$$\boxed{P = \frac{2\pi R_0}{V_0} = 230 \times 10^6\ \mathrm{yr}}\ . \tag{2.55}$$

Hence, during the lifetime of the Solar System, estimated to be $\sim 4.6 \times 10^9$ yr, it has completed about 20 orbits around the Galactic center.

2.4.2 The Rotation Curve of the Galaxy

From observations of the velocity of stars or gas around the Galactic center, the rotational velocity V can be determined as a function of the distance R from the Galactic center. In this section, we will describe methods to determine this *rotation curve* and discuss the result.

We consider an object at distance R from the Galactic center which moves along a circular orbit in the Galactic plane, has a distance D from the Sun, and is located at a Galactic longitude ℓ (see Fig. 2.16). In a Cartesian coordinate system with the Galactic center at the origin, the positional and velocity vectors (we only consider the two components in the Galactic plane because we assume a motion in the plane) are given by

$$\boldsymbol{r} = R\begin{pmatrix}\sin\theta\\ \cos\theta\end{pmatrix}, \quad \boldsymbol{V} = \dot{\boldsymbol{r}} = V(R)\begin{pmatrix}\cos\theta\\ -\sin\theta\end{pmatrix},$$

where θ denotes the angle between the Sun and the object as seen from the Galactic center. From the geometry shown in Fig. 2.16 it follows that

$$\boldsymbol{r} = \begin{pmatrix}D\sin\ell\\ R_0 - D\cos\ell\end{pmatrix}.$$

If we now identify the two expressions for the components of $\boldsymbol{r}$, we obtain

$$\sin\theta = (D/R)\sin\ell\,,$$
$$\cos\theta = (R_0/R) - (D/R)\cos\ell\,.$$

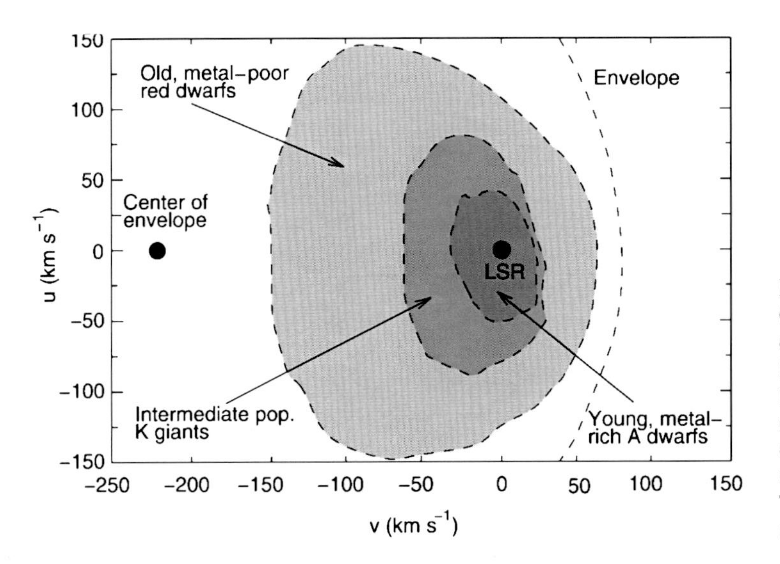

Fig. 2.15. The motion of the Sun around the Galactic center is reflected in the asymmetric drift: while young stars in the Solar vicinity have velocities very similar to the Solar velocity, i.e., small relative velocities, members of other populations (and of other Milky Way components) have different velocities – e.g., for halo objects $v = -220\,\mathrm{km/s}$ on average. Thus, different velocity ellipses show up in a $(u - v)$-diagram

If we disregard the difference between the velocities of the Sun and the LSR we get $\boldsymbol{V}_\odot \approx \boldsymbol{V}_{\rm LSR} = (V_0, 0)$ in this coordinate system. Thus the relative velocity between the object and the Sun is, in Cartesian coordinates,

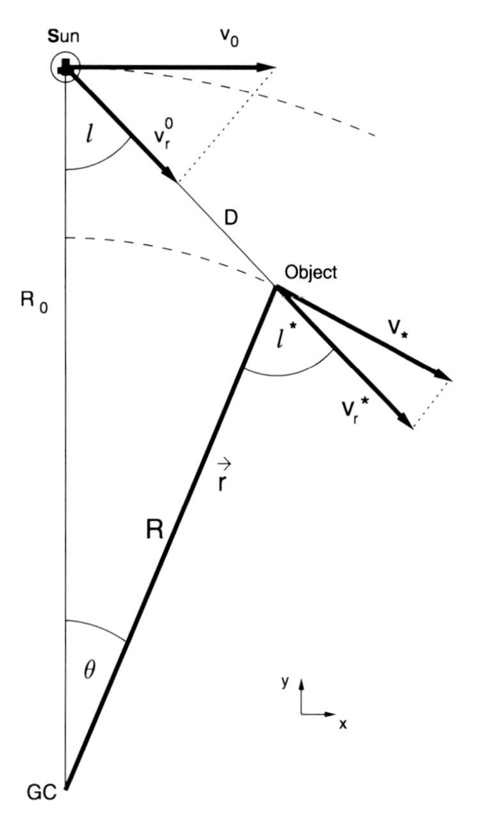

Fig. 2.16. Geometric derivation of the formalism of differential rotation:

$$\begin{aligned} v_{\rm r} &= v_{\rm r}^* - v_{\rm r}^\odot = v_* \sin \ell^* - v_\odot \sin \ell \,, \\ v_{\rm t} &= v_{\rm t}^* - v_{\rm t}^\odot = v_* \cos \ell^* - v_\odot \cos \ell \,. \end{aligned}$$

One has:

$$\begin{aligned} R \sin\theta &= D \sin\ell \,, \\ R\cos\theta + D\cos\ell &= R_0 \,, \end{aligned}$$

which implies

$$\begin{aligned} v_{\rm r} &= R_0 \left(\frac{v_*}{R} - \frac{v_\odot}{R_0} \right) \sin\ell \\ &= (\Omega - \Omega_0) R_0 \sin\ell \,, \\ v_{\rm t} &= R_0 \left(\frac{v_*}{R} - \frac{v_\odot}{R_0} \right) \cos\ell - D\frac{v_*}{R} \\ &= (\Omega - \Omega_0) R_0 \cos\ell - \Omega D \,. \end{aligned}$$

$$\begin{aligned} \Delta \boldsymbol{V} &= \boldsymbol{V} - \boldsymbol{V}_\odot \\ &= \begin{pmatrix} V\,(R_0/R) - V\,(D/R)\cos\ell - V_0 \\ -V\,(D/R)\sin\ell \end{pmatrix} . \end{aligned}$$

With the angular velocity defined as

$$\Omega(R) = \frac{V(R)}{R} \,, \tag{2.56}$$

we obtain for the relative velocity

$$\Delta \boldsymbol{V} = \begin{pmatrix} R_0(\Omega - \Omega_0) - \Omega\, D \cos\ell \\ -D\,\Omega\, \sin\ell \end{pmatrix} ,$$

where $\Omega_0 = V_0/R_0$ is the angular velocity of the Sun. The radial and tangential velocities of this relative motion then follow by projection of $\Delta \boldsymbol{V}$ along the direction parallel or perpendicular, respectively, to the separation vector,

$$\boxed{v_{\rm r} = \Delta \boldsymbol{V} \cdot \begin{pmatrix} \sin\ell \\ -\cos\ell \end{pmatrix} = (\Omega - \Omega_0) R_0 \sin\ell} \,, \tag{2.57}$$

$$\boxed{v_{\rm t} = \Delta \boldsymbol{V} \cdot \begin{pmatrix} \cos\ell \\ \sin\ell \end{pmatrix} = (\Omega - \Omega_0) R_0 \cos\ell - \Omega\, D} \,. \tag{2.58}$$

A purely geometric derivation of these relations is given in Fig. 2.16.

Rotation Curve near R_0; Oort Constants. Using (2.57) one can derive the angular velocity by means of measuring $v_{\rm r}$, but not the radius R to which it corresponds. Therefore, by measuring the radial velocity alone $\Omega(R)$ cannot be determined. If one measures $v_{\rm r}$ and, in addition, the proper motion $\mu = v_{\rm t}/D$ of stars, then Ω and D can be determined from the equations above, and from D and ℓ one obtains $R = \sqrt{R_0^2 + D^2 - 2R_0 D \cos\ell}$. The effects of extinction prohibits the use of this method for large distances D, since we have considered objects in the Galactic disk. For small distances $D \ll R_0$, which implies $|R - R_0| \ll R_0$, we can make a local approximation by evaluating the expressions above only up to first order in $(R - R_0)/R_0$. In this linear approximation we get

$$\Omega - \Omega_0 \approx \left(\frac{{\rm d}\Omega}{{\rm d}R} \right)_{|R_0} (R - R_0) \,, \tag{2.59}$$

where the derivative has to be evaluated at $R = R_0$. Hence

$$v_r = (R - R_0) \left(\frac{d\Omega}{dR}\right)_{|R_0} R_0 \sin \ell ,$$

and furthermore, with (2.56),

$$R_0 \left(\frac{d\Omega}{dR}\right)_{|R_0} \doteq \frac{R_0}{R} \left[\left(\frac{dV}{dR}\right)_{|R_0} - \frac{V}{R}\right] \approx \left(\frac{dV}{dR}\right)_{|R_0} - \frac{V_0}{R_0} ,$$

in zeroth order in $(R - R_0)/R_0$. Combining the last two equations yields

$$v_r = \left[\left(\frac{dV}{dR}\right)_{|R_0} - \frac{V_0}{R_0}\right] (R - R_0) \sin \ell ; \quad (2.60)$$

in analogy to this, we obtain for the tangential velocity

$$v_t = \left[\left(\frac{dV}{dR}\right)_{|R_0} - \frac{V_0}{R_0}\right] (R - R_0) \cos \ell - \Omega_0 D . \quad (2.61)$$

For $|R - R_0| \ll R_0$ it follows that $R_0 - R \approx D \cos \ell$; if we insert this into (2.60) and (2.61) we get

$$v_r \approx A D \sin 2\ell , \quad v_t \approx A D \cos 2\ell + B D , \quad (2.62)$$

where A and B are the *Oort constants*

$$A := -\frac{1}{2}\left[\left(\frac{dV}{dR}\right)_{|R_0} - \frac{V_0}{R_0}\right] ,$$
$$B := -\frac{1}{2}\left[\left(\frac{dV}{dR}\right)_{|R_0} + \frac{V_0}{R_0}\right] . \quad (2.63)$$

The radial and tangential velocity fields relative to the Sun show a sine curve with period π, where v_t and v_r are phase-shifted by $\pi/4$. This behavior of the velocity field in the Solar neighborhood is indeed observed (see Fig. 2.17). By fitting the data for $v_r(\ell)$ and $v_t(\ell)$ for stars of equal distance D one can determine A and B, and thus

$$\Omega_0 = \frac{V_0}{R_0} = A - B, \quad \left(\frac{dV}{dR}\right)_{|R_0} = -(A + B) . \quad (2.64)$$

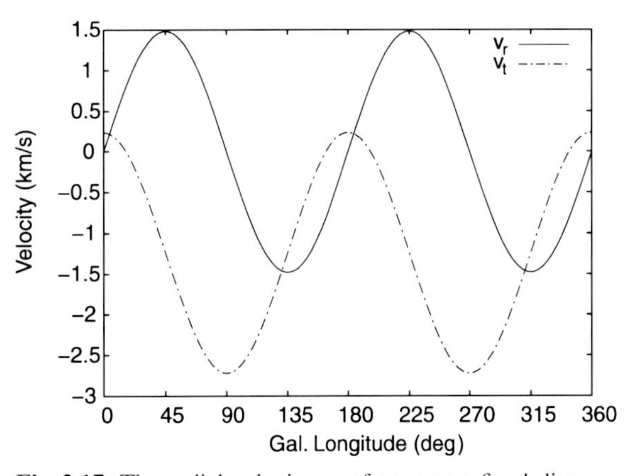

Fig. 2.17. The radial velocity v_r of stars at a fixed distance D is proportional to $\sin 2\ell$; the tangential velocity v_t is a linear function of $\cos 2\ell$. From the amplitude of the oscillating curves and from the mean value of v_t the Oort constants A and B can be derived, respectively (see (2.62))

The Oort constants thus yield the angular velocity of the Solar orbit and its derivative, and therefore the local kinematical information. If our Galaxy was rotating rigidly so that Ω was independent of the radius, $A = 0$ would follow. But the Milky Way rotates differentially, i.e., the angular velocity depends on the radius. Measurements yield the following values for A and B,

$$A = (14.8 \pm 0.8) \text{ km s}^{-1} \text{kpc}^{-1} ,$$
$$B = (-12.4 \pm 0.6) \text{ km s}^{-1} \text{kpc}^{-1} . \quad (2.65)$$

Galactic Rotation Curve for $R < R_0$; Tangent Point Method. To measure the rotation curve for radii that are significantly smaller than R_0, one has to turn to large wavelengths due to extinction in the disk. Usually the 21-cm emission line of neutral hydrogen is used, which can be observed over large distances, or the emission of CO in molecular gas. These gas components are found throughout the disk and are strongly concentrated towards the plane. Furthermore, the radial velocity can easily be measured from the Doppler effect. However, since the distance to a hydrogen cloud cannot be determined directly, a method is needed to link the measured radial velocities to the distance of the gas from the Galactic center. For this purpose the *tangent point method* is used.

Consider a line-of-sight at fixed Galactic longitude ℓ, with $\cos\ell > 0$ (thus "inwards"). The radial velocity v_r along this line-of-sight for objects moving on circular orbits is a function of the distance D, according to (2.57). If $\Omega(R)$ is a monotonically decreasing function, v_r attains a maximum where the line-of-sight is tangent to the local orbit, and thus its distance R from the Galactic center attains the minimum value $R_{\rm min}$. This is the case at

$$D = R_0 \cos\ell, \quad R_{\rm min} = R_0 \sin\ell \tag{2.66}$$

(see Fig. 2.18). The maximum radial velocity there, according to (2.57), is

$$\begin{aligned} v_{\rm r,max} &= [\Omega(R_{\rm min}) - \Omega_0]\; R_0 \sin\ell \\ &= V(R_{\rm min}) - V_0 \sin\ell\,, \end{aligned} \tag{2.67}$$

so that from the measured value of $v_{\rm r,max}$ as a function of direction ℓ, the rotation curve inside R_0 can be determined,

$$\boxed{V(R) = \left(\frac{R}{R_0}\right) V_0 + v_{\rm r,max}(\sin\ell = R/R_0)}\,. \tag{2.68}$$

In the optical regime of the spectrum this method can only be applied locally, i.e., for small D, due to extinction. This is the case if one observes in a direction nearly tangential to the orbit of the Sun, i.e., if

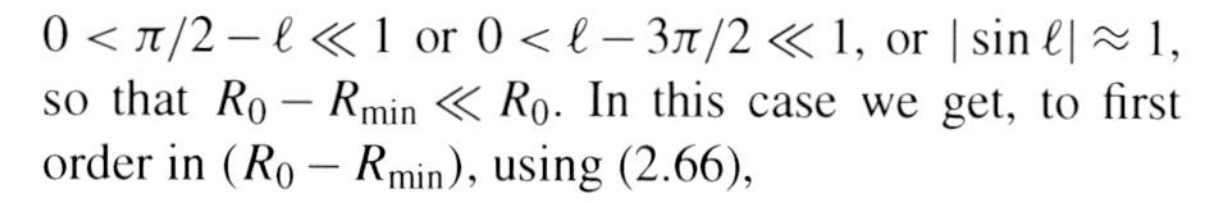
$0 < \pi/2 - \ell \ll 1$ or $0 < \ell - 3\pi/2 \ll 1$, or $|\sin\ell| \approx 1$, so that $R_0 - R_{\rm min} \ll R_0$. In this case we get, to first order in $(R_0 - R_{\rm min})$, using (2.66),

$$\begin{aligned} V(R_{\rm min}) &\approx V_0 + \left(\frac{{\rm d}V}{{\rm d}R}\right)_{|R_0} (R_{\rm min} - R_0) \\ &= V_0 - \left(\frac{{\rm d}V}{{\rm d}R}\right)_{|R_0} R_0\,(1 - \sin\ell)\,, \end{aligned} \tag{2.69}$$

so that with (2.67)

$$\boxed{\begin{aligned} v_{\rm r,max} &= \left[V_0 - \left(\frac{{\rm d}V}{{\rm d}R}\right)_{|R_0} R_0\right](1 - \sin\ell) \\ &= 2\,A\,R_0\,(1 - \sin\ell) \end{aligned}}\,, \tag{2.70}$$

where (2.63) was used in the last step. This relation can also be used for determining the Oort constant A.

To determine $V(R)$ for smaller R by employing the tangent point method, we have to observe in wavelength regimes in which the Galactic plane is transparent, using radio emission lines of gas. In Fig. 2.18, a typical intensity profile of the 21-cm line along a line-of-sight is sketched; according to the Doppler effect this can be converted directly into a velocity profile using $v_r = (\lambda - \lambda_0)/\lambda_0$. It consists of several maxima that originate in individual gas clouds. The radial velocity of each cloud is defined by its distance R from the Galactic

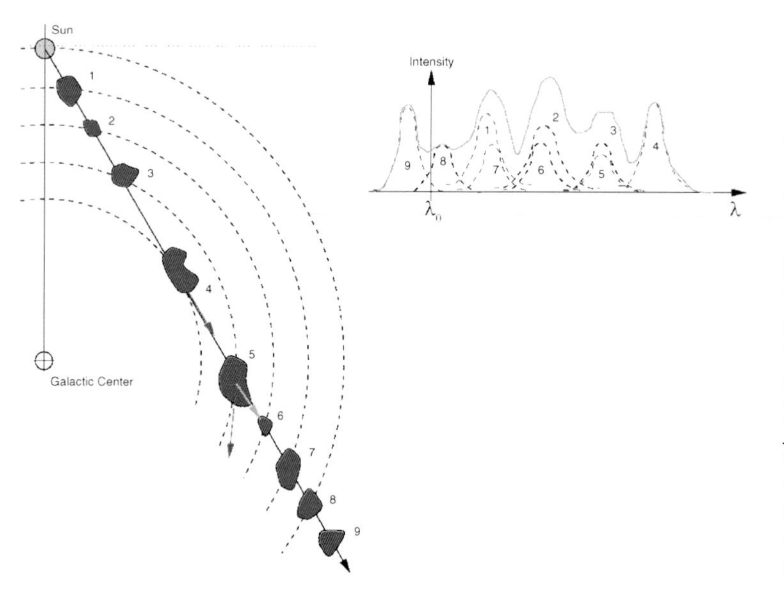

Fig. 2.18. The ISM is optically thin for 21-cm radiation, and thus we receive the 21-cm emission of HI regions from everywhere in the Galaxy. Due to the motion of an HI cloud relative to us, the wavelength is shifted. This can be used to measure the radial velocity of the cloud. With the assumption that the gas is moving on a circular orbit around the Galactic center, one expects that for the cloud in the tangent point (cloud 4), the full velocity is projected along the line-of-sight so that this cloud will therefore have the largest radial velocity. If the distance of the Sun to the Galactic center is known, the velocity of a cloud and its distance from the Galactic center can then be determined

center (if the gas follows the Galactic rotation), so that the largest radial velocity will occur for gas closest to the tangent point, which will be identified with $v_{r,\max}(\ell)$. Figure 2.19 shows the observed intensity profile of the ^{12}CO line as a function of the Galactic longitude, from which the rotation curve for $R < R_0$ can be read off.

> With the tangent point method, applied to the 21-cm line of neutral hydrogen or to radio emission lines of molecular gas, the rotation curve of the Galaxy inside the Solar orbit can be measured.

Rotation Curve for $R > R_0$. The tangent point method cannot be applied for $R > R_0$ because for lines-of-sight at $\pi/2 < \ell < 3\pi/2$, the radial velocity v_r attains no maximum. In this case, the line-of-sight is nowhere parallel to a circular orbit.

Measuring $V(R)$ for $R > R_0$ requires measuring v_r for objects whose distance can be determined directly, e.g., Cepheids, for which the period–luminosity relation (Sect. 2.2.7) is used, or O- and B-stars in HII regions. With ℓ and D known, R can then be calculated, and with (2.57) we obtain $\Omega(R)$ or $V(R)$, respectively. Any object with known D and v_r thus contributes one data point to the Galactic rotation curve. Since the distance estimates of individual objects are always affected by uncertainties, the rotation curve for large values of R is less accurately known than that inside the Solar circle.

It turns out that the rotation curve for $R > R_0$ does not decline outwards (see Fig. 2.20) as we would expect from the distribution of visible matter in the Milky Way. Both the stellar density and the gas density of the Galaxy decline exponentially for large R – e.g., see (2.34). This steep radial decline of the visible matter density should imply that $M(R)$, the mass inside R, is nearly constant for $R \gtrsim R_0$, from which a velocity profile like $V \propto R^{-1/2}$ would follow, according to Kepler's law. However, this is not the case: $V(R)$ is virtually constant for $R > R_0$, indicating that $M(R) \propto R$. Thus, to get a constant rotational velocity of the Galaxy much more matter has to be present than we observe in gas and stars.

> The Milky Way contains, besides stars and gas, an additional component of matter that dominates the mass at $R \gtrsim R_0$ but which has not yet been observed directly. Its presence is known only by its gravitational effect – hence, it is called dark matter.

In Sect. 3.3.3 we will see that this is a common phenomenon. The rotation curves of spiral galaxies are flat at large radii up to the maximum radius at which it can be measured; *spiral galaxies contain dark matter.*

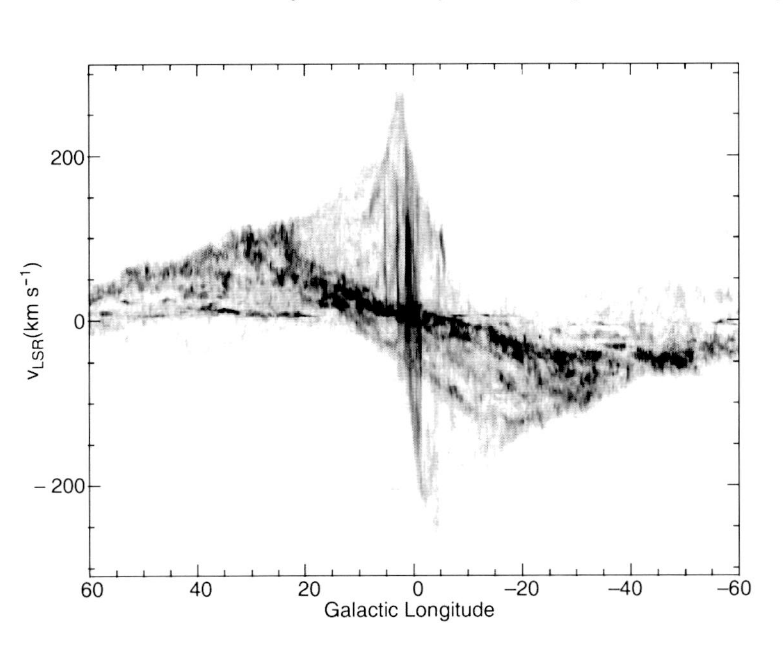

Fig. 2.19. ^{12}CO emission of molecular gas in the Galactic disk. For each ℓ, the intensity of the emission in the $\ell - v_r$ plane is plotted, integrated over the range $-2° \le b \le 2°$ (i.e., very close to the middle of the plane). Since v_r depends on the distance along each line-of-sight, characterized by ℓ, this diagram contains information on the rotation curve of the Galaxy as well as on the spatial distribution of the gas. The maximum velocity at each ℓ is rather well defined and forms the basis for the tangent point method

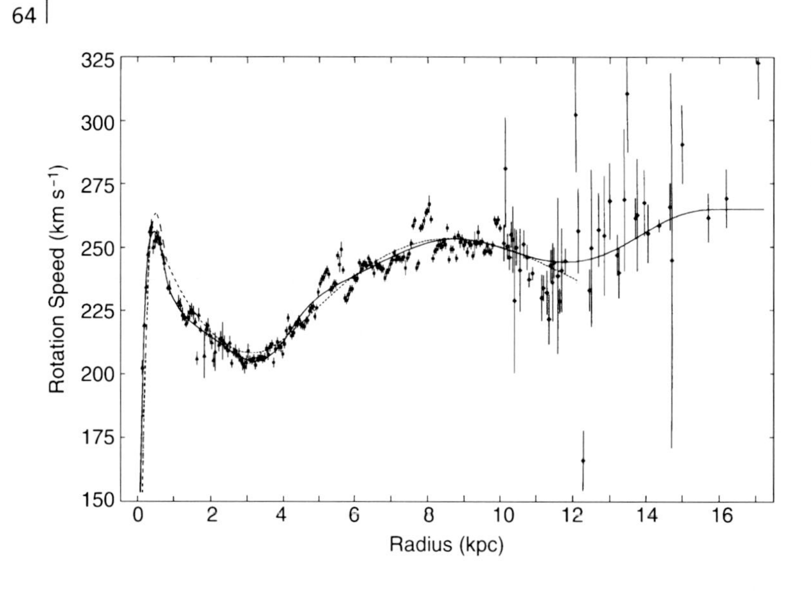

Fig. 2.20. Rotation curve of the Milky Way. Inside the "Solar circle", that is at $R < R_0$, the radial velocity is determined quite accurately using the tangent point method; the measurements outside have larger uncertainties

The nature of dark matter is thus far unknown; in principle, we can distinguish two totally different kinds of dark matter candidates:

- *Astrophysical dark matter*, consisting of compact objects – e.g., faint stars like white dwarfs, brown dwarfs, black holes, etc. Such objects were assigned the name MACHOs, which stands for "MAssive Compact Halo Objects".
- *Particle physics dark matter*, consisting of elementary particles which have thus far escaped detection in accelerator laboratories.

Although the origin of astrophysical dark matter would be difficult to understand (not least because of the baryon abundance in the Universe – see Sect. 4.4.4 – and because of the metal abundance in the ISM), a direct distinction between the two alternatives through observation would be of great interest. In the following section we will describe a method which is able to probe whether the dark matter in our Galaxy consists of MACHOs.

2.5 The Galactic Microlensing Effect: The Quest for Compact Dark Matter

In 1986, Bohdan Paczyński proposed to test the possible presence of MACHOs by performing microlensing experiments. As we will soon see, this was a daring idea at that time, but since then such experiments have been carried out. In this section we will mainly summarize and discuss the results of these searches for MACHOs. We will start with a description of the microlensing effect and then proceed with its specific application to the search for MACHOs.

2.5.1 The Gravitational Lensing Effect I

Einstein's Deflection Angle. *Light, just like massive particles, is deflected in a gravitational field.* This is one of the specific predictions by Einstein's theory of gravity, General Relativity. Quantitatively it predicts that a light beam which passes a point mass M at a distance ξ is deflected by an angle $\hat{\alpha}$, which amounts to

$$\boxed{\hat{\alpha} = \frac{4\,G\,M}{c^2\,\xi}} \; . \tag{2.71}$$

The deflection law (2.71) is valid as long as $\hat{\alpha} \ll 1$, which is the case for weak gravitational fields. If we now set $M = M_\odot$, $R = R_\odot$ in the foregoing equation, we obtain

$$\boxed{\hat{\alpha}_\odot \approx 1\overset{''}{.}74}$$

for the light deflection at the limb of the Sun. This deflection of light was measured during a Solar eclipse in 1919 from the shift of the apparent positions of stars close to the shaded Solar disk. Its agreement with the value predicted by Einstein made him world-famous over night, because this was the first real and challenging test of General Relativity. Although the precision of the measured value back then was only $\sim 30\%$, it was sufficient to confirm Einstein's theory. By now the law (2.71) has been measured in the Solar System with a precision of about 0.1%, and Einstein's prediction has been confirmed.

Not long after the discovery of gravitational light deflection at the Sun, the following scenario was considered. If the deflection of the light were sufficiently strong, light from a very distant source could be visible at two positions in the sky: one light ray could pass a mass concentration, located between us and the source, "to the right", and the second one "to the left", as sketched in Fig. 2.21. The astrophysical consequence of this gravitational light deflection is also called the *gravitational lens effect*. We will discuss various aspects of the lens effect in the course of this book, and we will review its astrophysical applications.

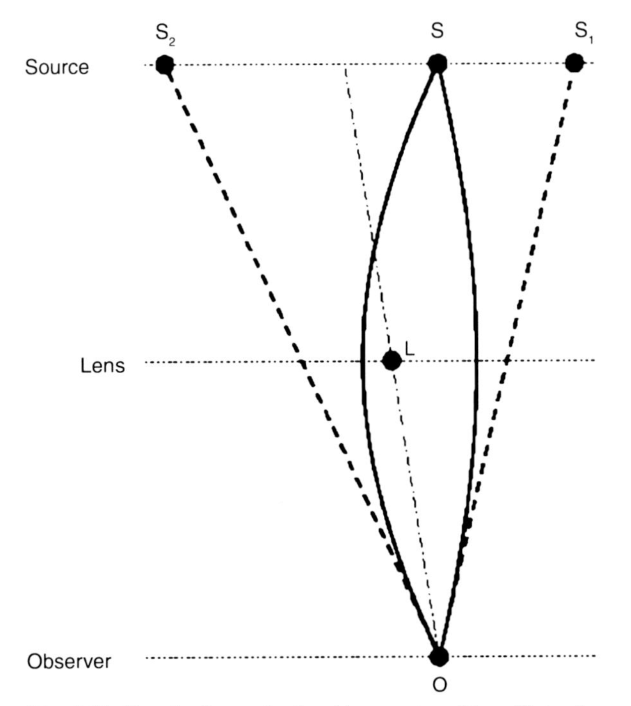

Fig. 2.21. Sketch of a gravitational lens system. If a sufficiently massive mass concentration is located between us and a distant source, it may happen that we observe this source at two different positions on the sphere

The Sun is not able to cause multiple images of distant sources. The maximum deflection angle $\hat{\alpha}_\odot$ is much smaller than the angular radius of the Sun, so that two beams of light that pass the Sun to the left and to the right cannot converge by light deflection at the position of the Earth. Given its radius, the Sun is too close to produce multiple images, since its angular radius is (far) larger than the deflection angle $\hat{\alpha}_\odot$. However, the light deflection by more distant stars (or other massive celestial bodies) can produce multiple images of sources located behind them.

Lens Geometry. The geometry of a gravitational lens system is depicted in Fig. 2.22. We consider light rays from a source at distance D_s from us that pass a mass concentration (called a lens or deflector) at a separation $\boldsymbol{\xi}$. The deflector is at a distance D_d from us. In Fig. 2.22 $\boldsymbol{\eta}$ denotes the true, two-dimensional position of the source in the source plane, and $\boldsymbol{\beta}$ is the true angular position of the source, that is the angular position at which it would be observed in the absence of light deflection,

$$\boldsymbol{\beta} = \frac{\boldsymbol{\eta}}{D_\mathrm{s}} . \tag{2.72}$$

The position of the light ray in the lens plane is denoted by $\boldsymbol{\xi}$, and $\boldsymbol{\theta}$ is the corresponding angular position,

$$\boldsymbol{\theta} = \frac{\boldsymbol{\xi}}{D_\mathrm{d}} . \tag{2.73}$$

Hence, $\boldsymbol{\theta}$ is the observed position of the source on the sphere relative to the position of the "center of the lens" which we have chosen as the origin of the coordinate system, $\boldsymbol{\xi} = 0$. D_ds is the distance of the source plane from the lens plane. As long as the relevant distances are much smaller than the "radius of the Universe" c/H_0, which is certainly the case within our Galaxy and in the Local Group, we have $D_\mathrm{ds} = D_\mathrm{s} - D_\mathrm{d}$. However, this relation is no longer valid for cosmologically distant sources and lenses; we will come back to this in Sect. 4.3.3.

Lens Equation. From Fig. 2.22 we can deduce the condition that a light ray from the source will reach us from the direction $\boldsymbol{\theta}$ (or $\boldsymbol{\xi}$),

$$\boldsymbol{\eta} = \frac{D_\mathrm{s}}{D_\mathrm{d}} \boldsymbol{\xi} - D_\mathrm{ds} \hat{\boldsymbol{\alpha}}(\boldsymbol{\xi}) , \tag{2.74}$$

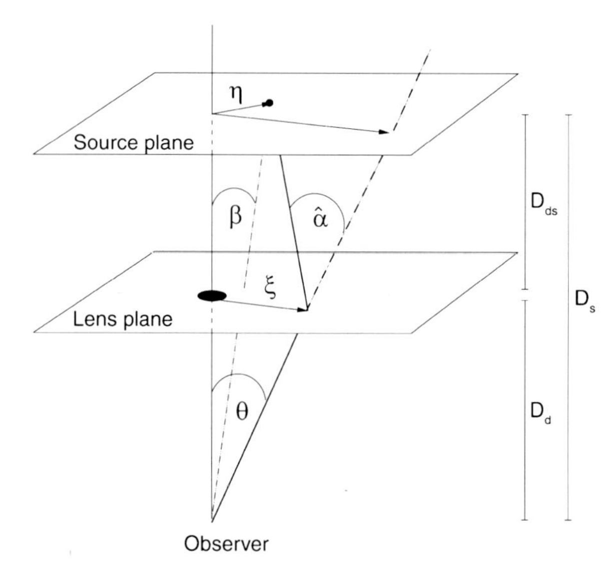

Fig. 2.22. Geometry of a gravitational lens system. Consider a source to be located at a distance D_s from us and a mass concentration at distance D_d. An optical axis is defined that connects the observer and the center of the mass concentration; its extension will intersect the so-called source plane, a plane perpendicular to the optical axis at the distance of the source. Accordingly, the lens plane is the plane perpendicular to the line-of-sight to the mass concentration at distance D_d from us. The intersections of the optical axis and the planes are chosen as the origins of the respective coordinate systems. Let the source be at the point $\boldsymbol{\eta}$ in the source plane; a light beam that encloses an angle $\boldsymbol{\theta}$ to the optical axis intersects the lens plane at the point $\boldsymbol{\xi}$ and is deflected by an angle $\hat{\boldsymbol{\alpha}}(\boldsymbol{\xi})$. All these quantities are two-dimensional vectors. The condition that the source is observed in the direction $\boldsymbol{\theta}$ is given by the lens equation (2.74) which follows from the theorem of intersecting lines

or, after dividing by D_s and using (2.72) and (2.73):

$$\boldsymbol{\beta} = \boldsymbol{\theta} - \frac{D_{ds}}{D_s}\,\hat{\boldsymbol{\alpha}}(D_d\boldsymbol{\theta})\,. \tag{2.75}$$

Due to the factor multiplying the deflection angle in (2.75), it is convenient to define the *reduced deflection angle*

$$\boxed{\boldsymbol{\alpha}(\boldsymbol{\theta}) := \frac{D_{ds}}{D_s}\,\hat{\boldsymbol{\alpha}}(D_d\boldsymbol{\theta})}\,, \tag{2.76}$$

so that the lens equation (2.75) attains the simple form

$$\boxed{\boldsymbol{\beta} = \boldsymbol{\theta} - \boldsymbol{\alpha}(\boldsymbol{\theta})}\,. \tag{2.77}$$

The deflection angle $\boldsymbol{\alpha}(\boldsymbol{\theta})$ depends on the mass distribution of the deflector. We will discuss the deflection angle for an arbitrary density distribution of a lens in Sect. 3.8. Here we will first concentrate on point masses, which is – in most cases – a good approximation for the lensing effect on stars.

For a point mass, we get – see (2.71)

$$|\boldsymbol{\alpha}(\boldsymbol{\theta})| = \frac{D_{ds}}{D_s}\,\frac{4\,G\,M}{c^2\,D_d\,|\boldsymbol{\theta}|}\,,$$

or, if we account for the direction of the deflection (the deflection angle always points towards the point mass),

$$\boxed{\boldsymbol{\alpha}(\boldsymbol{\theta}) = \frac{4\,G\,M}{c^2}\,\frac{D_{ds}}{D_s\,D_d}\,\frac{\boldsymbol{\theta}}{|\boldsymbol{\theta}|^2}}\,. \tag{2.78}$$

Multiple Images of a source occur if the lens equation (2.77) has multiple solutions $\boldsymbol{\theta}_i$ for a (true) source position $\boldsymbol{\beta}$ – in this case, the source is observed at the positions $\boldsymbol{\theta}_i$ on the sphere.

Explicit Solution of the Lens Equation for a Point Mass. The lens equation for a point mass is simple enough to be solved analytically which means that for each source position $\boldsymbol{\beta}$ the respective image positions $\boldsymbol{\theta}_i$ can be determined. If we define the so-called *Einstein angle* of the lens,

$$\boxed{\theta_E := \sqrt{\frac{4\,G\,M}{c^2}\,\frac{D_{ds}}{D_s\,D_d}}}\,, \tag{2.79}$$

then the lens equation (2.77) for the point-mass lens with a deflection angle (2.78) can be written as

$$\boldsymbol{\beta} = \boldsymbol{\theta} - \theta_E^2\,\frac{\boldsymbol{\theta}}{|\boldsymbol{\theta}|^2}\,.$$

Obviously, θ_E is a characteristic angle in this equation, so that for practical reasons we will use the scaling

$$\boldsymbol{y} := \frac{\boldsymbol{\beta}}{\theta_E}\,; \quad \boldsymbol{x} := \frac{\boldsymbol{\theta}}{\theta_E}\,.$$

Hence the lens equation simplifies to

$$\boldsymbol{y} = \boldsymbol{x} - \frac{\boldsymbol{x}}{|\boldsymbol{x}|^2}\,. \tag{2.80}$$

After multiplication with $\boldsymbol{x}$, this becomes a quadratic equation, whose solutions are

$$\boldsymbol{x} = \frac{1}{2}\left(|\boldsymbol{y}| \pm \sqrt{4+|\boldsymbol{y}|^2}\right)\frac{\boldsymbol{y}}{|\boldsymbol{y}|} . \quad (2.81)$$

From this solution of the lens equation one can immediately draw a number of conclusions:

- For each source position $\boldsymbol{y}$, the lens equation for a point-mass lens has two solutions – any source is (formally, at least) imaged twice. The reason for this is the divergence of the deflection angle for $\theta \to 0$. This divergence does not occur in reality because of the finite geometric extent of the lens (e.g., the radius of the star), as the solutions are of course physically relevant only if $\xi = D_\mathrm{d}\theta_\mathrm{E}|\boldsymbol{x}|$ is larger than the radius of the star. We need to point out again that we explicitly exclude the case of strong gravitational fields such as the light deflection near a black hole or a neutron star, for which the equation for the deflection angle has to be modified.
- The two images $\boldsymbol{x}_i$ are collinear with the lens and the source. In other words, the observer, lens, and source define a plane, and light rays from the source that reach the observer are located in this plane as well. One of the two images is located on the same side of the lens as the source ($\boldsymbol{x} \cdot \boldsymbol{y} > 0$), the second image is located on the other side – as is already indicated in Fig. 2.21.
- If $\boldsymbol{y} = 0$, so that the source is positioned exactly behind the lens, the full circle $|\boldsymbol{x}| = 1$, or $|\boldsymbol{\theta}| = \theta_\mathrm{E}$, is a solution of the lens equation (2.80) – the source is seen as a circular image. In this case, the source, lens, and observer no longer define a plane, and the problem becomes axially symmetric. Such a circular image is called an *Einstein ring*. Ring-shaped images have indeed been observed, as we will discuss in Sect. 3.8.3.
- The angular diameter of this ring is then $2\theta_\mathrm{E}$. From the solution (2.81), one can easily see that the distance between the two images is about $\Delta x = |\boldsymbol{x}_1 - \boldsymbol{x}_2| \gtrsim 2$ (as long as $|\boldsymbol{y}| \lesssim 1$), hence

  $$\Delta\theta \gtrsim 2\theta_\mathrm{E} ;$$

  the Einstein angle thus specifies the characteristic image separation. Situations with $|\boldsymbol{y}| \gg 1$, and hence angular separations significantly larger than $2\theta_\mathrm{E}$, are astrophysically of only minor relevance, as will be shown below.

Magnification: The Principle. Light beams are not only deflected as a whole, but they are also subject to differential deflection. For instance, those rays of a light beam (also called light bundle) that are closer to the lens are deflected more than rays at the other side of the beam. The differential deflection is an effect of the tidal component of the deflection angle; this is sketched in Fig. 2.23. By differential deflection, the solid angle which the image of the source subtends on the sky changes. Let ω_s be the solid angle the source would subtend if no lens were present, and ω the observed solid angle of the image of the source in the presence of a deflector. Since gravitational light deflection is not linked to emission or absorption of radiation, the sur-

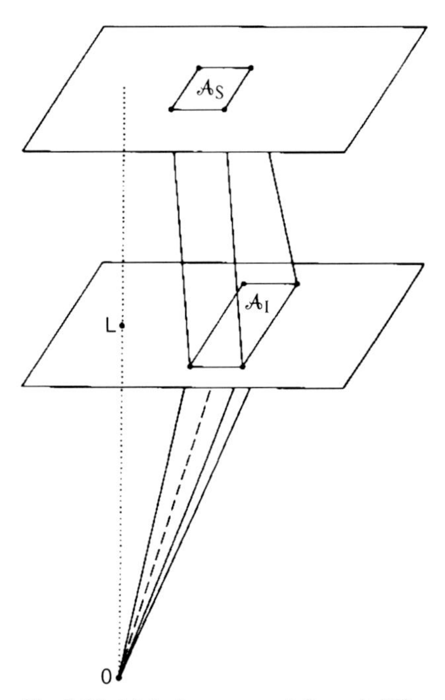

Fig. 2.23. Light beams are deflected differentially, leading to changes of the shape and the cross-sectional area of the beam. As a consequence, the observed solid angle subtended by the source, as seen by the observer, is modified by gravitational light deflection. In the example shown, the observed solid angle $\mathcal{A}_\mathrm{I}/D_\mathrm{d}^2$ is larger than the one subtended by the undeflected source, $\mathcal{A}_\mathrm{S}/D_\mathrm{s}^2$ – the image of the source is thus magnified

face brightness (or specific intensity) is preserved. The flux of a source is given as the product of surface brightness and solid angle. Since the former of the two factors is unchanged by light deflection, but the solid angle changes, the observed flux of the source is modified. If S_0 is the flux of the unlensed source and S the flux of an image of the source, then

$$\mu := \frac{S}{S_0} = \frac{\omega}{\omega_s} \qquad (2.82)$$

describes the change in flux that is caused by a magnification (or a diminution) of the image of a source. Obviously, the magnification is a purely geometrical effect.

Magnification for "Small" Sources. For sources and images that are much smaller than the characteristic scale of the lens, the magnification μ is given by the differential area distortion of the lens mapping (2.77),

$$\mu = \left|\det\left(\frac{\partial \boldsymbol{\beta}}{\partial \boldsymbol{\theta}}\right)\right|^{-1} \equiv \left|\det\left(\frac{\partial \beta_i}{\partial \theta_j}\right)\right|^{-1} . \qquad (2.83)$$

Hence for small sources, the ratio of solid angles of the lensed image and the unlensed source is described by the determinant of the local Jacobi matrix.[11]

The magnification can therefore be calculated for each individual image of the source, and the total magnification of a source, given by the ratio of the sum of the fluxes of the individual images and the flux of the unlensed source, is the sum of the magnifications for the individual images.

Magnification for the Point-Mass Lens. For a point-mass lens, the magnifications for the two images (2.81) are

$$\mu_\pm = \frac{1}{4}\left(\frac{y}{\sqrt{y^2+4}} + \frac{\sqrt{y^2+4}}{y} \pm 2\right) . \qquad (2.84)$$

From this it follows that for the "+"-image $\mu_+ > 1$ for all source positions $y = |\boldsymbol{y}|$, whereas the "−"-image can have magnification either larger or less than unity, depending on y. The magnification of the two images is illustrated in Fig. 2.24, while Fig. 2.25 shows the magnification for several different source positions y. For $y \gg 1$, one has $\mu_+ \gtrsim 1$ and $\mu_- \sim 0$, from which we draw the following conclusion: if the source and lens are not sufficiently well aligned, the secondary image is strongly demagnified and the primary image has magnification very close to unity. For this reason, situations with $y \gg 1$ are of little relevance since then essentially only one image is observed which has about the same flux as the unlensed source.

For $y \to 0$, the two magnifications diverge, $\mu_\pm \to \infty$. The reason for this is purely geometric: in this case, out of a zero-dimensional point source a one-dimensional image, the Einstein ring, is formed. This divergence is not physical, of course, since infinite magnifications do not occur in reality. The magnifications remain finite even for $y = 0$, for two reasons. First, real sources have a finite extent, and for these the magnifi-

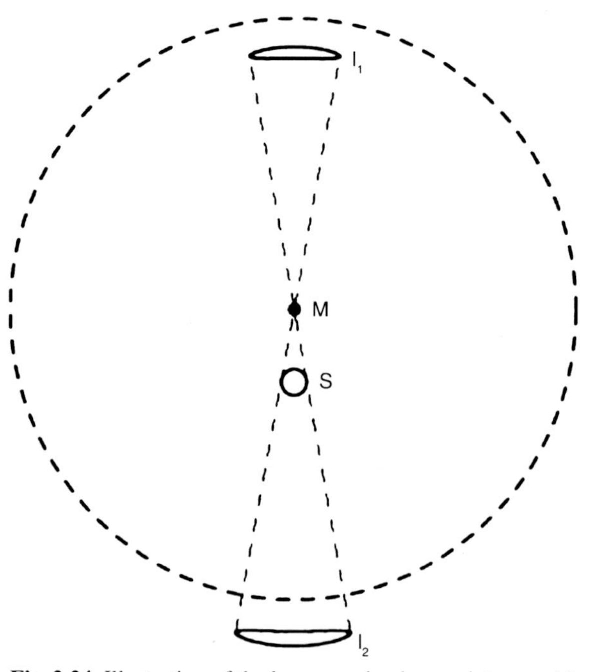

Fig. 2.24. Illustration of the lens mapping by a point mass M. The unlensed source S and the two images I_1 and I_2 of the lensed source are shown. We see that the two images have a solid angle different from the unlensed source, and they also have a different shape. The dashed circle shows the Einstein radius of the lens

[11] The determinant in (2.83) is a generalization of the derivative in one spatial dimension to higher dimensional mappings. Consider a scalar mapping $y = y(x)$; through this mapping, a "small" interval Δx is mapped onto a small interval Δy, where $\Delta y \approx (\mathrm{d}y/\mathrm{d}x)\,\Delta x$. The Jacobian determinant occurring in (2.83) generalizes this result for a two-dimensional mapping from the lens plane to the source plane.

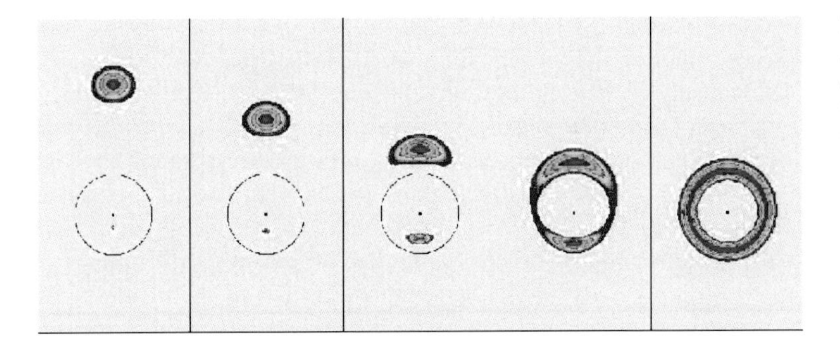

Fig. 2.25. Image of a circular source with a radial brightness profile – indicated by colors – for different relative positions of the lens and source. y decreases from left to right; in the rightmost figure $y = 0$ and an Einstein ring is formed

cation is finite. Second, even if one had a point source, wave effects of the light (interference) would lead to a finite value of μ. The total magnification of a point source by a point-mass lens follows from the sum of the magnifications (2.84),

$$\mu(y) = \mu_+ + \mu_- = \frac{y^2+2}{y\sqrt{y^2+4}} . \tag{2.85}$$

2.5.2 Galactic Microlensing Effect

After these theoretical considerations we will now return to the starting point of our discussion, employing the lensing effect as a potential diagnostic for dark matter in our Milky Way, if this dark matter were to consist of compact mass concentrations, e.g., very faint stars.

Image Splitting. Considering a star in our Galaxy as the lens, (2.79) yields the Einstein angle

$$\theta_{\rm E} = 0.902\,\text{mas}\left(\frac{M}{M_\odot}\right)^{1/2} \times \left(\frac{D_{\rm d}}{10\,\text{kpc}}\right)^{-1/2}\left(1-\frac{D_{\rm d}}{D_{\rm s}}\right)^{1/2} . \tag{2.86}$$

Since the angular separation $\Delta\theta$ of the two images is about $2\theta_{\rm E}$, the typical image splittings are about a milliarcsecond (mas) for lens systems including Galactic stars; such angular separations are as yet not observable with optical telescopes. This insight made Einstein believe in 1936, after he conducted a detailed quantitative analysis of gravitational lensing by point masses, that the lens effect will not be observable.[12]

Magnification. Bohdan Paczyński pointed out in 1986 that, although image splitting was unobservable, the magnification by the lens should nevertheless be measurable. To do this, we have to realize that the absolute magnification is observable only if the unlensed flux of the source is known – which is not the case, of course (for nearly all sources). However, the magnification, and therefore also the observed flux, changes with time by the relative motion of source, lens, and ourselves. Therefore, the flux is a function of time, caused by the time-dependent magnification.

Characteristic Time-Scale of the Variation. Let v be a typical transverse velocity of the lens, then its angular velocity is

$$\dot\theta = \frac{v}{D_{\rm d}} = 4.22\,\text{mas yr}^{-1}\left(\frac{v}{200\,\text{km/s}}\right)\left(\frac{D_{\rm d}}{10\,\text{kpc}}\right)^{-1} , \tag{2.87}$$

if we consider the source and the observer to be at rest. The characteristic time-scale of the variability is then given by

$$t_{\rm E} := \frac{\theta_{\rm E}}{\dot\theta} = 0.214\,\text{yr}\left(\frac{M}{M_\odot}\right)^{1/2}\left(\frac{D_{\rm d}}{10\,\text{kpc}}\right)^{1/2} \times\left(1-\frac{D_{\rm d}}{D_{\rm s}}\right)^{1/2}\left(\frac{v}{200\,\text{km/s}}\right)^{-1} . \tag{2.88}$$

This time-scale is of the order of a month for lenses with $M \sim M_\odot$ and typical Galactic velocities. Hence,

[12] The expression "microlens" has its origin in the angular scale (2.86) that was discussed in the context of the lens effect on quasars by stars at cosmological distances, for which one obtains image splittings of about one microarcsecond.

the effect is measurable in principle. In the general case that source, lens, and observer are all moving, v has to be considered as an effective velocity. Alternatively, the motion of the source in the source plane can be considered.

Light Curves. In most cases, the relative motion can be considered linear, so that the position of the source in the source plane can be written as

$$\boldsymbol{\beta} = \boldsymbol{\beta}_0 + \dot{\boldsymbol{\beta}}(t - t_0) .$$

Using the scaled position $\boldsymbol{y} = \boldsymbol{\beta}/\theta_{\rm E}$, for $y = |\boldsymbol{y}|$ we obtain

$$y(t) = \sqrt{p^2 + \left(\frac{t - t_{\max}}{t_{\rm E}}\right)^2} , \qquad (2.89)$$

where $p = y_{\min}$ specifies the minimum distance from the optical axis, and $t_{\max}$ is the time at which $y = p$ attains this minimum value, thus when the magnification $\mu = \mu(p) = \mu_{\max}$ is maximized. From this, and using (2.85), one obtains the light curve

$$S(t) = S_0\, \mu(y(t)) = S_0\, \frac{y^2(t) + 2}{y(t)\sqrt{y^2(t) + 4}} . \qquad (2.90)$$

Examples for such light curves are shown in Fig. 2.26. They depend on only four parameters: the flux of the unlensed source S_0, the time of maximum magnification $t_{\max}$, the smallest distance of the source from the optical axis p, and the characteristic time-scale $t_{\rm E}$. All these values are directly observable in a light curve. One obtains $t_{\max}$ from the time of the maximum of the light curve, S_0 is the flux that is measured for very large and small times, $S_0 = S(t \to \pm\infty)$, or $S_0 \approx S(t)$ for $|t - t_{\max}| \gg t_{\rm E}$. Furthermore, p follows from the maximum magnification $\mu_{\max} = S_{\max}/S_0$ by inversion of (2.85), and $t_{\rm E}$ from the width of the light curve.

Only $t_{\rm E}$ contains information of astrophysical relevance, because the time of the maximum, the unlensed flux of the source, and the minimum separation p provide no information about the lens. Since $t_{\rm E} \propto \sqrt{M\, D_{\rm d}}/v$, this time-scale contains the combined information on the lens mass, the distances to the lens and the source, and the transverse velocity: *Only the combination $t_{\rm E} \propto \sqrt{M\, D_{\rm d}}/v$ can be derived from the light curve, but not mass, distance, or velocity individually.*

Paczyński's idea can be expressed as follows: if the halo of our Milky Way consists (partially) of compact objects, a distant compact source should, from time to time, be lensed by one of these MACHOs and thus show characteristic changes in flux, corresponding to a light curve similar to those in Fig. 2.26. The number density of MACHOs is proportional to the probability or abundance of lens events, and the characteristic mass of the MACHOs is proportional to the square of the typical variation time-scale $t_{\rm E}$. All one has to do is measure the light curves of a sufficiently large number of background sources and extract all lens events from those light curves to obtain information on the population of potential MACHOs in the halo. A given halo model predicts the spatial density distribution and the distribution of velocities of the MACHOs and can therefore be compared to the observations in a statistical way. However, one faces the problem that the abundance of such lensing events is very small.

Probability of a Lens Event. In practice, a system of a foreground object and a background source is considered a lens system if $p < 1$ and hence $\mu_{\max} > 3/\sqrt{5} \approx 1.34$, i.e., if the relative trajectory of the source passes within the Einstein circle of the lens.

If the dark halo of the Milky Way consisted solely of MACHOs, the probability that a very distant source is lensed (in the sense of $|\boldsymbol{\beta}| \le \theta_{\rm E}$) would be $\sim 10^{-7}$, where the exact value depends on the direction to the source. At any one time, one of $\sim 10^7$ distant sources would be located within the Einstein radius of a MACHO in our halo. The immediate consequence of this is that the light curves of millions of sources have to be monitored to detect this effect. Furthermore, these sources have to be located within a relatively small region on the sphere to keep the total solid angle that has to be photometrically monitored relatively small. This condition is needed to limit the required observing time, so that many such sources should be present within the field-of-view of the camera used. The stars of the Magellanic Clouds are well suited for such an experiment: they are close together on the sphere, but can still be resolved into individual stars.

Problems, and their Solution. From this observational strategy, a large number of problems arise immediately; they were discussed in Paczyński's original paper. First,

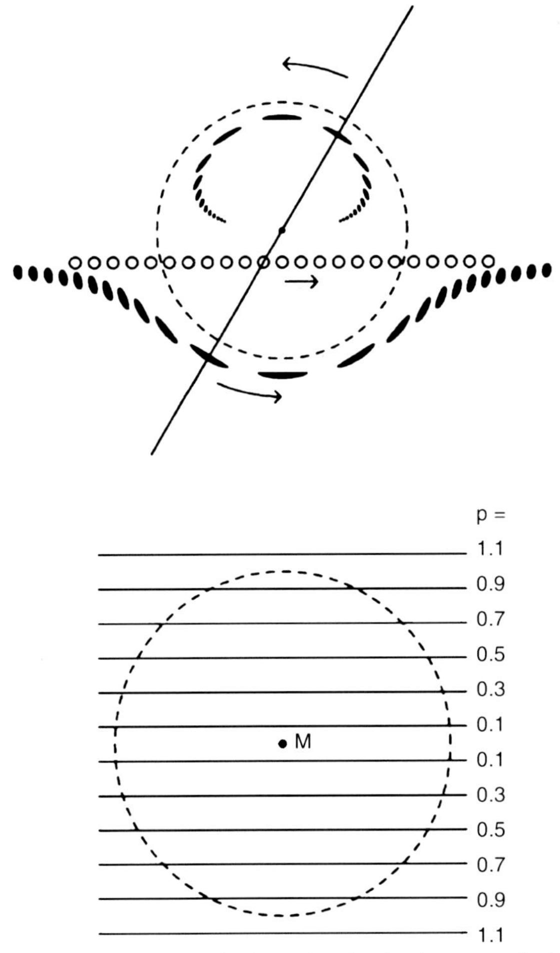

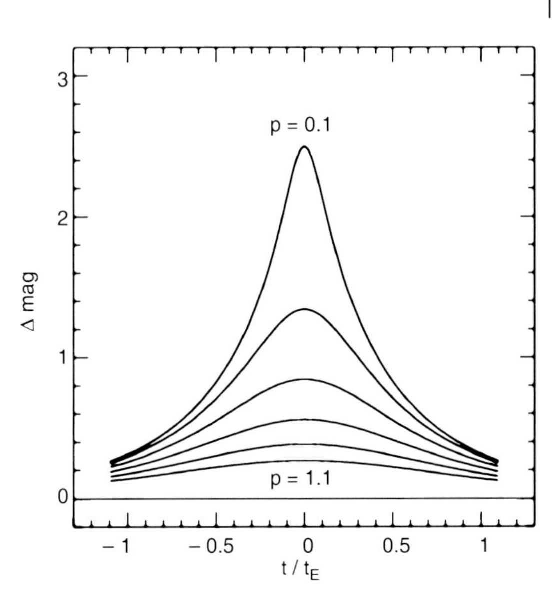

Fig. 2.26. Illustration of a Galactic microlensing event: In the upper left panel a source (depicted by the open circles) moves behind a point-mass lens; for each source position two images of the source are formed, which are indicated by the black ellipses. The identification of the corresponding image pair with the source position follows from the fact that, in projection, the source, the lens, and the two images are located on a straight line, which is indicated for one source position. The dashed circle represents the Einstein ring. In the lower left panel, different trajectories of the source are shown, each characterized by the smallest projected separation p to the lens. The light curves resulting from these relative motions, which can be calculated using equation (2.90), are then shown in the right-hand panel for different values of p. The smaller p is, the larger the maximum magnification will be, here measured in magnitudes

the photometry of so many sources over many epochs produces a huge amount of data that need to be handled; they have to be stored and reduced. Second, one has the problem of "crowding": the stars in the Magellanic Clouds are densely packed on the sky, which renders the photometry of individual stars difficult. Third, stars also show intrinsic variability – about 1% of all stars are variable. This intrinsic variability has to be distinguished from that due to the lens effect. Due to the small abundance of the latter, selecting the lens events is comparable to searching for a needle in a haystack. Finally, it should be mentioned that one has to ensure that the experiment is indeed sensitive enough to detect lens events. A "calibration experiment" would therefore be desirable.

Faced with these problems, it seemed daring to seriously think about the realization of such an observing program. However, a fortunate event helped, in the magnificent time of the easing of tension between the US and the Soviet Union, and their respective allies, at the end of the 1980s. Physicists and astrophysicists, partly occupied with issues concerning national security, then saw an opportunity to meet new challenges. In addition, scientists in national laboratories had much better access to sufficient computing power and storage capacity than those in other research institutes, attenuating some of the aforementioned problems. While the expected data volume was still a major problem in 1986, it could be handled a few years later. Also, wide-field cameras were constructed, with which large areas of the sky could be observed simultaneously. Software was developed which specializes in the photometry of objects in crowded fields, so that light curves could be measured even if individual stars in the image were no longer cleanly separated.

To distinguish between lensing events and intrinsic variablity of stars, we note that the microlensing light curves have a characteristic shape that is described by only four parameters. The light curves should be symmetric and achromatic because gravitational light deflection is independent of the frequency of the radiation. Furthermore, due to the small lensing probability, any source should experience at most one microlensing event and show a constant flux before and after, whereas intrinsic variations of stars are often periodic and in nearly all cases chromatic.

And finally a control experiment could be performed: the lensing probability in the direction of the Galactic bulge is known, or at least, we can obtain a lower limit for it from the observed density of stars in the disk. If a microlens experiment is carried out in the direction of the Galactic bulge, we *have to* find lens events if the experiment is sufficiently sensitive.

2.5.3 Surveys and Results

In the early 1990s, two collaborations (MACHO and EROS) began the search for microlensing events towards the Magellanic clouds. Another group (OGLE) started searching in an area of the Galactic bulge. Fields in the respective survey regions were observed regularly, typically once every night if weather conditions permitted. From the photometry of the stars in the fields, light curves for many millions of stars were generated and then checked for microlensing events.

First Detections. In 1993, all three groups reported their first results. The MACHO collaboration found one event in the Large Magellanic Cloud (LMC), the EROS group two events, and the OGLE group observed one event in the bulge. The light curve of the first MACHO event is plotted in Fig. 2.27. It was observed in two different filters, and the fit to the data, which corresponds to a standard light curve (2.90), is the same for both filters, proving that the event is achromatic. Together with the quality of the fit to the data, this is very strong evidence for the microlensing nature of the event.

Statistical Results. In the years since 1993, all three aforementioned teams have proceeded with their ob-

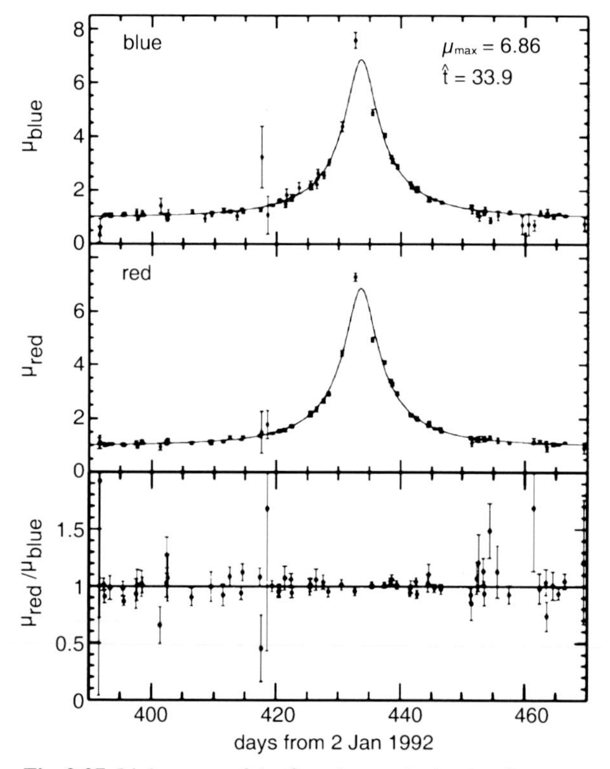

Fig. 2.27. Light curve of the first observed microlensing event in the LMC, in two broad-band filters. The solid curve is the best-fit microlens light curve as described by (2.90), with $\mu_{max} = 6.86$. The ratio of the magnifications in both filters is displayed at the bottom, and it is compatible with 1. Some of the data points deviate significantly from the curve; this means that either the errors in the measurements were underestimated, or this event is more complicated than one described by a point-mass lens – see Sect. 2.5.4

servations and analysis (Fig. 2.28), and more groups have begun the search for microlensing events, choosing various lines-of-sight. The most important results from these experiments can be summarized as follows:

About 20 events have been found in the direction of the Magellanic Clouds, and of the order a thousand in the direction of the bulge. The statistical analysis of the data revealed the lensing probability towards the bulge to be higher than originally expected. This can be explained by the fact that *our Galaxy features a bar* (see Chap. 3). This bar was also observed in IR maps such as those made by the COBE satellite. The events in the direction of the bulge are dominated by lenses that are part of the bulge themselves, and their column density is increased by the bar-like shape of the bulge. On the other hand, the lens probability in the direction of the Magellanic Clouds is *smaller* than expected for the case where the dark halo consists solely of MACHOs.

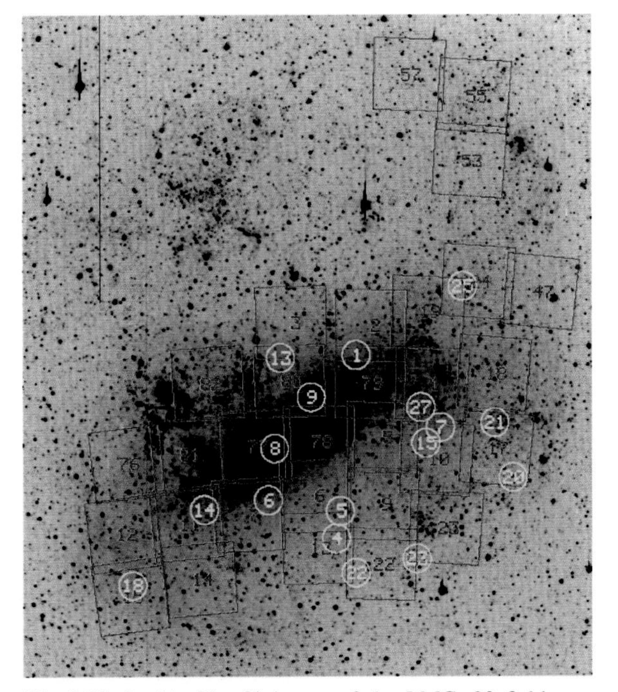

Fig. 2.28. In this 8° × 8° image of the LMC, 30 fields are marked in red which the MACHO group has searched for microlensing events during the ~ 5.5 years of their experiment; images were taken in two filters to test for achromaticity. The positions of 17 microlens events are marked by yellow circles; these have been subject to statistical analysis

Based on the analysis of the MACHO collaboration, the observed statistics of lensing events towards the Magellanic Clouds is best explained if about 20% of the halo mass consists of MACHOs, with a characteristic mass of about $M \sim 0.5 M_{\odot}$ (see Fig. 2.29).

Interpretation and Discussion. This latter result is not easy to interpret and came as a real surprise. If a result compatible with ~ 100% had been found, it would have been obvious to conclude that the dark matter in our Milky Way consists of compact objects. Otherwise, if very few lensing events had been found, it would have been clear that MACHOs do not contribute significantly to the dark matter. But a value of 20% does not allow any unambiguous interpretation. Taken at face value, the result from the MACHO group would imply that the total mass of MACHOs in the Milky Way halo is about the same as that in the stellar disk.

Furthermore, the estimated mass scale is hard to understand: what could be the nature of MACHOs with $M = 0.5 M_{\odot}$? Normal stars can be excluded, because they would be far too luminous not to be observed. White dwarfs are also unsuitable candidates, because to

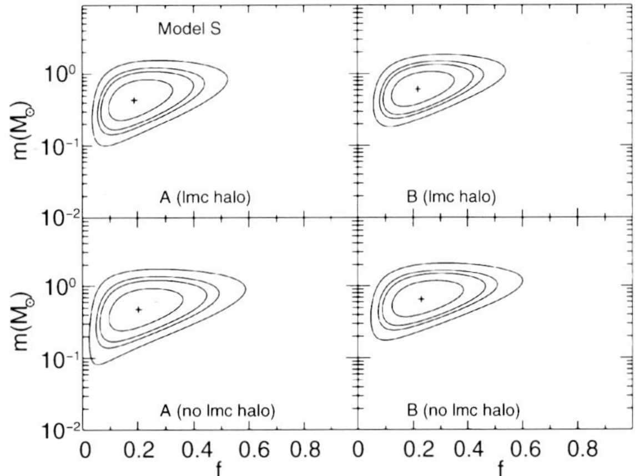

Fig. 2.29. Probability contours for a specific halo model as a function of the characteristic MACHO mass M (here denoted by m) and the mass fraction f of MACHOs in the halo. The halo of the LMC was either taken into account as an additional source for microlenses (lmc halo) or not (no lmc halo), and two different selection criteria (A,B) for the statistically complete microlensing sample have been used. In all cases, $M \sim 0.5 M_{\odot}$ and $f \sim 0.2$ are the best-fit values

produce such a large number of white dwarfs as a final stage of stellar evolution, the total star formation in our Milky Way, integrated over its lifetime, needs to be significantly larger than normally assumed. In this case, many more massive stars would also have formed, which would then have released the metals they produced into the ISM, both by stellar winds and in supernova explosions. In such a scenario, the metal content of the ISM would therefore be distinctly higher than is actually observed. The only possibility of escaping this argument is with the hypothesis that the mass function of newly formed stars (the initial mass function, IMF) was different in the early phase of the Milky Way compared to that observed today. The IMF that needs to be assumed in this case is such that for each star of intermediate mass which evolves into a white dwarf, far fewer high-mass stars, responsible for the metal enrichment of the ISM, must have formed in the past compared to today. However, we lack a plausible physical model for such a scenario, and it is in conflict with the star-formation history that we observe in the high-redshift Universe (see Chap. 9).

Neutron stars can be excluded as well, because they are too massive (typically $> 1 M_\odot$); in addition, they are formed in supernova explosions, implying that the aforementioned metallicity problem would be even greater for neutron stars. Would stellar-mass black holes be an alternative? The answer to this question depends on how they are formed. They could not originate in SN explosions, again because of the metallicity problem. If they had formed in a very early phase of the Universe (they are then called primordial black holes), this would be an imaginable, though perhaps quite exotic, alternative.

However, we have strong indications that the interpretation of the MACHO results is not as straightforward as described above. Some doubts have been raised as to whether all events reported as being due to microlensing are in fact caused by this effect. In fact, one of the microlensing source stars identified by the MACHO group showed another bump seven years after the first event. Given the extremely small likelihood of two microlensing events happening to a single source this is almost certainly a star with unusual variability.

As argued previously, by means of t_E we only measure a combination of lens mass, transverse velocity, and distance. The result given in Fig. 2.29 is therefore based on the statistical analysis of the lensing events in the framework of a halo model that describes the shape and the radial density profile of the halo. However, microlensing events have been observed for which more than just t_E can be determined – e.g., events in which the lens is a binary star, or those for which t_E is larger than a few months. In this case, the orbit of the Earth around the Sun, which is not a linear motion, has a noticeable effect, causing deviations from the standard curve. Such parallax events have indeed been observed.[13] Three events are known in the direction of the Magellanic Clouds in which more than just t_E could be measured. In all three cases the lenses are most likely located in the Magellanic Clouds themselves (an effect called self-lensing) and not in the halo of the Milky Way. If for those three cases, where the degeneracy between lens mass, distance, and transverse velocity can be broken, the respective lenses are not MACHOs in the Galactic halo, we might then suspect that in most of the other microlensing events the lens is not a MACHO either. Therefore, it is currently unclear how to interpret the results of the microlensing surveys. In particular, it is unclear to what extent self-lensing contributes to the results. Furthermore, the quantitative results depend on the halo model.

The EROS collaboration used an observation strategy which was sightly different from that of the MACHO group, by observing a number of fields in very short time intervals. Since the duration of a lensing event depends on the mass of the lens as $\Delta t \propto M^{1/2}$ – see (2.88) – they were also able to probe very small MACHO masses. The absence of lensing events of very short duration then allowed them to derive limits for the mass fraction of such low-mass MACHOs, as is shown in Fig. 2.30.

Despite this unsettled situation concerning the interpretation of the MACHO results, we have to emphasize that the microlensing surveys have been enormously successful experiments because they accomplished exactly what was expected at the beginning of the observations. They measured the lensing probability in the direction of the Magellanic Clouds and the Galactic bulge. The fact that the distribution of the lenses differs from that expected by no means diminishes the success of these surveys.

[13] These parallax events in addition prove that the Earth is in fact orbiting around the Sun – even though this is not really a new insight.

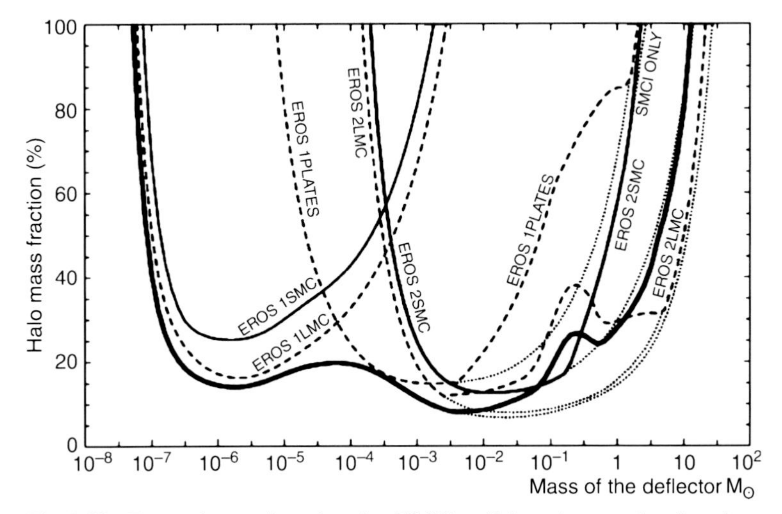

Fig. 2.30. From observations by the EROS collaboration, a large mass range for MACHO candidates can be excluded. The maximum allowed fraction of the halo mass contained in MACHOs is plotted as a function of the MACHO mass M, as an upper limit with 95% confidence. A standard model for the mass distribution in the Galactic halo was assumed which describes the rotation curve of the Milky Way quite well. The various curves show different phases of the EROS experiment. They are plotted separately for observations in the directions of the LMC and the SMC. The experiment EROS 1 searched for microlensing events on short time-scales but did not find any; this yields the upper limits at small masses. Upper limits at larger masses were obtained by the EROS 2 experiment. The thick solid curve represents the upper limit derived from combining the individual experiments. If not a single MACHO event had been found the upper limit would have been described by the dotted line

2.5.4 Variations and Extensions

Besides the search for MACHOs, microlensing surveys have yielded other important results and will continue to do so in the future. For instance, the distribution of stars in the Galaxy can be measured by analyzing the lensing probability as a function of direction. Thousands of variable stars have been newly discovered and accurately monitored; the extensive and publicly accessible databases of the surveys form an invaluable resource for stellar astrophysics. Furthermore, globular clusters in the LMC have been identified from these photometric observations.

For some lensing events, the radius and the surface structure of distant stars can be measured with very high precision. This is possible because the magnification μ depends on the position of the source. Situations can occur, for example where a binary star acts as a lens (see Fig. 2.31), in which the dependence of the magnification on the position in the source plane is very sensitive. Since the source – the star – is in motion relative to the line-of-sight between Earth and the lens, its different regions are subject to different magnification, depending on the time-dependent source position. A detailed analysis of the light curve of such events then enables us to reconstruct the light distribution on the surface of the star. The light curve of one such event is shown in Fig. 2.32.

For these lensing events the source can no longer be assumed to be a point source. Rather, the details of the light curve are determined by its light distribution. Therefore, another length-scale appears in the system, the radius of the star. This length-scale shows up in the corresponding microlensing light curve, as can be seen in Fig. 2.32, by the time-scale which characterizes the width of the peaks in the light curve – it is directly related to the ratio of the stellar radius and the transverse velocity of the lens. With this new scale, the degeneracy between M, v, and $D_{\rm d}$ is partially broken, so that these special events provide more information than the "classical" ones.

In fact, the light curve in Fig. 2.27 is probably not caused by a single lens star, but instead by additional slight disturbances from a companion star. This would

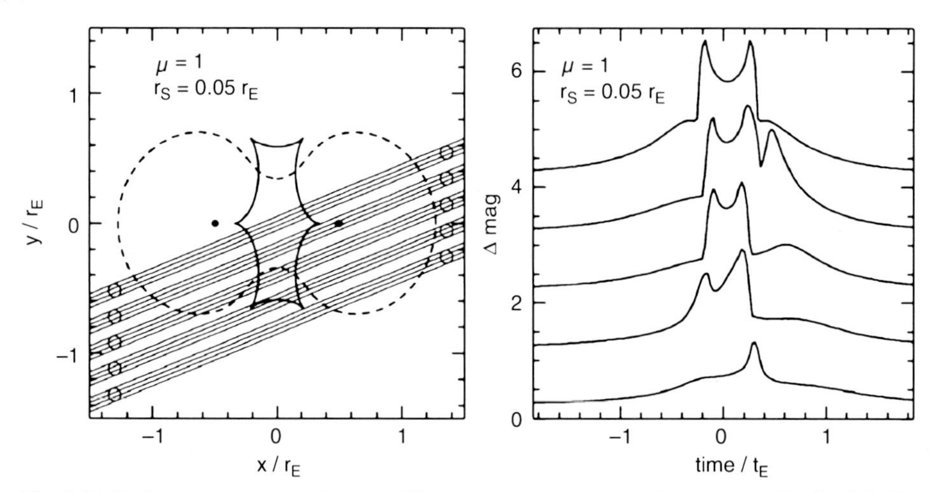

Fig. 2.31. If a binary star acts as a lens, significantly more complicated light curves can be generated. In the left-hand panel tracks are plotted for five different relative motions of a background source; the dashed curve is the so-called *critical curve*, formally defined by $\det(\partial\boldsymbol{\beta}/\partial\boldsymbol{\theta}) = 0$, and the solid line is the corresponding image of the critical curve in the source plane, called a *caustic*. Light curves corresponding to these five tracks are plotted in the right-hand panel. If the source crosses the caustic, the magnification μ becomes very large – formally infinite if the source was point-like. Since it has a finite extent, μ has to be finite as well; from the maximum μ during caustic crossing, the radius of the source can be determined, and sometimes even the variation of the surface brightness across the stellar disk, an effect known as limb darkening

explain the deviation of the observed light curve from a simple model light curve. However, the sampling in time of this particular light curve is not sufficient to determine the parameters of the binary system.

By now, detailed light curves with very good time coverage have been measured, which was made possible with an alarm system. The data from those groups searching for microlensing events are analyzed immediately after observations, and potential candidates for interesting events are published on the Internet. Other groups (such as the PLANET collaboration, for example) then follow-up these systems with very good time coverage by using several telescopes spread over a large range in geographical longitude. This makes around-the-clock observations of the event possible. Using this method light curves of extremely high quality have been measured. These groups hope to detect extra-solar planets by characteristic deviations in these light curves. Indeed, these microlens observations may be the most realistic (and cheapest) option for finding low-mass planets. Other methods for finding extra-solar planets, such as the search for small periodic changes of the radial velocity of stars which is caused by the gravitational pull of their orbiting planet, are mostly sensitive to high-mass planets. Whereas such surveys

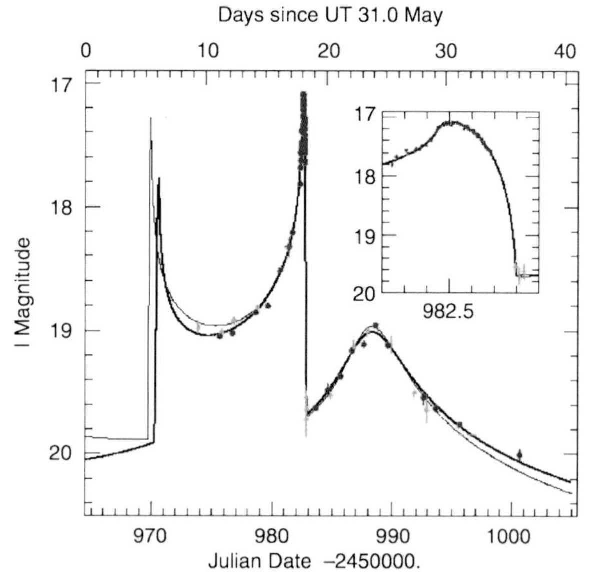

Fig. 2.32. Light curve of an event in which the lens was a binary star. The MACHO group discovered this "binary event". Members of the PLANET collaboration obtained this data using four different telescopes (in Chile, Tasmania, and South Africa). The second caustic crossing is highly resolved (displayed in the small diagram) and allows us to draw conclusions about the size and the brightness distribution of the source star. The two curves show the fits of a binary lens to the data

have been extremely successful in the past decade, having detected far more than one hundred planets around other stars, the characteristic mass of these planets is that of Jupiter, i.e., ~ 1000 times more massive than the Earth. At least two planet-mass companions to lens stars have already been discovered through microlensing, one of them having a mass of only six times that of the Earth.

Pixel Lensing. An extension to the microlensing search was suggested in the form of the so-called pixel lensing method. Instead of measuring the light curve of a single star one records the brightness of groups of stars that are positioned closely together on the sky. This method is applicable in situations where the density of source stars is very high, such as for the stars in the Andromeda galaxy (M31), which cannot be resolved individually. If one star is magnified by a microlensing event, the brightness in the corresponding region changes in a characteristic way, similar to that in the lensing events discussed above. To identify such events, the magnification needs to be relatively large, because only then can the light of the lensed star dominate over the local brightness in the region, so that the event can be recognized. On the other hand, the number of photometrically monitored stars (per solid angle) is larger than in surveys where single stars are observed, so that events of larger magnification are also more abundant. By now, several groups have successfully started to search for microlensing events in M31. The quantitative analysis of these surveys is more complicated than for the surveys targeting the Magellanic Clouds. However, the M31 experiments are equally sensitive to both MACHOs in the halo of our Milky Way and in that of M31. Therefore, these surveys promise to finally resolve the question of whether part of the dark matter consists of MACHOs.

Annihilation Radiation due to Dark Matter? The 511 keV annihilation radiation from the Galactic bulge, discussed in Sect. 2.3.4 above, has been suggested to be related to dark matter particles. Depending on the density of dark matter in the center of the Galaxy, as well as on the cross-section of the constituent particles of the dark matter (if it is indeed due to elementary particles), these particles can annihilate. In this process, positrons might be released which can then annihilate with the electrons of the interstellar medium. However, in order for this to be the source of the 511 keV line radiation, the dark matter particles must have rather "exotic" properties.

2.6 The Galactic Center

The Galactic center (GC, see Fig. 2.33) is not observable at optical wavelengths, because the extinction in the V-band is ~ 28 mag. Our information about the GC has

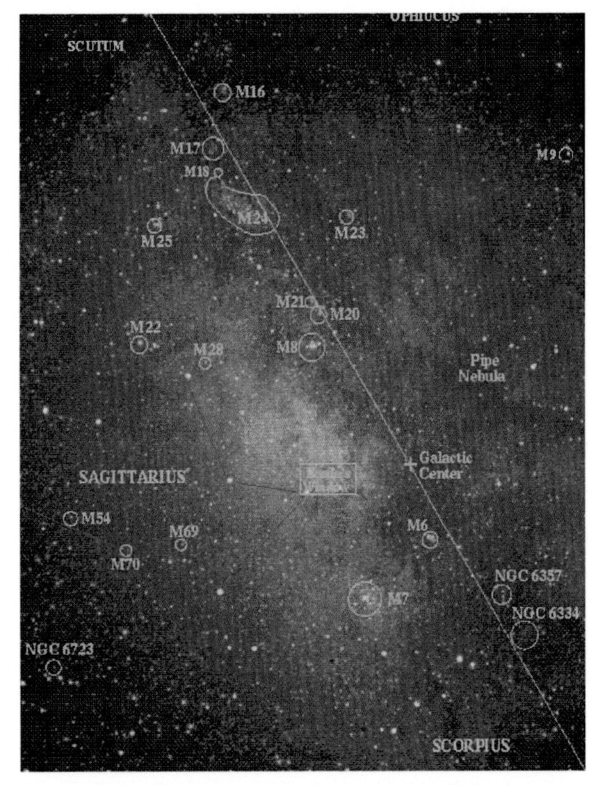

Fig. 2.33. Optical image in the direction of the Galactic center. Marked are some Messier objects: gas nebulae such as M8, M16, M17, M20; open star clusters such as M6, M7, M18, M21, M23, M24, and M25; globular clusters such as M9, M22, M28, M54, M69, and M70. Also marked is the Galactic center, as well as the Galactic plane, which is indicated by a line. Baade's Window can be easily recognized, a direction in which the extinction is significantly lower than in nearby directions, so that a clear increase in stellar density is visible there. This is the reason why the microlensing observations towards the Galactic center were preferably done in Baade's Window

been obtained from radio-, IR-, and X-ray radiation. Since the GC is nearby, and thus serves as a prototype of the central regions of galaxies, its observation is of great interest for our understanding of the processes taking place in the centers of galaxies.

2.6.1 Where is the Galactic Center?

The question of where the center of our Milky Way is located is by no means trivial, because the term "center" is in fact not well-defined. Is it the center of mass of the Galaxy, or the point around which the stars and the gas are orbiting? And how could we pinpoint this "center" accurately? Fortunately, the center can nevertheless be localized because, as we will see below, a distinct source exists that is readily identified as the Galactic center.

Radio observations in the direction of the GC show a relatively complex structure, as is displayed in Fig. 2.34.

A central disk of HI gas exists at radii from several 100 pc up to about 1 kpc. Its rotational velocity yields a mass estimate $M(R)$ for $R \gtrsim 100$ pc. Furthermore, radio filaments are observed which extend perpendicularly to the Galactic plane, and also a large number of supernova remnants are seen. Within about 2 kpc from the center, roughly $3 \times 10^7 M_\odot$ of atomic hydrogen is found. Optical images show regions close to the GC towards which the extinction is significantly lower. The best known of these is Baade's window – most of the microlensing surveys towards the bulge are conducted in this region. In addition, a fairly large number of globular clusters and gas nebulae are observed towards the central region. X-ray images (Fig. 2.35) show numerous X-ray binaries, as well as diffuse emission by hot gas.

The innermost 8 pc contain the radio source Sgr A (Sagittarius A), which itself consists of different components:

- A circumnuclear molecular ring, shaped like a torus, which extends between radii of $2\,\mathrm{pc} \lesssim R \lesssim 8$ pc and is inclined by about 20° relative to the Galactic disk. The rotational velocity of this ring is about ~ 110 km/s, nearly independent of R. This ring has a sharp inner boundary; this cannot be the result of an equilibrium flow, because internal turbulent motions would quickly (on a time-scale of $\sim 10^5$ yr) erase this boundary. Probably, it is evidence of an energetic event that occurred in the Galactic center within the past $\sim 10^5$ years. This interpretation is also supported by other observations, e.g., by a clumpiness in density and temperature.
- Sgr A East, a non-thermal (synchrotron) source of shell-like structure. Presumably this is a supernova remnant (SNR), with an age between 100 and 5000 years.
- Sgr A West is located about 1.′5 away from Sgr A East. It is a thermal source, an unusual HII region with a spiral-like structure.
- Sgr A* is a strong compact radio source close to the center of Sgr A West. Recent observations with mm-VLBI show that its extent is smaller than 3 AU. The radio luminosity is $L_{\rm rad} \sim 2 \times 10^{34}$ erg/s. Except for the emission in the mm and cm domain, Sgr A* is a weak source. Since other galaxies often have a compact radio source in their center, Sgr A* is an excellent candidate for being the center of our Milky Way.

Through observations of stars which contain a radio maser[14] source, the astrometry of the GC in the radio domain was matched to that in the IR, i.e., the position of Sgr A* is also known in the IR.[15] The uncertainty in the relative positions between radio and IR observations is only ~ 30 mas – at a presumed distance of the GC of 8 kpc, one arcsecond corresponds to 0.0388 pc, or about 8000 AU.

2.6.2 The Central Star Cluster

Density Distribution. Observations in the K-band ($\lambda \sim 2\,\mu$m) show a compact star cluster that is centered on Sgr A*. Its density behaves like $\propto r^{-1.8}$ in the distance range $0.1\,\mathrm{pc} \lesssim r \lesssim 1$ pc. The number density

[14] Masers are regions of stimulated non-thermal emission which show a very high surface brightness. The maser phenomenon is similar to that of lasers, except that the former radiate in the microwave regime of the spectrum. Masers are sometimes found in the atmospheres of active stars.

[15] One problem in the combined analysis of data taken in different wavelength bands is that astrometry in each individual wavelength band can be performed with a very high precision – e.g., individually in the radio and the IR band – however, the relative astrometry between these bands is less well known. To stack maps of different wavelength precisely "on top of each other", knowledge of exact relative astrometry is essential. This can be gained if a population of compact sources exists that is observable in both wavelength domains and for which accurate positions can be measured.

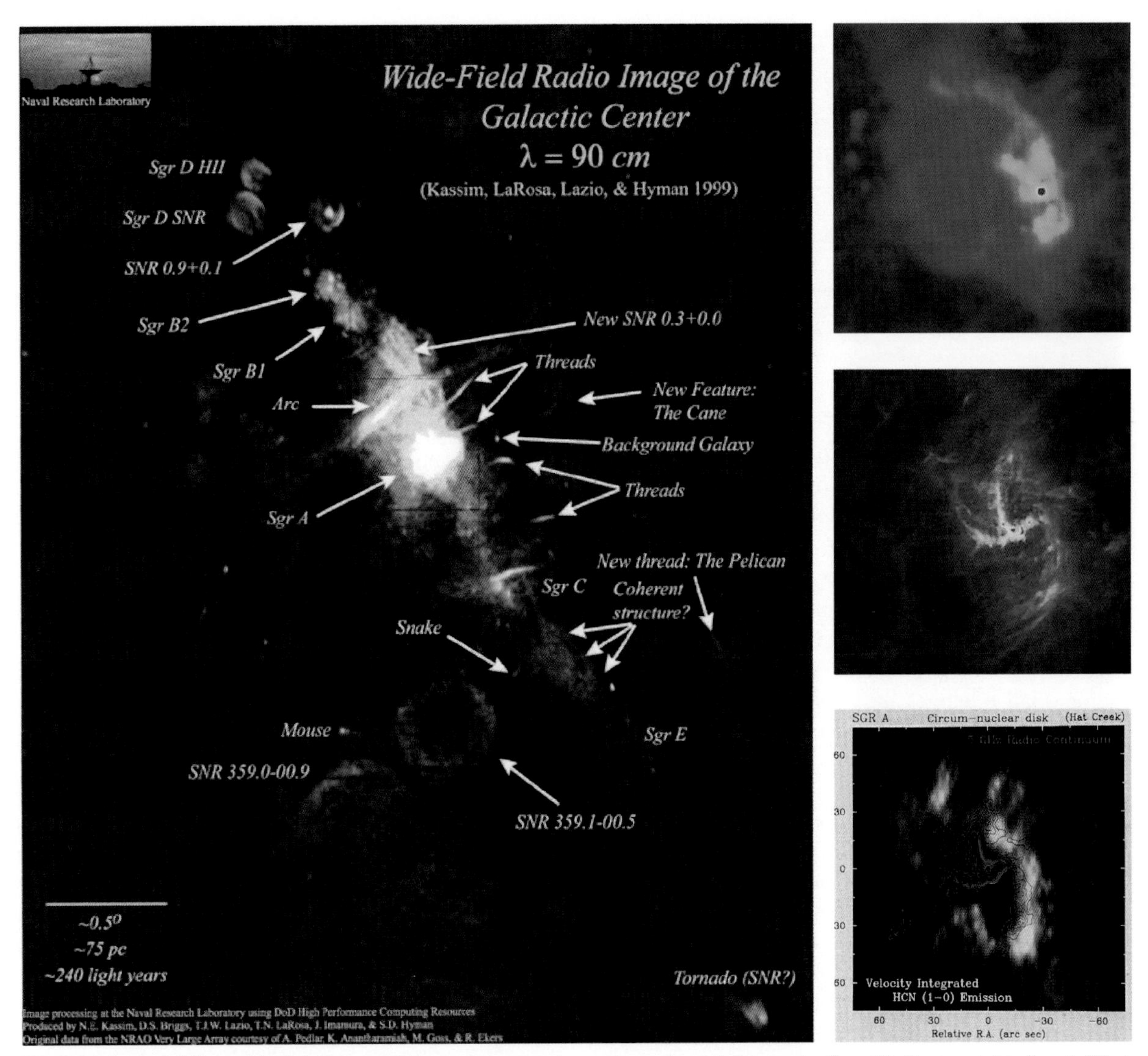

Fig. 2.34. Left: A VLA wide-field image of the region around the Galactic center, with a large number of sources identified. Upper right: a 20 cm continuum VLA image of Sgr A East, where the red dot marks Sgr A*. Center right: Sgr A West, as seen in a 6-cm continuum VLA image. Lower right: the circumnuclear ring in HCN line emission

of stars in its inner region is so large that close stellar encounters are common. It can be estimated that a star has a close encounter about every $\sim 10^6$ years. Thus, it is expected that the distribution of the stars is "thermalized", which means that the local velocity distribution of the stars is the same everywhere, i.e., it is close to a Maxwellian distribution with a constant velocity dispersion. For such an isothermal distribution we expect a density profile $n \propto r^{-2}$, which is in good agreement with the observation.

However, another observational result yields a striking and interesting discrepancy with respect to the idea of an isothermal distribution. Instead of the expected constant dispersion σ of the radial velocities of the stars, a strong radial dependence is observed: σ increases towards smaller r. For example, one finds $\sigma \sim 55\,\mathrm{km/s}$ at

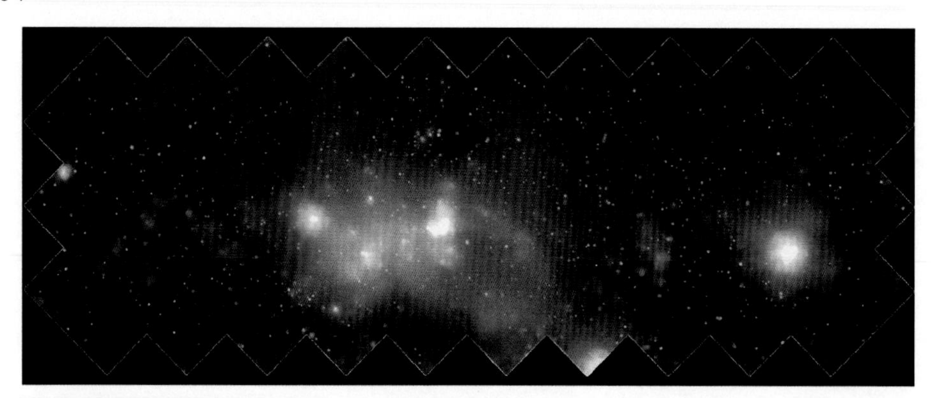

Fig. 2.35. Mosaic of X-ray images of the Galactic center, taken by the Chandra satellite. The image covers an area of about 130 pc × 300 pc (48′ × 120′). The actual GC, in which a supermassive black hole is suspected to reside, is located in the white region near the center of the image. Furthermore, on this image hundreds of white dwarfs, neutron stars, and black holes are visible that radiate in the X-ray regime due to accretion phenomena (accreting X-ray binaries). Colors code the photon energy, from low energy (red) to high energy (blue). The diffuse emission, predominantly red in this image, originates in diffuse hot gas with a temperature of about $T \sim 10^7$ K

$r = 5$ pc, but $\sigma \sim 180$ km/s at $r = 0.15$ pc. This discrepancy indicates that the gravitational potential in which the stars are moving is generated not only by themselves. According to the virial theorem, the strong increase of σ for small r implies the presence of a central mass concentration in the star cluster.

Proper Motions. Since the middle of the 1990s, proper motions of stars in this star cluster have also been measured, using the methods of speckle interferometry and adaptive optics. These produce images at diffraction-limited angular resolution, about $\sim 0.''15$ in the K-band at the ESO/NTT (3.5 m) and about $\sim 0.''05$ at the Keck (10 m). Proper motions are currently known for about 1000 stars within $\sim 10''$ of Sgr A*. This breakthrough was achieved independently by two groups, whose results are in excellent agreement. For more than 20 stars within $\sim 5''$ of Sgr A* both proper motions and radial velocities, and therefore their three-dimensional velocities are known. The radial and tangential velocity dispersions resulting from these measurements are in good mutual agreement. Thus, it can be concluded that a basically isotropic distribution of the stellar orbits exists, simplifying the study of the dynamics of this stellar cluster.

The Origin of Very Massive Stars near the Galactic Center. One of the unsolved problems is the presence of these massive stars close to the Galactic center. One finds that most of the innermost stars are main-sequence B-stars. Their small lifetime of $\sim 10^8$ yr probably implies that these stars were born close to the Galactic center. This, however, is very difficult to understand. Both the strong tidal gravitational field of the central black hole (see below) and the presumably strong magnetic field in this region will prevent the "standard" star-formation picture of a collapsing molecular cloud: the former effect tends to disrupt such a cloud while the latter stabilizes it against gravitational contraction. Several solutions to this problem have been suggested, such as a scenario in which the stars are born at larger distances from the Galactic center and then brought there by dynamical processes, involving strong gravitational scattering events. However, none of these models appears satisfactory at present.

2.6.3 A Black Hole in the Center of the Milky Way

Some stars within $0.''6$ of Sgr A* have a proper motion of more than 1000 km/s, as shown in Fig. 2.36. For instance, the star S1 has a separation of only $0.''1$ from Sgr A* and shows proper motion of 1470 km/s at the epoch displayed in Fig. 2.36. Combining the velocity dispersions in radial and tangential directions reveals it to be increasing according to the Kepler law for the presence of a point mass, $\sigma \propto r^{-1/2}$ down to $r \sim 0.01$ pc.

By now, the *acceleration* of some stars in the star cluster has also been measured, i.e., the change of proper

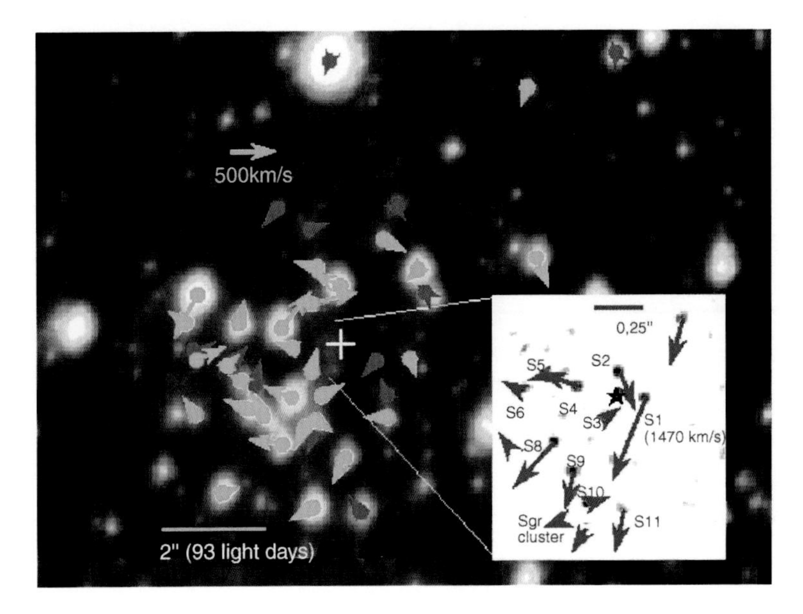

Fig. 2.36. Proper motions of stars in the central region of the GC. The differently colored arrows denote different types of stars. The small image shows the proper motions in the Sgr A* star cluster within half an arcsecond from Sgr A*; the fastest star (S1) has a proper motion of $\sim 1500\,\mathrm{km/s}$ (from Genzel, 2000, astro-ph/0008119

motion with time. From these measurements Sgr A* indeed emerges as the focus of the orbits and thus as the center of mass. Figure 2.37 shows the orbits of some stars around Sgr A*. The star S2 could be observed during a major fraction of its orbit, where a maximum velocity of more than 5000 km/s was found. The eccentricity of the orbit of S2 is 0.87, and its orbital period is ~ 15.7 yr. The minimum separation of this star from Sgr A* is only 6×10^{-4} pc, or about 100 AU.

From the observed kinematics, the enclosed mass $M(r)$ can be calculated, see Fig. 2.38. The corresponding analysis yields that $M(r)$ is basically constant over the range $0.01\,\mathrm{pc} \lesssim r \lesssim 0.5\,\mathrm{pc}$. This exciting result clearly indicates the presence of a point mass, for which a mass of

$$M = (3.6 \pm 0.4) \times 10^6 M_\odot \tag{2.91}$$

is determined. For larger radii, the mass of the star cluster dominates; it nearly follows an isothermal density distribution with a core radius of ~ 0.34 pc and a central density of $3.6 \times 10^6 M_\odot/\mathrm{pc}^3$. This result is also compatible with the kinematics of the gas in the center of the Galaxy. However, stars are much better kinematic indicators because gas can be affected by magnetic fields, viscosity, and various other processes besides gravity.

The kinematics of stars in the central star cluster of the Galaxy shows that our Milky Way contains a mass concentration in which $\sim 3 \times 10^6 M_\odot$ are concentrated within a region smaller than 0.01 pc. This is most probably a black hole in the center of our Galaxy at the position of the compact radio source Sgr A*.

Why a Black Hole? We have interpreted the central mass concentration as a black hole; this requires some further explanation:

- The energy for the central activity in quasars, radio galaxies, and other AGNs is produced by accretion of gas onto a supermassive black hole (SMBH); we will discuss this in more detail in Sect. 5.3. Thus we know that at least a subclass of galaxies contains a central SMBH. Furthermore, we will see in Sect. 3.5 that many "normal" galaxies, especially ellipticals, harbor a black hole in their center. The presence of a black hole in the center of our own Galaxy would therefore not be something unusual.
- To bring the radial mass profile $M(r)$ into accordance with an extended mass distribution, its density distribution must be very strongly concentrated, with a density profile steeper than $\propto r^{-4}$; otherwise the

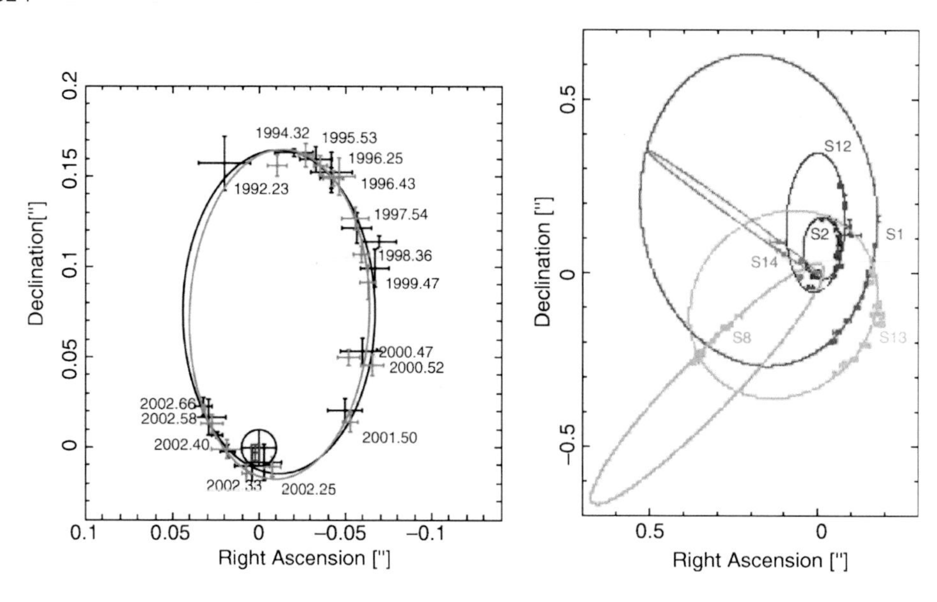

Fig. 2.37. At left, the orbit of the star S2 around Sgr A* is shown as determined by two different observing campaigns. The position of Sgr A* is indicated by the black circled cross. The individual points along the orbit are identified by the epoch of the observation. The right-hand image shows the orbits of some other stars for which accelerations have already been measured

mass profile $M(r)$ would not be as flat as observed in Fig. 2.38. Hence, this hypothetical mass distribution must be vastly different from the expected isothermal distribution which has a mass profile $\propto r^{-2}$, as discussed in Sect. 2.6.2. However, observations of the stellar distribution provide no indication of an inwardly increasing density of the star cluster with such a steep profile.

- Even if such an ultra-dense star cluster (with a central density of $\gtrsim 4 \times 10^{12} M_{\odot}/\mathrm{pc}^3$) were present it could not be stable, but instead would dissolve within $\sim 10^7$ years through frequent stellar collisions.
- Sgr A* itself has a proper motion of less than 20 km/s. It is therefore the dynamic center of the Milky Way. Due to the large velocities of its surrounding stars, one would derive a mass of $M \gg 10^3 M_{\odot}$ for the radio source, assuming equipartition of energy (see also Sect. 2.6.5). Together with the tight upper limits for its extent, a lower limit for the density of $10^{18} M_{\odot}/\mathrm{pc}^3$ can then be obtained.

Following the stellar orbits in forthcoming years will further complete our picture of the mass distribution in the GC.

We have to emphasize at this point that the gravitational effect of the black hole on the motion of stars and gas is constrained to the innermost region of the Milky Way. As one can see from Fig. 2.38, the gravitational field of the SMBH dominates the rotation curve of the Galaxy only for $R \lesssim 2$ pc – this is the very reason why the detection of the SMBH is so difficult. At larger radii, the presence of the SMBH is of no relevance for the rotation curve of the Milky Way.

2.6.4 Flares from the Galactic Center

In 2000, the X-ray satellite Chandra discovered a powerful X-ray flare from Sgr A*. This event lasted for about three hours, and the X-ray flux increased by a factor of 50 during this period. XMM-Newton confirmed the existence of X-ray flares, recording one where the luminosity increased by a factor of ~ 200. Combining the flare duration of a few hours with the short time-scale of variability of a few minutes indicates that the emission must originate from a very small source, not larger than $\sim 10^{13}$ cm in size.

Monitoring Sgr A* in the NIR, flare emission was also found in this spectral regime. These NIR flares are more frequent than in X-rays, occurring several times per day. Furthermore, the NIR emission seems to show some sort of periodicity of ~ 17 min, which is most likely to be identified with an orbital motion of the emitting material around the SMBH. Indeed, a reanalysis of the X-ray light curve shows some hint of

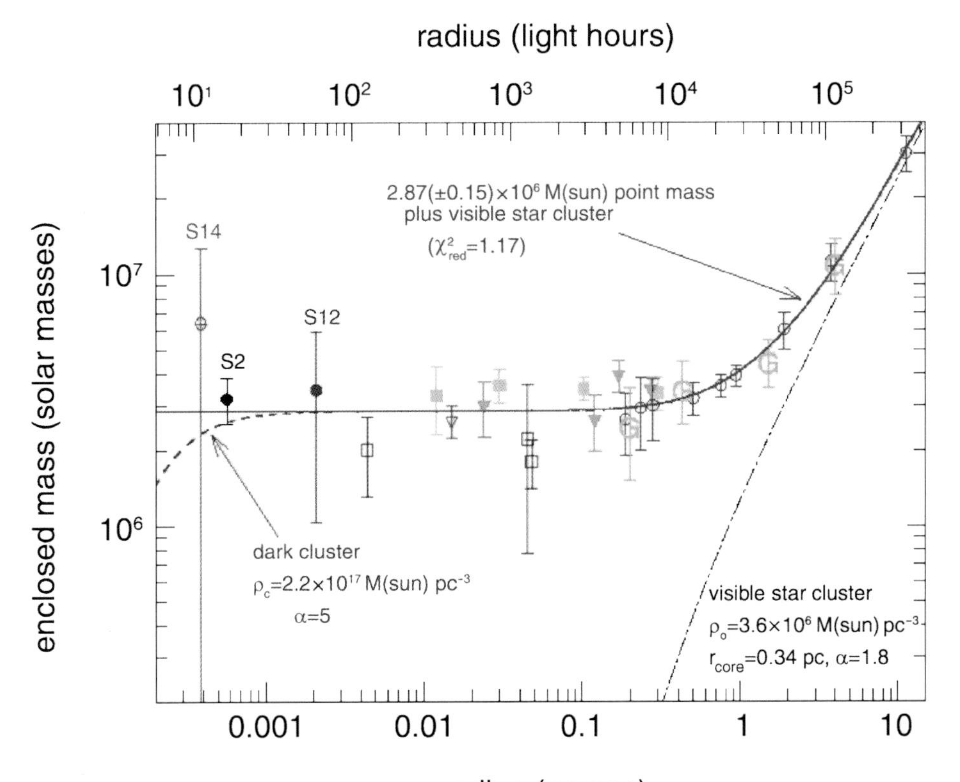

Fig. 2.38. Determination of the mass $M(r)$ within a radius r from Sgr A*, as measured by the radial velocities and proper motions of stars in the central cluster. Mass estimates obtained from individual stars (S14, S2, S12) are given by the points with error bars for small r. The other data points were derived from the kinematic analysis of the observed proper motions of the stars, where different methods have been applied. As can be seen, these methods produce results that are mutually compatible, so that the mass profile plotted here can be regarded to be robust. The solid curve is the best-fit model, representing a point mass of $2.9 \times 10^6 M_\odot$ plus a star cluster with a central density of $3.6 \times 10^6 M_\odot/\mathrm{pc}^3$ (the mass profile of this star cluster is indicated by the dash-dotted curve). The dashed curve shows the mass profile of a hypothetical cluster with a very steep profile, $n \propto r^{-5}$, and a central density of $2.2 \times 10^{17} M_\odot\,\mathrm{pc}^{-3}$

the same modulation time-scale. Observing the Galactic center simultaneously in the NIR and the X-rays revealed a clear correlation of the corresponding light curves; for example, simultaneous flares were found in these two wavelength regimes. These flares have similar light profiles, indicating a similar origin of their radiation. The consequences of these observations for the nature of the central black hole will be discussed in Sect. 5.4.6, after we have introduced the concept of black holes in a bit more detail. Flares were also observed at mm-wavelengths; their time-scale appears to be longer than that at higher frequencies, as expected if the emission comes from a more extended source component.

2.6.5 The Proper Motion of Sgr A*

From a series of VLBI observations of the position of Sgr A*, covering eight years, the proper motion of this compact radio source was measured with very high precision. To do this, the position of Sgr A* was determined relative to two compact extragalactic radio sources. Due to their large distances these are not expected to show any proper motion, and the VLBI measurements show that their separation vector is indeed constant over time. The position of Sgr A* over the observing period is plotted in Fig. 2.39.

From the plot, we can conclude that the observed proper motion of Sgr A* is essentially parallel to the Galactic plane. The proper motion perpendicular to the Galactic plane is about 0.2 mas/yr, compared to the proper motion in the Galactic plane of 6.4 mas/yr. If $R_0 = (8.0 \pm 0.5)$ kpc is assumed for the distance to the GC, this proper motion translates into an orbital velocity of (241 ± 15) km/s, where the uncertainty is dominated by the exact value of R_0 (the error in the measurement alone would yield an uncertainty of only 1 km/s). This proper motion is easily explained by the Solar orbital motion around the GC, i.e., this measurement contains no hint of a non-zero velocity of the radio source Sgr A* itself. In fact, the small deviation of the proper motion from the orientation of the Galactic plane can be explained by the peculiar velocity of the Sun relative to the LSR (see Sect. 2.4.1). If this is taken into account, a velocity perpendicular to the Galactic disk of $v_\perp = (-0.4 \pm 0.9)$ km/s is obtained for Sgr A*. The component of the orbital velocity within the disk has a much larger uncertainty because we know neither R_0 nor the rotational velocity V_0 of the LSR very precisely. The small upper limit for $v_\perp$ suggests, however, that the motion in the disk should also be very small. Under the (therefore plausible) assumption that Sgr A* has no peculiar velocity, the ratio R_0/V_0 can be determined from these measurements with an as yet unmatched precision.

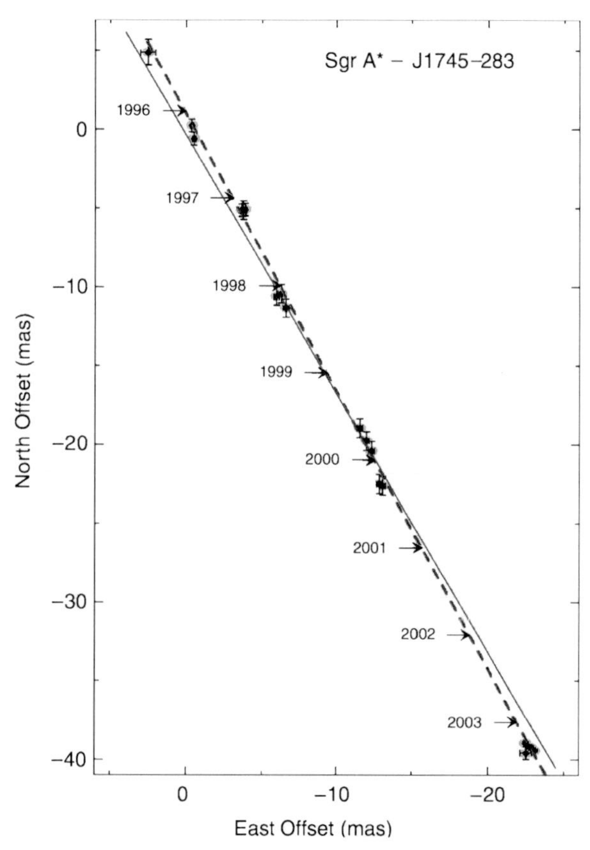

Fig. 2.39. The position of Sgr A* at different epochs, relative to the position in 1996. To a very good approximation the motion is linear, as indicated by the dashed best-fit straight line. In comparison, the solid line shows the orientation of the Galactic plane

What also makes this observation so impressive is that from it we can directly derive a lower limit for the mass of Sgr A*. Since this radio source is surrounded by $\sim 10^6$ stars within a sphere of radius ~ 1 pc, the net acceleration towards the center is not vanishing, even in the case of a statistically isotropic distribution of stars. Rather, due to the discrete nature of the mass distribution, a stochastic force exists that changes with time because of the orbital motion of the stars. The radio source is accelerated by this force, causing a motion of Sgr A* which becomes larger the smaller the mass of the source. The very strong limits to the velocity of Sgr A* enable us to derive a lower limit for its mass of $0.4 \times 10^6 M_\odot$. This mass limit is significantly lower than the mass of the SMBH that was derived from the stellar orbits, but it is the mass of the radio source itself. Although we have excellent reasons to assume that Sgr A* coincides with the SMBH, this new observation is the first proof for a large mass of the radio source itself.

2.6.6 Hypervelocity Stars in the Galaxy

Discovery. In 2005, a Galactic star was discovered which travels with a velocity of at least 700 km/s relative to the Galactic rest-frame. This B-star has a distance of ~ 110 kpc from the Galactic center, and its actual space velocity depends on its transverse motion which has not be yet been measured, due to the large distance of the object from us. The velocity of this star is so large that it exceeds the escape velocity from the Galaxy;

hence, this star is gravitationally unbound to the Milky Way. Within one year after this first discovery, four more such *hypervelocity stars* have been discovered, all of them early-type stars (O- or B-stars) with Galactic rest-frame velocities in excess of 500 km/s. They will all escape the gravitational potential of the Galaxy.

Acceleration of Hypervelocity Stars. The fact that the hypervelocity stars are gravitationally unbound to the Milky Way implies that they must have been accelerated very recently, i.e., less than a crossing time through the Galaxy ago. In addition, since they are early-type stars, they must have been accelerated within the lifetime of such stars. The acceleration mechanism must be of gravitational origin and is related to the *dynamical instability* of N-body systems, with $N > 2$. A pair of objects will orbit in their joint gravitational field, either on bound orbits (ellipses) or unbound ones (gravitational scattering on hyperbolic orbits); in the former case, the system is stable and the two masses will orbit around each other literally forever. If more than two masses are involved this is no longer the case – such a system is inherently unstable. Consider three masses, initially bound to each other, orbiting around their center-of-mass. In general, their orbits will not be ellipses but are more complicated; in particular, they are not periodic. Such a system is, mathematically speaking, chaotic. A chaotic system is characterized by the property that the state of a system at time t depends very sensitively on the initial conditions set at time $t_i < t$. Whereas for a dynamically stable system the positions and velocities of the masses at time t are changed only a little if their initial conditions are slightly varied (e.g., by giving one of the masses a slightly larger velocity), in a chaotic, dynamically unstable system even tiny changes in the initial conditions can lead to completely different states at later times. Any N-body system with $N > 2$ is dynamically unstable.

Back to our three-body system. The three masses may orbit around each other for an extended period of time, but their gravitational interaction may then change the state of the system suddenly, in that one of the three masses attains a sufficiently high velocity relative to the other two and may escape to infinity, whereas the other two masses form a binary system. What was a bound system initially may become an unbound system later on. This behavior may appear unphysical at first sight – where does the energy come from to eject one of the stars? Is this process violating energy conservation?

Of course not! The trick lies in the properties of gravity: a binary has *negative* binding energy, and the more negative, the tighter the binary orbit. By three-body interactions, the orbit of two masses can become tighter (one says that the binary "hardens"), and the corresponding excess energy is transferred to the third mass which may then become gravitationally unbound. In fact, a single binary of compact stars can in principle take up all the binding energy of a star cluster and "evaporate" all other stars.

This discussion then leads to the explanation of hypervelocity stars. The characteristic escape velocity of the "third mass" will be the orbital velocity of the three-body system before the escape. The only place in our Milky Way where orbital velocities are as high as that observed for the hypervelocity stars is the Galactic center. In fact, the travel time of a star with current velocity of ~ 600 km/s from the Galactic center to Galactrocentric distances of ~ 80 kpc is of order 10^8 yr, slightly shorter than the main-sequence lifetime of a B-star. Furthermore, most of the bright stars in the central $1''$ of the Galactic center region are B-stars. Therefore, the immediate environment of the central black hole is the natural origin for these hypervelocity stars. Indeed, long before their discovery the existence of such stars was predicted. When a binary star gets close to the black hole, this three-body interaction can lead to the ejection of one of the two stars into an unbound orbit. Thus, the existence of hypervelocity stars can be considered as an additional piece of evidence for the presence of a central black hole in our Galaxy.

3. The World of Galaxies

The insight that our Milky Way is just one of many galaxies in the Universe is less than 100 years old, despite the fact that many had already been known for a long time. The catalog by Charles Messier (1730–1817), for instance, lists 103 diffuse objects. Among them M31, the Andromeda galaxy, is listed as the 31st entry in the Messier catalog. Later, this catalog was extended to 110 objects. John Dreyer (1852–1926) published the *New General Catalog (NGC)* that contains nearly 8000 objects, most of them galaxies. In 1912, Vesto Slipher found that the spiral nebulae are rotating, using spectroscopic analysis. But the nature of these extended sources, then called nebulae, was still unknown at that time; it was unclear whether they are part of our Milky Way or outside it.

The year 1920 saw a public debate (the Great Debate) between Harlow Shapley and Heber Curtis. Shapley believed that the nebulae are part of our Milky Way, whereas Curtis was convinced that the nebulae must be objects located outside the Galaxy. The arguments which the two opponents brought forward were partly based on assumptions which later turned out to be invalid, as well as on incorrect data. We will not go into the details of their arguments which were partially linked to the assumed size of the Milky Way since, only a few years later, the question of the nature of the nebulae was resolved.

In 1925, Edwin Hubble discovered Cepheids in Andromeda (M31). Using the period-luminosity relation for these pulsating stars (see Sect. 2.2.7) he derived a distance of 285 kpc. This value is a factor of ~ 3 smaller than the distance of M31 known today, but it provided clear evidence that M31, and thus also other spiral nebulae, must be extragalactic. This then immediately implied that they consist of innumerable stars, like our Milky Way. Hubble's results were considered conclusive by his contemporaries and marked the beginning of extragalactic astronomy. It is not coincidental that at this time George Hale began to arrange the funding for an ambitious project. In 1928 he obtained six

Fig. 3.1. Galaxies occur in different shapes and sizes, and often they are grouped together in groups or clusters. This cluster, ACO 3341, at a redshift of $z = 0.037$, contains numerous galaxies of different types and luminosities

Peter Schneider, The World of Galaxies.
In: Peter Schneider, Extragalactic Astronomy and Cosmology. pp. 87–140 (2006)
DOI: 10.1007/11614371_3 © Springer-Verlag Berlin Heidelberg 2006

million dollars for the construction of the 5-m telescope on Mt. Palomar which was completed in 1949.

This chapter is about galaxies. We will confine the consideration here to "normal" galaxies in the local Universe; galaxies at large distances, some of which are in a very early evolutionary state, will be discussed in Chap. 9, and active galaxies, like quasars for example, will be discussed later in Chap. 5.

3.1 Classification

The classification of objects depends on the type of observation according to which this classification is made. This is also the case for galaxies. Historically, optical photometry was the method used to observe galaxies. Thus, the morphological classification defined by Hubble is still the best-known today. Besides morphological criteria, color indices, spectroscopic parameters (based on emission or absorption lines), the broad-band spectral distribution (galaxies with/without radio- and/or X-ray emission), as well as other features may also be used.

3.1.1 Morphological Classification: The Hubble Sequence

Figure 3.2 shows the classification scheme defined by Hubble. According to this, three main types of galaxies exist:

- *Elliptical galaxies* (E's) are galaxies that have nearly elliptical isophotes[1] without any clearly defined structure. They are subdivided according to their ellipticity $\epsilon \equiv 1 - b/a$, where a and b denote the semimajor and the semiminor axes, respectively. Ellipticals are found over a relatively broad range in ellipticity, $0 \le \epsilon \lesssim 0.7$. The notation E$n$ is commonly used to classify the ellipticals with respect to ϵ, with $n = 10\epsilon$; i.e., an E4 galaxy has an axis ratio of $b/a = 0.6$, and E0's have circular isophotes.
- *Spiral galaxies* consist of a disk with spiral arm structure and a central bulge. They are divided into two subclasses: *normal spirals* (S's) and *barred spirals* (SB's). In each of these subclasses, a sequence is defined that is ordered according to the brightness ratio of bulge and disk, and that is denoted by a, ab, b, bc, c, cd, d. Objects along this sequence are often referred to as being either an early-type or a late-type; hence, an Sa galaxy is an early-type spiral, and an SBc galaxy is a late-type barred spiral. We stress explicitly that this nomenclature is not a statement of the evolutionary stage of the objects but is merely a nomenclature of purely historical origin.
- *Irregular galaxies* (Irr's) are galaxies with only weak (Irr I) or no (Irr II) regular structure. The classification of Irr's is often refined. In particular, the sequence of spirals is extended to the classes Sdm, Sm, Im, and Ir (m stands for Magellanic; the Large Magellanic Cloud is of type SBm).
- *S0 galaxies* are a transition between ellipticals and spirals. They are also called lenticulars as they are lentil-shaped galaxies which are likewise subdivided into S0 and SB0, depending on whether or not they show a bar. They contain a bulge and a large enveloping region of relatively unstructured brightness which often appears like a disk without spiral arms. Ellipticals and S0 galaxies are referred to as early-type galaxies, spirals as late-type galaxies. As before,

[1] Isophotes are contours along which the surface brightness of a sources is constant. If the light profile of a galaxy is elliptical, then its isophotes are ellipses.

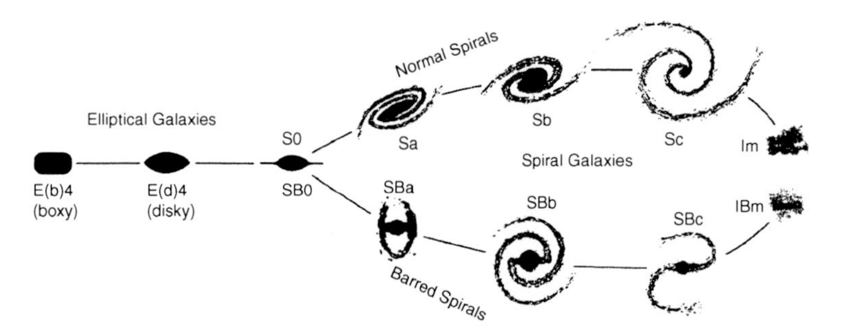

Fig. 3.2. Hubble's "tuning fork" for galaxy classification

these names are only historical and are not meant to describe an evolutionary track!

Obviously, the morphological classification is at least partially affected by projection effects. If, for instance, the spatial shape of an elliptical galaxy is a triaxial ellipsoid, then the observed ellipticity ϵ will depend on its orientation with respect to the line-of-sight. Also, it will be difficult to identify a bar in a spiral that is observed from its side ("edge-on").

Besides the aforementioned main types of galaxy morphologies, others exist which do not fit into the Hubble scheme. Many of these are presumably caused by interaction between galaxies (see below). Furthermore, we observe galaxies with radiation characteristics that differ significantly from the spectral behavior of "normal" galaxies. These galaxies will be discussed next.

3.1.2 Other Types of Galaxies

The light from "normal" galaxies is emitted mainly by stars. Therefore, the spectral distribution of the radiation from such galaxies is in principle a superposition of the spectra of their stellar population. The spectrum of stars is, to a first approximation, described by a Planck function (see Appendix A) that depends only on the star's surface temperature. A typical stellar population covers a temperature range from a few thousand Kelvin up to a few tens of thousand Kelvin. Since the Planck function has a well-localized maximum and from there steeply declines to both sides, most of the energy of such "normal" galaxies is emitted in a relatively narrow frequency interval that is located in the optical and NIR sections of the spectrum.

In addition to these, other galaxies exist whose spectral distribution cannot be described by a superposition of stellar spectra. One example is the class of active galaxies which generate a significant fraction of their luminosity from gravitational energy that is released in the infall of matter onto a supermassive black hole, as was mentioned in Sect. 1.2.4. The activity of such objects can be recognized in various ways. For example, some of them are very luminous in the radio and/or in the X-ray portion of the spectrum (see Fig. 3.3), or they show strong emission lines with a width of several thousand km/s if the line width is interpreted as due to Doppler broadening, i.e., $\Delta\lambda/\lambda = \Delta v/c$. In many cases, by far the largest fraction of luminosity is produced in a very small central region: the active galactic nucleus (AGN) that gave this class of galaxies its name. In quasars, the central luminosity can be of the order of $\sim 10^{13} L_\odot$, about a thousand times as luminous as the total luminosity of our Milky Way. We will discuss active galaxies, their phenomena, and their physical properties in detail in Chap. 5.

Another type of galaxy also has spectral properties that differ significantly from those of "normal" galaxies, namely the starburst galaxies. Normal spiral galaxies like our Milky Way form new stars at a star-formation rate of $\sim 3 M_\odot/\mathrm{yr}$ which can be derived, for instance, from the Balmer lines of hydrogen generated in the HII regions around young, hot stars. By contrast, elliptical galaxies show only marginal star formation or none at all. However, there are galaxies which have a much higher star-formation rate, reaching values of

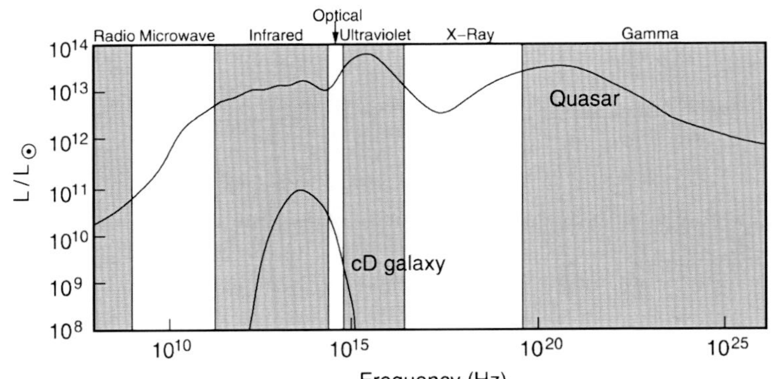

Fig. 3.3. The spectrum of a quasar (3C273) in comparison to that of an elliptical galaxy. While the radiation from the elliptical is concentrated in a narrow range spanning less than two decades in frequency, the emission from the quasar is observed over the full range of the electromagnetic spectrum, and the energy per logarithmic frequency interval is roughly constant. This demonstrates that the light from the quasar cannot be interpreted as a superposition of stellar spectra, but instead has to be generated by completely different sources and by different radiation mechanisms

$100 M_\odot$/yr and more. If many young stars are formed we would expect these starburst galaxies to radiate strongly in the blue or in the UV part of the spectrum, corresponding to the maximum of the Planck function for the most massive and most luminous stars. This expectation is not fully met though: star formation takes place in the interior of dense molecular clouds which often also contain large amounts of dust. If the major part of star formation is hidden from our direct view by layers of absorbing dust, these galaxies will not be very prominent in blue light. However, the strong radiation from the young, luminous stars heats the dust; the absorbed stellar light is then emitted in the form of thermal dust emission in the infrared and submillimeter regions of the electromagnetic spectrum – these galaxies can thus be extremely luminous in the IR. They are called ultra-luminous infrared galaxies (ULIRGs). We will describe the phenomena of starburst galaxies in more detail in Sect. 9.2.1. Of special interest is the discovery that the star-formation rate of galaxies seems to be closely related to interactions between galaxies – many ULIRGs are strongly interacting galaxies (see Fig. 3.4).

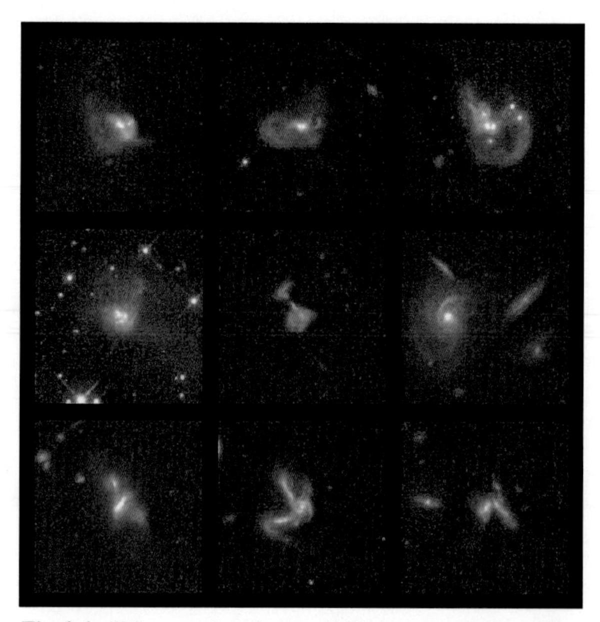

Fig. 3.4. This mosaic of nine HST images shows different ULIRGs in collisional interaction between two or more galaxies

3.2 Elliptical Galaxies

3.2.1 Classification

The general term "elliptical galaxies" (or ellipticals, for short) covers a broad class of galaxies which differ in their luminosities and sizes – some of them are displayed in Fig. 3.5. A rough subdivision is as follows:

- *Normal ellipticals.* This class includes giant ellipticals (gE's), those of intermediate luminosity (E's), and compact ellipticals (cE's), covering a range in absolute magnitudes from $M_B \sim -23$ to $M_B \sim -15$. In addition, S0 galaxies are often assigned to this class of early-type galaxies.
- *Dwarf ellipticals* (dE's). These differ from the cE's in that they have a significantly smaller surface brightness and a lower metallicity.
- *cD galaxies.* These are extremely luminous (up to $M_B \sim -25$) and large (up to $R \lesssim 1$ Mpc) galaxies that are only found near the centers of dense clusters of galaxies. Their surface brightness is very high close to the center, they have an extended diffuse envelope, and they have a very high M/L ratio.
- *Blue compact dwarf galaxies.* These "blue compact dwarfs" (BCD's) are clearly bluer (with $\langle B-V \rangle$ between 0.0 and 0.3) than the other ellipticals, and contain an appreciable amount of gas in comparison.
- *Dwarf spheroidals* (dSph's) exhibit a very low luminosity and surface brightness. They have been observed down to $M_B \sim -8$. Due to these properties, they have thus far only been observed in the Local Group.

Thus elliptical galaxies span an enormous range (more than 10^6) in luminosity and mass, as is shown by the compilation in Table 3.1.

3.2.2 Brightness Profile

The brightness profiles of normal E's and cD's follow a de Vaucouleurs profile (see (2.39) or (2.41), respectively) over a wide range in radius, as is demonstrated in Fig. 3.6. The effective radius R_e is strongly correlated with the absolute magnitude M_B, as can be seen in Fig. 3.7, with rather little scatter. In comparison, the dE's and the dSph's clearly follow a different distribution. Owing to the relation (2.42) between luminosity,

Fig. 3.5. Different types of elliptical galaxies. Upper left: the cD galaxy M87 in the center of the Virgo galaxy cluster; upper right: Centaurus A, a giant elliptical galaxy with a very distinct dust disk and an active galactic nucleus; lower left: the galaxy Leo I belongs to the nine known *dwarf spheroidals* in the Local Group; lower right: NGC 1705, a dwarf irregular, shows indications of massive star formation – a super star cluster and strong galactic winds

Table 3.1. Characteristic values for elliptical galaxies. D_{25} denotes the diameter at which the surface brightness has decreased to 25 B-mag/arcsec^2, S_N is the "specific frequency", a measure for the number of globular clusters in relation to the visual luminosity (see (3.13)), and M/L is the mass-to-light ratio in Solar units (the values of this table are taken from the book by Carroll & Ostlie, 1996)

	S0	cD	E	dE	dSph	BCD
M_B	−17 to −22	−22 to −25	−15 to −23	−13 to −19	−8 to −15	−14 to −17
$M(M_\odot)$	10^{10} to 10^{12}	10^{13} to 10^{14}	10^{8} to 10^{13}	10^{7} to 10^{9}	10^{7} to 10^{8}	$\sim 10^{9}$
D_{25} (kpc)	10–100	300–1000	1–200	1–10	0.1–0.5	< 3
$\langle M/L_B \rangle$	~ 10	> 100	10–100	1–10	5–100	0.1–10
$\langle S_N \rangle$	~ 5	~ 15	~ 5	4.8 ± 1.0	–	–

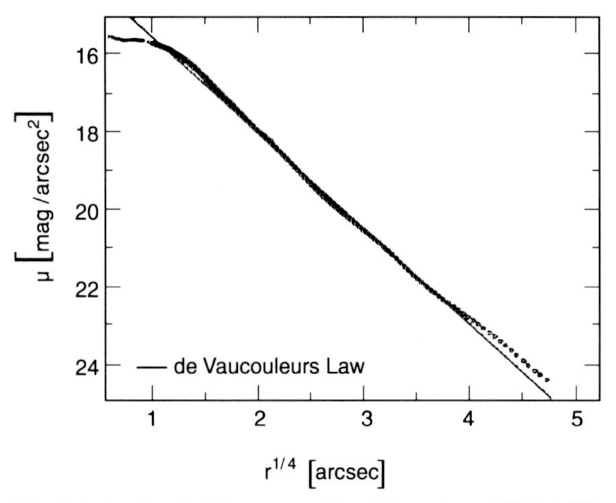

Fig. 3.6. Surface brightness profile of the galaxy NGC 4472, fitted by a de Vaucouleurs profile. The de Vaucouleurs profile describes a linear relation between the logarithm of the intensity (i.e., linear on a magnitude scale) and $r^{1/4}$; for this reason, it is also called an $r^{1/4}$-law

effective radius and central surface brightness, an analogous relation exists for the average surface brightness $\mu_{\rm ave}$ (unit: B − mag/arcsec^2) within $R_{\rm e}$ as a function of M_B. In particular, the surface brightness in normal E's decreases with increasing luminosity, while it increases for dE's and dSph's.

Yet another way of expressing this correlation is by eliminating the absolute luminosity, thus obtaining a relation between effective radius $R_{\rm e}$ and surface brightness $\mu_{\rm ave}$. This form is then called the Kormendy relation.

The de Vaucouleurs profile provides the best fits for normal E's, whereas for E's with exceptionally high (or low) luminosity the profile decreases more slowly (or rapidly) for larger radii. The profile of cD's extends much farther out and is not properly described by a de Vaucouleurs profile (Fig. 3.8), except in its innermost part. It appears that cD's are similar to E's but embedded in a very extended, luminous halo. Since cD's are only found in the centers of massive clusters of galaxies, a connection must exist between this morphology and the environment of these galaxies. In contrast to these classes of ellipticals, diffuse dE's are often better described by an exponential profile.

3.2.3 Composition of Elliptical Galaxies

Except for the BCD's, elliptical galaxies appear red when observed in the optical, which suggests an old stellar population. It was once believed that ellipticals contain neither gas nor dust, but these components have now been found, though at a much lower mass-fraction than in spirals. For example, in some ellipticals hot gas ($\sim 10^7$ K) has been detected by its X-ray emission. Furthermore, Hα emission lines of warm gas ($\sim 10^4$ K) have been observed, as well as cold gas ($\sim$ 100 K) in the HI (21-cm) and CO molecular lines. Many of the normal ellipticals contain visible amounts of dust, partially manifested as a dust disk. The metallicity of ellipticals and S0 galaxies increases towards the galaxy center, as derived from color gradients. Also in S0 galaxies the bulge appears redder than the disk. The Spitzer Space Telescope, launched in 2003, has detected a spatially extended distribution of warm dust in S0 galaxies, organized in some sort of spiral structure. Cold dust has also been found in ellipticals and S0 galaxies.

This composition of ellipticals clearly differs from that of spiral galaxies and needs to be explained by mod-

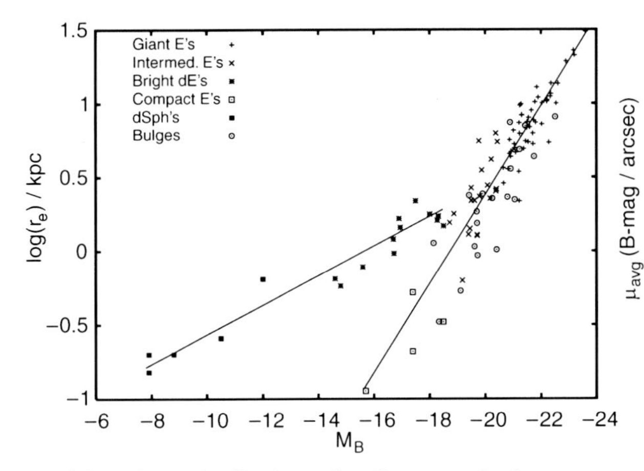

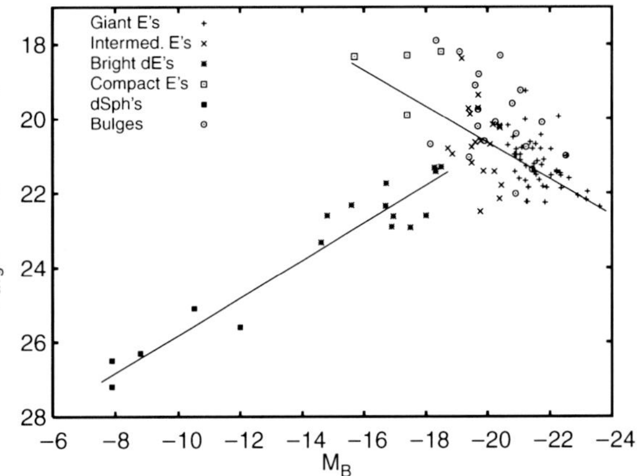

Fig. 3.7. Left panel: effective radius R_{e} versus absolute magnitude M_B; the correlation for normal ellipticals is different from that of dwarfs. Right panel: average surface brightness μ_{ave} versus M_B; for normal ellipticals, the surface brightness decreases with increasing luminosity while for dwarfs it increases

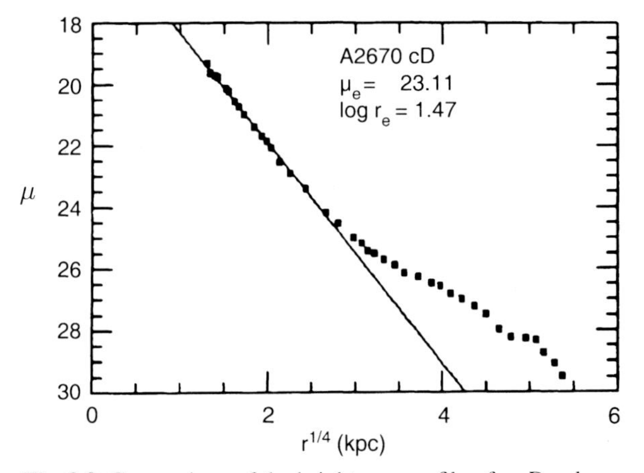

Fig. 3.8. Comparison of the brightness profile of a cD galaxy, the central galaxy of the cluster of galaxies Abell 2670, with a de Vaucouleurs profile. The light excess for large radii is clearly visible

els of the formation and evolution of galaxies. We will see later that the cosmic evolution of elliptical galaxies is also observed to be different from that of spirals.

3.2.4 Dynamics of Elliptical Galaxies

Analyzing the morphology of elliptical galaxies raises a simple question: *Why are ellipticals not round?* A simple explanation would be rotational flattening, i.e., as in a rotating self-gravitating gas ball, the stellar distribution bulges outwards at the equator due to centrifugal forces, as is also the case for the Earth. If this explanation were correct, the rotational velocity v_{rot}, which is measurable in the relative Doppler shift of absorption lines, would have to be of about the same magnitude as the velocity dispersion of the stars σ_v that is measurable through the Doppler broadening of lines. More precisely, by means of stellar dynamics one can show that for the rotational flattening of an axially symmetric, oblate[2] galaxy, the relation

$$\left(\frac{v_{\mathrm{rot}}}{\sigma_v}\right)_{\mathrm{iso}} \approx \sqrt{\frac{\epsilon}{1-\epsilon}} \tag{3.1}$$

has to be satisfied, where "iso" indicates the assumption of an isotropic velocity distribution of the stars. However, for luminous ellipticals one finds that, in general, $v_{\mathrm{rot}} \ll \sigma_v$, so that rotation cannot be the major cause of their ellipticity (see Fig. 3.9). In addition, many ellipticals are presumably triaxial, so that no unambiguous rotation axis is defined. Thus, luminous ellipticals are in general *not* rotationally flattened. For less luminous ellipticals and for the bulges of disk galaxies, however, rotational flattening can play an important role. The question remains of how to explain a stable elliptical distribution of stars without rotation.

[2] If $a \ge b \ge c$ denote the lengths of the major axes of an ellipsoid, then it is called an oblate spheroid (= rotational ellipsoid) if $a = b > c$, whereas a prolate spheroid is specified by $a > b = c$. If all three axes are different, it is called triaxial ellipsoid.

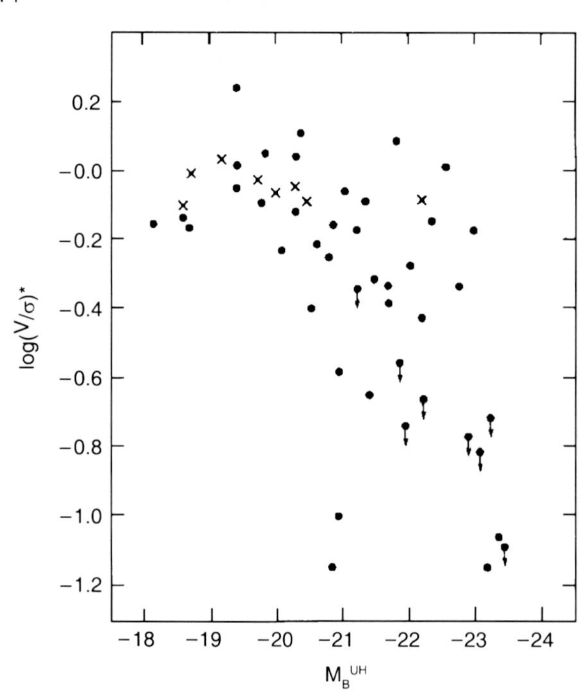

Fig. 3.9. The rotation parameter $\left(\frac{v_{\rm rot}}{\sigma_v}\right) / \left(\frac{v_{\rm rot}}{\sigma_v}\right)_{\rm iso}$ of elliptical galaxies, here denoted by $(V/\sigma)^*$, plotted as a function of absolute magnitude. Dots denote elliptical galaxies, crosses the bulges of disk galaxies

The brightness profile of an elliptical galaxy is defined by the distribution of its stellar orbits. Let us assume that the gravitational potential is given. The stars are then placed into this potential, with the initial positions and velocities following a specified distribution. If this distribution is not isotropic in velocity space, the resulting light distribution will in general not be spherical. For instance, one could imagine that the orbital planes of the stars have a preferred direction, but that an equal number of stars exists with positive and negative angular momentum L_z, so that the total stellar distribution has no angular momentum and therefore does not rotate. Each star moves along its orbit in the gravitational potential, where the orbits are in general not closed. If an initial distribution of stellar orbits is chosen such that the statistical properties of the distribution of the orbits are invariant in time, then one will obtain a stationary system. If, in addition, the distribution is chosen such that the respective mass distribution of the stars will generate exactly the originally chosen gravitational potential, one arrives at a self-gravitating equilibrium system. In general, it is a difficult mathematical problem to construct such self-gravitating equilibrium systems.

Relaxation Time-Scale. The question now arises whether such an equilibrium system can also be stable in time. One might expect that close encounters of pairs of stars would cause a noticeable disturbance in the distribution of orbits. These pair-wise collisions could then lead to a "thermalization" of the stellar orbits.[3] To examine this question we need to estimate the time-scale for such collisions and the changes in direction they cause.

For this purpose, we consider the relaxation time-scale by pair collisions in a system of N stars of mass m, total mass $M = Nm$, extent R, and a mean stellar density of $n = 3N/(4\pi R^3)$. We define the relaxation time $t_{\rm relax}$ as the characteristic time in which a star changes its velocity direction by $\sim 90°$ due to pair collisions with other stars. By simple calculation (see below), we find that

$$t_{\rm relax} \approx \frac{R}{v} \frac{N}{\ln N} , \tag{3.2}$$

or

$$\boxed{t_{\rm relax} = t_{\rm cross} \frac{N}{\ln N}} , \tag{3.3}$$

where $t_{\rm cross} = R/v$ is the crossing time-scale, i.e., the time it takes a star to cross the stellar system. If we now consider a typical galaxy, with $t_{\rm cross} \sim 10^8$ yr, $N \sim 10^{12}$ (thus $\ln N \sim 30$), then we find that the relaxation time is much longer than the age of the Universe. This means that *pair collisions do not play any role in the evolution of stellar orbits.* The dynamics of the orbits are determined solely by the large-scale gravitational field of the galaxy. In Sect. 7.5.1, we will describe a process called violent relaxation which most likely plays a central role in the formation of galaxies and which is probably also responsible for the stellar orbits establishing an equilibrium configuration.

The stars behave like a collisionless gas: elliptical galaxies are stabilized by (dynamical) pressure, and they are elliptical because the stellar distribution is

[3]Note that in a gas like air, scattering between molecules occurs frequently, which drives the velocity distribution of the molecules towards an isotropic Maxwellian, i.e., the thermal, distribution.

anisotropic in velocity space. This corresponds to an anisotropic pressure – where we recall that the pressure of a gas is nothing but the momentum transport of gas particles due to their thermal motions.

Derivation of the Collisional Relaxation Time-Scale. We consider a star passing by another one, with the impact parameter b being the minimum distance between the two. From gravitational deflection, the star attains a velocity component perpendicular to the incoming direction of

$$v_\perp^{(1)} \approx a\,\Delta t \approx \left(\frac{Gm}{b^2}\right)\left(\frac{2b}{v}\right) = \frac{2Gm}{bv}\,, \tag{3.4}$$

where a is the acceleration at closest separation and Δt the "duration of the collision", estimated as $\Delta t = 2b/v$ (see Fig. 3.10). Equation (3.4) can be derived more rigorously by integrating the perpendicular acceleration along the orbit. A star undergoes many collisions, through which the perpendicular velocity components will accumulate; these form two-dimensional vectors perpendicular to the original direction. After a time t we have $\boldsymbol{v}_\perp(t) = \sum_i \boldsymbol{v}_\perp^{(i)}$. The expectation value of this vector is $\langle \boldsymbol{v}_\perp(t)\rangle = \sum_i \left\langle \boldsymbol{v}_\perp^{(i)}\right\rangle = 0$ since the directions of the individual $\boldsymbol{v}_\perp^{(i)}$ are random. But the mean square velocity perpendicular to the incoming direction does not vanish,

$$\left\langle |\boldsymbol{v}_\perp|^2(t)\right\rangle = \sum_{ij}\left\langle \boldsymbol{v}_\perp^{(i)}\cdot\boldsymbol{v}_\perp^{(j)}\right\rangle = \sum_i \left\langle \left|\boldsymbol{v}_\perp^{(i)}\right|^2\right\rangle \neq 0\,, \tag{3.5}$$

where we set $\left\langle \boldsymbol{v}_\perp^{(i)}\cdot\boldsymbol{v}_\perp^{(j)}\right\rangle = 0$ for $i \neq j$ because the directions of different collisions are assumed to be uncorrelated. The velocity $\boldsymbol{v}_\perp$ performs a so-called *random walk*. To compute the sum, we convert it into an integral where we have to integrate over all collision parameters b. During time t, all collision partners with impact

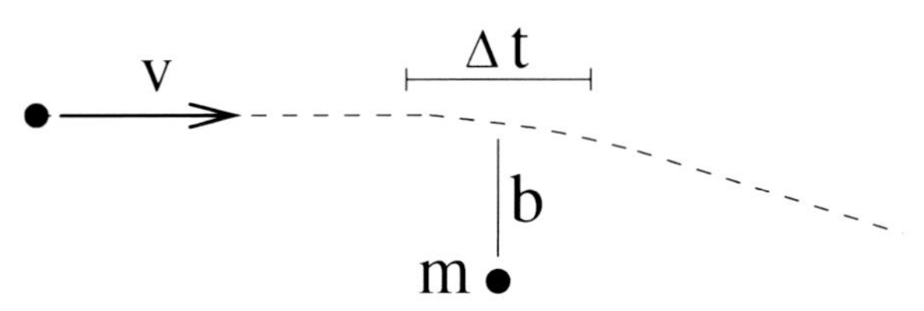

Fig. 3.10. Sketch related to the derivation of the dynamical time-scale

parameters within $\mathrm{d}b$ of b are located in a cylindrical shell of volume $(2\pi b\,\mathrm{d}b)\,(vt)$, so that

$$\begin{aligned}\left\langle |\boldsymbol{v}_\perp|^2(t)\right\rangle &= \int 2\pi\, b\,\mathrm{d}b\, v\, t\, n\, \left|v_\perp^{(1)}\right|^2 \\ &= 2\pi\left(\frac{2Gm}{v}\right)^2 v\, t\, n \int \frac{\mathrm{d}b}{b}\,.\end{aligned} \tag{3.6}$$

The integral cannot be performed from 0 to ∞. Thus, it has to be cut off at $b_{\rm min}$ and $b_{\rm max}$ and then yields $\ln(b_{\rm max}/b_{\rm min})$. Due to the finite size of the stellar distribution, $b_{\rm max} = R$ is a natural choice. Furthermore, our approximation which led to (3.4) will certainly break down if $v_\perp^{(1)}$ is of the same order of magnitude as v; hence we choose $b_{\rm min} = 2Gm/v^2$. With this, we obtain $b_{\rm max}/b_{\rm min} = Rv^2/(2Gm)$. The exact choice of the integration limits is not important, since $b_{\rm min}$ and $b_{\rm max}$ appear only logarithmically. Next, using the virial theorem, $|E_{\rm pot}| = 2E_{\rm kin}$, and thus $GM/R = v^2$ for a typical star, we get $b_{\rm max}/b_{\rm min} \approx N$. Thus,

$$\left\langle |\boldsymbol{v}_\perp|^2(t)\right\rangle = 2\pi\left(\frac{2Gm}{v}\right)^2 v\, t\, n\, \ln N\,. \tag{3.7}$$

We define the relaxation time $t_{\rm relax}$ by $\left\langle |\boldsymbol{v}_\perp|^2(t_{\rm relax})\right\rangle = v^2$, i.e., the time after which the perpendicular velocity roughly equals the infall velocity:

$$\begin{aligned} t_{\rm relax} &= \frac{1}{2\pi n v}\left(\frac{v^2}{2Gm}\right)^2 \frac{1}{\ln N} \\ &= \frac{1}{2\pi n v}\left(\frac{M}{2Rm}\right)^2 \frac{1}{\ln N} \approx \frac{R}{v}\frac{N}{\ln N}\,,\end{aligned} \tag{3.8}$$

from which we finally obtain (3.3).

3.2.5 Indicators of a Complex Evolution

The isophotes (that is, the curves of constant surface brightness) of many of the normal elliptical galaxies are well approximated by ellipses. These elliptical isophotes with different surface brightnesses are concentric to high accuracy, with the deviation of the isophote's center from the center of the galaxy being typically $\lesssim 1\%$ of its extent. However, in many cases the ellipticity varies with radius, so that the value for ϵ is not a constant. In addition, many ellipticals show a so-called isophote twist: the orientation of the semi-major axis of the isophotes changes with the radius.

This indicates that elliptical galaxies are not spheroidal, but triaxial systems (or that there is some intrinsic twist of their axes).

Although the light distribution of ellipticals appears rather simple at first glance, a more thorough analysis reveals that the kinematics can be quite complicated. For example, dust disks are not necessarily perpendicular to any of the principal axes, and the dust disk may rotate in a direction opposite to the galactic rotation. In addition, ellipticals may also contain (weak) stellar disks.

Boxiness and Diskiness. The so-called boxiness parameter describes the deviation of the isophotes' shape from that of an ellipse. Consider the shape of an isophote. If it is described by an ellipse, then after a suitable choice of the coordinate system, $\theta_1 = a\cos t$, $\theta_2 = b\sin t$, where a and b are the two semi-axes of the ellipse and $t \in [0, 2\pi]$ parametrizes the curve. The distance $r(t)$ of a point from the center is

$$r(t) = \sqrt{\theta_1^2 + \theta_2^2} = \sqrt{\frac{a^2+b^2}{2} + \frac{a^2-b^2}{2}\cos(2t)} .$$

Deviations of the isophote shape from this ellipse are now expanded in a Taylor series, where the term $\propto \cos(4t)$ describes the lowest-order correction that preserves the symmetry of the ellipse with respect to reflection in the two coordinate axes. The modified curve is then described by

$$\boldsymbol{\theta}(t) = \left(1 + \frac{a_4\cos(4t)}{r(t)}\right)\begin{pmatrix} a\cos t \\ b\sin t \end{pmatrix} , \tag{3.9}$$

with $r(t)$ as defined above. The parameter a_4 thus describes a deviation from an ellipse: if $a_4 > 0$, the isophote appears more disk-like, and if $a_4 < 0$, it becomes rather boxy (see Fig. 3.11). In elliptical galaxies we typically find $|a_4/a| \sim 0.01$, thus only a small deviation from the elliptical form.

Correlations of a_4 with Other Properties of Ellipticals. Surprisingly, we find that the parameter a_4/a is strongly correlated with other properties of ellipticals (see Fig. 3.12). The ratio $\left(\frac{v_{\rm rot}}{\sigma_v}\right) / \left(\frac{v_{\rm rot}}{\sigma_v}\right)_{\rm iso}$ (upper left in Fig. 3.12) is of order unity for disky ellipses ($a_4 > 0$) and, in general, significantly smaller than 1 for boxy ellipticals. From this we conclude that "diskies" are in part rotationally supported, whereas the flattening

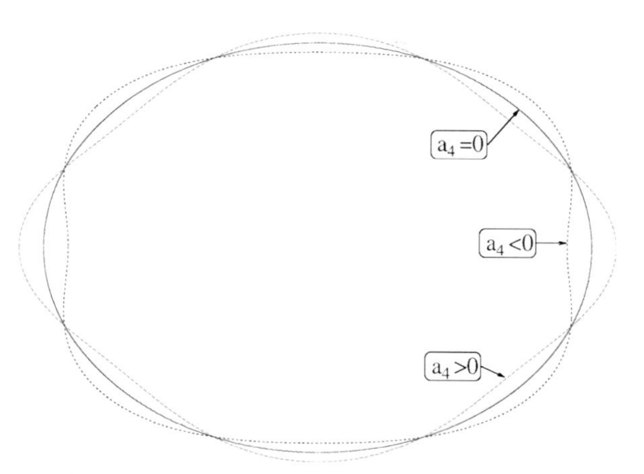

Fig. 3.11. Sketch to illustrate boxiness and diskiness. The solid red curve shows an ellipse ($a_4 = 0$), the green dashed curve a disky ellipse ($a_4 > 0$), and the blue dotted curve a boxy ellipse ($a_4 < 0$). In elliptical galaxies, the deviations in the shape of the isophotes from an ellipse are considerably smaller than in this sketch

of "boxies" is mainly caused by the anisotropic distribution of their stellar orbits in velocity space. The mass-to-light ratio is also correlated with a_4: boxies (diskies) have a value of M/L in their core which is larger (smaller) than the mean elliptical of comparable luminosity. A very strong correlation exists between a_4/a and the radio luminosity of ellipticals: while diskies are weak radio emitters, boxies show a broad distribution in $L_{\rm radio}$. These correlations are also seen in the X-ray luminosity, since diskies are weak X-ray emitters and boxies have a broad distribution in $L_{\rm X}$. This bimodality becomes even more obvious if the radiation contributed by compact sources (e.g., X-ray binary stars) is subtracted from the total X-ray luminosity, thus considering only the diffuse X-ray emission. Ellipticals with a different sign of a_4 also differ in the kinematics of their stars: boxies often have cores spinning against the general direction of rotation (counter-rotating cores), which is rarely observed in diskies.

About 70% of the ellipticals are diskies. The transition between diskies and S0 galaxies may be continuous along a sequence of varying disk-to-bulge ratio.

Shells and Ripples. In about 40% of the early-type galaxies that are not member galaxies of a cluster, sharp discontinuities in the surface brightness are found,

Fig. 3.12. Correlations of a_4/a with some other properties of elliptical galaxies. $100a(4)/a$ (corresponding to a_4/a) describes the deviation of the isophote shape from an ellipse in percent. Negative values denote boxy ellipticals, positive values disky ellipticals. The upper left panel shows the rotation parameter discussed in Sect. 3.2.4; at the lower left, the deviation from the average mass-to-light ratio is shown. The upper right panel shows the ellipticity, and the lower right panel displays the radio luminosity at 1.4 GHz. Obviously, there is a correlation of all these parameters with the boxiness parameter

a kind of shell structure ("shells" or "ripples"). They are visible as elliptical arcs curving around the center of the galaxy (see Fig. 3.13). Such sharp edges can only be formed if the corresponding distribution of stars is "cold", i.e., they must have a very small velocity dispersion, since otherwise such coherent structures would smear out on a very short time-scale. As a comparison, we can consider disk galaxies that likewise contain sharp structures, namely the thin stellar disk. Indeed, the stars in the disk have a very small velocity dispersion, ~ 20 km/s, compared to the rotational velocity of typically 200 km/s.

These peculiarities in ellipticals are not uncommon. Indicators for shells can be found in about half of the early-type galaxies, and about a third of them show boxy isophotes.

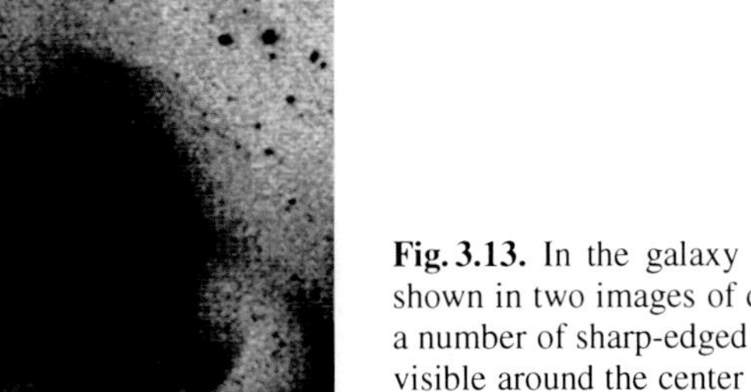

Fig. 3.13. In the galaxy NGC 474, here shown in two images of different contrast, a number of sharp-edged elliptical arcs are visible around the center of the galaxy, the so-called ripples or shells. The displayed image corresponds to a linear scale of about 90 kpc

Boxiness, counter-rotating cores, and shells and ripples are all indicators of a complex evolution that is probably caused by past mergers with other galaxies.

We will proceed with a discussion of this interpretation in Chap. 9.

3.3 Spiral Galaxies

3.3.1 Trends in the Sequence of Spirals

Looking at the sequence of early-type spirals (i.e., Sa's or SBa's) to late-type spirals, we find a number of differences that can be used for classification (see also Fig. 3.14):

- a decreasing luminosity ratio of bulge and disk, with $L_{\rm bulge}/L_{\rm disk} \sim 0.3$ for Sa's and ~ 0.05 for Sc's;
- an increasing opening angle of the spiral arms, from $\sim 6°$ for Sa's to $\sim 18°$ for Sc's;
- and an increasing brightness structure along the spiral arms: Sa's have a "smooth" distribution of stars along the spiral arms, whereas the light distribution in the spiral arms of Sc's is resolved into bright knots of stars and HII regions.

Compared to ellipticals, the spirals cover a distinctly smaller range in absolute magnitude (and mass). They are limited to $-16 \gtrsim M_B \gtrsim -23$ and $10^9 M_\odot \lesssim M \lesssim 10^{12} M_\odot$, respectively. Characteristic parameters of the various types of spirals are compiled in Table 3.2.

Bars are common in spiral galaxies, with $\sim 70\%$ of all disk galaxies containing a large-scale stellar bar. Such a bar perturbs the axial symmetry of the gravitational potential in a galaxy, which may have a number of consequences. One of them is that this perturbation can lead to a redistribution of angular momentum of the stars, gas, and dark matter. In addition, by perturbing the orbits, gas can be driven towards the center of the galaxy which may have important consequences for triggering nuclear activity (see Chap. 5).

3.3.2 Brightness Profile

The light profile of the bulge of spirals is described by a de Vaucouleurs profile to a good approximation – see (2.39) and (2.41) – while the disk follows an exponential brightness profile, as is the case for our Milky Way. Expressing these distributions of the surface brightness in $\mu \propto -2.5\log(I)$, measured in mag/arcsec^2, we obtain

$$\mu_{\rm bulge}(R) = \mu_{\rm e} + 8.3268\left[\left(\frac{R}{R_{\rm e}}\right)^{1/4} - 1\right] \quad (3.10)$$

and

$$\mu_{\rm disk}(R) = \mu_0 + 1.09\left(\frac{R}{h_r}\right) . \quad (3.11)$$

Table 3.2. Characteristic values for spiral galaxies. $V_{\rm max}$ is the maximum rotation velocity, thus characterizing the flat part of the rotation curve. The opening angle is the angle under which the spiral arms branch off, i.e., the angle between the tangent to the spiral arms and the circle around the center of the galaxy running through this tangential point. $S_{\rm N}$ is the specific abundance of globular clusters as defined in (3.13). The values in this table are taken from the book by Carroll & Ostlie (1996)

	Sa	Sb	Sc	Sd/Sm	Im/Ir
M_B	−17 to −23	−17 to −23	−16 to −22	−15 to −20	−13 to −18
$M\ (M_\odot)$	10^9–10^{12}	10^9–10^{12}	10^9–10^{12}	10^8–10^{10}	10^8–10^{10}
$\langle L_{\rm bulge}/L_{\rm tot}\rangle_B$	0.3	0.13	0.05	–	–
Diam. (D_{25}, kpc)	5–100	5–100	5–100	0.5–50	0.5–50
$\langle M/L_{\rm B}\rangle\ (M_\odot/L_\odot)$	6.2 ± 0.6	4.5 ± 0.4	2.6 ± 0.2	~ 1	~ 1
$\langle V_{\rm max}\rangle$(km s^{-1})	299	222	175	–	–
$V_{\rm max}$ range (km s^{-1})	163–367	144–330	99–304	–	50–70
Opening angle	$\sim 6°$	$\sim 12°$	$\sim 18°$	–	–
$\mu_{0,\rm B}$ (mag arcsec^{-2})	21.52 ± 0.39	21.52 ± 0.39	21.52 ± 0.39	22.61 ± 0.47	22.61 ± 0.47
$\langle B-V\rangle$	0.75	0.64	0.52	0.47	0.37
$\langle M_{\rm gas}/M_{\rm tot}\rangle$	0.04	0.08	0.16	0.25 (Scd)	–
$\langle M_{\rm H_2}/M_{\rm HI}\rangle$	2.2 ± 0.6 (Sab)	1.8 ± 0.3	0.73 ± 0.13	0.19 ± 0.10	–
$\langle S_{\rm N}\rangle$	1.2 ± 0.2	1.2 ± 0.2	0.5 ± 0.2	0.5 ± 0.2	–

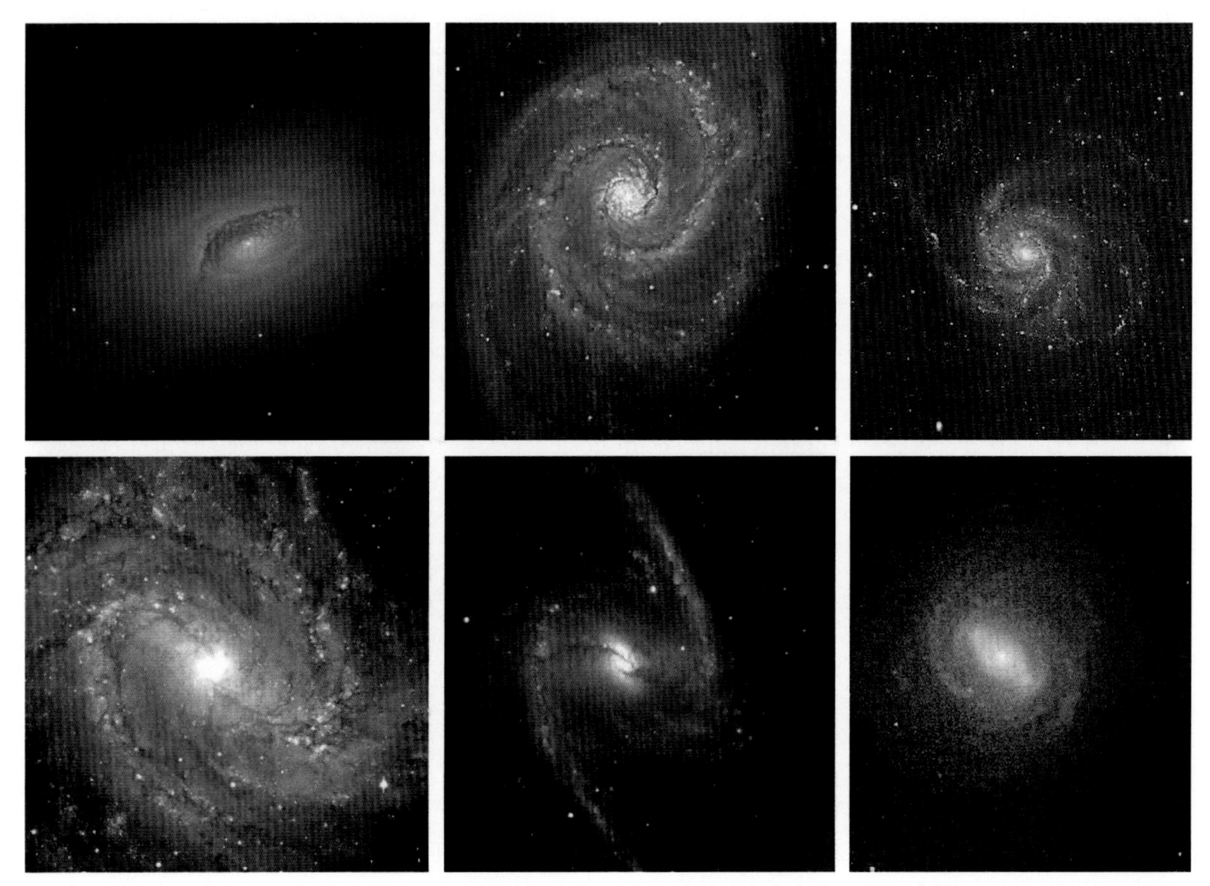

Fig. 3.14. Types of spiral galaxies. Top left: M94, an Sab galaxy. Top middle: M51, an Sbc galaxy. Top right: M101, an Sc galaxy. Lower left: M83, an SBa galaxy. Lower middle: NGC 1365, an SBb galaxy. Lower right: M58, an SBc galaxy

Here, μ_e is the surface brightness at the effective radius R_e which is defined such that half of the luminosity is emitted within R_e (see (2.40)). The central surface brightness and the scale-length of the disk are denoted by μ_0 and h_r, respectively. It has to be noted that μ_0 is not directly measurable since μ_0 is *not* the central surface brightness of the galaxy, only that of its disk component. To determine μ_0, the exponential law (3.11) is extrapolated from large R inwards to $R = 0$.

When Ken Freeman analyzed a sample of spiral galaxies, he found the remarkable result that the central surface brightness μ_0 of disks has a very low spread, i.e., it is very similar for different galaxies (*Freeman's law, 1970*). For Sa's to Sc's, a value of $\mu_0 = 21.52 \pm 0.39$ B-mag/arcsec^2 is observed, and for Sd spirals and later types, $\mu_0 = 22.61 \pm 0.47$ B-mag/arcsec^2. This result was critically discussed, for example with regard to its possible dependence on selection effects. Their importance is not implausible since the determination of precise photometry of galaxies is definitely a lot easier for objects with a high surface brightness. After accounting for such selection effects in the statistical analysis of galaxy samples, Freeman's law was confirmed for "normal" spiral galaxies. However, galaxies exist which have a significantly lower surface brightness, the *low surface brightness galaxies (LSBs)*. They seem to form a separate class of galaxies whose study is substantially more difficult compared to normal spirals because of their low surface brightness. In fact, the central surface brightness of LSBs is much lower than the brightness of the night sky, so that searching for these LSBs is problematic and requires very accurate data reduction and subtraction of the sky background.

Whereas the bulge and the disk can be studied in spirals even at fairly large distances, the stellar halo has too low a surface brightness to be seen in distant galaxies. However, our neighboring galaxy M31, the Andromeda galaxy, can be studied in quite some detail. In particular, the brightness profile of its stellar halo can be studied more easily than that of the Milky Way, taking advantage of our "outside" view. This galaxy should be quite similar to our Galaxy in many respects; for example, tidal streams from disrupted accreted galaxies were also clearly detected in M31.

A stellar halo of red giant branch stars was detected in M31, which extends out to more than 150 kpc from its center. The brightness profile of this stellar distribution indicates that for radii $r \lesssim 20$ kpc it follows the extrapolation from the brightness profile of the bulge, i.e., a de Vaucouleurs profile. However, for larger radii it exceeds this extrapolation, showing a power-law profile which corresponds to a radial density profile of approximately $\rho \propto r^{-3}$, not unlike that observed in our Milky Way. It thus seems that stellar halos form a generic property of spirals. Unfortunately, the corresponding surface brightness is so small that there is little hope of detecting such a halo in other spirals for which individual stars can no longer be resolved and classified.

The thick disk in other spirals can only be studied if they are oriented edge-on. In these cases, a thick disk can indeed be observed as a stellar population outside the plane of the disk and well beyond the scale-height of the thin disk. As is the case for the Milky Way, the scale-height of a stellar population increases with its age, increasing from young main-sequence stars to old asymptotic giant branch stars. For luminous disk galaxies, the thick disk does not contribute substantially to the total luminosity; however, in lower-mass disk galaxies with rotational velocities $\lesssim 120$ km/s, the thick disk stars can contribute nearly half the luminosity and may actually dominate the stellar mass. In this case, the dominant stellar population of these galaxies is old, despite the fact that they appear blue.

3.3.3 Rotation Curves and Dark Matter

The rotation curves of other spiral galaxies are easier to measure than that of the Milky Way because we are able to observe them "from outside". These measurements are achieved by utilizing the Doppler effect, where the inclination of the disk, i.e., its orientation with respect to the line-of-sight, has to be accounted for. The inclination angle is determined from the observed axis ratio of the disk, assuming that disks are intrinsically axially symmetric (except for the spiral arms). Mainly the stars and HI gas in the galaxies are used as luminous tracers, where the observable HI disk is in general significantly more extended than the stellar disk. Therefore, the rotation curves measured from the 21-cm line typically extend to much larger radii than those from optical stellar spectroscopy.

Like our Milky Way, other spirals also rotate considerably faster in their outer regions than one would expect from Kepler's law and the distribution of visible matter (see Fig. 3.15).

> The rotation curves of spirals do not decrease for $R \geq h_r$, as one would expect from the light distribution, but are basically flat. We therefore conclude that spirals are surrounded by a halo of dark matter. The density distribution of this dark halo can be derived from the rotation curves.

Indeed, the density distribution of the dark matter can be derived from the rotation curves. The force balance between gravitation and centrifugal acceleration yields the Kepler rotation law,

$$v^2(R) = G M(R)/R \, ,$$

from which one directly obtains the mass $M(R)$ within a radius R. The rotation curve expected from the visible matter distribution is[4]

$$v^2_{\rm lum}(R) = G M_{\rm lum}(R)/R \, .$$

$M_{\rm lum}(R)$ can be determined by assuming a constant, plausible value for the mass-to-light ratio of the luminous matter. This value is obtained either from the spectral light distribution of the stars, together with knowledge of the properties of stellar populations, or by fitting the innermost part of the rotation curve (where

[4]This consideration is strongly simplified insofar as the given relations are only valid in this form for spherical mass distributions. The rotational velocity produced by an oblate (disk-shaped) mass distribution is more complicated to calculate; for instance, for an exponential mass distribution in a disk, the maximum of $v_{\rm lum}$ occurs at $\sim 2.2 h_r$, with a Kepler decrease, $v_{\rm lum} \propto R^{-1/2}$, at larger radii.

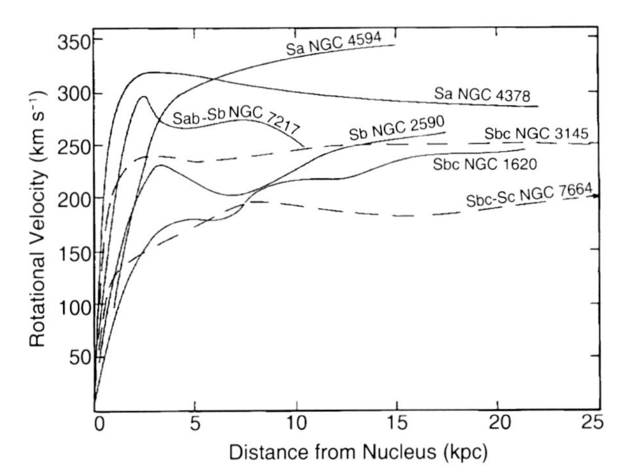

Fig. 3.15. Examples of rotation curves of spiral galaxies. They are all flat in the outer region and do not behave as expected from Kepler's law if the galaxy consisted only of luminous matter. Also striking is the fact that the amplitude of the rotation curve is higher for early types than for late types.

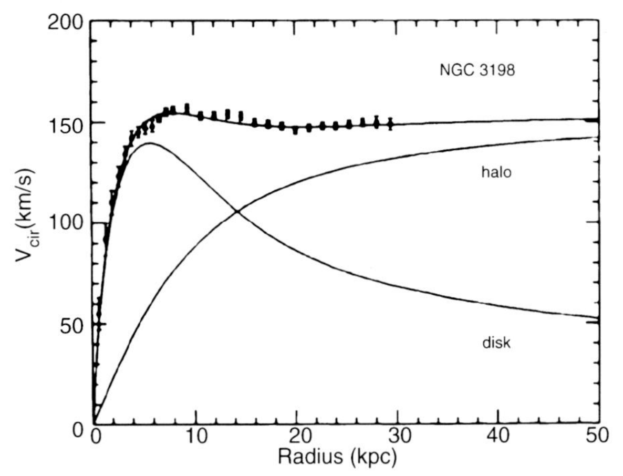

Fig. 3.16. The flat rotation curves of spiral galaxies cannot be explained by visible matter alone. The example of NGC 3198 demonstrates the rotation curve which would be expected from the visible matter alone (curve labeled "disk"). To explain the observed rotation curve, a dark matter component has to be present (curve labeled "halo"). However, the decomposition into disk and halo mass is not unambiguous because for it to be so it would be necessary to know the mass-to-light ratio of the disk. In the case considered here, a "maximum disk" was assumed, i.e., it was assumed that the innermost part of the rotation curve is produced solely by the visible matter in the disk

the mass contribution of dark matter can presumably be neglected), assuming that M/L is independent of radius for the stellar population. From this estimate of the mass-to-light ratio, the discrepancy between $v_{\rm lum}^2$ and v^2 yields the distribution of the dark matter, $v_{\rm dark}^2 = v^2 - v_{\rm lum}^2 = G M_{\rm dark}/R$, or

$$M_{\rm dark}(R) = \frac{R}{G}\left[v^2(R) - v_{\rm lum}^2(R)\right] . \tag{3.12}$$

An example of this decomposition of the mass contributions is shown in Fig. 3.16.

The corresponding density profiles of the dark matter halos seem to be flat in the inner region, and decreasing as R^{-2} at large radii. It is remarkable that $\rho \propto R^{-2}$ implies a mass profile $M \propto R$, i.e., the mass of the halo increases linearly with the radius for large R. As long as the extent of the halo is undetermined the total mass of a galaxy will be unknown. Since the observed rotation curves are flat out to the largest radius for which 21-cm emission can still be observed, a lower limit for the radius of the dark halo can be obtained, $R_{\rm halo} \gtrsim 30 h^{-1}$ kpc.

To derive the density profile out to even larger radii, other observable objects in an orbit around the galaxies are needed. Potential candidates for such luminous tracers are satellite galaxies – companions of other spirals, like the Magellanic Clouds are for the Milky Way. Because we cannot presume that these satellite galaxies move on circular orbits around their parent galaxy, conclusions can be drawn based only on a statistical sample of satellites. These analyses of the relative velocities of satellite galaxies around spirals still give no indication of an "edge" to the halo, leading to a lower limit for the radius of $R_{\rm halo} \gtrsim 100 h^{-1}$ kpc.

For elliptical galaxies the mass estimate, and thus the detection of a possible dark matter component, is significantly more complicated, since the orbits of stars are substantially more complex than in spirals. In particular, the mass estimate from measuring the stellar velocity dispersion via line widths depends on the anisotropy of the stellar orbits, which is a priori unknown. Nevertheless, in recent years it has been unambiguously proven that dark matter also exists in ellipticals. First, the degeneracy between the anisotropy of the orbits and the mass determination was broken by detailed kinematic analysis. Second, in some ellipticals hot gas was detected from its X-ray emission. As we will see in Sect. 6.3 in the context of clusters of galaxies, the temperature of the gas allows an estimate of the depth of the potential well, and therefore the mass. Both methods reveal that ellipticals are also surrounded by a dark halo.

The weak gravitational lens effect, which we will discuss in Sect. 6.5.2 in a different context, offers another way to determine the masses of galaxies up to very large radii. With this method we cannot study individual galaxies but only the mean mass properties of a galaxy population. The results of these measurements confirm the large size of dark halos in spirals and in ellipticals.

Correlations of Rotation Curves with Galaxy Properties. The form and amplitude of the rotation curves of spirals are correlated with their luminosity and their Hubble type. The larger the luminosity of a spiral, the steeper the rise of $v(R)$ in the central region, and the larger the maximum rotation velocity $v_{\rm max}$. This latter fact indicates that the mass of a galaxy increases with luminosity, as expected. For the characteristic values of the various Hubble types, one finds $v_{\rm max} \sim 300\,{\rm km/s}$ for Sa's, $v_{\rm max} \sim 175\,{\rm km/s}$ for Sc's, whereas Irr's have a much lower $v_{\rm max} \sim 70\,{\rm km/s}$. For equal luminosity, $v_{\rm max}$ is higher for earlier types of spirals. However, the shape (not the amplitude) of the rotation curves of different Hubble types is similar, despite the fact that they have a different brightness profile as seen, for instance, from the varying bulge-to-disk ratio. This point is another indicator that the rotation curves cannot be explained by visible matter alone.

These results leave us with a number of obvious questions. What is the nature of the dark matter? What are the density profiles of dark halos, how are they determined, and where is the "boundary" of a halo? Does the fact that galaxies with $v_{\rm rot} \lesssim 100\,{\rm km/s}$ have no prominent spiral structure mean that a minimum halo mass needs to be exceeded in order for spiral arms to form?

Some of these questions will be examined later, but here we point out that the major fraction of the mass of (spiral) galaxies consists of non-luminous matter. The fact that we do not know what this matter consists of leaves us with the question of whether this invisible matter is a new, yet unknown, form of matter. Or is the dark matter less exotic, normal (baryonic) matter that is just not luminous for some reason (for example, because it did not form any stars)? We will see in Chap. 4 that the problem of dark matter is not limited to galaxies, but is also clearly present on a cosmological scale; furthermore, the dark matter cannot be baryonic. A currently unknown form of matter is, therefore, revealing itself in the rotation curves of spirals.

3.3.4 Stellar Populations and Gas Fraction

The color of spiral galaxies depends on their Hubble type, with later types being bluer; e.g., one finds $B-V \sim 0.75$ for Sa's, 0.64 for Sb's, 0.52 for Sc's, and 0.4 for Irr's. This means that the fraction of massive young stars increases along the Hubble sequence towards later spiral types. This conclusion is also in agreement with the findings for the light distribution along spiral arms where we clearly observe active star-formation regions in the bright knots in the spiral arms of Sc's. Furthermore, this color sequence is also in agreement with the decreasing bulge fraction towards later types.

The formation of stars requires gas, and the mass fraction of gas is larger for later types, as can be measured, for instance, from the 21-cm emission of HI, from Hα and from CO emission. Characteristic values for the ratio $\langle M_{\rm gas}/M_{\rm tot} \rangle$ are about 0.04 for Sa's, 0.08 for Sb's, 0.16 for Sc's, and 0.25 for Irr's. In addition, the fraction of molecular gas relative to the total gas mass is smaller for later Hubble types. The dust mass is less than 1% of the gas mass.

Dust, in combination with hot stars, is the main source of far-infrared (FIR) emission from galaxies. Sc galaxies emit a larger fraction of FIR radiation than Sa's, and barred spirals have stronger FIR emission than normal spirals. The FIR emission arises due to dust being heated by the UV radiation of hot stars and then reradiating this energy in the form of thermal emission.

A prominent color gradient is observed in spirals: they are red in the center and bluer in the outer regions. We can identify at least two reasons for this trend. The first is a metallicity effect, as the metallicity is increasing inwards and metal-rich stars are redder than metal-poor ones, due to their higher opacity. Second, the color gradient can be explained by star formation. Since the gas fraction in the bulge is lower than in the disk, less star formation takes place in the bulge, resulting in a stellar population that is older and redder in general. Furthermore, it is found that the metallicity of spirals increases with luminosity.

Abundance of Globular Clusters. The number of globular clusters is higher in early types and in more luminous galaxies. The *specific abundance* of globular clusters in a galaxy is defined as their number, normalized to a galaxy of absolute magnitude $M_V = -15$. This can be done by scaling the observed number N_t of globular clusters in a galaxy of visual luminosity L_V or absolute magnitude M_V, respectively, to that of a hypothetical galaxy with $M_V = -15$:

$$S_N = N_t \frac{L_{15}}{L_V} = N_t \, 10^{0.4(M_V+15)} \, . \tag{3.13}$$

If the number of globular clusters were proportional to the luminosity (and thus roughly to the stellar mass) of a galaxy, then this would imply $S_N = \text{const}$. However, this is not the case: For Sa's and Sb's we find $S_N \sim 1.2$, whereas $S_N \sim 0.5$ for Sc's. S_N is larger for ellipticals and largest for cD galaxies.

3.3.5 Spiral Structure

The spiral arms are the bluest regions in spirals and they contain young stars and HII regions. For this reason, the brightness contrast of spiral arms increases as the wavelength of the (optical) observation decreases. In particular, the spiral structure is very prominent in a blue filter, as is shown impressively in Fig. 3.17.

Naturally, the question arises as to the nature of the spiral arms. Probably the most obvious answer would be that they are material structures of stars and gas, rotating around the galaxy's center together with the rest of the disk. However, this scenario cannot explain spiral arm structure since, owing to the differential rotation, they would wind up much more tightly than observed within only a few rotation periods.

Rather, it is suspected that spiral arms are a wave structure, the velocity of which does not coincide with the physical velocity of the stars. Spiral arms are quasi-stationary density waves, regions of higher density (possibly 10–20% higher than the local disk environment). If the gas, on its orbit around the center of the galaxy, enters a region of higher density, it is compressed, and this compression of molecular clouds results in an enhanced star-formation rate. This accounts for the blue color of spiral arms. Since low-mass (thus red) stars live longer, the brightness contrast of spiral arms is lower in red light, whereas massive blue stars are born in the spiral arms and soon after explode there as SNe. Indeed, only few blue stars are found outside spiral arms.

In order to better understand density waves we may consider, for example, the waves on the surface of a lake. Peaks at different times consist of different water particles, and the velocity of the waves is by no means the bulk velocity of the water.

3.3.6 Corona in Spirals?

Hot gas resulting from the evolution of supernova remnants may expand out of the disk and thereby be ejected to form a gaseous halo of a spiral galaxy. We might

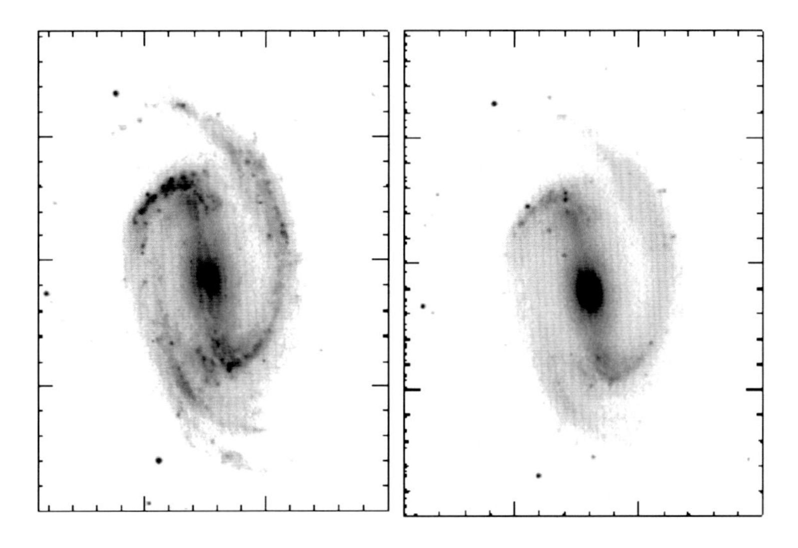

Fig. 3.17. The galaxy NGC 1300 in the B filter (left panel) and in the I filter (right panel). The spiral arms are much more prominent in the blue than in the red. Also, the tips of the bar are more pronounced in the blue – an indicator of an enhanced star-formation rate

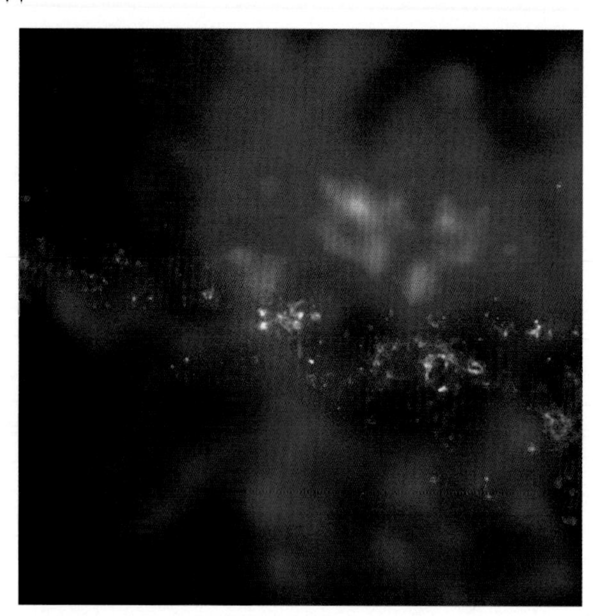

Fig. 3.18. The spiral galaxy NGC 4631. The optical (HST) image of the galaxy is shown in red; the many luminous areas are regions of very active star formation. The SN explosions of massive stars eject hot gas into the halo of the galaxy. This gas (at a temperature of $T \sim 10^6$ K) emits X-ray radiation, shown as the blue diffuse emission as observed by the Chandra satellite. The image has a size of 2.′5

therefore suspect that such a "coronal" gas exists outside the galactic disk. While the existence of this coronal gas has long been suspected, the detection of its X-ray emission was first made possible with the ROSAT satellite in the early 1990s. However, the limited angular resolution of ROSAT rendered the distinction between diffuse emission and clusters of discrete sources difficult. Finally, the Chandra observatory unambiguously detected the coronal gas in a number of spiral galaxies. As an example, Fig. 3.18 shows the spiral galaxy NGC 4631.

3.4 Scaling Relations

The kinematic properties of spirals and ellipticals are closely related to their luminosity. As we shall discuss below, spirals follow the *Tully–Fisher relation* (Sect. 3.4.1), whereas elliptical galaxies obey the *Faber–Jackson relation* (Sect. 3.4.2) and are located in the *fundamental plane* (Sect. 3.4.3). These scaling relations are a very important tool for distance estimations, as will be discussed in Sect. 3.6. Furthermore, these scaling relations express relations between galaxy properties which any successful model of galaxy evolution must be able to explain. Here we will describe these scaling relations and discuss their physical origin.

3.4.1 The Tully–Fisher Relation

Using 21-cm observations of spiral galaxies, in 1977 R. Brent Tully and J. Richard Fisher found that the maximum rotation velocity of spirals is closely related to their luminosity, following the relation

$$\boxed{L \propto v_{\rm max}^{\alpha}} \; , \tag{3.14}$$

where the slope of the Tully–Fisher relation is about $\alpha \sim 4$. The larger the wavelength of the filter in which the luminosity is measured, the smaller the dispersion of the Tully–Fisher relation (see Fig. 3.19). This is to be expected because radiation at larger wavelengths is less affected by dust absorption and by the current star-formation rate, which may vary to some extent between individual spirals. Furthermore, it is found that the value of α increases with the wavelength of the filter; the Tully–Fisher relation is steeper in the red. The dispersion of galaxies around the relation (3.14) in the near infrared (e.g., in the H-band) is about 10%.

Because of this close correlation, the luminosity of spirals can be estimated quite precisely by measuring the rotational velocity. The determination of the (maximum) rotational velocity is independent of the galaxy's distance. By comparing the luminosity, as determined from the Tully–Fisher relation, with the measured flux one can then estimate the distance of the galaxy without utilizing the Hubble relation!

The measurement of $v_{\rm max}$ is obtained either from a spatially resolved rotation curve, by measuring $v_{\rm rot}(\theta)$, which is possible for relatively nearby galaxies, or by observing an integrated spectrum of the 21-cm line of HI that has a Doppler width corresponding to about $2v_{\rm max}$ (see Fig. 3.20). The Tully–Fisher relation shown in Fig. 3.19 was determined by measuring the width of the 21-cm line.

Explaining the Tully–Fisher Relation. The shapes of the rotation curves of spirals are very similar to each

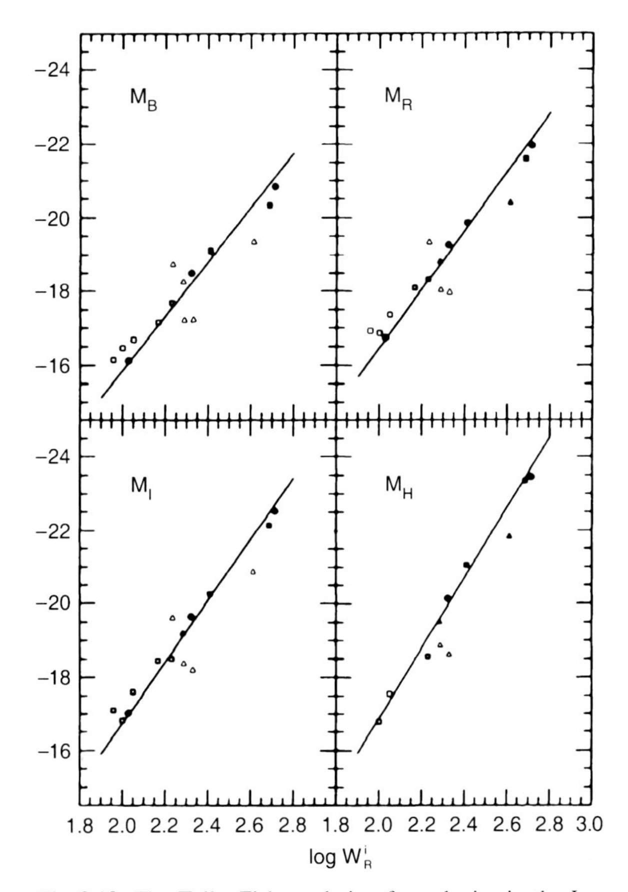

Fig. 3.19. The Tully–Fisher relation for galaxies in the Local Group (dots), in the Sculptor group (triangles), and in the M81 group (squares). The absolute magnitude is plotted as a function of the width of the 21-cm profile which indicates the maximum rotation velocity (see Fig. 3.20). Filled symbols represent galaxies for which independent distance estimates were obtained, either from RR Lyrae stars, Cepheids, or planetary nebulae. For galaxies represented by open symbols, the average distance of the respective group is used. The solid line is a fit to similar data for the Ursa-Major cluster, together with data of those galaxies for which individual distance estimates are available (filled symbols). The larger dispersion around the mean relation for the Sculptor group galaxies is due to the group's extent along the line-of-sight

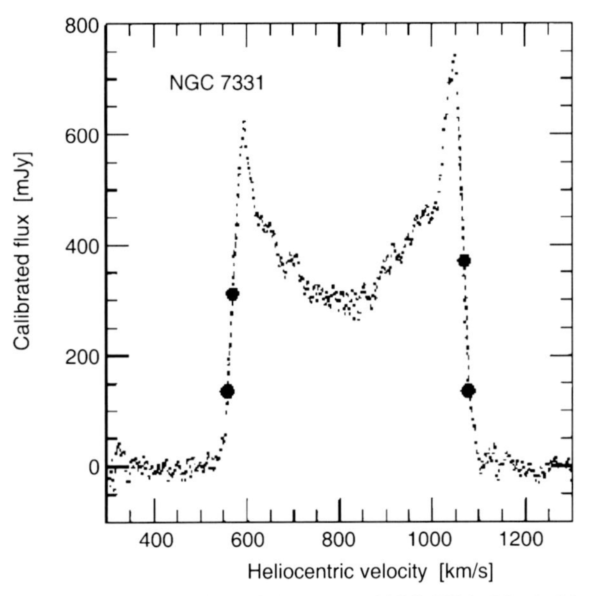

Fig. 3.20. 21 cm profile of the galaxy NGC 7331. The bold dots indicate 20% and 50% of the maximum flux; these are of relevance for the determination of the line width from which the rotational velocity is derived

other, in particular with regard to their flat behavior in the outer part. The flat rotation curve implies

$$\boxed{M = \frac{v_{\rm max}^2 R}{G}} \,, \tag{3.15}$$

where the distance R from the center of the galaxy refers to the flat part of the rotation curve. The exact value is not important, though, if only $v(R) \approx$ const. By re-writing (3.15),

$$L = \left(\frac{M}{L}\right)^{-1} \frac{v_{\rm max}^2 R}{G} \,, \tag{3.16}$$

and replacing R by the mean surface brightness $\langle I \rangle = L/R^2$, we obtain

$$L = \left(\frac{M}{L}\right)^{-2} \left(\frac{1}{G^2 \langle I \rangle}\right) v_{\rm max}^4 \,. \tag{3.17}$$

This is the Tully–Fisher relation *if* M/L and $\langle I \rangle$ are the same for all spirals. The latter is in fact suggested by Freeman's law (Sect. 3.3.2). Since the shapes of rotation curves for spirals seem to be very similar, the radial dependence of the ratio of luminous to dark matter may also be quite similar among spirals. Furthermore, since the red or infrared mass-to-light ratios of a stellar population do not depend strongly on its age, the constancy of M/L could also be valid if dark matter is included.

Although the line of argument presented above is far from a proper derivation of the Tully–Fisher-relation, it nevertheless makes the existence of such a scaling relation plausible.

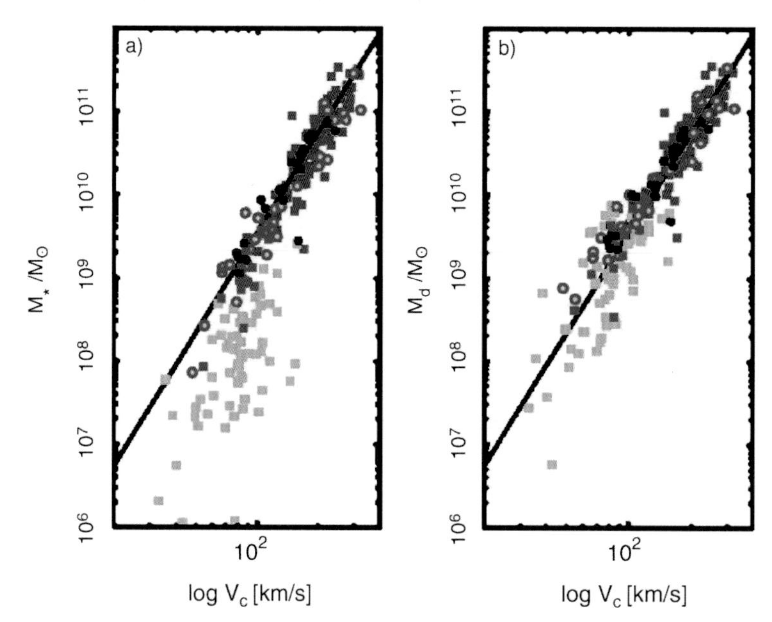

Fig. 3.21. Left panel: the mass contained in stars as a function of the rotational velocity V_c for spirals. This stellar mass is computed from the luminosity by multiplying it with a suitable stellar mass-to-light ratio which depends on the chosen filter and which can be calculated from stellar population models. This is the "classical" Tully–Fisher relation. Squares and circles denote galaxies for which V_c was determined from the 21-cm line width or from a spatially resolved rotation curve, respectively. The colors of the symbols indicate the filter band in which the luminosity was measured: H (red), K′ (black), I (green), B (blue). Right panel: instead of the stellar mass, here the sum of the stellar and gaseous mass is plotted. The gas mass was derived from the flux in the 21-cm line, $M_{\rm gas} = 1.4 M_{\rm HI}$, corrected for helium and metals. Molecular gas has no significant contribution to the baryonic mass. The line in both plots is the Tully–Fisher relation with a slope of $\alpha = 4$

Mass-to-Light Ratio of Spirals. We are unable to determine the total mass of a spiral because the extent of the dark halo is unknown. Thus we can measure M/L only within a fixed radius. We shall define this radius as R_{25}, the radius at which the surface brightness attains the value of 25 mag/arcsec^2 in the B-band;[5] then spirals follow the relation

$$\boxed{\log\left(\frac{R_{25}}{\rm kpc}\right) = -0.249 M_B - 4.00} \;, \tag{3.18}$$

independently of their Hubble type. Within R_{25} one finds $M/L_B = 6.2$ for Sa's, 4.5 for Sb's, and 2.6 for Sc's. This trend does not come as a surprise because late types of spirals contain more young, blue and luminous stars.

[5]We point out explicitly once more that the surface brightness does not depend on the distance of a source.

The Baryonic Tully–Fisher Relation. The above "derivation" of the Tully–Fisher relation is based on the assumption of a constant M/L value, where M is the total mass (i.e., including dark matter). Let us assume that (i) the ratio of baryons to dark matter is constant, and furthermore that (ii) the stellar populations in spirals are similar, so that the ratio of stellar mass to luminosity is a constant. Even under these assumptions we would expect the Tully–Fisher relation to be valid only if the gas does not, or only marginally, contribute to the baryonic mass. However, low-mass spirals contain a significant fraction of gas, so we should expect that the Tully–Fisher relation does not apply to these galaxies. Indeed, it is found that spirals with a small $v_{\max} \lesssim 100$ km/s deviate significantly from the Tully–Fisher relation – see Fig. 3.21(a).

Since the luminosity is approximately proportional to the stellar mass, $L \propto M_*$, the Tully–Fisher relation is a relation between $v_{\max}$ and M_*. Adding the mass of the gas, which can be determined from the strength of the 21-cm line, to the stellar mass a much tighter correlation is obtained, see Fig. 3.21(b). It reads

$$M_{\rm disk} = 2 \times 10^9\, h^{-2}\, M_\odot \left(\frac{v_{\rm max}}{100\,{\rm km/s}}\right)^4 , \qquad (3.19)$$

and is valid over five orders of magnitude in disk mass $M_{\rm disk} = M_* + M_{\rm gas}$. If no further baryons exist in spirals (such as, e.g., MACHOs), this close relation means that the ratio of baryons and dark matter in spirals is constant over a very wide mass range.

3.4.2 The Faber–Jackson Relation

A relation for elliptical galaxies, analogous to the Tully–Fisher relation, was found by Sandra Faber and Roger Jackson. They discovered that the velocity dispersion in the center of ellipticals, σ_0, scales with luminosity (see Fig. 3.22),

$$L \propto \sigma_0^4 ,$$

or

$$\log(\sigma_0) = -0.1 M_B + {\rm const} . \qquad (3.20)$$

"Deriving" the Faber–Jackson scaling relation is possible under the same assumptions as the Tully–Fisher relation. However, the dispersion of ellipticals about this relation is larger than that of spirals about the Tully–Fisher relation.

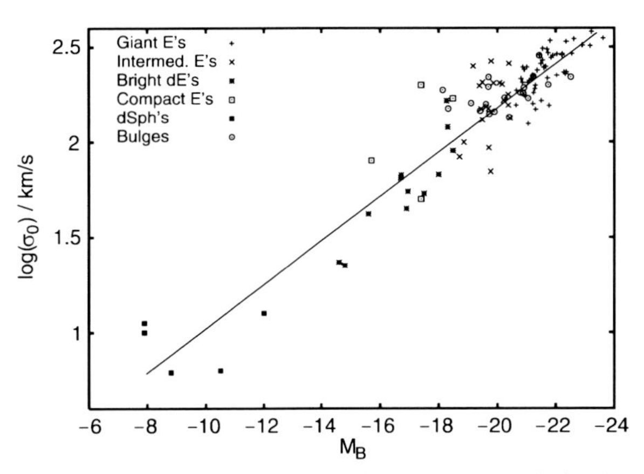

Fig. 3.22. The Faber–Jackson relation expresses a relation between the velocity dispersion and the luminosity of elliptical galaxies. It can be derived from the virial theorem

3.4.3 The Fundamental Plane

The Tully–Fisher and Faber–Jackson relations specify a connection between the luminosity and a kinematic property of galaxies. As we discussed previously, various relations exist between the parameters of elliptical galaxies. Thus one might wonder whether a relation exists between observables of elliptical galaxies for which the dispersion is smaller than that of the Faber–Jackson relation. Such a relation was indeed found and is known as the *fundamental plane*.

To explain this relation, we will consider the various relations between the parameters of ellipticals. In Sect. 3.2.2 we saw that the effective radius of normal ellipticals is related to the luminosity (see Fig. 3.7). This implies a relation between the surface brightness and the effective radius,

$$R_{\rm e} \propto \langle I \rangle_{\rm e}^{-0.83} , \qquad (3.21)$$

where $\langle I \rangle_{\rm e}$ is the average surface brightness within the effective radius, so that

$$L = 2\pi R_{\rm e}^2 \langle I \rangle_{\rm e} . \qquad (3.22)$$

From this, a relation between the luminosity and $\langle I \rangle_{\rm e}$ results,

$$L \propto R_{\rm e}^2 \langle I \rangle_{\rm e} \propto \langle I \rangle_{\rm e}^{-0.66}$$

or

$$\langle I \rangle_{\rm e} \propto L^{-1.5} . \qquad (3.23)$$

Hence, more luminous ellipticals have smaller surface brightnesses, as is also shown in Fig. 3.7. By means of the Faber–Jackson relation, L is related to σ_0, the central velocity dispersion, and therefore, σ_0, $\langle I \rangle_{\rm e}$, and $R_{\rm e}$ are related to each other. The distribution of elliptical galaxies in the three-dimensional parameter space ($R_{\rm e}$, $\langle I \rangle_{\rm e}$, σ_0) is located close to a plane defined by

$$R_{\rm e} \propto \sigma_0^{1.4} \langle I \rangle_{\rm e}^{-0.85} . \qquad (3.24)$$

Writing this relation in logarithmic form, we obtain

$$\log R_{\rm e} = 0.34 \langle \mu \rangle_{\rm e} + 1.4 \log \sigma_0 + {\rm const} , \qquad (3.25)$$

where $\langle\mu\rangle_e$ is the average surface brightness within R_e, measured in mag/arcsec^2. Equation (3.25) defines a plane in this three-dimensional parameter space that is known as the *fundamental plane (FP)*. Different projections of the fundamental plane are displayed in Fig. 3.23.

How can this be Explained? The mass within R_e can be derived from the virial theorem, $M \propto \sigma_0^2 R_e$. Combining this with (3.22) yields

$$R_e \propto \frac{L}{M} \frac{\sigma_0^2}{\langle I \rangle_e} , \qquad (3.26)$$

which agrees with the FP in the form of (3.24) if

$$\frac{L}{M} \frac{\sigma_0^2}{\langle I \rangle_e} \propto \frac{\sigma_0^{1.4}}{\langle I \rangle_e^{0.85}} ,$$

or

$$\frac{M}{L} \propto \frac{\sigma_0^{0.6}}{\langle I \rangle_e^{0.15}} \propto \frac{M^{0.3}}{R_e^{0.3}} \frac{R_e^{0.3}}{L^{0.15}} .$$

Hence, the FP follows from the virial theorem provided

$$\left(\frac{M}{L}\right) \propto M^{0.2} \quad \text{or} \quad \left(\frac{M}{L}\right) \propto L^{0.25} , \quad \text{respectively} , \qquad (3.27)$$

i.e., if the mass-to-light ratio of galaxies increases slightly with mass. Like the Tully–Fisher relation, the fundamental plane is an important tool for distance estimations. It will be discussed more thoroughly later.

3.4.4 The D_n–σ Relation

Another scaling relation for ellipticals which is of substantial importance in practical applications is the D_n–σ relation. D_n is defined as that diameter of an ellipse within which the average surface brightness I_n corresponds to a value of 20.75 mag/arcsec^2 in the B-band. If we now assume that all ellipticals have a self-similar brightness profile, $I(R) = I_e \, f(R/R_e)$, with $f(1) = 1$, then the luminosity within D_n can be written as

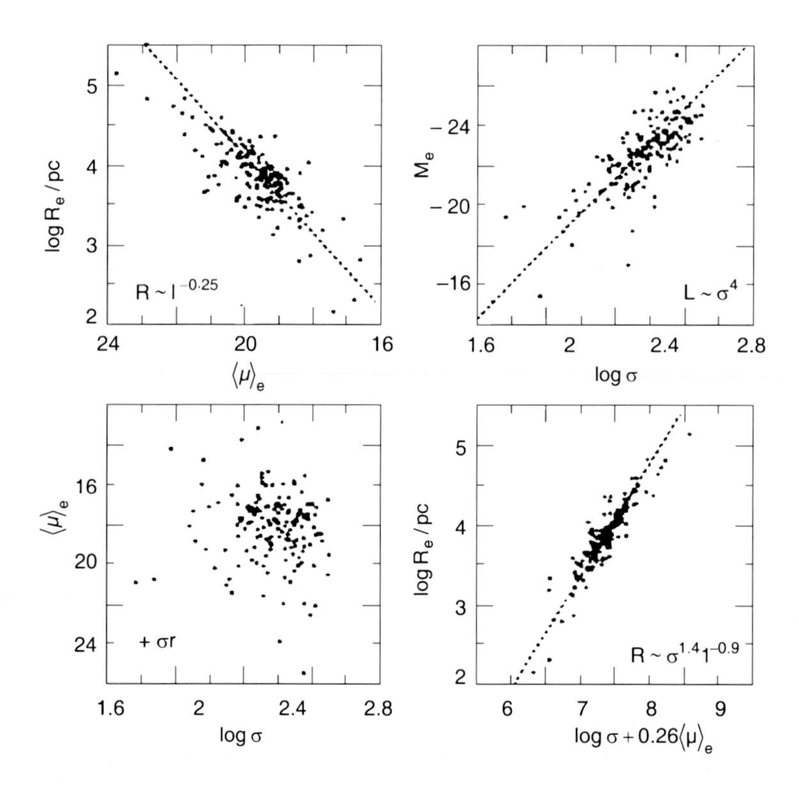

Fig. 3.23. Projections of the fundamental plane onto different two-parameter planes. Upper left: the relation between radius and mean surface brightness within the effective radius. Upper right: Faber–Jackson relation. Lower left: the relation between mean surface brightness and velocity dispersion shows the fundamental plane viewed from above. Lower right: the fundamental plane viewed from the side – the linear relation between radius and a combination of surface brightness and velocity dispersion

$$I_n \left(\frac{D_n}{2}\right)^2 \pi = 2\pi I_e \int_0^{D_n/2} dR\, R\, f(R/R_e)$$
$$= 2\pi I_e R_e^2 \int_0^{D_n/(2R_e)} dx\, x\, f(x) .$$

For a de Vaucouleurs profile we have approximately $f(x) \propto x^{-1.2}$ in the relevant range of radius. Computing the integral with this expression, we obtain

$$D_n \propto R_e\, I_e^{0.8} . \tag{3.28}$$

Replacing R_e by the fundamental plane (3.24) then results in

$$D_n \propto \sigma_0^{1.4}\, \langle I \rangle_e^{-0.85}\, I_e^{0.8} .$$

Since $\langle I \rangle_e \propto I_e$ due to the assumed self-similar brightness profile, we finally find

$$\boxed{D_n \propto \sigma_0^{1.4}\, I_e^{0.05}} . \tag{3.29}$$

This implies that D_n is nearly independent of I_e and only depends on σ_0. The D_n–σ relation (3.29) describes the properties of ellipticals considerably better than the Faber–Jackson relation and, in contrast to the fundamental plane, it is a relation between only two observables. Empirically, we find that ellipticals follow the normalized D_n–σ relation

$$\frac{D_n}{\text{kpc}} = 2.05 \left(\frac{\sigma_0}{100\,\text{km/s}}\right)^{1.33} , \tag{3.30}$$

and they scatter around this relation with a relative width of about 15%.

3.5 Black Holes in the Centers of Galaxies

As we have seen in Sect. 2.6.3, the Milky Way harbors a black hole in its center. Furthermore, it is generally accepted that the energy for the activity of AGNs is generated by accretion onto a black hole (see Sect. 5.3). Thus, the question arises as to whether all (or most) galaxies contain a supermassive black hole (SMBH) in their nuclei. We will pursue this question in this section and show that SMBHs are very abundant indeed.

This result then instigates further questions: what distinguishes a "normal" galaxy from an AGN if both have a SMBH in the nucleus? Is it the mass of the black hole, the rate at which material is accreted onto it, or the efficiency of the mechanism which is generating the energy?

We will start with a concise discussion of how to search for SMBHs in galaxies, then present some examples for the discovery of such SMBHs. Finally, we will discuss the very tight relationship between the mass of the SMBH and the properties of the stellar component of a galaxy.

3.5.1 The Search for Supermassive Black Holes

We will start with the question of what a black hole actually is. A technical answer is that a black hole is the simplest solution of Einstein's theory of General Relativity which describes the gravitational field of a point mass. Less technically – though sufficient for our needs – we may say that a black hole is a point mass, or a compact mass concentration, with an extent smaller than its Schwarzschild radius r_S (see below).

The Schwarzschild Radius. The first discussion of black holes can be traced back to Laplace in 1795, who considered the following: if one reduces the radius r of a celestial body of mass M, the escape velocity v_{esc} at its surface will change,

$$v_{esc} = \sqrt{\frac{2GM}{r}} .$$

As a thought experiment, one can now see that for a sufficiently small radius v_{esc} will be equal to the speed of light, c. This happens when the radius decreases to

$$\boxed{r_S := \frac{2GM}{c^2} = 2.95 \times 10^5\,\text{cm} \left(\frac{M}{M_\odot}\right)} . \tag{3.31}$$

The radius r_S is named the *Schwarzschild radius*, after Karl Schwarzschild who, in 1916, discovered the point-mass solution for Einstein's field equations. For our purpose we will define a black hole as a mass concentration with a radius smaller than r_S. As we can see, r_S is very small: about 3 km for the Sun, and $r_S \sim 10^{12}$ cm for the SMBH in the Galactic center. At a distance of $D = R_0 \approx 8$ kpc, this corresponds to an angular radius of $\sim 6 \times 10^{-6}$ arcsec. Current observing capabilities are still far from resolving scales of order r_S, but in the near future VLBI observations at very short radio wavelengths may achieve sufficient angular resolution to resolve the Schwarzschild radius for the Galactic black hole. The largest observed velocities of stars in the Galactic center, $\sim 5000\,\mathrm{km/s} \ll c$, indicate that they are still well away from the Schwarzschild radius. However, in the case of the SMBH in our Galactic center we can "look" much closer to the Schwarzschild radius: with VLBI observations at wavelengths of 3 mm the angular size of the compact radio source Sgr A* can be constrained to be less than 0.3 mas, corresponding to about $20 r_S$. We will show in Sect. 5.3.3 that relativistic effects are directly observed in AGNs and that velocities close to c do in fact occur there – which again is a very direct indication of the existence of a SMBH.

If even for the closest SMBH, the one in the GC, the Schwarzschild radius is significantly smaller than the achievable angular resolution, how can we hope to prove that SMBHs exist in other galaxies? Like in the GC, this proof has to be found indirectly by detecting a compact mass concentration incompatible with the mass concentration of the stars observed.

The Radius of Influence. We consider a mass concentration of mass $M_\bullet$ in the center of a galaxy where the characteristic velocity dispersion of stars (or gas) is σ. We compare this velocity dispersion with the characteristic velocity (e.g., the Kepler rotational velocity) around a SMBH at a distance r, given by $\sqrt{GM_\bullet/r}$. From this it follows that, for distances smaller than

$$r_{\rm BH} = \frac{GM_\bullet}{\sigma^2} \sim 0.4 \left(\frac{M_\bullet}{10^6 M_\odot}\right) \left(\frac{\sigma}{100\,\mathrm{km/s}}\right)^{-2} \mathrm{pc}\,, \tag{3.32}$$

the SMBH will significantly affect the kinematics of stars and gas in the galaxy. The corresponding angular scale is

$$\theta_{\rm BH} = \frac{r_{\rm BH}}{D} \sim 0\farcs1 \left(\frac{M_\bullet}{10^6 M_\odot}\right) \left(\frac{\sigma}{100\,\mathrm{km/s}}\right)^{-2} \left(\frac{D}{1\,\mathrm{Mpc}}\right)^{-1}, \tag{3.33}$$

where D is the distance of the galaxy. From this we immediately conclude that our success in finding SMBHs will depend heavily on the achievable angular resolution. The HST enabled scientists to make huge progress in this field. The search for SMBHs promises to be successful only in relatively nearby galaxies. In addition, from (3.33) we can see that for increasing distance D the mass $M_\bullet$ has to increase for a SMBH to be detectable at a given angular resolution.

Kinematic Evidence. The presence of a SMBH inside $r_{\rm BH}$ is revealed by an increase in the velocity dispersion for $r \lesssim r_{\rm BH}$, which should then behave as $\sigma \propto r^{-1/2}$ for $r \lesssim r_{\rm BH}$. If the inner region of the galaxy rotates, one expects, in addition, that the rotational velocity $v_{\rm rot}$ should also increase inwards $\propto r^{-1/2}$.

Problems in Detecting These Signatures. The practical problems in observing a SMBH have already been mentioned above. One problem is the angular resolution. To measure an increase in the velocities for small radii, the angular resolution needs to be better than $\theta_{\rm BH}$. Furthermore, projection effects play a role because only the velocity dispersion of the projected stellar distribution, weighted by the luminosity of the stars, is measured. Added to this, the kinematics of stars can be rather complicated, so that the observed values for σ and $v_{\rm rot}$ depend on the distribution of orbits and on the geometry of the distribution.

Despite these difficulties, the detection of SMBHs has been achieved in recent years, largely due to the much improved angular resolution of optical telescopes (like the HST) and to improved kinematic models.

3.5.2 Examples for SMBHs in Galaxies

Figure 3.24 shows an example for the kinematical method discussed in the previous section. A long-slit spectrum across the nucleus of the galaxy M84 clearly shows that, near the nucleus, both the rotational velocity and the velocity dispersion change; both increase dramatically towards the center. Figure 3.25 illustrates how strongly the measurability of the kinematical evidence for a SMBH depends on the achievable angular resolution of the observation. For this example of NGC 3115, observing with the resolution offered by space-based spectroscopy yields much higher measured velocities than is possible from the ground. Particularly interesting is the observation of the rotation curve very close to the center. Another impressive example is the central region of M87, the central galaxy of the Virgo Cluster. The increase of the rotation curve and the broadening of the [OII]-line (a spectral line of singly-ionized oxygen) at $\lambda = 3727$ Å towards the center are displayed in Fig. 3.26 and argue very convincingly for a SMBH with $M_\bullet \approx 3 \times 10^9 M_\odot$.

The mapping of the Kepler rotation in the center of the Seyfert galaxy NGC 4258 is especially spectacular. This galaxy contains water masers – very compact sources whose position can be observed with very high precision using VLBI techniques (Fig. 3.27). In this case, the deviation from a Kepler rotation in the gravitational field of a point mass of $M_\bullet \sim 3.5 \times 10^7 M_\odot$ is much less than 1%. The maser sources are embedded in an accretion disk having a thickness of less than 0.3% of its radius, of which also a warping is detected. Changes in the radial velocities and the proper motions of these maser sources have already been measured, so that the model of a Kepler accretion disk has been confirmed in detail.

All these observations are of course no proof of the existence of a SMBH in these galaxies because the sources from which we obtain the kinematic evidence are still too far away from the Schwarzschild radius. The conclusion of the presence of SMBHs is rather that of a missing alternative, as was already explained for the case of the GC (Sect. 2.6.3). We have no other plausible model for the mass concentrations detected. As for the case of the SMBH in the Milky Way, an ultra-compact star cluster might be postulated, but such a cluster would not be stable over a long period of time. Based on the existence of a SMBH in our Galaxy and in AGNs, the SMBH hypothesis is the only plausible explanation for these mass concentrations.

3.5.3 Correlation Between SMBH Mass and Galaxy Properties

Currently, strong indications of SMBHs have been found in about 35 normal galaxies, and their masses have been estimated. This permits us to examine whether, and in what way, $M_\bullet$ is related to the properties of the host galaxy. This leads us to the discovery of a remarkable correlation; it is found that $M_\bullet$ is correlated with the absolute magnitude of the bulge component (or the spheroidal component) of the galaxy in which

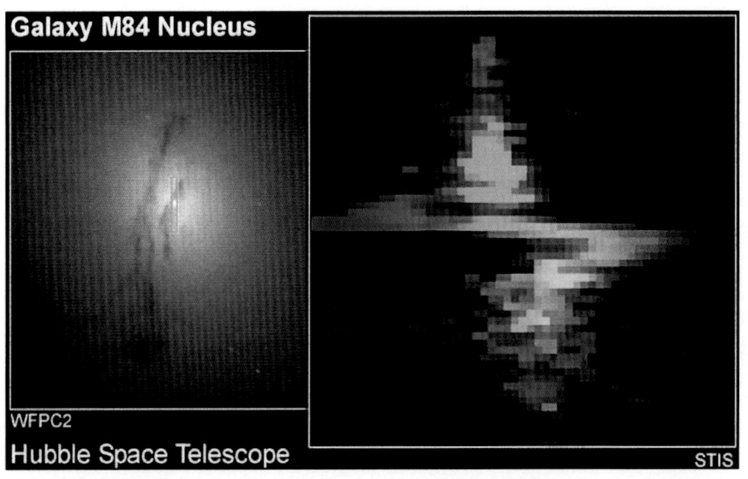

Fig. 3.24. An HST image of the nucleus of the galaxy M84 is shown in the left-hand panel. M84 is a member of the Virgo Cluster, about 15 Mpc away from us. The small rectangle depicts the position of the slit used by the STIS (Space Telescope Imaging Spectrograph) instrument on-board the HST to obtain a spectrum of the central region. This long-slit spectrum is shown in the right-hand panel; the position along the slit is plotted vertically, the wavelength of the light horizontally, also illustrated by colors. Near the center of the galaxy the wavelength suddenly changes because the rotational velocity steeply increases inwards and then changes sign on the other side of the center. This shows the Kepler rotation in the central gravitational field of a SMBH, whose mass can be estimated as $M_\bullet \sim 3 \times 10^8 M_\odot$

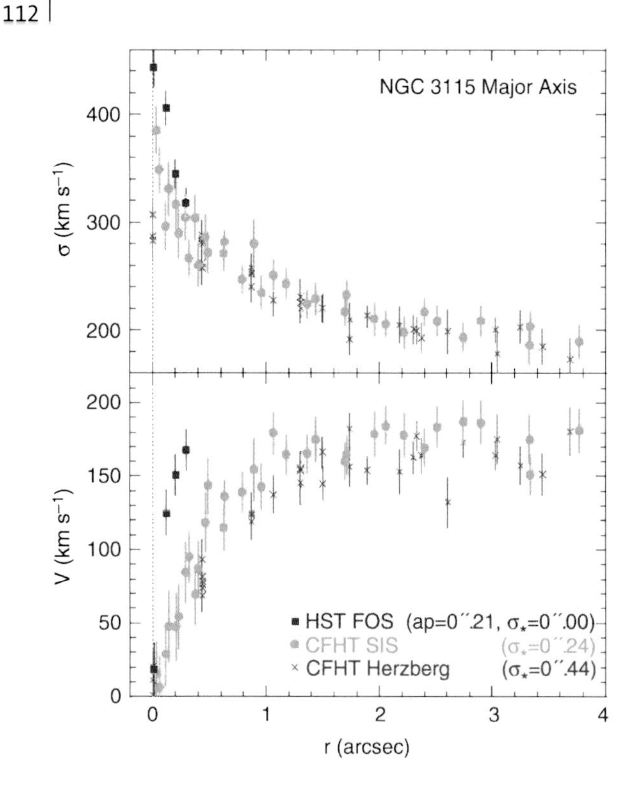

Fig. 3.25. Rotational velocity (bottom) and velocity dispersion (top), as functions of the distance from the center along the major axis of the galaxy NGC 3115. Colors of the symbols mark observations with different instruments. Results from CFHT data which have an angular resolution of 0.″44 are shown in blue. The SIS instrument at the CFHT uses active optics to achieve roughly twice this angular resolution; corresponding results are plotted in green. Finally, the red symbols show the result from HST observations using the Faint Object Spectrograph (FOS). As expected, with improved angular resolution an increase in the velocity dispersion is seen towards the center. Even more dramatic is the impact of resolution on measurements of the rotational velocity. Due to projection effects, the measured central velocity dispersion is smaller than the real one; this effect can be corrected for. After correction, a central value of $\sigma \sim 600\,\mathrm{km/s}$ is found. This value is much higher than the escape velocity from the central star cluster if it were to consist solely of stars – it would dissolve within $\sim 2\times10^4$ years. Therefore, an additional compact mass component of $M_\bullet \sim 10^9 M_\odot$ must exist

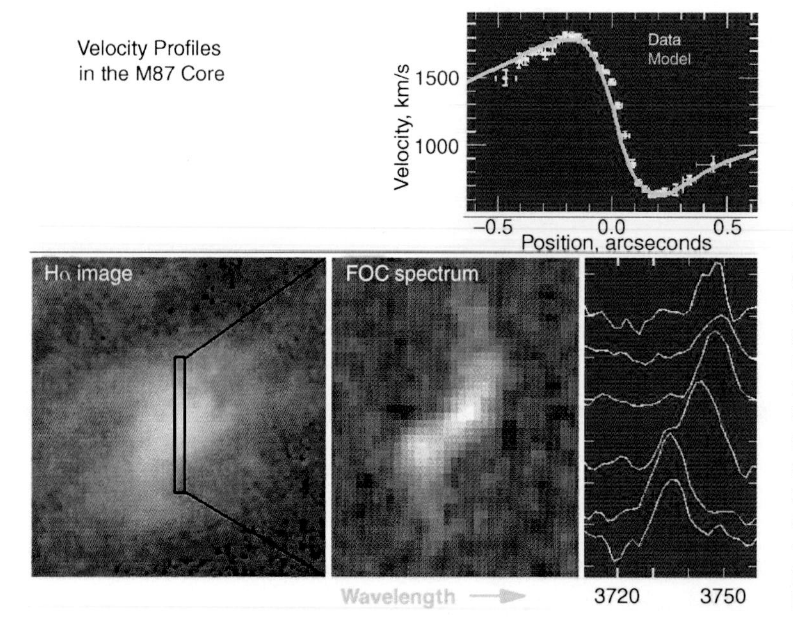

Fig. 3.26. M87 has long been one of the most promising candidates for harboring an SMBH in its center. In this figure, the position of the slit is shown superimposed on an Hα image of the galaxy (lower left) together with the spectrum of the [OII] line along this slit (bottom, center), and six spectra corresponding to six different positions along the slit, separated by 0.″14 each (lower right). In the upper right panel the rotation curve extracted from the data using a kinematical model is displayed. These results show that a central mass concentration with $\sim 3\times10^9 M_\odot$ must be present, confined to a region less than 3 pc across – indeed leaving basically no alternative but a SMBH

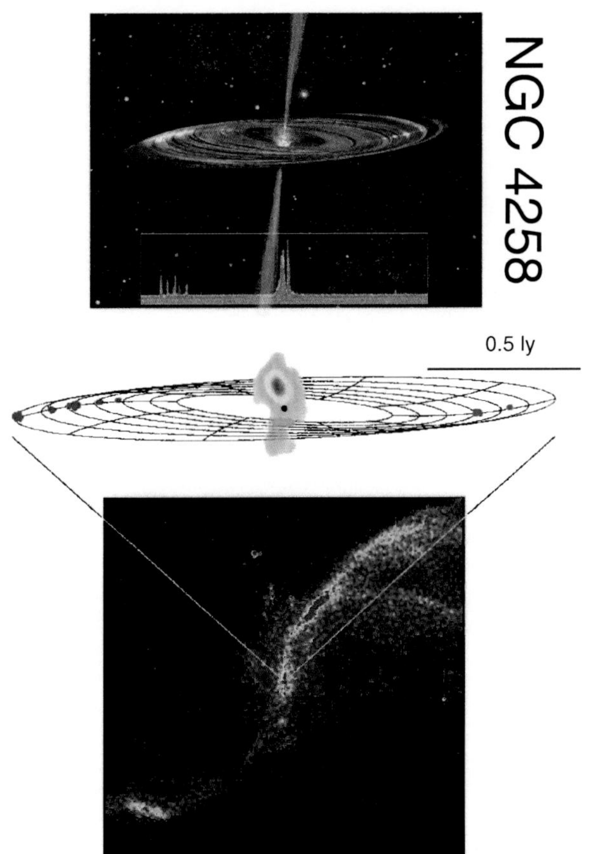

Fig. 3.27. The Seyfert galaxy NGC 4258 contains an accretion disk in its center in which several water masers are embedded. In the top image, an artist's impression of the hidden disk and the jet is displayed, together with the line spectrum of the maser sources. Their positions (center image) and velocities have been mapped by VLBI observations. From these measurements, the Kepler law for rotation in the gravitational field of a point mass of $M_\bullet = 25 \times 10^6 M_\odot$ in the center of this galaxy was verified. The best-fitting model of the central disk is also plotted. The bottom image is a 20-cm map showing the large-scale radio structure of the Seyfert galaxy

the SMBH is located (see Fig. 3.28, left). Here, the bulge component is either the bulge of a spiral galaxy or an elliptical galaxy as a whole. This correlation is described by

$$M_\bullet = 0.93 \times 10^8 \, M_\odot \left(\frac{L_{\mathrm{B,bulge}}}{10^{10} L_{\mathrm{B}\odot}} \right)^{1.11} ; \qquad (3.34)$$

it is statistically highly significant, but the deviations of the data points from this power law are considerably larger than their error bars. An alternative way to express this correlation is provided by the relation $M/L \propto L^{0.25}$ found previously – see (3.27) – by which we can also write $M_\bullet \propto M_{\mathrm{bulge}}^{0.9}$.

An even better correlation exists between $M_\bullet$ and the velocity dispersion in the bulge component, as can be seen in the right-hand panel of Fig. 3.28. This relation is best described by

$$\boxed{M_\bullet = 1.35 \times 10^8 \, M_\odot \left(\frac{\sigma_{\mathrm{e}}}{200 \, \mathrm{km/s}} \right)^4} , \qquad (3.35)$$

where the exact value of the exponent is still subject to discussion, and where a slightly higher value $M_\bullet \propto \sigma^{4.5}$ might better describe the data. The difference in the results obtained by different groups can partially be traced back to different definitions of the velocity dispersion, especially concerning the choice of the spatial region across which it is measured. It is remarkable that the deviations of the data points from the correlation (3.35) are compatible with the error bars for the measurements of $M_\bullet$. Thus, we have at present no indication of an intrinsic dispersion of the $M_\bullet$-σ relation.

In fact, there have been claims in the literature that even globular clusters contain a black hole; however, these claims are not undisputed. In addition, there may be objects that appear like globular clusters, but are in fact the stripped nucleus of a former dwarf galaxy. In this case, the presence of a central black hole is not unexpected, provided the scaling relation (3.35) holds down to very low velocity dispersion.

To date, the physical origin of this very close relation has not been understood in detail. The most obvious apparent explanation – that in the vicinity of a SMBH with a very large mass the stars are moving faster than around a smaller-mass SMBH – is not conclusive: the mass of the SMBH is significantly less than one percent of the mass of the bulge component. We can therefore disregard its contribution to the gravitational field in which the stars are orbiting. Instead, this correlation has to be linked to the fact that the spheroidal component of a galaxy evolves together with the SMBH. A better understanding of this relation can only be found from models of galaxy evolution. We will continue with this topic in Sect. 9.6.

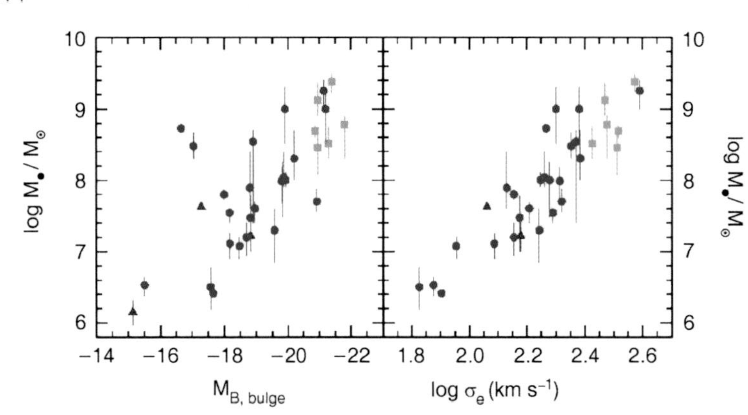

Fig. 3.28. Correlation of SMBH mass $M_\bullet$ with the absolute magnitude $M_{\rm B,bulge}$ (left) and the velocity dispersion $\sigma_{\rm e}$ (right) in the bulge component of the host galaxy. Circles (squares, triangles) indicate measurements that are based on stellar kinematics (gas kinematics, maser disks)

3.6 Extragalactic Distance Determination

In Sect. 2.2 we discussed methods for distance determination within our own Galaxy. We will now proceed with the determination of distances to other galaxies. It should be noted that the Hubble law (1.2) specifies a relation between the redshift of an extragalactic object and its distance. The redshift z is easily measured from the shift in spectral lines. For this reason, the Hubble law (and its generalization – see Sect. 4.3.3) provides a simple method for determining distance. However, to apply this law, the Hubble constant H_0 must first be known, i.e., the Hubble law must be calibrated. Therefore, in order to determine the Hubble constant, distances have to be measured independently from redshift.

Furthermore, it has to be kept in mind that besides the general cosmic expansion, which is expressed in the Hubble law, objects also show *peculiar motion*, like the velocities of galaxies in clusters of galaxies or the motion of the Magellanic Clouds around our Milky Way. These peculiar velocities are induced by gravitational acceleration resulting from the locally inhomogeneous mass distribution in the Universe. For instance, our Galaxy is moving towards the Virgo Cluster of galaxies, a dense accumulation of galaxies, due to the gravitational attraction caused by the cluster mass. The measured redshift, and therefore the Doppler shift, is always a superposition of the cosmic expansion velocity and peculiar velocities.

CMB Dipole Anisotropy. The peculiar velocity of the Galaxy is very precisely known. The radiation of the cosmic microwave background is not completely isotropic but instead shows a dipole component. This component originates in the velocity of the Solar System relative to the rest-frame in which the CMB appears isotropic (see Fig. 1.17). Due to the Doppler effect, the CMB appears hotter than average in the direction of our motion and cooler in the opposite direction. Analyzing this CMB dipole allows us to determine our peculiar velocity, which yields the result that the Sun moves at a velocity of (368 ± 2) km/s relative to the CMB rest-frame. Furthermore, the Local Group of galaxies (see Sect. 6.1) is moving at $v_{\rm LG} \approx 600$ km/s relative to the CMB rest-frame.

Distance Ladder. For the redshift of a source to be dominated by the Hubble expansion, the cosmic expansion velocity $v = cz = H_0 D$ has to be much larger than typical peculiar velocities. This means that in order to determine H_0 we have to consider sources at large distances for the peculiar velocities to be negligible compared to $H_0\,D$.

Making a direct estimate of the distances of distant galaxies is very difficult. Traditionally one uses a *distance ladder*: at first, the *absolute distances* to nearby galaxies are measured directly. If methods to measure *relative distances* (that is, distance ratios) with sufficient precision are utilized, the distances to galaxies further away are then determined relative to those nearby. In this way, by means of relative methods, distances are estimated for galaxies that are sufficiently far away for their redshift to be dominated by the Hubble flow.

3.6.1 Distance of the LMC

The distance of the Large Magellanic Cloud (LMC) can be estimated using various methods. For example, we can resolve and observe individual stars in the LMC, which forms the basis of the MACHO experiments (see Sect. 2.5.2). Because the metallicity of the LMC is significantly lower than that of the Milky Way, some of the methods discussed in Sect. 2.2 are only applicable after correcting for metallicity effects, e.g., the photometric distance determination or the period–luminosity relation for pulsating stars.

Perhaps the most precise method of determining the distance to the LMC is a purely geometrical one. The supernova SN 1987A that exploded in 1987 in the LMC illuminates a nearly perfectly elliptical ring (see Fig. 3.29). This ring consists of material that was once ejected by the stellar winds of the progenitor star of the supernova and that is now radiatively excited by energetic photons from the supernova explosion. The corresponding recombination radiation is thus emitted only when photons from the SN hit this gas. Because the observed ring is almost certainly intrinsically circular and the observed ellipticity is caused only by its inclination with respect to the line-of-sight, the distance to SN 1987A can be derived from observations of the ring. First, the inclination angle is determined from its observed ellipticity. The gas in the ring is excited by photons from the SN a time R/c after the original explosion, where R is the radius of the ring. We do not observe the illumination of the ring instantaneously because light from the section of the ring closer to us reaches us earlier than light from the more distant part. Thus, its illumination was seen sequentially along the ring. Combining the time delay in the illumination between the nearest and farthest part of the ring with its inclination angle, we then obtain the physical diameter of the ring. When this is compared to the measured angular diameter of $\sim 1\overset{''}{.}7$, the ratio yields the distance to SN 1987A,

$$D_{\rm SN1987A} \approx 51.8\ {\rm kpc} \pm 6\% .$$

If we now assume the extent of the LMC along the line-of-sight to be small, this distance can be identified with the distance to the LMC. The value is also compatible with other distance estimates (e.g., as derived by using photometric methods based on the properties of main-sequence stars – see Sect. 2.2.4).

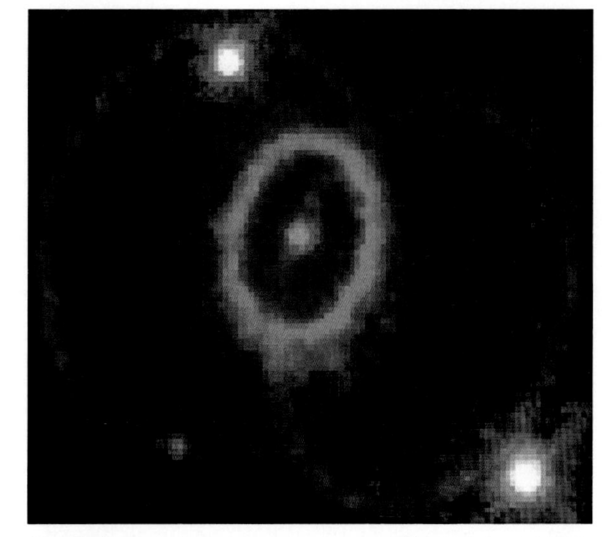

Fig. 3.29. The ring around supernova 1987A in the LMC is illuminated by photons from the explosion which induce the radiation from the gas in the ring. It is inclined towards the line-of-sight; thus it appears to be elliptical. Lighting up of the ring was not instantaneous, due to the finite speed of light: those sections of the ring closer to us lit up earlier than the more distant parts. From the time shift in the onset of radiation across the ring, its diameter can be derived. Combining this with the measured angular diameter of the ring, the distance to SN 1987A – and thus the distance to the LMC – can be determined

3.6.2 The Cepheid Distance

In Sect. 2.2.7, we discussed the period–luminosity relation of pulsating stars. Due to their high luminosity, Cepheids turn out to be particularly useful since they can be observed out to large distances.

For the period–luminosity relation of the Cepheids to be a good distance measure, it must first be calibrated. This calibration has to be done with as large a sample of Cepheids as possible at a known distance. Cepheids in the LMC are well-suited for this purpose because we believe we know the distance to the LMC quite precisely, see above. Also, due to the relatively small extent of the LMC along the line-of-sight, all Cepheids in the

LMC should be located at approximately the same distance. For this reason, the period–luminosity relation is calibrated in the LMC. Due to the large number of Cepheids available for this purpose (many of them have been found in the microlensing surveys), the resulting statistical errors are small. Uncertainties remain in the form of systematic errors related to the metallicity dependence of the period–luminosity relation; however, these can be corrected for since the color of Cepheids depends on the metallicity as well.

With the high angular resolution of the HST, individual Cepheids in galaxies are visible at distances up to that of the Virgo cluster of galaxies. In fact, determining the distance to Virgo as a central step in the determination of the Hubble constant was one of the major scientific aims of the HST. In the *Hubble Key Project*, the distances to numerous spiral galaxies in the Virgo Cluster were determined by identifying Cepheids and measuring their periods.

3.6.3 Secondary Distance Indicators

The Virgo Cluster, at a measured distance of about 16 Mpc, is not sufficiently far away from us to directly determine the Hubble constant from its distance and redshift, because peculiar velocities still contribute considerably to the measured redshift at this distance. To get to larger distances, a number of relative distance indicators are used. They are all based on measuring the distance *ratio* of galaxies. If the distance to one of the two is known, the distance to the other is then obtained from the ratio. By this procedure, distances to more remote galaxies can be measured. Below, we will review some of the most important secondary distance indicators.

SN Ia. Supernovae of Type Ia are to good approximation standard candles, as will be discussed more thoroughly in Sect. 8.3.1. This means that the absolute magnitudes of SNe Ia are all within a very narrow range. To measure the value of this absolute magnitude, distances must be known for galaxies in which SN Ia explosions have been observed and accurately measured. Therefore, the Cepheid method was applied especially to such galaxies, in this way calibrating the brightness of SNe Ia. SNe Ia are visible over very large distances, so that they also permit distance estimates at such large redshifts that the simple Hubble law (1.6) is no longer valid, but needs to be generalized based on a cosmological model (Sect. 4.3.3). As we will see later, these measurements belong to the most important pillars on which our standard model of cosmology rests.

Surface Brightness Fluctuations of Galaxies. Another method of estimating distance ratios is surface brightness fluctuations. It is based on the fact that the number of bright stars per area element in a galaxy fluctuates – purely by Poisson noise: If N stars are expected in an area element, relative fluctuations of $\sqrt{N}/N = 1/\sqrt{N}$ of the number of stars will occur. These are observed in fluctuations of the local surface brightness. To demonstrate that this effect can be used to estimate distances, we consider a solid angle $\mathrm{d}\omega$. The corresponding area element $\mathrm{d}A = D^2\,\mathrm{d}\omega$ depends quadratically on the distance D of the galaxy; the larger the distance, the larger the number of stars N in this solid angle, and the smaller the relative fluctuations of the surface brightness. By comparing the surface brightness fluctuations of different galaxies, one can then estimate relative distances. This method also has to be calibrated on the galaxies for which Cepheid distances are available.

Planetary Nebulae. The brightness distribution of planetary nebulae in a galaxy seems to have an upper limit which is the nearly the same for each galaxy (see Fig. 3.30). If a sufficient number of planetary nebulae are observed and their brightnesses measured, it enables us to determine their luminosity function from which the maximum apparent magnitude is then derived. By calibration on galaxies of known Cepheid distance, the corresponding maximum absolute magnitude can be determined, which then allows the determination of the distance modulus for other galaxies, thus their distances.

Scaling Relations. The scaling relations for galaxies – fundamental plane for ellipticals, Tully–Fisher relation for spirals (see Sect. 3.4) – can be calibrated on local groups of galaxies or on the Virgo Cluster, the distances of which have been determined from Cepheids. Although the scatter of these scaling relations can be 15% for individual galaxies, the statistical fluctuations are reduced when observing several galaxies at about the same distance (such as in clusters and groups). This

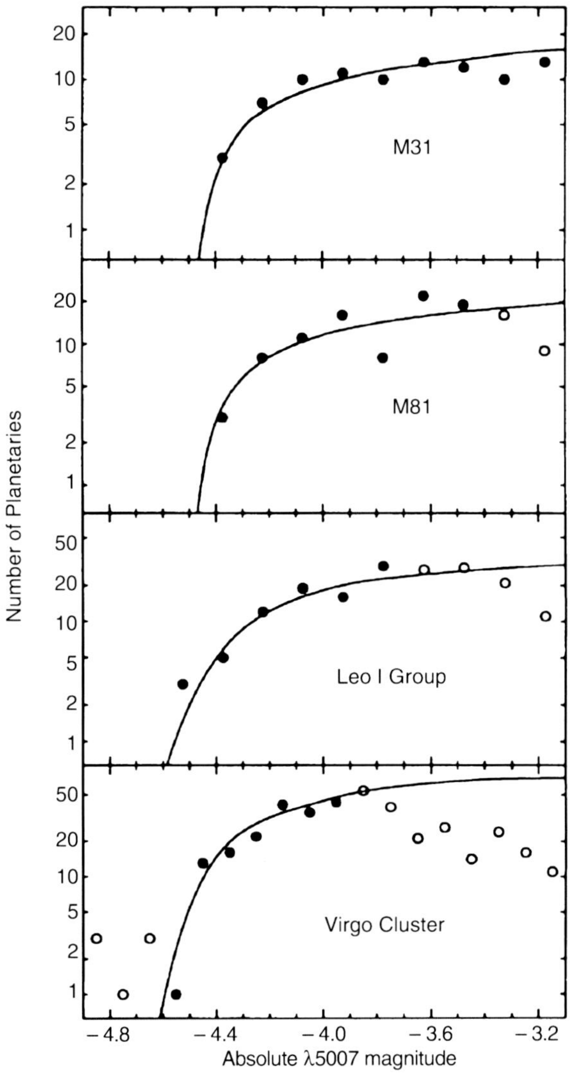

Fig. 3.30. Brightness distribution of planetary nebulae in Andromeda (M31), M81, three galaxies in the Leo I group, and six galaxies in the Virgo Cluster. The plotted absolute magnitude was measured in the emission line of double-ionized oxygen at $\lambda = 5007\,\text{Å}$ in which a large fraction of the luminosity of a planetary nebula is emitted. This characteristic property is also used in the identification of such objects in other galaxies. In all cases, the distribution is described by a nearly identical luminosity function; it seems to be a universal function in galaxies. Therefore, the brightness distribution of planetary nebulae can be used to estimate the distance of a galaxy. In the fits shown, the data points marked by open symbols were disregarded: at these magnitudes, the distribution function is probably not complete

enables us to estimate the distance ratio of two clusters of galaxies.

The Hubble Constant. In particular, the ratio of distances to the Virgo and the Coma clusters of galaxies is estimated by means of these various secondary distance measures. Together with the distance to the Virgo Cluster as determined from Cepheids, we can then derive the distance to Coma. Its redshift ($z \approx 0.023$) is large enough for its peculiar velocity to make no significant contribution to its redshift, so that it is dominated by the Hubble expansion. By combining the various methods we obtain a distance to the Coma cluster of about 90 Mpc, resulting in a Hubble constant of

$$\boxed{H_0 = 72 \pm 8\ \text{km/s/Mpc}}\ . \tag{3.36}$$

The error given here denotes the statistical uncertainty in the determination of H_0. Besides this uncertainty, possible systematic errors of the same order of magnitude may exist. In particular, the distance to the LMC plays a crucial role. As the lowest rung in the distance latter, it has an effect on all further distance estimates. We will see later (Sect. 8.7.1) that the Hubble constant can also be measured by a completely different method, based on tiny small-scale anisotropies of the cosmic microwave background, and that this method results in a value which is in impressively good agreement with the one in (3.36).

3.7 Luminosity Function of Galaxies

Definition of the Luminosity Function. The luminosity function specifies the way in which the members of a class of objects are distributed with respect to their luminosity. More precisely, the luminosity function is the number density of objects (here galaxies) of a specific luminosity. $\Phi(M)\,\mathrm{d}M$ is defined as the number density of galaxies with absolute magnitude in the interval $[M, M + \mathrm{d}M]$. The total density of galaxies is then

$$\nu = \int_{-\infty}^{\infty} \mathrm{d}M\ \Phi(M)\ . \tag{3.37}$$

Accordingly, $\Phi(L)\,\mathrm{d}L$ is defined as the number density of galaxies with a luminosity between L and $L + \mathrm{d}L$. It should be noted here explicitly that both definitions of the luminosity function are denoted by the same symbol, although they represent different mathematical functions, i.e., they describe different functional relations. It is therefore important (and in most cases not difficult) to deduce from the context which of these two functions is being referred to.

Problems in Determining the Luminosity Function. At first sight, the task of determining the luminosity function of galaxies does not seem very difficult. The history of this topic shows, however, that we encounter a number of problems in practice. As a first step, the determination of galaxy luminosities is required, for which, besides measuring the flux, distance estimates are also necessary. For very distant galaxies redshift is a sufficiently reliable measure of distance, whereas for nearby galaxies the methods discussed in Sect. 3.6 have to be applied.

Another problem occurs for nearby galaxies, namely the large-scale structure of the galaxy distribution. To obtain a representative sample of galaxies, a sufficiently large volume has to be surveyed because the galaxy distribution is heavily structured on scales of $\sim 100\,h^{-1}$ Mpc. On the other hand, galaxies of particularly low luminosity can only be observed locally, so the determination of $\Phi(L)$ for small L always needs to refer to local galaxies. Finally, one has to deal with the so-called *Malmquist bias*; in a flux-limited sample luminous galaxies will always be overrepresented because they are visible at larger distances (and therefore are selected from a larger volume). A correction for this effect is always necessary.

3.7.1 The Schechter Luminosity Function

The global galaxy distribution is well approximated by the *Schechter luminosity function*

$$\boxed{\Phi(L) = \left(\frac{\Phi^*}{L^*}\right)\left(\frac{L}{L^*}\right)^{\alpha} \exp\left(-L/L^*\right)}\;, \tag{3.38}$$

where L^* is a characteristic luminosity above which the distribution decreases exponentially, α is the slope of the luminosity function for small L, and Φ^* specifies the normalization of the distribution. A schematic plot of this function is shown in Fig. 3.31.

Expressed in magnitudes, this function appears much more complicated. Considering that an interval $\mathrm{d}L$ in luminosity corresponds to an interval $\mathrm{d}M$ in absolute magnitude, with $\mathrm{d}L/L = -0.4\,\ln 10\,\mathrm{d}M$, and using $\Phi(L)\,\mathrm{d}L = \Phi(M)\,\mathrm{d}M$, i.e., the number of sources in these intervals are of course the same, we obtain

$$\Phi(M) = \Phi(L)\left|\frac{\mathrm{d}L}{\mathrm{d}M}\right| = \Phi(L)\,0.4\,\ln 10\,L \tag{3.39}$$

$$= (0.4\,\ln 10)\Phi^*\,10^{0.4(\alpha+1)(M^*-M)} \times \exp\left(-10^{0.4(M^*-M)}\right)\;. \tag{3.40}$$

As mentioned above, the determination of the parameters entering the Schechter function is difficult; a set of parameters in the blue band is

$$\begin{aligned} \Phi^* &= 1.6\times 10^{-2}\,h^3\,\mathrm{Mpc}^{-3}\;, \\ M_B^* &= -19.7 + 5\log h\;, \qquad \text{or} \\ L_B^* &= 1.2\times 10^{10}\,h^{-2}\,L_\odot\;, \\ \alpha &= -1.07\;. \end{aligned} \tag{3.41}$$

While the blue light of galaxies is strongly affected by star formation, the luminosity function in the red bands measures the typical stellar distribution. In the K-band, we have

$$\begin{aligned} \Phi^* &= 1.6\times 10^{-2}\,h^3\,\mathrm{Mpc}^{-3}\;, \\ M_K^* &= -23.1 + 5\log h\;, \\ \alpha &= -0.9\;. \end{aligned} \tag{3.42}$$

The total number density of galaxies is formally infinite if $\alpha \le -1$, but the validity of the Schechter function does of course not extend to arbitrarily small L. The luminosity density

$$l_{\rm tot} = \int_0^\infty \mathrm{d}L\;L\,\Phi(L) = \Phi^*\,L^*\,\Gamma(2+\alpha) \tag{3.43}$$

is finite for $\alpha \ge -2$.[6] The integral in (3.43), for $\alpha \sim -1$, is dominated by $L \sim L^*$, and $n = \Phi^*$ is thus a good estimate for the mean density of L^*-galaxies.

[6] $\Gamma(x)$ is the Gamma function, defined by

$$\Gamma(x) = \int_0^\infty \mathrm{d}y\;y^{(x-1)}\,\mathrm{e}^{-y}\;. \tag{3.44}$$

For positive integers, $\Gamma(n+1) = n!$. We have $\Gamma(0.7) \approx 1.30$, $\Gamma(1) = 1$, $\Gamma(1.3) \approx 0.90$. Since these values are all close to unity, $l_{\rm tot} \sim \Phi^* L^*$ is a good approximation for the luminosity density.

Deviations of the galaxy luminosity function from the Schechter form are common. There is also no obvious reason why such a simple relation for describing the luminosity distribution of galaxies should exist. Although the Schechter function seems to be a good representation of the total distribution, each type of galaxy has its own luminosity function, with each function having a form that strongly deviates from the Schechter function – see Fig. 3.32. For instance, spirals are relatively narrowly distributed in L, whereas the distribution of ellipticals is much broader if we account for the full L-range, from giant ellipticals to dwarf ellipticals. E's dominate in particular at large L; the low end of the luminosity function is likewise dominated by dwarf ellipticals and Irr's. In addition, the luminosity distribution of cluster and group galaxies differs from that of field galaxies. The fact that the total luminosity function can be described by an equation as simple as (3.38) is, at least partly, a coincidence ("cosmic conspiracy") and cannot be modeled easily.

3.7.2 The Bimodal Color Distribution of Galaxies

The classification of galaxies by morphology, given by the Hubble classification scheme (Fig. 3.2), has the disadvantage that morphologies of galaxies are not easy to quantify. Traditionally, this was done by visual inspection but of course this method bears some subjectivity of the researcher doing it. Furthermore, this visual inspection is time consuming and cannot be performed on large samples of galaxies. Various techniques were developed to perform such a classification automatically, including brightness profile fitting – a de Vaucouleurs profile indicates an elliptical galaxy whereas an exponential brightness profile corresponds to a spiral.

Even these methods cannot be applied to galaxy samples for which the angular resolution of the imaging is not much better than the angular size of galaxies – since then, no brightness profiles can be fitted. An alternative to classify galaxies is provided by their color. We expect that early-type galaxies are red, whereas late-type galaxies are considerably bluer. Colors are much eas-

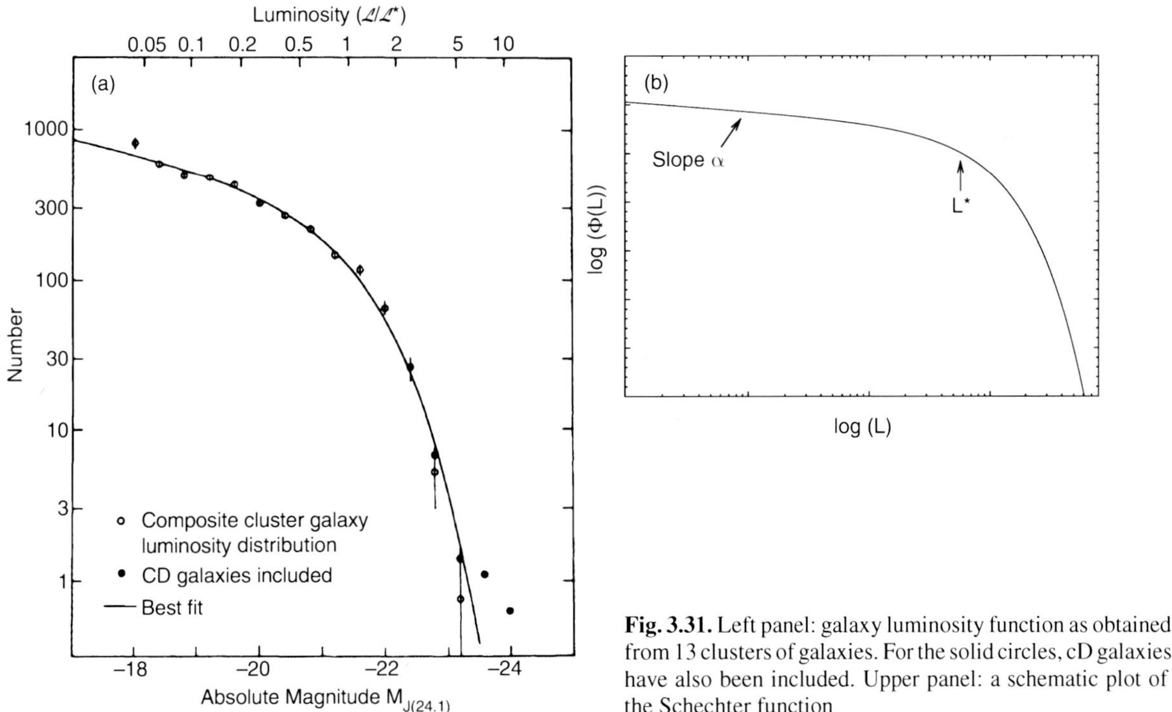

Fig. 3.31. Left panel: galaxy luminosity function as obtained from 13 clusters of galaxies. For the solid circles, cD galaxies have also been included. Upper panel: a schematic plot of the Schechter function

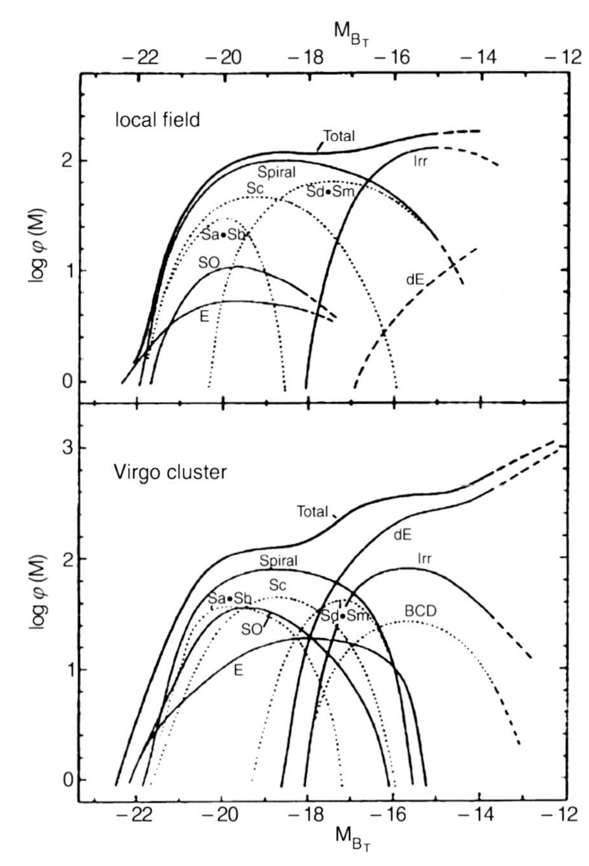

Fig. 3.32. The luminosity function for different Hubble types of field galaxies (top) and galaxies in the Virgo Cluster of galaxies (bottom). Dashed curves denote extrapolations. In contrast to Fig. 3.31, the more luminous galaxies are plotted towards the left. The Schechter luminosity function of the total galaxy distribution is compiled from the sum of the luminosity distributions of individual galaxy types that all deviate significantly from the Schechter function. One can see that in clusters the major contribution at faint magnitudes comes from the dwarf ellipticals (dEs), and that at the bright end ellipticals and S0's contribute much more strongly to the luminosity function than they do in the field. This trend is even more prominent in regular clusters of galaxies

ier to measure than morphology, in particular for very small galaxies. Therefore, one can study the luminosity function of galaxies, classifying them by their color.

Using photometric measurements and spectroscopy from the Sloan Digital Sky Survey (see Sect. 8.1.2), the colors and absolute magnitudes of $\sim 70\,000$ low-redshift galaxies has been studied; their density distribution in a color–magnitude diagram are plotted in the left-hand side of Fig. 3.33. From this figure we see immediately that there are two density peaks of the galaxy distribution in this diagram: one at high luminosities and red color, the other at significantly fainter absolute magnitudes and much bluer color. It appears that the galaxies are distributed at and around these two density peaks, hence galaxies tend to be either luminous and red, or less luminous and blue. We can also easily see from this diagram that the luminosity function of red galaxies is quite different from that of blue galaxies, which is another indication for the fact that the simple Schechter luminosity function (3.38) for the whole galaxy population most likely is a coincidence.

We can next consider the color distribution of galaxies at a fixed absolute magnitude M_r. This is obtained by plotting the galaxy number density along vertical cuts through the left-hand side of Fig. 3.33. When this is done for different M_r, it turns out that the color distribution of galaxies is bimodal: over a broad range in absolute magnitude, the color distribution has two peaks, one at red, the other at blue $u-r$. Again, this fact can be seen directly from Fig. 3.33. For each value of M_r, the color distribution of galaxies can be very well fitted by the sum of two Gaussian functions. The central colors of the two Gaussians is shown by the two dashed curves in the left panel of Fig. 3.33. They become redder the more luminous the galaxies are. This luminosity-dependent reddening is considerably more pronounced for the blue population than for the red galaxies.

To see how good this fit indeed is, the right-hand side of Fig. 3.33 shows the galaxy density as obtained from the two-Gaussian fits, with solid contours corresponding to the red galaxies and dashed contours to the blue ones. We thus conclude that the local galaxy population can be described as a bimodal distribution in $u-r$ color, where the characteristic color depends slightly on absolute magnitude. The galaxy distribution at bright absolute magnitudes is dominated by red galaxies, whereas for less luminous galaxies the blue population dominates. The luminosity function of both populations can be described by Schechter functions; however these two are quite different. The characteristic luminosity is about one magnitude brighter for the red galaxies than for the blue ones, whereas the faint-end slope α is significantly steeper for the blue galaxies. This

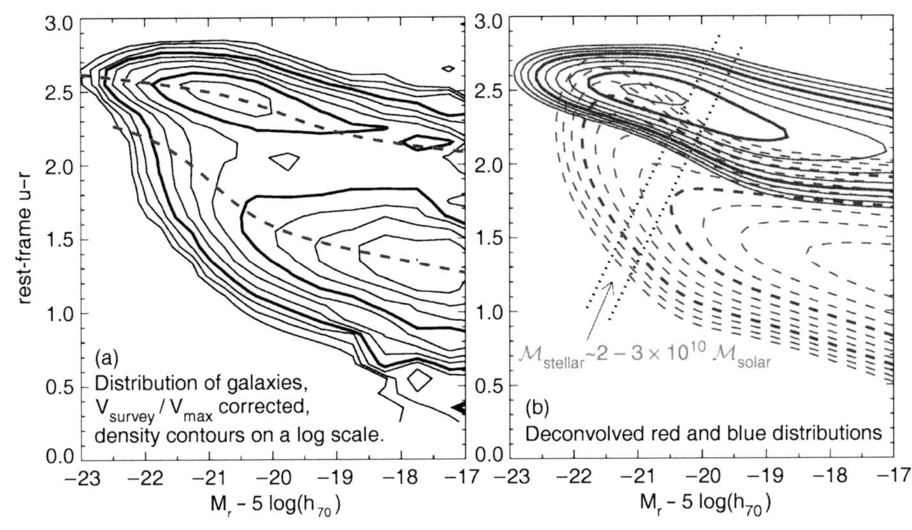

Fig. 3.33. The density of galaxies in color–magnitude space. The color of $\sim 70\,000$ galaxies with redshifts $0.01 \le z \le 0.08$ from the Sloan Digital Sky Survey is measured by the rest-frame $u-r$, i.e., after a (small) correction for their redshift was applied. The density contours, which were corrected for selection effects like the Malmquist bias, are logarithmically spaced, with a factor of $\sqrt{2}$ between consecutive contours. In the left-hand panel, the measured distribution is shown. Obviously, two peaks of the galaxy density are clearly visible, one at a red color of $u-r \sim 2.5$ and an absolute magnitude of $M_r \sim -21$, the other at a bluer color of $u-r \sim 1.3$ and significantly fainter magnitudes. The right-hand panel corresponds to the modeled galaxy density, as is described in the text

again is in agreement of what we just learned: for high luminosities, the red galaxies clearly dominate, whereas at small luminosities, the blue galaxies are much more abundant.

The mass-to-light ratio of a red stellar population is larger than that of a blue population, since the former no longer contains massive luminous stars. The difference in the peak absolute magnitude between the red and blue galaxies therefore corresponds to an even larger difference in the stellar mass of these two populations. Red galaxies in the local Universe have on average a much higher stellar mass than blue galaxies. This fact is illustrated by the two dotted lines in the right-hand panel of Fig. 3.33 which correspond to lines of constant stellar mass of ~ 2–$3 \times 10^{10}\,M_\odot$. This seems to indicate a very characteristic mass scale for the galaxy distribution: most galaxies with a stellar mass larger than this characteristic mass scale are red, whereas most of those with a lower stellar mass are blue.

Obviously, these statistical properties of the galaxy distribution must have an explanation in terms of the evolution of galaxies; we will come back to this issue in Chap. 9.

3.8 Galaxies as Gravitational Lenses

In Sect. 2.5 the gravitational lens effect was discussed, where we concentrated on the deflection of light by point masses. The lensing effect by stars leads to image separations too small to be resolved by any existing telescope. Since the separation angle is proportional to the square root of the lens mass (2.79), the angular separation of the images will be about a million times larger if a galaxy acts as a gravitational lens. In this case it should be observable, as was predicted in 1937 by Fritz Zwicky. Indeed, multiple images of very distant sources have been found, together with the galaxy responsible for the image splitting. In this section we will first describe this effect by continuing the discussion we began in Sect. 2.5.1. Examples of the lens effect and its various applications will then be discussed.

3.8.1 The Gravitational Lensing Effect – Part II

The geometry of a typical gravitational lens system is sketched in Fig. 2.21 and again in Fig. 3.34. The phys-

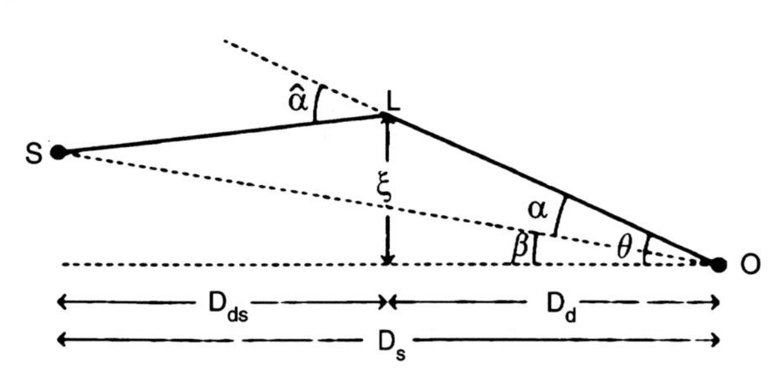

Fig. 3.34. As a reminder, another sketch of the lens geometry

ical description of such a lens system for an arbitrary mass distribution of the deflector is obtained from the following considerations.

If the gravitational field is weak (which is the case in all situations considered here), the gravitational effects can be linearized.[7] Hence, the deflection angle of a lens that consists of several mass components can be described by a linear superposition of the deflection angles of the individual components,

$$\hat{\boldsymbol{\alpha}} = \sum_i \hat{\boldsymbol{\alpha}}_i \, . \tag{3.45}$$

We assume that the deflecting mass has a small extent along the line-of-sight, as compared to the distances between observer and lens (D_d) and between lens and source (D_ds), $L \ll D_\mathrm{d}$ and $L \ll D_\mathrm{ds}$. All mass elements can then be assumed to be located at the same distance D_d. This physical situation is called a *geometrically thin lens*. If a galaxy acts as the lens, this condition is certainly fulfilled – the extent of galaxies is typically $\sim 100\,h^{-1}$ kpc while the distances of lens and source are typically $\sim$ Gpc. We can therefore write (3.45) as a superposition of Einstein angles of the form (2.71),

$$\hat{\boldsymbol{\alpha}}(\boldsymbol{\xi}) = \sum_i \frac{4Gm_i}{c^2} \frac{\boldsymbol{\xi} - \boldsymbol{\xi}_i}{|\boldsymbol{\xi} - \boldsymbol{\xi}_i|^2} \, , \tag{3.46}$$

[7]To characterize the strength of a gravitational field, we refer to the gravitational potential Φ. The ratio Φ/c^2 is dimensionless and therefore well suited to distinguishing between strong and weak gravitational fields. For weak fields, $\Phi/c^2 \ll 1$. Another possible way to quantify the field strength is to apply the virial theorem: if a mass distribution is in virial equilibrium, then $v^2 \sim \Phi$, and weak fields are therefore characterized by $v^2/c^2 \ll 1$. Because the typical velocities in galaxies are ~ 200 km/s, for galaxies $\Phi/c^2 \lesssim 10^{-6}$. The typical velocities of galaxies in a cluster of galaxies are ~ 1000 km/s, so that in clusters $\Phi/c^2 \lesssim 10^{-5}$. Thus the gravitational fields occurring are weak in both cases.

where $\boldsymbol{\xi}_i$ is the projected position vector of the mass element m_i, and $\boldsymbol{\xi}$ describes the position of the light ray in the lens plane, also called the impact vector.

For a continuous mass distribution we can imagine subdividing the lens into mass elements of mass $\mathrm{d}m = \Sigma(\boldsymbol{\xi})\mathrm{d}^2\xi$, where $\Sigma(\boldsymbol{\xi})$ describes the *surface mass density* of the lens at the position $\boldsymbol{\xi}$, obtained by projecting the spatial (three-dimensional) mass density ρ along the line-of-sight to the lens. With this definition the deflection angle (3.46) can be transformed into an integral,

$$\boxed{\hat{\boldsymbol{\alpha}}(\boldsymbol{\xi}) = \frac{4G}{c^2} \int \mathrm{d}^2\xi' \; \Sigma(\boldsymbol{\xi}') \frac{\boldsymbol{\xi} - \boldsymbol{\xi}'}{|\boldsymbol{\xi} - \boldsymbol{\xi}'|^2}} \, . \tag{3.47}$$

This deflection angle is then inserted into the lens equation (2.75),

$$\boxed{\boldsymbol{\beta} = \boldsymbol{\theta} - \frac{D_\mathrm{ds}}{D_\mathrm{s}} \hat{\boldsymbol{\alpha}}(D_\mathrm{d}\boldsymbol{\theta})} \, , \tag{3.48}$$

where $\boldsymbol{\xi} = D_\mathrm{d}\boldsymbol{\theta}$ describes the relation between the position $\boldsymbol{\xi}$ of the light ray in the lens plane and its apparent direction $\boldsymbol{\theta}$. We define the scaled deflection angle as in (2.76),

$$\boldsymbol{\alpha}(\boldsymbol{\theta}) = \frac{D_\mathrm{ds}}{D_\mathrm{s}} \hat{\boldsymbol{\alpha}}(D_\mathrm{d}\boldsymbol{\theta}) \, ,$$

so that the lens equation (3.48) can be written in the simple form (see Fig. 3.34)

$$\boxed{\boldsymbol{\beta} = \boldsymbol{\theta} - \boldsymbol{\alpha}(\boldsymbol{\theta})} \, . \tag{3.49}$$

A more convenient way to write the scaled deflection is as follows,

$$\boldsymbol{\alpha}(\boldsymbol{\theta}) = \frac{1}{\pi} \int \mathrm{d}^2\theta' \; \kappa(\boldsymbol{\theta}') \frac{\boldsymbol{\theta} - \boldsymbol{\theta}'}{|\boldsymbol{\theta} - \boldsymbol{\theta}'|^2} \, , \tag{3.50}$$

where

$$\kappa(\boldsymbol{\theta}) = \frac{\Sigma(D_\mathrm{d}\boldsymbol{\theta})}{\Sigma_\mathrm{cr}} \tag{3.51}$$

is the *dimensionless surface mass density*, and the so-called *critical surface mass density*

$$\Sigma_\mathrm{cr} = \frac{c^2 \, D_\mathrm{s}}{4\pi G \, D_\mathrm{d} \, D_\mathrm{ds}} \tag{3.52}$$

depends only on the distances to the lens and to the source. Although Σ_{cr} incorporates a combination of cosmological distances, it is of a rather "human" order of magnitude,

$$\Sigma_{cr} \approx 0.35 \left(\frac{D_d \, D_{ds}}{D_s \; 1\,\text{Gpc}} \right)^{-1} \text{g cm}^{-2} \, .$$

A source is visible at several positions $\boldsymbol{\theta}$ on the sphere, or multiply imaged, if the lens equation (3.49) has several solutions $\boldsymbol{\theta}$ for a given source position $\boldsymbol{\beta}$. A more detailed analysis of the properties of this lens equation yields the following general result:

If $\Sigma \geq \Sigma_{cr}$ in at least one point of the lens, then source positions $\boldsymbol{\beta}$ exist such that a source at $\boldsymbol{\beta}$ has multiple images. It immediately follows that κ is a good measure for the strength of the lens. A mass distribution with $\kappa \ll 1$ at all points is a weak lens, unable to produce multiple images, whereas one with $\kappa \gtrsim 1$ for certain regions of $\boldsymbol{\theta}$ is a strong lens.

For sources that are small compared to the characteristic scales of the lens, the magnification μ of an image, caused by the differential light deflection, is given by (2.83), i.e.,

$$\mu = \left| \det \left(\frac{\partial \boldsymbol{\beta}}{\partial \boldsymbol{\theta}} \right) \right|^{-1} \, . \tag{3.53}$$

The importance of the gravitational lens effect for extragalactic astronomy stems from the fact that gravitational light deflection is independent of the nature and the state of the deflecting matter. Therefore, it is equally sensitive to both dark and baryonic matter and independent of whether or not the matter distribution is in a state of equilibrium. The lens effect is thus particularly suitable for probing matter distributions, without requiring any further assumptions about the state of equilibrium or the relation between dark and luminous matter.

3.8.2 Simple Models

Axially Symmetric Mass Distributions. The simplest models for gravitational lenses are those which are axially symmetric, for which $\Sigma(\boldsymbol{\xi}) = \Sigma(\xi)$, where $\xi = |\boldsymbol{\xi}|$ denotes the distance of a point from the center of the lens. In this case, the deflection angle is directed radially inwards, and we obtain

$$\hat{\alpha} = \frac{4GM(\xi)}{c^2 \, \xi} \, , \tag{3.54}$$

where $M(\xi)$ is the mass within radius ξ. Accordingly, for the scaled deflection angle we have

$$\alpha(\theta) = \frac{m(\theta)}{\theta} := \frac{1}{\theta} \, 2 \int_0^\theta \mathrm{d}\theta' \, \theta' \, \kappa(\theta') \, , \tag{3.55}$$

where, in the last step, $m(\theta)$ was defined as the dimensionless mass within θ. Since $\boldsymbol{\alpha}$ and $\boldsymbol{\theta}$ are collinear, the lens equation becomes one-dimensional because only the radial coordinate needs to be considered,

$$\beta = \theta - \alpha(\theta) = \theta - \frac{m(\theta)}{\theta} \, . \tag{3.56}$$

An illustration of this one-dimensional lens mapping is shown in Fig. 3.35.

Example: Point-Mass Lens. For a point mass M, the dimensionless mass becomes

$$m(\theta) = \frac{4GM}{c^2} \frac{D_{ds}}{D_d \, D_s} \, ,$$

reproducing the lens equation from Sect. 2.5.1 for a point-mass lens.

Example: Isothermal Sphere. We saw in Sect. 2.4.2 that the rotation curve of our Milky Way is flat for large radii, and we know from Sect. 3.3.3 that the rotation curves of other spiral galaxies are flat as well. This indicates that the mass of a galaxy increases proportional to r, thus $\rho(r) \propto r^{-2}$, or more precisely,

$$\rho(r) = \frac{\sigma_v^2}{2\pi G r^2} \, . \tag{3.57}$$

Here, σ_v is the one-dimensional velocity dispersion of stars in the potential of the mass distribution if the distribution of stellar orbits is isotropic. In principle, σ_v is therefore measurable spectroscopically from the line width. The mass distribution described by (3.57) is called a *singular isothermal sphere* (SIS). Because this mass model is of significant importance not only for the analysis of the lens effect, we will discuss its properties in a bit more detail.

The density (3.57) diverges for $r \to 0$ as $\rho \propto r^{-2}$, so that the mass model cannot be applied up to the very center of a galaxy. However, the steep central increase of the rotation curve shows that the core region of the mass distribution, in which the density function will deviate considerably from the r^{-2}-law, must be small for galaxies. Furthermore, the mass diverges for large r such that $M(r) \propto r$. The mass profile thus has to be cut off at some radius in order to get a finite total mass. This cut-off radius is probably very large ($\gtrsim 100$ kpc for L^*-galaxies) because the rotation curves are flat to at least the outermost point at which they are observable.

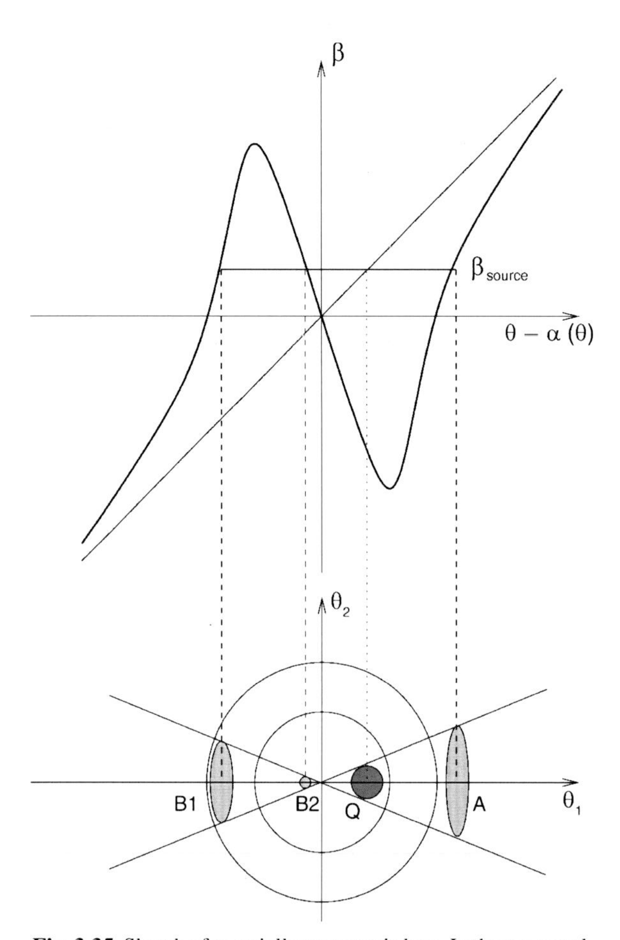

Fig. 3.35. Sketch of an axially symmetric lens. In the top panel, $\theta - \alpha(\theta)$ is plotted as a function of the angular separation θ from the center of the lens, together with the straight line $\beta = \theta$. The three intersection points of the horizontal line at fixed β with the curve $\theta - \alpha(\theta)$ are the three solutions of the lens equation. The bottom image indicates the positions and sizes of the images on the observer's sky. Here, Q is the unlensed source (which is not visible itself in the case of light deflection, of course!), and A, B1, B2 are the observed images of the source. The sizes of the images, and thus their fluxes, differ considerably; the inner image B2 is particularly weak in the case depicted here. The flux of B2 relative to that of image A depends strongly on the core radius of the lens; it can be so low as to render the third image unobservable. In the special case of a singular isothermal sphere, the innermost image is in fact absent

The SIS is an appropriate simple model for gravitational lenses over a wide range in radius since it seems to reproduce the basic properties of lens systems (such as image separation) quite well. The surface mass density is obtained from the projection of (3.57) along the line-of-sight,

$$\Sigma(\xi) = \frac{\sigma_v^2}{2G\xi} , \tag{3.58}$$

which yields the projected mass $M(\xi)$ within radius ξ

$$M(\xi) = 2\pi \int_0^\xi d\xi' \, \xi' \, \Sigma(\xi') = \frac{\pi \sigma_v^2 \xi}{G} . \tag{3.59}$$

With (3.54) the deflection angle can be obtained,

$$\hat{\alpha}(\xi) = 4\pi \left(\frac{\sigma_v}{c}\right)^2 ,$$

$$\boxed{\alpha(\theta) = 4\pi \left(\frac{\sigma_v}{c}\right)^2 \left(\frac{D_{\rm ds}}{D_{\rm s}}\right) \equiv \theta_{\rm E}} . \tag{3.60}$$

Thus the deflection angle for an SIS is constant and equals $\theta_{\rm E}$, and it depends quadratically on σ_v. $\theta_{\rm E}$ is called the *Einstein angle* of the SIS. The characteristic scale of the Einstein angle is

$$\theta_{\rm E} = 1''\!.15 \left(\frac{\sigma_v}{200\,{\rm km/s}}\right)^2 \left(\frac{D_{\rm ds}}{D_{\rm s}}\right) , \tag{3.61}$$

from which we conclude that the angular scale of the lens effect in galaxies is about an arcsecond for massive galaxies. The lens equation (3.56) for an SIS is

$$\beta = \theta - \theta_E \frac{\theta}{|\theta|} , \qquad (3.62)$$

where we took into account the fact that the deflection angle is negative for $\theta < 0$ since it is always directed inwards.

Solution of the Lens Equation for the Singular Isothermal Sphere. If $|\beta| < \theta_E$, two solutions of the lens equation exist,

$$\theta_1 = \beta + \theta_E , \quad \theta_2 = \beta - \theta_E . \qquad (3.63)$$

Without loss of generality, we assume $\beta \geq 0$; then $\theta_1 > \theta_E > 0$ and $0 > \theta_2 > -\theta_E$: one image of the source is located on either side of the lens center, and the separation of the images is

$$\boxed{\Delta\theta = \theta_1 - \theta_2 = 2\theta_E = 2''\!.3 \left(\frac{\sigma_v}{200\,\text{km/s}}\right)^2 \left(\frac{D_{ds}}{D_s}\right)} . \qquad (3.64)$$

Thus, the angular separation of the images does not depend on the position of the source. For massive galaxies acting as lenses it is of the order of somewhat more than one arcsecond. For $\beta > \theta_E$ only one image of the source exists, at θ_1, meaning that it is located on the same side of the center of the lens as the unlensed source.

For the magnification, we find

$$\mu(\theta) = \frac{|\theta/\theta_E|}{||\theta/\theta_E| - 1|} . \qquad (3.65)$$

If $\theta \approx \theta_E$, μ is very large. Such solutions of the lens equation exist for $|\beta| \ll \theta_E$, so that sources close to the center of the source plane may be highly magnified. If $\beta = 0$, the image of the source will be a ring of radius $\theta = \theta_E$, a so-called *Einstein ring*.

More Realistic Models. Mass distributions occurring in nature are not expected to be truly symmetric. The ellipticity of the mass distribution or external shear forces (caused, for example, by the tidal gravitational field of neighboring galaxies) will disturb the symmetry. The lensing properties of the galaxy will change by this symmetry breaking. For example, more than two images may be generated. Figure 3.36 illustrates the lens properties of such elliptical mass distributions. One can see, for example, that pairs of images, which are both heavily magnified, may be observed with a separation significantly smaller than the Einstein radius of the lens. Nevertheless, the characteristic image separation is still of the order of magnitude given by (3.64).

3.8.3 Examples for Gravitational Lenses

Currently, about 70 gravitational lens systems are known in which a galaxy acts as the lens. Some of them were discovered serendipitously, but most were found in systematic searches for lens systems. Amongst the most important lens surveys are: (1) *The HST Snapshot Survey.* The ~ 500 most luminous quasars have been imaged with the HST, and six lens systems have been identified. (2) *JVAS.* About 2000 bright radio sources with a flat radio spectrum (these often contain compact radio components, see Sect. 5.1.3) were scanned for multiple components with the VLA. Six lens systems have been found. (3) *CLASS.* Like in JVAS, radio sources with a flat spectrum were searched with the VLA for multiple components, but the flux limit was lower than in JVAS, which form a subset of the CLASS sources. The survey contains 15 000 sources, of which, to data, 22 have been identified as lenses. In this section we will discuss some examples of identified lens systems.

QSO 0957+561: The First Double Quasar. The first lens system was discovered in 1979 by Walsh, Carswell & Weymann when the optical identification of a radio source showed two point-like optical sources (see Fig. 3.37). Both could be identified as quasars located at the same redshift of $z_s = 1.41$ and having very similar spectra (see Fig. 3.38). Deep optical images of the field show an elliptical galaxy situated between the two quasar images, at a redshift of $z_d = 0.36$. The galaxy is so massive and so close to image B of the source that it *has to* produce a lens effect. However, the observed image separation of $\Delta\theta = 6''\!.1$ is considerably larger than expected from the lens effect by a single galaxy (3.64). The explanation for this is that the lens galaxy is located in a cluster of galaxies; the additional lens effect of the cluster adds to that of the galaxy,

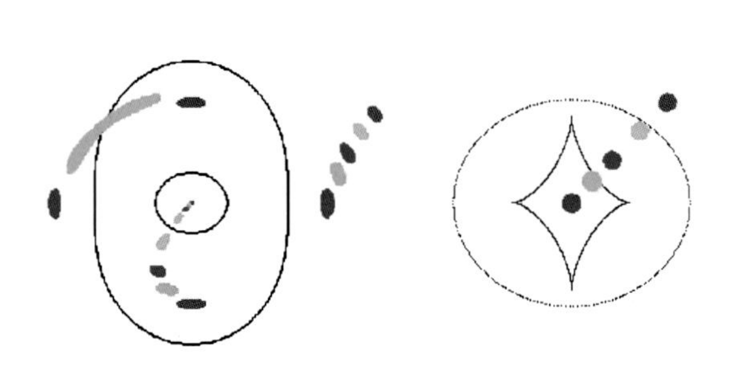

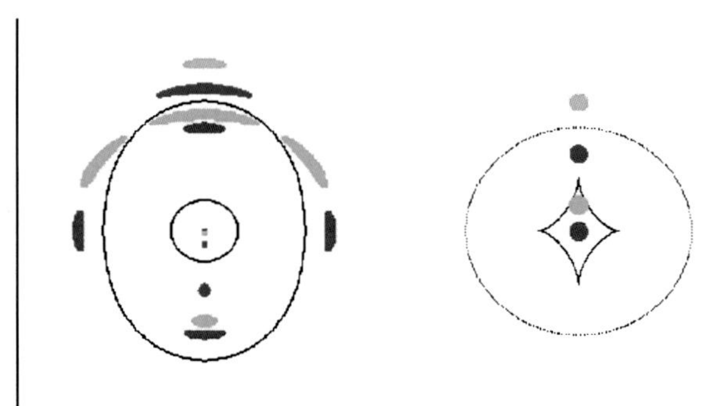

Fig. 3.36. Geometry of an "elliptical" lens, whereby it is of little importance whether the surface mass density Σ is constant on ellipses (i.e., the mass distribution has elliptical isodensity contours) or whether an originally spherical mass distribution is distorted by an external tidal field. On the right-hand side in both panels, several different source positions in the source plane are displayed, each corresponding to a different color. The origin in the source plane is chosen as the intersection point of the line connecting the center of symmetry in the lens and the observer with the source plane (see also Fig. 2.22). Depending on the position of the source, one, three, or five images may appear in the lens plane (i.e., the observer's sky); they are shown on the left-hand side of each panel. The curves in the lens plane are the *critical curves*, the location of all points for which $\mu \to \infty$. The curves in the source plane (i.e., on the right-hand side of each panel) are *caustics*, obtained by mapping the critical curves onto the source plane using the lens equation. Obviously, the number of images of a source depends on the source location relative to the location of the caustics. Strongly elongated images of a source occur close to the critical curves

boosting the image separation to a large value. The lens system QSO 0957+561 was observed in all wavelength ranges, from the radio to the X-ray. The two images of the quasar are very similar at all λ, including the VLBI structure (Fig. 3.38) – as would be expected since the lens effect is independent of the wavelength, i.e., achromatic.

QSO PG1115+080. In 1980, the so-called triple quasar was discovered, composed of three optical quasars at a maximum angular separation of just below $3''$. Component (A) is significantly brighter than the other two images (B, C; see Fig. 3.39, left). In high-resolution images it was found that the brightest image is in fact a double image: A is split into A1 and A2. The angular separation of the two roughly equally bright images is $\sim 0\overset{''}{.}5$, which is considerably smaller than all other angular separations in this system. The four quasar images have a redshift of $z_{\rm s} = 1.72$, and the lens is located at $z_{\rm d} = 0.31$. The image configuration is one of those that are expected for an elliptical lens, see Fig. 3.36.

With the NIR camera NICMOS on-board HST, not only were the quasar images and the lens galaxy observed, but also a nearly complete Einstein ring (Fig. 3.39, right). The source of this ring is the host galaxy of the quasar (see Sect. 5.4.5) which is substantially redder than the active galactic nucleus itself.

From the image configuration in such a quadruple system, the mass of the lens within the images can be estimated very accurately. The four images of the lens system trace a circle around the center of the lens galaxy, the radius of which can be identified with the Einstein radius of the lens. From this, the mass of the lens within the Einstein radius follows immediately because the Einstein radius is obtained from the lens equation (3.56) by setting $\beta = 0$. Therefore, the Einstein radius is the solution of the equation

$$\theta = \alpha(\theta) = \frac{m(\theta)}{\theta} ,$$

or

$$m(\theta_{\rm E}) = \frac{4G\,M(\theta_{\rm E})}{c^2} \frac{D_{\rm ds}}{D_{\rm d}\,D_{\rm s}} = \theta_{\rm E}^2 .$$

This equation is best written as

$$\boxed{M(\theta_{\rm E}) = \pi\,(D_{\rm d}\theta_{\rm E})^2\,\Sigma_{\rm cr}} , \tag{3.66}$$

which is readily interpreted:

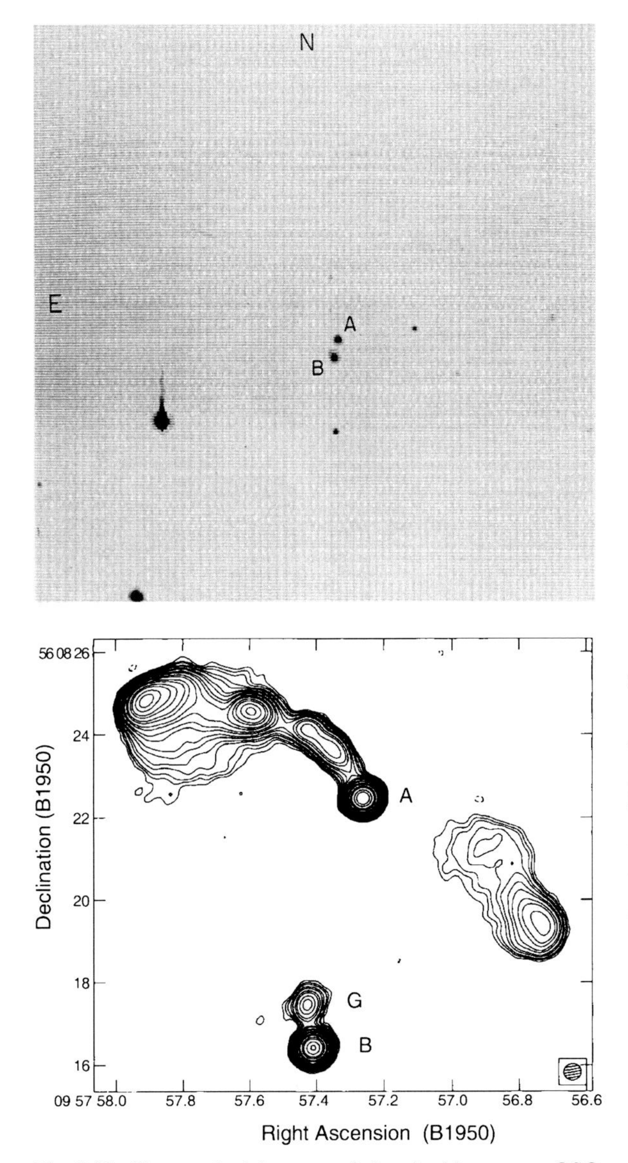

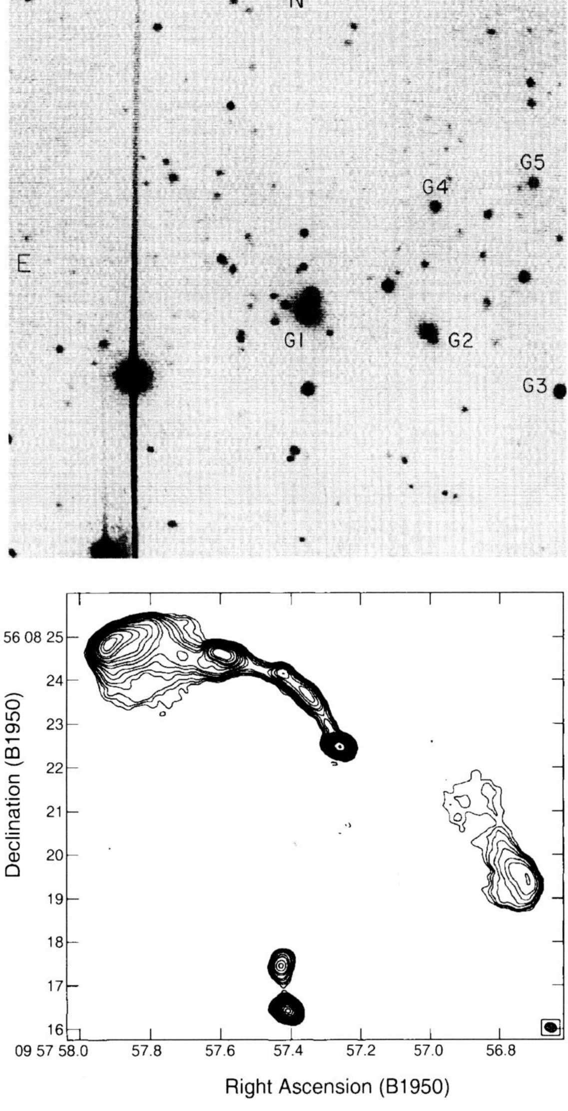

Fig. 3.37. Top: optical images of the double quasar QSO 0957+561. The image on the left has a short exposure time; here, the two point-like images A,B of the quasar are clearly visible. In contrast, the image on the right has a longer exposure time, showing the lens galaxy G1 between the two quasar images. Several other galaxies (G2-G5) are visible as well. The lens galaxy is a member of a cluster of galaxies at $z_d = 0.36$. Bottom: two radio maps of QSO 0957+561, observed with the VLA at 6 cm (left) and 3.6 cm (right), respectively. The two images of the quasar are denoted by A,B; G is the radio emission of the lens galaxy. The quasar has a radio jet, which is a common property of many quasars (see Sect. 5.3.1). On small angular scales, the jet can be observed by VLBI techniques in both images (see Fig. 3.38). On large scales only a single image of the jet exists, seen in image A; this property should be compared with Fig. 3.36 where it was demonstrated that the number of images of a source (component) depends on its position in the source plane

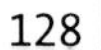

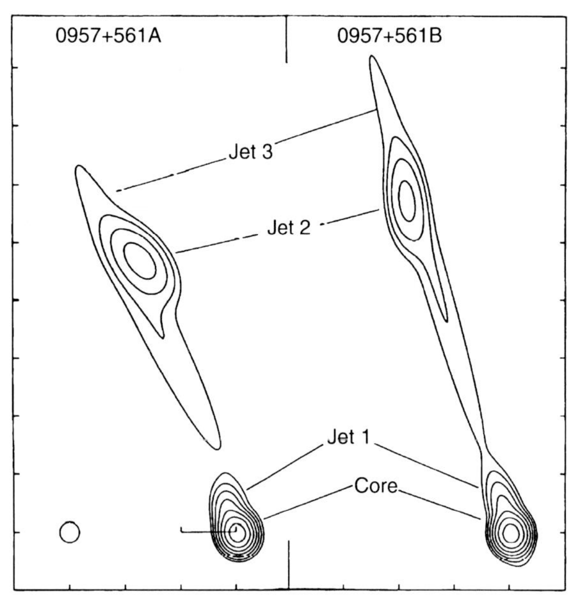

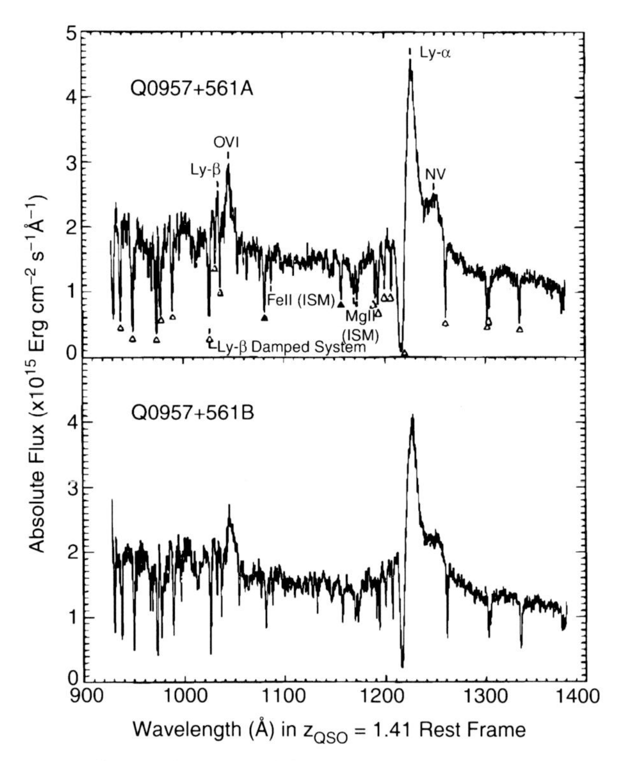

Fig. 3.38. Left: milliarcsecond structure of the two images of the quasar QSO 0957+561, a VLBI map at 13 cm wavelength by Gorenstein et al. Both quasar images show a core-jet structure, and it is clearly seen that they are mirror-symmetric, as predicted by lens models. right: spectra of the two quasar images QSO 0957+561A,B, observed by the Faint Object Camera (FOC) on-board HST. The similarity of the spectra, in particular the identical redshift, is a clear indicator of a common source of the two quasar images. The broad Lyα line, in the wings of which an Nv line is visible, is virtually always the strongest emission line in quasars

The mass within θ_E of a lens follows from the fact that the mean surface mass density within θ_E equals the critical surface mass density Σ_{cr}. A more accurate determination of lens masses is possible by means of detailed lens models. For quadruple image systems, the masses can be derived with a precision of a few percent – these are the most precise mass determinations in (extragalactic) astronomy.

QSO 2237+0305: The Einstein Cross. A spectroscopic survey of galaxies found several unusual emission lines in the nucleus of a nearby spiral galaxy which cannot originate from this galaxy itself. Instead, they are emitted by a background quasar at redshift $z_s = 1.7$ situated exactly behind this spiral. High-resolution images show four point sources situated around the nucleus of this galaxy, with an image separation of $\Delta\theta \approx 1''.8$ (Fig. 3.40). The spectroscopic analysis of these point sources revealed that all four are images of the same quasar (Fig. 3.41).

The images in this system are positioned nearly symmetrically around the lens center; this is also a typical lens configuration which may be caused by an elliptical lens (see Fig. 3.36). The Einstein radius of this lens is $\theta_E \approx 0''.9$, and we can determine the mass within this radius with a precision of $\sim 3\%$.

Einstein Rings. More examples of Einstein rings are displayed in Figs. 3.42 and 3.43. The first of these is a radio galaxy, with its two radio components be-

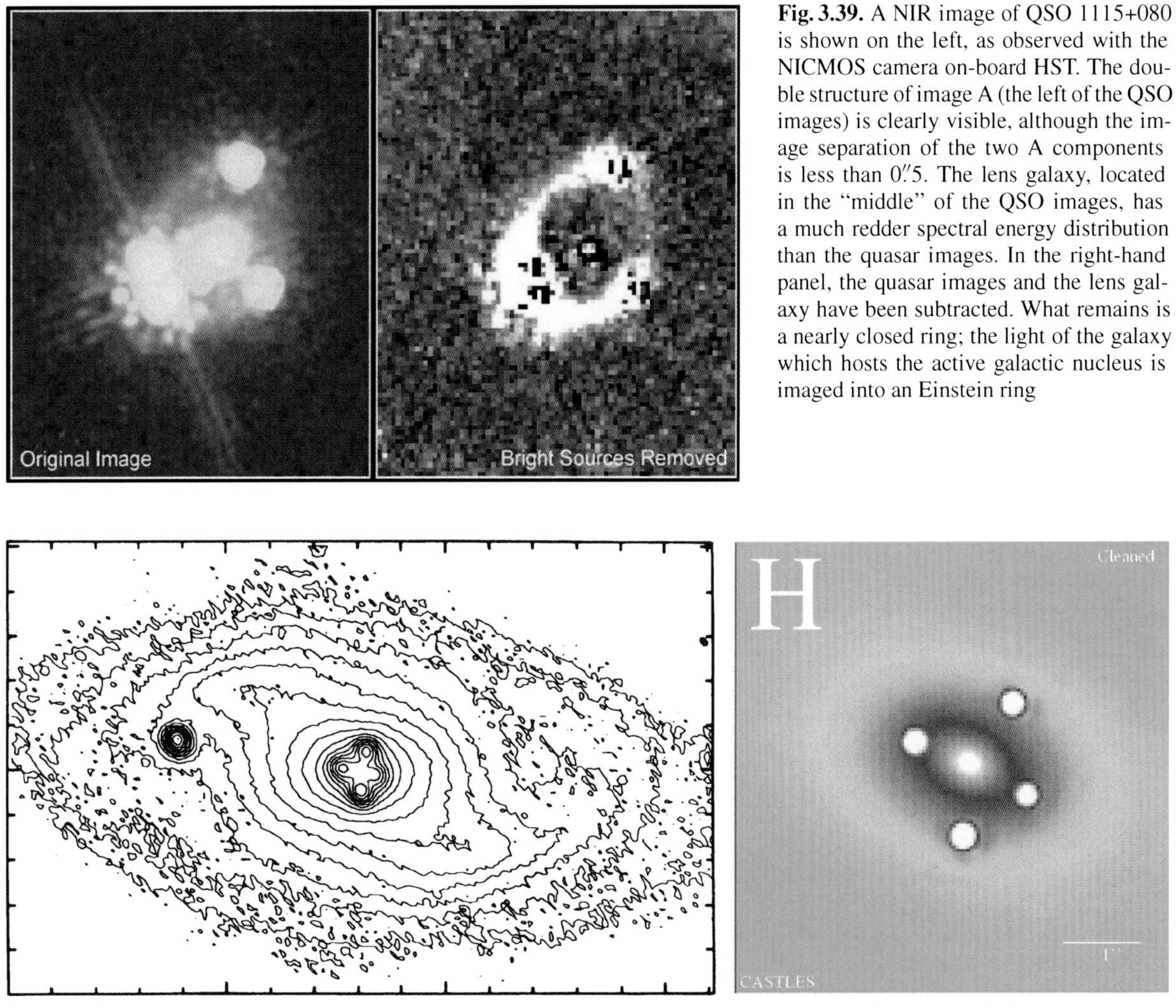

Fig. 3.39. A NIR image of QSO 1115+080 is shown on the left, as observed with the NICMOS camera on-board HST. The double structure of image A (the left of the QSO images) is clearly visible, although the image separation of the two A components is less than 0.″5. The lens galaxy, located in the "middle" of the QSO images, has a much redder spectral energy distribution than the quasar images. In the right-hand panel, the quasar images and the lens galaxy have been subtracted. What remains is a nearly closed ring; the light of the galaxy which hosts the active galactic nucleus is imaged into an Einstein ring

Fig. 3.40. Left: in the center of a nearby spiral galaxy, four point-like sources were found whose spectra show strong emission lines. This image from the CFHT clearly shows the bar structure in the core of the lens galaxy. An HST/NICMOS image of the center of QSO 2237+0305 is shown on the right. The central source is not a fifth quasar image but rather the bright nucleus of the lens galaxy

ing multiply imaged by a lens galaxy – one of the two radio sources is imaged into four components, the other mapped into a double image. In the NIR the radio galaxy is visible as a complete Einstein ring. This example shows very clearly that the appearance of the images of a source depends on the source size: to obtain an Einstein ring a sufficiently extended source is needed.

At radio wavelengths, the quasar MG 1654+13 consists of a compact central source and two radio lobes. As we will discuss in Sect. 5.1.3, this is a very typical radio morphology for quasars. One of the two lobes has a ring-shaped structure, which prior to this observation had never been observed before. An optical image of the field shows the optical quasar at the position of the compact radio component and, in addition, a bright elliptical galaxy right in the center of the ring-shaped radio lobe. This galaxy has a significantly lower redshift than the quasar and hence is the gravitational lens responsible for imaging the lobe into an Einstein ring.

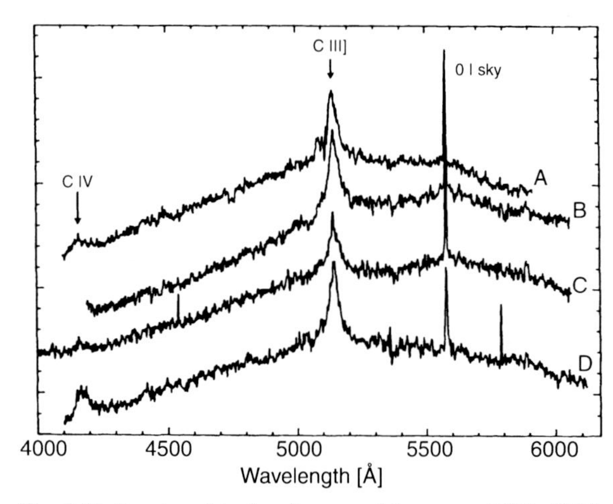

Fig. 3.41. Spectra of the four images of the quasar 2237+0305, observed with the CFHT. As is clearly visible, the spectral properties of these four images are very similar; this is the final proof that we are dealing with a lens system here. Measuring the individual spectra of these four very closely spaced sources is extremely difficult and can only be performed under optimum observing conditions

3.8.4 Applications of the Lens Effect

Mass Determination. As mentioned previously, the mass within a system of multiple images can be determined directly, sometimes very precisely. Since the length-scale in the lens plane (at given angular scale) and Σ_{cr} depend on H_0, these mass estimates scale with H_0. For instance, for QSO 2237+0305, a mass within $0\farcs9$ of $(1.08 \pm 0.02)h^{-1} \times 10^{10} M_\odot$ is derived. An even more precise determination of the mass was obtained for the lens galaxy of the Einstein ring in the system MG 1654+13 (Fig. 3.43). The dependence on the other cosmological parameters is comparatively weak, especially at low redshifts of the source and the lens. Most lens galaxies are early-type galaxies (ellipticals), and from the determination of their mass it can be concluded that ellipticals also contain dark matter.

Environmental Effects. Detailed lens models show that the light deflection of most gravitational lenses is affected by an external tidal field. This is due to the fact that lens galaxies are often members of galaxy groups which contribute to the light deflection as well. In some cases the members of the group have been identified. Mass properties of the corresponding group can be derived from the strength of this external influence.

Determination of the Hubble Constant. The light travel times along the different paths (according to the multiple images) are not the same. On the one hand the paths have different geometrical lengths, and on the other hand the light rays traverse different depths of the gravitational potential of the lens, resulting in a (general relativistic) time dilation effect. The difference in the light travel times Δt is measurable because luminosity variations of the source are observed at different times in the individual images. Δt can be measured from this difference in arrival time, called the time delay.

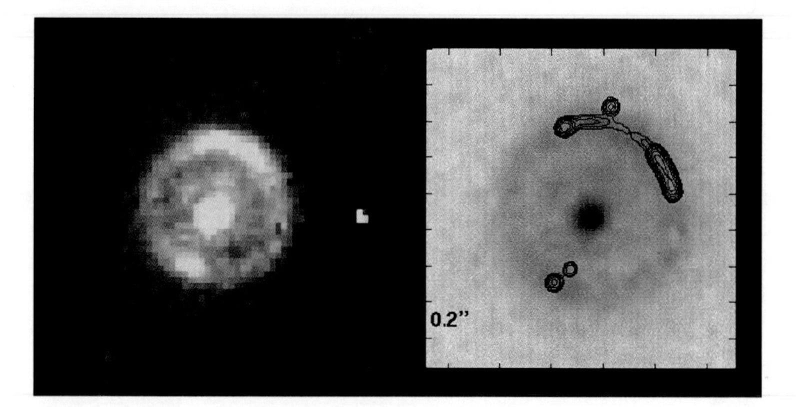

Fig. 3.42. The radio source 1938+666 is seen to be multiply imaged (contours in the right-hand figure); here, the radio source consists of two components, one of which is imaged four-fold, the other two-fold. A NIR image taken with the NICMOS camera onboard the HST (left-hand figure, also shown on the right in gray-scale) shows the lens galaxy in the center of an Einstein ring that originates from the stellar light of the host galaxy of the active galactic nucleus

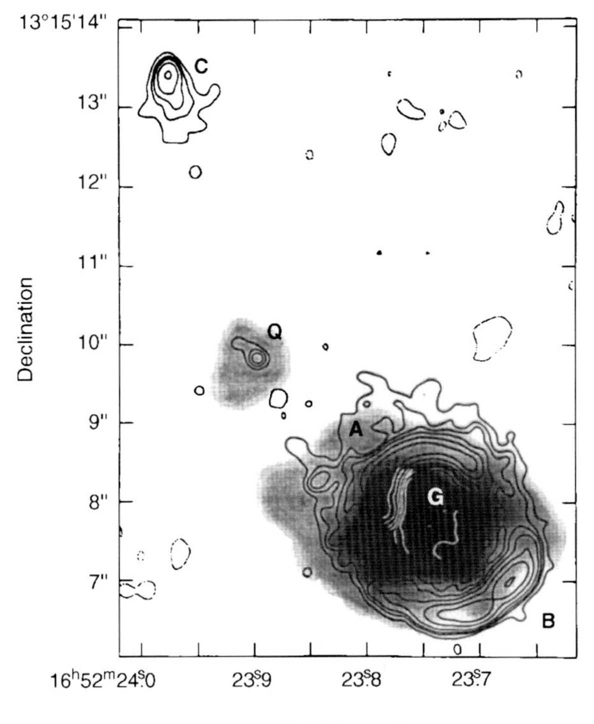

Fig. 3.43. The quasar MG1654+13 shows, in addition to the compact radio core (Q), two radio lobes; the northern lobe is denoted by C, whereas the southern lobe is imaged into a ring. An optical image is displayed in gray-scales, showing not only the quasar at Q ($z_s = 1.72$) but also a massive foreground galaxy at $z_d = 0.25$ that is responsible for the lensing of the lobe into an Einstein ring. The mass of this galaxy within the ring can be derived with a precision of $\sim 1\%$

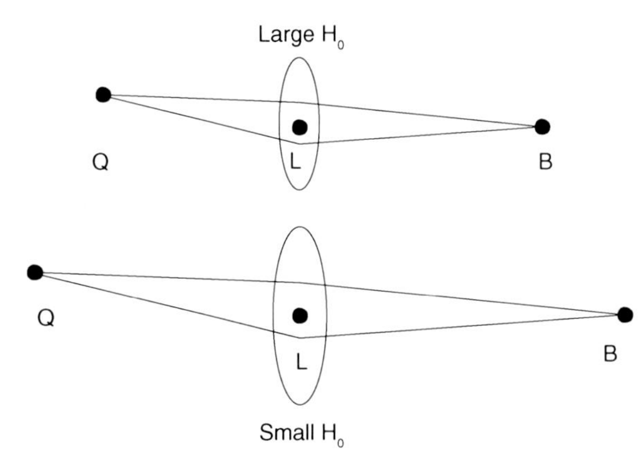

Fig. 3.44. Lens geometry in two universes with different Hubble constant. All observables are dimensionless – angular separations, flux ratios, redshifts – except for the difference in the light travel time. This is larger in the universe at the bottom than in the one at the top; hence, $\Delta t \propto H_0^{-1}$. If the time delay Δt can be measured, and if one has a good model for the mass distribution of the lens, then the Hubble constant can be derived from measuring Δt

It is easy to see that Δt depends on the Hubble constant, or in other words, on the size of the Universe. If a universe is twice the size of our own, Δt would be twice as large as well – see Fig. 3.44. Thus if the mass distribution of the lens can be modeled sufficiently well, by modeling the geometry of the image configuration, then the Hubble constant can be derived from measuring the difference in the light travel time. To date, Δt has been measured in about 10 lens systems (see Fig. 3.45 for an example). Based on "plausible" lens models we can derive values for the Hubble constant that are compatible with other measurements (see Sect. 3.6), but which tend towards slightly smaller values of H_0 than that determined from the HST Key Project (3.36). The main difficulty here is that the mass distribution in lens galaxies cannot unambiguously be derived from the positions of the multiple images. Therefore, these determinations of H_0 are currently not considered to be precision measurements. On the other hand, we can draw interesting conclusions about the radial mass profile of lens galaxies from Δt if we assume H_0 is known. In Sect. 6.3.4 we will discuss the value of H_0 determinations from lens time delays in a slightly different context.

The ISM in Lens Galaxies. Since the same source is seen along different sight lines passing through the lens galaxy, the comparison of the colors and spectra of the individual images provides information on reddening and on dust extinction in the ISM of the lens galaxy. From such investigations it was shown that the extinction in ellipticals is in fact very low, as is to be expected from the small amount of interstellar medium they contain, whereas the extinction is considerably higher for spirals. These analyses also enable us to study the relation between extinction and reddening, and from this to search for deviations from the Galactic reddening law (2.21). In fact, the constant of proportionality R_V is different in other galaxies, indicating a different composition of the dust, e.g., with respect to the chemical composition and to the size distribution of the dust grains.

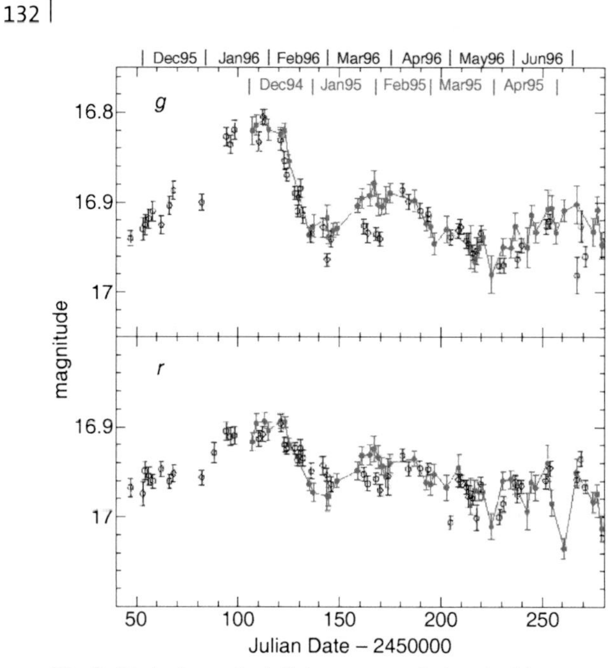

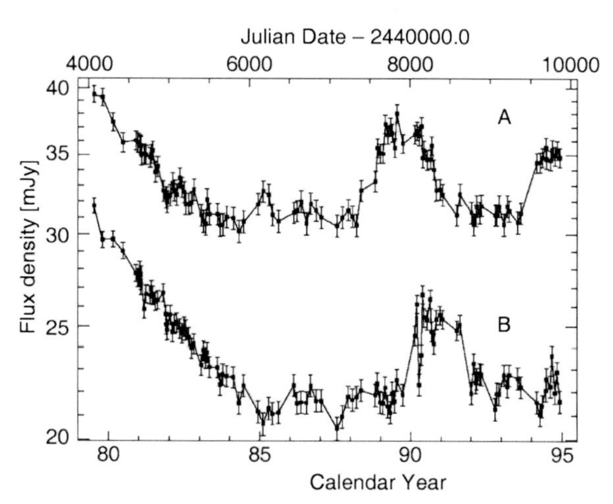

Fig. 3.45. Left: optical light curves of the double quasar 0957+561 in two broad-band filters. The light curve of image A is displayed in red and that of image B in blue, where the latter is shifted in time by 417 days. With this shift, the two light curves are made to coincide – this light travel time difference of 417 days is determined with an accuracy of $\sim \pm 3$ days. Right: radio light curves of QSO 0957+561A,B at 6 cm. From these radio measurements Δt can also be measured, and the corresponding value is compatible with that obtained from optical data

3.9 Population Synthesis

The light of normal galaxies originates from stars. Stellar evolution is largely understood, and the spectral radiation of stars can be calculated from the theory of stellar atmospheres. If the distribution of the number density of stars is known as a function of their mass, chemical composition, and evolutionary stage, we can compute the light emitted by them. The *theory of population synthesis* aims at interpreting the spectrum of galaxies as a superposition of stellar spectra. We have to take into account the fact that the distribution of stars changes over time; e.g., massive stars leave the main sequence after several 10^6 years, the number of luminous blue stars thus decreases, which means that the spectral distribution of the population also changes in time. The spectral energy distribution of a galaxy thus reflects its history of star formation and stellar evolution. For this reason, simulating different star-formation histories and comparing them with observed galaxy spectra provides important clues to understanding the evolution of galaxies. In this section, we will discuss some aspects of the theory of population synthesis; this subject is of tremendous importance for our understanding of galaxy spectra.

3.9.1 Model Assumptions

The processes of star formation are not understood in detail; for instance, it is currently impossible to compute the mass spectrum of a group of stars that jointly formed in a molecular cloud. Obviously, high-mass and low-mass stars are born together and form young (open) star clusters. The mass spectra of these stars are determined empirically from observations.

The *initial mass function* (IMF) is defined as the initial mass distribution at the time of birth of the stars, such that $\phi(m)\,\mathrm{d}m$ specifies the fraction of stars in the mass interval of width $\mathrm{d}m$ around m, where the distribution is normalized,

$$\int_{m_L}^{m_U} \mathrm{d}m\; m\, \phi(m) = 1 M_\odot \; .$$

The integration limits are not well defined. Typically, one puts $m_L \sim 0.1 M_\odot$ because less massive stars do not ignite their hydrogen (and are thus brown dwarfs), and $m_U \sim 100 M_\odot$, because more massive stars have not been observed. Such very massive stars would be difficult to observe because of their very short lifetime; furthermore, the theory of stellar structure tells us that more massive stars can probably not form a stable configuration due to excessive radiation pressure. The shape of the IMF is also subject to uncertainties; in most cases, the *Salpeter-IMF* is used,

$$\boxed{\phi(m) \propto m^{-2.35}} \,, \tag{3.67}$$

as obtained from investigating the stellar mass spectrum in young star clusters. It is by no means clear whether a universal IMF exists, or whether it depends on specific conditions like metallicity, the mass of the galaxy, or other parameters. The Salpeter IMF seems to be a good description for stars with $M \gtrsim 1 M_\odot$, whereas the IMF for less massive stars is less steep.

The *star-formation rate* is the gas mass that is converted into stars per unit time,

$$\psi(t) = -\frac{\mathrm{d}M_{\mathrm{gas}}}{\mathrm{d}t} \,.$$

The metallicity Z of the ISM defines the metallicity of the newborn stars, and the stellar properties in turn depend on Z. During stellar evolution, metal-enriched matter is ejected into the ISM by stellar winds, planetary nebulae, and SNe, so that $Z(t)$ is an increasing function of time. This chemical enrichment must be taken into account in population synthesis studies in a self-consistent form.

Let $S_{\lambda,Z}(t')$ be the emitted energy per wavelength and time interval, normalized to an initial total mass of $1 M_\odot$, emitted by a group of stars of initial metallicity Z and age t'. The function $S_{\lambda,Z(t-t')}(t')$, which describes this emission at any point t in time, accounts for the different evolutionary tracks of the stars in the Hertzsprung–Russell diagram (HRD) – see Appendix B.2. It also accounts for their initial metallicity (i.e., at time $t-t'$), where the latter follows from the chemical evolution of the ISM of the corresponding galaxy. Then the total spectral luminosity of this galaxy at a time t is given by

$$F_\lambda(t) = \int_0^t \mathrm{d}t' \, \psi(t-t') \, S_{\lambda,Z(t-t')}(t') \,, \tag{3.68}$$

thus by the convolution of the star-formation rate with the spectral energy distribution of the stellar population. In particular, $F_\lambda(t)$ depends on the star-formation history.

3.9.2 Evolutionary Tracks in the HRD; Integrated Spectrum

In order to compute $S_{\lambda,Z(t-t')}(t')$, models for stellar evolution and stellar atmospheres are needed. As a reminder, Fig. 3.46(a) displays the evolutionary tracks in the HRD. Each track shows the position of a star with specified mass in the HRD and is parametrized by the time since its formation. Positions of equal time in the HRD are called *isochrones* and are shown in Fig. 3.46(b). As time proceeds, fewer and fewer massive stars exist because they quickly leave the main sequence and end up as supernovae or white dwarfs. The number density of stars along the isochrones depends on the IMF. The spectrum $S_{\lambda,Z(t-t')}(t')$ is then the sum over all spectra of the stars on an isochrone – see Fig. 3.47(b).

In the beginning, the spectrum and luminosity of a stellar population are dominated by the most massive stars, which emit intense UV radiation. But after $\sim 10^7$ years, the flux below 1000 Å is diminished significantly, and after $\sim 10^8$ years, it hardly exists any more. At the same time, the flux in the NIR increases because the massive stars evolve into red supergiants.

For $10^8\,\mathrm{yr} \lesssim t \lesssim 10^9\,\mathrm{yr}$, the emission in the NIR remains high, whereas short-wavelength radiation is more and more diminished. After $\sim 10^9$ yr, red giant stars (RGB stars) account for most of the NIR production. After $\sim 3 \times 10^9$ yr, the UV radiation increases again due to blue stars on the horizontal branch into which stars evolve after the AGB phase, and due to white dwarfs which are hot when they are born. Between an age of 4 and 13 billion years, the spectrum of a stellar population evolves fairly little.

Of particular importance is the spectral break located at about 4000 Å which becomes visible in the spectrum after a few 10^7 years. This break is caused by a strongly changing opacity of stellar atmospheres at this wavelength, mainly due to strong transitions of singly ionized calcium and the Balmer lines of hydrogen. This *4000 Å-break* is one of the most important spectral properties of galaxies; as we will discuss in Sect. 9.1.2, it allows us to estimate the redshifts of early-type galaxies from their

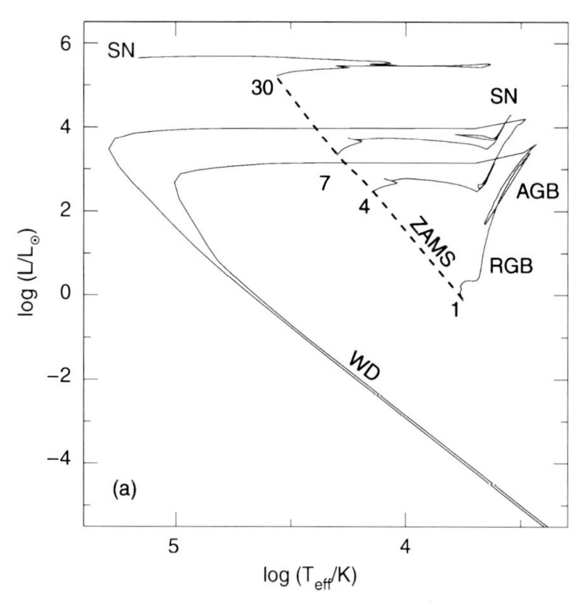

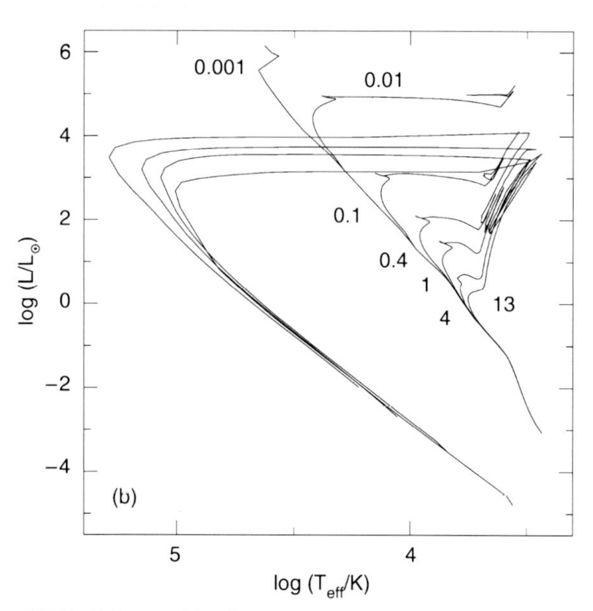

Fig. 3.46. **a**) Evolutionary tracks in the HRD for stars of different masses, as indicated by the numbers near the tracks (in units of $M_{\odot}$). The ZAMS (zero age main sequence) is the place of birth in the HRD; evolution moves stars away from the main sequence. Depending on the mass, they explode as a core-collapse SN (for $M \geq 8M_{\odot}$) or end as a white dwarf (WD). Prior to this, they move along the red giant branch (RGB) and the asymptotic giant branch (AGB). **b**) Isochrones at different times, indicated in units of 10^9 years. The upper main sequence is quickly depopulated by the rapid evolution of massive stars, whereas the red giant branch is populated over time

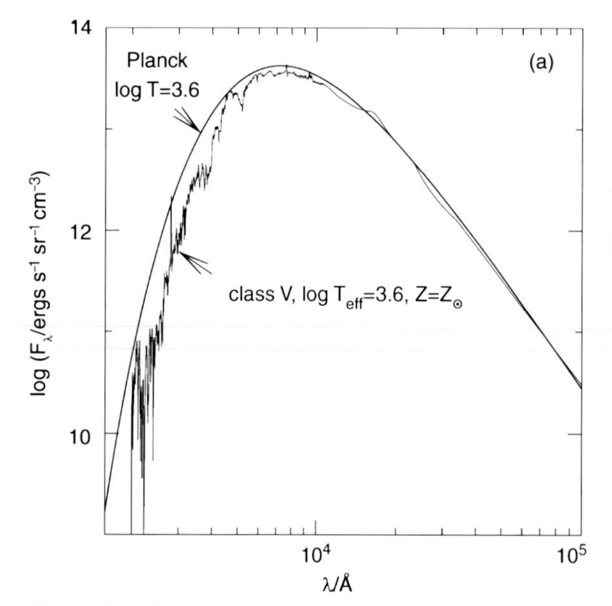

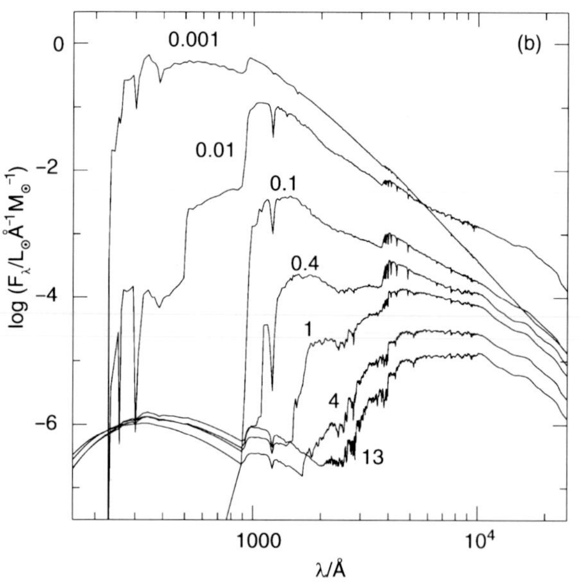

Fig. 3.47. **a**) Comparison of the spectrum of a main-sequence star with a blackbody spectrum of equal effective temperature. The opacity of the stellar atmosphere causes clear deviations from the Planck spectrum in the UV/optical. **b**) Spectrum of a stellar population with solar metallicity that was instantaneously born a time t ago; t is given in units of 10^9 years

photometric properties – so-called photometric redshift estimates.

3.9.3 Color Evolution

Detailed spectra of galaxies are often not available. Instead we have photometric images in different broad-band filters, since the observing time required for spectroscopy is substantially larger than for photometry. In addition, modern wide-field cameras can obtain photometric data of numerous galaxies simultaneously. From the theory of population synthesis we can derive photometric magnitudes by multiplying model spectra with the filter functions, i.e., the transmission curves of the color filters used in observations, and then integrating over wavelength (A.25). Hence the spectral evolution implies a color evolution, as is illustrated in Fig. 3.48(a).

For a young stellar population the color evolution is rapid and the population becomes redder, again because the hot blue stars have a higher mass and thus evolve quickly in the HRD. For the same reason, the evolution is faster in $B-V$ than in $V-K$. It should be mentioned that this color evolution is also observed in star clusters of different ages. The mass-to-light ratio M/L also increases with time because M remains constant while L decreases.

As shown in Fig. 3.48(b), the blue light of a stellar population is always dominated by main-sequence stars, although at later stages a noticeable contribution also comes from horizontal branch stars. The NIR radiation is first dominated by stars burning helium in their center (this class includes the supergiant phase of massive stars), later by AGB stars, and after $\sim 10^9$ yr by red giants. Main sequence stars never contribute more than 20% of the light in the K-band. The fact that M/L_K varies only little with time implies that the NIR luminosity is a good indicator for the total stellar mass: the NIR mass-to-light ratio is much less dependent on the age of the stellar population than that for bluer filters.

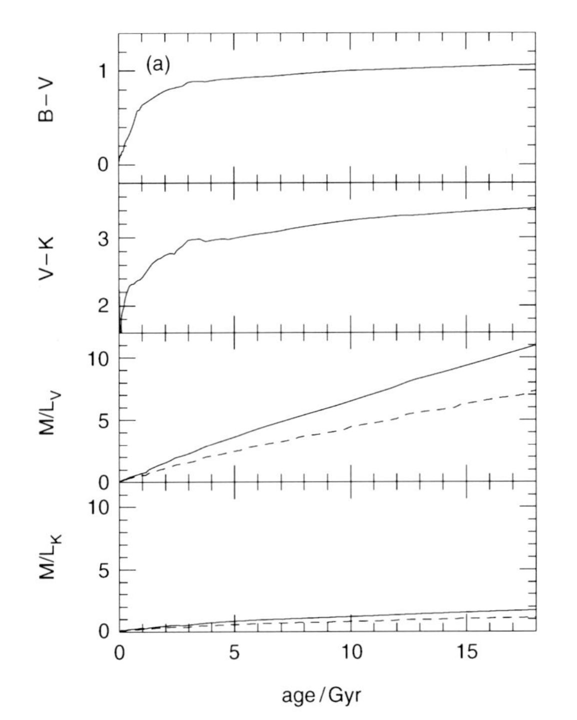

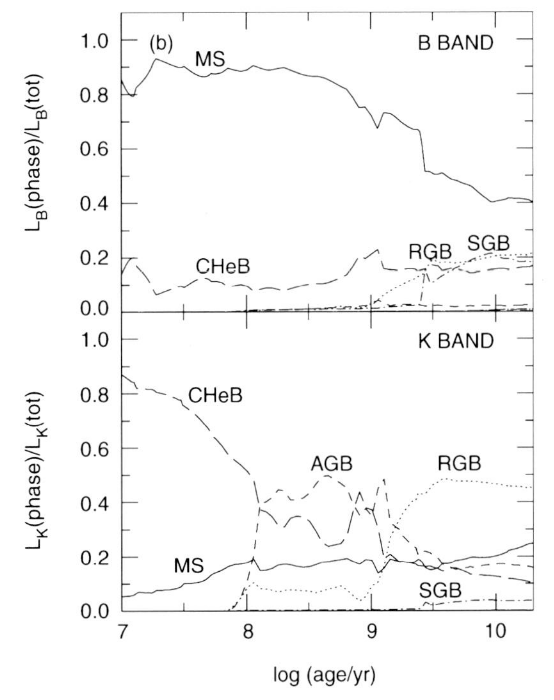

Fig. 3.48. a) For the same stellar population as in Fig. 3.47(b), the upper two graphs show the colors $B-V$ and $V-K$ as a function of age. The lower two graphs show the mass-to-light ratio M/L in two color bands in Solar units. The solid curves show the total M/L (i.e., including the mass that is later returned into the ISM), whereas the dashed curves show the M/L of the stars itself. **b**) The fraction of B- (top) and K-luminosity (bottom) contributed by stars in their different phases of stellar evolution (CHeB: core helium burning stars; SGB: subgiant branch)

3.9.4 Star Formation History and Galaxy Colors

Up to now, we have considered the evolution of a stellar population of a common age (called an *instantaneous burst of star formation*). However, star formation in a galaxy takes place over a finite period of time. We expect that the star-formation rate decreases over time because more and more matter is bound in stars and thus no longer available to form new stars. Since the star-formation history of a galaxy is a priori unknown, it needs to be parametrized in a suitable manner. A "standard model" of an exponentially decreasing star-formation rate was established for this,

$$\psi(t) = \tau^{-1} \exp\left[-(t - t_{\rm f})/\tau\right] \, {\rm H}(t - t_{\rm f}) \,, \quad (3.69)$$

where τ is the characteristic duration and $t_{\rm f}$ the onset of star formation. The last factor in (3.69) is the Heaviside step function, ${\rm H}(x) = 1$ for $x \geq 0$, ${\rm H}(x) = 0$ for $x < 0$. This Heaviside step function accounts for the fact that $\psi(t) = 0$ for $t < t_{\rm f}$. We may hope that this simple model describes the basic aspects of a stellar population. Results of this model are plotted in Fig. 3.49(a) in a color–color diagram.

From the diagram we find that the colors of the population depend strongly on τ. Specifically, galaxies do not become very red if τ is large because their star-formation rate, and thus the fraction of massive blue stars, does not decrease sufficiently. The colors of Sc spirals, for example, are not compatible with a constant star-formation rate – except if the total light of spirals is strongly reddened by dust absorption (but there are good reasons why this is not the case). To explain the colors of early-type galaxies we need $\tau \lesssim 4 \times 10^9$ yr. In general, one deduces from these models that a substantial evolution to redder colors occurs for $t \gtrsim \tau$. Since the luminosity of a stellar population in the blue spectral range decreases quickly with the age of the population, whereas increasing age affects the red luminosity much less, we conclude:

> The spectral distribution of galaxies is mainly determined by the ratio of the star-formation rate today to the mean star-formation rate in the past, $\psi(\text{today})/\langle\psi\rangle$.

One of the achievements of this standard model is that it explains the colors of present day galaxies, which have an age of $\gtrsim 10$ billion years. However, this model is not unambiguous because other star-formation histories $\psi(t)$ can be constructed with which the colors of galaxies can be modeled as well.

3.9.5 Metallicity, Dust, and HII Regions

Predictions of the model depend on the metallicity Z – see Fig. 3.49(b). A small value of Z results in a bluer color and a smaller M/L ratio. The age and metallicity of a stellar population are degenerate in the sense that an increase in the age by a factor X is nearly equivalent to an increase of the metallicity by a factor $0.65X$ with respect to the color of a population. The age estimate of a population from color will therefore strongly depend on the assumed value for Z. However, this degeneracy can be broken by taking several colors, or information from spectroscopy, into account.

Intrinsic dust absorption will also change the colors of a population. This effect cannot be easily accounted for in the models because it depends not only on the properties of the dust but also on the geometric distribution of dust and stars. For example, it makes a difference whether the dust in a galaxy is homogeneously distributed or concentrated in a thin disk. Empirically, it is found that galaxies show strong extinction during their active phase of star formation, whereas normal galaxies are presumably not much affected by extinction, with early-type galaxies (E/S0) affected the least.

Besides stellar light, the emission by HII regions also contributes to the light of galaxies. It is found, though, that after $\sim 10^7$ yr the emission from gas nebulae only marginally contributes to the broad-band colors of galaxies. However, this nebular radiation is the origin of emission lines in the spectra of galaxies. Therefore, emission lines are used as diagnostics for the star-formation rate and the metallicity in a stellar population.

3.9.6 Summary

After this somewhat lengthy section, we shall summarize the most important results of population synthesis here:

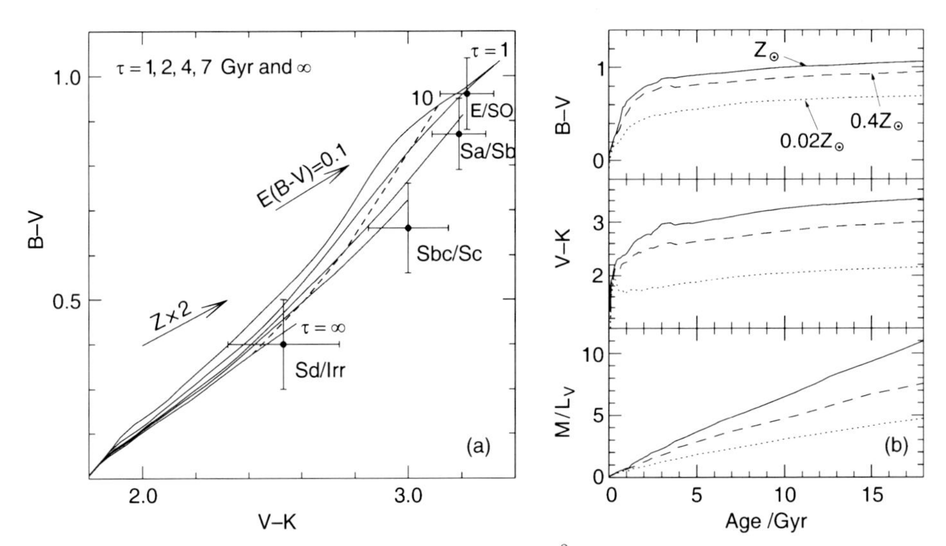

Fig. 3.49. a) Evolution of colors between $0 \leq t \leq 17 \times 10^9$ yr for a stellar population with star-formation rate given by (3.69), for five different values of the characteristic time-scale τ ($\tau = \infty$ is the limiting case for a constant star-formation rate) –Galactic center see solid curves. The typical colors for four different morphological types of galaxies are plotted. For each τ, the evolution begins at the lower left, i.e., as a blue population in both color indices. In the case of constant star formation, the population never becomes redder than Irr's; to achieve redder colors, τ has to be smaller. The dashed line connects points of $t = 10^{10}$ yr on the different curves. Here, a Salpeter IMF and Solar metallicity was assumed. The shift in color obtained by doubling the metallicity is indicated by an arrow, as well as that due to an extinction coefficient of $E(B-V) = 0.1$; both effects will make galaxies appear redder. **b**) The dependence of colors and M/L on the metallicity of the population

- A simple model of star-formation history reproduces the colors of today's galaxies fairly well.
- (Most of) the stars in elliptical and S0 galaxies are old – the earlier the Hubble type, the older the stellar population.
- Detailed models of population synthesis provide information about the star-formation history, and predictions by the models can be compared with observations of galaxies at high redshift (and thus smaller age).

We will frequently refer to results from population synthesis in the following chapters. For example, we will use them to interpret the colors of galaxies at high redshifts and the different spatial distributions of early-type and late-type galaxies (see Chap. 6). Also, we will present a method of estimating the redshift of galaxies from their broad-band colors (photometric redshifts). As a special case of this method, we will discuss the efficient selection of galaxies at very high redshift (Lyman-break galaxies, LBGs, see Chap. 9). Because the color and luminosity of a galaxy are changing even when no star formation is taking place, tracing back such a *passive evolution* allows us to distinguish this passive aging process from episodes of star formation and other processes.

3.9.7 The Spectra of Galaxies

At the end of this section we shall consider the typical spectra of different galaxy types. They are displayed for six galaxies of different Hubble types in Fig. 3.50. To make it easier to compare them, they are all plotted in a single diagram where the logarithmic flux scale is arbitrarily normalized (since this normalization does not affect the shape of the spectra).

It is easy to recognize the general trends in these spectra: the later the Hubble type, (1) the bluer the overall spectral distribution, (2) the stronger the emission lines, (3) the weaker the absorption lines, and (4) the smaller the 4000-Å break in the spectra. From the above discussion, we would also expect these trends if the Hubble sequence is considered an ordering of galaxy types

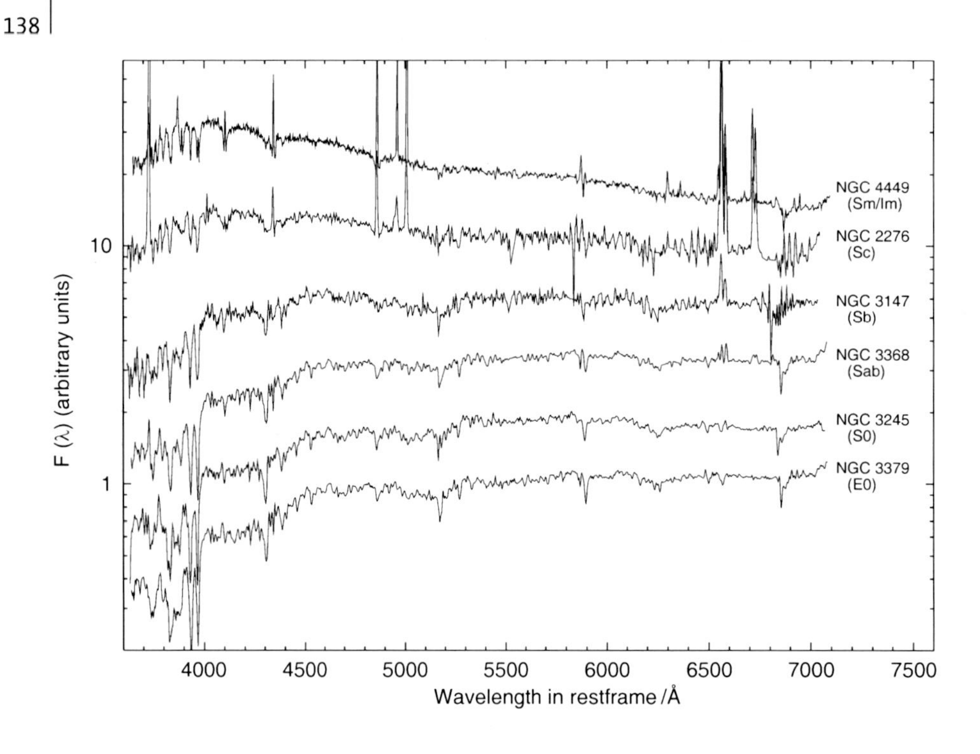

Fig. 3.50. Spectra of galaxies of different types, where the spectral flux is plotted logarithmically in arbitrary units. The spectra are ordered according to the Hubble sequence, with early types at the bottom and late-type spectra at the top

according to the characteristic age of their stellar population or according to their star-formation rate. Elliptical and S0 galaxies essentially have no star-formation activity, which renders their spectral energy distribution dominated by red stars. Furthermore, in these galaxies there are no HII regions where emission lines could be generated. The old stellar population produces a pronounced 4000-Å break, which corresponds to a jump by a factor of ~ 2 in the spectra of early-type galaxies. It should be noted that the spectra of ellipticals and S0 galaxies are quite similar.

By contrast, Sc spirals and irregular galaxies have a spectrum which is dominated by emission lines, where the Balmer lines of hydrogen as well as nitrogen and oxygen lines are most pronounced. The relative strength of these emission lines are characteristic for HII regions, implying that most of this line emission is produced in the ionized regions surrounding young stars. For irregular galaxies, the spectrum is nearly totally dominated by the stellar continuum light of hot stars and the emission lines from HII regions, whereas clear contributions by cooler stars can be identified in the spectra of Sc spiral galaxies.

The spectra of Sa and Sb galaxies form a kind of transition between those of early-type galaxies and Sc galaxies. Their spectra can be described as a superposition of an old stellar population generating a red continuum and a young population with its blue continuum and its emission lines. This can be seen in connection with the decreasing contribution of the bulge to the galaxy luminosity towards later spiral types.

The properties of the spectral light distribution of different galaxy types, as briefly discussed here, is described and interpreted in the framework of population synthesis. This gives us a detailed understanding of stellar populations as a function of the galaxy type. Extending these studies to spectra of high-redshift galaxies allows us to draw conclusions about the evolutionary history of their stellar populations.

3.10 Chemical Evolution of Galaxies

During its evolution, the chemical composition of a galaxy changes. Thus the observed metallicity yields information about the galaxy's star-formation history. We expect the metallicity Z to increase with star-formation rate, integrated over the lifetime of the galaxy. We will now discuss a simple model of the chemical evolution of a galaxy, which will provide insight into some of the principal aspects.

We assume that at the formation epoch of the stellar population of a galaxy, at time $t = 0$, no metals were present; hence $Z(0) = 0$. Furthermore, the galaxy did not contain any stars at the time of its birth, so that all baryonic matter was in the form of gas. In addition, we consider the galaxy as a closed system out of which no matter can escape or be added later on by processes of accretion or merger. Finally, we assume that the time-scales of the stellar evolution processes that lead to the metal enrichment of the galaxy are small compared to the evolutionary time-scale of the galaxy. Under these assumptions, we can now derive a relation between the metallicity and the gas content of a galaxy.

Of the total mass of a newly formed stellar population, part of it is returned to the ISM by supernova explosions and stellar winds. We define this fraction as R, so that the fraction $\alpha = (1 - R)$ of a newly-formed stellar population remains enclosed in stars, i.e., it no longer takes part in the further chemical evolution of the ISM. The value of α depends on the IMF of the stellar population and can be computed from models of population synthesis. Furthermore, let q be the ratio of the mass in metals, which is produced by a stellar population and then returned into the ISM, and the initial total mass of the population. The *yield* $y = q/\alpha$ is defined as the ratio of the mass in metals that is produced by a stellar population and returned into the ISM, and the mass that stays enclosed in the stellar population. The yield can also be calculated from population synthesis models. If $\psi(t)$ is the star-formation rate as a function of time, then the mass of all stars formed in the history of the galaxy is given by

$$S(t) = \int_0^t \mathrm{d}t' \, \psi(t') ,$$

and the total mass that remains enclosed in stars is $s(t) = \alpha S(t)$. Since we have assumed a closed system for the baryons, the sum of gas mass $g(t)$ and stellar mass $s(t)$ is a constant, namely the baryon mass of the galaxy,

$$g(t) + s(t) = M_\mathrm{b} \;\Rightarrow\; \frac{\mathrm{d}g}{\mathrm{d}t} + \frac{\mathrm{d}s}{\mathrm{d}t} = 0 . \qquad (3.70)$$

The mass of the metals in the ISM is gZ; it changes when stars are formed. Through this formation, the mass of the ISM and thus also that of its metals decreases. On the other hand, metals are also returned into the ISM by processes of stellar evolution. Under the above assumption that the time-scales of stellar evolution are small, this return occurs virtually instantaneously. The metals returned to the ISM are composed of metals that were already present at the formation of the stellar population – a fraction R of these will be returned – and newly formed metals. Together, the total mass of the metals in the ISM obeys the evolution equation

$$\frac{\mathrm{d}(gZ)}{\mathrm{d}t} = \psi\,(RZ + q) - Z\psi ,$$

where the last term specifies the rate of the metals extracted from the ISM in the process of star formation and the first term describes the return of metals to the ISM by stellar evolution processes. Since $\mathrm{d}S/\mathrm{d}t = \psi$, this can also be written as

$$\frac{\mathrm{d}(gZ)}{\mathrm{d}S} = (R-1)Z + q = q - \alpha Z .$$

Dividing this equation by α and using $s = \alpha S$ and the definition of the yield, $y = q/\alpha$, we obtain

$$\frac{\mathrm{d}(gZ)}{\mathrm{d}s} = \frac{\mathrm{d}g}{\mathrm{d}s} Z + g\frac{\mathrm{d}Z}{\mathrm{d}s} = y - Z . \qquad (3.71)$$

From (3.70) it follows that $\mathrm{d}g/\mathrm{d}s = -1$ and $\mathrm{d}Z/\mathrm{d}s = -\mathrm{d}Z/\mathrm{d}g$, and so we obtain a simple equation for the metallicity,

$$g\frac{\mathrm{d}Z}{\mathrm{d}g} = \frac{\mathrm{d}Z}{\mathrm{d}\ln g} = -y$$

$$\Rightarrow\; Z(t) = -y\,\ln\left(\frac{g(t)}{M_\mathrm{b}}\right) = -y\,\ln(\mu_\mathrm{g}) , \qquad (3.72)$$

where $\mu_\mathrm{g} = g/M_\mathrm{b}$ is the fraction of baryons in the ISM, and where we chose the integration constant such that at the beginning, when $\mu_\mathrm{g} = 1$, the metallicity was $Z = 0$. From this relation, we can now see that with decreasing gas content in a galaxy, the metallicity will increase; in our simple model this increase depends only on the yield y. Since y can be calculated from population synthesis models, (3.72) is a well-defined relation.

If (3.72) is compared with observations of galaxies, rather strong deviations from this relation are found which are particularly prominent for low-mass galaxies. While the assumption of an instantaneous evolution of the ISM is fairly well justified, we know from structure formation in the Universe (Chap. 7) that galaxies are by no means isolated systems: their mass continuously changes through accretion and merging processes. In addition, the kinetic energy transferred to the ISM by supernova explosions causes an outflow of the ISM, in particular in low-mass galaxies where the gas is not strongly gravitationally bound. Therefore, the observed deviations from relation (3.72) allow us to draw conclusions about these processes.

Also, from observations in our Milky Way we find indications that the model of the chemical evolution sketched above is too simplified. This is known as the *G-dwarf problem*. The model described above predicts that about half of the F- and G-main-sequence stars should have a metallicity of less than a quarter of the Solar value. These stars have a long lifetime on the main sequence, so that many of those observed today should have been formed in the early stages of the Galaxy. Thus, in accordance with our model they should have very low metallicity. However, a low metallicity is in fact observed in only very few of these stars. The discrepancy is far too large to be explained by selection effects. Rather, observations show that the chemical evolution of our Galaxy must have been substantially more complicated than described by our simple model.

4. Cosmology I: Homogeneous Isotropic World Models

We will now begin to consider the Universe as a whole. Individual objects such as galaxies and stars will no longer be the subject of discussion, but instead we will turn our attention to the space and time in which these objects are embedded. These considerations will then lead to a world model, the model of our cosmos.

This chapter will deal with aspects of homogeneous cosmology. As we will see, the Universe can, to first approximation, be considered as being homogeneous. At first sight this fact obviously seems to contradict observations because the world around us is highly inhomogeneous and structured. Thus the assumption of homogeneity is certainly not valid on small scales. But observations are compatible with the assumption that the Universe is homogeneous when averaged over large spatial scales. Aspects of inhomogeneous cosmology, and thus the formation and evolution of structures in the Universe, will be considered later in Chap. 7.

4.1 Introduction and Fundamental Observations

Cosmology is a very special science indeed. To be able to appreciate its peculiar role we should recall the typical way of establishing knowledge in natural sciences. It normally starts with the observation of some regular patterns, for instance the observation that the height h a stone falls through is related quadratically to the time t it takes to fall, $h = (g/2)t^2$. This relation is then also found for other objects and observed at different places on Earth. Therefore, this relation is formulated as the "law" of free fall. The constant of proportionality $g/2$ in this law is always the same. This law of physics is tested by its prediction of how an object falls, and wherever this prediction is tested it is confirmed – disregarding the resistance of air in this simple example, of course.

Relations become physical laws if the predictions they make are confirmed again and again; the validity of such a law is considered more secure the more diverse the tests have been. The law of free fall was tested only on the surface of the Earth and it is only valid there with this constant of proportionality.[1] In contrast to this, Newton's law of gravity contains the law of free fall as a special case, but it also describes the free fall on the surface of the Moon, and the motion of planets around the Sun. If only a single stone were available, we would not know whether the law of free fall is a property of this particular stone or whether it is valid more generally.

In some ways, cosmology corresponds to the latter example: we have only one single Universe available for observation. Relations that are found in our cosmos cannot be verified in other universes. Thus it is not possible to consider any property of our Universe as "typical" – we have no statistics on which we could base a statement like this. Despite this special situation, enormous progress has been made in understanding our Universe, as we will describe here and in subsequent chapters.

Cosmological observations are difficult in general, simply because the majority of the Universe (and with it most of the sources it contains) is very far away from us. Distant sources are very dim. This explains why our knowledge of the Universe runs in parallel with the development of large telescopes and sensitive detectors. Much of today's knowledge of the distant Universe became available only with the new generation of optical telescopes of the 8-m class, as well as new and powerful telescopes in other wavelength regimes.

The most important aspect of cosmological observations is the finite speed of light. We observe a source at distance D in an evolutionary state at which it was $\Delta t = (D/c)$ younger than today. Thus we can observe the current state of the Universe only very locally. Another consequence of this effect, however, is of even greater importance: due to the finite speed of light, it is possible to look back into the past. At a distance of 10 billion light years we observe galaxies in an evolutionary state when the Universe had only a third of its current age. Although we cannot observe the past of our own Milky Way, we can study that of other galaxies. If we are able to identify among them the ones that will form objects similar to our Galaxy in the course of cos-

[1] Strictly speaking, the constant of proportionality g depends slightly on the location.

Peter Schneider, Cosmology I: Homogeneous Isotropic World Models.
In: Peter Schneider, Extragalactic Astronomy and Cosmology. pp. 141–174 (2006)
DOI: 10.1007/11614371_4

mic evolution, we will be able to learn a great deal about the typical evolutionary history of such spirals.

The finite speed of light in a Euclidean space, in which we are located at the origin $\boldsymbol{r} = 0$ today ($t = t_0$), implies that we can only observe points in spacetime for which $|\boldsymbol{r}| = c(t_0 - t)$; an arbitrary point $(\boldsymbol{r}, t)$ in spacetime is not observable. The set of points in spacetime which satisfy the relation $|\boldsymbol{r}| = c(t_0 - t)$ is called our *backward light cone*.

The fact that our astronomical observations are restricted to sources which are located on our backward light cone implies that our possibilities to observe the Universe are fundamentally limited. If somewhere in spacetime there would be a highly unusual event, we will not be able to observe it unless it happens to lie on our backward light cone. Only if the Universe has an essentially "simple" structure we will be able to understand it, by combining astronomical observations with theoretical modeling. Luckily, our Universe seems to be basically simple in this sense.

Fig. 4.1. The APM survey: galaxy distribution in a $\sim 100 \times 50\,\text{degree}^2$ field around the South Galactic Pole. The intensities of the pixels are scaled with the number of galaxies per pixel, i.e., the projected galaxy number density on the sphere. The "holes" are regions around bright stars, globular clusters, etc., that have not been surveyed

4.1.1 Fundamental Cosmological Observations

We will begin with a short list of key observations that have proven to be of particular importance for cosmology. Using these observational facts we will then be able to draw a number of immediate conclusions; other observations will be explained later in the context of a cosmological model.

1. The sky is dark at night (Olbers' paradox).
2. Averaged over large angular scales, faint galaxies (e.g., those with $R > 20$) are uniformly distributed on the sky (see Fig. 4.1).
3. With the exception of a very few very nearby galaxies (e.g., Andromeda = M31), a redshift is observed in the spectra of galaxies – most galaxies are moving away from us, and their escape velocity increases linearly with distance (Hubble law; see Fig. 1.10).
4. In nearly all cosmic objects (e.g., gas nebulae, main-sequence stars), the mass fraction of helium is 25–30%.
5. The oldest star clusters in our Galaxy have an age of $\sim 12\,\text{Gyr} = 12 \times 10^9\,\text{yr}$ (see Fig. 4.2).
6. A microwave radiation (cosmic microwave background radiation, CMB) is observed, reaching us from all directions. This radiation is isotropic except for very small, but immensely important, fluctuations with relative amplitude $\sim 10^{-5}$.
7. The spectrum of the CMB corresponds, within the very small error bars that were obtained by the measurements with COBE, to that of a perfect blackbody, i.e., a Planck radiation of a temperature of $T_0 = 2.728 \pm 0.004\,\text{K}$ – see Fig. 4.3.
8. The number counts of radio sources at high Galactic latitude does *not* follow the simple law $N(> S) \propto S^{-3/2}$ (see Fig. 4.4).

4.1.2 Simple Conclusions

We will next draw a number of simple conclusions from the observational facts listed above. These will then serve as a motivation and guideline for developing the cosmological model. We will start with the assumption of an infinite, Euclidean, static Universe, and show that these assumptions are in direct contradiction to observations (1) and (8).

Olbers' Paradox (1): We can show that the night sky would be bright in such a universe – uncomfortably bright, in fact. Let n_* be the mean number density of stars, constant in space and time according to the assumptions, and let R_* be their mean radius. A spherical

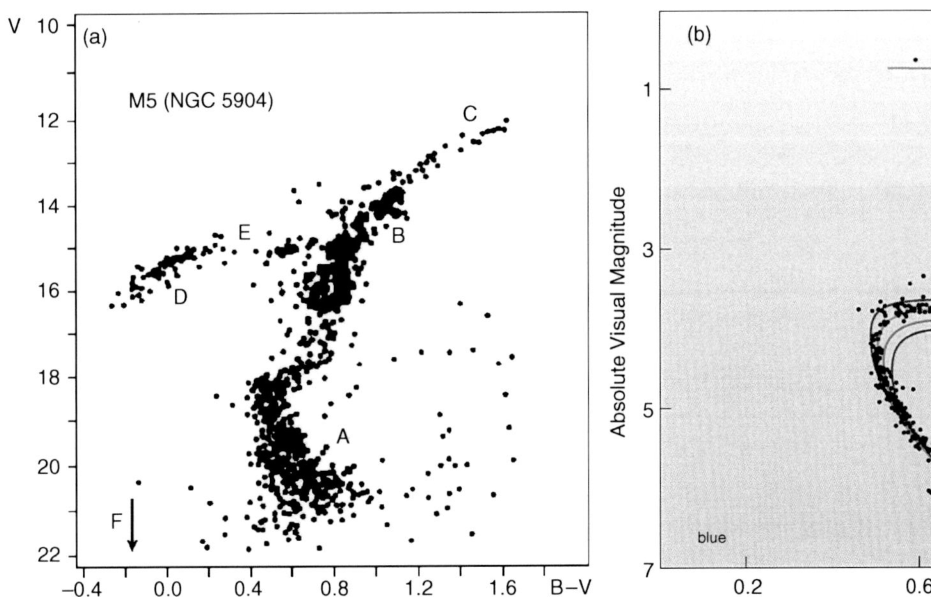

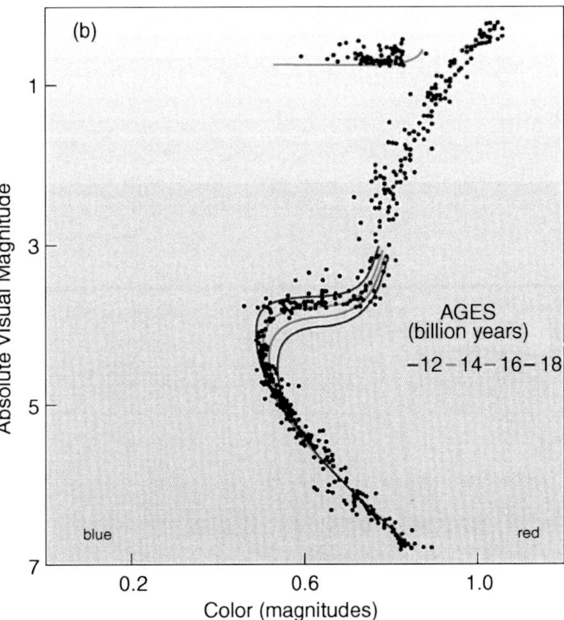

Fig. 4.2. **(a)** Color–magnitude diagram of the globular cluster M 5. The different sections in this diagram are labeled. A: main sequence; B: red giant branch; C: point of helium flash; D: horizontal branch; E: Schwarzschild gap in the horizontal branch; F: white dwarfs, below the arrow. At the point where the main sequence turns over to the red giant branch (called the "turn-off point"), stars have a mass corresponding to a main-sequence lifetime which is equal to the age of the globular cluster (see Appendix B.3). Therefore, the age of the cluster can be determined from the position of the turn-off point by comparing it with models of stellar evolution. **(b)** Isochrones, i.e., curves connecting the stellar evolutionary position in the color–magnitude diagram of stars of equal age, are plotted for different ages and compared to the stars of the globular cluster 47 Tucanae. Such analyses reveal that the oldest globular clusters in our Milky Way are about 13 billion years old, where different authors obtain slightly differing results – details of stellar evolution may play a role here. The age thus obtained also depends on the distance of the cluster. A revision of these distances by the HIPPARCOS satellite led to a decrease of the estimated ages by about 2 billion years

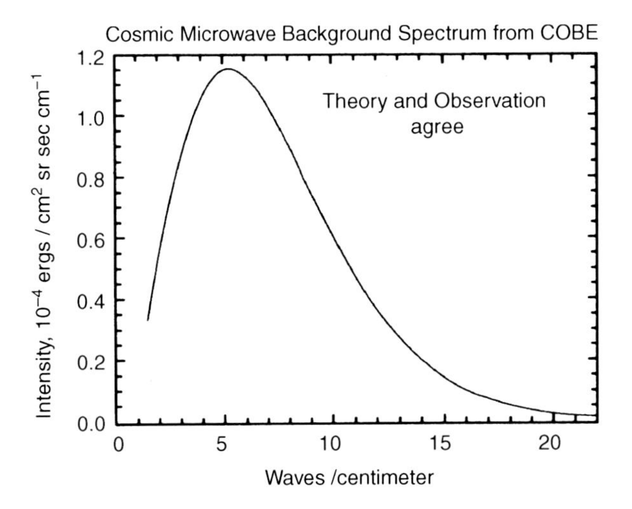

Fig. 4.3. CMB spectrum, plotted as intensity vs. frequency, measured in waves per centimeter. The solid line shows the expected spectrum of a blackbody of temperature $T = 2.728$ K. The error bars of the data, observed by the FIRAS instrument on-board COBE, are so small that the data points with error bars cannot be distinguished from the theoretical curve

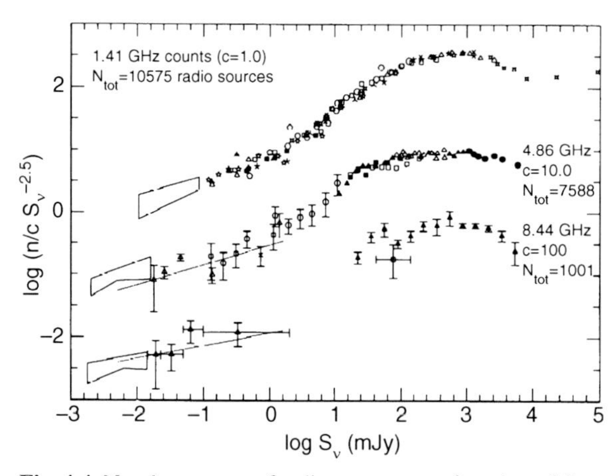

Fig. 4.4. Number counts of radio sources as a function of flux, normalized by the Euclidean expectation $N(S) \propto S^{-5/2}$, corresponding to the integrated counts $N(>S) \propto S^{-3/2}$. Counts are displayed for three different frequencies; they clearly deviate from the Euclidean expectation

shell of radius r and thickness $\mathrm{d}r$ around us contains $n_* \,\mathrm{d}V = 4\pi r^2 \,\mathrm{d}r\, n_*$ stars. Each of these stars subtends a solid angle of $\pi R_*^2/r^2$ on our sky, so the stars in the shell cover a total solid angle of

$$\mathrm{d}\omega = 4\pi r^2 \,\mathrm{d}r\, n_* \frac{R_*^2 \pi}{r^2} = 4\pi^2 n_* R_*^2 \,\mathrm{d}r \,. \tag{4.1}$$

We see that this solid angle is independent of the radius r of the spherical shell because the solid angle of a single star $\propto r^{-2}$ just compensates the volume of the shell $\propto r^2$. To compute the total solid angle of all stars in a static Euclidean universe, (4.1) has to be integrated over all distances r, but the integral

$$\omega = \int_0^\infty \mathrm{d}r \,\frac{\mathrm{d}\omega}{\mathrm{d}r} = 4\pi^2 n_* R_*^2 \int_0^\infty \mathrm{d}r$$

diverges. Formally, this means that the stars cover an infinite solid angle, which of course makes no sense physically. The reason for this divergence is that we disregarded the effect of overlapping stellar disks on the sphere. However, these considerations demonstrate that the sky would be completely filled with stellar disks, i.e., from any direction, along any line-of-sight, light from a star would reach us. Since the specific intensity I_ν is independent of distance – the surface brightness of the Sun as observed from Earth is the same as seen by an observer who is much closer to the Solar surface – the sky would have a temperature of $\sim 10^4$ K; fortunately, this is not the case!

Source Counts (8): Consider now a population of sources with a luminosity function that is constant in space and time, i.e., let $n(>L)$ be the spatial number density of sources with luminosity larger than L. A spherical shell of radius r and thickness $\mathrm{d}r$ around us contains $4\pi r^2 \,\mathrm{d}r\, n(>L)$ sources with luminosity larger than L. Because the observed flux S is related to the luminosity via $L = 4\pi r^2 S$, the number of sources with flux $> S$ in this spherical shell is given as $\mathrm{d}N(>S) = 4\pi r^2 \,\mathrm{d}r\, n(>4\pi r^2 S)$, and the total number of sources with flux $> S$ results from integration over the radii of the spherical shells,

$$N(>S) = \int_0^\infty \mathrm{d}r\, 4\pi r^2\, n(>4\pi r^2 S) \,.$$

Changing the integration variable to $L = 4\pi r^2 S$, or $r = \sqrt{L/(4\pi S)}$, with $\mathrm{d}r = \mathrm{d}L/(2\sqrt{4\pi L S})$, yields

$$N(>S) = \int_0^\infty \frac{\mathrm{d}L}{2\sqrt{4\pi L S}} \frac{L}{4\pi S} n(>L)$$
$$= \frac{1}{16\pi^{3/2}} S^{-3/2} \int_0^\infty \mathrm{d}L\, \sqrt{L}\, n(>L) \,. \tag{4.2}$$

From this result we deduce that the source counts in such a universe is $N(>S) \propto S^{-3/2}$, independent of the luminosity function. This is in contradiction to the observations.

From these two contradictions – Olbers' paradox and the non-Euclidean source counts – we conclude that at least one of the assumptions must be wrong. Our Universe cannot be all three of Euclidean, infinite, and static. The Hubble flow, i.e., the redshift of galaxies, indicates that the assumption of a static Universe is wrong.

The **age of globular clusters (5)** requires that the Universe is at least 12 Gyr old because it cannot be younger than the oldest objects it contains. Interestingly, the age estimates for globular clusters yield values which are very close to the *Hubble time* $H_0^{-1} = 9.78\, h^{-1}$ Gyr. This similarity suggests that the Hubble expansion may be directly linked to the evolution of the Universe.

The apparently isotropic **distribution of galaxies (2)**, when averaged over large scales, and the **CMB isotropy (6)** suggest that the Universe around us is isotropic on large angular scales. Therefore we will first consider a world model that describes the Universe around us as isotropic. If we assume, in addition, that our place in the cosmos is not privileged over any other place, then the assumption of isotropy around us implies that the Universe appears isotropic as seen from any other place. The homogeneity of the Universe follows immediately from the isotropy around every location, as explained in Fig. 4.5. The combined assumption of homogeneity and isotropy of the Universe is also known as the *cosmological principle*. We will see that a world model based on the cosmological principle in fact provides an excellent description of numerous observational facts.

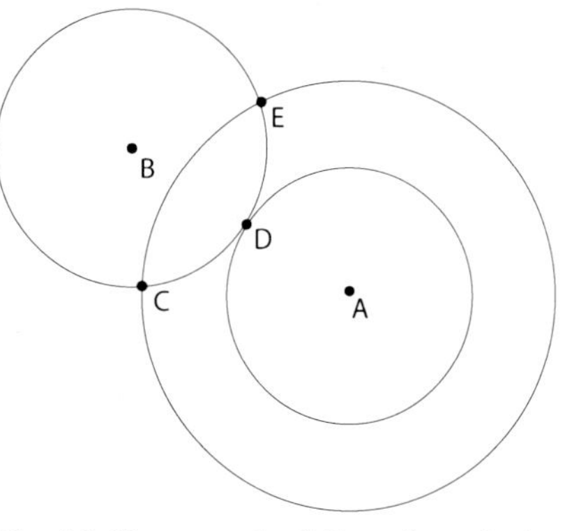

Fig. 4.5. Homogeneity follows from the isotropy around two points. If the Universe is isotropic around observer B, the densities at C, D, and E are equal. Drawing spheres of different radii around observer A, it is seen that the region within the spherical shell around A has to be homogeneous. By varying the radius of the shell, we can conclude the whole Universe must be homogeneous

However, homogeneity is in principle unobservable because observations of distant objects show those at an earlier epoch. If the Universe evolves in time, as the aforementioned observations suggest, evolutionary effects cannot directly be separated from spatial variations.

The assumption of homogeneity of course breaks down on small scales. We observe structures in the Universe, like galaxies and clusters of galaxies, and even accumulations of clusters of galaxies, so-called superclusters. Structures have been found in redshift surveys that extend over $\sim 100\,h^{-1}$ Mpc. However, we have no indication of the existence of structures in the Universe with scales $\gg 100$ Mpc. This length-scale can be compared to a characteristic length of the Universe, which is obtained from the Hubble constant. If H_0^{-1} specifies the characteristic age of the Universe, then light will travel a distance c/H_0 in this time. With this, one obtains the *Hubble radius* as a characteristic length-scale of the Universe (or more precisely, of the observable Universe),

$$\boxed{R_{\rm H} := \frac{c}{H_0} = 2997\,h^{-1}\,{\rm Mpc}\ :\ \text{Hubble length}}\ . \tag{4.3}$$

The Hubble volume $\sim R_{\rm H}^3$ can contain a very large number of structures of size $\sim 100\,h^{-1}$ Mpc, so that it still makes sense to assume an on-average homogeneous Universe. In this homogeneous Universe we then have density fluctuations that are identified with the observed large-scale structures; these will be discussed in detail in Chap. 7. To a first approximation we can neglect these density perturbations in a description of the Universe as a whole. We will therefore consider next world models that are based on the cosmological principle, i.e., in which the Universe looks the same for all observers (or, in other words, if observed from any point).

Homogeneous and isotropic world models are the simplest cosmological solutions of the equations of General Relativity (GR). We will examine how far such simple models are compatible with observations. As we shall see, the application of the cosmological principle results in the observational facts which were mentioned in Sect. 4.1.1.

4.2 An Expanding Universe

Gravitation is the fundamental force in the Universe. Only gravitational forces and electromagnetic forces can act over large distance. Since cosmic matter is electrically neutral on average, electromagnetic forces do not play any significant role on large scales, so that gravity has to be considered as the driving force in the

Universe. The laws of gravity are described by the theory of General Relativity, formulated by A. Einstein in 1915. It contains Newton's theory of gravitation as a special case for weak gravitational fields and small spatial scales. Newton's theory of gravitation has been proven to be eminently successful, e.g., in describing the motion of planets. Thus it is tempting to try to design a cosmological model based on Newtonian gravity. We will proceed to do that as a first step because not only is this Newtonian cosmology very useful from a didactic point of view, but one can also argue why the Newtonian cosmos correctly describes the major aspects of a relativistic cosmology.

4.2.1 Newtonian Cosmology

The description of a gravitational system necessitates the application of GR if the length-scales in the system are comparable to the radius of curvature of spacetime; this is certainly the case in our Universe. Even if we cannot explain at this point what exactly the "curvature radius of the Universe" is, it should be plausible that it is of the same order of magnitude as the Hubble radius $R_{\rm H}$. We will discuss this more thoroughly further below. Despite this fact, one can expect that a Newtonian description is essentially correct: in a homogeneous universe, any small spatial region is characteristic for the whole universe. If the evolution of a small region in space is known, we also know the history of the whole universe, due to homogeneity. However, on small scales, the Newtonian approach is justified. We will therefore, based on the cosmological principle, first consider spatially homogeneous and isotropic world models in the framework of Newtonian gravity.

4.2.2 Kinematics of the Universe

Comoving Coordinates. We consider a homogeneous sphere which may be radially expanding (or contracting); however, we require that the density $\rho(t)$ remains spatially homogeneous. The density may vary in time due to expansion or contraction. We choose a point $t = t_0$ in time and introduce a coordinate system $\boldsymbol{x}$ at this instant with the origin coinciding with the center of the sphere. A particle in the sphere which is located at position $\boldsymbol{x}$ at time t_0 will be located at some other time t at the position $\boldsymbol{r}(t)$ which results from the expansion of the sphere. Since the expansion is radial or, in other words, the velocity vector of a particle at position $\boldsymbol{r}(t)$ is parallel to $\boldsymbol{r}$, the direction of $\boldsymbol{r}(t)$ is constant. Because $\boldsymbol{r}(t_0) = \boldsymbol{x}$, this means that

$$\boldsymbol{r}(t) = a(t)\,\boldsymbol{x} \; . \tag{4.4}$$

The function $a(t)$ can depend only on time. Although requiring radial expansion alone could make a depend on $|\boldsymbol{x}|$ as well, the requirement that the density remains homogeneous implies that a must be spatially constant. The function $a(t)$ is called the *cosmic scale factor*; due to $\boldsymbol{r}(t_0) = \boldsymbol{x}$, it obeys

$$a(t_0) = 1 \; . \tag{4.5}$$

The value of t_0 is arbitrary; we choose $t_0 =$ today. Particles (or observers) which move according to (4.4) are called *comoving particles (observers)*, and $\boldsymbol{x}$ is the comoving coordinate. The world line $(\boldsymbol{r}, t)$ of a comoving observer is unambiguously determined by $\boldsymbol{x}$, $(\boldsymbol{r}, t) = [a(t)\boldsymbol{x}, t]$.

Expansion Rate. The velocity of such a comoving particle is obtained from the time derivative of its position,

$$\boldsymbol{v}(\boldsymbol{r}, t) = \frac{\mathrm{d}}{\mathrm{d}t}\boldsymbol{r}(t) = \frac{\mathrm{d}a}{\mathrm{d}t}\,\boldsymbol{x} \equiv \dot{a}\,\boldsymbol{x} = \frac{\dot{a}}{a}\,\boldsymbol{r} \equiv H(t)\,\boldsymbol{r} \; , \tag{4.6}$$

where in the last step we defined the *expansion rate*

$$\boxed{H(t) := \frac{\dot{a}}{a}} \; . \tag{4.7}$$

The choice of this notation is not accidental, since H is closely related to the Hubble constant. To see this, we consider the relative velocity vector of two comoving particles at positions $\boldsymbol{r}$ and $\boldsymbol{r} + \Delta\boldsymbol{r}$, which follows directly from (4.6):

$$\Delta\boldsymbol{v} = \boldsymbol{v}(\boldsymbol{r} + \Delta\boldsymbol{r}, t) - \boldsymbol{v}(\boldsymbol{r}, t) = H(t)\,\Delta\boldsymbol{r} \; . \tag{4.8}$$

Hence, the relative velocity is proportional to the separation vector, and the constant of proportionality $H(t)$ depends only on time but not on the position of the two particles. Obviously, (4.8) is very similar to the Hubble law

$$v = H_0\,D \; , \tag{4.9}$$

in which v is the radial velocity of a source at distance D from us. Therefore, setting $t = t_0$ and $H_0 \equiv H(t_0)$, (4.8) is simply the Hubble law, in other words, (4.8) is a generalization of (4.9) for arbitrary time. It expresses the fact that any observer expanding with the sphere will observe an isotropic velocity field that follows the Hubble law. Since we are observing an expansion today – sources are moving away from us – we have $H_0 > 0$, and $\dot{a}(t_0) > 0$.

4.2.3 Dynamics of the Expansion

The above discussion describes the kinematics of the expansion. However, to obtain the behavior of the function $a(t)$ in time, and thus also the motion of comoving observers and the time evolution of the density of the sphere, it is necessary to consider the dynamics. The evolution of the expansion rate is determined by self-gravity of the sphere, from which it is expected that it will cause a deceleration of the expansion.

Equation of Motion. We therefore consider a spherical surface of radius x at time t_0 and, accordingly, a radius $r(t) = a(t)\,x$ at arbitrary time t. The mass $M(x)$ enclosed in this comoving surface is constant in time, and is given by

$$M(x) = \frac{4\pi}{3}\,\rho_0\,x^3 = \frac{4\pi}{3}\,\rho(t)\,r^3(t) = \frac{4\pi}{3}\,\rho(t)\,a^3(t)\,x^3\,, \tag{4.10}$$

where ρ_0 must be identified with the mass density of the Universe today ($t = t_0$). The density is a function of time and, due to mass conservation, it is inversely proportional to the volume of the sphere,

$$\rho(t) = \rho_0\,a^{-3}(t)\,. \tag{4.11}$$

The gravitational acceleration of a particle on the spherical surface is $GM(x)/r^2$, directed towards the center. This then yields the *equation of motion* of the particle,

$$\ddot{r}(t) \equiv \frac{\mathrm{d}^2 r}{\mathrm{d}t^2} = -\frac{G\,M(x)}{r^2} = -\frac{4\pi G}{3}\,\frac{\rho_0\,x^3}{r^2}\,, \tag{4.12}$$

or, after substituting $r(t) = x\,a(t)$, an equation for a,

$$\ddot{a}(t) = \frac{\ddot{r}(t)}{x} = -\frac{4\pi G}{3}\,\frac{\rho_0}{a^2(t)} = -\frac{4\pi G}{3}\,\rho(t)\,a(t)\,. \tag{4.13}$$

It is important to note that this equation of motion does not dependent on x. The dynamics of the expansion, described by $a(t)$, is determined solely by the matter density.

"Conservation of Energy". Another way to describe the dynamics of the expanding shell is based on the law of energy conservation: the sum of kinetic and potential energy is constant in time. This conservation of energy is derived directly from (4.13). To do this, (4.13) is multiplied by $2\dot{a}$, and the resulting equation can be integrated with respect to time since $\mathrm{d}(\dot{a}^2)/\mathrm{d}t = 2\dot{a}\ddot{a}$, and $\mathrm{d}(-1/a)/\mathrm{d}t = \dot{a}/a^2$:

$$\dot{a}^2 = \frac{8\pi G}{3}\,\rho_0\,\frac{1}{a} - Kc^2 = \frac{8\pi G}{3}\,\rho(t)\,a^2(t) - Kc^2\,; \tag{4.14}$$

here, Kc^2 is a constant of integration that will be interpreted later. After multiplication with $x^2/2$, (4.14) can be written as

$$\frac{v^2(t)}{2} - \frac{G\,M}{r(t)} = -Kc^2\,\frac{x^2}{2}\,,$$

which is interpreted such that the kinetic + potential energy (per unit mass) of a particle is a constant on the spherical surface. Thus (4.14) in fact describes the conservation of energy. The latter equation also immediately suggests an interpretation of the integration constant: K is proportional to the total energy of a comoving particle, and thus the history of the expansion depends on K. The sign of K characterizes the qualitative behavior of the cosmic expansion history.

- If $K < 0$, the right-hand side of (4.14) is always positive. Since $\mathrm{d}a/\mathrm{d}t > 0$ today, $\mathrm{d}a/\mathrm{d}t$ remains positive for all times or, in other words, the Universe will expand forever.
- If $K = 0$, the right-hand side of (4.14) is always positive, i.e., $\mathrm{d}a/\mathrm{d}t > 0$ for all times, and the Universe will also expand forever, but in a way that $\mathrm{d}a/\mathrm{d}t \to 0$ for $t \to \infty$ – the limiting expansion velocity for $t \to \infty$ is zero.
- If $K > 0$, the right-hand side of (4.14) vanishes if $a = a_{\max} = (8\pi G\rho_0)/(3Kc^2)$. For this value of a, $\mathrm{d}a/\mathrm{d}t = 0$, and the expansion will come to a halt. After that, the expansion will turn into a contraction, and the Universe will re-collapse.

In the special case of $K = 0$, which separates eternally expanding world models from those that will re-collapse in the future, the Universe has a current density called *critical density* which can be inferred from (4.14) by setting $t = t_0$ and $H_0 = \dot{a}(t_0)$:

$$\rho_{\rm cr} := \frac{3H_0^2}{8\pi G} = 1.88 \times 10^{-29}\, h^2\, {\rm g/cm^3} \; . \tag{4.15}$$

Obviously, $\rho_{\rm cr}$ is a characteristic density of the current Universe. As in many situations in physics, it is useful to express physical quantities in terms of dimensionless parameters, for instance the real density of the Universe today. We therefore define the *density parameter*

$$\Omega_0 := \frac{\rho_0}{\rho_{\rm cr}} \; ; \tag{4.16}$$

where $K > 0$ corresponds to $\Omega_0 > 1$, and $K < 0$ corresponds to $\Omega_0 < 1$. Thus, Ω_0 is one of the central cosmological parameters. Its determination was made possible only quite recently, and we shall discuss this in detail later. However, we should mention here that matter which is visible as stars contributes only a small fraction to the density of the Universe, $\Omega_* \lesssim 0.01$. But, as we already discussed in the context of rotation curves of spiral galaxies, we find clear indications of the presence of dark matter which can in principle dominate the value of Ω_0. We will see that this is indeed the case.

4.2.4 Modifications due to General Relativity

The Newtonian approach contains nearly all essential aspects of homogeneous and isotropic world models, otherwise we would not have discussed it in detail. Most of the above equations are also valid in relativistic cosmology, although the interpretation needs to be altered. In particular, the image of an expanding sphere needs to be revised – this picture implies that a "center" of the Universe exists. Such a picture implicitly contradicts the cosmological principle in which no point is singled out over others – our Universe neither has a center, nor is it expanding away from a privileged point. However, the image of a sphere does not show up in any of the relevant equations: Eq. (4.11) for the evolution of the density of the Universe and Eqs. (4.13) and (4.14) for the evolution of the scale factor $a(t)$ contain no quantities that refer to a sphere.

General Relativity modifies the Newtonian model in several respects:

- We know from the theory of Special Relativity that mass and energy are equivalent, according to Einstein's famous relation $E = m\,c^2$. This implies that it is not only the matter density that contributes to the equations of motion. For example, a radiation field like the CMB has an energy density and, due to the equivalence above, this has to enter the expansion equations. We will see below that such a radiation field can be characterized as matter with pressure. The pressure will then explicitly appear in the equation of motion for $a(t)$.
- The field equation of GR as originally formulated by Einstein did not allow a solution which corresponds to a homogeneous, isotropic, and static cosmos. But since Einstein, like most of his contemporaries, believed the Universe to be static, he modified his field equations by introducing an additional term, the cosmological constant.
- The interpretation of the expansion is changed completely: it is not the particles or the observers that are expanding away from each other, nor is the Universe an expanding sphere. Instead, it is space itself that expands. In particular, the redshift is no Doppler redshift, but is itself a property of expanding spacetimes. However, we may still visualize redshift locally as being due to the Doppler effect without making a substantial conceptual error.

In the following, we will explain the first two aspects in more detail.

First Law of Thermodynamics. When air is compressed, for instance when pumping up a tire, it heats up. The temperature increases and accordingly so does the thermal energy of the air. In the language of thermodynamics, this fact is described by the first law: the change in internal energy $\mathrm{d}U$ through an (adiabatic) change in volume $\mathrm{d}V$ equals the work $\mathrm{d}U = -P\,\mathrm{d}V$, where P is the pressure in the gas. From the equations of GR as applied to a homogeneous isotropic cosmos, a relation is derived which reads

$$\frac{\mathrm{d}}{\mathrm{d}t}\left(c^2\,\rho\,a^3\right) = -P\,\frac{\mathrm{d}a^3}{\mathrm{d}t} \; , \tag{4.17}$$

in full analogy to this law. Here, $\rho\,c^2$ is the energy density, i.e., for "normal" matter, ρ is the mass density, and P is the pressure of the matter. If we now consider a constant comoving volume element V_x, then its physical volume $V = a^3(t)\, V_x$ will change due to expansion. Thus, $a^3 = V/V_x$ is the volume, and $c^2\, \rho\, a^3$ the energy contained in the volume, each divided by V_x. Taken together, (4.17) corresponds to the first law of thermodynamics in an expanding universe.

The Friedmann–Lemaître Expansion Equations. Next, we will present equations for the scale factor $a(t)$ which follow from GR for a homogeneous isotropic universe. Afterwards, we will derive these equations from the relations stated above – as we shall see, the modifications by GR are in fact only minor, as expected from the argument that a small section of a homogeneous universe characterizes the cosmos as a whole. The field equations of GR yield the equations of motion

$$\boxed{\left(\frac{\dot a}{a}\right)^2 = \frac{8\pi G}{3}\rho - \frac{Kc^2}{a^2} + \frac{\Lambda}{3}} \tag{4.18}$$

and

$$\boxed{\frac{\ddot a}{a} = -\frac{4\pi G}{3}\left(\rho + \frac{3P}{c^2}\right) + \frac{\Lambda}{3}}\,, \tag{4.19}$$

where Λ is the aforementioned cosmological constant introduced by Einstein. Compared to equations (4.13) and (4.14), these two equations have been changed in two places. First, the cosmological constant occurs in both equations, and second, the equation of motion (4.19) now contains a pressure term. The pair of equations (4.18) and (4.19) are called the *Friedmann equations.*

The Cosmological Constant. When Einstein introduced the Λ-term into his equations, he did this solely for the purpose of obtaining a static solution for the resulting expansion equations. We can easily see that (4.18) and (4.19), without the Λ-term, have no solution for $\dot a \equiv 0$. However, if the Λ-term is included, such a solution can be found (which is irrelevant, however, as we now know that the Universe is expanding). Einstein had no real physical interpretation for this constant, and after the expansion of the Universe was discovered he discarded it again. But with the genie out of the bottle, the cosmological constant remained in the minds of cosmologists, and their attitude towards Λ has changed frequently in the past 90 years. Around the turn of the millennium, observations were made which strongly suggest a non-vanishing cosmological constant, i.e., we believe today that $\Lambda \neq 0$.

But the physical interpretation of the cosmological constant has also been modified. In quantum mechanics even completely empty space, the so-called vacuum, may have a finite energy density, the vacuum energy density. For physical measurements not involving gravity, the value of this vacuum energy density is of no relevance since those measurements are only sensitive to energy *differences*. For example, the energy of a photon that is emitted in an atomic transition equals the energy difference between the two corresponding states in the atom. Thus the absolute energy of a state is measurable only up to a constant. Only in gravity does the absolute energy become important, because $E = m\, c^2$ implies that it corresponds to a mass.

It is now found that the cosmological constant is equivalent to a finite vacuum energy density – the equations of GR, and thus also the expansion equations, are not affected by this new interpretation. We will explain this fact in the following.

4.2.5 The Components of Matter in the Universe

Starting from the equation of energy conservation (4.14), we will now derive the relativistically correct expansion equations (4.18) and (4.19). The only change with respect to the Newtonian approach in Sect. 4.2.3 will be that we introduce other forms of matter. The essential components of the Universe can be described as pressure-free matter, radiation, and vacuum energy.

Pressure-Free Matter. The pressure in a gas is determined by the thermal motion of its constituents. At room temperature, molecules in the air move at a speed comparable to the speed of sound, $c_{\rm s} \sim 300\,{\rm m/s}$. For such a gas, $P \sim \rho\, c_{\rm s}^2 \ll \rho c^2$, so that its pressure is of course gravitationally completely insignificant. In cosmology, a substance with $P \ll \rho c^2$ is denoted as (pressure-free) matter, also called cosmological dust.[2] We approximate $P_{\rm m} = 0$, where the index "m" stands for matter. The constituents of the (pressure-free) matter move with velocities much smaller than c.

[2] The notation "dust" should not be confused with the dust that is responsible for the extinction of light – "dust" in cosmology only denotes matter with $P = 0$.

Radiation. If this condition is no longer satisfied, thus if the thermal velocities are no longer negligible compared to the speed of light, then the pressure will also no longer be small compared to ρc^2. In the limiting case that the thermal velocity equals the speed of light, we denote this component as "radiation". One example of course is electromagnetic radiation, in particular the CMB photons. Another example would be other particles of vanishing rest mass. Even particles of finite mass can have a thermal velocity very close to c if the thermal energy of the particles is much larger than the rest mass energy, i.e., $k_B T \gg mc^2$. In these cases, the pressure is related to the density via the equation of state for radiation,

$$P_r = \frac{1}{3} \rho_r c^2 . \tag{4.20}$$

Vacuum Energy. The equation of state for vacuum energy takes a very unusual form which results from the first law of thermodynamics. Because the energy density ρ_v of the vacuum is constant in space and time, (4.17) immediately yields the relation

$$P_v = -\rho_v c^2 . \tag{4.21}$$

Thus the vacuum energy has a negative pressure. This unusual form of an equation of state can also be made plausible as follows: consider the change of a volume V that contains only vacuum. Since the internal energy is $U \propto V$, and thus a growth by $\mathrm{d}V$ implies an increase in U, the first law $\mathrm{d}U = -P\,\mathrm{d}V$ demands that P be negative.

4.2.6 "Derivation" of the Expansion Equation

Beginning with the equation of energy conservation (4.14), we are now able to derive the expansion equations (4.18) and (4.19). To achieve this, we differentiate both sides of (4.14) with respect to t and obtain

$$2\,\dot a\,\ddot a = \frac{8\pi G}{3}\left(\dot\rho\, a^2 + 2\,a\,\dot a\,\rho\right) .$$

Next, we carry out the differentiation in (4.17), thereby obtaining $\dot\rho a^3 + 3\rho a^2 \dot a = -3 P a^2 \dot a / c^2$. This relation is then used to replace the term containing $\dot\rho$ in the previous equation, yielding

$$\frac{\ddot a}{a} = -\frac{4\pi G}{3}\left(\rho + \frac{3P}{c^2}\right) . \tag{4.22}$$

This derivation therefore reveals that the pressure term in the equation of motion results from the combination of energy conservation and the first law of thermodynamics. However, we point out that the first law in the form (4.17) is based explicitely on the equivalence of mass and energy, resulting from Special Relativity. When assuming this equivalence, we indeed obtain the Friedmann equations from Newtonian cosmology, as expected from the discussion at the beginning of Sect. 4.2.1.

Next we consider the three aforementioned components of the cosmos and write the density and pressure as the sum of dust, radiation, and vacuum energy,

$$\rho = \rho_m + \rho_r + \rho_v = \rho_{m+r} + \rho_v , \qquad P = P_r + P_v ,$$

where ρ_{m+r} combines the density in matter and radiation. In the second equation, the pressureless nature of matter, $P_m = 0$, was used so that $P_{m+r} = P_r$. By inserting the first of these equations into (4.14), we indeed obtain the first Friedmann equation (4.18) if the density ρ there is identified with ρ_{m+r} (the density in "normal matter"), and if

$$\rho_v = \frac{\Lambda}{8\pi G} . \tag{4.23}$$

Furthermore, we insert the above decomposition of density and pressure into the equation of motion (4.22) and immediately obtain (4.19) if we identify ρ and P with ρ_{m+r} and $P_{m+r} = P_r$, respectively. Hence, this approach yields both Friedmann equations; the density and the pressure in the Friedmann equations refer to normal matter, i.e., all matter except the contribution by Λ. Alternatively, the Λ-terms in the Friedmann equations may be discarded if instead the vacuum energy density and its pressure are explicitly included in P and ρ.

4.2.7 Discussion of the Expansion Equations

Following the "derivation" of the expansion equations, we will now discuss their consequences. First we consider the density evolution of the various cosmic components resulting from (4.17). For pressure-free matter, we immediately obtain $\rho_m \propto a^{-3}$ which is in agreement with (4.11). Inserting the equation of state (4.20) for radiation into (4.17) yields the behavior $\rho_r \propto a^{-4}$; the vacuum energy density is a constant in time. Hence

$$\rho_{\rm m}(t) = \rho_{\rm m,0}\, a^{-3}(t)\,;\quad \rho_{\rm r}(t) = \rho_{\rm r,0}\, a^{-4}(t)\,;$$
$$\rho_{\rm v}(t) = \rho_{\rm v} = {\rm const}\,, \tag{4.24}$$

where the index "0" indicates the current time, $t = t_0$. The physical origin of the a^{-4} dependence of the radiation density is seen as follows: as for matter, the number density of photons changes $\propto a^{-3}$ because the number of photons in a comoving volume is unchanged. However, photons are redshifted by the cosmic expansion. Their wavelength λ changes proportional to a (see Sect. 4.3.2). Since the energy of a photon is $E = h_{\rm P}\,\nu$ ($h_{\rm P}$: Planck constant) and $\nu = c/\lambda$, the energy of a photon changes as a^{-1} due to cosmic expansion so that the photon energy density changes $\propto a^{-4}$.

Analogous to (4.16), we define the dimensionless density parameters for matter, radiation, and vacuum,

$$\boxed{\Omega_{\rm m} = \frac{\rho_{\rm m,0}}{\rho_{\rm cr}}\,;\quad \Omega_{\rm r} = \frac{\rho_{\rm r,0}}{\rho_{\rm cr}}\,;\quad \Omega_\Lambda = \frac{\rho_{\rm v}}{\rho_{\rm cr}} = \frac{\Lambda}{3H_0^2}}\,, \tag{4.25}$$

so that $\Omega_0 = \Omega_{\rm m} + \Omega_{\rm r} + \Omega_\Lambda$.[3]

By now we know the current composition of the Universe quite well. The matter density of galaxies (including their dark halos) corresponds to $\Omega_{\rm m} \gtrsim 0.02$, depending on the – largely unknown – extent of their dark halos. This value therefore provides a lower limit for $\Omega_{\rm m}$. Studies of galaxy clusters, which will be discussed in Chap. 6, yield a lower limit of $\Omega_{\rm m} \gtrsim 0.1$. Finally, we will show in Chap. 8 that $\Omega_{\rm m} \sim 0.3$.

In comparison to matter, the radiation energy density today is much smaller. It is dominated by photons of the cosmic background radiation and by neutrinos from the early Universe, as will be explained below. For the density parameter of radiation we obtain

$$\Omega_{\rm r} \sim 4.2 \times 10^{-5}\, h^{-2}\,, \tag{4.26}$$

so that today, the energy density of radiation in the Universe can be neglected when compared to that of matter. However, equations (4.24) reveal that the ratio between matter and radiation density was different at earlier epochs since $\rho_{\rm r}$ evolves faster with a than $\rho_{\rm m}$,

$$\frac{\rho_{\rm r}(t)}{\rho_{\rm m}(t)} = \frac{\rho_{\rm r,0}}{\rho_{\rm m,0}}\,\frac{1}{a(t)} = \frac{\Omega_{\rm r}}{\Omega_{\rm m}}\,\frac{1}{a(t)}\,. \tag{4.27}$$

Thus radiation and dust had the same energy density at an epoch when the scaling factor was

$$\boxed{a_{\rm eq} = \frac{\Omega_{\rm r}}{\Omega_{\rm m}} = 4.2 \times 10^{-5}\left(\Omega_{\rm m} h^2\right)^{-1}}\,. \tag{4.28}$$

This value of the scaling factor and the corresponding epoch in cosmic history play a very important role in structure evolution in the Universe, as we will see in Chap. 7.

With $\rho = \rho_{\rm m+r} = \rho_{\rm m,0}\, a^{-3} + \rho_{\rm r,0}\, a^{-4}$ and (4.25), the expansion equation (4.18) can be written as

$$H^2(t) = \tag{4.29}$$
$$H_0^2\left[a^{-4}(t)\Omega_{\rm r} + a^{-3}(t)\Omega_{\rm m} - a^{-2}(t)\frac{Kc^2}{H_0^2} + \Omega_\Lambda\right].$$

Evaluating this equation at the present epoch, with $H(t_0) = H_0$ and $a(t_0) = 1$, yields the value of the integration constant K,

$$\boxed{K = \left(\frac{H_0}{c}\right)^2(\Omega_0 - 1) \approx \left(\frac{H_0}{c}\right)^2(\Omega_{\rm m} + \Omega_\Lambda - 1)}\,. \tag{4.30}$$

Hence the constant K is obtained from the density parameters of matter and vacuum and from the aforementioned fact that $\Omega_{\rm r} \ll \Omega_{\rm m}$, and has the dimension of (length)$^{-2}$. In the context of GR, K is interpreted as the curvature scalar of the Universe today, or more precisely, the homogeneous, isotropic three-dimensional space at time $t = t_0$ has a curvature K. Depending on the sign of K, we can distinguish the following cases:

- If $K = 0$, the three-dimensional space for any fixed time t is Euclidean, i.e., flat.
- If $K > 0$, $1/\sqrt{K}$ can be interpreted as the curvature radius of the spherical 3-space – the two-dimensional analogy would be the surface of a sphere. As already

[3] In the literature, different definitions for Ω_0 are used. Often the notation Ω_0 is used for $\Omega_{\rm m}$.

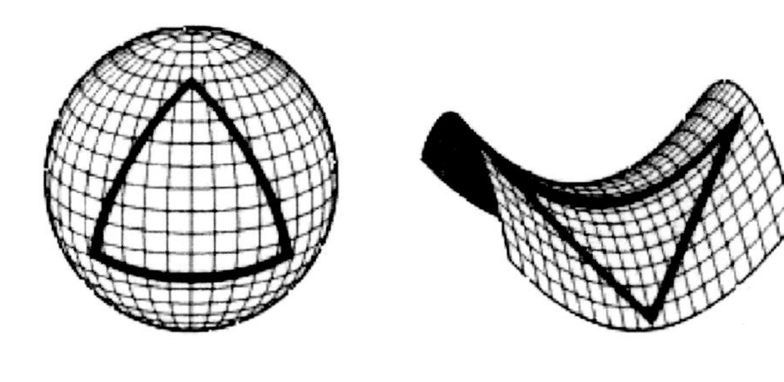
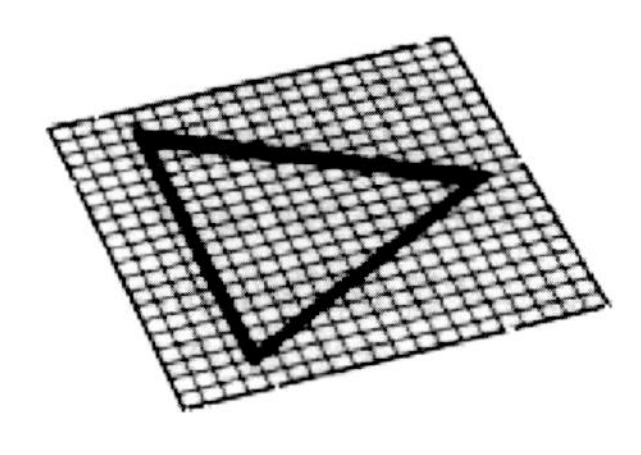

Closed Open Flat

Fig. 4.6. Two-dimensional analogies for the three possible curvatures of space. In a universe with positive curvature ($K > 0$) the sum of the angles in a triangle is larger than 180°, in a universe of negative curvature it is smaller than 180°, and in a flat universe the sum of angles is exactly 180°

speculated in Sect. 4.2.1, the order of magnitude of the curvature radius is c/H_0 according to (4.30).
- If $K < 0$, the space is called hyperbolic – the two-dimensional analogy would be the surface of a saddle (see Fig. 4.6).

Hence GR provides a relation between the curvature of space and the density of the Universe. In fact, this is the central aspect of GR which links the geometry of spacetime to its matter content. However, Einstein's theory makes no statement about the topology of spacetime and, in particular, says nothing about the topology of the Universe.[4] If the Universe has a simple topology, it is finite in the case of $K > 0$, whereas it is infinite if $K \le 0$. However, in both cases it has no boundary (compare: the surface of a sphere is a finite space without boundaries).

With (4.29) and (4.30), we finally obtain the expansion equation in the form

$$\left(\frac{\dot{a}}{a}\right)^2 = H^2(t) = H_0^2 \left[a^{-4}(t)\Omega_\mathrm{r} + a^{-3}(t)\Omega_\mathrm{m} + a^{-2}(t)(1 - \Omega_\mathrm{m} - \Omega_\Lambda) + \Omega_\Lambda\right] . \tag{4.31}$$

[4]The surface of a cylinder is also considered a flat space, like a plane, because the sum of angles in a triangle on a cylinder is also 180°. But the surface of a cylinder obviously has a topology different from a plane; in particular, closed straight lines do exist – walking on a cylinder in a direction perpendicular to its axis, one will return to the starting point after a finite amount of time.

4.3 Consequences of the Friedmann Expansion

The cosmic expansion equations imply a number of immediate consequences, some of which will be discussed next. In particular, we will first demonstrate that the early Universe must have evolved out of a very dense and hot state called the *Big Bang*. We will then link the scaling factor a to an observable, the redshift, and explain what the term "distance" means in cosmology.

4.3.1 The Necessity of a Big Bang

The terms on the right-hand side of (4.31) each have a different dependence on a:

- For very small a, the first term dominates and the Universe is radiation dominated then.
- For slightly larger $a \gtrsim a_\mathrm{eq}$, the dust (or matter) term dominates.
- If $K \neq 0$, the third term, also called the curvature term, dominates for larger a.
- For very large a, the cosmological constant dominates if it is different from zero.

The differential equation (4.31) in general cannot be solved analytically. However, its numerical solution for $a(t)$ poses no problems. Nevertheless, we can analyze the qualitative behavior of the function $a(t)$ and thereby understand the essential aspects of the expansion history. From the Hubble law, we conclude that $\dot{a}(t_0) > 0$, i.e., a is currently an increasing function of time. Equation (4.31) shows that $\dot{a}(t) > 0$ for all times, unless the right-hand side of (4.31) vanishes for some value of a: the sign of $\dot{a}$ can only switch when the right-hand side of (4.31) is zero. If $H^2 = 0$ for a value of $a > 1$, the expansion will come to a halt and the Universe will recollapse afterwards. On the other hand, if $H^2 = 0$ for a value $a = a_{\min}$ with $0 < a_{\min} < 1$, then the sign of $\dot{a}$ switches at $a_{\min}$. At this epoch, a collapsing Universe changes into an expanding one.

Which of these alternatives describes our Universe depends on the density parameters. We find the following classification (also see Fig. 4.7):

- If $\Lambda = 0$, then $H^2 > 0$ for all $a \le 1$, whereas the behavior for $a > 1$ depends on Ω_m:
  - if $\Omega_m \le 1$ (or $K \le 0$, respectively), $H^2 > 0$ for all a: the Universe will expand for all times. This behavior is expected from the Newtonian approach because if $K \le 0$, the kinetic energy in any spherical shell is larger than the modulus of the potential energy, i.e., the expansion velocity exceeds the escape velocity and the expansion will never come to a halt.
  - If $\Omega_m > 1$ $(K > 0)$, H^2 will vanish for $a = a_{\max} = \Omega_m/(\Omega_m - 1)$. The Universe will have its maximum expansion when the scale factor is $a_{\max}$ and will recollapse thereafter. In Newtonian terms, the total energy of any spherical shell is negative, so that it is gravitationally bound.
- In the presence of a cosmological constant $\Lambda > 0$, the discussion becomes more complicated:
  - If $\Omega_m < 1$, the Universe will expand for all $a > 1$.
  - However, for $\Omega_m > 1$ the future behavior of $a(t)$ depends on Ω_Λ: if Ω_Λ is sufficiently small, a value $a_{\max}$ exists at which the expansion comes to a halt and reverses. In contrast, if Ω_Λ is large enough the Universe will expand forever.
  - If $\Omega_\Lambda < 1$, then $H^2 > 0$ for all $a \le 1$.
  - However, if $\Omega_\Lambda > 1$, it is in principle possible that $H^2 = 0$ for an $a = a_{\min} < 1$. Such models, in which a minimum value for a existed in the past, can be excluded by observations (see Sect. 4.3.2).

With the exception of the last case, which can be excluded, we come to the conclusion that a must have attained the value $a = 0$ at some point in the past, at least formally. At this instant the "size of the Universe" formally vanished. As $a \to 0$, both matter and radiation densities diverge so that the density in this state must have been singular. The epoch at which $a = 0$ and the evolution away from this state is called the *Big Bang*. It is useful to define this epoch ($a = 0$) as the origin of time, so that t is identified with the age of the Universe, the time since the Big Bang. As we will show, the predictions of the Big Bang model are in impressive agreement with observations. The expansion history for the special case of a vanishing vacuum energy density is sketched in Fig. 4.8 for three values of the curvature.

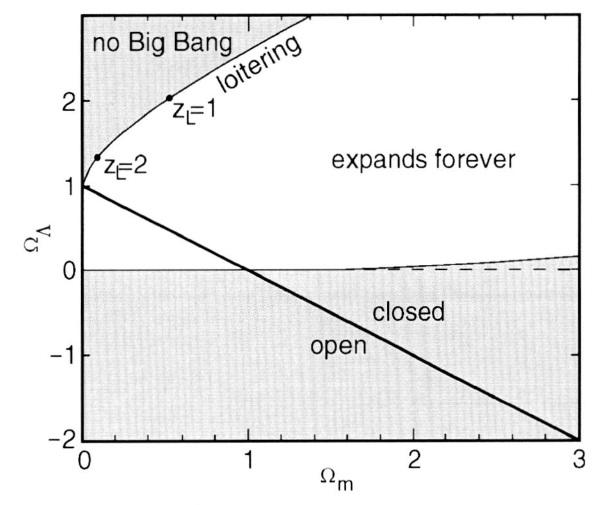

Fig. 4.7. Classification of cosmological models. The straight solid line connects flat models (i.e., those without spatial curvature, $\Omega_m + \Omega_\Lambda = 1$) and separates open ($K < 0$) and closed ($K > 0$) models. The nearly horizontal curve separates models that will expand forever from those that will recollapse in the distant future. Models in the upper left corner have an expansion history where a has never been close to zero and thus did not experience a Big Bang. In those models, a maximum redshift for sources exists, which is indicated for two cases. Since we know that $\Omega_m > 0.1$, and sources at redshift > 6 have been observed, these models can be excluded

To characterize whether the current expansion of the Universe is decelerated or accelerated, the *deceleration parameter*

$$q_0 := -\ddot{a}\, a/\dot{a}^2 \tag{4.32}$$

is defined where the right-hand side has to be evaluated at $t = t_0$. With (4.19) and (4.31) it follows that

$$q_0 = \Omega_m/2 - \Omega_\Lambda \, . \tag{4.33}$$

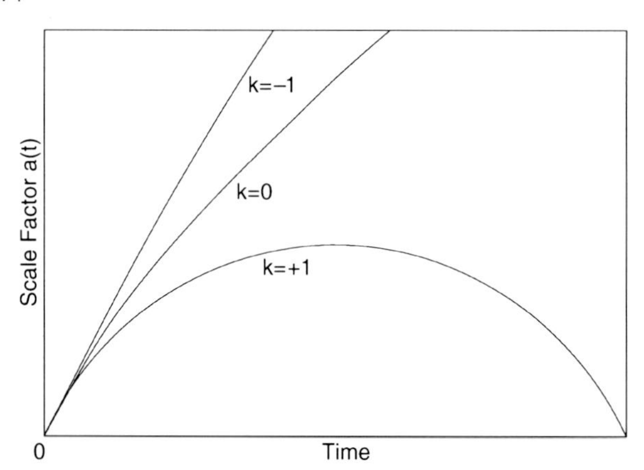

Fig. 4.8. The scale factor $a(t)$ as a function of cosmic time t for three models with a vanishing cosmological constant, $\Omega_\Lambda = 0$. Closed models ($K > 0$) attain a maximum expansion and then recollapse. In contrast, open models ($K \le 0$) expand forever, and the Einstein–de Sitter model of $K = 0$ separates these two cases. In all models, the scale factor tends towards zero in the past; this time is called the Big Bang and defines the origin of the time axis

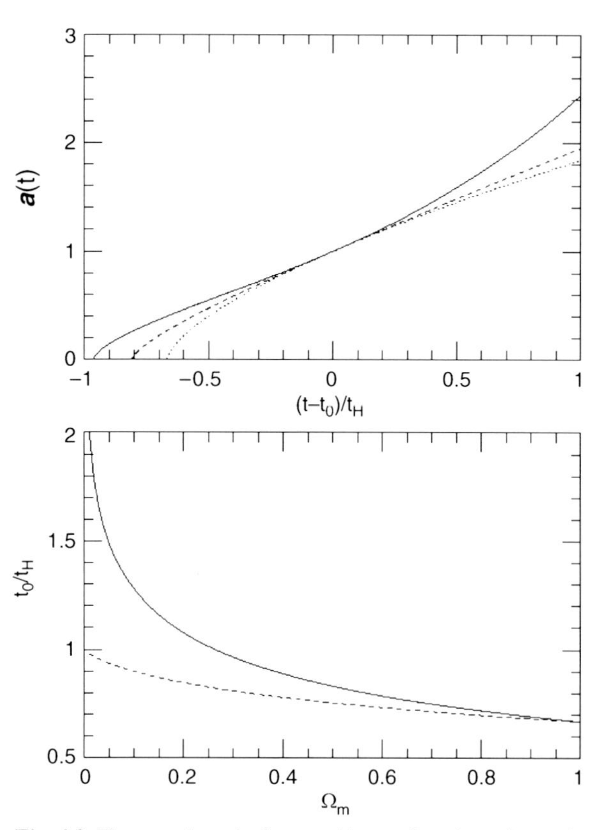

Fig. 4.9. Top panel: scale factor $a(t)$ as a function of cosmic time, here scaled as $(t - t_0)H_0$, for an Einstein–de Sitter model ($\Omega_m = 1$, $\Omega_\Lambda = 0$; dotted curve), an open universe ($\Omega_m = 0.3$, $\Omega_\Lambda = 0$; dashed curve), and a flat universe of low density ($\Omega_m = 0.3$, $\Omega_\Lambda = 0.7$; solid curve). At the current epoch, $t = t_0$ and $a = 1$. Bottom panel: age of the universe in units of the Hubble time $t_H = H_0^{-1}$ for flat world models with $K = 0$ ($\Omega_m + \Omega_\Lambda = 1$; solid curve) and models with a vanishing cosmological constant (dashed curve). We see that for a flat universe with small Ω_m (thus large $\Omega_\Lambda = 1 - \Omega_m$), t_0 may be considerably larger than H_0^{-1}

If $\Omega_\Lambda = 0$ then $q_0 > 0$, $\ddot{a} < 0$, i.e., the expansion decelerates, as expected due to gravity. However, if Ω_Λ is sufficiently large the deceleration parameter may become negative, corresponding to an accelerated expansion of the Universe. The reason for this behavior, which certainly contradicts intuition, is seen in the vacuum energy. Only a negative pressure can cause an accelerated expansion – more precisely, as seen from (4.22), $P < -\rho c^2/3$ is needed for $\ddot{a} > 0$. Indeed, we believe today that the Universe is currently undergoing an accelerated expansion and thus that the cosmological constant differs significantly from zero.

Age of the Universe. The age of the Universe at a given scale factor a follows from $\mathrm{d}t = \mathrm{d}a(\mathrm{d}a/\mathrm{d}t)^{-1} = \mathrm{d}a/(aH)$. This relation can be integrated,

$$t(a) = \frac{1}{H_0} \int_0^a \mathrm{d}a \left[a^{-2}\Omega_r + a^{-1}\Omega_m + (1 - \Omega_m - \Omega_\Lambda) + a^2\Omega_\Lambda\right]^{-1/2}, \tag{4.34}$$

where the contribution from radiation for $a \gg a_{eq}$ can be neglected because it is relevant only for very small a and thus only for a very small fraction of cosmic time. To obtain the current age t_0 of the Universe, (4.34) is calculated for $a = 1$. For models of vanishing spatial curvature $K = 0$ and for those with $\Lambda = 0$, Fig. 4.9 displays t_0 as a function of Ω_m.

The qualitative behavior of the cosmological models is characterized by the density parameters Ω_m and Ω_Λ, whereas the Hubble constant H_0 determines "only" the overall length- or time-scale. Today, mainly two families of models are considered:

- Models without a cosmological constant, $\Lambda = 0$. The difficulties in deriving a "sensible" value for Λ from particle physics is often taken as an argument for neglecting the vacuum energy density. However, there are very strong observational indications that in fact $\Lambda > 0$.

- Models with $\Omega_\mathrm{m} + \Omega_\Lambda = 1$, i.e., $K = 0$. Such flat models are preferred by the so-called inflationary models, which we will briefly discuss further below.

A special case is the Einstein–de Sitter model, $\Omega_\mathrm{m} = 1$, $\Omega_\Lambda = 0$. For this model, $t_0 = 2/(3H_0) \approx 6.7\,h^{-1} \times 10^9$ yr.

For many world models, t_0 is larger than the age of the oldest globular clusters, so they are compatible with this age determination. The Einstein–de Sitter model, however, is compatible with stellar ages only if H_0 is very small, considerably smaller than the value of H_0 derived from the HST Key Project discussed in Sect. 3.6. Hence, this model is ruled out by observations.

It is believed that the values of the cosmological parameters are now quite well known. We list them here for later reference without any further discussion. Their determination will be described in the course of this chapter and in Chap. 8. The values are approximately

$$\boxed{\Omega_\mathrm{m} \sim 0.3\,;\;\; \Omega_\Lambda \sim 0.7\,;\;\; h \sim 0.7}\,. \tag{4.35}$$

4.3.2 Redshift

The Hubble law describes a relation between the redshift, or the radial component of the relative velocity, and the distance of an object from us. Furthermore, (4.6) specifies that any observer is experiencing a local Hubble law with an expansion rate $H(t)$ which depends on the cosmic epoch. We will now derive a relation between the redshift of a source, which is directly observable, and the cosmic time t or the scaling factor $a(t)$, respectively, at which the source emitted the light we receive today.

To do this, we consider a light ray that reaches us today. Along this light ray we imagine fictitious comoving observers. The light ray is parametrized by the cosmic time t, and is supposed to have been emitted by the source at epoch t_e. Two comoving observers along the light ray with separation $\mathrm{d}r$ from each other see their relative motion due to the cosmic expansion according to (4.6), $\mathrm{d}v = H(t)\,\mathrm{d}r$, and they measure it as a redshift of light, $\mathrm{d}\lambda/\lambda = \mathrm{d}z = \mathrm{d}v/c$. It takes a time $\mathrm{d}t = \mathrm{d}r/c$ for the light to travel from one observer to the other. Furthermore, from the definition of the Hubble parameter, $\dot{a} = \mathrm{d}a/\mathrm{d}t = H\,a$, we obtain the relation $\mathrm{d}t = \mathrm{d}a/(H\,a)$. Combining these relations, we find

$$\frac{\mathrm{d}\lambda}{\lambda} = \frac{\mathrm{d}v}{c} = \frac{H}{c}\,\mathrm{d}r = H\,\mathrm{d}t = \frac{\mathrm{d}a}{a}\,. \tag{4.36}$$

The relation $\mathrm{d}\lambda/\lambda = \mathrm{d}a/a$ is now easily integrated since the equation $\mathrm{d}\lambda/\mathrm{d}a = \lambda/a$ obviously has the solution $\lambda = Ca$, where C is a constant. That constant is determined by the wavelength λ_obs of the light as observed today (i.e., at $a = 1$), so that

$$\lambda(a) = a\,\lambda_\mathrm{obs} \tag{4.37}$$

(see Fig. 4.10). The wavelength at emission was therefore $\lambda_\mathrm{e} = a(t_\mathrm{e})\lambda_\mathrm{obs}$. On the other hand, the redshift z is defined as $(1+z) = \lambda_\mathrm{obs}/\lambda_\mathrm{e}$. From this, we finally obtain the relation

$$\boxed{1 + z = \frac{1}{a}} \tag{4.38}$$

between the observable z and the scale factor a which is linked via (4.34) to the cosmic time. The same relation can also be derived by considering light rays in GR.

The relation between redshift and the scale factor is of immense importance for cosmology because, for most sources, redshift is the only distance information that we are able to measure. If the scale factor is a monotonic function of time, i.e., if the right-hand side of (4.31) is different from zero for all $a \in [0, 1]$, then z is also a monotonic function of t. In this case, which corresponds to the Universe we happen to live in, a, t, and z are equally good measures of the distance of a source from us.

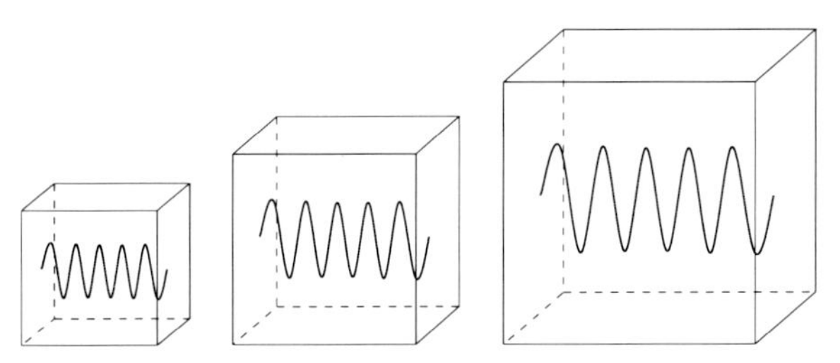

Fig. 4.10. Due to cosmic expansion, photons are redshifted, i.e., their wavelength, as measured by a comoving observer, increases with the scale factor a. This sketch visualizes this effect as a standing wave in an expanding box

Local Hubble Law. The Hubble law applies for nearby sources: with (4.8) and $v \approx zc$ it follows that

$$z = \frac{H_0}{c} D \approx \frac{h\,D}{3000\,\text{Mpc}} \quad \text{for } z \ll 1 \;, \tag{4.39}$$

where D is the distance of a source with redshift z. This corresponds to a light travel time of $\Delta t = D/c$. On the other hand, due to the definition of the Hubble parameter, we have $\Delta a = (1-a) \approx H_0\,\Delta t$, where a is the scale factor at time $t_0 - \Delta t$, and we used $a(t_0) = 1$ and $H(t_0) = H_0$. This implies $D = (1-a)c/H_0$. Utilizing (4.39), we then find $z = 1-a$, or $a = 1-z$, which agrees with (4.38) in linear approximation since $(1+z)^{-1} = 1 - z + \mathcal{O}(z^2)$. Hence we conclude that the general relation (4.38) contains the local Hubble law as a special case.

Energy Density in Radiation. A further consequence of (4.38) is the dependence of the energy density of radiation on the scale parameter. As mentioned previously, the number density of photons is $\propto a^{-3}$ if we assume that photons are neither created nor destroyed. In other words, the number of photons in a comoving volume element is conserved. According to (4.38), the frequency ν of a photon changes due to cosmic expansion. Since the energy of a photon is $\propto \nu$, $E_\gamma = h_P \nu \propto 1/a$, the energy density of photons decreases, $\rho_r \propto n\,E_\gamma \propto a^{-4}$. Therefore (4.38) implies (4.24).

Cosmic Microwave Background. Assuming that, at some time t_1, the Universe contained a blackbody radiation of temperature T_1, we can examine the evolution of this photon population in time by means of relation (4.38). We should recall that the Planck function B_ν (A.13) specifies the radiation energy of blackbody radiation that passes through a unit area per unit time, per unit area, per unit frequency interval, and per unit solid angle. Using this definition, the number density dN_ν of photons in the frequency interval between ν and $\nu + d\nu$ is obtained as

$$\frac{dN_\nu}{d\nu} = \frac{4\pi\,B_\nu}{c\,h_P\nu} = \frac{8\pi\nu^2}{c^3} \frac{1}{\exp\left(\frac{h_P\nu}{k_B T_1}\right) - 1} \;. \tag{4.40}$$

At a later time $t_2 > t_1$, the Universe has expanded by a factor $a(t_2)/a(t_1)$. An observer at t_2 therefore observes the photons redshifted by a factor $(1+z) = a(t_2)/a(t_1)$, i.e., a photon with frequency ν at t_1 will then be measured to have frequency $\nu' = \nu/(1+z)$. The original frequency interval is transformed accordingly as $d\nu' = d\nu/(1+z)$. The number density of photons decreases with the third power of the scale factor, so that $dN'_{\nu'} = dN_\nu/(1+z)^3$. Combining these relations, we obtain for the number density $dN'_{\nu'}$ of photons in the frequency interval between ν' and $\nu' + d\nu'$

$$\begin{aligned}
\frac{dN'_{\nu'}}{d\nu'} &= \frac{dN_\nu/(1+z)^3}{d\nu/(1+z)} \\
&= \frac{1}{(1+z)^2} \frac{8\pi(1+z)^2\nu'^2}{c^3} \frac{1}{\exp\left(\frac{h_P(1+z)\nu'}{k_B T_1}\right) - 1} \\
&= \frac{8\pi\nu'^2}{c^3} \frac{1}{\exp\left(\frac{h_P\nu'}{k_B T_2}\right) - 1} \;,
\end{aligned} \tag{4.41}$$

where we used $T_2 = T_1/(1+z)$ in the last step. The distribution (4.41) has the same form as (4.40) except that the temperature is reduced by a factor $(1+z)^{-1}$. If a Planck distribution of photons had been established at an earlier time, it will persist during cosmic expansion. As we have seen above, the CMB is such a blackbody radiation, with a current temperature of $T_0 = T_{\rm CMB} \approx 2.73$ K. We will show in Sect. 4.4 that this radiation originates in the early phase of the cosmos. Thus it is meaningful to consider the temperature of the CMB as the "temperature of the Universe" which is a function of redshift,

$$T(z) = T_0(1+z) = T_0\,a^{-1} \;, \tag{4.42}$$

i.e., the Universe was hotter in the past than it is today. The energy density of the Planck spectrum is

$$\rho_r = a_{SB}\,T^4 = \left(\frac{\pi^2\,k_B^4}{15\hbar^3\,c^3}\right) T^4 \;, \tag{4.43}$$

so ρ_r behaves like $(1+z)^4 = a^{-4}$ in accordance with (4.24).

Generally, it can be shown that the specific intensity I_ν changes due to redshift according to

$$\frac{I_\nu}{\nu^3} = \frac{I'_{\nu'}}{(\nu')^3} \;. \tag{4.44}$$

Here, I_ν is the specific intensity today at frequency ν and $I'_{\nu'}$ is the specific intensity at redshift z at frequency $\nu' = (1+z)\nu$.

Finally, it should be stressed again that (4.38) allows all relations to be expressed as functions of a as well as of z. For example, the age of the Universe as a function of z is obtained by replacing the upper integration limit, $a \to (1+z)^{-1}$, in (4.34).

The Necessity of a Big Bang. We discussed in Sect. 4.3.1 that the scale factor must have attained the value $a = 0$ at some time in the past. One gap in our argument that inevitably led to the necessity of a Big Bang still remains, namely the possibility that at sometime in the past $\dot{a} = 0$ occurred, i.e., that the Universe underwent a transition from a contracting to an expanding state. This is possible only if $\Omega_\Lambda > 1$ and if the matter density parameter is sufficiently small (see Fig. 4.7). In this case, a attained a minimum value in the past. This minimum value depends on both $\Omega_{\rm m}$ and Ω_Λ. For instance, for $\Omega_{\rm m} > 0.1$, the value is $a_{\rm min} > 0.3$. But a minimum value for a implies a maximum redshift $z_{\rm max} = 1/a_{\rm min} - 1$. However, since we have observed quasars and galaxies with $z > 6$ and the density parameter is known to be $\Omega_{\rm m} > 0.1$, such a model without a Big Bang can be excluded.

4.3.3 Distances in Cosmology

In the previous sections, different distance measures were discussed. Because of the monotonic behavior of the corresponding functions, each of a, t, and z provide the means to sort objects according to their distance. An object at higher redshift z_2 is more distant than one at $z_1 < z_2$ such that light from a source at z_2 may become absorbed by gas in an object at redshift z_1, but not vice versa. The object at redshift z_1 is located between us and the object at z_2. The more distant a source is from us, the longer the light takes to reach us, the earlier it was emitted, the smaller a was at emission, and the larger z is. Since z is the only observable of these parameters, distances in extragalactic astronomy are nearly always expressed in terms of redshift.

But how can a redshift be translated into a distance that has the dimension of a length? Or, phrasing this question differently, how many Megaparsecs away from us is a source with redshift $z = 2$? The corresponding answer is more complicated than the question suggests. For very small redshifts, the local Hubble relation (4.39) may be used, but this is valid only for $z \ll 1$.

In Euclidean space, the separation between two points is unambiguously defined, and several prescriptions exist for measuring a distance. We will give two examples here. A sphere of radius R situated at distance D subtends a solid angle of $\omega = \pi R^2/D^2$ on our sky. If the radius is known, D can be measured using this relation. As a second example, we consider a source of luminosity L at distance D which then has a measured flux $S = L/(4\pi D^2)$. Again, if the luminosity is known, the distance can be computed from the observed flux. If we use these two methods to determine, for example, the distance to the Sun, we would of course obtain identical results for the distance (within the range of accuracy), since these two prescriptions for distance measurements are defined to yield equal results.

In a non-Euclidean space like, for instance, our Universe this is no longer the case. The equivalence of different distance measures is only ensured in Euclidean space, and we have no reason to expect this equivalence to also hold in a curved spacetime. In cosmology, the same measuring prescriptions as in Euclidean space are used for defining distances, but the different definitions lead to different results. The two most important definitions of distance are:

- *Angular-diameter distance*: As above, we consider a source of radius R observed to cover a solid angle ω. The angular-diameter distance is defined as

$$D_{\rm A}(z) = \sqrt{\frac{R^2\,\pi}{\omega}} \; . \tag{4.45}$$

- *Luminosity distance*: We consider a source with luminosity L and flux S and define its luminosity distance as

$$D_{\rm L}(z) = \sqrt{\frac{L}{4\pi\, S}} \; . \tag{4.46}$$

These two distances agree locally, i.e., for $z \ll 1$; on small scales, the curvature of spacetime is not noticeable. In addition, they are *unique* functions of redshift. They can be computed explicitly. However, to do this some tools of GR are required. Since we have not discussed GR in this book, these tools are not available to us here. The distance–redshift relations depend on the cosmological parameters; Fig. 4.11 shows the angular-diameter distance for different models. For $\Lambda = 0$, the

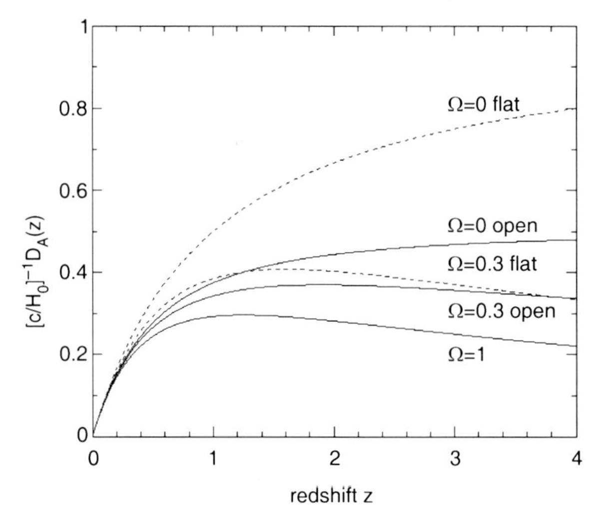

Fig. 4.11. Angular-diameter distance vs. redshift for different cosmological models. Solid curves display models with no vacuum energy; dashed curves show flat models with $\Omega_m + \Omega_\Lambda = 1$. In both cases, results are plotted for $\Omega_m = 1$, 0.3, and 0

famous Mattig relation applies,

$$D_A(z) = \frac{c}{H_0} \frac{2}{\Omega_m^2 (1+z)^2} \times \left[\Omega_m z + (\Omega_m - 2) \left(\sqrt{1 + \Omega_m z} - 1 \right) \right] . \quad (4.47)$$

In particular, D_A is not necessarily a monotonic function of z. To better comprehend this, we consider the geometry on the surface of a sphere. Two great circles on Earth are supposed to intersect at the North Pole enclosing an angle $\varphi \ll 1$ – they are therefore meridians. The separation L between these two great circles, i.e., the length of the connecting line perpendicular to both great circles, can be determined as a function of the distance D from the North Pole, which is measured as the distance along one of the two great circles. If θ is the geographical latitude ($\theta = \pi/2$ at the North Pole, $\theta = -\pi/2$ at the South Pole), $L = R\varphi \cos\theta$ is found, where R is the radius of the Earth. L vanishes at the North Pole, attains its maximum at the equator (where $\theta = 0$), and vanishes again at the South Pole; this is because both meridians intersect there again. Furthermore, $D = R(\pi/2 - \theta)$, e.g., the distance to the equator $D = R\pi/2$ is a quarter of the Earth's circumference. Solving the last relation for θ, the distance is then given by $L = R\varphi \cos(\pi/2 - D/R) = R\varphi \sin(D/R)$. For the angular-diameter distance on the Earth's surface, we define $D_A(D) = L/\varphi = R \sin(D/R)$, in analogy to the definition (4.45). For values of D that are considerably smaller than the curvature radius R of the sphere, we therefore obtain that $D_A \approx D$, whereas for larger D, D_A deviates considerably from D. In particular, D_A is not a monotonic function of D, rather it has a maximum at $D = \pi R/2$.

There exists a general relation between angular-diameter distance and luminosity distance,

$$\boxed{D_L(z) = (1+z)^2 \, D_A(z)} \, . \quad (4.48)$$

The reader might now ask which of these distances is the *correct one*? Well, this question does not make sense since there is no unique definition of *the* distance in a curved spacetime like our Universe. Instead, the aforementioned measurement prescriptions must be used. The choice of a distance definition depends on the desired application of this distance. For example, if we want to compute the linear diameter of a source with observed angular diameter, the angular-diameter distance must be employed because it is defined just in this way. On the other hand, to derive the luminosity of a source from its redshift and observed flux, the luminosity distance needs to be applied. Due to the definition of the angular-diameter distance (length/angular diameter), those are the relevant distances that appear in the gravitational lens equation (3.48). A statement that a source is located "at a distance of 3 billion light years" away from us is meaningless unless it is mentioned which type of distance is meant. Again, in the low-redshift Universe ($z \ll 1$), the differences between different distance definitions are very small, and thus it *is* meaningful to state, for example, that the Coma cluster of galaxies lies at a distance of ~ 90 Mpc.

In Fig. 4.12 a Hubble diagram extending to high redshifts is shown, where the brightest galaxies in clusters of galaxies have been used as approximate standard candles. With an assumed constant intrinsic luminosity for these galaxies, the apparent magnitude is a measure of their distance, where the luminosity distance $D_L(z)$ must be applied to compute the flux as a function of redshift.

Without derivation, we compile several expressions that are required to compute distances in general Friedmann–Lemaître models. To do this, we need to

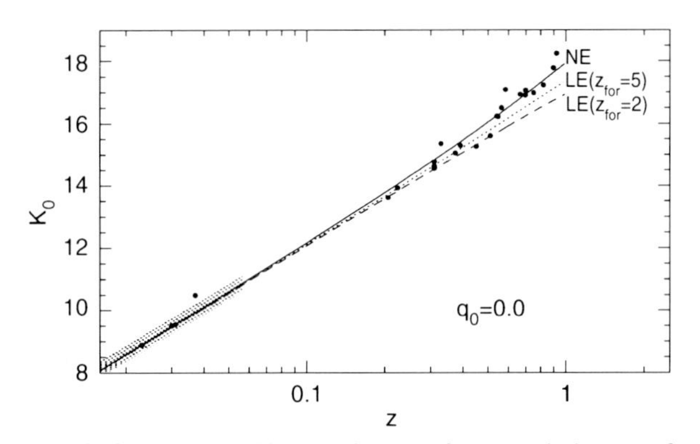

Fig. 4.12. A modern Hubble diagram: for several clusters of galaxies, the K-band magnitude of the brightest cluster galaxy is plotted versus the escape velocity, measured as redshift $z = \Delta\lambda/\lambda$ (symbols). If these galaxies all had the same luminosity, the apparent magnitude would be a measure of distance. For low redshifts, the curves follow the linear Hubble law (4.9), with $z \approx v/c$, whereas for higher redshifts modifications to this law are necessary. The solid curve corresponds to a constant galaxy luminosity at all redshifts, whereas the two other curves take evolutionary effects of the luminosity into account according to models of population synthesis (Sect. 3.9). Two different epochs of star formation were assumed for these galaxies. The diagram is based on a cosmological model with a deceleration parameter of $q_0 = 0$ (see Eq. 4.33)

define the function

$$f_K(x) = \begin{cases} 1/\sqrt{K}\ \sin(\sqrt{K}x) & K > 0 \\ x & K = 0 \\ 1/\sqrt{-K}\ \sinh(\sqrt{-K}x) & K < 0 \end{cases} ,$$

where K is the curvature scalar (4.30). The comoving radial distance x of a source at redshift z can be computed using $\mathrm{d}x = a^{-1}\,\mathrm{d}r = -a^{-1}\,c\,\mathrm{d}t = -c\,\mathrm{d}a/(a^2 H)$. Hence with (4.31)

$$x(z) = \frac{c}{H_0} \times \int_{(1+z)^{-1}}^{1} \frac{\mathrm{d}a(c/H_0)}{\sqrt{a\Omega_\mathrm{m} + a^2(1-\Omega_\mathrm{m}-\Omega_\Lambda) + a^4\Omega_\Lambda}} . \tag{4.49}$$

The angular-diameter distance is then given as

$$D_\mathrm{A}(z) = \frac{1}{1+z}\, f_K\left[x(z)\right] , \tag{4.50}$$

and thus can be computed for all redshifts and cosmological parameters by (in general numerical) integration of (4.49). The luminosity distance then follows from (4.48). The angular-diameter distance of a source at redshift z_2, as measured by an observer at redshift $z_1 < z_2$, reads

$$D_\mathrm{A}(z_1, z_2) = \frac{1}{1+z_2}\, f_K\left[x(z_2) - x(z_1)\right] . \tag{4.51}$$

This is the distance that is required in equations of gravitational lens theory for D_ds. In particular, $D_\mathrm{A}(z_1, z_2) \neq D_\mathrm{A}(z_2) - D_\mathrm{A}(z_1)$.

4.3.4 Special Case: The Einstein–de Sitter Model

As a final note in this section, we will briefly examine one particular cosmological model more closely, namely the model with $\Omega_\Lambda = 0$ and vanishing curvature, $K = 0$, and hence $\Omega_\mathrm{m} = 1$. We disregard the radiation component, which contributes to the expansion only at very early times and thus for very small a. For a long time, this Einstein–de Sitter (EdS) model was the preferred model among cosmologists because inflation (see Sect. 4.5.3) predicts $K = 0$ and because a finite value for the cosmological constant was considered "unnatural". In fact, as late as the mid-1990s, this model was termed the "standard model". In the meantime we have learned that $\Lambda \neq 0$; thus we are not living in an EdS universe. But there is at least one good reason to examine this model a bit more, since the expansion equations become much simpler for these parameters and we can formulate simple explicit expressions for the quantities introduced above. These then yield estimates which for other model parameters are only possible by means of numerical integration.

The resulting expansion equation $\dot{a} = H_0\, a^{-1/2}$ is easily solved by making the ansatz $a = (Ct)^\beta$ which, when inserted into the equation, yields the solution

$$a(t) = \left(\frac{3\,H_0\,t}{2}\right)^{2/3} . \tag{4.52}$$

Setting $a = 1$, we obtain the age of the Universe, $t_0 = 2/(3H_0)$. The same result also follows immediately from (4.34) if the parameters of an EdS model are inserted there. Using $H_0 \approx 70\,\mathrm{km\,s^{-1}\,Mpc^{-1}}$ results in an age of about 10 Gyr, which is slightly too low to be compatible with the age of the oldest star clusters. The angular-diameter distance (4.45) in an EdS universe is obtained by considering the Mattig relation (4.38) for the case $\Omega_m = 1$:

$$D_A(z) = \frac{2c}{H_0} \frac{1}{(1+z)} \left(1 - \frac{1}{\sqrt{1+z}}\right) ,$$

$$D_L(z) = \frac{2c}{H_0} (1+z) \left(1 - \frac{1}{\sqrt{1+z}}\right) , \tag{4.53}$$

where we used (4.48) to obtain the second relation from the first.

4.3.5 Summary

We shall summarize the most important points of the two preceeding lengthy sections:

- Observations are compatible with the fact that the Universe around us is isotropic and homogeneous on large scales. The cosmological principle postulates this homogeneity and isotropy of the Universe.
- General Relativity allows homogeneous and isotropic world models, the Friedmann–Lemaître models. In the language of GR, the cosmological principle reads as follows: "A family of solutions of Einstein's field equations exists such that a set of comoving observers see the same history of the Universe; for each of them, the Universe appears isotropic."
- The *shape* of these Friedmann–Lemaître world models is characterized by the density parameter Ω_m and by the cosmological constant Ω_Λ, the *size* by the Hubble constant H_0. The cosmological parameters determine the expansion rate of the Universe as a function of time.
- The scale factor $a(t)$ of the Universe is a monotonically increasing function from the beginning of the Universe until now; at earlier times the Universe was smaller, denser, and hotter. There must have been an instant when $a \to 0$, which is called the Big Bang. The future of the expansion depends on Ω_m and Ω_Λ.
- The expansion of the Universe causes a redshift of photons. The more distant a source is from us, the more its photons are redshifted.

4.4 Thermal History of the Universe

Since $T \propto (1+z)$ the Universe was hotter at earlier times. For example, at a redshift of $z = 1100$ the temperature was about $T \sim 3000$ K. And at an even higher redshift, $z = 10^9$, it was $T \sim 3 \times 10^9$ K, hotter than in a stellar interior. Thus we might expect energetic processes like nuclear fusion to have taken place in the early Universe.

In this section we shall describe the essential processes in the early Universe. To do so we will assume that the laws of physics have not changed over time. This assumption is by no means trivial – we have no guarantee whatsoever that the cross-sections in nuclear physics were the same 13 billion years ago as they are today. But if they have changed in the course of time the only chance of detecting this is through cosmology. Based on this assumption of time-invariant physical laws, we will study the consequences of the Big Bang model developed in the previous section and then compare them with observations. Only this comparison can serve as a test of the success of the model. A few comments should serve as preparation for the discussion in this section.

1. Temperature and energy may be converted into each other since $k_B T$ has the dimension of energy. We use the electron volt (eV) to measure temperatures and energies, with the conversion $1\,\text{eV} = 1.1605 \times 10^4\, k_B\,\text{K}$.
2. Elementary particle physics is very well understood for energies below ~ 1 GeV. For much higher energies our understanding of physics is a lot less certain. Therefore, we will begin the consideration of the thermal history of the cosmos at energies below 1 GeV.
3. Statistical physics and thermodynamics of elementary particles are described by quantum mechanics. A distinction has to be made between *bosons*, which are particles of integer spin (like the photon), and *fermions*, particles of half-integer spin (like, for instance, electrons, protons, or neutrinos).
4. If particles are in thermodynamical and chemical equilibrium, their number density and their energy distribution are specified solely by the temperature – e.g., the Planck distribution (A.13), and thus the en-

ergy density of the radiation (4.43), is a function of T only.

The necessary condition for establishing chemical equilibrium is the possibility for particles to be be created and destroyed, such as in electron–positron pair production and annihilation.

4.4.1 Expansion in the Radiation-Dominated Phase

As mentioned above (4.28), the energy density of radiation dominates in the early Universe, at redshifts $z \gg z_{\rm eq}$ where

$$z_{\rm eq} = a_{\rm eq}^{-1} - 1 \approx 23\,900\,\Omega_{\rm m}\,h^2 \,, \quad (4.54)$$

The radiation density behaves like $\rho_{\rm r} \propto T^4$, where the constant of proportionality depends on the number of species of relativistic particles (these are the ones for which $k_{\rm B}T \gg mc^2$). Since $T \propto 1/a$ and thus $\rho_{\rm r} \propto a^{-4}$, radiation then dominates in the expansion equation (4.18). The latter can be solved by a power law, $a(t) \propto t^\beta$, which after insertion into the expansion equation yields $\beta = 1/2$ and thus

$$a \propto t^{1/2}\,, \quad t = \sqrt{\frac{3}{32\pi G\rho}}\,, \quad t \propto T^{-2}$$
$$\text{in radiation-dominated phase}\,, \quad (4.55)$$

where the constant of proportionality depends again on the number of relativistic particle species. Since the latter is known, assuming thermodynamical equilibrium, the time dependence of the early expansion is uniquely specified by (4.55). This is reasonable because for early times neither the curvature term nor the cosmological constant contribute significantly to the expansion dynamics.

4.4.2 Decoupling of Neutrinos

We start our consideration of the Universe at a temperature of $T \approx 10^{12}$ K which corresponds to ~ 100 MeV. This energy can be compared to the rest mass of various particles:

proton, $m_{\rm p} = 938.3\,{\rm MeV}/c^2$,
neutron, $m_{\rm n} = 939.6\,{\rm MeV}/c^2$,
electron, $m_{\rm e} = 511\,{\rm keV}/c^2$,
muon, $m_\mu = 140\,{\rm MeV}/c^2$.

Protons and neutrons (i.e., the baryons) are too heavy to be produced at the temperature considered. Thus all baryons that exist today must have already been present at this early time. Also, the production of muon[5] pairs, according to the reaction $\gamma + \gamma \to \mu^+ + \mu^-$, is not efficient because the temperature, and thus the typical photon energy, is not sufficiently high. Hence, at the temperature considered the following relativistic particle species are present: electrons and positrons, photons and neutrinos. These species contribute to the radiation density $\rho_{\rm r}$. The mass of the neutrinos is not accurately known, though we recently learned that they have a small but finite rest mass. As will be explained in Sect. 8.7, cosmology allows us to obtain a very strict limit on the neutrino mass, which is currently below 1 eV. For the purpose of this discussion they may be considered as massless.

In addition to relativistic particles, non-relativistic particles also exist. These are the protons and neutrons, and probably also the constituents of dark matter. We assume that the latter consists of weakly interacting massive particles (WIMPs), with rest mass larger than ~ 100 GeV because up to these energies no WIMP candidates have been found in terrestrial particle accelerator laboratories. With this assumption, WIMPs are non-relativistic at the energies considered. Thus, like the baryons, they virtually do not contribute to the energy density in the early Universe.

Apart from the WIMPs, all the aforementioned particle species are in equilibrium, e.g., by the following reactions:

$e^\pm + \gamma \leftrightarrow e^\pm + \gamma$: Compton scattering,
$e^+ + e^- \leftrightarrow \gamma + \gamma$: pair-production and annihilation,
$\nu + \overline{\nu} \leftrightarrow e^+ + e^-$: neutrino–antineutrino scattering,
$\nu + e^\pm \leftrightarrow \nu + e^\pm$: neutrino–electron scattering.

[5]Muons are particles which behave in many respects as electrons, except that they are much heavier. Furthermore, muons are unstable and decay on a time-scale of $\sim 2 \times 10^{-6}$ s into an electron (or positron) and two neutrinos.

Reactions involving baryons will be discussed later. The energy density at this epoch is

$$\rho = \rho_{\rm r} = 10.75\,\frac{\pi^2}{30}\,\frac{(k_{\rm B}T)^4}{\hbar c^3}\;,$$

which yields – see (4.55) –

$$t \approx 0.3\,{\rm s}\left(\frac{T}{1\,{\rm MeV}}\right)^{-2}\;. \tag{4.56}$$

Hence, about one second after the Big Bang the temperature of the Universe was about 10^{10} K. For the particles to maintain equilibrium, the reactions above have to occur at a sufficient rate. The equilibrium state, specified by the temperature, continuously changes, so that the particle distribution needs to continually adjust to this changing equilibrium. This is possible only if the mean time between two reactions is much shorter than the time-scale on which equilibrium conditions change. The latter is given by the expansion. This means that the reaction rates (the number of reactions per particle per unit time) must be larger than the cosmic expansion rate $H(t)$ in order for the particles to maintain equilibrium.

The reaction rates Γ are proportional to the product of the number density n of the particles and the cross-section σ of the corresponding reaction. Both decrease with time: the number density decreases as $n \propto a^{-3} \propto t^{-3/2}$ because of the expansion. Furthermore, the cross-sections for weak interaction, which is responsible for the reactions involving neutrinos, depend on energy, approximately as $\sigma \propto E^2 \propto T^2 \propto a^{-2}$. Together this yields $\Gamma \propto n\sigma \propto a^{-5} \propto t^{-5/2}$, whereas the expansion rate decreases only as $H \propto t^{-1}$. At sufficiently early times, the reaction rates were larger than the expansion rate, and thus particles remained in equilibrium. Later, however, the reactions no longer took place fast enough to maintain equilibrium. The time or temperature, respectively, of this transition can be calculated from the cross-section of weak interaction,

$$\frac{\Gamma}{H} \approx \left(\frac{T}{1.6\times 10^{10}\,{\rm K}}\right)^3\;,$$

so that for $T \lesssim 10^{10}$ K neutrinos are no longer in equilibrium with the other particles. This process of decoupling from the other particles is also called *freeze-out*; neutrinos freeze out at $T \sim 10^{10}$ K. At the time of freeze-out, they had a thermal distribution with the same temperature as the other particle species which stayed in mutual equilibrium. From this time on neutrinos propagate without further interactions, and so have kept their thermal distribution up to the present day, with a temperature decreasing as $T \propto 1/a$. This consideration predicts that these neutrinos, which decoupled from the rest of the matter about one second after the Big Bang, are still around in the Universe today. They have a number density of $113\,{\rm cm}^{-3}$ per neutrino family and are at a temperature of 1.9 K (this value will be explained in more detail below). However, these neutrinos are currently undetectable because of their extremely low cross-section.

The expansion behavior is unaffected by the neutrino freeze-out and continues to proceed according to (4.56).

4.4.3 Pair Annihilation

At temperatures smaller than $\sim 5\times 10^9$ K, or $k_{\rm B}T \sim$ 500 keV, electron–positron pairs can no longer be produced efficiently since the number density of photons with energies above the pair production threshold of 511 keV is becoming too small. However, the annihilation $e^+ + e^- \to \gamma + \gamma$ continues to proceed and, due to its large cross-section, the density of e^+e^--pairs decreases rapidly.

Pair annihilation injects additional energy into the photon gas, originally present as kinetic and rest mass energy of the e^+e^- pairs. This changes the energy distribution of photons, which continues to be a Planck distribution but now with a modified temperature relative to that it would have had without annihilation. The neutrinos, already decoupled at this time, do not benefit from this additional energy. This means that after the annihilation the photon temperature exceeds that of the neutrinos. From the thermodynamics of this process, the change in photon temperature is computed as

$$\begin{aligned} &T\ (\text{after annihilation}) \\ &= \left(\frac{11}{4}\right)^{1/3} T\ (\text{before annihilation}) \\ &= \left(\frac{11}{4}\right)^{1/3} T_\nu\;. \end{aligned} \tag{4.57}$$

This temperature ratio is preserved afterwards, so that neutrinos have a temperature lower than that of the photons by $(11/4)^{1/3} \sim 1.4$ – until the present epoch. This result has already been mentioned and taken into account in the estimate of $\rho_{r,0}$ in (4.26); we find $\rho_{r,0} = 1.68\rho_{CMB,0}$.

After pair annihilation, the expansion law

$$t = 0.55\,\mathrm{s}\left(\frac{T}{1\,\mathrm{MeV}}\right)^{-2} \tag{4.58}$$

applies. This means that, as a result of annihilation, the constant in this relation changes compared to (4.56) because the number of relativistic particles species has decreased. Furthermore, the ratio η of baryon to photon density remains constant after pair annihilation. The former is characterized by the density parameter $\Omega_b = \rho_{b,0}/\Omega_{cr}$ in baryons (today), and the latter is determined by T_0:

$$\eta := \left(\frac{n_b}{n_\gamma}\right) = 2.74 \times 10^{-8}\,\left(\Omega_b h^2\right) \,. \tag{4.59}$$

Before pair annihilation there were about as many electrons and positrons as there were photons. After annihilation *nearly* all electrons were converted into photons – but not entirely because there was a very small excess of electrons over positrons to compensate for the positive electrical charge density of the protons. Therefore, the number density of electrons that survive the pair annihilation is exactly the same as the number density of protons, for the Universe to remain electrically neutral. Thus, the ratio of electrons to photons is also given by η, or more precisely by about 0.8η, since η includes both protons and neutrons.

4.4.4 Primordial Nucleosynthesis

Protons and neutrons can fuse to form atomic nuclei if the temperature and density of the plasma are sufficiently high. In the interior of stars, these conditions for nuclear fusion are provided. The high temperatures in the early phases of the Universe suggest that atomic nuclei may also have formed then. As we will discuss below, in the first few minutes after the Big Bang some of the lightest atomic nuclei were formed. The quantitative discussion of this primordial nucleosynthesis (Big Bang nucleosynthesis, BBN) will explain observation (4) of Sect. 4.1.1.

Proton-to-Neutron Abundance Ratio. As already discussed, the baryons (or nucleons) do not play any role in the expansion dynamics in the early Universe because of their low density. The most important reactions through which they maintain chemical equilibrium with the rest of the particles are

$$\mathrm{p} + \mathrm{e} \leftrightarrow \mathrm{n} + \nu \,,$$
$$\mathrm{p} + \overline{\nu} \leftrightarrow \mathrm{n} + \mathrm{e}^+ \,,$$
$$\mathrm{n} \rightarrow \mathrm{p} + \mathrm{e} + \overline{\nu} \,.$$

The latter is the decay of free neutrons, with a time-scale for the decay of $\tau_n = 887$ s. The first two reactions maintain the equilibrium proton-to-neutron ratio as long as the corresponding reaction rates are large compared to the expansion rate. The equilibrium distribution is specified by the Boltzmann factor,

$$\frac{n_n}{n_p} = \exp\left(-\frac{\Delta m\, c^2}{k_B T}\right) \,, \tag{4.60}$$

where $\Delta m = m_n - m_p = 1.293\,\mathrm{MeV}/c^2$ is the mass difference between neutrons and protons. Hence, neutrons are slightly heavier than protons; otherwise the neutron decay would not be possible. After neutrino freeze-out equilibrium reactions become rare because the above reactions are based on weak interactions, the same as those which kept the neutrinos in chemical equilibrium. At the time of neutrino decoupling, we have $n_n/n_p \approx 1/3$. After this, protons and neutrons are no longer in equilibrium, and their ratio is no longer described by (4.60). Instead, it changes only by the decay of free neutrons on the time-scale τ_n. To have neutrons survive at all until the present day, they must quickly become bound in atomic nuclei.

Deuterium Formation. The simplest compound nucleus is that of deuterium (D), consisting of a proton and a neutron and formed in the reaction

$$\mathrm{p} + \mathrm{n} \rightarrow \mathrm{D} + \gamma \,.$$

The binding energy of D is $E_b = 2.225$ MeV. This energy is only slightly larger than $m_e c^2$ and Δm – all these energies are of comparable size. The formation of deuterium is based on strong interactions and therefore occurs very efficiently. However, at the time of neutrino decoupling and pair annihilation, T is not much smaller than E_b. This has an important consequence: because photons are so much more abundant than baryons, a sufficient number of highly energetic photons, with $E_\gamma \geq E_b$, exist in the Wien tail of the Planck distribution to instantly destroy newly formed D by photo-dissociation. Only when the temperature has decreased considerably, $k_B T \ll E_b$, can the deuterium abundance become appreciable. With the corresponding balance equations we can calculate that the formation rate exceeds the photo-dissociation rate of deuterium at about $T_D \approx 8 \times 10^8$ K, corresponding to $t \sim 3$ min. Up to then, a fraction of the neutrons has thus decayed, yielding a neutron–proton ratio at T_D of $n_n/n_p \approx 1/7$.

After that time, everything happens very rapidly. Owing to the strong interaction, virtually all neutrons first become bound in D. Once the deuterium density has become appreciable, helium (He^4) forms, which is a nucleus with high binding energy (~ 28 MeV) which can therefore not be destroyed by photo-dissociation. Except for a small (but, as we will later see, very important) remaining fraction, all deuterium is quickly transformed into He^4. For this reason, the dependence of helium formation on the small binding energy of D is known as the "bottleneck of nucleosynthesis".

Helium Abundance. The number density of helium nuclei can now be calculated since virtually all neutrons present are bound in He^4. First, $n_{He} = n_n/2$, since every helium nucleus contains two neutrons. Second, the number density of protons after the formation of helium is $n_H = n_p - n_n$, since He^4 contains an equal number of protons and neutrons. From this, the *mass fraction Y of* He^4 *of the baryon density* follows,

$$Y = \frac{4n_{He}}{4n_{He} + n_H} = \frac{2n_n}{n_p + n_n} = \frac{2(n_n/n_p)}{1 + (n_n/n_p)} \approx 0.25 \quad , \tag{4.61}$$

where in the last step we used the above ratio of $n_n/n_p \approx 1/7$ at T_D. This consideration thus leads to the following:

About 1/4 of the baryonic mass in the Universe should be in the form of He^4. This is a robust prediction of Big Bang models, and it is in excellent agreement with observations.

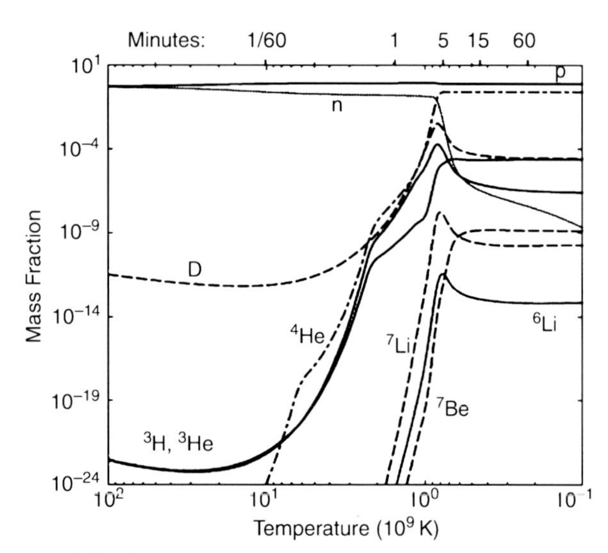

Fig. 4.13. The evolution of abundances of the light elements formed in BBN, as a function of temperature (lower axis) and cosmic time t (upper axis). The decrease in neutron abundance in the first ~ 3 min is due to neutron decay. The density of deuterium increases steeply – linked to the steep decrease in neutron density – and reaches a maximum at $t \sim 3$ min because then its density becomes sufficiently large for efficient formation of He^4 to set in. Only a few deuterium nuclei do not find a reaction partner and remain, with a mass fraction of $\sim 10^{-5}$. Only a few other light nuclei are formed in the Big Bang, mainly He^3 and Li^7

The helium content in the Universe may change later by nuclear fusion in stars, which also forms heavier nuclei ("metals"). Observations of fairly unprocessed material (i.e., that which has a low metal content) reveal that in fact $Y \approx 0.25$. Figure 4.13 shows the result of a quantitative model of BBN where the mass fraction of several species is plotted as a function of time or temperature, respectively.

Dependence of the Primordial Abundances on the Baryon Density. At the end of the first 3 min, the composition of the baryonic component of the Universe is about as follows: 25% of the baryonic mass is bound in helium nuclei, 75% in hydrogen nuclei (i.e., protons), with traces of D, He^3 and Li^7. Heavier nuclei cannot form because no stable nucleus of mass number 5 or 8 exists and thus no new, stable nuclei can be formed in collisions of two helium nuclei or of a proton with a helium nucleus. Collisions between three nuclei are far too rare to contribute to nucleosynthesis. The density in He^4 and D depends on the baryon density in the Universe, as can be seen in Fig. 4.14 and through the following considerations:

- The larger the baryon density Ω_b, thus the larger the baryon-to-photon ratio η (4.59), the earlier D can form, i.e., the fewer neutrons have decayed, which then results in a larger n_n/n_p ratio. From this and (4.61) it follows that Y increases with increasing Ω_b.
- A similar argument is valid for the abundance of deuterium: the larger Ω_b is, the higher the baryon density during the conversion of D into He^4. Thus the conversion will be more efficient and more complete. This means that fewer deuterium nuclei remain without a reaction partner for helium formation. Thus fewer of them are left over in the end, so the fraction of D will be lower.

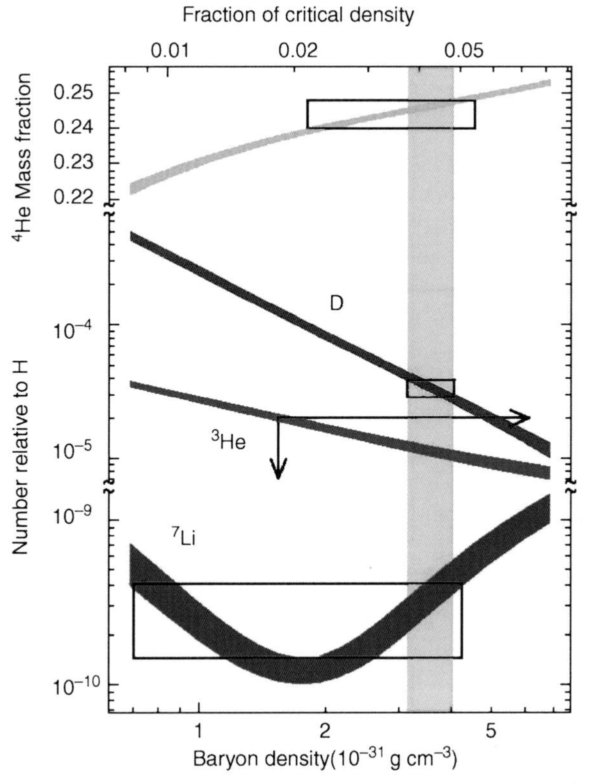

Fig. 4.14. BBN predictions of the primordial abundances of light elements as a function of today's baryon density ($\rho_{b,0}$, lower axis) and the corresponding density parameter Ω_b where $h = 0.65$ is assumed. The vertical extent of the rectangles marks the measured values of the abundances (top: He^4, center: D, bottom: Li^7). The horizontal extent results from the overlap of these intervals with curves computed from theoretical models. The ranges in Ω_b that are allowed by these three species do overlap, as is indicated by the vertical strip. The deuterium measurements yield the most stringent constraints for Ω_b

Baryon Content of the Universe. From measurements of the primordial abundances of He^4 and D and their comparison with detailed models of nucleosynthesis in the early Universe, η or Ω_b, respectively, can be determined (see Fig. 4.14). The abundance of deuterium is a particularly sensitive measure for Ω_b. Measurements of the relative strength of the Lyα lines of H and D, which have slightly different transition frequencies due to the different masses of their nuclei, in QSO absorption lines (see Sect. 5.6.3) yield $D/H \approx 3.4 \times 10^{-5}$. Since the intergalactic gas producing these absorption lines is very metal-poor and thus presumably barely affected by nucleosynthesis in stars, its D/H ratio should be close to the primordial value. Combining the quoted value of D/H with the model curves shown in Fig. 4.14 we find

$$\boxed{\Omega_b h^2 \approx 0.02} \; . \tag{4.62}$$

With a Hubble constant of $H_0 \sim 70\,\mathrm{km\,s^{-1}\,Mpc^{-1}}$, thus $h \sim 0.7$, we have $\Omega_b \approx 0.04$. But since $\Omega_m > 0.1$, this result implies that baryons represent only a small fraction of the matter in the Universe. *The major fraction of matter is non-baryonic dark matter.*

WIMPs as Dark Matter. Big Bang nucleosynthesis therefore provides a clear indication of the existence of non-baryonic dark matter on cosmological scales. Whereas our discussion of rotation curves of spiral galaxies could not fully exclude the possibility that the dark matter consists of baryons, BBN shows that this cannot be the case.

As mentioned previously, the most promising candidate for a dark matter constituent is an as yet unknown elementary particle, a WIMP. In fact, from the above considerations, constraints on the properties of such a particle can be derived. If the WIMP is weakly interacting, it will decouple in a similar way to the neutrinos. If its mass m_{WIMP} is smaller than the decoupling temperature ($T \sim 1$ MeV), the WIMP was relativistic at the epoch of freeze-out and thus its number density is the same as that of the neutrinos, $n = 113\,\mathrm{cm^{-3}}$ today. Hence, we can compute the corresponding density parameter,

$$\Omega_{\rm WIMP}\, h^2 = \frac{m_{\rm WIMP}}{91.5\ {\rm eV}}\,, \tag{4.63}$$

provided $m_{\rm WIMP} \lesssim 1$ MeV. This equation is of course also valid for neutrinos with $m_\nu \lesssim 1$ MeV. Since $\Omega_{\rm m} < 2$ certainly, we conclude from (4.63) that no stable weakly interacting particle can exist in the mass range of $100\ {\rm eV} \lesssim m \lesssim 1$ MeV. In particular, none of the three neutrinos can have a mass in this range. Until recently, these constraints on the mass of the μ- and τ-neutrinos were many orders of magnitude better than those from laboratory experiments. Only measurements of neutrino oscillations, in connection with laboratory constraints on the $\nu_{\rm e}$-mass, provided better mass constraints. From structure formation in the Universe, which will be discussed in Chap. 7, we now know that neutrinos cannot make a dominant contribution to the dark matter, and therefore an upper limit on their mass $m_\nu \lesssim 1$ eV can be obtained.

If the WIMP is heavier than 1 MeV, it decouples at a time when it is already non-relativistic. Then the estimate of the number density, and with it the relation (4.63), needs to be modified. In particular, if the WIMP mass exceeds that of the Z-boson[6] ($m_{\rm Z} = 91$ GeV), then $\Omega_{\rm WIMP}\, h^2 \simeq (m_{\rm WIMP}/1\ {\rm TeV})^2$. This means that a WIMP mass of several hundred GeV would provide a density of $\Omega_{\rm WIMP} \approx \Omega_{\rm m} \sim 0.3$. The next generation of particle accelerators, in particular the Large Hadron Collider at CERN which is supposed to start operations in 2007, will most likely be able to detect such a particle in the laboratory if it really exists. In fact, arguably the most promising extension to the standard model of particle physics – the model of supersymmetry – predicts a stable particle with a mass of several hundred GeV, the neutralino.

In the analysis of BBN we implicitly assumed that not more than three (relativistic, i.e., with $m_\nu < 1$ MeV) neutrino families exist. If $N_\nu > 3$, the quantitative predictions of BBN will change. In this case, the expansion would occur faster (see Eq. 4.55) because $\rho(T)$ would be larger, leaving less time until the Universe has cooled down to $T_{\rm D}$ – thus, fewer neutrons would decay and the resulting helium abundance would be higher. Even before 1990, it was concluded from BBN (with relatively large uncertainties, however) that $N_\nu = 3$. In 1990, the value of $N_\nu = 3$ was then confirmed in the laboratory from Z-boson decay.

To circumvent the conclusion of a dominant fraction of non-baryonic matter, inhomogeneous models of BBN have been investigated, but these also yield values for $\Omega_{\rm b}$ which are too low and therefore do not provide a viable alternative.

4.4.5 Recombination

About 3 min after the Big Bang, BBN comes to an end. At this time, the Universe has a temperature of roughly $T \sim 8 \times 10^8$ K and consists of photons, protons, helium nuclei, traces of other light elements, and electrons. In addition, there are neutrinos that dominate, together with photons, the energy density and thus also the expansion rate, and there are (probably) WIMPs. Except for the neutrinos and the WIMPs, all particle species have the same temperature, which is established by interactions of charged particles with the photons, which resemble some kind of heat bath.

At $z = z_{\rm eq} \approx 23\,900\, \Omega_{\rm m}\, h^2$, pressureless matter (i.e., the so-called dust) begins to dominate the energy density in the Universe and thus the expansion rate. The second term in (4.31) then becomes largest, i.e., $H^2 \approx H_0^2\, \Omega_{\rm m}/a^3$. If a power-law ansatz for the scale factor, $a \propto t^\beta$, is inserted into the expansion equation, we find that $\beta = 2/3$, and hence

$$\boxed{a(t) = \left(\frac{3}{2}\sqrt{\Omega_{\rm m}}\, H_0\, t\right)^{2/3} \quad \text{for} \quad a_{\rm eq} \ll a \ll 1}\ . \tag{4.64}$$

This describes the expansion behavior until either the curvature term or, if this is zero, the Λ-term starts to dominate.

After further cooling, the free electrons can combine with the nuclei and form neutral atoms. This process is called *recombination*, although this expression is misleading: since the Universe was fully ionized until then,

[6]The Z-boson is one of the exchange particles in weak interactions; the other two are the charged $W^\pm$ particles. They were predicted by the model of electroweak interactions, and finally being found in particle accelerators, thus beautifully supporting the validity of this electroweak unification model which lies at the heart of the standard model of particle physics. The Z-boson plays about the same role in weak interactions as the photon does in electromagnetic interactions.

it is not a *re*combination but rather the (first) transition to a neutral state – however the expression recombination has now long been established. The recombination of electrons and nuclei is in competition with the ionization of neutral atoms by energetic photons (photoionization), whereas collisional ionization can be disregarded completely since η – (4.59) – is so small. Because photons are so much more numerous than electrons, cooling has to proceed to well below the ionization temperature, corresponding to the binding energy of an electron in hydrogen, before neutral atoms become abundant. This happens for the same reasons as apply in the context of deuterium formation: there are plenty of ionizing photons in the Wien tail of the Planck distribution, even if the temperature is well below the ionization temperature. The ionization energy of hydrogen is $\chi = 13.6\,\mathrm{eV}$, corresponding to a temperature of $T > 10^5\,\mathrm{K}$, but T has to first decrease to $\sim 3000\,\mathrm{K}$ before the ionization fraction

$$x = \frac{\text{number density of free electrons}}{\text{total number density of existing protons}} \tag{4.65}$$

falls considerably below 1, for the reason mentioned above. At temperatures $T > 10^4\,\mathrm{K}$ we have $x \approx 1$, i.e., virtually all electrons are free. Only at $z \sim 1300$ does x deviate significantly from unity.

The onset of recombination can be described by an equilibrium consideration which leads to the so-called Saha equation,

$$\frac{1-x}{x^2} \approx 3.84\,\eta \left(\frac{k_\mathrm{B}T}{m_\mathrm{e}c^2}\right)^{3/2} \exp\left(\frac{\chi}{k_\mathrm{B}T}\right),$$

which describes the ionization fraction x as a function of temperature. However, once recombination occurs, the assumption of thermodynamical equilibrium is no longer justified. This can be seen from the following consideration.

Any recombination directly to the ground state leads to the emission of a photon with energy $E_\gamma > \chi$. However, these photons can ionize other, already recombined (thus neutral), atoms. Because of the large cross-section for photoionization, this happens very efficiently. Thus for each recombination to the ground state, one neutral atom will become ionized, yielding a vanishing net effect. But recombination can also happen in steps, first into an excited state and then evolving into the ground state by radiative transitions. Each of these recombinations will yield a Lyα photon in the transition from the first excited state into the ground state. This Lyα photon will then immediately excite another atom from the ground state into the first excited state, which has an ionization energy of only $\chi/4$. This yields no net production of atoms in the ground state. Since the density of photons with $E_\gamma > \chi/4$ is very much larger than of those of $E_\gamma > \chi$, the excited atoms are more easily ionized, and this indeed happens. Stepwise recombination thus also provides no route towards a lower ionization fraction.

The processes described above cause a small distortion of the Planck spectrum due to recombination radiation (in the range $\chi \gg k_\mathrm{B}T$) which affects recombination. One cannot get rid of these energetic photons – in contrast to gas nebulae like HII regions, in which the Lyα photons may escape due to the finite geometry.

Ultimately, recombination takes place by means of a very rare process, the two-photon decay of the first excited level. This process is less probable than the direct Lyα transition by a factor of $\sim 10^8$. However, it leads to the emission of two photons, both of which are not sufficiently energetic to excite an atom from the ground state. This 2γ-transition is therefore a net sink for energetic photons.[7] Taking into account all relevant processes and using a rate equation, which describes the evolution of the distribution of particles and photons even in the absence of thermodynamic equilibrium, gives for the ionization fraction in the relevant redshift range $800 \lesssim z \lesssim 1200$

$$\boxed{x(z) = 2.4 \times 10^{-3} \frac{\sqrt{\Omega_\mathrm{m} h^2}}{\Omega_\mathrm{b} h^2} \left(\frac{z}{1000}\right)^{12.75}}\,. \tag{4.66}$$

The ionization fraction is thus a very strong function of redshift since x changes from 1 (complete ionization)

[7] The recombination of hydrogen – and also that of helium which occurred at higher redshifts – perturbed the exact Planck shape of the photon distribution, adding to it the Lyman-alpha photons and the photon pairs from the two-photon transition. This slight perturbation in the CMB spectrum should in principle still be present today. Unfortunately, it lies in a wavelength range ($\sim 200\,\mu\mathrm{m}$) where the dust emission from the Galaxy is very strong; in addition, the wavelength range coincides with the peak of the far-infrared background radiation (see Sect. 9.3.1). Therefore, the detection of this spectral distortion will be extremely difficult.

to $x \sim 10^{-4}$ (where essentially all atoms are neutral) within a relatively small redshift range. The recombination process is not complete, however. A small ionization fraction of $x \sim 10^{-4}$ remains since the recombination rate for small x becomes smaller than the expansion rate – some nuclei do not find an electron fast enough before the density of the Universe becomes too low. From (4.66), the optical depth for Thomson scattering (scattering of photons by free electrons) can be computed,

$$\tau(z) = 0.37 \left(\frac{z}{1000}\right)^{14.25} , \tag{4.67}$$

which is virtually independent of cosmological parameters. Equation (4.67) implies that photons can propagate from $z \sim 1000$ (the *last-scattering surface*) until the present day essentially without any interaction with matter – provided the wavelength is larger than 1216 Å. For photons of smaller wavelength, the absorption cross-section of neutral atoms is large. Disregarding these highly energetic photons here – their energies are $\gtrsim 10\,\mathrm{eV}$, compared to $T_{\rm rec} \sim 0.3\,\mathrm{eV}$, so they are far out in the Wien tail of the Planck distribution – we conclude that the photons present in the early Universe have been able to propagate without further interactions until the present epoch. Before recombination they followed a Planck spectrum. As was discussed in Sect. 4.3.2, the distribution will remain a Planck spectrum with only its temperature changing. Thus these photons from the early Universe should still be observable today, redshifted into the microwave regime of the electromagnetic spectrum.

Our consideration of the early Universe predicts thermal radiation from the Big Bang, as was first realized by George Gamow in 1946 – the cosmic microwave background. The CMB is therefore a visible relic of the Big Bang.

The CMB was detected in 1965 by Arno Penzias & Robert Wilson (see Fig. 4.15), who were awarded the 1978 Nobel prize in physics for this very important discovery. At the beginning of the 1990s, the COBE satellite measured the spectrum of the CMB with a very high precision – it is the most perfect blackbody ever measured (see Fig. 4.3). From upper limits of deviations from the Planck spectrum, very tight limits for possible later energy injections into the photon gas, and thus on energetic processes in the Universe, can be obtained.[8]

A MEASUREMENT OF EXCESS ANTENNA TEMPERATURE
AT 4080 Mc/s

Measurements of the effective zenith noise temperature of the 20-foot horn-reflector antenna (Crawford, Hogg, and Hunt 1961) at the Crawford Hill Laboratory, Holmdel, New Jersey, at 4080 Mc/s have yielded a value about 3.5° K higher than expected. This excess temperature is, within the limits of our observations, isotropic, unpolarized, and

Fig. 4.15. The first lines of the article by Penzias & Wilson, 1965, ApJ, **142**, 419

We have only discussed the recombination of hydrogen. Since helium has a higher ionization energy it recombines earlier than hydrogen. Although recombination defines a rather sharp transition, (4.67) tells us that we receive photons from a recombination layer of finite thickness ($\Delta z \sim 60$). This aspect will be of importance later.

The gas in the intergalactic medium at lower redshift is highly ionized. If this were not the case we would not be able to observe any UV photons from sources at high redshift ("Gunn–Peterson test", see Sect. 8.5.1). Sources with redshifts $z > 6$ have been observed, and we also observe photons with wavelengths shorter than the Lyα line of these objects. Thus at least at the epoch corresponding to redshift $z \sim 6$, the Universe must have been nearly fully ionized or else these photons would have been absorbed by photoionization of neutral hydrogen. This means that at some time between $z \sim 1000$ and $z \sim 6$, a reionization of the intergalactic medium must have occurred, presumably by a first generation of stars or by the first AGNs. The results from the new CMB satellite WMAP suggest a reionization at redshift $z \sim 15$; this will be discussed more thoroughly in Sect. 8.7.

[8] For instance, there exists an X-ray background (XRB) which is radiation that appeared isotropic in early measurements. For a long time, a possible explanation for this was suggested to be a hot intergalactic medium with temperature of $k_{\rm B}T \sim 40\,\mathrm{keV}$ emitting bremsstrahlung radiation. But such a hot intergalactic gas would modify the spectrum of the CMB via the scattering of CMB photons to higher frequencies by energetic electrons (inverse Compton scattering). This explanation for the source of the XRB was excluded by the COBE measurements. From observations by the X-ray satellites ROSAT, Chandra, and XMM-Newton, with their high angular resolution, we know today that the XRB is a superposition of radiation from discrete sources, mostly AGNs.

4.4.6 Summary

We will summarize this somewhat long section as follows:

- Our Universe originated from a very dense, very hot state, the so-called *Big Bang*. Shortly afterwards, it consisted of a mix of various elementary particles, all interacting with each other.
- We are able to examine the history of the Universe in detail, starting at an early epoch where it cooled down by expansion such as to leave only those particle species known to us (electrons, protons, neutrons, neutrinos, and photons).
- Because of their weak interaction and the decreasing density, the neutrinos experience only little interaction at temperatures below $\sim 10^{10}$ K, the decoupling temperature.
- At $T \sim 5 \times 10^9$ K, electrons and positrons annihilate into photons. At this low temperature, pair production ceases to take place.
- Protons and neutrons interact and form deuterium nuclei. As soon as $T \sim 10^9$ K, deuterium is no longer efficiently destroyed by energetic photons. Further nuclear reactions produce mainly helium nuclei. About 25% of the mass in nucleons is transformed into helium, and traces of lithium are produced, but no heavier elements.
- At about $T \sim 3000$ K, some 400 000 years after the Big Bang, the protons and helium nuclei combine with the electrons, and the Universe becomes essentially neutral (we say that it "recombines"). From then on, photons can travel without further interactions. At recombination, the photons follow a blackbody distribution (i.e., a thermal spectrum, or a Planck distribution). By the ongoing cosmic expansion, the temperature of the spectral distribution decreases, $T \propto (1+z)^{-1}$, though its Planck property remains.
- After recombination, the matter in the Universe is almost completely neutral. However, we know from the observation of sources at very high redshift that the intergalactic medium is essentially fully ionized at $z \lesssim 6$. Before $z > 6$, the Universe must therefore have experienced a phase of reionization. This effect cannot be explained in the context of the *strictly homogeneous* world models; rather it must be examined in the context of structure formation in the Universe and the formation of the first stars and AGNs. These aspects will be discussed in Sect. 9.4.

4.5 Achievements and Problems of the Standard Model

To conclude this chapter, we will evaluate the cosmological model which has been presented. We will review its achievements and successes, but also apparent problems, and point out the route by which those might be understood. As is always the case in natural sciences, problems with an otherwise very successful model are often the key to a new and deeper understanding.

4.5.1 Achievements

The standard model of the Friedmann–Lemaître Universe described above has been extremely successful in numerous ways:

- It predicts that gas which has not been subject to much chemical processing (i.e., metal-poor gas) should have a helium content of $\sim 25\%$. This is in extraordinarily good agreement with observations.
- It predicts that sources of lower redshift are closer to us than sources of higher redshift.[9] Therefore, modulo any peculiar velocities, the absorption of radiation from sources at high redshift must happen at smaller redshifts. Not a single counter-example has been found yet.
- It predicts the existence of a microwave background, which has indeed been found.
- It predicts the correct number of neutrino families, which was confirmed in laboratory experiments of the decay of the Z-boson.

Further achievements will be discussed in the context of structure evolution in the Universe.

A good physical model is one that can also be falsified. In this respect, the Friedmann–Lemaître Universe is also an excellent model: a single observation could ei-

[9] We ignore peculiar motions here which may cause an additional (Doppler-)redshift. These are typically $\lesssim 1000\,\mathrm{km/s}$ and are thus small compared to cosmological redshifts.

ther cause a lot of trouble for this model or even disprove it. To wit, it would be incompatible with the model

1. if the helium content of a gas cloud or of a low-metallicity star were significantly below 25%;
2. if it were found that one of the neutrinos has a rest mass $\gtrsim 100$ eV;
3. if the Wien part of the CMB had a smaller amplitude compared to the Planck spectrum;
4. if a source with emission lines at z_e were found to show absorption lines at $z_a \gg z_e$;
5. if the cosmological parameters were such that $t_0 \lesssim$ 10 Gyr.

On (1): While the helium content may increase by stellar evolution due to fusion of hydrogen into helium, only a small fraction of helium is burned in stars. In this process, heavier elements are of course produced. A gas cloud or a star with low metallicity therefore cannot consist of material in which helium has been destroyed; it must contain at least the helium abundance from BBN. On (2): Such a neutrino would lead to $\Omega_m > 2$, which is in strict contradiction to the derived model parameters. On (3): Though it is possible to generate additional photons by energetic processes in the Universe, thereby increasing the Wien part of the coadded spectrum compared to that of a Planck function, it is thermodynamically impossible to extract photons from the Wien part. On (4): Such an observation would question the role of redshift as a monotonic measure of relative distances and thus remove one of the pillars of the model. On (5): Our knowledge of stellar evolution allows us to determine the age of the oldest stars with a precision of better than $\sim 20\%$. An age of the Universe below ~ 10 Gyr would be incompatible with the age of the globular clusters – naturally, these have to be younger than the age of the Universe, i.e., the time after the Big Bang.

Although these predictions have been known for more than 30 years, no observation has yet been made which disproves the standard model. Indeed, at any given time there have been astronomers who disagree with the standard model. These astronomers have tried to make a discovery, like the examples above, which would pose great difficulties for the model. So far, they have not succeeded; this does not mean that such results cannot be found in the literature, but rather such results did not withstand closer examination. The simple opportunities to falsify the model and the lack of any corresponding observation, together with the achievements listed above have made the Friedmann–Lemaître model the standard model of cosmology. Alternative models have either been excluded by observation (such as steady-state cosmology) or have been unable to make any predictions. Currently, there is no serious alternative to the standard model.

4.5.2 Problems of the Standard Model

Despite these achievements, there are some aspects of the model which require further consideration. Here we will describe two conceptual problems with the standard model more thoroughly – the horizon problem and the flatness problem.

Horizons. The finite speed of light implies that we are only able to observe a finite part of the Universe, namely those regions from which light can reach us within a time t_0. Since $t_0 \approx 13.5$ Gyr, our visible Universe has – roughly speaking – a radius of 13.5 billion light years. More distant parts of the Universe are at the present time unobservable for us. This means that there exists a horizon beyond which we cannot see. Such horizons do not only exist for us today: at an earlier time t, the size of the horizon was about ct, hence smaller than today. We will now describe this aspect quantitatively.

In a time interval $\mathrm{d}t$, light travels a distance $c\,\mathrm{d}t$, which corresponds to a comoving distance interval $\mathrm{d}x = c\,\mathrm{d}t/a$ at scale factor a. From the Big Bang to a time t (or redshift z) the light traverses a comoving distance of

$$r_{\mathrm{H,com}}(z) = \int_0^t \frac{c\,\mathrm{d}t}{a(t)} .$$

From $\dot{a} = \mathrm{d}a/\mathrm{d}t$ we get $\mathrm{d}t = \mathrm{d}a/\dot{a} = \mathrm{d}a/(aH)$, so that

$$\boxed{r_{\mathrm{H,com}}(z) = \int_0^{(1+z)^{-1}} \frac{c\,\mathrm{d}a}{a^2\,H(a)}} . \tag{4.68}$$

If $z_{\mathrm{eq}} \gg z \gg 0$, the main contribution to the integral comes from times (or values of a) in which the so-called dust dominates the expansion rate H. Then with

(4.31) we find $H(a) \approx H_0 \sqrt{\Omega_m} a^{-3/2}$, and (4.68) yields

$$r_{H,com}(z) \approx 2 \frac{c}{H_0} \frac{1}{\sqrt{(1+z)\Omega_m}} \quad \text{for} \quad z_{eq} \gg z \gg 0 \; . \tag{4.69}$$

In earlier phases, $z \gg z_{eq}$, H is radiation-dominated, $H(a) \approx H_0 \sqrt{\Omega_r}/a^2$, and (4.68) becomes

$$r_{H,com}(z) \approx \frac{c}{H_0 \sqrt{\Omega_r}} \frac{1}{(1+z)} \quad \text{for} \quad z \gg z_{eq} \; . \tag{4.70}$$

The earlier the cosmic epoch, the smaller the comoving horizon length, as was to be expected. In particular, we will now consider the recombination epoch, $z_{rec} \sim 1000$, for which (4.69) applies (see Fig. 4.16). The comoving length $r_{H,com}$ corresponds to a physical proper length $r_{H,prop} = a\, r_{H,com}$, and thus

$$r_{H,prop}(z_{rec}) = 2 \frac{c}{H_0} \Omega_m^{-1/2} (1+z_{rec})^{-3/2} \tag{4.71}$$

is the horizon length at recombination. We can then calculate the angular size on the sky that this length corresponds to,

$$\theta_{H,rec} = \frac{r_{H,prop}(z_{rec})}{D_A(z_{rec})} \; ,$$

where D_A is the angular-diameter distance (4.45) to the last scattering surface of the CMB. Using (4.47), we find that in the case of $\Omega_\Lambda = 0$

$$D_A(z) \approx \frac{c}{H_0} \frac{2}{\Omega_m z} \quad \text{for} \quad z \gg 1 \; ,$$

and hence

$$\theta_{H,rec} \approx \sqrt{\frac{\Omega_m}{z_{rec}}} \sim \frac{\sqrt{\Omega_m}}{30} \sim \sqrt{\Omega_m}\, 2^\circ \quad \text{for } \Omega_\Lambda = 0 \; . \tag{4.72}$$

This means that the horizon length at recombination subtends an angle of about one degree on the sky.

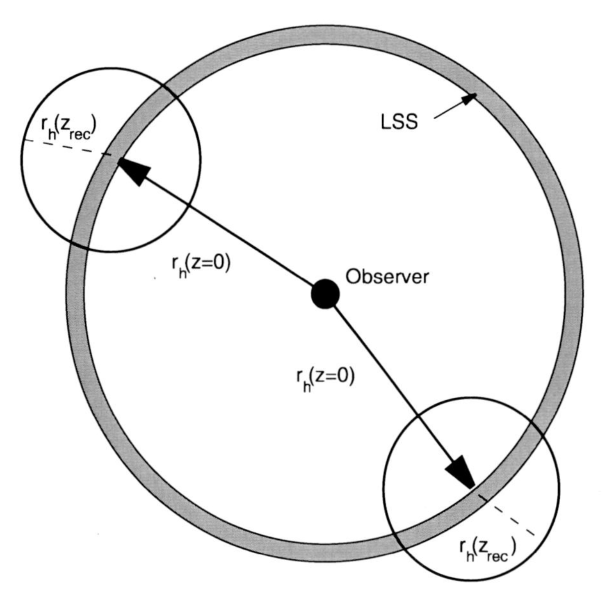

Fig. 4.16. The horizon problem: the region of space which was in causal contact before recombination has a much smaller radius than the spatial separation between two regions from which we receive the CMB photons. Thus the question arises how these two regions may "know" of each other's temperature

The Horizon Problem: Since no signal can travel faster than light, (4.72) means that CMB radiation from two directions separated by more than about one degree originates in regions that were not in causal contact before recombination, i.e., the time when the CMB photons interacted with matter the last time. Therefore, these two regions have never been able to exchange information, for example about their temperature. Nevertheless their temperature is the same, as seen from the high degree of isotropy of the CMB, which shows relative fluctuations of only $\Delta T/T \sim 10^{-5}$!

Redshift-Dependent Density Parameter. We have defined the density parameters Ω_m and Ω_Λ as the current density divided by the critical mass density ρ_{cr} today. These definitions can be generalized. If we existed at a different time, the densities and the Hubble constant would have had different values and consequently we would obtain different values for the density parameters. Thus we define the total density parameter for an arbitrary redshift

$$\Omega_0(z) = \frac{\rho_{\rm m}(z) + \rho_{\rm r}(z) + \rho_{\rm v}}{\rho_{\rm cr}(z)} , \qquad (4.73)$$

where the critical density $\rho_{\rm cr}$ is also a function of redshift,

$$\rho_{\rm cr}(z) = \frac{3H^2(z)}{8\pi G} . \qquad (4.74)$$

Then by inserting (4.24) into (4.73), we find

$$\Omega_0(z) = \left(\frac{H_0}{H}\right)^2 \left(\frac{\Omega_{\rm m}}{a^3} + \frac{\Omega_{\rm r}}{a^4} + \Omega_\Lambda\right) .$$

Using (4.31), this yields

$$1 - \Omega_0(z) = F\,[1 - \Omega_0(0)] , \qquad (4.75)$$

where $\Omega_0(0)$ is the total density parameter today, and

$$F = \left(\frac{H_0}{a\,H(a)}\right)^2 . \qquad (4.76)$$

From (4.75) we can now draw two important conclusions. Since $F > 0$ for all a, the sign of $\Omega_0 - 1$ is preserved and thus is the same at all times as today. Since the sign of $\Omega_0 - 1$ is the same as that of the curvature – see (4.30) – the sign of the curvature is preserved in cosmic evolution: a flat Universe will be flat at all times, a closed Universe with $K > 0$ will always have a positive curvature.

The second conclusion follows from the analysis of the function F at early cosmic epochs, e.g., at $z \gg z_{\rm eq}$, thus in the radiation-dominated Universe. Back then, with (4.31), we have

$$F = \frac{1}{\Omega_{\rm r}(1+z)^2} ,$$

so that for very early times, F becomes very small. For instance, at $z \sim 10^{10}$, the epoch of the neutrino freeze-out, $F \sim 10^{-15}$. Today, Ω_0 is of order unity but not necessarily exactly 1. From observations, we know that $0.1 \lesssim \Omega_0(0) \lesssim 2$, where this is a very conservative estimate (from the more recent CMB measurements we are able to constrain this interval to [0.97, 1.04]), so that $|1 - \Omega_0(0)| \lesssim 1$. Since F is so small at large redshifts, this means that $\Omega_0(z)$ must have been very, very close to 1; for example at $z \sim 10^{10}$ it is required that $|\Omega_0 - 1| \lesssim 10^{-15}$.

The Flatness Problem: For the total density parameter to be of order unity today, it must have been extremely close to 1 at earlier times, which means that a very precise "fine tuning" of this parameter was necessary.

This aspect can be illustrated very well by another physical example. If we throw an object up into the air, it takes several seconds until it falls back to the ground. The higher the initial velocity, the longer it takes to hit the ground. To increase the time of flight we need to increase the initial velocity, for instance by using a cannon. In this way, the time of flight may be extended to up to about a minute. Assume that we want the object to be back only after one day; in this case we must use a rocket. But we know that if the initial velocity of a rocket exceeds the escape velocity $v_{\rm esc} \sim 11.2$ km/s, it will leave the gravitational field of the Earth and never fall back. On the other hand, if the initial velocity is too much below $v_{\rm esc}$, the object will be back in significantly less than a day. So the initial velocity must be *very* well chosen for the object to return after being up for at least a day.

The flatness problem is completely analogous to this. If Ω_0 had not been so extremely close to 1 at $z \sim 10^{10}$, the Universe would have recollapsed long ago, or it would have expanded significantly more than the Universe we live in. In either case, the consequences for the evolution of life in the Universe would have been catastrophic. In the first case, the total lifetime of the Universe would have been much shorter than is needed for the formation of the first stars and the first planetary systems, so that in such a world no life could be formed. In the second case, extreme expansion would have prevented the formation of structure in the Universe. In such a Universe no life could have evolved either.

This consideration can be interpreted as follows: we live in a Universe which had, at a very early time, a very precisely tuned density parameter, because only in such a Universe can life evolve and astronomers exist to examine the flatness of the Universe. In all other conceivable universes this would not be possible. This approach is meaningful only if a large number of universes existed – in this case we should not be too surprised about living in one of those where this initial fine-tuning took place – in the other ones, we, and the question about the cosmological parameters, would just not exist. This approach is called the *anthropic principle*. It may either be seen as an "explanation" for the flatness of *our* Universe, or as a capitulation – where we give up attempting to solve the question of the origin of the flatness of the Universe.

The example of the rocket given above is helpful in understanding another aspect of cosmic expansion. If the rocket is supposed to have a long time of flight but not escape the gravitational field of the Earth, its initial velocity must be very, very close to, but a tiny little bit smaller than $v_{\rm esc}$. In other words, the absolute value of the sum of kinetic and potential energy has to be very much smaller than either of these two components. This is also true for a large part of the initial trajectory. Independent of the exact value of the time of flight, the initial trajectory can be approximated by the limiting case $v_0 = v_{\rm esc}$ at which the total energy is exactly zero. Transferred to the Hubble expansion, this reads as follows: independent of the exact values of the cosmological parameters, the curvature term can be disregarded in the early phases of expansion (as we have already seen above). This is because our Universe can reach its current age only if at early times the modulus of potential and kinetic energy were nearly exactly equal, i.e., the curvature term in (4.14) must have been a lot smaller than the other two terms.

4.5.3 Extension of the Standard Model: Inflation

We will consider the horizon and flatness problems from a different, more technical point of view. Einstein's field equations of GR, one solution of which has been described as our world model, are a system of coupled partial differential equations. As is always the case for differential equations, their solutions are determined by (1) the system of equations itself and (2) the initial conditions. If the initial conditions at e.g., $t = 1$ s were as they have been described, the two aforementioned problems would not exist. But why are the conditions at $t = 1$ s such that they allow a homogeneous, isotropic, nearly flat model? The set of homogeneous and isotropic solutions to the Einstein equation is of measure zero (i.e., nearly all solutions of the Einstein equation are not homogeneous and isotropic); thus these particular solutions are *very* special. Taking the line of reasoning that the initial conditions "just happened to be so" is not satisfying because it does not explain anything. Besides the anthropic principle, the answer to this question can only be that processes must have taken place even earlier, due to known or as yet unknown physics, which have produced these "initial conditions" at $t = 1$ s. The initial conditions of the normal Friedmann–Lemaître expansion thus have a physical origin. Cosmologists believe they have found such a physical reason: the inflationary model.

Inflation. In the early 1980s, a model was developed which was able to solve the flatness and horizon problems (and some others as well). As a motivation for this model, we first recall that the physical laws and properties of elementary particles are well known up to energies of ~ 100 GeV because they were experimentally tested in particle accelerators. For higher energies, particles and their interactions are unknown. This means that the history of the Universe, as sketched above, can be considered secure only up to energies of 100 GeV. The extrapolation to earlier times, up to the Big Bang, is considerably less certain. From particle physics we expect new phenomena to occur at an energy scale of the Grand Unified Theories (GUTs), at about 10^{14} GeV, corresponding to $t \sim 10^{-34}$ s.

In the inflationary scenario it is presumed that at very early times the vacuum energy density was much higher than today, so that Ω_Λ dominated the Hubble expansion. Then from (4.18) we find that $\dot a/a \approx \sqrt{\Lambda/3}$. This implies an exponential expansion of the Universe,

$$a(t) = C \exp\left(\sqrt{\frac{\Lambda}{3}}\,t\right) . \tag{4.77}$$

Obviously, this exponential expansion (or inflationary phase) cannot last forever. We assume that a phase transition took place in which the vacuum energy density is transformed into normal matter and radiation (a process called reheating), which ends the exponential expansion and after which the normal Friedmann evolution of the Universe begins. Figure 4.17 sketches the expansion history of the Universe in an inflationary model.

Inflation Solves the Horizon Problem. During inflation, $H(a) = \sqrt{\Lambda/3}$ is constant so that the integral (4.68) for the comoving horizon length formally diverges. This implies that the horizon may become arbitrarily large in the inflationary phase, depending on the duration of

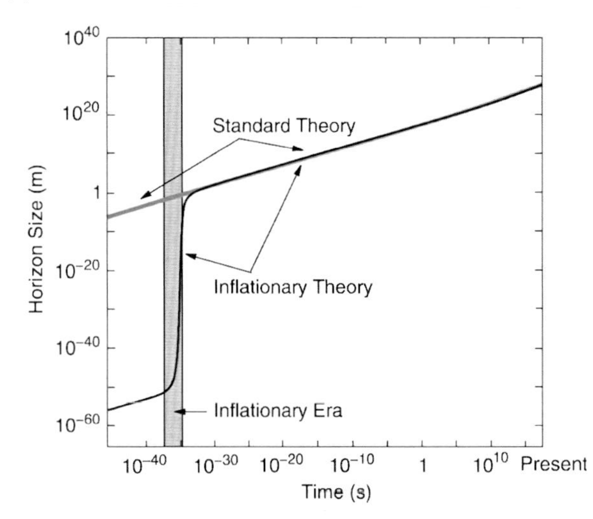

Fig. 4.17. During an inflationary phase, indicated here by the gray bar, the Universe expands exponentially; see (4.77). This phase comes to an end when a phase transition transforms the vacuum energy into matter and radiation, after which the Universe follows the normal Friedmann expansion

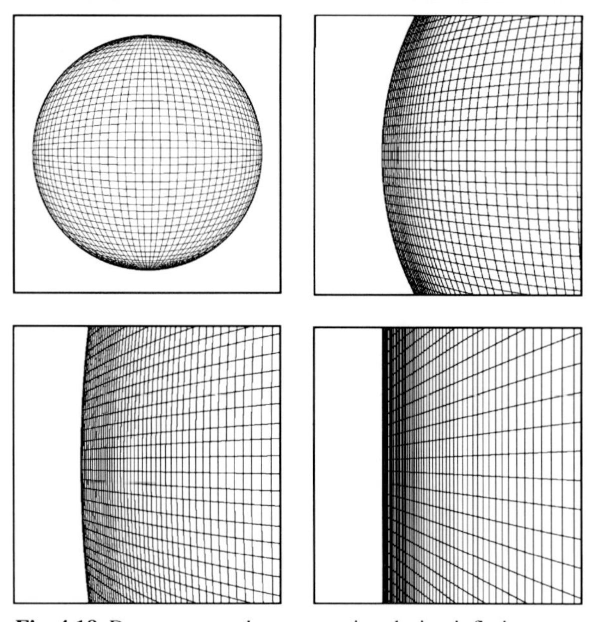

Fig. 4.18. Due to tremendous expansion during inflation, even a Universe with initial curvature will appear to be a flat Universe by the end of the inflationary phase

the exponential expansion. For illustration we consider a very small region in space of size $L < ct_{\rm i}$ at a time $t_{\rm i} \sim 10^{-34}$ s prior to inflation which is in causal contact. Through inflation, it expands tremendously, e.g., by a factor $\sim 10^{40}$; the original $L \sim 10^{-24}$ cm inflate to about 10^{16} cm by the end of the inflationary phase, at $t_{\rm f} \sim 10^{-32}$ s. By today, this spatial region will have expanded by another factor of $\sim 10^{25}$ by following (for $t > t_{\rm f}$) the normal cosmic expansion, to $\sim 10^{41}$ cm. This scale is considerably larger than the size of the currently visible Universe, c/H_0. According to this scenario, the whole Universe visible today was in causal contact prior to inflation, so that the homogeneity of the physical conditions at recombination, and with it the nearly perfect isotropy of the CMB, is provided by causal processes.

Inflation Solves the Flatness Problem as well. Due to the tremendous expansion, any initial curvature is straightened out (see Fig. 4.18). Formally this can be seen as follows: during the inflationary phase we have

$$\Omega_\Lambda = \frac{\Lambda}{3H^2} = 1 ,$$

and since it is assumed that the inflationary phase lasts long enough for the vacuum energy to be completely dominant, when it ends we then have $\Omega_0 = 1$. Hence the Universe is flat to an extremely good approximation.

> The inflationary model of the very early Universe predicts that today $\Omega_0 = 1$ is valid to very high precision; any other value of Ω_0 would require another fine-tuning. Thus the Universe is flat.

The physical details of the inflationary scenario are not very well known. In particular it is not yet understood how the phase transition at the end of the inflationary phase took place and why it did not occur earlier. But the two achievements presented above (and some others) make an inflationary phase appear a very plausible scenario. As we will see below (Chap. 8), the prediction of a flat Universe was recently accurately tested and it was indeed confirmed. Furthermore, the inflationary model provides a natural explanation for the origin of density fluctuations in the Universe which must have been present at very early epochs as the seeds of structure formation. We will discuss these aspects further in Chap. 7.

5. Active Galactic Nuclei

The light of normal galaxies in the optical and near infrared part of the spectrum is dominated by stars, with small contributions by gas and dust. This is thermal radiation since the emitting plasma in stellar atmospheres is basically in thermodynamical equilibrium. To a first approximation, the spectral properties of a star can be described by a Planck spectrum whose temperature depends on the stellar mass and the evolutionary state of the star. As we have seen in Sect. 3.9, the spectrum of galaxies can be described quite well as a superposition of stellar spectra. The temperature of stars varies over a relatively narrow range. Only few stars are found with $T \gtrsim 40\,000$ K, and those with $T \lesssim 3000$ K hardly contribute to the spectrum of a galaxy, due to their low luminosity. Therefore, as a rough approximation, the light distribution of a galaxy can be described by a superposition of Planck spectra from a temperature range that covers about one decade. Since the Planck spectrum has a very narrow energy distribution around its maximum at $h_{\rm P}\nu \sim 3k_{\rm B}T$, the spectrum of a galaxy is basically confined to a range between ~ 4000 Å and $\sim 20\,000$ Å. If the galaxy is actively forming stars, young hot stars extend this frequency range to higher frequency, and the thermal radiation from dust, heated by these new-born stars, extends the emission to the far-infrared.

However, there are galaxies which have a much broader energy distribution. Some of these show significant emission in the full range from radio wavelengths to the X-ray and even Gamma range (see Fig. 3.3). This emission originates mainly from a very small central region of such an *active galaxy* which is called the *active galactic nucleus* (*AGN*). Active galaxies form a family of many different types of AGN which differ in their spectral properties, their luminosities and their ratio of nuclear luminosity to that of the stellar light. The optical spectra of three AGNs are presented in Fig. 5.1.

Some classes of AGNs, in particular the quasars, belong to the most luminous sources in the Universe, and they have been observed out to the highest measured redshifts ($z \sim 6$). The luminosity of quasars can exceed the luminosity of normal galaxies by a factor of a thousand. This luminosity originates from a very small region in space, $r \leq 1$ pc. The optical/UV spectra

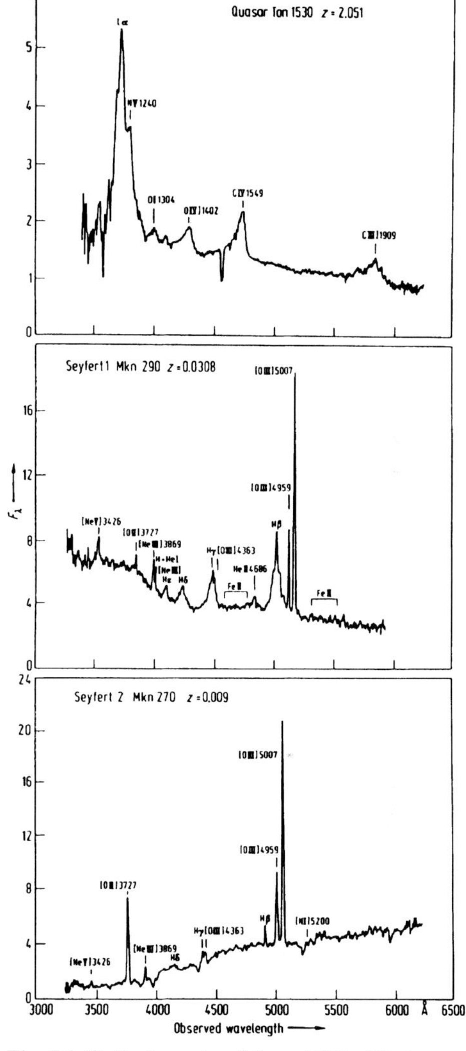

Fig. 5.1. Optical spectra of three AGNs. The top panel displays the spectrum of a quasar at redshift $z \sim 2$, which shows the characteristic broad emission lines. The strongest are Lyα of hydrogen, and the CIV-line and CIII]-line of triple and double ionized carbon, respectively (where the squared bracket means that this is a semi-forbidden transition, as will be explained in Sect. 5.4.2). The middle panel shows the spectrum of a nearby Seyfert galaxy of Type 1. Here both very broad emission lines and narrow lines, in particular of double ionized oxygen, are visible. In contrast, the spectrum in the bottom panel, of a Seyfert galaxy of Type 2, shows only relatively narrow emission lines

Peter Schneider, Active Galactic Nuclei.
In: Peter Schneider, Extragalactic Astronomy and Cosmology. pp. 175–222 (2006)
DOI: 10.1007/11614371_5

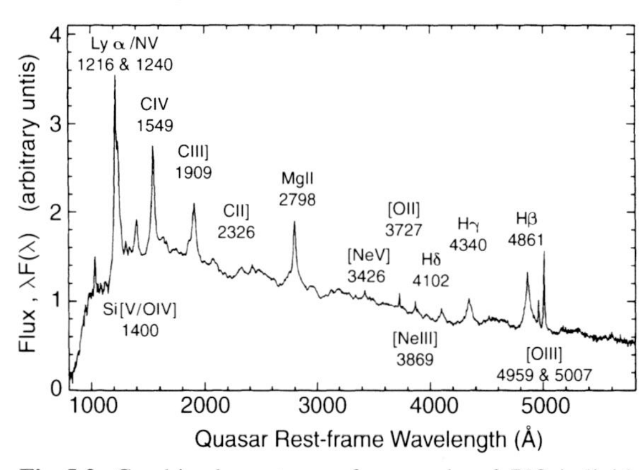

Fig. 5.2. Combined spectrum of a sample of 718 individual QSOs, taken from the Large Bright Quasar Survey. This "mean" spectrum has a considerably better signal-to-noise ratio and a larger wavelength coverage than individual spectra. It was combined from the individual quasar spectra by transforming their wavelengths into the sources' rest-frames. The most prominent lines are marked

of quasars are dominated by numerous strong and very broad emission lines, some of them emitted by highly ionized atoms (see Figs. 5.2 and 5.3). The processes in AGNs are among the most energetic in astrophysics. The enormous bandwidth of AGN spectra suggests that the radiation is nonthermal. As we will discuss later, AGNs host processes which produce highly energetic particles and which are the origin of the nonthermal radiation.

After an introduction in which we will briefly present the history of the discovery of AGNs and their basic properties, in Sect. 5.2 we will describe the most important subgroups of the AGN family. In Sect. 5.3, we will discuss several arguments which lead to the conclusion that the energy source of an AGN originates in accretion of matter onto a supermassive black hole (SMBH). In particular, we will learn about the phenomenon of superluminal motion, where apparent velocities of source components are larger than the speed of light. We will then consider the different components of an AGN where radiation in different wavelength regions is produced.

Of particular importance for understanding the phenomenon of active galaxies are the unified models of AGNs that will be discussed next. We will see that the seemingly quite different appearances of AGNs can all be explained by geometric or projection effects. Finally, we will consider AGNs as cosmological probes. Due to their enormous luminosity they are observable up to very high redshifts. These observations allow us to draw conclusions about the properties of the early Universe.

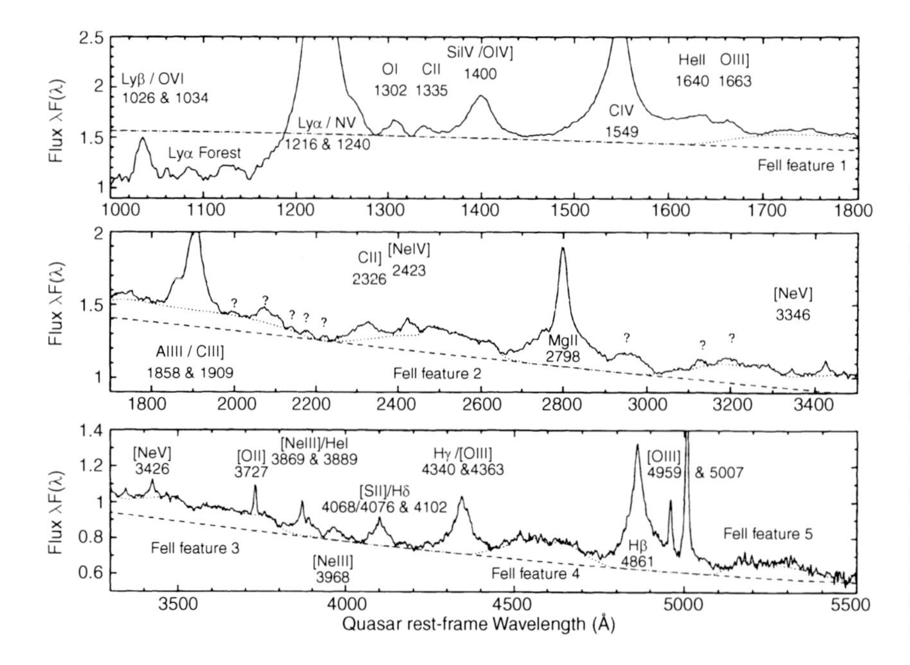

Fig. 5.3. An enlargement of the composite QSO spectrum shown in Fig. 5.2. Here, weaker lines are also visible. Also clearly visible is the break in the spectral flux bluewards of the Lyα line which is caused by the Lyα forest (Sect. 5.6.3), absorption by intergalactic hydrogen along the line-of-sight. The dashed line indicates the average continuum, whereas the dotted line marks line complexes of singly-ionized iron that has such a high line density that they blend into a quasi-continuum at the spectral resolution shown here

5.1 Introduction

5.1.1 Brief History of AGNs

As long ago as 1908, strong and broad emission lines were discovered in the galaxy NGC 1068. However, only the systematic analysis by Carl Seyfert in 1943 drew the focus of astronomers to this new class of galaxies. The cores of these *Seyfert galaxies* have an extremely high surface brightness, as demonstrated in Fig. 5.4, and the spectrum of their central region is dominated by emission lines of very high excitation. Some of these lines are extremely broad (see Fig. 5.1). The line width, when interpreted as Doppler broadening, $\Delta\lambda/\lambda = \Delta v/c$, yields values of up to $\Delta v \sim 8500\,\mathrm{km/s}$ for the full line width. The high excitation energy of some of the line-emitting atoms shows that they must have been excited by photons that are more energetic than photons from young stars that are responsible for the ionization of HII regions. The hydrogen lines are often broader than other spectral lines. Most of the Seyfert galaxies are spirals, but one cD galaxy is also found in his original catalog.

In 1959, Lodewijk Woltjer argued that the extent of the cores of Seyfert galaxies cannot be larger than $r \lesssim 100\,\mathrm{pc}$ because they appear point-like on optical images, i.e., they are spatially not resolved. If the line-emitting gas is gravitationally bound, the relation

$$\frac{GM}{r} \simeq v^2$$

between the central mass $M(<r)$, the separation r of the gas from the center, and the typical velocity v must be satisfied. The latter is obtained from the line width: typically $v \sim 1000\,\mathrm{km/s}$. Therefore, with $r \lesssim 100\,\mathrm{pc}$ a mass estimate is immediately obtained,

$$M \gtrsim 10^{10}\left(\frac{r}{100\,\mathrm{pc}}\right) M_\odot\,. \tag{5.1}$$

Thus, either $r \sim 100\,\mathrm{pc}$, which implies an enormous mass concentration in the center of these galaxies, or r is much smaller than the estimated upper limit, which then implies an enormous energy density inside AGNs.

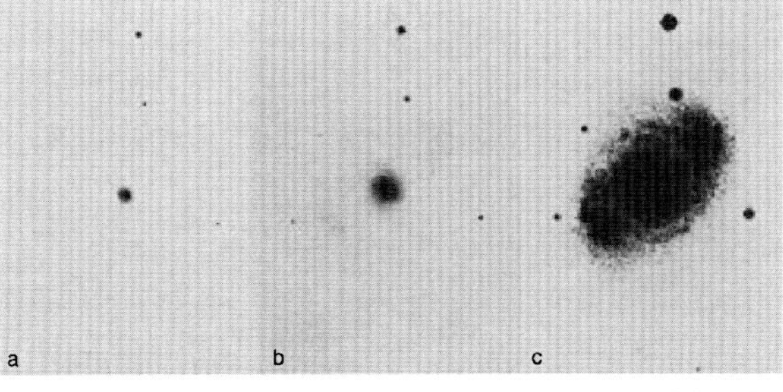

Fig. 5.4a–c. Three images of the Seyfert galaxy NGC 4151, with the exposure time increasing to the right. In short exposures, the source appears point-like, with longer exposures displaying the galaxy

An important milestone in the history of AGNs was made with the 3C and 3CR radio catalogs which were completed around 1960. These are surveys of the northern ($\delta > -22°$) sky at 158 MHz and 178 MHz, with a flux limit of $S_{\rm min} = 9\,\mathrm{Jy}$ (a Jansky is the flux unit used by radio astronomers, where $1\,\mathrm{Jy} = 10^{-23}\,\mathrm{erg\,s^{-1}\,cm^{-2}\,Hz^{-1}}$). Many of these 3C sources could be identified with relatively nearby galaxies, but the low angular resolution of radio telescopes at these low frequencies and the resulting large positional uncertainty of the respective sources rendered the identification with optical counterparts very difficult. If no striking nearby galaxy was found on optical photoplates within the positional uncertainty, the source was at first marked as unidentified.[1]

In 1963, Thomas Matthews and Allan Sandage showed that 3C48 is a point-like ("stellar-like") source of $m = 16\,\mathrm{mag}$. It has a complex optical spectrum consisting of a blue continuum and strong, broad emission lines which could not be assigned to any atomic transition, and thus could not be identified. In the same year, Maarten Schmidt succeeded in identifying the radio source 3C273 with a point-like optical source which also showed strong and broad emission lines at unusual wavelengths. This was achieved by a lunar eclipse: the Moon passed in front of the radio source and eclipsed it. From the exact measurement of the time when the radio emission was blocked and became visible again, the position of the radio source was pinned down accurately. Schmidt could identify the emission lines of the source with those of the Balmer series of hydrogen, but at an, for that time, extremely high redshift of $z = 0.158$. Presuming the validity of the Hubble law and interpreting

[1] The complete optical identification of the 3CR catalog, which was made possible by the enormously increased angular resolution of interferometric radio observations and thus by a considerably improved positional accuracy, was finalized only in the 1990s – some of these luminous radio sources are very faint optically.

the redshift as cosmological redshift, 3C273 is located at the large distance of $D \sim 500\,h^{-1}$ Mpc. This huge distance of the source then implies an absolute magnitude of $M_B = -25.3 + 5 \log h$, i.e., it is about ~ 100 times brighter than normal (spiral) galaxies. Since the optical source had not been resolved but appeared point-like, this enormous luminosity must originate from a small spatial region. With the improving determination of radio source positions, many such *quasars* (quasi-stellar radio sources = quasars) were identified in quick succession, the redshifts of some being significantly higher than that of 3C273.

5.1.2 Fundamental Properties of Quasars

In the following, we will review some of the most important properties of quasars. Although quasars are not the only class of AGNs, we will at first concentrate on them because they incorporate most of the properties of the other types of AGNs.

As already mentioned, quasars were discovered by identifying radio sources with point-like optical sources. Quasars emit at all wavelengths, from the radio to the X-ray domain of the spectrum. The flux of the source varies at nearly all frequencies, where the variability time-scale differs among the objects and also depends on the wavelength. In general, it is found that the variability time-scale is smaller and its amplitude larger when going to higher frequencies of the observed radiation. The optical spectrum is very blue; most quasars at redshifts $z \lesssim 2$ have $U - B < -0.3$ (for comparison: only hot white dwarfs have a similarly blue color index). Besides this blue continuum, very broad emission lines are characteristic of the optical spectrum. Some of them correspond to transitions of very high ionization energy (see Fig. 5.3).

The continuum spectrum of a quasar can often be described, over a broad frequency range, by a power law of the form

$$S_\nu \propto \nu^{-\alpha} , \tag{5.2}$$

where α is the spectral index. $\alpha = 0$ corresponds to a flat spectrum, whereas $\alpha = 1$ describes a spectrum in which the same energy is emitted in every logarithmic frequency interval. Finally, we shall point out again the high redshift of many quasars.

5.1.3 Quasars as Radio Sources: Synchrotron Radiation

The morphology of quasars in the radio regime depends on the observed frequency and can often be very complex, consisting of several extended source components and one compact central one. In most cases, the extended source is observed as a double source in the form of two radio lobes situated more or less symmetrically around the optical position of the quasar. These lobes are frequently connected to the central core by jets, which are thin emission structures probably related to the energy transport from the core into the lobes. The observed length-scales are often impressive, in that the total extent of the radio source can reach values of up to 1 Mpc. The position of the optical quasar coincides with the compact radio source, which has an angular extent of $\ll 1''$ and is in some cases not resolvable even with VLBI methods. Thus the extent of these sources is $\lesssim 1$ mas, corresponding to $r \lesssim 1$ pc. This dynamical range in the extent of quasars is thus extremely large.

Classification of Radio Sources. Extended radio sources are often divided into two classes. *Fanaroff–Riley Type I* (FR I) are brightest close to the core, and the surface brightness decreases outwards. They typically have a luminosity of $L_\nu(1.4\,\text{GHz}) \lesssim 10^{32}\,\text{erg}\,\text{s}^{-1}\,\text{Hz}^{-1}$. In contrast, the surface brightness of *Fanaroff–Riley Type II* sources (FR II) increases outwards, and their luminosity is in general higher than that of FR I sources, $L_\nu(1.4\,\text{GHz}) \gtrsim 10^{32}\,\text{erg}\,\text{s}^{-1}\,\text{Hz}^{-1}$. One example for each of the two classes is shown in Fig. 5.5. FR II radio sources often have *jets*; they are extended linear structures that connect the compact core with a radio lobe. Jets often show internal structure such as knots and kinks. Their appearance indicates that they transport energy from the core out into the radio lobe. One of the most impressive examples of this is displayed in Fig. 5.6.

The jets are not symmetric. Often only one jet is observed, and in most sources where two jets are found one of them (the "counter-jet") is much weaker than the other. The relative intensity of core, jet, and extended components varies with frequency, for sources as a whole and also within a source, because the components have different spectral indices. For this reason, radio catalogs of AGNs suffer from strong selection ef-

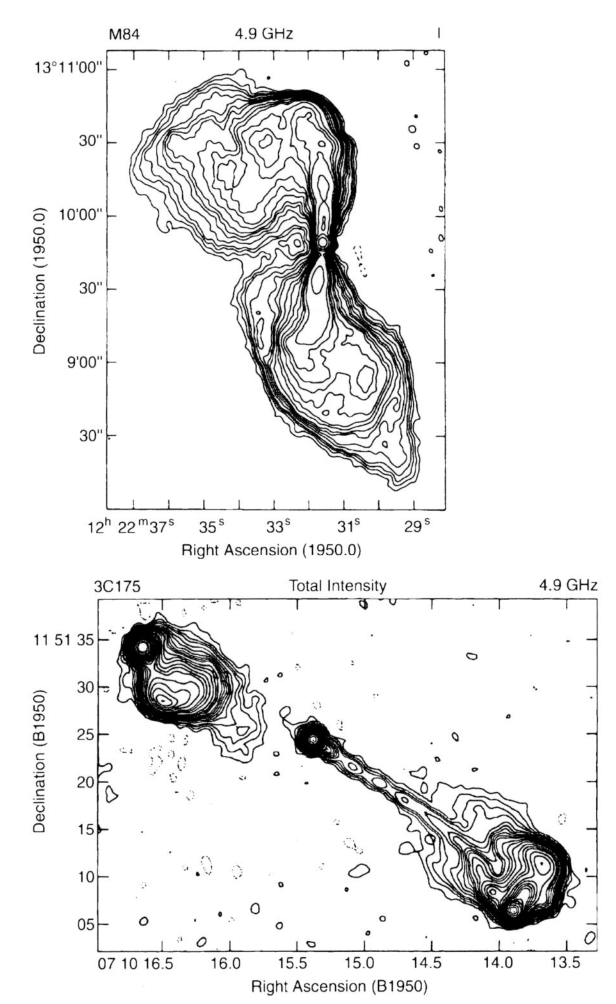

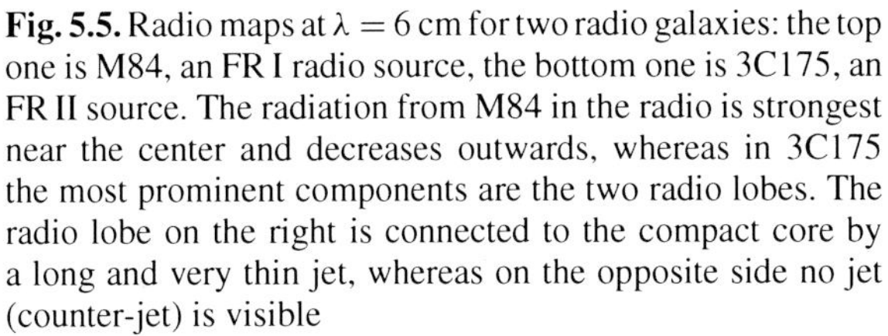

Fig. 5.5. Radio maps at $\lambda = 6$ cm for two radio galaxies: the top one is M84, an FR I radio source, the bottom one is 3C175, an FR II source. The radiation from M84 in the radio is strongest near the center and decreases outwards, whereas in 3C175 the most prominent components are the two radio lobes. The radio lobe on the right is connected to the compact core by a long and very thin jet, whereas on the opposite side no jet (counter-jet) is visible

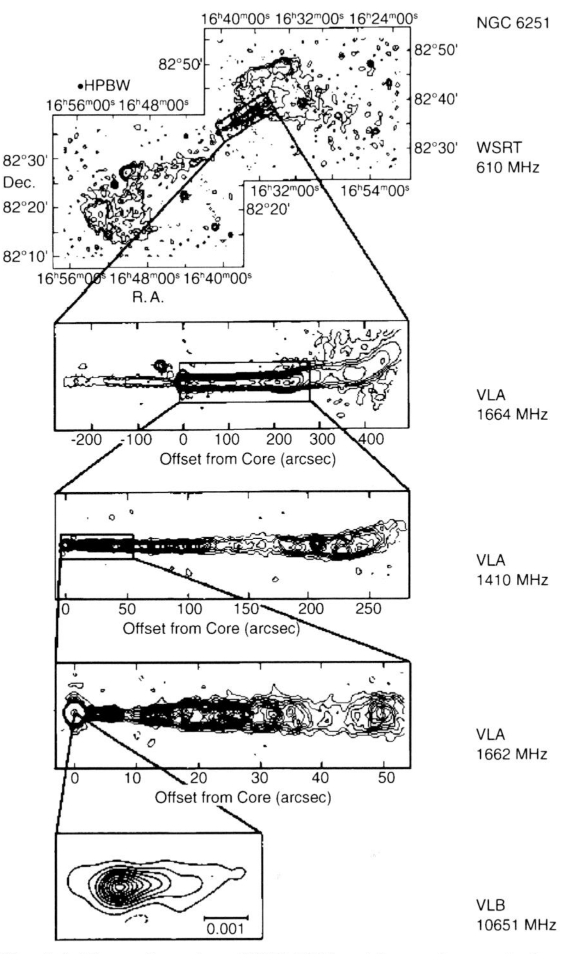

Fig. 5.6. The radio galaxy NGC 6251, with angular resolution increasing towards the bottom. On large scales (and at low frequencies), the two radio lobes dominate, while the core and the jets are clearly prominent at higher frequencies. NGC 6251 has a counter-jet, but with significantly lower luminosity than the main jet. Even at the highest resolution obtained by VLBI, structure can still be seen. The jets have a very small opening angle and are therefore strongly collimated

fects. Catalogs that are sampled at low frequencies will predominantly select sources that have a steep spectrum, i.e., in which the extended structures dominate, whereas high-frequency samples will preferentially contain core-dominated sources with a flat spectrum.[2]

[2]For this reason, radio surveys for gravitational lens systems, which have been mentioned in Sect. 3.8.3, concentrate on sources with a flat spectral index because these are dominated by the compact nucleus. Multiple image systems are thus more easily recognized as such.

Synchrotron Radiation. Over a broad range in wavelengths, the radio spectrum of AGNs follows a power law of the form (5.2), with $\alpha \sim 0.7$ for the extended components and $\alpha \sim 0$ for the compact core components. Radiation in the radio is often linearly polarized, where the extended radio source may reach a degree of polarization up to 30% or even more. The spectral form and the high degree of polarization are interpreted such that the radio emission is produced by *synchrotron radiation* of relativistic electrons. Electrons in a magnetic field propagate along a helical, i.e., corkscrew-shaped path, so that they are continually accelerated by the Lorentz force. Since accelerated charges emit electromagnetic radiation, this motion of the electrons leads to the emission of synchrotron radiation. Because of its importance for our understanding of the radio emission of AGNs, we will review some aspects of synchrotron radiation next.

The radiation can be characterized as follows. If an electron has energy $E = \gamma\, m_e\, c^2$, the characteristic frequency of the emission is

$$\nu_c = \frac{3\gamma^2 eB}{4\pi m_e c} \sim 4.2 \times 10^6\, \gamma^2 \left(\frac{B}{1\,\mathrm{G}}\right) \mathrm{Hz}\,, \qquad (5.3)$$

where B denotes the magnetic field strength, e the electron charge, and $m_e = 511\,\mathrm{keV}/c^2$ the mass of the electron. The *Lorentz factor* γ, and thus the energy of an electron, is related to its velocity v via

$$\gamma := \frac{1}{\sqrt{1-(v/c)^2}}\,. \qquad (5.4)$$

For frequencies considerably lower than ν_c, the spectrum of a single electron is $\propto \nu^{1/3}$, whereas it decreases exponentially for larger frequencies. To a first approximation, the spectrum of a single electron can be considered as quasi-monochromatic, i.e., the width of the spectral distribution is small compared to the characteristic emission frequency ν_c. The synchrotron radiation of a single electron is linearly polarized, where the polarization direction depends on the direction of the magnetic field projected onto the sky. The degree of polarization of the radiation from an ensemble of electrons depends on the complexity of the magnetic field. If the magnetic field is homogeneous in the spatial region from which the radiation is measured, the observed polarization may reach values of up to 75%. However, if the spatial region that lies within the telescope beam contains a complex magnetic field, with the direction changing strongly within this region, the polarizations partially cancel each other out and the observed degree of linear polarization is significantly reduced.

To produce radiation at cm wavelengths ($\nu \sim 10\,\mathrm{GHz}$) in a magnetic field of strength $B \sim 10^{-4}\,\mathrm{G}$, $\gamma \sim 10^5$ is required, i.e., the *electrons need to be highly relativistic!* To obtain particles at such high energies, very efficient processes of particle acceleration must occur in the inner regions of quasars. It should be mentioned in this context that comic ray particles of considerably higher energies are observed (see Sect. 2.3.4). The majority of cosmic rays are presumably produced in the shock fronts of supernova remnants. Thus, it is supposed that the energetic electrons in quasars (and other AGNs) are also produced by "diffusive shock acceleration", where here the shock fronts are not caused by supernova explosions but rather by other hydrodynamical phenomena. As we will see later, we find clear indications in AGNs for outflow velocities that are considerably higher than the speed of sound in the plasma, so that the conditions for the formation of shock fronts are satisfied.

Synchrotron radiation will follow a power law if the energy distribution of relativistic electrons also behaves like a power law (see Fig. 5.7). If $N(E)\,\mathrm{d}E \propto E^{-s}\,\mathrm{d}E$ represents the number density of electrons with energies between E and $E+\mathrm{d}E$, the power-law index of the resulting radiation will be $\alpha = (s-1)/2$, i.e., the slope in the power law of the electrons defines the spectral shape of the resulting synchrotron emission. In particular, an index of $\alpha = 0.7$ results for $s = 2.4$. An electron distribution with $N(E) \propto E^{-2.4}$ is very similar to the energy distribution of the cosmic rays in our Galaxy, which may be another indicator for the same or at least a similar mechanism being responsible for the generation of this energy spectrum.

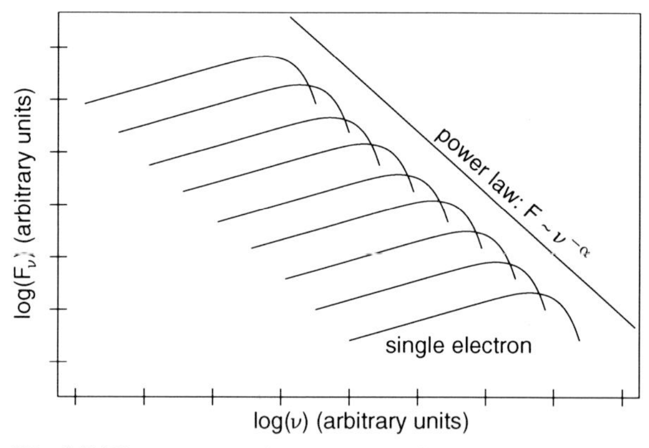

Fig. 5.7. Electrons at a given energy emit a synchrotron spectrum which is indicated by the individual curves; the maximum of the radiation is at ν_c (5.3), which depends on the electron energy. The superposition of many such spectra, corresponding to an energy distribution of the electrons, results in a power-law spectrum provided the energy distribution of the electrons follows a power law

The synchrotron spectrum is self-absorbed at low frequencies, i.e., the optical depth for absorption due to the synchrotron process is close to or larger than unity. In this case, the spectrum becomes flatter and, for small ν, it may even rise. In the limiting case of a high optical depth for self-absorption, we obtain $S_\nu \propto \nu^{2.5}$ for $\nu \to 0$. The extended radio components are optically thin at cm wavelength, so that $\alpha \sim 0.7$, whereas the compact core component is often optically thick and thus self-absorbed, which yields $\alpha \sim 0$, or even inverted so that $\alpha < 0$.

Through emission, the electrons lose energy. Thus, the electrons cool and for only a limited time can they radiate at the frequency described by (5.3). The power emitted by an electron of Lorentz factor γ, integrated over all frequencies, is

$$P = -\frac{\mathrm{d}E}{\mathrm{d}t} = \frac{4}{9}\frac{e^4 B^2 \gamma^2}{m_\mathrm{e}^2 c^3} \; . \tag{5.5}$$

The characteristic time in which an electron loses its energy is then obtained from its energy $E = \gamma m_\mathrm{e} c^2$ and its energy loss rate $\dot{E} = -P$ as

$$t_\mathrm{cool} = \frac{E}{P} = 2.4 \times 10^5 \left(\frac{\gamma}{10^4}\right)^{-1} \left(\frac{B}{10^{-4}\,\mathrm{G}}\right)^{-2} \mathrm{yr} \; . \tag{5.6}$$

For relatively low-frequency radio emission, this lifetime is longer than or comparable to the age of radio sources. But as we will see later, high-frequency synchrotron emission is also observed for which t_cool is considerably shorter than the age of a source component. The corresponding relativistic electrons can then only be generated locally. This means that the processes of particle acceleration are not confined to the inner core of an AGN, but also occur in the extended source components.

Since the characteristic frequency (5.3) of synchrotron radiation depends on a combination of the Lorentz factor γ and the magnetic field B, we cannot measure these two quantities independently. Therefore, it is difficult to estimate the magnetic field of a synchrotron source. In most cases, the (plausible) assumption of an equipartition of the energy density in the magnetic field and the relativistic particles is made, i.e., one assumes that the energy density $B^2/(8\pi)$ of the magnetic field roughly agrees with the energy density

$$\int \mathrm{d}\gamma \; n_\mathrm{e}(\gamma) \, \gamma m_\mathrm{e} c^2$$

of the relativistic electrons. Such approximate equipartition holds for the cosmic rays in our Galaxy and its magnetic field. Another approach is to estimate the magnetic field such that the total energy of relativistic electrons and magnetic field is minimized for a given source luminosity. The resulting value for B basically agrees with that derived from the assumption of equipartition.

5.1.4 Broad Emission Lines

The UV and optical spectra of quasars feature strong and very broad emission lines. Typically, lines of the Balmer series and Lyα of hydrogen, and metal lines of ions like MgII, CIII, CIV[3] are observed – these are found in virtually all quasar spectra. In addition, a large number of other emission lines occur which are not seen in every spectrum (Fig. 5.2).

To characterize the *strength* of an emission line, we define the *equivalent width* of a line W_λ as

$$W_\lambda = \int \mathrm{d}\lambda \, \frac{S_\mathrm{l}(\lambda) - S_\mathrm{c}(\lambda)}{S_\mathrm{c}(\lambda)} \approx \frac{F_\mathrm{line}}{S_\mathrm{c}(\lambda_0)} \; , \tag{5.7}$$

where $S_\mathrm{l}(\lambda)$ is the total spectral flux, and $S_\mathrm{c}(\lambda)$ is the spectral flux of the continuum radiation interpolated across the wavelength range of the line. F_line is the total flux in the line and λ_0 its wavelength. Hence, W_λ is the width of the wavelength interval over which the continuum needs to be integrated to obtain the same flux as measured in the line. Therefore, the equivalent width is a measure of the strength of a line relative to the continuum intensity.

The *width* of a line is characterized as follows: after subtracting the continuum, interpolated across the wavelength range of the line, the width is measured at half of the maximum line intensity. This width $\Delta\lambda$ is called the FWHM (full width at half maximum); it may be specified either in Å, or in km/s if the line width is interpreted as Doppler broadening, with $\Delta\lambda/\lambda_0 = \Delta v/c$.

Broad emission lines in quasars often have a FWHM of $\sim 10\,000$ km/s, while narrower emission lines still have widths of several 100 km/s. Thus the "narrow"

[3]The ionization stages of an element are distinguished by Roman numbers. A neutral atom is denoted by "I", a singly ionized atom by "II", and so on. So, CIV is three times ionized carbon.

emission lines are still broad compared to the typical velocities in normal galaxies.

Redshift. Quasar surveys are always flux limited, i.e., one tries to find all quasars in a certain sky region with a flux above a predefined threshold. Only with such a selection criterion are the samples obtained of any statistical value. In addition, the selection of sources may include further criteria such as color, variability, radio or X-ray flux. For instance, radio surveys are defined by $S_\nu > S_{\rm lim}$ at a specific wavelength. The optical identification of such radio sources reveals that quasars have a very broad redshift distribution. For decades, quasars have been the only sources known at $z > 3$. Below we will discuss different kinds of AGN surveys.

In the 1993 issue of the quasar catalog by Hewitt & Burbidge, 7236 sources are listed. This catalog contains a broad variety of different AGNs. Although it is statistically not well-defined, this catalog provides a good indication of the width of the redshift and brightness distribution of AGNs (see Fig. 5.8).

The luminosity function of quasars extends over a very large range in luminosity, nearly three orders of magnitude in L. It is steep at its bright end and has a significantly flatter slope at lower luminosities (see Sect. 5.6.2). We can compare this to the luminosity function of galaxies which is described by a Schechter function (see Sect. 3.7). While the faint end of the distribution is also described here by a relatively shallow power law, the Schechter function decreases exponentially for large L, whereas that of quasars decreases as a power law. For this reason, one finds quasars whose luminosity is much larger than the value of L where the break in the luminosity function occurs.

5.2 AGN Zoology

Quasars are the most luminous members of the class of AGNs. Seyfert galaxies are another type of AGN and were mentioned previously. In fact, a wide range of objects are subsumed under the name AGN, all of which

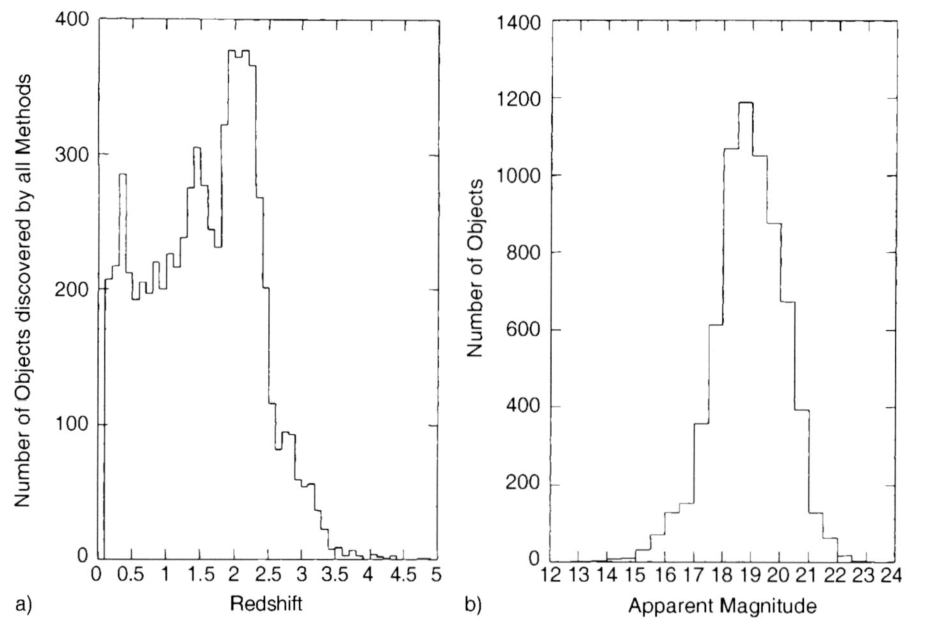

Fig. 5.8. The redshift (left) and brightness distribution (right) of QSOs in the 1993 Hewitt & Burbidge catalog. These distributions provide no proper statistical information, but they clearly show the width of the distributions. The decrease in abundances for $z \geq 2.3$ is a selection effect: many QSO surveys start with a color selection, typically $U - B < -0.3$. If $z \geq 2.3$, the strong Lyα emission line moves into the B-filter and hence the quasar becomes redder in this color index and drops out of the color selection

have in common strong non-thermal emission in the core of a galaxy (*host galaxy*). We will mention the most important types of AGN in this section. It is important to keep in mind that the frequency range in which sources are studied affects the source classification. We shall return to this point at the end of this section.

The classification of AGNs described below is very confusing at first glance. Different classes refer to different appearances of AGNs but do not necessarily correspond to the physical nature of these sources. As we will discuss in Sect. 5.5, the appearance of an AGN in the context of unified models depends very strongly on the orientation of the source with respect to its line-of-sight. We will then be able to better organize the variety of classes.

5.2.1 Quasi-Stellar Objects

The unusually blue color of quasars suggested the possibility of searching for them not only with radio observations but also at optical wavelengths, namely to look for point-like sources with a very blue $U-B$ color index. These photometric surveys were very successful. In fact, many more such sources were found than expected from radio counts. Most of these sources are (nearly) invisible in the radio domain of the spectrum; such sources are called radio-quiet. Their optical properties are virtually indistinguishable from those of quasars. In particular, they have a blue optical energy distribution (of course, since this was the search criterion!), strong and broad emission lines, and in general a high redshift.

Hence, apart from their radio properties, these sources appear to be like quasars. Therefore they were called *radio-quiet quasars*, or quasi-stellar objects, QSOs. Today this terminology is no longer very common because the clear separation between sources with and without radio emission is not considered valid any more. Radio-quiet quasars also show radio emission if they are observed at sufficiently high sensitivity. In modern terminology, the expression QSO encompasses both the quasars and the radio-quiet QSOs. About 10 times more radio-quiet QSOs than quasars are thought to exist.

The QSOs are the most luminous AGNs. Their core luminosity can be as high as a thousand times that of an L^*-galaxy. Therefore they outshine their host galaxy and appear point-like on optical images. For QSOs of lower L, their host galaxies were identified and spatially resolved with the HST (see Fig. 1.11). According to our current understanding, AGNs are the active cores of galaxies. These galaxies are supposed to be fairly normal galaxies, except for their intense nuclear activity, and we will discuss possible reasons for the onset of this activity further below.

5.2.2 Seyfert Galaxies

Seyfert galaxies are the AGNs which were detected first. Their luminosity is considerably lower than that of QSOs. On optical images they are identified as spiral galaxies which have an extraordinarily bright core (Fig. 5.4) whose spectrum shows strong and broad emission lines.

We distinguish Seyfert galaxies of Type 1 and Type 2: Seyfert 1 galaxies have both very broad and also narrower emission lines, where "narrow" still means several hundred km/s and thus a significantly larger width than characteristic velocities (like rotational velocities) found in normal galaxies. Seyfert 2 galaxies show only the narrower lines. Later, it was discovered that intermediate variants exist – one now speaks of Seyfert 1.5 and Seyfert 1.8 galaxies, for instance – in which very broad lines exist but which are much less prominent than they are in Seyfert 1 galaxies. The archetype of a Seyfert 1 galaxy is NGC 4151 (see Fig. 5.4), while NGC 1068 is a typical Seyfert 2 galaxy.

The optical spectrum of Seyfert 1 galaxies is very similar to that of QSOs. A smooth transition exists between (radio-quiet) QSOs and Seyfert 1 galaxies. Formally, these two classes of AGNs are separated at an absolute magnitude of $M_B = -21.5 + 5 \log h$. The separation of Seyfert 1 galaxies and QSOs is historical since these two categories were introduced only because of the different methods of discovering them. However, except for the different core luminosity, no fundamental physical difference seems to exist. Often both classes are combined under the name Type 1-AGNs.

5.2.3 Radio Galaxies

Radio galaxies are elliptical galaxies with an active nucleus. They were the first sources that were identified with optical counterparts in the early radio surveys. Characteristic radio galaxies are Cygnus A and Centaurus A.

In a similar fashion to Seyfert galaxies, for radio galaxies we also distinguish between those with and without broad emission lines: broad-line radio galaxies (BLRG) and narrow-line radio galaxies (NLRG), respectively. In principle, the two types of radio galaxy can be considered as radio-loud Seyfert 1 and Seyfert 2 galaxies but with a different morphology of the host galaxy. A smooth transition between BLRG and quasars also seems to exist, again separated by optical luminosity as for Seyfert galaxies.

Besides the classification of radio galaxies into BLRG and NLRG with respect to the optical spectrum, they are distinguished according to their radio morphology. As was discussed in Sect. 5.1.2, radio sources are divided into FR I and FR II sources.

5.2.4 Optically Violently Variables

One subclass of QSOs is characterized by the very strong and rapid variability of its optical radiation. The flux of these sources, which are known as Optically Violently Variables (OVVs), can vary by a significant fraction on time-scales of days (see Fig. 5.9). Besides this strong variability, OVVs also stand out because of their relatively high polarization of optical light, typi-

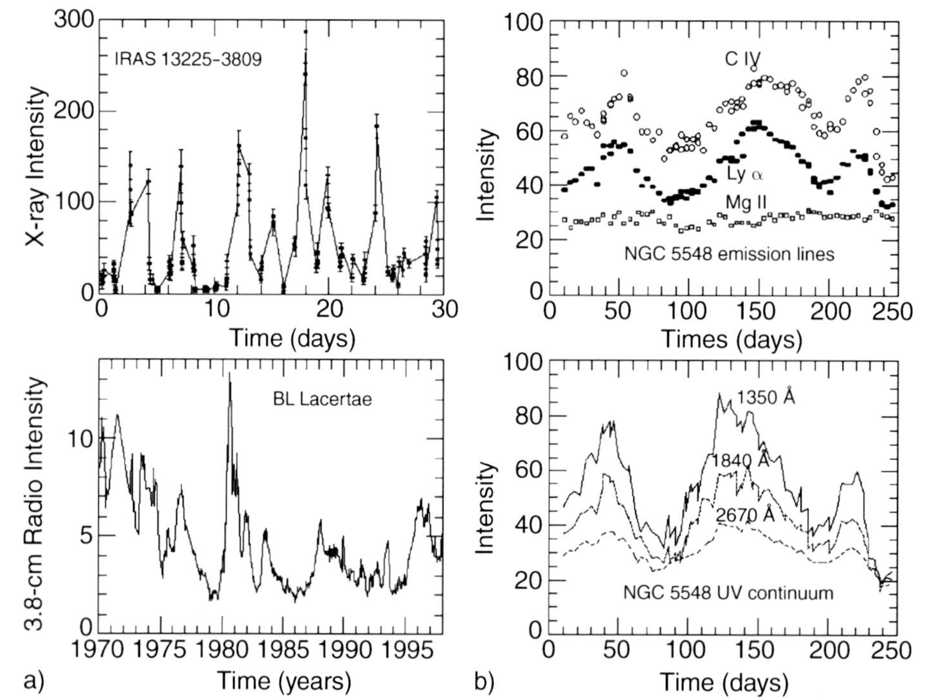

Fig. 5.9. Quasars, BL Lac objects, and Seyfert galaxies all show clear variability at many different wavelengths. In the upper left panel, the X-ray light curve of the Seyfert 1 galaxy IRAS 13225–3809 is plotted (observed by ROSAT); on time-scales of days, the source frequently varies by more than a factor of 20. The radio light curve of BL Lacertae at $\lambda =$ 3.8 cm covering a period of 28 years is shown in the lower left panel. Variations of such blazars are observed in a number of bursts, some overlapping (see, e.g., the burst in 1981). The UV variability of NGC 5548, a Seyfert 1 galaxy, observed by the IUE satellite is plotted for three wavelengths in the lower right panel. Variations at these frequencies appear to be in phase, but the amplitude becomes larger towards smaller wavelengths. Simultaneously, the line strengths of three broad emission lines of this Seyfert 1 galaxy have been measured and are plotted in the upper right panel. It is found that lines of high ionization potentials, like CIV, have higher variability amplitudes than those of low ionization potentials, like MgII. From the relative temporal shift in the line variability and the continuum flux, the size of the broad-line region can be estimated – see Sect. 5.4.2

cally a few percent, whereas the polarization of normal QSOs is below $\sim 1\%$. OVVs are usually strong radio emitters. Their radiation also varies in other wavelength regions besides the optical, with shorter time-scales and larger amplitudes as one moves to higher frequencies.

5.2.5 BL Lac Objects

The class of AGNs called BL Lac objects (or short: BL Lacs) is named after its prototypical source BL Lacertae. They are AGNs with very strongly varying radiation, like the OVVs, but without strong emission and absorption lines. As for OVVs, the optical radiation of BL Lacs is highly polarized. Since no emission lines are observed in the spectra of BL Lacs, the determination of their redshift is often difficult and sometimes impossible. In some cases, absorption lines are detected in the spectrum which are presumed to derive from the host galaxy of the AGN and are then identified with the redshift of the BL Lac.

The optical luminosity of some BL Lacs varies by several magnitudes if observed over a sufficiently long time period. Particularly remarkable is the fact that in epochs of low luminosity, emission lines are sometimes observed and then a BL Lac appears like an OVV. For this reason, OVVs and BL Lacs are collectively called *blazars*. All known blazars are radio sources. Besides the violent variability, blazars also show highly energetic and strongly variable γ-radiation (Fig. 5.10). Table 5.1 summarizes the fundamental properties of the different classes of AGNs.

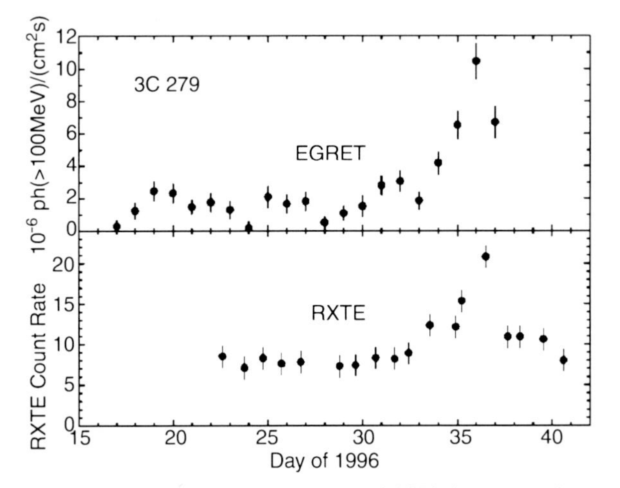

Fig. 5.10. Variability of the blazar 3C279 in X-ray (bottom) and in γ-radiation at photon energies above 100 MeV (top). On time-scales of a few days, the luminosity varies by a factor ~ 10

5.3 The Central Engine: A Black Hole

We have previously mentioned that the energy production in AGNs must be related to a supermassive black hole (SMBH) in its center. We will present arguments for this conclusion in this section. To do this, we will first summarize some of the relevant observational facts for AGNs.

- The extent of some radio sources in AGNs may reach $\gtrsim 1$ Mpc. From this length-scale a minimum lifetime for the activity in the nucleus of these objects can be derived, since even if the radio source expands outwards from the core with the speed of light, the age of such a source would be $\tau \gtrsim 10^7$ yr.
- Luminous QSOs have a luminosity of up to $L_{\rm bol} \sim 10^{47}$ erg/s. Assuming that the luminosity does not

Table 5.1. Overview of the classification of active galactic nuclei

	Normal galaxy	Radio galaxy	Seyfert galaxy	Quasar	Blazar
Example	Milky Way	M87, Cygnus A	NGC 4151	3C273	BL Lac, 3C279
Galaxy type	spiral	elliptical, irregular	spiral	irregular	elliptical?
$L/L_\odot$	$< 10^4$	$10^6 - 10^8$	$10^8 - 10^{11}$	$10^{11} - 10^{14}$	$10^{11} - 10^{14}$
$M_{\rm BH}/M_\odot$	3×10^6	3×10^9	$10^6 - 10^9$	$10^6 - 10^9$	$10^6 - 10^9$
Radio emission	weak	core, jets, lobes	only $\approx 5\%$ radio-loud	only $\approx 5\%$ radio-loud	strong, short-time variable
Radiation in optical / NIR	fully absorbed	old stars, continuum	broad emission lines	broad emission lines	weak or no lines
X-ray emission	weak	strong	strong	strong	strong
Gamma emission	weak	weak	medium	strong	strong
Variability	unknown	months-years	hours-months	weeks-years	hours-years

change substantially over the lifetime of the source, a total energy can be estimated from the luminosity and the minimum age,

$$E \gtrsim 10^{47}\,\mathrm{erg/s} \times 10^7\,\mathrm{yr} \sim 3 \times 10^{61}\,\mathrm{erg}\,, \quad (5.8)$$

however, the assumption of an essentially constant luminosity is not necessarily justified.

- The luminosity of some AGNs varies by more than 50% on time-scales of a day. From this variability time-scale, an upper limit for the spatial extent of the source can be determined, because the source luminosity can change substantially only on such time-scales where the source as a whole, or at least a major part of the emitting region, is in causal contact. Otherwise "one end" of the source does not know that the "other end" is about to vary. This yields a characteristic extent of the central source of $R \lesssim 1$ lightday $\sim 3 \times 10^{15}$ cm.

5.3.1 Why a Black Hole?

We will now combine the aforementioned observations and derive from them that the basic energy production in AGNs has to be of a gravitational nature. To do this, we note that the most efficient "classical" method of energy production is nuclear fusion, as is taking place in stars. We will therefore make the provisional assumption (which will soon lead to a contradiction) that the energy production in AGNs is based on thermonuclear processes.

By burning hydrogen into iron – the nucleus with the highest binding energy per nucleon – 8 MeV/nucleon are released, or $0.008\, m_\mathrm{p} c^2$ per nucleon. The maximum efficiency of nuclear fusion is therefore $\epsilon \lesssim 0.8\%$, where ϵ is defined as the mass fraction of "fuel" that is converted into energy, according to

$$\boxed{E = \epsilon\, m c^2}\,. \quad (5.9)$$

To generate the energy of $E = 3 \times 10^{61}$ erg by nuclear fusion, a total mass m of fuel would be needed, where m is given by

$$m = \frac{E}{\epsilon c^2} \sim 4 \times 10^{42}\,\mathrm{g} \sim 2 \times 10^9 M_\odot\,, \quad (5.10)$$

where we used the energy estimate from (5.8). If the energy of an AGN was produced by nuclear fusion, burnt-out matter of mass m [more precisely, $(1-\epsilon)m$] must be present in the core of the AGN.

However, the Schwarzschild radius of this mass is (see Sect. 3.5.1)

$$r_\mathrm{S} = \frac{2Gm}{c^2} = \frac{2GM_\odot}{c^2}\frac{m}{M_\odot} = 3 \times 10^5\,\mathrm{cm}\,\frac{m}{M_\odot} \sim 6 \times 10^{14}\,\mathrm{cm}\,,$$

i.e., the Schwarzschild radius of the "nuclear cinder" is of the same order of magnitude as the above estimate of the extent of the central source. This argument demonstrates that gravitational effects *must* play a crucial role – the assumption of thermonuclear energy generation has been disproven because its efficiency ϵ is too low. The only known mechanism yielding larger ϵ is gravitational energy production.

Through the infall of matter onto a central black hole, potential energy is converted into kinetic energy. If it is possible to convert part of this inward-directed kinetic energy into internal energy (heat) and subsequently emit this in the form of radiation, ϵ can be larger than that of thermonuclear processes. From the theory of accretion onto black holes, a maximum efficiency of $\epsilon \sim 6\%$ for accretion onto a non-rotating black hole (also called a Schwarzschild hole) is derived. A black hole with the maximum allowed angular momentum can have an efficiency of $\epsilon \sim 29\%$.

5.3.2 Accretion

Due to its broad astrophysical relevance beyond the context of AGNs, we will consider the accretion process in somewhat more detail.

The Principle of Accretion. Gas falling onto a compact object loses its potential energy, which is first converted into kinetic energy. If the infall is not prevented, the gas will fall into the black hole without being able to radiate this energy. In general one can expect that the gas has finite angular momentum. Thus it cannot fall straight onto the compact object, since this is prevented by the angular momentum barrier. Through friction with other gas particles and by the resulting momentum transfer, the gas will assemble in a disk oriented perpendicular to the direction of the angular momentum vector. The frictional forces in the gas are expected to be much smaller than the gravitational force. Hence the disk will locally rotate with approximately the Kepler velocity. Since a Kepler disk rotates differentially, in the sense as the angular velocity depends on radius, the gas in the disk will be heated by internal friction. In addition, the same friction causes a slight deceleration of the rotational velocity, whereby the gas will slowly move inwards. The energy source for heating the gas in the disk is provided by this inward motion – namely the conversion of potential energy into kinetic energy, which is then converted into internal energy (heat) by friction.

According to the virial theorem, half of the potential energy released is converted into kinetic energy; in the situation considered here, this is the rotational energy of the disk. The other half of the potential energy can be converted into internal energy. We now present an approximately quantitative description of this process, specifically for accretion onto a black hole.

Temperature Profile of a Geometrically Thin, Optically Thick Accretion Disk. When a mass m falls from radius $r+\Delta r$ to r, the energy

$$\Delta E = \frac{GM_\bullet m}{r} - \frac{GM_\bullet m}{r+\Delta r} \approx \frac{GM_\bullet m}{r}\frac{\Delta r}{r}$$

is released. Here $M_\bullet$ denotes the mass of the SMBH, assumed to dominate the gravitational potential, so that self-gravity of the disk can be neglected. Half of this energy is converted into heat, $E_{\rm heat} = \Delta E/2$. If we assume that this energy is emitted locally, the corresponding luminosity is

$$\Delta L = \frac{GM_\bullet \dot m}{2r^2}\Delta r \,, \tag{5.11}$$

where $\dot m$ denotes the accretion rate, which is the mass that falls into the black hole per unit time. In the stationary case, $\dot m$ is independent of radius since otherwise matter would accumulate at some radii. Hence the same amount of matter per unit time flows through any cylindrical radius.

If the disk is optically thick, the local emission corresponds to that of a black body. The ring between r and $r+\Delta r$ then emits a luminosity

$$\Delta L = 2\times 2\pi r\,\Delta r\,\sigma_{\rm SB} T^4(r) \,, \tag{5.12}$$

where the factor 2 originates from the fact that the disk has two sides. Combining (5.11) and (5.12) yields the radial dependence of the disk temperature,

$$T(r) = \left(\frac{GM_\bullet \dot m}{8\pi\sigma_{\rm SB} r^3}\right)^{1/4} .$$

A more accurate derivation explicitly considers the dissipation by friction and accounts for the fact that part of the generated energy is used for heating the gas, where the corresponding thermal energy is also partially advected inwards. Except for a numerical correction factor, the same result is obtained,

$$T(r) = \left(\frac{3GM_\bullet \dot m}{8\pi\sigma_{\rm SB} r^3}\right)^{1/4} , \tag{5.13}$$

which is valid in the range $r \gg r_{\rm S}$. Scaling r with the Schwarzschild radius $r_{\rm S}$, we obtain

$$T(r) = \left(\frac{3GM_\bullet \dot m}{8\pi\sigma_{\rm SB} r_{\rm S}^3}\right)^{1/4} \left(\frac{r}{r_{\rm S}}\right)^{-3/4} .$$

By replacing $r_{\rm S}$ with (3.31) in the first factor, this can be written as

$$T(r) = \left(\frac{3c^6}{64\pi\sigma_{\rm SB} G^2}\right)^{1/4} \dot m^{1/4} M_\bullet^{-1/2} \left(\frac{r}{r_{\rm S}}\right)^{-3/4} . \tag{5.14}$$

From this analysis, we can immediately draw a number of conclusions. The most surprising one may be the independence of the temperature profile of the disk from the detailed mechanism of the dissipation because the equations do not explicitly contain the viscosity. This fact allows us to obtain quantitative predictions based on the model of a *geometrically thin, optically thick accretion disk*.[4] The temperature in the disk increases inwards $\propto r^{-3/4}$, as expected. Therefore, the total emission of the disk is, to a first approximation, a superposition of black bodies consisting of rings with

[4]The physical mechanism that is responsible for the viscosity is unknown. The molecular viscosity is far too small to be considered as the primary process. Rather, the viscosity is probably produced by turbulent flows in the disk or by magnetic fields, which become spun up by differential rotation and thus amplified, so that these fields may act as an effective friction. In addition, hydrodynamic instabilities may act as a source of viscosity. Although the properties of the accretion disk presented here – luminosity and temperature profile – are independent of the specific mechanism of the viscosity, other disk properties definitely depend on it. For example, the temporal behavior of a disk in the presence of a perturbation, which is responsible for the variability in some binary systems, depends on the magnitude of the viscosity, which therefore can be estimated from observations of such systems.

different radii at different temperatures. For this reason, the resulting spectrum does not have a Planck shape but instead shows a much broader energy distribution.

For any fixed ratio r/r_S, the temperature increases with the accretion rate. This again was expected: since the local emission is $\propto T^4$ and the locally dissipated energy is $\propto \dot{m}$, it must be $T \propto \dot{m}^{1/4}$. Furthermore, at fixed ratio r/r_S, the temperature decreases with increasing mass $M_\bullet$ of the black hole. This implies that the maximum temperature attained in the disk is lower for more massive black holes. This may be unexpected, but it is explained by a decrease of the tidal forces, at fixed r/r_S, with increasing $M_\bullet$. In particular, it implies that the maximum temperature of the disk in an AGN is much lower than in accretion disks around stellar sources. Accretion disks around neutron stars and stellar-mass black holes emit in the hard X-ray part of the spectrum and are known as X-ray binaries. In contrast, the thermal radiation of the disk of an AGN extends to the UV range only (see below).

5.3.3 Superluminal Motion

Besides the generation of energy, another piece of evidence for the existence of SMBHs in the centers of AGNs results from observing relative motions of source components at *superluminal* velocities. These observations of central radio components in AGNs are mainly made using VLBI methods since they provide the highest available angular resolution. They measure a time dependence of the angular separation of source components, which often leads to values $> c$ if the angular velocity is translated into a transverse spatial velocity (Fig. 5.11). These superluminal motions caused some discomfort upon their discovery. In particular, they at first raised concerns that the redshift of QSOs may not originate from cosmic expansion. Only if the QSO redshifts are interpreted as being of cosmological origin can they be translated into a distance, which is needed to convert the observed angular velocity into a spatial velocity.

We consider two source components (e.g., the radio core and a component in the jet) which are observed to have a time-dependent angular separation $\theta(t)$. If D denotes the distance of the source, then the apparent relative transverse velocity of the two components is

$$v_{\rm app} = \frac{{\rm d}r}{{\rm d}t} = D\,\frac{{\rm d}\theta}{{\rm d}t}\,, \tag{5.15}$$

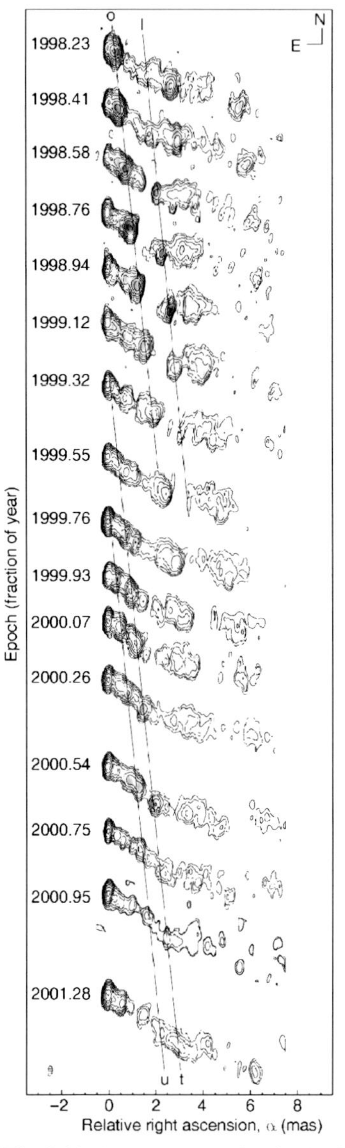

Fig. 5.11. Apparent superluminal velocities of source components in the radio jet of the source 3C120. VLBA observations of this source are presented for 16 different epochs (indicated by the numbers at the left of the corresponding radio map), observed at 7 mm wavelength. The ellipse at the lower left indicates the beam of the VLBA interferometer and thus the angular resolution of these observations. At the distance of 3C120 of 140 Mpc, a milliarcsecond corresponds to a linear scale of 0.70 pc. The four straight lines, denoted by l, o, t, and u, connect the same source components at different epochs. The linear motion of these components is clearly visible. The observed angular velocities of the components yield apparent transverse velocities in the range of 4.1c to 5c

where $r = D\theta$ is the transverse separation of the two components. The final expression in (5.15) shows that $v_{\rm app}$ is directly observable if the distance D is assumed to be known.

Frequently, VLBI observations of compact radio sources yield values for $v_{\rm app}$ that are larger than c! Characteristic values for sources with a dominant core component are $v_{\rm app} \sim 5c$ (see Fig. 5.11). But according to the theory of Special Relativity, velocities $> c$ do not exist. Thus it is not surprising that the phenomenon of superluminal motion engendered various kinds of explanations upon its discovery. By now, superluminal motion has also been seen in optical observations of jets, as is displayed in Fig. 5.12.

One possible explanation is that the cosmological interpretation of the redshifts may be wrong, because for a sufficiently small D velocities smaller than the speed of light would result from (5.15). However, no plausible alternative explanations for the observed redshifts of QSOs exist, and more than 40 years of QSO observations have consistently confirmed that redshift is an excellent measure for their distances – see Sect. 4.5.1.

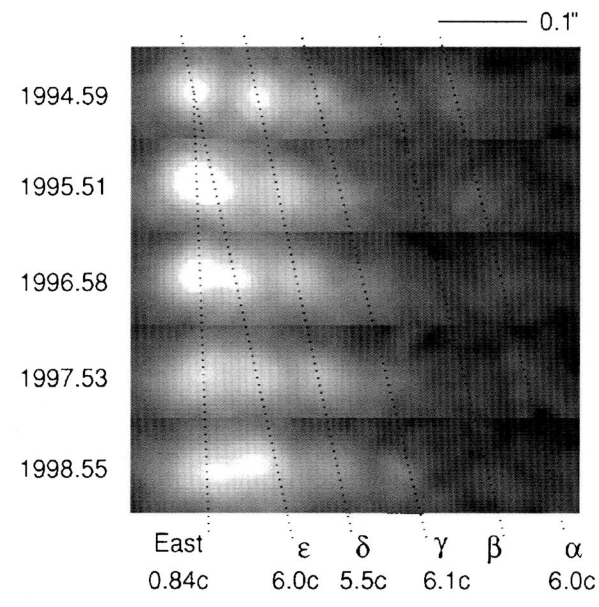

Fig. 5.12. Also at optical wavelengths, apparent superluminal motion was observed. The figure shows the optical jet in M87, based on HST images taken over a period of about four years. The angular velocity of the components is up to 23 mas/yr. Assuming a distance of M87 of $D = 16$ Mpc, velocities of up to $\sim 6c$ are obtained for the components

However, relativity only demands that no *signal* may propagate with velocities $> c$. It is easy to construct a thought-experiment in which superluminal velocities occur. For instance, consider a laser beam or a flashlight that is rotating perpendicular to its axis of symmetry. The corresponding light point on a screen changes its position with a speed proportional to the angular velocity and to the distance of the screen from the light source. If we make the latter sufficiently large, it is "easy" to obtain a superluminal light point on the screen. But this light point does not carry a signal along its track. Therefore, the superluminal motions in compact radio sources may be explained by such a screen effect, but what is the screen and what is the laser beam?

The generally accepted explanation of apparent superluminal motion combines very fast motions of source components with the finite speed of light. For this, we consider a source component moving at speed v at an angle ϕ with respect to the line-of-sight (see Fig. 5.13). We arbitrarily choose the origin of time $t = 0$ to be the time at which the moving component is close to the core component. At time $t = t_{\rm e}$, the source has a distance $v\,t_{\rm e}$ from the original position. The observed separation is the transverse component of this distance,

$$\Delta r = v\,t_{\rm e} \sin\phi \; .$$

Since at time $t_{\rm e}$ the source has a smaller distance from Earth than at $t = 0$, the light will accordingly take slightly less time to reach us. Photons emitted at times $t = 0$ and $t = t_{\rm e}$ will reach us with a time difference of

$$\Delta t = t_{\rm e} - \frac{v\,t_{\rm e}\cos\phi}{c} = t_{\rm e}\,(1 - \beta\cos\phi) \; ,$$

where we define

$$\beta := \frac{v}{c} \tag{5.16}$$

as the velocity in units of the speed of light. Equation (5.15) then yields the apparent velocity,

$$\boxed{v_{\rm app} = \frac{\Delta r}{\Delta t} = \frac{v\sin\phi}{1 - \beta\cos\phi}} \; . \tag{5.17}$$

We can directly draw some conclusions from this equation. The apparent velocity $v_{\rm app}$ is a function of the direction of motion relative to the line-of-sight and of

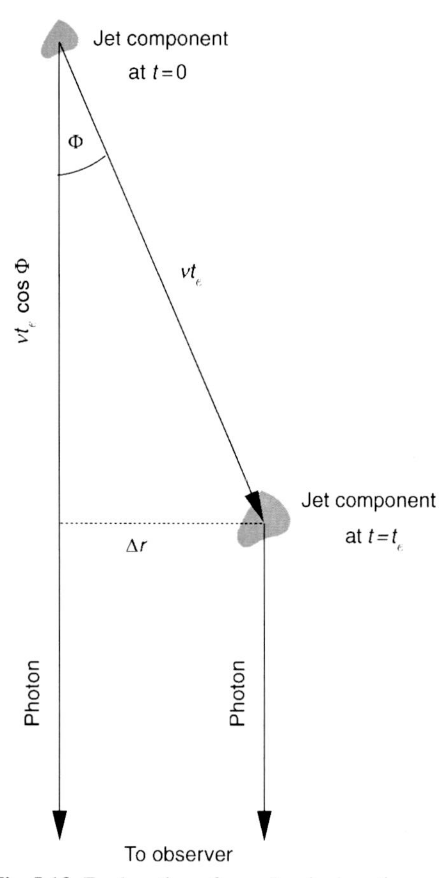

Fig. 5.13. Explanation of superluminal motion: a source component is moving at velocity v and at an angle ϕ relative to the line-of-sight. We consider the emission of photons at two different times $t=0$ and $t=t_e$. Photons emitted at $t=t_e$ will reach us by $\Delta t = t_e(1-\beta\cos\phi)$ later than those emitted at $t=0$. The apparent separation of the two source components then is $\Delta r = vt_e \sin\phi$, yielding an apparent velocity on the sky of $v_{app} = \Delta r/\Delta t = v\sin\phi/(1-\beta\cos\phi)$

the true velocity of the component. For a given value of v, the maximum velocity v_{app} is obtained if

$$(\sin\phi)_{max} = \frac{1}{\gamma} , \tag{5.18}$$

where the *Lorentz factor* $\gamma = (1-\beta^2)^{-1/2}$ was already defined in (5.4). The corresponding value for the maximum apparent velocity is then

$$\left(v_{app}\right)_{max} = \gamma\, v . \tag{5.19}$$

Since γ may become arbitrarily large for values of $v \to c$, the apparent velocity can be much larger than c,

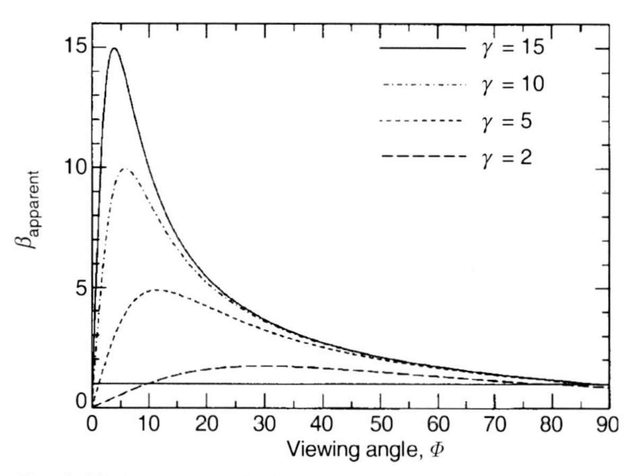

Fig. 5.14. Apparent velocity $\beta_{app} = v_{app}/c$ of a source component moving with Lorentz factor γ at an angle ϕ with respect to the line-of-sight, for four different values of γ. Over a wide range in θ, $\beta_{app} > 1$, thus apparent superluminal motion occurs. The maximum values for β_{app} are obtained if $\sin\theta = 1/\gamma$

even if the true velocity v is – as required by Special Relativity – smaller than c. In Fig. 5.14, v_{app} is plotted as a function of ϕ for different values of the Lorentz factor γ. To get $v_{app} > c$ for an angle ϕ, we need

$$\beta > \frac{1}{\sin\phi + \cos\phi} \geq \frac{1}{\sqrt{2}} \approx 0.707 .$$

Hence, superluminal motion is a consequence of the finiteness of the speed of light. Its occurrence implies that source components in the radio jets of AGNs are accelerated to velocities close to the speed of light.

In various astrophysical situations we find that the outflow speeds are of the same order as the escape velocities from the corresponding sources. Examples are the Solar wind, stellar winds in general, or the jets of neutron stars, such as in the famous example of SS433 (in which the jet velocity is $0.26\,c$). Therefore, if the outflow velocity of the jets in AGNs is close c, the jets should originate in a region where the escape velocity has a comparable value. The only objects compact enough to be plausible candidates for this are neutron stars and black holes. And since the central mass in AGNs is considerably larger than the maximum mass of a neutron star, a SMBH is the only option left for the central object. This argument, in addition, yields the conclusion that jets in AGNs must be formed and accelerated very close to the Schwarzschild radius of the SMBH.

The processes that lead to the formation of jets are still subject to intensive research. Most likely magnetic fields play a central role. Such fields may be anchored in the accretion disk, and then spun up and thereby amplified. The wound-up field lines may then act as a kind of spring, accelerating plasma outwards along the rotation axis of the disk. In addition, it is possible that rotational energy is extracted from a rotating black hole, a process in which magnetic fields again play a key role. As is always the case in astrophysics, detailed predictions in situations where magnetic fields dominate the dynamics of a system (like, e.g., in star formation) are extremely difficult to obtain because the corresponding coupled equations for the plasma and the magnetic field are very hard to solve.

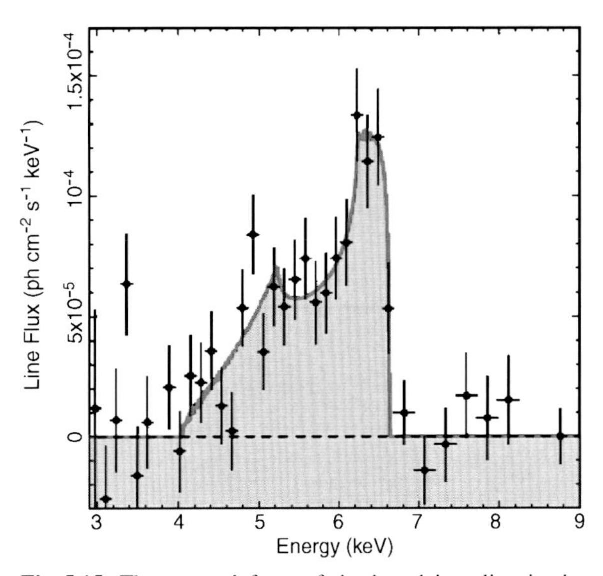

Fig. 5.15. The spectral form of the broad iron line in the Seyfert 1 galaxy MCG-6-30-15 as observed with the ASCA satellite. If the material emitting the line were at rest we would observe a narrow line at $h_P\nu = 6.35$ keV. We see that the line is (a) broad, (b) strongly asymmetric, and (c) shifted to smaller energies. A model for the shape of the line, based on a disk around a black hole that is emitting in the radius range $r_S \le r \le 20 r_S$, is sketched in Fig. 5.16

5.3.4 Further Arguments for SMBHs

A black hole is not only the simplest solution of the equations of Einstein's General Relativity, it is also the natural final state of a very compact mass distribution. The occurrence of SMBHs is thus highly plausible from a theoretical point of view. The evidence for the existence of SMBHs in the center of galaxies that has been detected in recent years (see Sect. 3.5) provides an additional argument for the presence of SMBHs in AGNs.

Furthermore, we find that the direction of the jets on a milliarcsecond scale, as observed by VLBI, is essentially identical to the direction of jets on much larger scales and to the direction of the corresponding radio lobes. These lobes often have a huge distance from the core, indicating a long lifetime of the source. Hence, the central engine must have some long-term memory because the outflow direction is stable over $\sim 10^7$ yr. A rotating SMBH is an ideal gyroscope, with a direction being defined by its angular momentum vector.

X-ray observations of an iron line of rest energy $h_P\nu = 6.35$ keV in Seyfert galaxies clearly indicate that the emission must be produced in the inner region of an accretion disk, within only a few Schwarzschild radii of a SMBH. An example for this is given in Fig. 5.15. The shape of the line is caused by a combination of a strong Doppler effect due to high rotation velocities in the disk and by the strong gravitational field of the black hole, as is explained in Fig. 5.16.

This iron line is not only detected in individual AGNs, but also in the average spectrum of an ensemble of AGNs. In a deep ($\sim 7.7 \times 10^5$ s) XMM-Newton exposure of the Lockman hole, a region of very low column density of Galactic hydrogen, a large number of AGNs were identified and spectroscopically verified. The X-ray spectrum of these AGNs in the energy ranges of 0.2 to 3 keV and of 8 to 20 keV (each in the AGN rest-frame) was modeled by a power law plus intrinsic absorption. The ratio of the measured spectrum of each individual AGN and the fitted model spectrum was then averaged over the AGN population, after transforming the spectra into the rest-frame of the individual sources. As shown in Fig. 5.17, this ratio clearly shows the presence of a strong and broad emission line. The shape of this average emission line can be very well modeled by emission from an accretion disk around a black hole where the radiation originates from a region lying between ~ 3 and ~ 400 Schwarzschild radii. The strength of the iron line indicates a high metallicity of the gas in these AGNs.

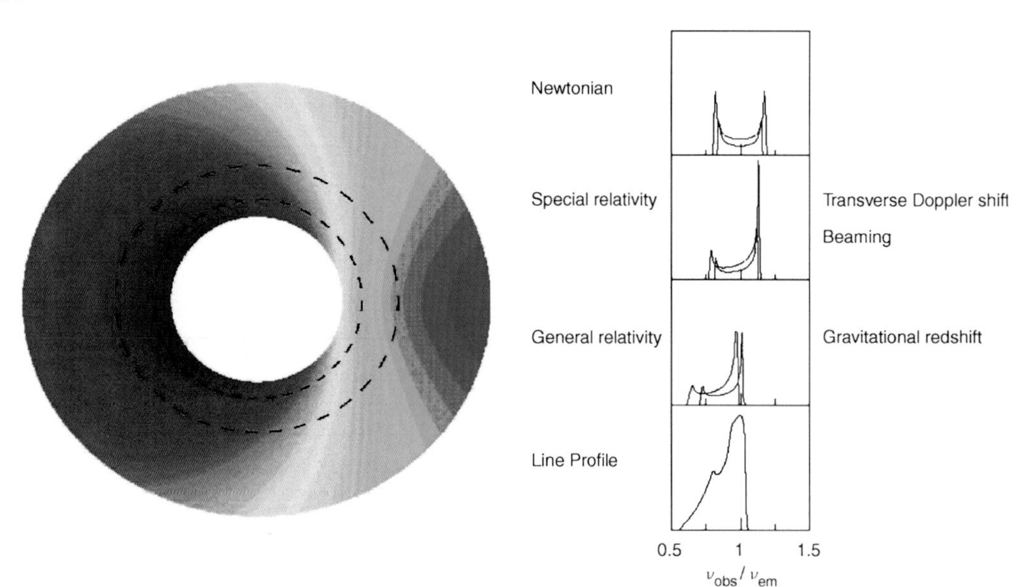

Fig. 5.16. The profile of the broad iron line is caused by a combination of Doppler shift, relativistic beaming, and gravitational redshift. On the left, the observed energy of the line as a function of position on a rotating disk is indicated by colors. Here, the energy in the right part of the disk which is moving towards us is blueshifted, whereas the left part of the disk emits redshifted radiation. Besides this Doppler effect, all radiation is redshifted because the photons must escape from the deep potential well. The smaller the radius of the emitting region, the larger this gravitational redshift. The line profile we would obtain from a ring-shaped section of the disk (dashed ellipses) is plotted in the panels on the right. The uppermost panel shows the shape of the line we would obtain if no relativistic effects occurred besides the non-relativistic Doppler effect. Below, the line profile is plotted taking the relativistic Doppler effect and beaming (see Eq. 5.31) into account. This line profile is shifted towards smaller energies by gravitational redshift so that, in combination, the line profile shown at the bottom results

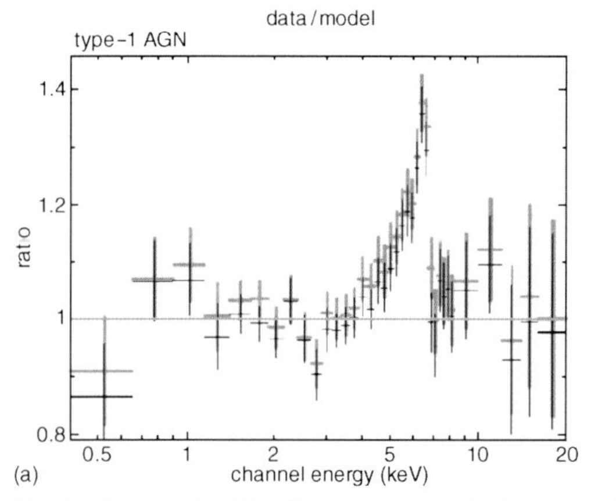

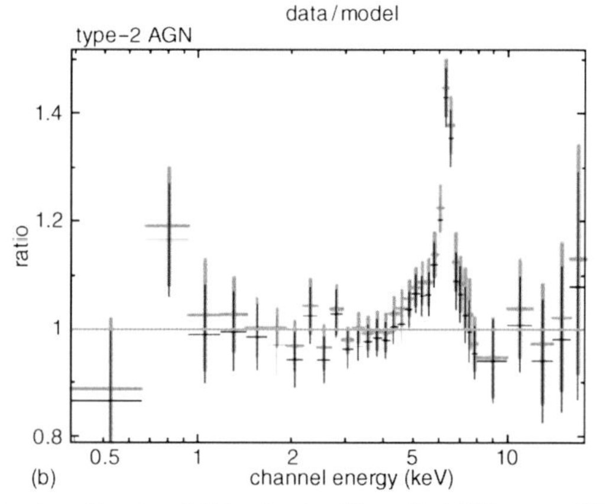

Fig. 5.17. The ratio of the X-ray spectrum of AGNs and a fitted power law averaged over 53 Type 1 AGNs (left panel) and 41 Type 2 AGNs (right panel). The gray and black data points are from two different detectors on-board the XMM-Newton observatory. In both AGN samples, a broad relativistic iron line is visible; in the Type 2 AGNs, an additional narrow line component at 6.4 keV can be identified. The line strength indicates that the average iron abundance in these sources is about three times the Solar value

5.3.5 A First Mass Estimate for the SMBH: The Eddington Luminosity

Radiation Force. As we have seen, the primary energy production in AGNs occurs through accretion of matter onto a SMBH, where the largest part of the energy is produced in the innermost region, close to the Schwarzschild radius. The energy produced in the central region then propagates outwards and can interact with infalling matter by absorption or scattering. Through this interaction of outward-directed radiation with matter, the momentum of the radiation is transferred to the matter, i.e., the infalling matter experiences an outwards-directed radiation force. In order for matter to fall onto the SMBH at all, this radiation force needs to be smaller than the gravitational force. This condition can be translated into a minimum mass of the SMBH, required for its gravity to dominate the total force at a given luminosity.

We consider a fully ionized gas, so that the interaction of radiation with this infalling plasma is basically due to scattering of photons by free electrons. This is called *Thomson scattering*. The mean radiation force on an electron at radius r is then

$$F_{\rm rad} = \sigma_{\rm T} \, \frac{L}{4\pi \, r^2 \, c} \, , \tag{5.20}$$

where

$$\sigma_{\rm T} = \frac{8\pi}{3} \left(\frac{e^2}{m_{\rm e} c^2} \right)^2 = 6.65 \times 10^{-25} \, {\rm cm}^2 \tag{5.21}$$

denotes the *Thomson cross-section* (in cgs units). This cross-section is independent of photon frequency.[5] To derive (5.20), we note that the flux $S = L/(4\pi r^2)$ is the radiation energy which flows through a unit area at distance r from the central source per unit time. Then S/c is the momentum of photons flowing through this unit area per time, or the radiation pressure, because the momentum of a photon is given by its energy divided by the speed of light. Thus the momentum transfer to an electron per unit time, or the radiation force, is given by $\sigma_{\rm T} S/c$. From (5.20), we can see that the radiation force has the same dependence on radius as the gravitational force, $\propto r^{-2}$, so that the ratio of the two forces is independent of radius.

[5]When a photon scatters off an electron at rest, this process is called Thomson scattering. To a first approximation, the energy of the photon is unchanged in this process, only its direction is different after scattering. This is not really true, though. Due to the fact that a photon with energy E_γ carries a momentum E_γ/c, scattering will impose a recoil on the electron. After the scattering event the electron will thus have a non-zero velocity and a corresponding kinetic energy. Owing to energy conservation the photon energy after scattering is therefore slightly smaller than before. This energy loss of the photon is very small as long as $E_\gamma \ll m_{\rm e} c^2$. When this energy loss becomes appreciable, this scattering process is then called Compton scattering. If the electron is not at rest, the scattering can also lead to net energy transfer to the photon, such as it happens when low-frequency photons propagate through a hot gas (as we will discuss in Sect. 6.3.4) or through a distribution of relativistic electrons. In this case one calls it the inverse Compton effect. The physics of all these effects is the same, only their kinematics are different.

Eddington Luminosity. For matter to be able to fall in – the condition for energy production – the radiation force must be smaller than the gravitational force. For each electron there is a proton, and these two kinds of particles are electromagnetically coupled. The gravitational force per electron–proton pair is given by

$$F_{\rm grav} = \frac{G M_\bullet m_{\rm p}}{r^2} \, ,$$

where we have neglected the mass of the electron since it is nearly a factor of 2000 smaller than the proton mass $m_{\rm p}$. Hence, the condition

$$F_{\rm rad} < F_{\rm grav} \tag{5.22}$$

for the dominance of gravity can be written as

$$\frac{\sigma_{\rm T} L}{4\pi \, r^2 \, c} < \frac{G M_\bullet m_{\rm p}}{r^2} \, ,$$

or

$$\boxed{L < L_{\rm edd} := \frac{4\pi G c m_{\rm p}}{\sigma_{\rm T}} M_\bullet \approx 1.3 \times 10^{38} \left(\frac{M_\bullet}{M_\odot} \right) {\rm erg/s}} \, , \tag{5.23}$$

where we have defined the *Eddington luminosity* $L_{\rm edd}$ *of a black hole of mass* $M_\bullet$. Since $\sigma_{\rm T}$ is independent of photon frequency, the luminosity referred to above is the bolometric luminosity.

For accretion to occur at all, we need $L < L_{\rm edd}$. Remembering that the Eddington luminosity is proportional to $M_\bullet$ we can turn the above argument around: if a luminosity L is observed, we conclude $L_{\rm edd} > L$, or

$$\boxed{M_\bullet > M_{\rm edd} := \frac{\sigma_{\rm T}}{4\pi G c m_{\rm p}} L \approx 8 \times 10^7 \left(\frac{L}{10^{46}\,{\rm erg/s}} \right) M_\odot} \;. \tag{5.24}$$

Therefore, a lower limit for the mass of the SMBH can be derived from the luminosity. For luminous AGNs, like QSOs, typical masses are $M_\bullet \gtrsim 10^8 M_\odot$, while Seyfert galaxies have lower limits of $M_\bullet \gtrsim 10^6 M_\odot$. Hence, the SMBH in our Galaxy could in principle provide a Seyfert galaxy with the necessary energy.

In the above definition of the Eddington luminosity we have implicitly assumed that the emission of radiation is isotropic. In principle, the above argument of a maximum luminosity can be avoided, and thus luminosities exceeding the Eddington luminosity can be obtained, if the emission is highly anisotropic. A geometrical concept for this would be, for example, accretion through a disk in the equatorial plane and the emission of a major part of the radiation along the polar axes (see Fig. 5.18). Models of this kind have indeed been constructed. It was shown that the Eddington limit may be exceeded by this, but not by a large factor. However, the possibility of anisotropic emission has another very important consequence. To derive a value for the luminosity from the observed flux of a source, the relation $L = 4\pi D_{\rm L}^2 \, S$ is applied, which is explicitly based on the assumption of isotropic emission. But if this emission is anisotropic and thus depends on the direction to the observer, the true luminosity may differ considerably from that which is derived under the assumption of isotropic emission. Later we will discuss the evidence for anisotropic emission in more detail.

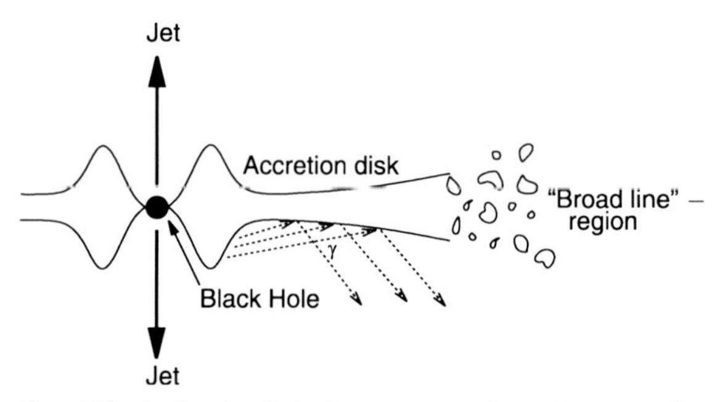

Fig. 5.18. A sketch of the innermost region of an accretion disk. Because of high temperatures in this region, radiation pressure can dominate the gas pressure inside the disk; this leads to an inflation into a thick disk. Radiation from the thick part of the disk can then hit the thin parts and be partially reflected. This reflection is a plausible explanation of the X-ray spectra of AGNs

Eddington Accretion Rate. If the conversion of infalling mass into energy takes place with an efficiency ϵ, the accretion rate can be determined,

$$\dot{m} = \frac{L}{\epsilon\, c^2} \approx 0.18\, \frac{1}{\epsilon} \left(\frac{L}{10^{46}\,{\rm erg/s}} \right) \left(\frac{M_\odot}{1\,{\rm yr}} \right) . \tag{5.25}$$

Since the maximum efficiency is of order $\epsilon \sim 0.1$, this implies accretion rates of typically several Solar masses per year for very luminous QSOs. If L is measured in units of the Eddington luminosity, we obtain with (5.23)

$$\dot{m} = \frac{L}{L_{\rm edd}} \left(\frac{1.3 \times 10^{38}\,{\rm erg/s}}{\epsilon c^2} \right) \left(\frac{M_\bullet}{M_\odot} \right) \equiv \frac{L}{L_{\rm edd}}\, \dot{m}_{\rm edd} \,, \tag{5.26}$$

where in the last step the Eddington accretion rate has been defined,

$$\dot{m}_{\rm edd} = \frac{L_{\rm edd}}{\epsilon c^2} \approx \frac{1}{\epsilon}\, 2 \times 10^{-9}\, M_\bullet\, {\rm yr}^{-1} \,. \tag{5.27}$$

Growth Rate of the SMBH Mass. The Eddington accretion rate is the maximum accretion rate if isotropic emission is assumed, and it depends on the assumed efficiency ϵ. We can now estimate a characteristic time in which the mass of the SMBH will significantly increase,

$$t_{\rm evo} := \frac{M_\bullet}{\dot{m}} \approx \epsilon \left(\frac{L}{L_{\rm edd}} \right)^{-1} 5 \times 10^8\,{\rm yr} \,, \tag{5.28}$$

i.e., even with efficient energy production ($\epsilon \sim 0.1$), the mass of a SMBH can increase greatly on cosmologically short time-scales by accretion. However, this is not the only mechanism which can produce SMBHs of large mass. They can also be formed through the merger of two black holes, each of smaller mass, as would be expected after the merger of two galaxies if both partners hosted a SMBH in its center. This aspect will be discussed more extensively later.

5.4 Components of an AGN

In contrast to stars, which have a simple geometry, we expect several source components in AGNs with different, sometimes very complex geometric configurations to produce the various components of the spectrum; this is sketched in Fig. 5.19. Accretion disks and jets in AGNs are clear indicators for a significant deviation from spherical symmetry in these sources. The relation between source components and the corresponding spectral components is not always obvious. However, combining theoretical arguments with detailed observations has led to quite satisfactory models.

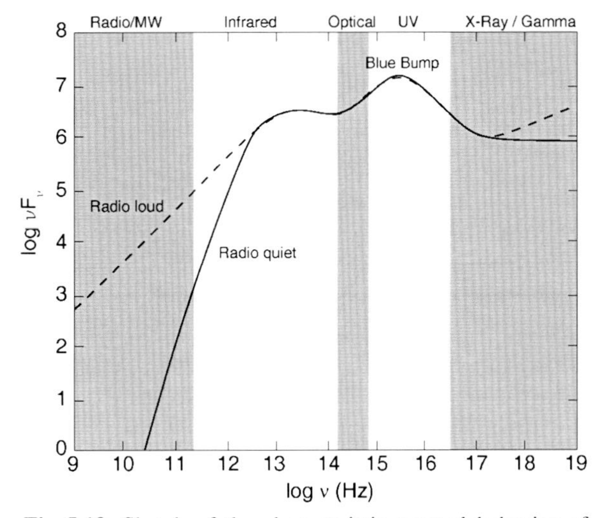

Fig. 5.19. Sketch of the characteristic spectral behavior of a QSO. We distinguish between radio-loud (dashed curve) and radio-quiet (solid curve) QSOs. Plotted is νS_ν (in arbitrary units), so that flat sections in the spectrum correspond to equal energy per logarithmic frequency interval. The most prominent feature is the big blue bump, a broad maximum in the UV up to the soft X-ray domain of the spectrum. Besides this maximum, a less prominent secondary maximum is found in the IR. The spectrum increases towards higher energies in the X-ray domain of the spectrum – typically $\sim 10\%$ of the total energy is emitted as X-rays

5.4.1 The IR, Optical, and UV Continuum

In Sect. 5.3.2 we considered an accretion disk with a characteristic temperature, following from (5.14), of

$$T(r) \approx 6.3 \times 10^5\ \mathrm{K} \left(\frac{\dot{m}}{\dot{m}_{\mathrm{edd}}}\right)^{1/4} \times \left(\frac{M_\bullet}{10^8 M_\odot}\right)^{-1/4} \left(\frac{r}{r_\mathrm{S}}\right)^{-3/4} . \tag{5.29}$$

The thermal emission of an accretion disk with this radial temperature profile produces a broad spectrum with its maximum in the UV. The continuum spectrum of QSOs indeed shows an obvious increase towards UV wavelengths, up to the limit of observable wavelengths, $\lambda \gtrsim 1000$ Å. (This is the observed wavelength; QSOs at high redshifts can be observed at significantly shorter wavelengths in the QSO rest-frame.) At wavelengths $\lambda \le 912$ Å, photoelectric absorption by neutral hydrogen in the ISM of the Galaxy sets in, so that the Milky Way is opaque for this radiation. Only at considerably higher frequencies, namely in the soft X-ray band ($h_\mathrm{P}\nu \gtrsim 0.2$ keV), does the extragalactic sky become observable again.

If the UV radiation of a QSO originates mainly from an accretion disk, which can be assumed because of the observed increase of the spectrum towards the UV, the question arises whether the thermal emission of the disk is also visible in the X-ray domain. In this case, the spectrum in the range hidden from observation, at $13\ \mathrm{eV} \lesssim h_\mathrm{P}\nu \lesssim 0.2$ keV, could be interpolated by such an accretion disk spectrum. This seems indeed to be the case. The X-ray spectrum of QSOs often shows a very simple spectral shape in the form of a power law, $S_\nu \propto \nu^{-\alpha}$, where $\alpha \sim 0.7$ is a characteristic value. However, the spectrum follows this power law only at energies down to ~ 0.5 keV. At lower energies, the spectral flux is higher than predicted by the extrapolation of the power-law spectrum observed at higher energies. One interpretation of this finding is that the (non-thermal) source of the X-ray emission produces a simple power law, and the additional flux at lower X-ray energies is thermal emission from the accretion disk (see Fig. 5.19).

Presumably, these two spectral properties – the increase of the spectrum towards the UV and the radiation excess in the soft X-ray – have the same origin, being two wings of a broad maximum in the energy distribution, which itself is located in the spectral range unobservable for us. This maximum is called the *big blue bump* (BBB). A description of the BBB is possible using detailed models of accretion disks (Fig. 5.20). For this modeling, however, the assumption of a local

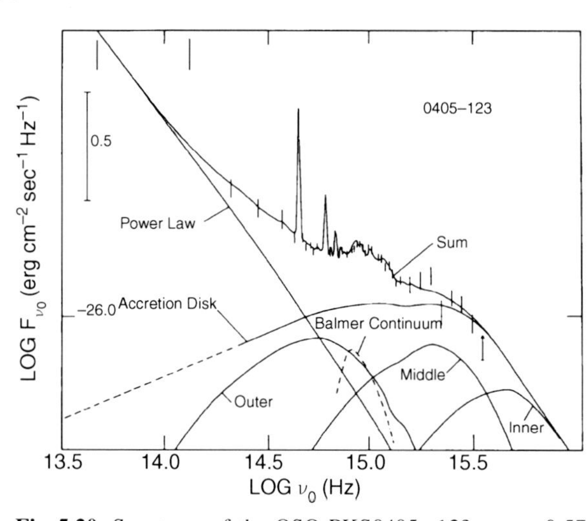

Fig. 5.20. Spectrum of the QSO PKS0405−123 at $z = 0.57$ (data points with error bars) from the NIR and the optical up to the UV spectral region, plus a model for this spectrum (solid curve). The latter combines various components: (1) the radiation from an accretion disk that causes the big blue bump and whose spectrum is also shown for three individual radius ranges, (2) the Balmer continuum, and (3) an underlying power law which may have its origin in synchrotron emission

Planck spectrum at all radii of the disk is too simple because the structure of the accretion disk is more complicated. The spectral properties of an accretion disk have to be modeled by an "atmosphere" for each radius, similar to that in stars.

Besides the BBB, an additional maximum exists in the MIR (IR-bump). This can be described by thermal emission of warm dust ($T \lesssim 2000$ K). Later in this chapter we will discuss other observations which provide additional evidence for this dust component.

The optical continuum of blazars is different from that of Seyfert galaxies and QSOs. It often features a spectral pattern that follows, to very good approximation, a power law and is strongly variable and polarized. This indicates that the radiation is predominantly non-thermal. The origin of this radiation thus probably does not lie in an accretion disk. Rather, the radiation presumably has its origin in the relativistic jets which we already discussed for the radio domain, with their synchrotron radiation extending up to optical wavelengths. This assumption was strongly supported by many sources where (HST) observations discovered optical emission from jets (see Fig. 5.12 and Sect. 5.5.4).

5.4.2 The Broad Emission Lines

Characteristics of the Broad Line Region. One of the most surprising characteristics of AGNs is the presence of very broad emission lines. Interpreted as Doppler velocities, the corresponding width of the velocity distribution of the components in the emitting region is of order $\Delta v \lesssim 10\,000$ km/s (or $\Delta\lambda/\lambda \lesssim 0.03$). These lines cannot be due to thermal line broadening because that would imply $k_B T \sim m_p(\Delta v)^2/2 \sim 1$ MeV, or $T \sim 10^{10}$ K – no emission lines would be produced at such high temperatures because all atoms would be fully ionized (plus the fact that at such temperatures a plasma would efficiently produce e^+e^--pairs, and the corresponding annihilation line at 511 keV should be observable in Gamma radiation). Therefore, the observed line width is interpreted as Doppler broadening. The gas emitting these lines then has large-scale velocities of order $\sim 10\,000$ km/s. Velocities this high are indicators of the presence of a strong gravitational field, as would occur in the vicinity of a SMBH. If the emission of the lines occurs in gas at a distance r from a SMBH, we expect characteristic velocities of

$$v_{\rm rot} \sim \sqrt{\frac{GM_\bullet}{r}} = \frac{c}{\sqrt{2}}\left(\frac{r}{r_S}\right)^{-1/2} ,$$

so for velocities of $v \sim c/30$, we obtain a radial distance of

$$\left(\frac{r}{r_S}\right) \sim 500 .$$

Hence, the Doppler broadening of the broad emission lines can be produced by Kepler rotation at radii of about $1000\, r_S$. Although this estimate is based on the assumption of a rotational motion, the infall velocity for free fall does not differ by more than a factor $\sqrt{2}$ from this rotational velocity. Thus the kinematic state of the emitting gas is of no major relevance for this rough estimate if only gravity is responsible for the occurrence of high velocities.

The region in which the broad emission lines are produced is called the *broad-line region* (BLR). The density of the gas in the BLR can be estimated from the lines that are observed. To see this, it must be pointed out that allowed and semi-forbidden transitions are found among the broad lines. Examples of the former are Lyα, MgII, and CIV, whereas CIII] and NIV] are semi-forbidden transitions. However, no forbidden transitions are observed among the broad lines. The classification into allowed, semi-forbidden, and forbidden transitions is done by means of quantum mechanical transition probabilities, or the resulting mean time for a spontaneous radiational transition. Allowed transitions correspond to electric dipole radiation, which has a large transition probability, and the lifetime of the excited state is then typically only 10^{-8} s. For forbidden transitions, the time-scales are considerably larger, typically 1 s, because their quantum mechanical transition probability is substantially lower. Semi-forbidden transitions have a lifetime between these two values. To mark the different kinds of transitions, a double square bracket is used for forbidden transitions, like in [OIII], while semi-forbidden lines are marked by a single square bracket, like in CIII].

An excited atom can transit into its ground state (or another lower-lying state) either by spontaneous emission of a photon or by losing energy through collisions with other atoms. The probability for a radiational transition is defined by the atomic parameters, whereas the collisional de-excitation depends on the gas density. If the density of the gas is high, the mean time between two collisions is much shorter than the average lifetime of forbidden or semi-forbidden radiational transitions. Therefore the corresponding line photons are not observed.[6] The absence of forbidden lines is then used to derive a lower limit for the gas density, and the occurrence of semi-forbidden lines yields an upper bound for the density. To minimize the dependence of this argument on the chemical composition of the gas, transitions of the same element are preferentially used for these estimates. However, this is not always possible. From the presence of the CIII] line and the non-existence of the [OIII] line in the BLR, combined with model calculations, a density estimate of $n_e \sim 3 \times 10^9\,\mathrm{cm}^{-3}$ is obtained.

Furthermore, from the ionization stages of the line-emitting elements, a temperature can be estimated, typically yielding $T \sim 20\,000$ K. Detailed photoionization models for the BLR are very successful and are able to reproduce details of line ratios very well.

[6]To make forbidden transitions visible, the gas density needs to be very low. Densities this low cannot be produced in the laboratory. Forbidden lines are in fact not observed in laboratory spectra; they are "forbidden".

From the density of the gas and its temperature, the emission measure can then be calculated (i.e., the number of line photons per volume element). From the observed line strength and the distance to the AGN, the total number of emitted line photons can be calculated, and by dividing through the emission measure, the volume of the line-emitting gas can be determined. This estimated volume of the gas is much smaller than the total volume ($\sim r^3$) of the BLR. We therefore conclude that the BLR is not homogeneously filled with gas; rather, the gas has a very small filling factor. The gas in which the broad lines originate fills only $\sim 10^{-7}$ of the total volume of the BLR; hence, it must be concentrated in clouds.

Geometrical Picture of the BLR. From the previous considerations, a picture of the BLR emerges in which it contains gas clouds with a characteristic particle density of $n_e \sim 10^9\,\mathrm{cm}^{-3}$. In these clouds, heating and cooling processes take place. Probably the most important cooling process is the observed emission in the form of broad emission lines. Heating of the gas is provided by energetic continuum radiation from the AGN which photoionizes the gas, similar to processes in Galactic gas clouds. The difference between the energy of a photon and the ionization energy yields the energy of the released electron, which is then thermalized by collisions and leads to gas heating. In a stationary state, the heating rate equals the cooling rate, and this equilibrium condition defines the temperature the clouds will attain.

The comparison of continuum radiation and line emission yields the fraction of ionizing continuum photons which are absorbed by the BLR clouds; a value of about 10% is obtained. Since the clouds are optically thick to ionizing radiation, the fraction of absorbed continuum photons is also the fraction of the solid angle subtended by the clouds, as seen from the central continuum source. From the filling factor and this solid angle, the characteristic size of the clouds can be estimated, from which we obtain typical values of $\sim 10^{11}$ cm. In addition, based on these argument, the number of clouds in the BLR can be estimated. This yields a typical value of $\sim 10^{10}$.

The characteristic velocity of the clouds corresponds to the line width, hence several thousand km/s. However, the kinematics of the clouds are unknown. We do not know whether they are rotating around the SMBH, whether they are infalling or streaming outwards, or whether their motion is rather chaotic. It is also possible that different regions within the BLR exist with different kinematic properties.

Reverberation Mapping. A direct method to examine the extent of the BLR is provided by *reverberation mapping*. This observational technique utilizes the fact that heating and ionization of the gas in the BLR are both caused by the central continuum source of the AGN. Since the UV radiation of AGNs varies, we expect corresponding variations of the physical conditions in the BLR. In this picture, a decreasing continuum flux should then lead to a lower line flux, as is demonstrated in Fig. 5.21. Due to the finite extent of the BLR, the observed variability in the lines will be delayed in time compared to the ionizing continuum. This delay Δt can be identified with the light travel time across the BLR, $\Delta t \sim r/c$. In other words, the BLR feels the variation in the continuum source only after a delay of Δt. From the observed correlated variabilities of continuum and line emission, Δt can be determined for different line transitions, and so the corresponding values of r can be estimated.

Such analyses of reverberation mapping are extremely time-consuming and complex because one needs to continuously monitor the continuum light and, simultaneously, the line fluxes of an AGN over a long period. The relevant time-scales are typically months for Seyfert 1 galaxies (see Fig. 5.22). To perform such measurements, coordinated campaigns involving many observatories are necessary because the light curves have to be observed without any gaps, and one should not depend on the local weather conditions at any observatory. From the results of such campaigns and the correlation of the light curves in the UV continuum and the different line fluxes (Fig. 5.23), the picture of an inhomogeneous BLR is obtained which extends over a large range in r and which consists of different "layers". The extent of the BLR scales with the luminosity of the AGN. Its ionization structure varies with r; the higher the ionization energy of a transition, the smaller the corresponding radius r. For the

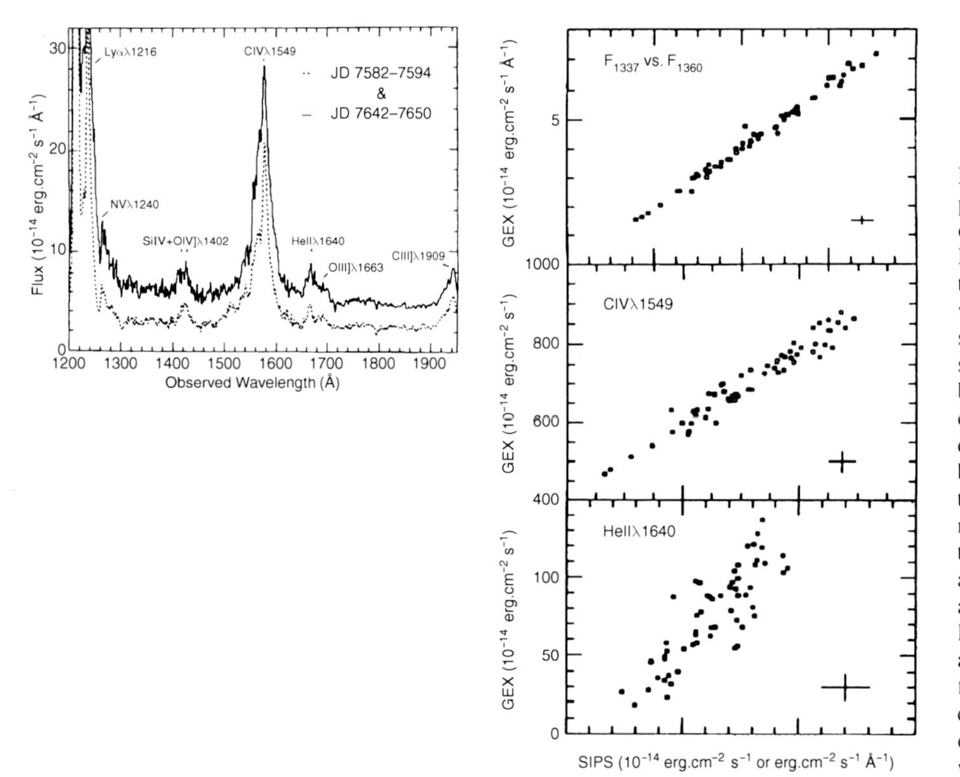

Fig. 5.21. In the left-hand panel, the UV spectrum of the Seyfert 1 galaxy NGC 5548 is plotted for two different epochs in which the source radiated strongly and weakly, respectively. It can clearly be seen that not only does the continuum radiation of the source vary but also the strength of the emission lines. The right-hand panels show the flux of the continuum at ~ 1300 Å, the C IV line at $\lambda = 1549$ Å, and the He II line at $\lambda = 1640$ Å, as a function of the near-UV flux at different epochs during an eight-month observational campaign with the IUE

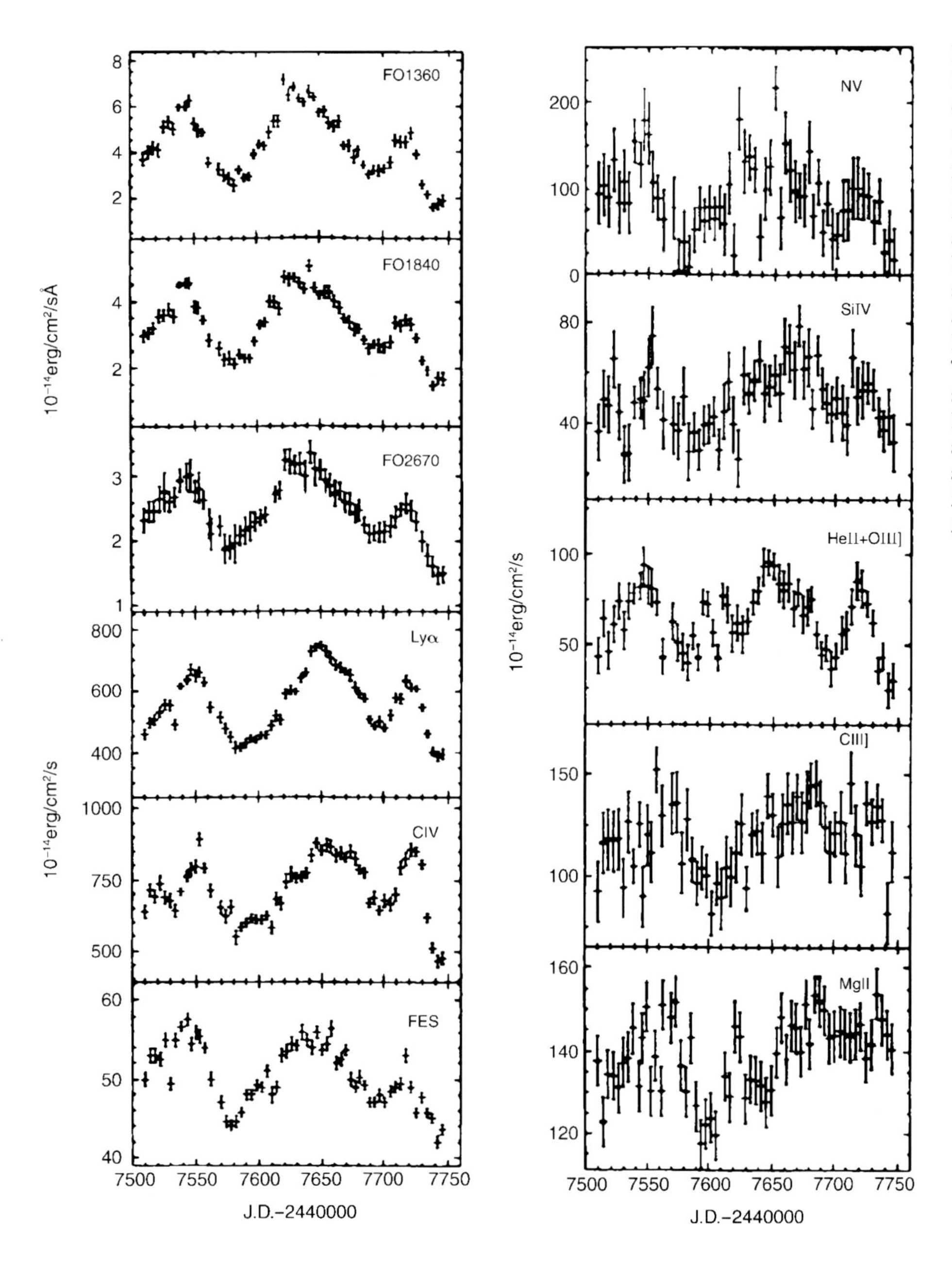

Fig. 5.22. Light curve of NGC 5548 over a period of 8 months at different wavelengths. In the left-hand panels, from top to bottom, the continuum at $\lambda = 1350$ Å, $\lambda = 1840$ Å, and $\lambda = 2670$ Å, the broad and strong emission lines Lyα and CIV, as well as the optical light curve are plotted. The right-hand panels shows the weaker lines NV at $\lambda = 1240$ Å, SiIV at $\lambda = 1402$ Å, HeII+OIII] at $\lambda = 1640$ Å, CIII] at $\lambda = 1909$ Å, and MgII at $\lambda = 2798$ Å

Seyfert 1 galaxy NGC 5548, one obtains $\Delta t \sim 12$ d for Lyα, about $\Delta t \sim 26$ d for CIII], and about 50 d for MgII. This may not come as a surprise because the ionizing flux increases for smaller r. Furthermore, the relative flux variations in lines of higher ionization energy are larger, as can also be seen in Fig. 5.9. This picture is also consistent with the different width of the various lines, in that lines of higher ionization energy are broader. In addition, lines of higher ionization energy have a mean redshift systematically shifted bluewards compared to narrower emission lines. This hints at a radial outward motion of the clouds together with an intrinsic absorption of part of the BLR radiation by intervening absorbing material. For a simple picture, but not the

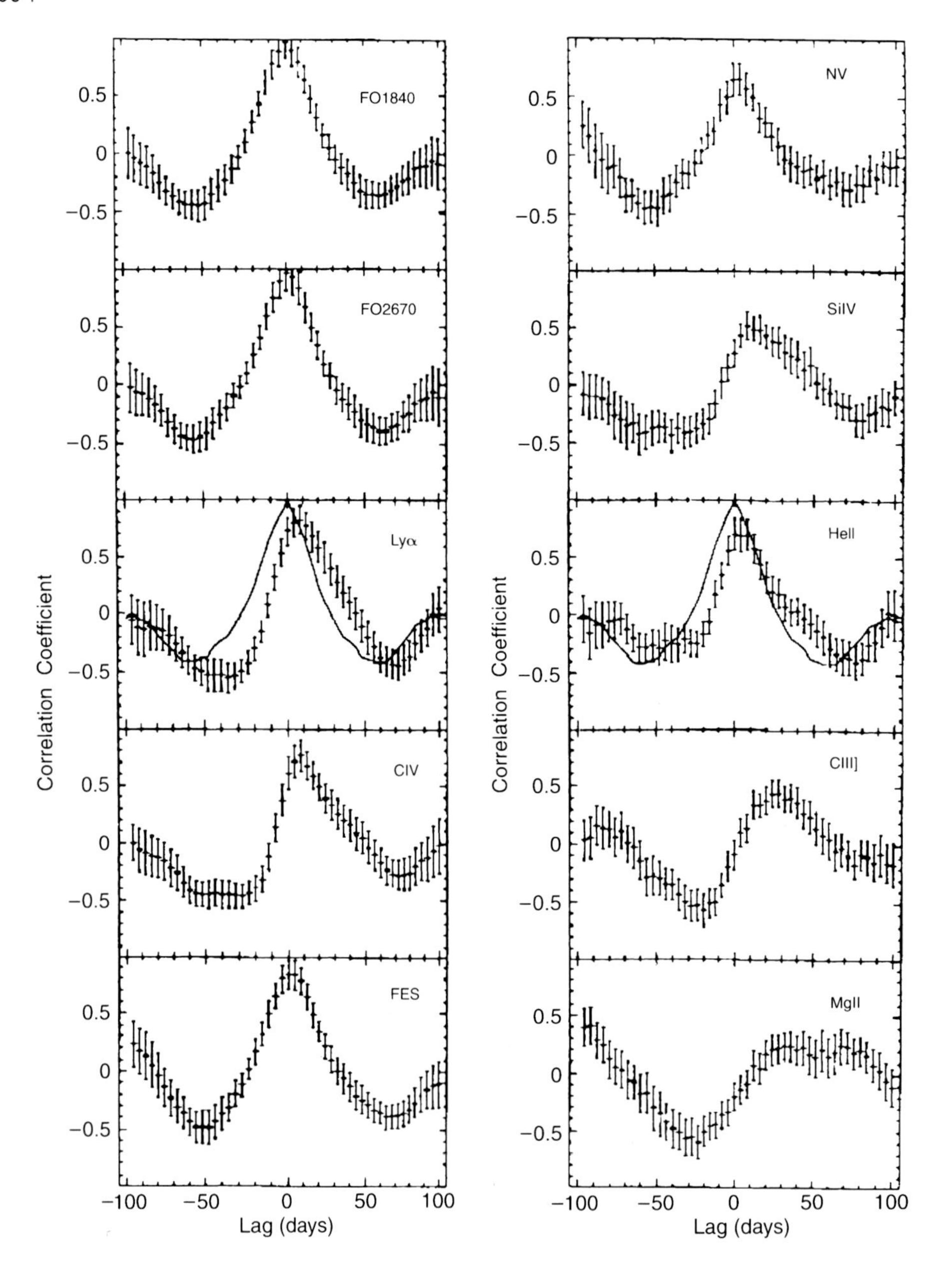

Fig. 5.23. The different light curves from Fig. 5.22 are correlated with the continuum flux at $\lambda = 1350$ Å. The autocorrelation function is shown by the solid line in the central panels, the others are cross-correlation functions. We can see that the maximum of the correlation is shifted towards positive times – variations in the continuum flux are not simultaneously followed by the emission lines but appear only after a delay. This delay corresponds to the light travel time from the center of the AGN to the clouds of the BLR where the lines are emitted. The smaller the ionization level of the respective ion, the longer the delay. For example, we obtain a delay of 12 days for Lyα, 26 days for CIII], and about 50 days for MgII, where the latter value could not be measured exactly because the relative flux variations of this line are small and thus the correlation function does not show a very prominent maximum

only plausible one, an outflow motion of the clouds in the BLR can be assumed, where we see those clouds that are, from our point of view, located behind the accretion disk partly absorbed by the disk material. The received line radiation is therefore dominated by the clouds that are in front of the disk and moving towards us, so that it is systematically blueshifted.

The nature of the clouds in the BLR is unknown. Their small extent and high temperature imply that they should vaporize on very small time-scales unless they are somehow stabilized. Therefore these clouds need to be either permanently replenished or they have to be stabilized, either by external pressure, e.g., from a very hot but thin medium in the BLR in between the clouds, by magnetic fields, or even gravitationally. One possibility is that the clouds are the extended atmospheres of stars; this would, however, imply a very high (too high?) total mass of the BLR.

5.4.3 Narrow Emission Lines

Besides the broad emission lines that occur in QSOs, Seyfert 1 galaxies, and broad-line radio galaxies, most AGNs (with exception of the BL Lacs) show narrow emission lines. Their typical width is $\sim 400\,\mathrm{km/s}$. This is considerably narrower than lines of the BLR, but still significantly broader than characteristic velocities in normal galaxies. In analogy to the BLR, the region in which these lines are produced is known as the *narrow line region* (NLR). The strongest line from the NLR is, besides Lyα and CIV, the forbidden [OIII] line at $\lambda = 5007\,\text{Å}$. The existence of forbidden lines implies that the gas density in the NLR is significantly lower than in the BLR. From estimates analogous to those for the BLR, the characteristic properties of the NLR are determined. It should be noted that no reverberation mapping can be applied, since the extent of the NLR is $\sim 100\,\mathrm{pc}$. Because of this large extent, no variability of the narrow line intensities is expected on time-scales accessible to observation, and none has been found. The line ratios of allowed and forbidden lines yield $n_\mathrm{e} \sim 10^3\,\mathrm{cm}^{-3}$ for the typical density of the gas in which the lines originate. The characteristic temperature of the gas is likewise obtained from line ratios, $T \sim 16\,000\,\mathrm{K}$, which is slightly lower than in the BLR. The filling factor here is also significantly smaller than one, about 10^{-2}. Hence, the geometrical picture of clouds in the NLR also emerges. Like in the BLR, the properties of the NLR are not homogeneous but vary with r.

Since the extent of the NLR in Seyfert galaxies is of the order of $r \sim 100\,\mathrm{pc}$, it can be spatially resolved for nearby Seyfert galaxies. The morphology of the NLR is very interesting: it is not spherical, but appears as two cone-shaped regions (Fig. 5.24). It seems as if the ionization of the NLR by the continuum radiation of the AGN is not isotropic, but instead depends strongly on the direction.

5.4.4 X-Ray Emission

The most energetic radiation of an AGN is expected to be produced in the immediate vicinity of the SMBH.

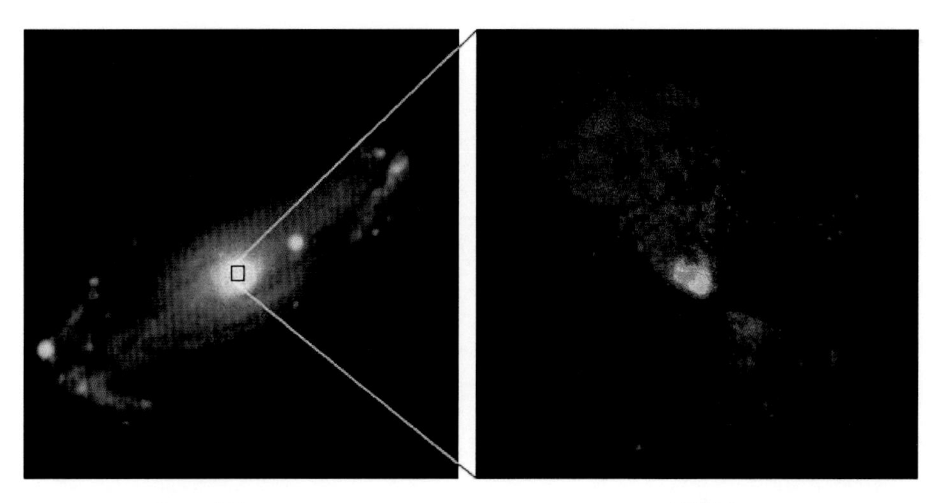

Fig. 5.24. Image of the Seyfert galaxy NGC 5728. Left: a large-scale image showing the disk galaxy; right: an HST image of its central region taken through a filter with a small bandwidth (narrow-band filter) centered on a narrow emission line. This image shows the spatially resolved NLR. We can see that it is not spherical but consists of two cones. From this, it is concluded that the ionizing radiation of the AGN is not isotropic, but is emitted in two preferred directions perpendicular to the disk of the Galaxy (and thus probably perpendicular to the central accretion disk)

Therefore, the X-ray emission of AGNs is of special interest for probing the innermost regions of these objects, as we have already seen from the relativistic iron line shown in Fig. 5.15. In fact, the variability on very short time-scales (see Fig. 5.9) is a clear indicator of a small extent of the X-ray source.

To a first approximation, the X-ray spectrum is characterized by a power law, $S_\nu \propto \nu^{-\alpha}$, with slope $\alpha \sim 0.7$. At energies $h_P\nu \gtrsim 10$ keV, the spectrum exceeds the extrapolation of this power law, i.e., it becomes flatter. Towards lower X-ray energies, the spectrum seems to be steeper than the power law, which presumably results from the blue part of the BBB, as was mentioned previously.

Besides this continuum radiation, emission and absorption lines are also found in the X-ray domain, the strongest lines being those of highly ionized iron. The improved sensitivity and spectral resolution of the X-ray telescopes Chandra and XMM-Newton compared to earlier X-ray observatories have greatly advanced the X-ray spectroscopy of AGNs. Figure 5.25 shows an example of the quality of these spectra.

The X-ray emission of Seyfert 1 and Seyfert 2 galaxies is very different. In the energy range of the ROSAT X-ray satellite (0.1 keV $\le h_P\nu \le 2.4$ keV), significantly more Seyfert 1 galaxies were discovered than Seyfert 2 galaxies. The origin of this was later uncovered by Chandra and XMM-Newton. In contrast to ROSAT, these two satellites are sensitive up to energies of $h_P\nu \sim 10$ keV and they have found large numbers of Seyfert 2 galaxies. However, their spectrum differs from that of Seyfert 1 galaxies because it is cut off towards lower X-ray energies. The spectrum indicates the presence of an absorber with a hydrogen column density of $\gtrsim 10^{22}\,\mathrm{cm}^{-2}$ and in some cases even orders of magnitude higher. This fact will be used in the context of unified models (Sect. 5.5) of AGNs.

5.4.5 The Host Galaxy

As the term "active galactic nuclei" already implies, AGNs are considered the central engine of otherwise quite normal galaxies. This nuclear activity is nourished by accretion of matter onto a SMBH. Since it seems that all galaxies (at least those with a spheroidal component) harbor a SMBH, the question of activity is rather one of accretion rate. What does it take to turn on a Seyfert galaxy, and why are most SMBHs virtually inactive? And by what mechanism is matter brought into the vicinity of the SMBH to serve as fuel?

For a long time it was not clear as to whether QSOs are also hosted in a galaxy. Their high luminosity renders it difficult to identify the surrounding galaxy on images taken from the ground, with their resolution being limited by seeing to $\sim 1''$. In the 1980s, the surrounding galaxies of some QSOs were imaged for the first time, but only with the HST did it become pos-

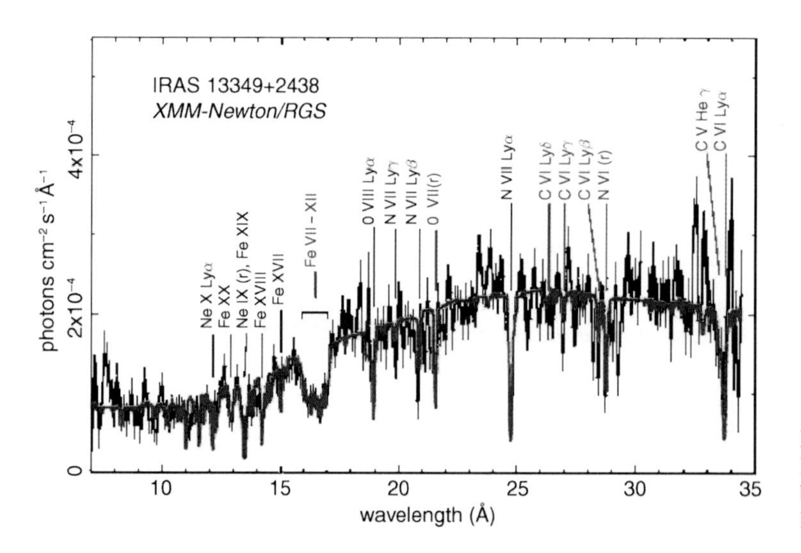

Fig. 5.25. X-ray spectrum of the quasar IRAS 13349+2438 ($z = 0.108$), observed by the XMM satellite. Various absorption lines are marked

sible to obtain detailed images of QSO host galaxies (see Fig. 5.26) and thus to include them in the class of galactic nuclei. In these investigations, it was also found that the host galaxies of QSOs are often heavily disturbed, e.g., by tidal interaction with other galaxies or even by merging processes. These disturbances of the gravitational potential are considered essential for the gas to overcome the angular momentum barrier and to flow towards the center of the galaxy. At the same time, such disturbances seem to increase the star-formation rate enormously, because starburst galaxies are also often characterized by disturbances and interactions. A close connection seems to exist between AGN activity and starbursts. Optical and NIR images of QSOs (see Fig. 5.26) cannot unambiguously answer the question of whether QSO hosts are spirals or ellipticals.

Today it seems established that the hosts of low-redshift QSOs are predominantly massive and bulge-dominated galaxies. This finding is in good agreement with the fact that the black hole mass in "normal" galaxies scales with the mass of the spheroidal component of the galaxies. It was recently found that higher-redshift QSOs are also hosted by massive elliptical galaxies. Furthermore, the host galaxies of radio-loud QSOs seem to be systematically more luminous than that of radio-quiet QSOs.

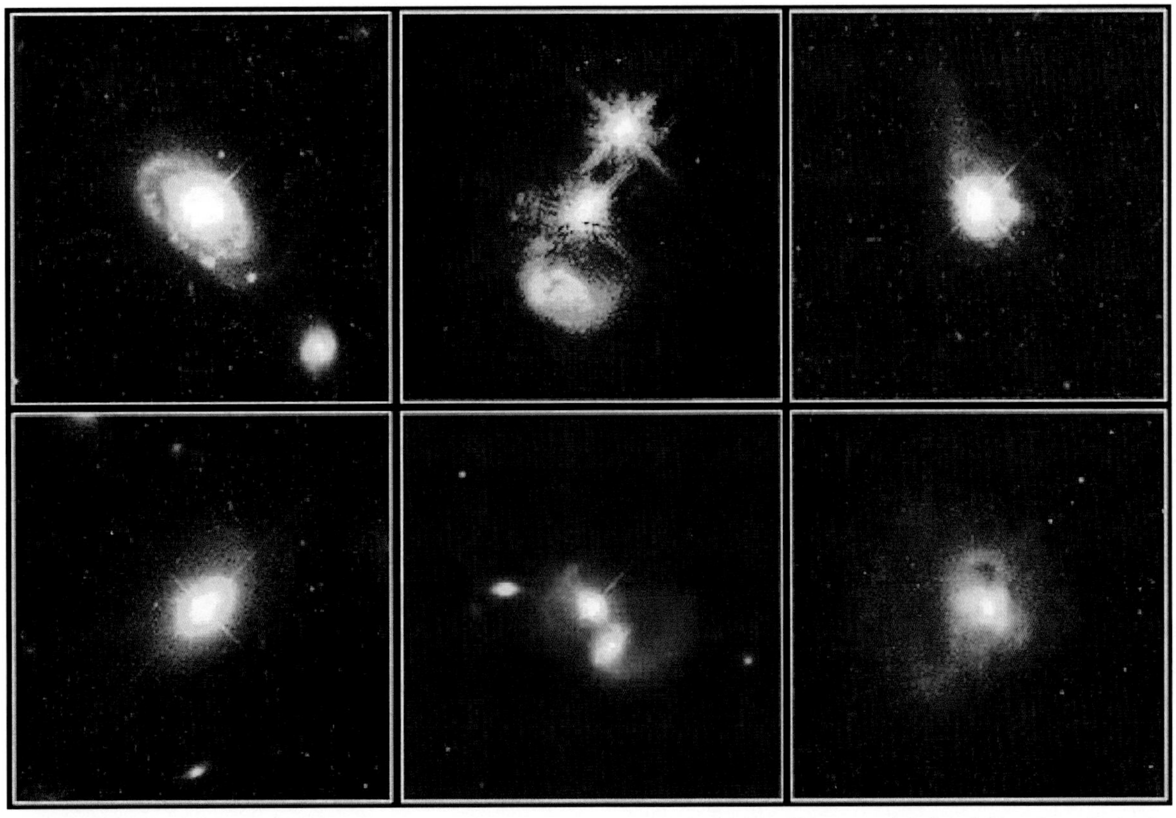

Fig. 5.26. HST images of QSOs. In all cases the host galaxy can clearly be identified, with the QSO itself being visible as a (central) point source in these images. Top left: PG 0052+251 is located in the center of an apparently normal spiral galaxy. Lower left: PHL 909 seems to be located in the center of a normal elliptical galaxy. Top center: the QSO IRAS 04505−2958 is obviously part of a collision of two galaxies and may be provided with "fuel" by material ripped from the galaxies by tidal forces. Surrounding the QSO core, a region of active star formation is visible. PG 1012+008 (lower center) is also part of a pair of merging galaxies. Top right: the host galaxy of QSO 0316−346 seems to be about to capture a tidal tail. Lower right: the QSO IRAS 13218+0552 seems to be located in a galaxy which just went through a merger process

Binary QSOs. The connection between the activity of galaxies and the presence of close neighbors is also seen from the clustering properties of QSOs. In surveys for gravitational lens systems, pairs of QSO images have been detected which have angular separations of a few arcseconds and very similar redshifts, but sufficiently different spectra to exclude them being gravitationally lensed images of the same source. The number of binary QSOs thus found is considerably larger than the expectation from the large-scale correlation function of QSOs. This conclusion was further strengthened by an extensive analysis from the QSOs in the Sloan Digital Sky Survey (see Sect. 8.1.2). The correlation function of QSOs at separations below $\sim 30h^{-1}$ kpc exceeds that of the extrapolation of the correlation function from larger scales by a factor of 10 or more. Hence it seems that the small-scale clustering of QSOs is very much enhanced, say compared to normal galaxies, which could be due to the triggering of activity by the proximity of the neighbor: in this case, both galaxies attain a perturbed gravitational potential and start to become active.

5.4.6 The Black Hole Mass in AGNs

We now return to the determination of the mass of the central black hole in AGNs. In Sect. 5.3.5, a lower limit on the mass was derived, based on the fact that the luminosity of an AGN cannot exceed the Eddington luminosity. However, this estimate cannot be very precise, for at least two reasons. The first is related to the anisotropic appearance of an AGN. The observed flux can be translated into a luminosity only on the assumption that the emission from the AGN is isotropic, and we have discussed several reasons why this assumption is not justified in many cases. Second, we do not have a clear idea what the ratio of AGN luminosity to its Eddington luminosity is. It is clear that this ratio can vary a lot between different black holes. For example, the black hole at the center of our Galaxy could power a luminosity of several 10^{44} erg/s if radiating with the Eddington luminosity – and we know that the true luminosity is several order of magnitudes below this value.

$M_\bullet$ from Reverberation Mapping. A far more accurate method for estimating the black hole mass in AGNs comes from reverberation mapping which we described in Sect. 5.4.2. The principal quantity that is derived from this technique is the size r of the BLR for a given atomic line or for a given ionization state of a chemical element. Furthermore, the relative line width $\Delta\lambda/\lambda$ can be measured, and can be related to the characteristic velocity dispersion σ in the BLR, $\sigma = c\,\Delta\lambda/\lambda$. Assuming that the gas is virialized, or moving on Keplerian orbits around the black hole, the mass of the latter can be estimated to be

$$M_\bullet \approx r\,\sigma^2/G\,, \tag{5.30}$$

where the difference between random motion and circular orbits corresponds to a factor of order 2 in this estimate. Thus, once reverberation mapping has been conducted, the black hole mass can be estimated with very reasonable accuracy.

However, this is a fairly expensive observing technique, requiring the photometric and spectroscopic monitoring of sources over long periods of time, and it can therefore be applied only to relatively small samples of sources. Furthermore, this technique is restricted to low-luminosity AGNs, since the size of the BLR, and thus the time delay and the necessary length of the monitoring campaign, increases with the black hole mass. We might therefore want to look at alternative methods for estimating $M_\bullet$.

$M_\bullet$ from Scaling Relations. When applied to a set of nearby Seyfert 1 galaxies, for which reverberation mapping has been carried out, one finds that the black hole mass in these AGNs satisfies the same relation (3.35) between $M_\bullet$ and the velocity dispersion $\sigma_{\rm e}$ of the bulge as has been obtained for inactive galaxies. This scaling relation then yields a useful estimate of the black hole mass from the stellar velocity dispersion. Unfortunately, even this method cannot be applied to a broad range of AGNs, since the velocity dispersion of stars cannot be measured in AGNs which are either too luminous – since then the nuclear emission outshines the stellar light, rendering spectroscopy of the latter impossible – or too distant, so that a spatial separation of nuclear light from stellar light is no longer possible.

However, another scaling relation was found which turns out to be very useful and which can be extended to luminous and high-redshift sources. The size of the BLR correlates strongly with the continuum luminosity of an AGN. This behavior can be understood by the following argument. As we have seen, the BLR covers a broad range in radii around the center, and the physical conditions in the BLR are "layered": ions of higher ionization energy are closer to the continuum source than those with lower ionization energy. The gas in the BLR is subject to photoionization, and hence the distribution of ionization states will depend on the flux of energetic photons. This flux is $\propto L/r^2$, thus it depends on the luminosity L of the ionizing radiation and the distance r for the central source. For a given ionization state, and thus for a given broad emission line, the value of the *ionization parameter* $\Xi = L/r^2$ should be very similar in all AGNs. Hence, this argument yields $r \propto L^{1/2}$. In fact, direct estimates from sources where the radius was determined with reverberation mapping confirmed such a relation, which might be slightly steeper, $r \propto L^{\sim 0.6}$. The value of Ξ can be obtained for those sources for which reverberation mapping yields a determination of the size r. These sources are then used to calibrate the L-r relation for a given line transition. Once this is done, (5.30) can be applied again, with the radius now determined from the calibrated value of Ξ and the continuum luminosity of the AGN. This method can be extended to high redshifts, if the value of Ξ can be determined for emission lines which are located in the optical window at a given redshift, where the MgII and CIV lines are the most important transitions.

The Eddington Efficiency. Once an estimate for $M_\bullet$ is obtained, the Eddington luminosity can be calculated and compared with the observed luminosity. The ratio of these two, $\epsilon_{\rm Edd} \equiv L/L_{\rm Edd}$, is called the *Eddington efficiency*. If one can ignore strongly beamed emission, $\epsilon_{\rm Edd}$ should be smaller than unity. For the estimate of $\epsilon_{\rm Edd}$, the observed luminosity in the optical band needs to be translated into a bolometric luminosity, which can be done with the help of the average spectral energy distribution of AGNs of a given class. All of these steps involve statistical errors of a factor ~ 2 in any individual object, but when averaged over an ensemble of sources, they should yield approximately the correct mean values.

We find that $\epsilon_{\rm Edd}$ varies between a few percent to nearly unity among QSOs. Hence, once a black hole becomes sufficiently active as to radiate like a QSO, its luminosity approaches the Eddington luminosity. There might be a trend that radio-loud QSOs have a somewhat larger $\epsilon_{\rm Edd}$, but these correlations are controversial and might be based on selection effects. The fact that $\epsilon_{\rm Edd}$ is confined to a fairly narrow interval implies that the luminosity of a QSO can be used to estimate $M_\bullet$, just by setting $M_\bullet = \epsilon_{\rm Edd} M_{\rm Edd}(L)$. This mass estimate has a statistical uncertainty of about a factor of ~ 3 in individual sources.

The Galactic Black Hole. The Eddington efficiency of the SMBH in the Galactic center is many orders of magnitude smaller than unity; in fact, with its total luminosity of 5×10^{36} erg/s, $\epsilon_{\rm Edd} \sim 10^{-8}$. Such a small value indicates that the SMBH in our Galaxy is starved; the accretion rate must be very small. However, one can estimate a minimum mass rate with which the SMBH in the Galactic center is fed, by considering the mass-loss rate of the stars near the Galactic center. This amounts to $\sim 10^{-4}\, M_\odot$/yr, enough material to power an accretion flow with $L \sim 10^{-2} L_{\rm Edd}$. The fact that the observed luminosity is so much smaller than this value leads to two implications. The first of these is that there must be other modes of accretion which are far less efficient than that of the geometrically thin, optically thick accretion disk described in Sect. 5.3.2. Such models for accretion flows were indeed developed. In these models, the generated internal energy (heat) is not radiated away locally, but instead advected with the flow towards the black hole. The second conclusion is that the central mass concentration must indeed be a black hole – a black hole is the only object which does not have a surface. If, for example, one would postulate a hypothetical object with $M \sim 3 \times 10^6\, M_\odot$ which has a hard surface (like a scaled-up version of a neutron star), the accreted material would fall onto the surface, and its kinetic and inner energy would be deposited there. Hence, this surface would heat up and radiate thermally. Since we have strict upper limits on the radius of the object, coming from mm-VLBI observations, we can estimate the minimum luminosity such a source would have. This estimate is again several orders of magnitude larger than the observed luminosity from Sgr A*, firmly ruling out the existence of such a solid surface.

The observed flaring activity of Sgr A* (see Sect. 2.6.4) yields further information about the properties of the Galactic SMBH. In particular, the quasi-periodicity of ~ 17 min most likely must be identified with a source component orbiting the SMBH. From the theory of black holes it follows that objects can have stable orbits around a black hole only if the orbital radius is larger than some threshold. For a black hole without rotation, this last stable orbit has a radius of $3r_S$, whereas it can be smaller for spinning black holes. Since we know the mass of the SMBH in our Galaxy, and thus its Schwarzschild radius r_S, we can calculate the orbital period for this last stable orbit. This turns out to be larger than 17 min for a non-rotating black hole. In fact, assuming that the material which emits the flared radiation orbits the black hole at or near the last stable orbit, one concludes that the SMBH in Sgr A* spins at about half the maximally allowed rate.

Recently, flaring activity from other low-luminosity AGNs has been detected. Since their corresponding black hole mass is estimated to be larger than that of Sgr A*, the time-scale of variability is accordingly longer.

Black Hole Mass Scaling Relations at High Redshifts. As we have seen in Sect. 3.5.3, the black hole mass in normal, nearby galaxies is correlated with the bulge (or spheroidal) luminosity. As this component of galaxies consists of an old stellar population, its luminosity is very closely related to its stellar mass. Estimating the black hole mass from the continuum luminosity of the QSOs, and observing the spheroidal luminosity of their host galaxies (which requires the high angular resolution of HST), we can now investigate whether such a scaling relation already existed at earlier epochs, i.e., at high redshifts. When the evolution of the stellar population is taken into account in determining the mass of the stellar spheroidal component – stars at high redshift are necessarily younger than the old stellar population in local ellipticals or bulges – essentially the same relation between $M_\bullet$ and the spheroidal stellar mass is obtained for $z \lesssim 2$ QSOs as for local galaxies. This means that, whatever causes the close correlation between these two quantities in the local Universe, these processes must have already occurred in the early Universe. Needless to say, this observational result places strong constraints on the joint evolution of galaxies and their central supermassive black holes. At even higher redshifts, there are indications that the ratio of black hole mass and stellar mass was larger than today.

Black Hole Demography. Given that supermassive black holes grow by accretion,[7] and that this accretion is related to the energy release in AGNs, one might ask whether the total mass density of black holes at the present epoch is compatible with the integrated AGN luminosity. In other words, can the mass density of black holes be accounted for by the total accretion luminosity over cosmic time, as seen in the AGN population?

The first of these numbers is obtained from the scaling relation between SMBH mass and the properties of the spheroidal components in galaxies, as discussed in Sect. 3.5.3. This yields a value of the spatial mass density of SMBHs in the mass range $10^6 \le M_\bullet/M_\odot \le 5 \times 10^9$ of $\sim 4 \times 10^5 M_\odot/\mathrm{Mpc}^3$, with about a 30% uncertainty. About a quarter of this mass is contributed by SMBHs in the bulges of late-type galaxies; hence, the total SMBH mass density is dominated by ellipticals.

The overall accreted mass is obtained from the redshift-dependent luminosity function of AGNs (see Sect. 5.6.2), by assuming an efficiency ϵ of the conversion of mass into energy. Indeed, the local mass density of SMBHs is matched if the accretion efficiency is $\epsilon \sim 0.10$, as is expected from standard accretion disk models. It therefore seems that the population of SMBHs located in normal galaxies at the present epoch have undergone an active phase in their past, causing their mass growth. However, it may be that the efficiency here is underestimated, as some fraction of the energy released during the accretion process is converted into kinetic energy, as seen by powerful jets in AGN. This fraction is largely undetermined at present, but may not be negligible. In this case, the true ϵ needs to be higher than 0.1, which is only possible for black holes which rotate rapidly. In fact, the observed profile of the iron emission line from AGNs indicates black hole rotation.

A more detailed comparison between the SMBH and AGN populations reveals that the characteristic Eddington efficiency is $\epsilon_{\mathrm{Edd}} \sim 0.3$. With this value, combined with (5.28), one can estimate the mean time-scale over which a typical SMBH was active in the past, yield-

[7] The population of supermassive black holes can also be changed by merging processes, i.e., as the result of merging black holes when their host galaxies merge. However, in this case the total black hole mass is largely conserved, modulo some general relativistic effects.

ing $t_{act} \sim 2 \times 10^8$ yr. Hence, the SMBH of a current day massive galaxy was active during about 2% of its lifetime.

5.5 Family Relations of AGNs

5.5.1 Unified Models

In Sect. 5.2, different types of AGNs were listed. We saw that many of their properties are common to all types, but also that there are considerable differences. Why are some AGNs seen as broad-line radio galaxies, others as BL Lac objects? The obvious question arises as to whether the different classes of AGNs consist of rather similar objects which differ in their appearance due to geometric or light propagation effects, or whether more fundamental differences exist. In this section we will discuss differences and similarities of the various classes of AGNs and show that they presumably all derive from the same physical model.

Common Properties. Common to all AGNs is a SMBH in the center of the host galaxy, the supposed central engine, and also an accretion disk that is feeding the black hole. This suggests that a classification can be based on $M_\bullet$ and the accretion rate $\dot{m}$, or perhaps more relevantly the ratio $\dot{m}/\dot{m}_{edd}$. $M_\bullet$ defines the maximum (isotropic) luminosity of the SMBH in terms of the Eddington luminosity, and the ratio $\dot{m}/\dot{m}_{edd}$ describes the accretion rate relative to its maximum value. Furthermore, the observed properties, in particular the seemingly smooth transition between the different classes, suggest that radio-quiet quasars and Seyfert 1 galaxies basically differ only in their central luminosity. From this, we would then deduce that they have a similar value of $\dot{m}/\dot{m}_{edd}$ but differ in $M_\bullet$. An analogous argument may be valid for the transition from BLRGs to radio-loud quasars.

The difference between these two classes may be due to the nature of the host galaxy. Radio galaxies (and maybe radio-loud quasars?) are situated in elliptical galaxies, Seyfert nuclei (and maybe radio-quiet quasars?) in spirals. A correlation between the luminosity of the AGN and that of the host galaxy also seems to exist. This is to be expected if the luminosity of the AGN is strongly correlated with the respective Eddington luminosity, because of the correlation between the SMBH mass in normal galaxies and the properties of the galaxy (Sect. 3.5.3). Another question is how to fit blazars and Seyfert 2 galaxies into this scheme.

Anisotropic Emission. In the context of the SMBH plus accretion disk model, another parameter exists that will affect the observed characteristics of an AGN, namely the angle between the rotation axis of the disk and the direction from which we observe the AGN. We should mention that in fact there are many indications that the radiation of an AGN is not isotropic and thus its appearance is dependent on this direction. Among these are the observed ionization cones in the NLR (see Fig. 5.24) and the morphology of the radio emission, as the radio lobes define a preferred direction. Furthermore, our discussion of superluminal motion has shown that the observed superluminal velocities are possible only if the direction of motion of the source component is close to the direction of the line-of-sight. The X-ray spectrum of many AGNs shows intrinsic (photoelectric) absorption caused by high column density gas, where this effect is mainly observed in Seyfert 2 galaxies. Because of these clear indications it seems obvious to examine the dependence of the appearance of an AGN on the viewing direction. For example, the observed difference between Seyfert 1 and Seyfert 2 galaxies may simply be due to a different orientation of the AGN relative to the line-of-sight.

Broad Emission Lines in Polarized Light. In fact, another observation of anisotropic emission provides a key to understanding the relation between AGN types, which supports the above idea. The galaxy NGC 1068 has no visible broad emission lines and is therefore classified as a Seyfert 2 galaxy. Indeed, it is considered an archetype of this kind of AGN. However, the optical spectrum of NGC 1068 in polarized light shows broad emission lines (Fig. 5.27) such as one would find in a Seyfert 1 galaxy. Obviously the galaxy must have a BLR, but it is only visible in polarized light. The photons that are emitted by the BLR are initially unpolarized. Polarization may be induced through scattering of the light, however, where the direction perpendicular to the directions of incoming and scattered photons define a preferred direction, which then defines the polarization direction.

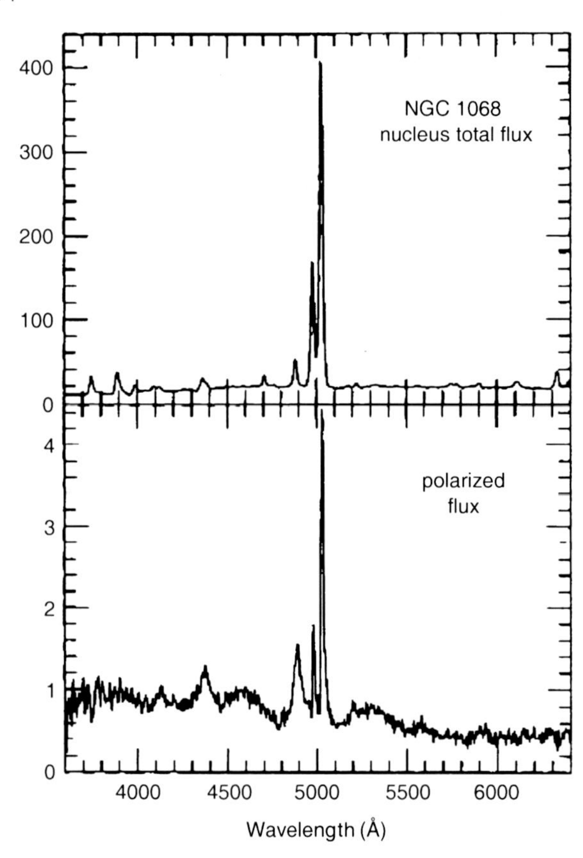

Fig. 5.27. Spectrum of the Seyfert 2 galaxy NGC 1068. The top panel displays the total flux which, besides the continuum, also shows narrow emission lines, in particular [OIII] at $\lambda = 5007$ Å and $\lambda = 4959$ Å. However, in polarized light (bottom panel), broad emission lines (like Hβ and Hγ) typical of a Seyfert 1 galaxy are also visible. Therefore, it is concluded that the BLR becomes visible in light polarized via scattering; the BLR is thus visible only indirectly

The interpretation of this observation (see Fig. 5.28) now is that NGC 1068 has a BLR but our direct view of it is obscured by absorbing material. However, this absorber does not fully engulf the BLR in all directions but only within a solid angle of $< 4\pi$ as seen from the central core. If photons from the BLR are scattered by dust or electrons in a way that we are able to observe the scattered radiation, then the BLR would be visible in this scattered light. Direct light from the AGN completely outshines the scattered light, which is the reason why we cannot identify the latter in the total flux. By scattering, however, this radiation is also polarized. Thus in observations made in polarized light, the (unpolarized) direct radiation is suppressed and the BLR becomes visible in the scattered light.

This interpretation is additionally supported by a strong correlation of the spatial distribution of the polarization and the color of the radiation in NGC 1068 (see Fig. 5.29). We can conclude from this that the differences between Seyfert 1 and Seyfert 2 galaxies originate in the orientation of the accretion disk and thus of the absorbing material relative to the line-of-sight.

From the abundance ratio of Seyfert 1 to Seyfert 2 galaxies (which is about 1:2), the fraction of solid angle in which the view to the BLR is obscured, as seen from the AGN, can be estimated. This ratio then tells us that about 2/3 of the solid angle is covered by an absorber. Such a blocking of light may be caused by dust. It is assumed that the dust is located in the plane of the accretion disk in the form of a thick torus (see Fig. 5.28 and Fig. 5.30 for a view of this geometry).

Search for Type 2 QSOs. If the difference between Seyfert galaxies of Type 1 and Type 2 is caused merely by their orientation, and if likewise the difference between Seyfert 1 galaxies and QSOs is basically one of absolute luminosity, then the question arises as to whether a luminous analog for Seyfert 2 galaxies exists, a kind of Type 2 QSO. Until a few years ago such Type 2 QSOs had not been observed, from which it was concluded that either no dust torus is present in QSOs due to the high luminosity (and therefore no Type 2 QSOs exist) or that Type 2 QSOs are not easy to identify.

This question has finally now been settled: the current X-ray satellites Chandra and XMM-Newton have identified the population of Type 2 QSOs. Due to the high column density of hydrogen which is distributed in the torus together with the dust, low-energy X-ray radiation is almost completely absorbed by the photoelectric effect if the line-of-sight to the center of these sources passes through the obscuring torus. These sources were therefore not visible for ROSAT ($E \leq 2.4$ keV), but the energy ranges of Chandra and XMM-Newton finally allowed the X-ray detection and identification of these Type 2 QSOs.

Another candidate for Type 2 QSOs are the ultra-luminous infrared galaxies (ULIRGs), in which extreme IR-luminosity is emitted by large amounts of warm dust which is heated either by very strong star formation or by an AGN. Since ULIRGs have total luminosi-

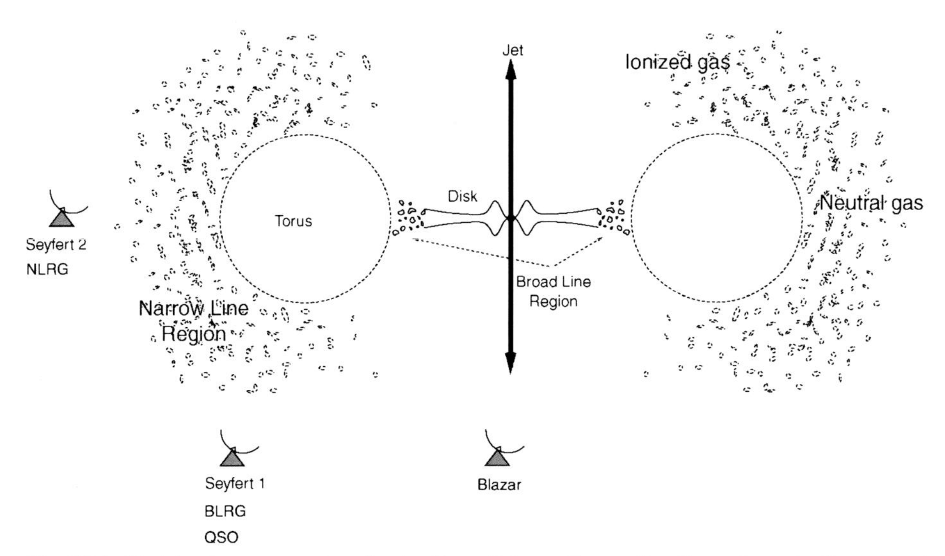

Fig. 5.28. Sketch of our current understanding of the unification of AGN types. The accretion disk is surrounded by a thick torus containing dust which thus obscures the view to the center of the AGN. When looking from a direction near the plane of the disk, a direct view of the continuum source and the BLR is blocked, whereas it is directly visible from directions closer to the symmetry axis of the disk. The difference between Seyfert 1 (and BLRG) and Seyfert 2 (and NLRG) is therefore merely a matter of orientation relative to the line-of-sight. If an AGN is seen exactly along the jet axis, it appears as a blazar

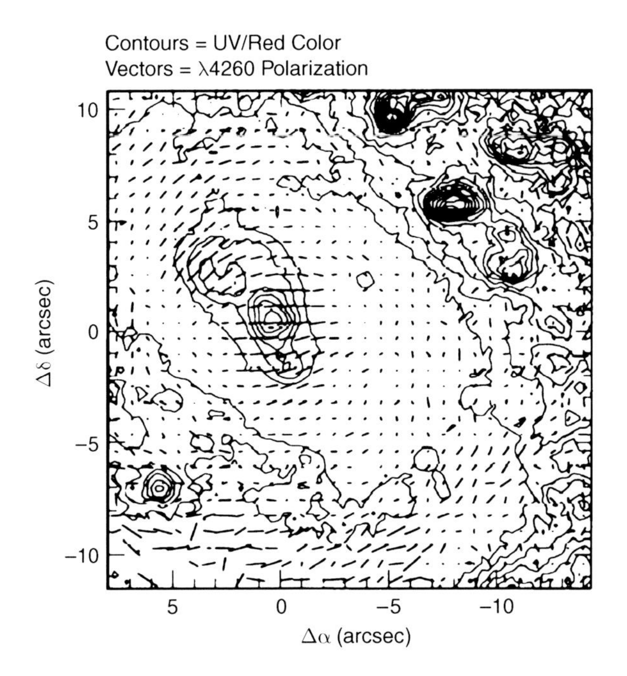

Fig. 5.29. The contours show the color of the optical emission in the Seyfert 2 galaxy NGC 1068, namely the flux ratio in the U- and R-bands. The sticks indicate the strength and orientation of the polarization in B-band light. The center of the galaxy is located at $\Delta\alpha = 0 = \Delta\delta$. At its bluest (center left), the polarization of the optical emission is strongest and is perpendicular to the direction to the center of the galaxy; this is the direction of polarization expected for local scattering by electrons. Hence, where the scattering is strongest, the largest fraction of direct light from the AGN is also observed, and the optical spectrum of AGNs is considerably bluer than the stellar light from galaxies

ties comparable to QSOs, the latter interpretation is possible. In fact, distinguishing between the two possibilities is not easy for individual ULIRGs, and in many sources indicators of both strong star formation and non-thermal emission (e.g., in the form of X-ray emission) are found. This discovery indicates that in many

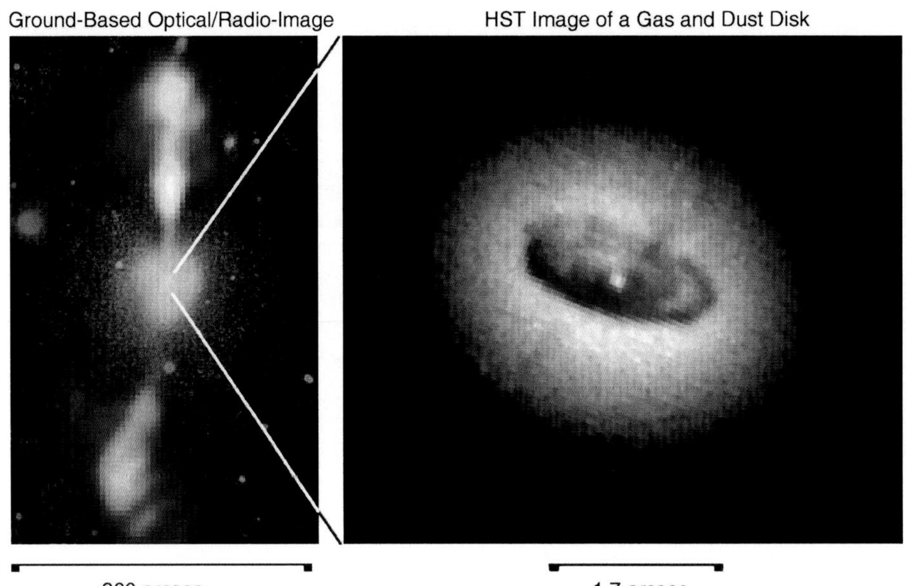

Fig. 5.30. The elliptical galaxy NGC 4261. The left-hand panel shows an optical image of this galaxy together with the radio emission (shown in orange). An HST image showing the innermost region of the galaxy is shown on the right. The jet is virtually perpendicular to the central disk of gas and dust, which is in agreement with the theoretical picture in the context of a unification model

objects, the processes of strong star formation and accretion onto a SMBH are linked. For both processes, large amounts of gas are necessary, and the fact that both starburst galaxies and AGNs are often found in interacting galaxies, where the disturbance in the gravitational field provides the conditions for a gas flow into the center of the galaxy, suggests a link between the two phenomena.

Next we will examine how blazars fit into this unified scheme. A first clue comes from the fact that all blazars are radio sources. Furthermore, in our interpretation of superluminal motion (Sect. 5.3.3) we saw that the appearance and apparent velocity of the central source components depend on the orientation of the source with respect to us, and that it requires relativistic velocities of the source components. To obtain an interpretation of the blazar phenomenon that fits into the above scheme, we first need to discuss an effect that results from Special Relativity.

5.5.2 Beaming

Due to relativistic motion of the source components relative to us, another effect occurs, known as *beaming*. Due to beaming, the relation between source luminosity and observed flux from a moving source depends on its velocity with respect to the observer. One aspect of this phenomenon is the Doppler shift in frequency space: the measured flux at a given frequency is different from that of a non-moving source because the measured frequency corresponds to a Doppler-shifted frequency in the rest-frame of the source. Another effect described by Special Relativity is that a moving source which emits isotropically in its rest-frame has an anisotropic emission pattern, with the angular distribution depending on its velocity. The radiation is emitted preferentially in the direction of the velocity vector of the source (thus, in the forward direction), so that a source will appear brighter if it is moving towards the observer. In Sect. 4.3.2, we already mentioned the relation (4.44) between the radiation intensity in the rest-frame of a source and in the system of the observer. Due to the strong Doppler shift, this implies that a source moving towards us appears brighter by a factor

$$\mathcal{D}_+ = \left(\frac{1}{\gamma(1-\beta\cos\phi)} \right)^{2+\alpha} \tag{5.31}$$

than the source at rest, where α is the spectral index. Furthermore, $\beta = v/c$, ϕ is the angle between the velocity vector of the source component and the line-of-sight to the source, and the Lorentz factor $\gamma = (1-\beta^2)^{-1/2}$ has already been defined in Sect. 5.3.3. Even at weakly relativistic velocities ($\beta \sim 0.9$) this can already be a considerable factor, i.e., the radiation from the relativistic jet may appear highly amplified. Another consequence of beaming is that if a second jet exists which is moving away from us (the so-called counter-jet), its radiation will be weakened by a factor

$$\mathcal{D}_- = \left(\frac{1}{\gamma(1+\beta\cos\phi)} \right)^{2+\alpha} \tag{5.32}$$

relative to the stationary source. Obviously, $\mathcal{D}_-$ can be obtained from $\mathcal{D}_+$ by replacing ϕ by $\phi+\pi$, since the counter-jet is moving in the opposite direction. In particular, the flux ratio of jet and counter-jet is

$$\frac{\mathcal{D}_+}{\mathcal{D}_-} = \left(\frac{1+\beta\cos\phi}{1-\beta\cos\phi} \right)^{2+\alpha} , \tag{5.33}$$

and this factor may easily be a hundred or more (Fig. 5.31). The large flux ratio (5.33) for relativistic jets is the canonical explanation for VLBI jets being virtually always only one-sided. This effect is also denoted as "Doppler favoritism" – the jet pointing towards us is observed preferentially because of the beaming effect and the resulting amplification of its flux.

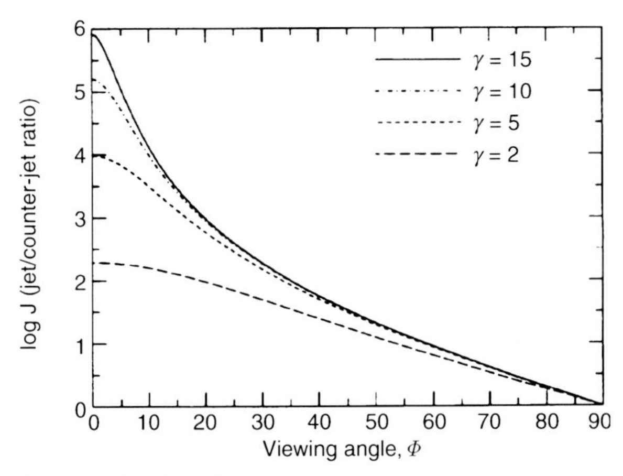

Fig. 5.31. The logarithm of the flux ratio of jet and counter-jet (5.33) is plotted as a function of the angle ϕ for different values of the Lorentz factor γ. Even at relatively small values of γ, this ratio is large if ϕ is close to 0, but even at $\phi \sim 30°$ the ratio is still appreciable. Hence the plot shows the Doppler favoritism and explains why, in most compact radio AGNs, one jet is visible but the counter-jet is not

Beaming and the Blazar Phenomenon. If we observe a source from a direction very close to the jet axis and if the jet is relativistic, its radiation can outshine all other radiation from the AGN because $\mathcal{D}_+$ can become very large in this case. Especially if the beamed radiation extends into the optical/UV part of the spectrum, the line emission may also become invisible relative to the jet emission, and the source will appear to us as a BL Lac object. If the line radiation is not outshined completely, the source may appear as an OVV. The synchrotron nature of the optical light is also the explanation for the optical polarization of blazars since synchrotron emission can be polarized, in contrast to thermal emission.

The strong beaming factor also provides an explanation for the rapid variability of blazars. If the velocity of the emitting component is close to the speed of light, $\beta \lesssim 1$, even small changes in the jet velocity or its direction may noticeably change the Doppler factor $\mathcal{D}_+$. Such small changes in the direction are expected because there is no reason to expect a smooth outflow of material along the jet at constant velocity. In addition, we argued that, very probably, magnetic fields play an important role in the generation and collimation of jets. These magnetic fields are toroidally spun-up, and emitting plasma can, at least partially, follow the field lines along helical orbits (see Fig. 5.32).

Hence beaming can explain the dominance of radiation from the jet components if the gas is relativistic, and also the absence or relative weakness of emission lines. At the same time, it provides a plausible scenario for the strong variability of blazars. The relative strength of the core emission and the extended radio emission depends heavily on the viewing direction. In blazars, a dominance of the core emission is expected, which is exactly what we observe.

5.5.3 Beaming on Large Scales

A consequence of this model is that the jets on kpc scales, which are mainly observed by the VLA, also need to be at least semi-relativistic: kiloparsec-scale jets are in most cases also one-sided, and they are always on the same side of the core as the VLBI jet on pc scales. Thus, if the one-sidedness of the VLBI jet is caused by beaming and the corresponding Doppler fa-

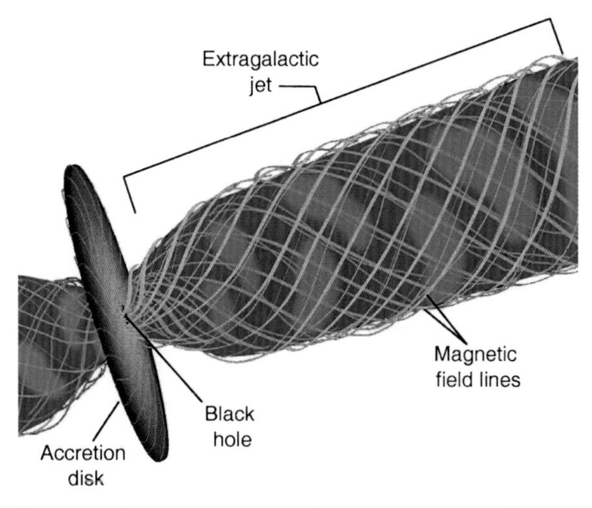

Fig. 5.32. Illustration of the relativistic jet model. The acceleration of the jet to velocities close to the speed of light is probably caused by a combination of very strong gravitational fields in the vicinity of the SMBH and strong magnetic fields which are rotating rapidly because they are anchored in the accretion disk. Shock fronts within the jet lead to acceleration processes of relativistic electrons, which then strongly radiate and become visible as "blobs" in the jets. By rotation of the accretion disk in which the magnetic field lines are anchored, the field lines obtain a characteristic helical shape. It is supposed that this process is responsible for the focusing (collimation) of the jet

voritism of an otherwise intrinsically symmetric source, the one-sidedness of large-scale jets should have the same explanation, implying relativistic velocities for them as well. These do not need to be as close to c as those of the components that show superluminal motion, but their velocity should also be at least a few tenths of the speed of light. In addition, it follows that the kiloparsec-scale jet is moving towards us and is therefore closer to us than the core of the AGN; for the counter-jet we have the opposite case. This prediction can be tested empirically, and it was confirmed in polarization measurements. Radiation from the counter-jet crosses the ISM of the host galaxy, where it experiences additional Faraday rotation (see Sect. 2.3.4). It is in fact observed that the Faraday rotation of counter-jets is systematically larger than that of jets. This can be explained by the fact that the counter-jet is located behind the host galaxy and we are thus observing it through the gas of that galaxy.

5.5.4 Jets at Higher Frequencies

Optical Jets. In Sect. 5.1.2, we discussed the radio emission of jets, and Sect. 5.3.3 described how their relativistic motion is detected from their structural changes, i.e., superluminal motion. However, jets are not only observable at radio frequencies; they also emit at much shorter wavelengths. Indeed, the first two jets were detected in optical observations, namely in QSO 3C273 (Fig. 5.33) and in the radio galaxy M87 (Fig. 5.34), as a linear source structure pointing radially away from the core of the respective galaxy. With the commissioning of the VLA (Fig. 1.21) as a sensitive and high-resolution radio interferometer, the discovery and examination of hundreds of jets at radio frequencies became possible.

The HST, with its unique angular resolution, has detected numerous jets in the optical (see also Fig. 5.12). They are situated on the same side of the corresponding AGNs as the main radio jet. Optical counterparts of radio counter-jets have not been detected thus far. Optical jets are always shorter, narrower, and show more structure than the corresponding radio jets. The spectrum of optical jets follows a power law (5.2) similar to that in the radio domain, with an index α that describes, in general, a slightly steeper spectrum. In some cases, linear polarization in the optical jet radiation of $\sim 10\%$ was also detected. If we also take into account that the positions of the knots in the optical and in the radio jets agree very well, we inevitably come to the conclusion that the optical radiation is also synchrotron emission. This conclusion is further supported by a nearly constant flux ratio of radio and optical radiation along the jets.

As was mentioned in Sect. 5.1.3, the relativistic electrons that produce the synchrotron radiation lose energy by emission. In many cases, the cooling time (5.6) of the electrons responsible for the radio emission is longer than the time of flow of the material from the central core along the jet, in particular if the flow is (semi-)relativistic. It is thus possible that relativistic electrons are produced or accelerated in the immediate vicinity of the AGN and are then transported away by the jet. This is not the case for those electrons producing the optical synchrotron radiation, however, because the cooling time for emission at optical wavelengths is only $t_{\rm cool} \sim 10^3\,(B/10^{-4}\,{\rm G})\,{\rm yr}$. Even if the relativistic

Fig. 5.33. Jets are visible not only in the radio domain but in some cases also at other wavelengths. Left: an HST image of the quasar 3C273 is shown, with the point-like quasar in the center and (displayed in blue) jet-shaped optical emission that spatially coincides with the radio jet (displayed in red). Right: an X-ray image of this quasar taken by the Chandra satellite. The jet is also visible at very high energies

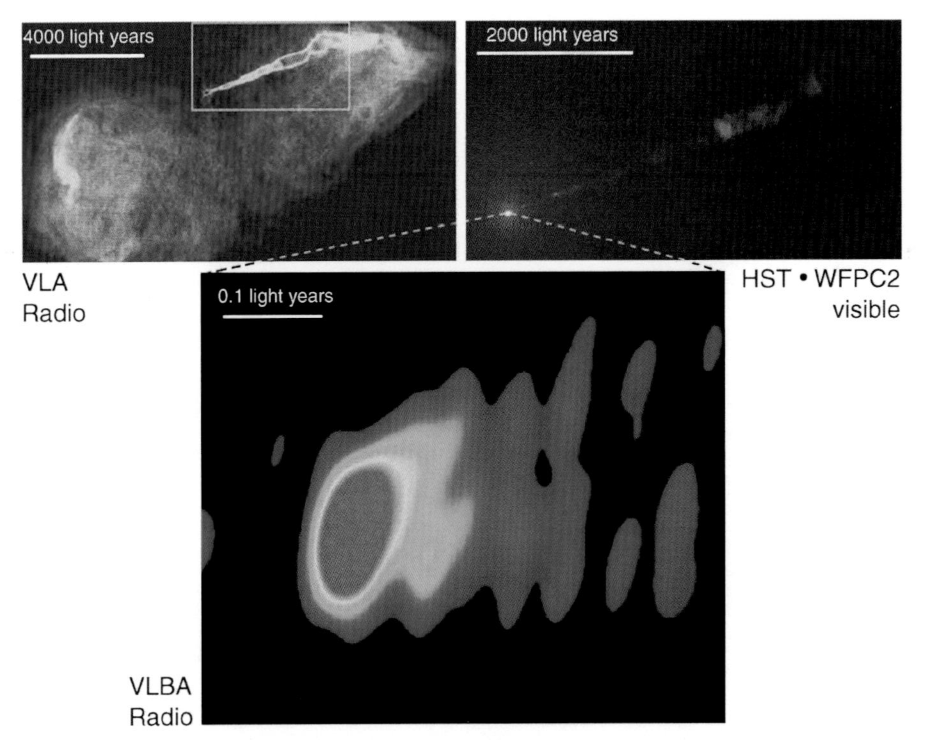

Fig. 5.34. Top left: a radio map of M87, the central galaxy in the Virgo Cluster of galaxies. Top right: an HST image of the region shown in the inset of the left-hand panel. The radio jet is also visible at optical wavelengths. The lower image shows a VLBI map of the region around the galaxy core; the jet is formed within a few 10^{17} cm from the core of the galaxy, which presumably contains a black hole of $M_\bullet \sim 3 \times 10^9 M_\odot$. Very close to the center the opening angle of the jet is significantly larger than further out. This indicates that the jet only becomes collimated at a larger distance

electrons are transported in a (semi-)relativistic jet, they cannot travel more than a distance of ~ 1 kpc before losing their energy. The observed length of optical jets is much larger, though. For this reason, the corresponding electrons cannot be originating in the AGN itself but instead must be produced locally in the jet. The knots in the jets, which are probably shock fronts in the outflow, are thought to represent the location of the acceleration of relativistic particles. Quantitative estimates of the cooling time are hampered by the unknown beaming factor (5.31). Since optical jets are all one-sided, and in most cases observed in radio sources with a flat spectrum, a very large beaming factor is generally assumed. Transforming back into the rest-frame of the electrons yields a lower frequency and a lower luminosity. Since the latter is utilized for estimating the strength of the magnetic fields (by assuming equipartition of energy, for instance), this also changes the estimated cooling time.

X-Ray Radiation of Jets. The Chandra satellite discovered that many of the jets which had been identified in the radio are also visible in X-ray light (Fig. 5.35). This came as a real surprise. This discovery and the strong correlation of the spatial distribution of radio, optical, and X-ray emission imply that they must all originate from the same regions in the jets, i.e., that the origins of the emission must be linked to each other. As we have discussed, radio and optical radiation originate from synchrotron emission, the emission by relativistic electrons moving in a magnetic field. The same electrons that are responsible for the radio emission can also produce X-ray photons by inverse Compton scattering. In this process, low-energy photons are scattered to much higher energies by collisions with relativistic electrons – a photon of frequency ν may have a frequency $\nu' \approx \gamma^2 \nu$ after being scattered by an electron of energy $\gamma m_e c^2$. Since the characteristic Lorentz factors of electrons causing the synchrotron radiation of radio jets may reach values of $\gamma \sim 10^4$, these electrons may scatter, by inverse Compton scattering, radio photons into the X-ray domain of the spectrum. This effect is also called synchrotron self Compton radiation. Alternatively, relativistic electrons can also scatter optical photons from the AGN, for which less energetic electrons are required. The omnipresent CMB may also be considered as a photon source for the inverse Compton effect, and in many cases the observed X-ray radiation is probably Compton-scattered CMB radiation.

The inverse Compton model cannot, however, be applied to all X-ray jets without serious problems occurring. For instance, variability in X-ray emission was observed in the knots of M87, indicating a very short cooling time for the electrons. Since the electrons must have a much larger Lorentz factor γ if the radiation, at

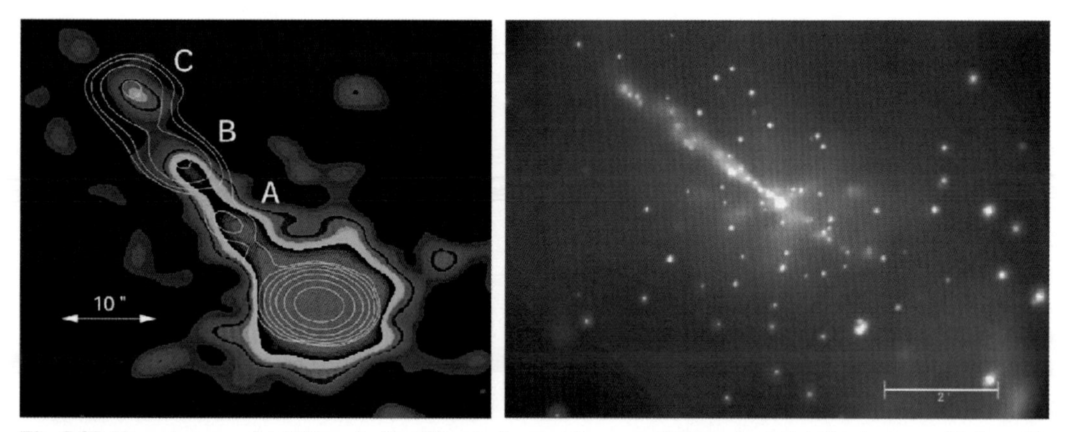

Fig. 5.35. X-ray images of AGN jets. Left: a Chandra image of the jet in the QSO PKS 1127−145, with overlaid contours of radio emission (1.4 cm, VLA). The direction of the jet and its substructure are very similar at both wavelengths, suggesting an interpretation in which the radiation is caused by the same population of relativistic electrons. Right: a Chandra image of the active galaxy Centaurus A. Here the jet is visible, as well as a large number of compact sources interpreted to be X-ray binaries

a given frequency, originates from synchrotron emission instead of by inverse Compton scattering, their cooling time $t_{\rm cool}$ (5.6) would be much shorter as well. In such sources, which are typically FR I radio sources, the synchrotron process itself probably accounts for the X-ray emission. This implies, on the one hand, very short cooling time-scales and therefore the increased necessity for a local acceleration of the electrons. On the other hand, the required energies for the electrons are very high, $\sim 100\,\mathrm{TeV}$. It is currently unclear which acceleration processes may account for these high energies.

Detecting radio jets at X-ray frequencies seems to be a frequent phenomenon: about half of the flat-spectrum radio QSO with jet-like extended radio emission also show an X-ray jet. All of those are one-sided, although the corresponding radio images often show lobes opposite the X-ray jets, reinforcing the necessity for Doppler favoritism also in the X-ray waveband.

Finally, it should be mentioned that our attempts at finding a unification scheme for the different classes of AGNs have been quite successful. The scheme of unification is generally accepted, even though some aspects are still subject to discussion. One particular model is sketched in Fig. 5.36.

5.6 AGNs and Cosmology

AGNs, and QSOs in particular, are visible out to very high redshifts. Since their discovery in 1963, QSOs have held the redshift record nearly without interruption. Only in recent years have QSOs and galaxies been taking turns in holding the record. Today, several hundred QSOs are known with $z \geq 4$, and the number of those with $z > 5$ continues to grow since a criterion was found to identify these objects. This leads to the possibility that QSOs could be used as cosmological probes, and thus to the question of what we can learn about the Universe from QSOs. For example, one of the most exciting questions is how does the QSO population evolve with redshift – was the abundance of QSO at high redshifts, i.e., at early epochs of the cosmos, similar to that today, or does it evolve over time?

5.6.1 The K-Correction

To answer this question, we must know the luminosity function of QSOs, along with its redshift dependence. As we did for galaxies, we define the luminosity function $\Phi(L, z)\,\mathrm{d}L$ as the spatial number density of

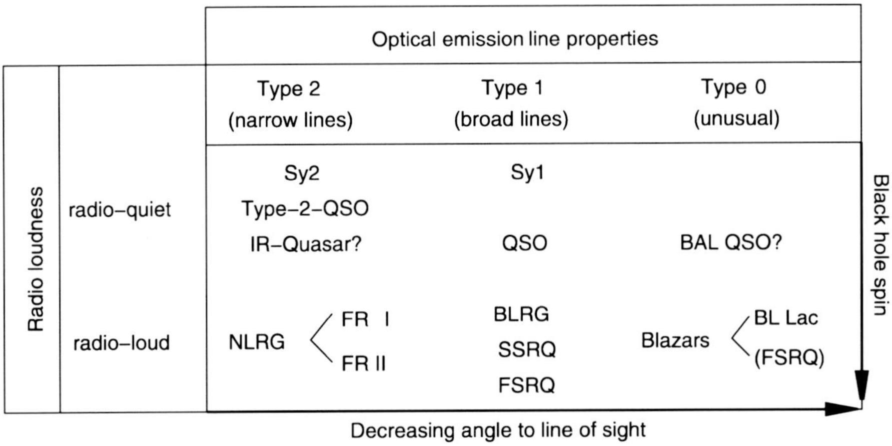

Radio loudness		Optical emission line properties: Type 2 (narrow lines)	Type 1 (broad lines)	Type 0 (unusual)
	radio–quiet	Sy2 Type–2–QSO IR–Quasar?	Sy1 QSO	BAL QSO?
	radio–loud	NLRG < FR I / FR II	BLRG SSRQ FSRQ	Blazars < BL Lac / (FSRQ)

Black hole spin (arrow pointing down)

Decreasing angle to line of sight (arrow pointing right)

Fig. 5.36. This table presents a unification scheme for AGNs via the angular momentum of the central black hole and the orientation of the accretion disk with respect to the line-of-sight. The closer the direction of the jet is to the line-of-sight, the more the jet component dominates. Furthermore, the relative strength of the radio emission in this particular unified scheme is linked to the angular momentum of the black hole. The classification pattern shown here is only one of several possibilities, but the dependence of the AGN class on orientation is generally considered to be accepted

QSOs with luminosity between L and $L+\mathrm{d}L$. Φ normally refers to a comoving volume element, so that a non-evolving QSO population would correspond to a z-independent Φ. One of the problems in determining Φ is related to the question of which kind of luminosity is meant here. For a given observed frequency band, the corresponding rest-frame radiation of the sources depends on their redshift. For optical observations, the measured flux of nearby QSOs corresponds to the rest-frame optical luminosity, whereas it corresponds to the UV luminosity for higher-redshift QSOs. In principle, using the bolometric luminosity would be a possible solution; however, this is not feasible since it is *very* difficult to measure the bolometric luminosity (if at all possible) due to the very broad spectral distribution of AGNs. Observations at all frequencies, from the radio to the gamma domain, would be required, and obviously, such observations can only be obtained for selected individual sources.

Of course, the same problem occurs for all sources at high redshift. In comparing the luminosity of galaxies at high redshift with that of nearby galaxies, for instance, it must always be taken into account that, at given observed wavelength, different spectral ranges in the galaxies' rest-frames are measured. This means in order to investigate the optical emission of galaxies at $z \sim 1$, observations in the NIR region of the spectrum are necessary.

Frequently the only possibility is to use the luminosity in some spectral band and to compensate for the above effect as well as possible by performing observations in several bands. For instance, one picks as a reference the blue filter which has its maximum efficiency at ~ 4500 Å and measures the blue luminosity for nearby objects in this filter, whereas for objects at redshift $z \sim 1$ the intrinsic blue luminosity is obtained by observing with the I-band filter, and for even larger redshifts observations need to be extended into the near-IR. The observational problems with this strategy, and the corresponding corrections for the different sensitivity profiles of the filters, must not be underestimated and are always a source of systematic uncertainties. An alternative is to perform the observation in only one (or a few) filters and to approximately correct for the redshift effect.

In Sect. 4.3.3, we defined various distance measures in cosmology. In particular, the relation $S = L/(4\pi D_\mathrm{L}^2)$ between the observed flux S and the luminosity L of a source defines the luminosity distance D_L. Here both the flux and the luminosity refer to bolometric quantities, i.e., flux and luminosity integrated over all frequencies. Due to the redshift, the measured spectral flux S_ν is related to the spectral luminosity $L_{\nu'}$ at a frequency $\nu' = \nu(1+z)$, where one finds

$$S_\nu = \frac{(1+z)L_{\nu'}}{4\pi D_\mathrm{L}^2} . \tag{5.34}$$

We write this relation in a slightly different form,

$$S_\nu = \frac{L_\nu}{4\pi D_\mathrm{L}^2} \left[\frac{L_{\nu'}}{L_\nu}(1+z) \right] , \tag{5.35}$$

where the first factor is of the same form as in the relation between the bolometric quantities while the second factor corrects for the spectral shift. This factor is denoted the *K-correction*. It obviously depends on the spectrum of the source, i.e., to determine the K-correction for a source its spectrum needs to be known. Furthermore, this factor depends on the filter used. Since in optical astronomy magnitudes are used as a measure for brightness, (5.35) is usually written in the form

$$m_\mathrm{int} = m_\mathrm{obs} + K(z)$$
$$\text{with} \quad K(z) = -2.5 \log \left[\frac{L_{\nu'}}{L_\nu}(1+z) \right] , \tag{5.36}$$

where m_int is the magnitude that would be measured in the absence of redshift, and m_obs describes the brightness actually observed. The K-correction is not only relevant for QSOs but for all objects at high redshift, in particular also for galaxies.

5.6.2 The Luminosity Function of Quasars

By counting QSOs, we obtain the number density $N(>S)$ of QSOs with a flux larger than S. We find a relation of roughly $N(>S) \propto S^{-2}$ for large fluxes, whereas the source counts are considerably flatter for smaller fluxes. The flux at which the transition from steep counts to flatter ones occurs corresponds to an apparent magnitude of about $B \sim 19.5$. Up to this magnitude, about 10 QSOs per square degree are found.

From QSO number counts, combined with measurements of QSO redshifts, the luminosity function $\Phi(L, z)$ can be determined. As already defined above,

$\Phi(L,z)\,\mathrm{d}L$ is the number density *in a comoving volume element* of QSOs at redshift z with a luminosity between L and $L+\mathrm{d}L$.

Two fundamental problems exist in determining the luminosity function. The first is related to the above discussion of wavelength shift due to cosmological redshift: a fixed wavelength range in which the brightness is observed corresponds to different wavelength intervals in the intrinsic QSO spectra, depending on their redshift. We need to correct for this effect if the number density of QSOs above a given luminosity in a certain frequency interval is to be compared for local and distant QSOs. One way to achieve this is by assuming a universal spectral shape for QSOs; over a limited spectral range (e.g., in the optical and the UV ranges), this assumption is indeed quite well satisfied. This universal spectrum is obtained by averaging over the spectra of a larger number of QSOs (Fig. 5.2). By this means, a useful K-correction of QSOs as a function of redshift can then be derived.

The second difficulty in determining $\Phi(L,z)$ is to construct QSO samples that are "complete". Since QSOs are point-like they cannot be distinguished from stars by morphology on optical images, but rather only by their color properties and subsequent spectroscopy. However, with the star density being much higher than that of QSOs, this selection of QSO candidates by color criteria, and subsequent spectroscopic verification, is very time-consuming. Only more recent surveys, which image large areas of the sky in several filters, were sufficiently successful in their color selection and subsequent spectroscopic verification, so that very large QSO samples could be compiled. An enormous increase in statistically well-defined QSO samples was achieved by two large surveys with the 2dF spectrograph and the Sloan Digital Sky Survey. We will discuss this in the context of galaxy redshift surveys in Sect. 8.1.2.

The luminosity function that results from such analyses is typically parametrized as

$$\Phi(L,z)=\frac{\Phi^*}{L^*(z)}\left[\left(\frac{L}{L^*(z)}\right)^{\alpha}+\left(\frac{L}{L^*(z)}\right)^{\beta}\right]^{-1}; \tag{5.37}$$

i.e., for fixed z, Φ is a double power law in L. At $L \gg L^*(z)$, the first term in the square brackets in (5.37) dominates if $\alpha > \beta$, yielding $\Phi \propto L^{-\alpha}$. On the other hand, the second term dominates for $L \ll L^*(z)$, so that $\Phi \propto L^{-\beta}$. Typical values for the exponents are $\alpha \approx 3.9$, $\beta \approx 1.5$. The characteristic luminosity $L^*(z)$ where the L-dependence changes, strongly depends on redshift. A good fit to the data for $z \lesssim 2$ is achieved by

$$L^*(z) = L_0^*(1+z)^k\,, \tag{5.38}$$

with $k \approx 3.45$, where the value of k depends on the assumed density parameters Ω_m and Ω_Λ. This approximation is valid for $z \lesssim 2$, whereas for larger redshifts $L^*(z)$ seems to vary less with z. The normalization constant is determined to be $\Phi^* \approx 5.2 \times 10^3\,h^3\,\mathrm{Gpc}^{-3}$, and L_0^* corresponds to roughly $M_B = -20.9 + 5\log h$. The luminosity function as determined from an extensive QSO survey is plotted in Fig. 5.37.

With this form of the luminosity function, a number of conclusions can be drawn. The luminosity function of QSOs is considerably broader than that of galaxies, which we found to decrease exponentially for large L. The strong dependence of the characteristic luminosity $L^*(z)$ on redshift clearly shows a very significant cosmological evolution of the QSO luminosity func-

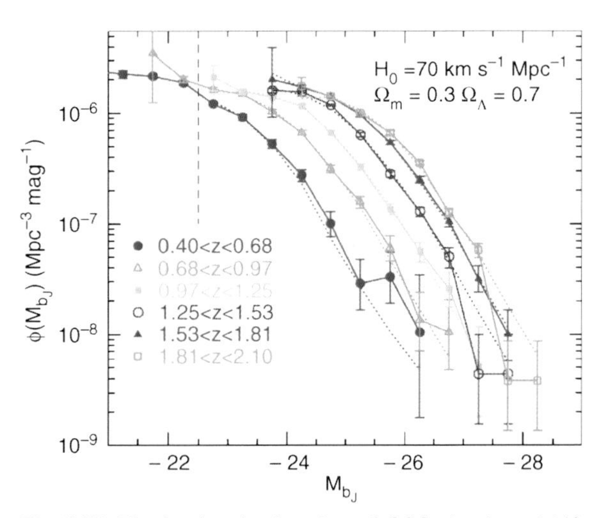

Fig. 5.37. The luminosity function of QSOs in six redshift intervals in the range $0.4 \le z \le 2.1$, determined from the 2dF QSO Redshift Survey by spectroscopy of more than 23 000 QSOs. The dotted curves represent the best fit to the data that is achieved by a double power law as in (5.37) where the data have been corrected for the selection function of the survey. The increase in QSO density with increasing redshift is clearly visible. The dashed line denotes the formal separation between Seyfert galaxies and QSOs

tion. For example, at $z \sim 2$, $L^*(z)$ is about 50 times larger than today. Furthermore, for high luminosities, $\Phi \propto [L^*(z)]^{\alpha-1} L^{-\alpha}$. This means that the spatial number density of luminous QSOs was more than 1000 times larger at $z \sim 2$ than it is today (see Fig. 5.38). Another way of seeing this is that the low-redshift luminosity function in Fig. 5.37 does not extend to the very bright luminosities for which the luminosity function at high redshifts was measured. The reason for this is that the number density of very luminous QSOs at low redshifts is so small that essentially none of them are contained in the survey volume from which the results in Fig. 5.37 were derived.

For redshifts $z \gtrsim 3$, the evolution of the QSO population seems to turn around, i.e., the spatial density apparently decreases again. The exact value of z at which the QSO density attains its maximum is still somewhat uncertain because of the difficulties in obtaining a complete sample of QSOs at high redshift. Since a redshift $z \sim 3$ corresponds to an epoch where the Universe had only about 20% of its current age (the exact value depends on the cosmological parameters), a kind of "QSO epoch" seems to have occurred, in the sense that the QSO population seem to have quickly formed and then largely became extinct again.

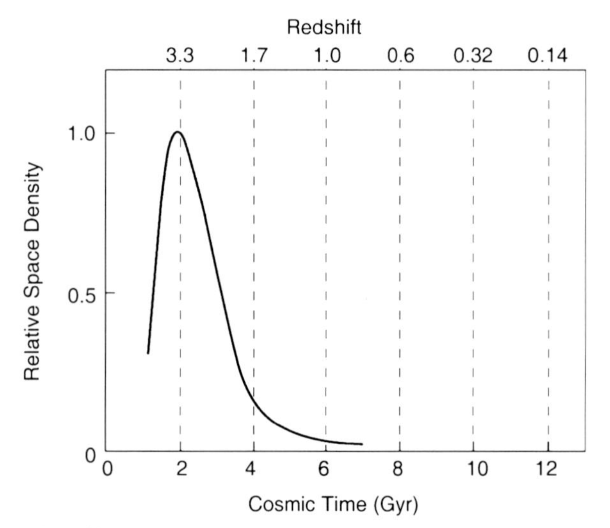

Fig. 5.38. The relative spatial density of QSOs as a function of the age of the Universe. We can see that the QSO density has a well-defined narrow maximum which corresponds to a redshift of about $z \sim 2.5$; towards even larger redshifts, the density seems to decrease again. This plot suggests the notion of a "QSO epoch"

There are several possible interpretations of the QSO luminosity function and its redshift dependence. One of them is that the luminosity of any one QSO varies in time, parallel to the evolution of $L^*(z)$. Most likely this interpretation is wrong because it implies that a luminous QSO will always remain luminous. Although the efficiency of energy conversion into radiation is much higher for accretion than for thermonuclear burning, an extremely high mass would nevertheless accumulate in this case. This would then be present as the mass of the SMBH in local QSOs.[8] However, estimates of $M_\bullet$ in QSOs rarely yield values larger than $\sim 3 \times 10^9 M_\odot$.

However, it is by no means clear that a given source will be a QSO throughout its lifetime: a source may be active as a QSO for a limited time, and later appear as a normal galaxy again. For instance, it is possible that virtually any massive galaxy hosts a potential AGN. This supposition is clearly supported by the fact that apparently all massive galaxies harbor a central SMBH. If the SMBH is fed by accreting matter, this galaxy will then host an AGN. However, if no more mass is provided, the nucleus will cease to radiate and the galaxy will no longer be active. Our Milky Way may serve as an example of this effect, since although the mass of the SMBH in the center of the Galaxy would be sufficient to power an AGN luminosity of more than 10^{44} erg/s considering its Eddington luminosity (5.23), the observed luminosity is lower by many orders of magnitude.

AGNs are often found in the vicinity of other galaxies. One possible interpretation is that the neighboring galaxy disturbs the gravitational field of the QSO's host, such that it allows its interstellar medium to flow into the center of the host galaxy where it accretes onto the central black hole – and "the monster starts to shine". If this is the case, the luminosity function (5.37) does not provide information about individual AGNs, but only about the population as a whole.

Interpreting the redshift evolution then becomes obvious. The increase in QSO density with redshift in the scenario described above originates from the fact that at earlier times in the Universe interactions between galaxies and merger processes were significantly more frequent than today. On the other hand, the decrease

[8] Compare the mass estimate in Sect. 5.3.1 where, instead of 10^7 yr, the lifetime to be inserted here is the age of the Universe, $\sim 10^{10}$ yr.

at very high z is to be expected because the SMBHs in the center of galaxies first need to form, and this obviously happens in the first $\sim 10^9$ years after the Big Bang.

5.6.3 Quasar Absorption Lines

The optical/UV spectra of quasars are characterized by strong emission lines. In addition, they also show absorption lines, which we have not mentioned thus far. Depending on the redshift of the QSOs, the wavelength range of the spectrum, and the spectral resolution, QSO spectra may contain a large variety of absorption lines. In principle, several possible explanations exist. They may be caused by absorbing material in the AGN itself or in its host galaxy, so they have an intrinsic origin. Alternatively, they may arise during the long journey between the QSO and us due to intervening gas along the line-of-sight. We will see that different kinds of absorption lines exist, and that both of these possibilities indeed occur. The analysis of those absorption lines which do not have their origin in the QSO itself provides information about the gas in the Universe. For this purpose, a QSO is basically a very distant bright light source used for probing the intervening gas.

This gas can be either in intergalactic space or is correlated with foreground galaxies. In the former case, we expect that this gas is metal-poor and thus consists mainly of hydrogen and helium. Furthermore, in order to cause absorption, the intergalactic medium must not be fully ionized, but needs to contain a fraction of neutral hydrogen. Gas located closer to galaxies may be expected to also contain appreciable amounts of metals which can give rise to absorption lines.

The identification of a spectral line with a specific line transition and a corresponding redshift is, in general, possible only if at least two lines occur at the same redshift. For this reason, doublet transitions are particularly valuable, such as those of MgII ($\lambda = 2795$ Å and $\lambda = 2802$ Å), and CIV ($\lambda = 1548$ Å and $\lambda = 1551$ Å). The spectrum of virtually any QSO at high (emission line) redshift z_{em} shows narrow absorption lines by CIV and MgII at absorption line redshifts $z_{abs} < z_{em}$. If the spectral coverage extends to shorter wavelengths than the observed Lyα emission line of the QSO, numerous narrow absorption lines exist at $\lambda_{obs} \lesssim \lambda_{obs}(\text{Ly}\alpha) = (1 + z_{em})\,1216$ Å. The set of these absorption lines is denoted as the *Lyman-α forest*. In about 15% of all QSOs, very broad absorption lines are found, the width of which may even considerably exceed that of the broad emission lines.

Classification of QSO Absorption Lines. The different absorption lines in QSOs are distinguished by classes according to their wavelength and width.

- *Metal systems:* In general these are narrow absorption lines, of which MgII and CIV most frequently occur (and which are the easiest to identify). However, in addition, a number of lines of other elements exist (Fig. 5.39). The redshift of these absorption lines is $0 < z_{abs} < z_{em}$; therefore they are caused by intervening matter along the line-of-sight and are not associated with the QSO. Normally a metal system consists of many different lines of different ions,

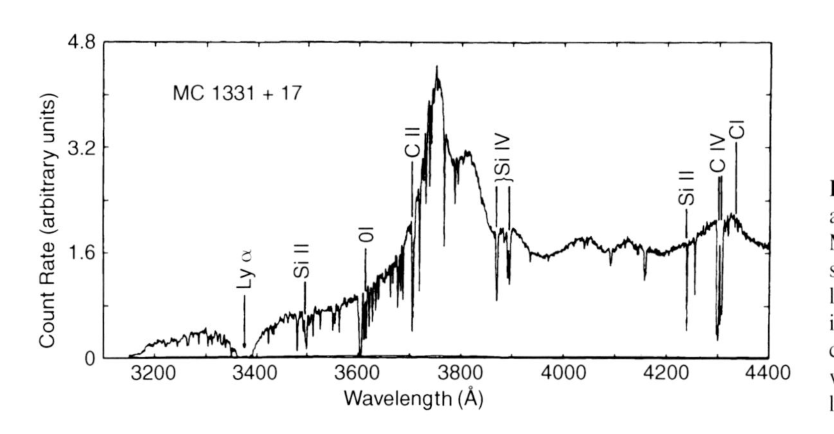

Fig. 5.39. Spectrum of the QSO 1331+17 at $z_{em} = 2.081$ observed by the Multi-Mirror Telescope in Arizona. In the spectrum, a whole series of absorption lines can be seen which have all been identified with gas at $z_{abs} = 1.776$. The corresponding Lyα line at $\lambda \approx 3400$ Å is very broad; it belongs to the damped Lyα lines

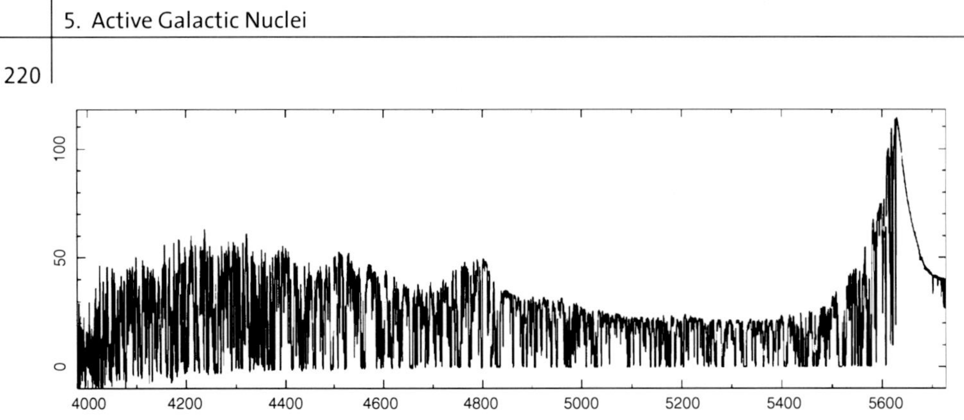

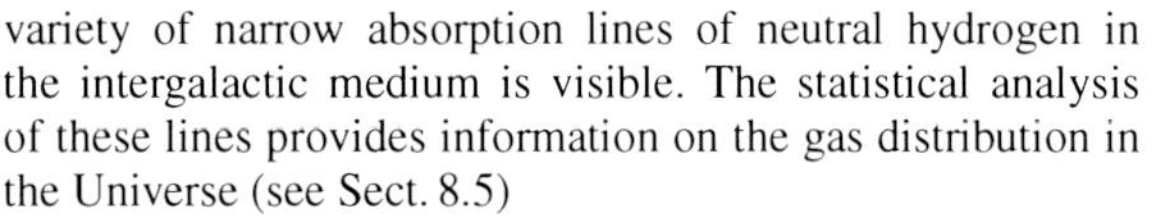

Fig. 5.40. Keck spectrum of the Lyman-α forest towards QSO 1422+231, a QSO at $z = 3.62$. As an aside, this is a quadruply-imaged lensed QSO. The wavelength resolution is about 7 km/s. On the blue side of the Lyα emission line, a large variety of narrow absorption lines of neutral hydrogen in the intergalactic medium is visible. The statistical analysis of these lines provides information on the gas distribution in the Universe (see Sect. 8.5)

all at the same redshift. From the line strength, the column density of the absorbing ions can be derived. For an assumed chemical composition and degree of ionization of the gas, the corresponding column density of hydrogen can then be determined. Estimates for such metal systems yield typical values of $10^{17}\,\mathrm{cm}^{-2} \lesssim N_\mathrm{H} \lesssim 10^{21}\,\mathrm{cm}^{-2}$, where the lower limit depends on the sensitivity of the spectral observation.

- *Associated metal systems:* These systems have characteristics very similar to those of the aforementioned intervening metal systems, but their redshift is $z_\mathrm{abs} \sim z_\mathrm{em}$. Since such systems are over-abundant compared to a statistical z-distribution of the metal systems, these systems are interpreted as belonging to the QSO itself. Thus the absorber is physically associated with the QSO and may be due, for example, to absorption in the QSO host galaxy or in a galaxy associated with it.
- *Lyα forest:* The large set of lines at $\lambda < (1 + z_\mathrm{em})\,1216\,\text{Å}$, as shown in Fig. 5.40, is interpreted to be Lyα absorption by hydrogen along the line-of-sight to the QSO. The statistical properties of these lines are essentially the same for all QSOs and seem to depend only on the redshift of the Lyα lines, but not on z_em. This interpretation is confirmed by the fact that for nearly any line in the Lyα forest, the corresponding Lyβ line is found if the quality and the wavelength range of the observed spectra permit this. The Lyα forest is further subdivided, according to the strength of the absorption, into narrow lines, Lyman-limit systems, and damped Lyα systems. Narrow Lyα lines are caused by absorbing gas of neutral hydrogen column densities of $N_\mathrm{H} \lesssim 10^{17}\,\mathrm{cm}^{-2}$. Lyman-limit systems derive their name from the fact that at column densities of $N_\mathrm{H} \gtrsim 10^{17}\,\mathrm{cm}^{-2}$, neutral hydrogen almost totally absorbs all radiation at $\lambda \lesssim 912\,\text{Å}$ (in the hydrogen rest-frame), where photons ion-

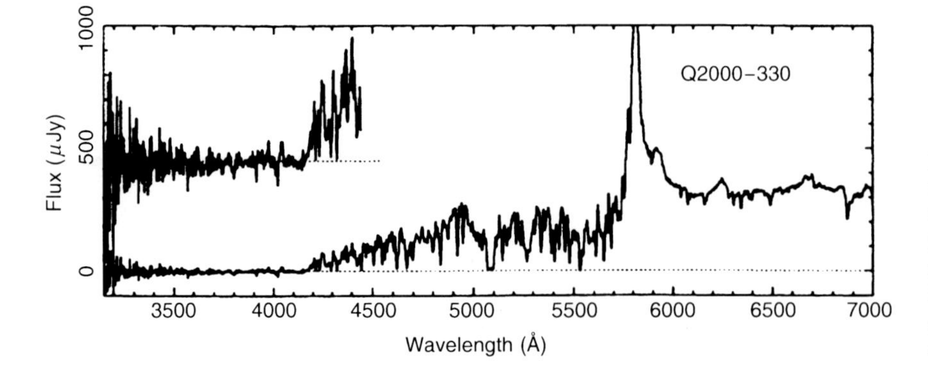

Fig. 5.41. A Lyman-limit system along the line-of-sight towards the QSO 2000−330 is absorbing virtually all radiation at wavelengths $\lambda \leq 912\,\text{Å}$ in the rest-frame of the absorber, here redshifted to about 4150 Å

ize hydrogen (Fig. 5.41). If such a system is located at $z_{\rm limit}$ in the spectrum of a QSO, the spectrum at $\lambda < (1 + z_{\rm limit})\,912\,\text{Å}$ is almost completely suppressed. Damped Lyα systems occur if the column density of neutral hydrogen is $N_{\rm H} \gtrsim 2 \times 10^{20}\,{\rm cm}^{-2}$. In this case, the absorption line becomes very broad due to the extended damping wings of the Voigt profile.[9]

- *Broad absorption lines:* For about 15% of the QSOs, very broad absorption lines are found in the spectrum at redshifts slightly below $z_{\rm em}$ (Fig. 5.42). The lines show a profile which is typical for sources with outflowing material, as seen, for instance, in stars with stellar winds. However, in contrast to the latter, the Doppler width of the lines in the *broad absorption line* (BAL) QSOs is a significant fraction of the speed of light.

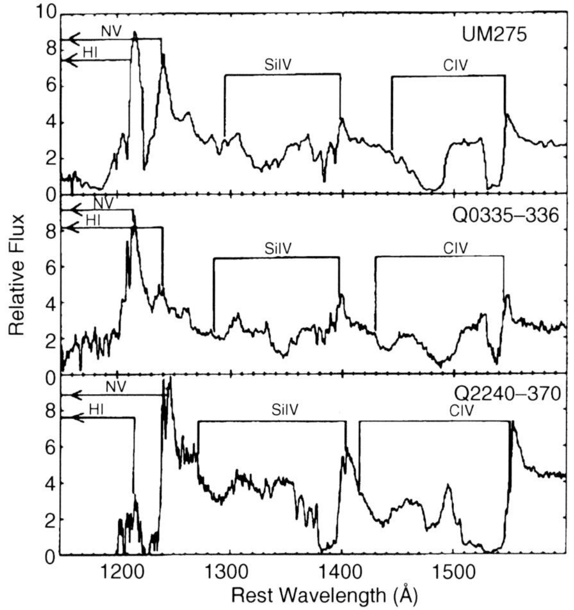

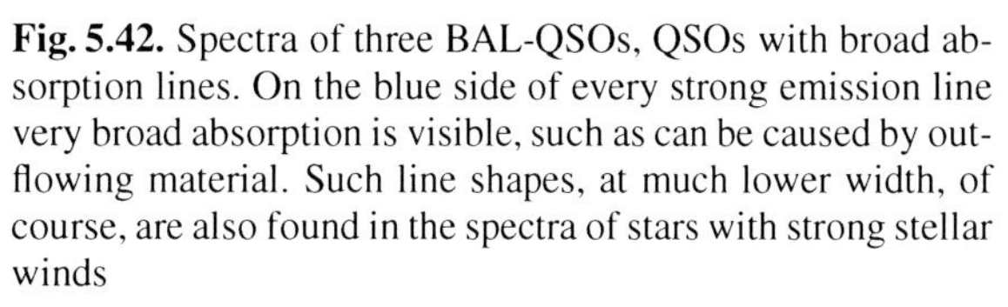
Fig. 5.42. Spectra of three BAL-QSOs, QSOs with broad absorption lines. On the blue side of every strong emission line very broad absorption is visible, such as can be caused by outflowing material. Such line shapes, at much lower width, of course, are also found in the spectra of stars with strong stellar winds

Interpretation. The metal systems with a redshift significantly smaller than $z_{\rm em}$ originate either in overdense regions in intergalactic space or they are associated with galaxies (or more specifically, galaxy halos) located along the line-of-sight. In fact, MgII systems always seem to be correlated with a galaxy at the same redshift as the absorbing gas. From the statistics of the angular separations of these associated galaxies to the QSO sight-line and from their redshifts, we obtain a characteristic extent of the gaseous halos of such galaxies of $\sim 25h^{-1}$ kpc. For CIV systems, the extent seems to be even larger, $\sim 40h^{-1}$ kpc.

The Lyα forest is caused by the diffuse intergalactic distribution of gas. In Sect. 8.5, we will discuss models of the Lyα forest and its relevance for cosmology more thoroughly (see also Fig. 5.43).

Broad absorption lines originate from material in the AGN itself, as follows immediately from their redshift and their enormous width. Since the redshift of the broad absorption lines is slightly lower than that of the corresponding emission lines, the absorbing gas must be moving towards us. The idea is that this is material flowing out at a very high velocity. BAL-QSOs (broad absorption line QSOs) are virtually always radio-quiet. The role of BAL-QSOs in the AGN family is unclear. A plausible interpretation is that the BAL property also depends on the orientation of the QSO. In this case, any

[9] The Voigt profile $\phi(\nu)$ of a line, which specifies the spectral energy distribution of the photons around the central frequency ν_0 of the line, is the convolution of the intrinsic line profile, described by a Lorentz profile,

$$\phi_{\rm L}(\nu) = \frac{\Gamma/4\pi^2}{(\nu-\nu_0)^2 + (\Gamma/4\pi)^2}\;,$$

and the Maxwellian velocity distribution of atoms in a thermal gas of temperature T. From this, the Voigt profile follows,

$$\phi(\nu) = \frac{\Gamma}{4\pi^2}\int\limits_{-\infty}^{\infty} {\rm d}v\; \frac{\sqrt{m/2\pi k_{\rm B}T}\,\exp\left(-mv^2/2k_{\rm B}T\right)}{(\nu-\nu_0-\nu_0 v/c)^2 + (\Gamma/4\pi)^2}\;, \qquad (5.39)$$

where the integral extends over the velocity component along the line-of-sight. In these equations, Γ is the intrinsic line width which results from the natural line width (related to the lifetime of the atomic states) and pressure broadening. m is the mass of the atom, which defines, together with the temperature T of the gas, the Maxwellian velocity distribution. If the natural line width is small compared to the thermal width, the Doppler profile dominates in the center of the line, that is for frequencies close to ν_0. The line profile is then well approximated by a Gaussian. In the wings of the line, the Lorentz profile dominates. For the wings of the line, where $\phi(\nu)$ is small, to become observable the optical depth needs to be high. This is the case in damped Lyα systems.

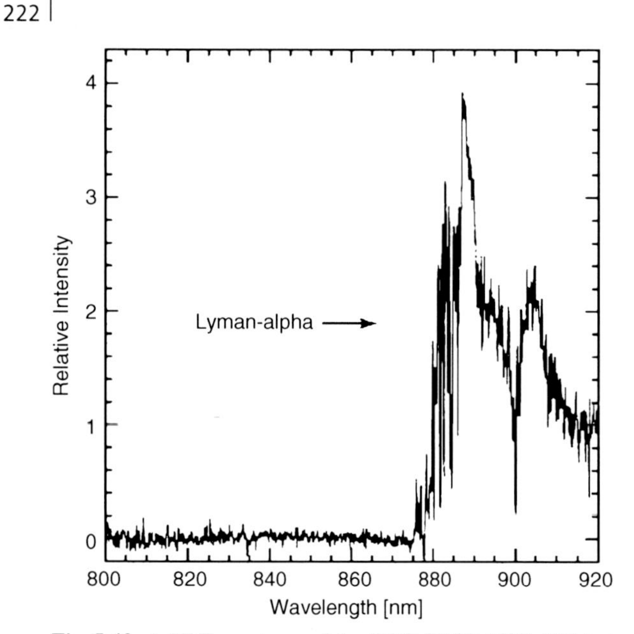

Fig. 5.43. A VLT spectrum of the QSO SDSS 1030+0524 at $z = 6.28$, currently one of the highest known QSO redshifts. The blue side of the Lyα emission line and the adjacent continuum are almost completely devoured by the dense Lyα forest

QSO would be a BAL if observed from the direction into which the absorbing material streams out.

Discussion. Most absorption lines in QSO spectra are not physically related to the AGN phenomenon. Rather, they provide us with an opportunity to probe the matter along the line-of-sight to the QSO. The Lyα forest will be discussed in relation to this aspect in Sect. 8.5. Furthermore, absorption line spectroscopy of QSOs carried out with UV satellites has proven the existence of very hot gas in the halo of our Milky Way. Such UV spectroscopy provides one of the very few opportunities to analyze the intergalactic medium if its temperature is of the order of $\sim 10^6$ K – gas at this temperature is very difficult to detect since it emits in the extreme UV which is unobservable from our location inside the Milky Way, and since almost all atoms are fully ionized and therefore cause no absorption. Only absorption lines from very highly ionized metals (such as the five times ionized oxygen) can still be observed. Since the majority of the baryons should be found in this hot gas phase today, this test is of great interest for cosmology.

6. Clusters and Groups of Galaxies

Galaxies are not uniformly distributed in space, but instead show a tendency to gather together in *galaxy groups* and *clusters of galaxies*. This effect can be clearly recognized in the projection of bright galaxies on the sky (see Figs. 6.1 and 6.2). The Milky Way itself is a member of a group, called the Local Group (Sect. 6.1), which implies that we are living in a locally overdense region of the Universe.

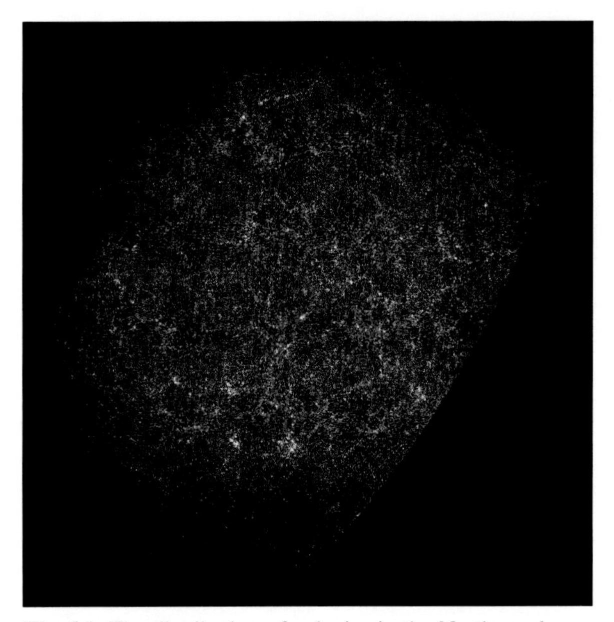

Fig. 6.1. The distribution of galaxies in the Northern sky, as compiled in the Lick catalog. This catalog contains the galaxy number counts for "pixels" of $10' \times 10'$ each. It is clearly seen that the distribution of galaxies on the sphere is far from being homogeneous. Instead it is distinctly structured

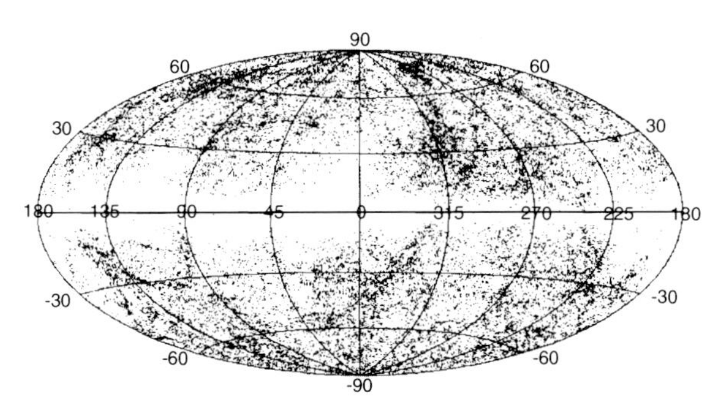

Fig. 6.2. The distribution of all galaxies brighter than $B < 14.5$ on the sphere, plotted in Galactic coordinates. The Zone of Avoidance is clearly seen as the region near the Galactic plane

The transition between groups and clusters of galaxies is smooth. The distinction is made by the number of their member galaxies. Roughly speaking, an accumulation of galaxies is called a group if it consists of $N \lesssim 50$ members within a sphere of diameter $D \lesssim 1.5h^{-1}$ Mpc. Clusters have $N \gtrsim 50$ members and diameters $D \gtrsim 1.5h^{-1}$ Mpc. A formal definition of a cluster is presented further below. An example of a group and a cluster of galaxies is displayed in Fig. 6.3.

Clusters of galaxies are the most massive gravitationally bound structures in the Universe. Typical values for the mass are $M \gtrsim 3 \times 10^{14} M_\odot$ for massive clusters, whereas for groups $M \sim 3 \times 10^{13} M_\odot$ is characteristic, with the total mass range of groups and clusters extending over $10^{12} M_\odot \lesssim M \lesssim 10^{15} M_\odot$.

Originally, clusters of galaxies were characterized as such by the observed spatial concentration of galaxies. Today we know that, although the galaxies determine the optical appearance of a cluster, the mass contained in galaxies contributes only a small fraction to the total mass of a cluster. Through advances in X-ray astronomy, it was discovered that galaxy clusters are intense sources of X-ray radiation which is emitted by a hot gas ($T \sim 3 \times 10^7$ K) located between the galaxies. This intergalactic gas (*intracluster medium*, ICM) contains more baryons than the stars seen in the member galaxies. From the dynamics of galaxies, from the properties of the X-ray emission of the clusters, and from the gravitational lens effect we deduce the existence of dark matter in galaxy clusters, dominating the cluster mass like it does for galaxies.

Clusters of galaxies play a very important role in observational cosmology. They are the most massive bound and relaxed (i.e., in a state of approximate dynamical equilibrium) structures in the Universe, as mentioned before, and therefore mark the most prominent density peaks of the large-scale structure in the Universe. Their cosmological evolution is therefore directly related to the growth of cosmic structures. Due to their high galaxy density, clusters and groups are also ideal laboratories for studying interactions between

Peter Schneider, Clusters and Groups of Galaxies.
In: Peter Schneider, Extragalactic Astronomy and Cosmology. pp. 223–275 (2006)
DOI: 10.1007/11614371_6

Fig. 6.3. Left: HCG40, a compact group of galaxies, observed with the Subaru telescope on Mauna-Kea. Right: the cluster of galaxies Cl 0053−37, observed with the WFI at the ESO/MPG 2.2-m telescope

galaxies and their effect on the galaxy population. For instance, the fact that elliptical galaxies are preferentially found in clusters indicates the impact of the local galaxy density on the morphology and evolution of galaxies.

6.1 The Local Group

The galaxy group of which the Milky Way is a member is called the *Local Group*. Within a distance of ~ 1 Mpc around our Galaxy, about 35 galaxies are currently known; they are listed in Table 6.1. A sketch of their spatial distribution is given in Fig. 6.4.

6.1.1 Phenomenology

The Milky Way (MW), M31 (Andromeda), and M33 are the three spiral galaxies in the Local Group, and they are also its most luminous members. The Andromeda galaxy is located at a distance of 770 kpc from us. The Local Group member next in luminosity is the Large Magellanic Cloud (LMC, see Fig. 6.5), which is orbiting around the Milky Way, together with the Small Magellanic Cloud (SMC), at a distance of ~ 50 kpc (~ 60 kpc, respectively, for the SMC). Both are satellite galaxies of the Milky Way and belong to the class of irregular galaxies (like about 11 other Local Group members). The other members of the Local Group are dwarf galaxies, which are very small and faint. Because of their low luminosity and their low surface brightness, many of the known members of the Local Group have been detected only in recent years. For example, the Antlia galaxy, a dwarf spheroidal galaxy, was found in 1997. Its luminosity is about 10^4 times smaller than that of the Milky Way.

Many of the dwarf galaxies are grouped around the Galaxy or around M31; these are known as *satellite*

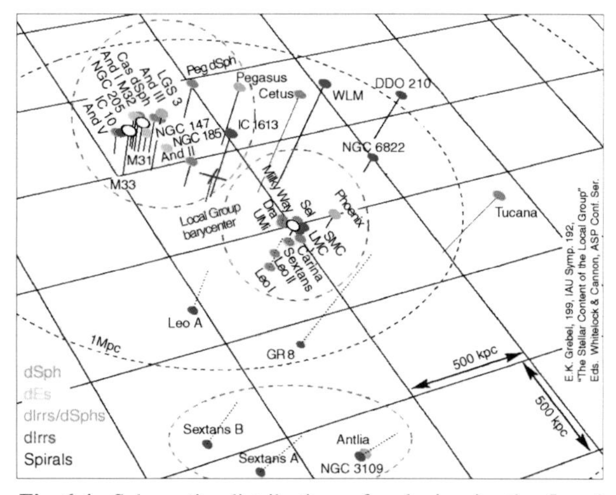

Fig. 6.4. Schematic distribution of galaxies in the Local Group, with the Milky Way at the center of the figure

Table 6.1. Members of the Local Group. Listed are the name of the galaxy, its morphological type, the absolute B-band magnitude, its position on the sphere in both right ascension/declination and in Galactic coordinates, its distance from the Sun, and its radial velocity. A sketch of the spatial configuration is displayed in Fig. 6.4

Galaxy	Type	M_B	RA/Dec.	ℓ, b	D (kpc)	v_r (km/s)
Milky Way	Sbc I-II	−20.0	1830−30	0, 0	8	0
LMC	Ir III-IV	−18.5	0524−60	280, −33	50	270
SMC	Ir IV-V	−17.1	0051−73	303, −44	63	163
Sgr I	dSph?		1856−30	6, −14	20	140
Fornax	dE0	−12.0	0237−34	237,−65	138	55
Sculptor Dwarf	dSph	−9.8	0057−33	286, −84	88	110
Leo I	dSph	−11.9	1005+12	226, +49	790	168
Leo II	dSph	−10.1	1110+22	220, +67	205	90
Ursa Minor	dSph	−8.9	1508+67	105, +45	69	−209
Draco	dSph	−9.4	1719+58	86, +35	79	−281
Carina	dSph	−9.4	0640−50	260, −22	94	229
Sextans	dSph	−9.5	1010−01	243, +42	86	230
M31	Sb I-II	−21.2	0040+41	121, −22	770	−297
M32=NGC 221	dE2	−16.5	0039+40	121, −22	730	−200
M110=NGC 205	dE5p	−16.4	0037+41	121, −21	730	−239
NGC 185	dE3p	−15.6	0036+48	121, −14	620	−202
NGC 147	dE5	−15.1	0030+48	120, −14	755	−193
And I	dSph	−11.8	0043+37	122, −25	790	—
And II	dSph	−11.8	0113+33	129, −29	680	—
And III	dSph	−10.2	0032+36	119, −26	760	—
Cas = And VII	dSph		2326+50	109, −09	690	—
Peg = DDO 216	dIr/dSph	−12.9	2328+14	94, −43	760	—
Peg II = And VI	dSph	−11.3	2351+24	106, −36	775	—
LGS 3	dIr/dSph	−9.8	0101+21	126, −41	620	−277
M33	Sc II-III	−18.9	0131+30	134, −31	850	−179
NGC 6822	dIr IV-V	−16.0	1942−15	025, −18	500	−57
IC 1613	dIr V	−15.3	0102+01	130, −60	715	−234
Sagittarius	dIr V	−12.0	1927−17	21, +16	1060	−79
WLM	dIr IV-V	−14.4	2359−15	76, −74	945	−116
IC 10	dIr IV	−16.0	0017+59	119, −03	660	−344
DDO 210, Aqr	dIr/dSph	−10.9	2044−13	34, −31	950	−137
Phoenix Dwarf	dIr/dSph	−9.8	0149−44	272, 68	405	56
Tucana	dSph	−9.6	2241−64	323, −48	870	—
Leo A = DDO 69	dIr V	−11.7	0959+30	196, 52	800	—
Cetus Dwarf	dSph	−10.1	0026−11	101, −72	775	—

galaxies. Distributed around the Milky Way are the LMC, the SMC, and nine dwarf galaxies, several of them in the so-called *Magellanic Stream* (see Fig. 6.6), a long, extended band of neutral hydrogen which was stripped from the Magellanic Clouds about 2×10^8 yr ago by tidal interactions with the Milky Way. The Magellanic Stream contains about $2 \times 10^8 M_\odot$ of neutral hydrogen.

The spatial distribution of satellite galaxies around the Milky Way shows a pronounced peculiarity, in that these 11 satellites form a highly flattened system. These satellites appear to lie essentially in a plane which is oriented perpendicular to the Galactic plane. The satellites around M31 also seem to be distributed in an anisotropic way around their host. In fact, satellites galaxies around spirals seem to be preferentially located near the short axes of the projected light distribution, which has been termed the Holmberg effect, although the statistical significance of this alignment has been questioned.

6.1.2 Mass Estimate

We will present a simple estimate of the mass of the Local Group, from which we will find that it is considerably more massive than one would conclude from the observed luminosity of the associated galaxies.

Fig. 6.5. An image of the Large Magellanic Cloud (LMC), taken with the CTIO 4-m telescope

M31 is one of the very few galaxies with a blueshifted spectrum. Hence, Andromeda and the Milky Way are approaching each other at a relative velocity of $v \approx 120$ km/s. This value results from the velocity of M31 relative to the Sun of $v \approx 300$ km/s, and from the motion of the Sun around the Galactic center. Together with the distance to M31 of $D \sim 770$ kpc, we conclude that both galaxies will collide on a time-scale of $\sim 6 \times 10^9$ yr (if we disregard the transverse component of the relative velocity).

The luminosity of the Local Group is dominated by the Milky Way and by M31, which together produce about 90% of the total luminosity. If the mass density follows the light distribution, the dynamics of the Local Group should also be dominated by these two galaxies. Therefore, one can try to estimate the mass of the two galaxies from their relative motion, and with this also the mass of the Local Group.

In the early phases of the Universe, the Galaxy and M31 were close together and both took part in the Hubble expansion. By their mutual gravitational attraction, their relative motion was decelerated until it came to a halt – at a time t_{max} at which the two galaxies had their maximum separation r_{max} from each other. From this time on, they have been moving towards each other. The relative velocity $v(t)$ and the separation $r(t)$ follow from the conservation of energy,

$$\frac{v^2}{2} = \frac{GM}{r} - C\,, \tag{6.1}$$

where M is the sum of the masses of the Milky Way and M231, and C is an integration constant. The latter can be

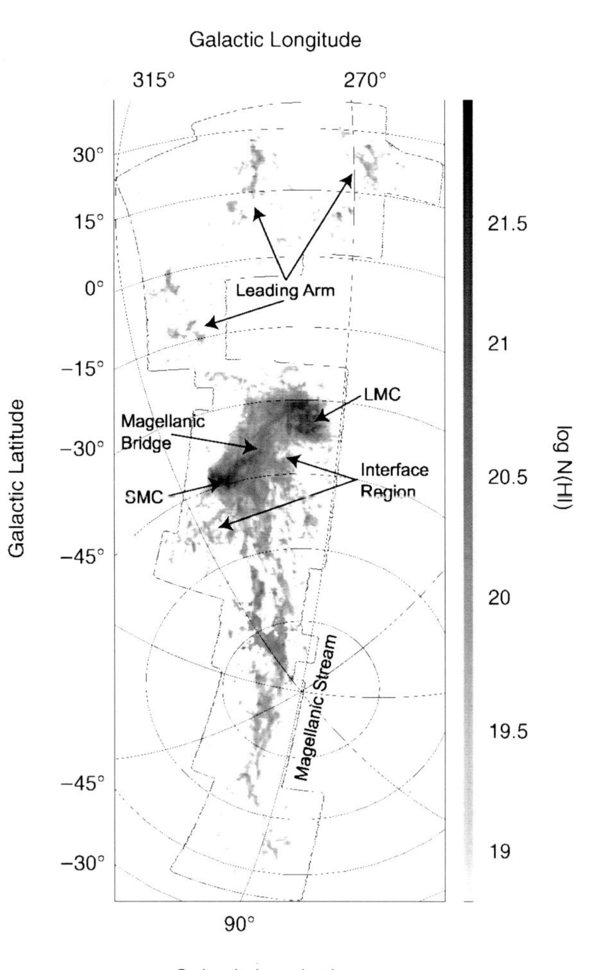

Fig. 6.6. HI map of a large region in the sky containing the Magellanic Clouds. This map is part of a large survey of HI, observed through its 21-cm line emission, that was performed with the Parkes telescope in Australia, and which maps about a quarter of the Southern sky with a pixel size of 5′ and a velocity resolution of ~ 1 km/s. The emission from gas at Galactic velocities has been removed in this map. Besides the HI emission by the Magellanic Clouds themselves, gas between them is visible, the Magellanic Bridge and the Magellanic Stream, the latter connected to the Magellanic Clouds by an "interface region". Gas is also found in the direction of the orbital motion of the Magellanic Clouds around the Milky Way, forming the "leading arm"

determined by considering (6.1) at the time of maximum separation, when $r = r_{max}$ and $v = 0$. With this,

$$C = \frac{GM}{r_{max}}$$

follows immediately. Since $v = \mathrm{d}r/\mathrm{d}t$, (6.1) is a differential equation for $r(t)$,

$$\frac{1}{2}\left(\frac{\mathrm{d}r}{\mathrm{d}t}\right)^2 = GM\left(\frac{1}{r} - \frac{1}{r_{\max}}\right) .$$

It can be solved using the initial condition $r = 0$ at $t = 0$. For our purpose, an approximate consideration is sufficient. Solving the equation for $\mathrm{d}t$ we obtain, by integration, a relation between $r_{\max}$ and $t_{\max}$,

$$t_{\max} = \int_0^{t_{\max}} \mathrm{d}t = \int_0^{r_{\max}} \frac{\mathrm{d}r}{\sqrt{2GM}\sqrt{1/r - 1/r_{\max}}} = \frac{\pi\, r_{\max}^{3/2}}{2\sqrt{2GM}} . \qquad (6.2)$$

Since the differential equation is symmetric with respect to changing $v \to -v$, the collision will happen at $2t_{\max}$. Estimating the time from today to the collision, by assuming the relative velocity to be constant during this time, then yields $r(t_0)/v(t_0) = D/v = 770\,\mathrm{kpc}/120\,\mathrm{km/s}$, and one obtains $2t_{\max} \approx t_0 + D/v$, or

$$t_{\max} \approx \frac{t_0}{2} + \frac{D}{2v} , \qquad (6.3)$$

where t_0 is the current age of the Universe. Hence, together with (6.2) this yields

$$\frac{v^2}{2} = \frac{GM}{r} - \frac{GM}{r_{\max}} = \frac{GM}{r} - \frac{1}{2}\left(\frac{\pi GM}{t_{\max}}\right)^{2/3} . \qquad (6.4)$$

Now by inserting the values $r(t_0) = D$ and $v = v(t_0)$, we obtain the mass M,

$$M \sim 3 \times 10^{12} M_\odot , \qquad (6.5)$$

where we have assumed $t_0 \approx 14 \times 10^9$ yr. This mass is much larger than the mass of the two galaxies as observed in stars and gas. The mass estimate yields a mass-to-light ratio for the Local Group of $M/L \sim 70\, M_\odot/L_\odot$. This is therefore another indication of the presence of dark matter because we can see only about 5% of the estimated mass in the Milky Way and Andromeda. Another mass estimate follows from the kinematics of the Magellanic Stream, which also yields $M/L \gtrsim 80 M_\odot/L_\odot$.

6.1.3 Other Components of the Local Group

One of the most interesting galaxies in the Local Group is the *Sagittarius dwarf galaxy* which was only discovered in 1994. Since it is located in the direction of the Galactic bulge, it is barely visible on optical images, if at all, as an overdensity of stars. Furthermore, it has a very low surface brightness. It was discovered in an analysis of stellar kinematics in the direction of the bulge, in which a coherent group of stars was found with a velocity distinctly different from that of bulge stars. In addition, the stars belonging to this overdensity have a much lower metallicity, reflected in their colors. The Sagittarius dwarf galaxy is located close to the Galactic plane, at a distance of about 16 kpc from the Galactic center and nearly in the direct extension of our line-of-sight to the GC. This proximity implies that it must be experiencing strong tidal gravitational forces on its orbit around the Milky Way; over the course of time, these will have the effect that the Sagittarius dwarf galaxy will be slowly disrupted. In fact, in recent years a relatively narrow band of stars has been found around the Milky Way. These stars are located along the orbit of the Sagittarius galaxy. Their chemical composition supports the interpretation that they are stars stripped from the Sagittarius dwarf galaxy by tidal forces. In addition, globular clusters have been identified which presumably once belonged to the Sagittarius dwarf galaxy, but which have also been removed from it by tidal forces and are now part of the globular cluster population in the Galactic halo.

Compact high-velocity clouds (CHVCs) are high-velocity clouds (see Sect. 2.3.6) with an angular diameter of $\lesssim 1°$. The distance of these clouds is difficult to determine, since they do not seem to contain any stars, and hence the methods of distance determination based on stellar properties cannot be applied. In those cases where the spectrum of a background object shows an absorption line at the same radial velocity as determined for the cloud from measurements of the 21-cm line, an upper limit for the cloud distance is obtained; namely the distance of the object whose spectrum displays the absorption line.

Indirect arguments sometimes yield rather large estimates, of several hundred kpc, for the distance of the CHVCs. If their distance is indeed this large, the rotation curves of CHVCs, i.e., their differential infall velocities, suggest high masses for the clouds. In this model, CHVCs would contain a large fraction of dark matter, $M \sim 10^7 M_\odot$, and hence much more dark matter than their neutral hydrogen mass. CHVCs would then be additional members of the Local Group, having a mass not very different from that of dwarf galaxies, but in which star formation was suppressed for some reason so that they contain no, or only very few, stars.

This model of CHVCs is controversial, however, and its verification or falsification would be of considerable interest for cosmology, as we will discuss later. If a concentration of CHVCs exists around the Milky Way at distances like the ones assumed in this model, a similar concentration should also exist around our sister galaxy M31. Currently, an intensive search for these systems is in progress. While HVCs have been found around M31, the search for CHVCs has been without success thus far. Therefore, one concludes a relatively low characteristic Galacto-centric distance for Galactic CHVCs of ~ 50 kpc. In this case, they would not be high-mass objects.

The Neighborhood of the Local Group. The Local Group is indeed a concentration of galaxies: while it contains about 35 members within ~ 1 Mpc, the next neighboring galaxies are found only in the Sculptor Group, which contains about six members and is located at a distance of $D \sim 1.8$ Mpc. The next galaxy group after this is the M81 group of about eight galaxies at $D \sim 3.1$ Mpc, the two most prominent galaxies of which are displayed in Fig. 6.7.

The other nearby associations of galaxies within 10 Mpc from us shall also be mentioned: the Centaurus group with 17 members and $D \sim 3.5$ Mpc, the M101 group with five members and $D \sim 7.7$ Mpc, the M66 and M96 group with together 10 members located at $D \sim 9.4$ Mpc, and the NGC 1023 group with six members at $D = 9.6$ Mpc. The numbers given here are those of currently known galaxies. Dwarf galaxies like Sagittarius would be very difficult to detect at the distances of these groups.

Most galaxies are members of a group. Many more dwarf galaxies exist than luminous galaxies, and dwarf galaxies are located preferentially in the vicinity of larger galaxies. Some members of the Local Group are so under-luminous that they would hardly be observable outside the Local Group.

Fig. 6.7. M81 (left) and M82 (right), two galaxies of the M81 group, about 3.1 Mpc away. These two galaxies are moving around each other, and the gravitational interaction taking place may be the reason for the violent star formation in M82. M82 is an archetypical starburst galaxy

One large concentration of galaxies was already known in the eighteenth century (W. Herschel) – the *Virgo Cluster*. Its galaxies extend over a region of about $10° \times 10°$ in the sky, and its distance is $D \sim 16$ Mpc. The Virgo Cluster consists of about 250 large galaxies and more than 2000 smaller ones. In the classification scheme of galaxy clusters, Virgo is considered an irregular cluster. The closest regular massive galaxy cluster is the *Coma cluster* (see Fig. 1.14), at a distance of about $D \sim 90$ Mpc.

6.2 Galaxies in Clusters and Groups

6.2.1 The Abell Catalog

George Abell compiled a catalog of galaxy clusters, published in 1958, in which he identified regions in the sky that show an overdensity of galaxies. This identification was performed by eye on photoplates from the *Palomar Observatory Sky Survey* (POSS), a photographic atlas of the Northern ($\delta > -30°$) sky.[1] He

[1]The POSS, or more precisely the first Palomar Sky Survey, consists of 879 pairs of photoplates observed in two color bands, and covers the Northern sky at declinations $\gtrsim -30°$. It was completed in 1960. The coverage of the southern part of the sky was completed in 1980 in the ESO/SERC Southern Sky Surveys, where this survey is about two magnitudes deeper ($B \lesssim 23$, $R \lesssim 22$) than POSS. The photoplates from both surveys have been digitized, forming the Digitized

omitted the Galactic disk region because the observation of galaxies is considerably more problematic there, due to extinction and the high stellar density (see also Fig. 6.2).

Abell's Criteria and his Catalog. The criteria Abell applied for the identification of clusters refer to an overdensity of galaxies within a specified solid angle. According to these criteria, a cluster contains ≥ 50 galaxies in a magnitude interval $m_3 \leq m \leq m_3 + 2$, where m_3 is the apparent magnitude of the third brightest galaxy in the cluster.[2] These galaxies must be located within a circle of angular radius

$$\theta_{\rm A} = \frac{1\overset{\prime}{.}7}{z} \tag{6.6}$$

where z is the estimated redshift. The latter is determined by the assumption that the luminosity of the tenth brightest galaxy in a cluster is the same for all clusters. A calibration of this distance estimate is performed on clusters of known redshift. $\theta_{\rm A}$ is called the *Abell radius* of a cluster, and corresponds to a physical radius of $R_{\rm A} \approx 1.5h^{-1}$ Mpc.

The so-determined redshift should be within the range $0.02 \leq z \leq 0.2$ for the selection of Abell clusters. The lower limit is chosen such that a cluster can be found on a single POSS photoplate ($\sim 6° \times 6°$) and does not extend over several plates, which would make the search more difficult, e.g., because the photographic sensitivity may differ for individual plates. The upper redshift bound is chosen due to the sensitivity limit of the photoplates.

The Abell catalog contains 1682 clusters which all fulfill the above criteria. In addition, it lists 1030 clusters that have been found in the search, but which do not fulfill all of the criteria (most of these contain between 30 and 49 galaxies). An extension of the catalog to the Southern sky was published by Abell, Corwin & Olowin in 1989. This ACO catalog contains 4076 clusters, including the members of the original catalog. Another important catalog of galaxy clusters is the Zwicky catalog (1961–68), which contains more clusters, but for which the applied selection criteria are considered less reliable.

Problems in the Optical Search for Clusters. The selection of galaxy clusters from an overdensity of galaxies on the sphere is not without problems, in particular if these catalogs are to be used for statistical purposes. An ideal catalog ought to fulfill two criteria: first it should be complete, in the sense that all objects which fulfill the selection criteria are contained in the catalog. Second it should be reliable, i.e., it should not contain any objects that do not belong in the catalog because they do not fulfill the criteria (so-called false positives). The Abell catalog is neither complete, nor is it reliable. We will briefly discuss why completeness and reliability cannot be expected in a catalog compiled in this way.

A galaxy cluster is a three-dimensional object, whereas galaxy counts on images are necessarily based on the projection of galaxy positions onto the sky. Therefore, projection effects are inevitable. Random overdensities on the sphere caused by line-of-sight projection may easily be classified as clusters. The reverse effect is likewise possible: due to fluctuations in the number density of foreground galaxies, a cluster at high redshift may be classified as an insignificant fluctuation – and thus remain undiscovered.

Of course, not all members of a cluster classified as such are in fact galaxies in the cluster, as here projection effects also play an important role. Furthermore, the redshift estimate is relatively coarse. In the meantime, spectroscopic analyses have been performed for many of the Abell clusters, and it has been found that Abell's redshift estimates have an error of about 30% – surprisingly accurate, considering the coarseness of his assumptions.

The Abell catalog is based on visual inspection of photographic plates. It is therefore partly subjective. Today, the Abell criteria can be applied to digitized images in an objective manner, using automated searches.

Sky Survey (DSS) that covers the full sky. Sections from the DSS can be obtained directly via the Internet, with the full DSS having a data volume of some 600 GB. Currently, the second Palomar Sky Survey (POSS-II) is in progress, which will be about one magnitude deeper compared to the first one and will contain data from three (instead of two) color filters. This will probably be the last photographic atlas of the sky because, with the development of large CCD cameras, we will soon be able to perform such surveys digitally. The most prominent example of this is the Sloan Digital Sky Survey, which we will discuss in a different context in Sect. 8.1.2.

[2] The reason for choosing the third brightest galaxy is that the luminosity of the brightest galaxy may vary considerably among clusters. Even more important is the fact that there is a finite probability for the brightest galaxy in a sky region under consideration to not belong to the cluster, but to be located at some smaller distance from us.

From these, it has been found that the results are not much different. The visual search must have been performed with great care and has to be recognized as a great accomplishment. For this reason, and in spite of the potential problems discussed above, the Abell and the ACO catalogs are still frequently used.

The clusters in the catalog are ordered by right ascension and are numbered. For example, Abell 851 is the 851st entry in the catalog, also denoted as A851. With a redshift of $z = 0.41$, A851 is the most distant Abell cluster.

Abell Classes. The Abell and ACO catalogs divide clusters into so-called richness and distance classes. Table 6.2 lists the criteria for the richness classes, while Table 6.3 lists those for the distance classes.

There are six *richness classes*, denoted from 0 to 5, according to the number of cluster member galaxies. Richness class 0 contains between 30 and 49 members and therefore does not belong to the cluster catalog proper. One can see from Table 6.2 that the number of clusters rapidly decreases with increasing richness class, so only very few clusters exist with a very large number of cluster galaxies. As a reminder, the region of the sky from where the Abell clusters were detected is about 2/3 of the total sphere. Thus, only a few very rich clusters do indeed exist (at redshift $\lesssim 0.2$).

Table 6.2. Definition of Abell's richness classes. N is the number of cluster galaxies with magnitudes between m_3 and m_3+2 inside the Abell radius (6.6), where m_3 is the brightness of the third brightest cluster galaxy.

Richness class R	N	Number in Abell's catalog
(0)	(30–49)	(≥ 1000)
1	50–79	1224
2	80–129	383
3	130–199	68
4	200–299	6
5	≥ 300	1

Table 6.3. Definition of Abell's distance classes. m_{10} is the magnitude of the tenth brightest cluster galaxy.

Distance class	m_{10}	Estimated average redshift	Number in Abell's catalog with $R \geq 1$
1	13.3–14.0	0.0283	9
2	14.1–14.8	0.0400	2
3	14.9–15.6	0.0577	33
4	15.7–16.4	0.0787	60
5	16.5–17.2	0.131	657
6	17.3–18.0	0.198	921

The subdivision into six *distance classes* is based on the apparent magnitude of the tenth brightest galaxy, in accordance with the redshift estimate for the cluster. Hence, the distance class provides a coarse measure of the distance.

6.2.2 Luminosity Function of Cluster Galaxies

The luminosity function of galaxies in a cluster is defined as in Sect. 3.7 for the total galaxy population. In many clusters, the Schechter luminosity function (3.38) represents a very good fit to the data if the brightest galaxy is disregarded in each cluster (see Fig. 3.32 for the Virgo Cluster of galaxies). The slope α at the faint end is not easy to determine, since projection effects become increasingly important for fainter galaxies. The value of α seems to vary between clusters, but it is not entirely clear whether this result may also be affected by projection effects in different clusters of differing strength. Thus, no final conclusion has been reached as to whether the luminosity function has a steep increase at $L \ll L^*$ or not, i.e., whether many more faint galaxies exist than luminous $\sim L^*$-galaxies (compare the galaxy content in the Local Group, Sect. 6.1.1, where even in our close neighborhood it is difficult to obtain a complete census of the galaxy population). L^* is very similar for many clusters, which is the reason why the distance estimate by apparent brightness of cluster members is quite reliable. However, a number of clusters exists with a clearly deviating value of L^*.

Many clusters contain *cD galaxies* at their centers; these differ from large ellipticals in several respects. They have a very extended stellar envelope, whose size may exceed $R \sim 100$ kpc and whose surface brightness profile is much broader than that of a de Vaucouleurs profile (see Fig. 3.8). cD galaxies are found only in the centers of clusters or groups, thus only in regions of strongly enhanced galaxy density. Many cD galaxies have multiple cores, which is a rather rare phenomenon among the other cluster members.

6.2.3 Morphological Classification of Clusters

Clusters are also classified by the morphology of their galaxy distribution. Several classifications are used, one of which is displayed in Fig. 6.8. Since this is a description of the visual impression of the galaxy distribution, the exact class of a cluster is not of great interest. However, a rough classification can provide an idea of the state of a cluster, i.e., whether it is currently in dynamical equilibrium or whether it has been heavily disturbed by a merger process with another cluster. Therefore, one distinguishes in particular between regular and irregular clusters, and also those which are intermediate; the transition between classes is of course continuous. Regular clusters are compact whereas, in contrast, irregular clusters are "open" (Zwicky's classification criteria).

This morphological classification indeed points at physical differences between clusters, as correlations between morphology and other properties of galaxy clusters show. For example, it is found that regular clusters are completely dominated by early-type galaxies, whereas irregular clusters have a fraction of spirals nearly as large as in the general distribution of field galaxies. Very often, regular clusters are dominated by a cD galaxy at the center, and their central galaxy density is very high. In contrast, irregular clusters are significantly less dense in the center. Irregular clusters often show strong substructure, which is rarely found in regular clusters. Furthermore, regular clusters have a high richness, whereas irregular clusters have fewer cluster members. To summarize, regular clusters can be said to be in a relaxed state, whereas irregular clusters are still in the process of evolution.

6.2.4 Spatial Distribution of Galaxies

Most regular clusters show a centrally condensed number density distribution of cluster galaxies, i.e., the galaxy density increases strongly towards the center. If the cluster is not very elliptical, this density distribution can be assumed, to a first approximation, as being spherically symmetric. Only the projected density distribution $N(R)$ is observable. This is related to the three-dimensional number density $n(r)$ through

$$N(R) = \int_{-\infty}^{\infty} \mathrm{d}z\, n\left(\sqrt{R^2+z^2}\right) = 2\int_R^{\infty} \frac{\mathrm{d}r\, r\, n(r)}{\sqrt{r^2-R^2}}\,, \tag{6.7}$$

where in the second step a simple transformation of the integration variable from the line-of-sight coordinate z to the three-dimensional radius $r=\sqrt{R^2+z^2}$ was made.

Of course, no function $N(R)$ can be observed, but only points (the positions of the galaxies) that are distributed in a certain way. If the number density of galaxies is sufficiently large, $N(R)$ is obtained by smoothing the point distribution. Alternatively, one considers parametrized forms of $N(R)$ and fits the parameters to the observed galaxy positions. In most cases, the second approach is taken because its results are more robust. A parametrized distribution needs to contain at least five parameters to be able to describe at least the basic characteristics of a cluster. Two of these parameters describe the position of the cluster center on the sky. One parameter is used to describe the amplitude of the density, for which, e.g., the central density $N_0 = N(0)$

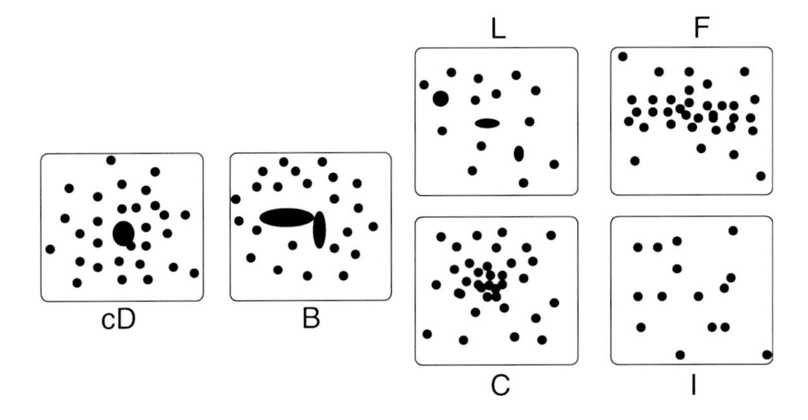

Fig. 6.8. Rough morphological classification of clusters by Rood & Sastry: cDs are those which are dominated by a central cD galaxy, Bs contain a pair of bright galaxies in the center. Ls are clusters with a nearly linear alignment of the dominant galaxies, Cs have a single core of galaxies, Fs are clusters with an oblate galaxy distribution, and Is are clusters with an irregular distribution

may be used. A forth parameter is a characteristic scale of a cluster, often taken to be the core radius r_c, defined such that at $R = r_c$, the projected density has decreased to half the central value, $N(r_c) = N_0/2$. Finally, one parameter is needed to describe "where the cluster ends"; the Abell radius is a first approximation for such a parameter.

Parametrized cluster models can be divided into those which are physically motivated, and those which are of a purely mathematical nature. One example for the latter is the de Vaucouleurs profile which is not derived from dynamical models. Next, we will consider a class of distributions that are based on a dynamical model.

Isothermal Distributions. These models are based on the assumption that the velocity distribution of the massive particles (this may be both galaxies in the cluster or dark matter particles) of a cluster is locally described by a Maxwell distribution, i.e., they are thermalized. As shown from spectroscopic analyses of the distribution of the radial velocities of cluster galaxies, this is not a bad assumption. Assuming, in addition, that the mass profile of the cluster follows that of the galaxies (or vice versa), and that the temperature (or equivalently the velocity dispersion) of the distribution does not depend on the radius (so that one has an isothermal distribution of galaxies), then one obtains a one-parameter set of models, the so-called *isothermal spheres*. These can be described physically as follows.

In dynamical equilibrium, the pressure gradient must be equal to the gravitational acceleration, so that

$$\boxed{\frac{\mathrm{d}P}{\mathrm{d}r} = -\rho \frac{GM(r)}{r^2}}\,, \tag{6.8}$$

where $\rho(r)$ denotes the density of the distribution, e.g., the density of galaxies. By $\rho(r) = \langle m \rangle\, n(r)$, this mass density is related to the number density $n(r)$, where $\langle m \rangle$ is the average particle mass. $M(r) = 4\pi \int_0^r \mathrm{d}r'\, r'^2\, \rho(r')$ is the mass of the cluster within a radius r. By differentiation of (6.8), we obtain

$$\frac{\mathrm{d}}{\mathrm{d}r}\left(\frac{r^2}{\rho}\frac{\mathrm{d}P}{\mathrm{d}r}\right) + 4\pi G r^2 \rho = 0\,. \tag{6.9}$$

The relation between pressure and density is $P = nk_{\mathrm{B}}T$. On the other hand, the temperature is related to the velocity dispersion of the particles,

$$\frac{3}{2}k_{\mathrm{B}}T = \frac{\langle m \rangle}{2}\left\langle v^2 \right\rangle\,, \tag{6.10}$$

where $\left\langle v^2 \right\rangle$ is the mean squared velocity, i.e., the velocity dispersion, provided the average velocity vector is set to zero. The latter assumption means that the cluster does not rotate, or contract or expand. If T (or $\left\langle v^2 \right\rangle$) is independent of r, then

$$\frac{\mathrm{d}P}{\mathrm{d}r} = \frac{k_{\mathrm{B}}T}{\langle m \rangle}\frac{\mathrm{d}\rho}{\mathrm{d}r} = \frac{\left\langle v^2 \right\rangle}{3}\frac{\mathrm{d}\rho}{\mathrm{d}r} = \sigma_v^2 \frac{\mathrm{d}\rho}{\mathrm{d}r}\,, \tag{6.11}$$

where σ_v^2 is the one-dimensional velocity dispersion, e.g., the velocity dispersion along the line-of-sight, which can be measured from the redshift of the cluster galaxies. If the velocity distribution corresponds to an isotropic (Maxwell) distribution, the one-dimensional velocity dispersion is exactly 1/3 times the three-dimensional velocity dispersion, because of $\left\langle v^2 \right\rangle = \sigma_x^2 + \sigma_y^2 + \sigma_z^2$, or

$$\sigma_v^2 = \frac{\left\langle v^2 \right\rangle}{3}\,. \tag{6.12}$$

With (6.9), it then follows that

$$\frac{\mathrm{d}}{\mathrm{d}r}\left(\frac{\sigma_v^2\, r^2}{\rho}\frac{\mathrm{d}\rho}{\mathrm{d}r}\right) + 4\pi G r^2 \rho = 0\,. \tag{6.13}$$

Singular Isothermal Sphere. In general, the differential equation (6.13) for $\rho(r)$ cannot be solved analytically. Physically reasonable boundary conditions are $\rho(0) = \rho_0$, the central density, and $(\mathrm{d}\rho/\mathrm{d}r)_{|r=0} = 0$, for the density profile to be flat at the center. One particular analytical solution of the differential equation exists, however: By substitution, we can easily show that

$$\boxed{\rho(r) = \frac{\sigma_v^2}{2\pi G r^2}} \tag{6.14}$$

solves (6.13). This density distribution is called *singular isothermal sphere*; we have encountered it before, in the discussion of gravitational lens models in Sect. 3.8.2. This distribution has a diverging density as $r \to 0$ and an infinite total mass $M(r) \propto r$. It is remarkable that this density distribution is just what is needed to explain the flat rotation curves of galaxies at large radii.

Numerical solutions of (6.13) with the initial conditions specified above (thus, with a flat core) reveal that the central density and the core radius are related to each other by

$$\rho_0 = \frac{9\sigma_v^2}{4\pi G r_c^2} \; . \tag{6.15}$$

Hence, these physical solutions of (6.13) avoid the infinite density of the singular isothermal sphere. However, these solutions also decrease outwards with $\rho \propto r^{-2}$, so they have a diverging mass as well. The origin of this mass divergence is easily understood because these isothermal distributions are based on the assumption that the velocity distribution is isothermal, thus Maxwellian with a spatially constant temperature. A Maxwell distribution has wings, hence it (formally) contains particles with arbitrarily high velocities. Since the distribution is assumed stationary, such particles must not escape, so their velocity must be lower than the escape velocity from the gravitational well of the cluster. But for a Maxwell distribution this is only achievable for an infinite total mass.

King Models. To remove the problem of the diverging total mass, self-gravitating dynamical models with an upper cut-off in the velocity distribution of their constituent particles are introduced. These are called *King models* and cannot be expressed analytically. However, an analytical approximation exists for the central region of these mass profiles,

$$\rho(r) = \rho_0 \left[1 + \left(\frac{r}{r_c}\right)^2\right]^{-3/2} \; . \tag{6.16}$$

Using (6.7), we obtain from this the projected surface mass density

$$\Sigma(R) = \Sigma_0 \left[1 + \left(\frac{R}{r_c}\right)^2\right]^{-1} \quad \text{with} \quad \Sigma_0 = 2\rho_0 r_c \; . \tag{6.17}$$

The analytical fit (6.16) of the King profile also has a diverging total mass, but this divergence is "only" logarithmic.

These analytical models for the density distribution of galaxies in clusters are only approximations, of course, because the galaxy distribution in clusters is often heavily structured. Furthermore, these dynamical models are applicable to a galaxy distribution only if the galaxy number density follows the matter density. However, one finds that the distribution of galaxies in a cluster often depends on the galaxy type. The fraction of early-type galaxies (Es and S0s) is often largest near the center. Therefore, one should consider the possibility that the distribution of galaxies in a cluster may be different from that of the total matter. A typical value for the core radius is about $r_c \sim 0.25 h^{-1}$ Mpc.

6.2.5 Dynamical Mass of Clusters

The above argument relates the velocity distribution of cluster galaxies to the mass profile of the cluster, and from this we obtain physical models for the density distribution. This implies the possibility of deriving the mass, or the mass profile, respectively, of a cluster from the observed velocities of cluster galaxies. We will briefly present this method of mass determination here. For this, we consider the dynamical time-scale of clusters, defined as the time a typical galaxy needs to traverse the cluster once,

$$t_{\rm cross} \sim \frac{R_A}{\sigma_v} \sim 1.5 h^{-1} \times 10^9\,{\rm yr} \; , \tag{6.18}$$

where a (one-dimensional) velocity dispersion $\sigma_v \sim 1000$ km/s was assumed. The dynamical time-scale is shorter than the age of the Universe. One therefore concludes that clusters of galaxies are gravitationally bound systems. If this were not the case they would dissolve on a timescale $t_{\rm cross}$. Since $t_{\rm cross} \ll t_0$ one assumes a *virial equilibrium*, hence that the virial theorem applies, so that in a time-average sense,

$$2E_{\rm kin} + E_{\rm pot} = 0 \; , \tag{6.19}$$

where

$$E_{\rm kin} = \frac{1}{2} \sum_i m_i v_i^2 \; ; \quad E_{\rm pot} = -\frac{1}{2} \sum_{i \neq j} \frac{G m_i m_j}{r_{ij}} \tag{6.20}$$

are the kinetic and the potential energy of the cluster galaxies, m_i is the mass of the i-th galaxy, v_i is the absolute value of its velocity, and r_{ij} is the spatial separation between the i-th and the j-th galaxy. The factor $1/2$ in the definition of E_{pot} occurs since each pair of galaxies occurs twice in the sum.

We define the total mass of the cluster,

$$M := \sum_i m_i \, , \tag{6.21}$$

the velocity dispersion, weighted by mass,

$$\langle v^2 \rangle := \frac{1}{M} \sum_i m_i \, v_i^2 \tag{6.22}$$

and the gravitational radius,

$$r_G := 2M^2 \left(\sum_{i \neq j} \frac{m_i \, m_j}{r_{ij}} \right)^{-1} \, . \tag{6.23}$$

With this, we obtain

$$E_{kin} = \frac{M}{2} \langle v^2 \rangle \; ; \quad E_{pot} = -\frac{G\,M^2}{r_G} \, , \tag{6.24}$$

for the kinetic and potential energy. Applying the virial theorem (6.19) yields the mass estimate

$$\boxed{M = \frac{r_G \langle v^2 \rangle}{G}} \, . \tag{6.25}$$

Transition to Projected Quantities. The above derivation uses the three-dimensional separations r_i of the galaxies from the cluster center, which are, however, not observable. To be able to apply these equations to observations, they need to be transformed to projected separations. If the galaxy positions and the directions of their velocity vectors are uncorrelated, as it is the case, e.g., for an isotropic velocity distribution, then

$$\langle v^2 \rangle = 3\sigma_v^2 \, , \quad r_G = \frac{\pi}{2} R_G$$
$$\text{with} \quad R_G = 2M^2 \left(\sum_{i \neq j} \frac{m_i m_j}{R_{ij}} \right)^{-1} \, , \tag{6.26}$$

where R_{ij} denotes the projected separation between the galaxies i and j. The parameters σ_v and R_G are direct observables; thus, the total mass of the cluster can be determined. One obtains

$$\boxed{\begin{aligned} M &= \frac{3\pi R_G \sigma_v^2}{2G} \\ &= 1.1 \times 10^{15} M_\odot \left(\frac{\sigma_v}{1000\,\text{km/s}} \right)^2 \left(\frac{R_G}{1\,\text{Mpc}} \right) \end{aligned}} \, . \tag{6.27}$$

We explicitly point out that this mass estimate no longer depends on the masses m_i of the individual galaxies – rather the galaxies are now test particles in the gravitational potential. With $\sigma_v \sim 1000$ km/s and $R_G \sim 1$ Mpc as typical values for rich clusters of galaxies, one obtains a characteristic mass of $\sim 10^{15} M_\odot$ for rich clusters.

The "Missing Mass" Problem in Clusters of Galaxies. With M and the number N of galaxies, one can now derive a characteristic mass $m = M/N$ for the luminous galaxies. This mass is found to be very high, $m \sim 10^{13} M_\odot$. Alternatively, M can be compared with the total optical luminosity of the cluster galaxies, $L_{tot} \sim 10^{12}$–$10^{13}\,L_\odot$, and hence the mass-to-light ratio can be calculated; typically

$$\boxed{\left(\frac{M}{L_{tot}} \right) \sim 300\,h \left(\frac{M_\odot}{L_\odot} \right)} \, . \tag{6.28}$$

This value exceeds the M/L ratio of early-type galaxies by at least a factor of 10. Realizing this discrepancy, Fritz Zwicky concluded as early as 1933, from an analysis of the Coma cluster, that clusters of galaxies must contain considerably more mass than is visible in galaxies – the dawn of the *missing mass problem*. As we will see further below, this problem has by now been firmly established, since other methods for the mass determination of clusters also yield comparable values and indicate that a major fraction of the mass in galaxy clusters consists of (non-baryonic) dark matter. *The stars visible in galaxies contribute less than about 5% to the total mass in clusters of galaxies.*

6.2.6 Additional Remarks on Cluster Dynamics

Given the above line of argument, the question of course arises as to whether the application of the virial theorem is still justified if the main fraction of mass is not contained in galaxies. The derivation remains valid in this form as long as the spatial distribution of galaxies follows the total mass distribution. The dynamical mass determination can be affected by an anisotropic velocity distribution of the cluster galaxies and by the possibly non-spherical cluster mass distribution. In both cases, projection effects, which are dealt with relatively easily in the spherically-symmetric case, obviously become more complicated. This is also one of the reasons for the necessity to consider alternative methods of mass determination.

Two-body collisions of galaxies in clusters are of no importance dynamically, as is easily seen from the corresponding relaxation time-scale (3.3),

$$t_{\rm relax} = t_{\rm cross} \frac{N}{\ln N} ,$$

which is much larger than the age of the Universe. The motion of galaxies is therefore governed by the collective gravitational potential of the cluster. The velocity dispersion is approximately the same for the different types of galaxies, and also only a weak tendency exists for a dependence of σ_v on galaxy luminosity, restricted to the brightest ones (see below in Sect. 6.2.9). From this, we conclude that the galaxies in a cluster are not "thermalized" because this would mean that they all have the same mean kinetic energy, implying $\sigma_v \propto m^{-1/2}$. Furthermore, the independence of σ_v from L implies that collisions of galaxies with each other are not dynamically relevant; rather, the velocity distribution of galaxies is defined by collective processes during cluster formation.

Violent Relaxation. One of the most important of the aforementioned processes is known as *violent relaxation*. This process very quickly establishes a virial equilibrium in the course of the gravitational collapse of a mass concentration. The reason for it are the small-scale density inhomogeneities within the collapsing matter distribution which generate, via Poisson's equation, corresponding fluctuations in the gravitational field. These then scatter the infalling particles and, by this, the density inhomogeneities are further amplified. The fluctuations of the gravitational field act on the matter like scattering centers. In addition, these field fluctuations change over time, yielding an effective exchange of energy between the particles. In a statistical average, all galaxies obtain the same velocity distribution by this process. As confirmed by numerical simulations, this process takes place on a time-scale of $t_{\rm cross}$, i.e., roughly as quickly as the collapse itself.

Dynamical Friction. Another important process for the dynamics of galaxies in a cluster is *dynamical friction*. The simplest picture of dynamical friction is obtained by considering the following. If a massive particle of mass m moves through a statistically homogeneous distribution of massive particles, the gravitational force on this particle vanishes due to homogeneity. But since the particle itself has a mass, it will attract other massive particles and thus cause the distribution to become inhomogeneous. As the particle moves, the surrounding "background" particles will react to its gravitational field and slowly start moving towards the direction of the particle trajectory. Due to the inertia of matter, the resulting density inhomogeneity will be such that an overdensity of mass will be established along the track of the particle, where the density will be higher on the side opposite to the direction of motion (thus, behind the particle) than in the forward direction (see Fig. 6.9). By this process, a gravitational field will form that causes an acceleration of the particle against the direction of motion, so that the particle will be slowed down. Because this "polarization" of the medium is caused by the gravity of the particle, which is proportional to its mass, the deceleration will also be proportional to m. Furthermore, a fast-moving particle will cause less polarization in the medium than a slow-moving one because each mass element in the medium is experiencing the gravitational attraction of the particle for a shorter time, thus the medium becomes less polarized. In addition, the particle is on average farther away from the density accumulation on its backward track, and thus will experience a smaller acceleration if it is faster. Combining these arguments, one obtains for the dependence of this dynamical friction

$$\frac{{\rm d}\boldsymbol{v}}{{\rm d}t} \propto -\frac{m\,\rho\,\boldsymbol{v}}{|\boldsymbol{v}|^3} , \tag{6.29}$$

where ρ is the mass density in the medium. Applied to clusters of galaxies, this means that the most mas-

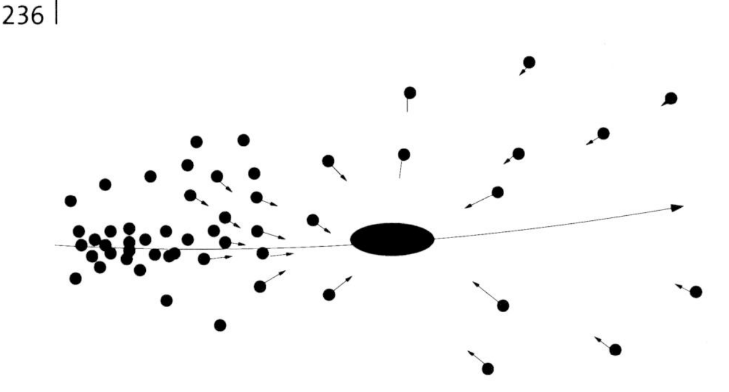

Fig. 6.9. The principle of dynamical friction. The gravitational field of a massive particle (here indicated by the large symbol) accelerates the surrounding matter towards its track. Through this, an overdensity establishes on the backward side of its orbit, the gravitational force of which decelerates the particle

sive galaxies will experience the strongest dynamical friction, so that they are subject to a significant deceleration through which they move deeper into the potential well. The most massive cluster galaxies should therefore be concentrated around the cluster center, so that a spatial separation of galaxy populations with respect to their masses occurs (mass segregation). If dynamical friction acts over a sufficiently long time, the massive cluster galaxies in the center may merge into a single one. This is one possible explanation for the formation of cD galaxies.

Dynamical friction also plays an important role in other dynamical processes in astrophysics. For example, the Magellanic Clouds experience dynamical friction on their orbit around the Milky Way and thereby lose kinetic energy. Consequently, their orbit will become smaller over the course of time and, in a distant future, these two satellite galaxies will merge with our Galaxy. In fact, dynamical friction is of vital importance in galaxy merger processes which occur in the evolution of the galaxy population, a subject we will return to in Sect. 9.6.

6.2.7 Intergalactic Stars in Clusters of Galaxies

The space between the galaxies in a cluster is filled with hot gas, as visible from X-ray observations. In recent years, it has been found that besides hot gas there are also stars in between the galaxies. The detection of such an intergalactic stellar population comes as a surprise at first sight, because our understanding of star formation implies that they can only form in the dense centers of molecular clouds. Hence, one expects that stars cannot form in intergalactic space. This is not necessarily implied by the presence of intergalactic stars, however, since they can also be stripped from galaxies in the course of gravitational interactions between galaxies in the cluster, and so form an intergalactic population. The fate of these stars is thus comparable to that of the interstellar medium, which is metal-enriched by the processes of stellar evolution in galaxies before it is removed from these galaxies and becomes part of the intergalactic medium in clusters; otherwise, the substantial metallicity of the ICM could not be explained.

The observation of diffuse optical light in clusters of galaxies and, related to this, the detection of the intracluster stellar population, is extremely difficult. Although first indications have already been found with photographic plate measurements, the surface brightness of this cluster component is so low that even with CCD detectors the observation is extraordinarily challenging. To quantify this, we note that the surface brightness of this diffuse light component is about 30 mag arcsec^{-2} at a distance of several hundred kpc from the cluster center. This value needs to be compared with the brightness of the night sky, which is about 21 mag arcsec^{-2} in the V-band. One therefore needs to correct for the effects of the night sky to better than a tenth of a percent for the intergalactic stellar component to become visible in a cluster. Furthermore, cluster galaxies and objects in the foreground and background need to be masked out in the images, in order to measure the radial profile of this diffuse component. This is possible only up to a certain limiting magnitude, of course, up to which individual objects can be identified. The existence of weaker sources has to be accounted for with statistical methods, which in turn use the luminosity function of galaxies.

The diffuse light component is best investigated in a statistical superposition of the images of several galaxy clusters. Statistical fluctuations in the sky background and uncertainties in the flatfield[3] determination

[3]The flatfield of an image (or, more precisely, of the system consisting of telescope, filter, and detector) is defined as the image of a uniformly illuminated field, so that in the ideal case each pixel of the detector produces the same output signal. This is not the case in reality, however, as the sensitivity differs for individual pixels. For this reason, the flatfield measures the sensitivity distribution of the pixels, which is then accounted for in the image analysis.

are in this case averaged out. In these analyses an $r^{-1/4}$-law is found for the light distribution in the inner region of clusters, i.e., the (de Vaucouleurs) brightness profile of the central galaxy is measured. For radii larger than about ~ 50 kpc, the brightness profile exceeds the extrapolation of the de Vaucouleurs profile, and has been detected out to very large distances from the cluster center. This fact needs to be considered in the context of the existence of cD galaxies, which are defined by exactly this light excess. The separation between the diffuse light component and the extended light profile of a cD galaxy is not easily performed, but at large distances from the cluster center, one can exclude the possibility that the corresponding stars are gravitationally bound to the central cluster galaxy.

Besides this diffuse component, individual stars and planetary nebulae have been detected in some neighboring galaxy clusters which cannot be assigned to any cluster galaxy. The diffuse cluster component accounts for about 10% of the total optical light in a cluster. Therefore, models of galaxy evolution in clusters should provide an explanation for these observations.

Hydrogen Clouds in the Virgo Cluster, and a Dark Galaxy? The interaction of galaxies in clusters cannot only lead to the stripping of stars from galaxies, but in the case of gas-rich galaxies its ISM can be (partly) removed. Indeed, in the Virgo Cluster several large clouds of gas have been found, through their HI 21-cm emission, which are not centered on optically luminous galaxies. For one of them, with a neutral hydrogen mass of $\sim 10^8 M_\odot$, the rotational velocity has been measured, yielding the result that this cloud is dominated by dark matter – an optically dark galaxy. Combining the rotational velocity with the size of the HI distribution, a lower bound of the dynamical mass of $\sim 10^{11} M_\odot$ is inferred – the lack of any visible counterpart in the optical then yields a lower bound on the mass-to-light ratio of about 500 in Solar units. However, this conclusion is based on the assumption that the gas is in dynamical equilibrium. Rather, if it has been recently stripped from a galaxy, an equilibrium state may not have been established, and therefore the mass estimate from the measured velocity field may be in error. In other cases, the HI clouds can be identified with the galaxy from which they originated, owing to the tidal tails connecting the cloud with the corresponding galaxy. In any case, the possibility of having identified a "dark galaxy" of this large mass is exceedingly exciting, as such objects are not expected from our current understanding of galaxy formation.

6.2.8 Galaxy Groups

Accumulations of galaxies that do not satisfy Abell's criteria are in most cases galaxy groups. Hence, groups are the continuation of clusters towards fewer member galaxies and are therefore presumably of lower mass, lower velocity dispersion, and smaller extent. The distinction between groups and clusters is at least partially arbitrary. It was defined by Abell mainly to be not too heavily affected by projection effects in the identification of clusters. Groups are of course more difficult to detect, since the overdensity criterion for them is more sensitive to projection effects by foreground and background galaxies than for clusters.

A special class of groups are the *compact groups*, assemblies of (in most cases, few) galaxies with very small projected separations. The best known examples for compact groups are Stephan's Quintet and Seyfert's Sextet (see Fig. 6.10). In 1982, a catalog of 100 compact groups (Hickson Compact Groups, HCGs) was published, where a group consists of four or more bright members. These were also selected on POSS photoplates, again solely by an overdensity criterion. The median redshift of the HCGs is about $z = 0.03$. Examples of optical images of HCGs are given in Figs. 6.3 and 1.16.

Follow-up spectroscopic studies of the HCGs have verified that 92 of them have at least three galaxies with conforming redshifts, defined such that the corresponding recession velocities lie within 1000 km/s of the median velocity of group members. Of course, the similarity in redshift does not necessarily imply that these groups form a gravitationally bound and relaxed system. For instance, the galaxies could be tracers of an overdense structure which we happen to view from a direction where the galaxies are projected near each other on the sky. However, more than 40% of the galaxies in HCGs show evidence of interactions, indicating that these galaxies have near neighbors in three-dimensional

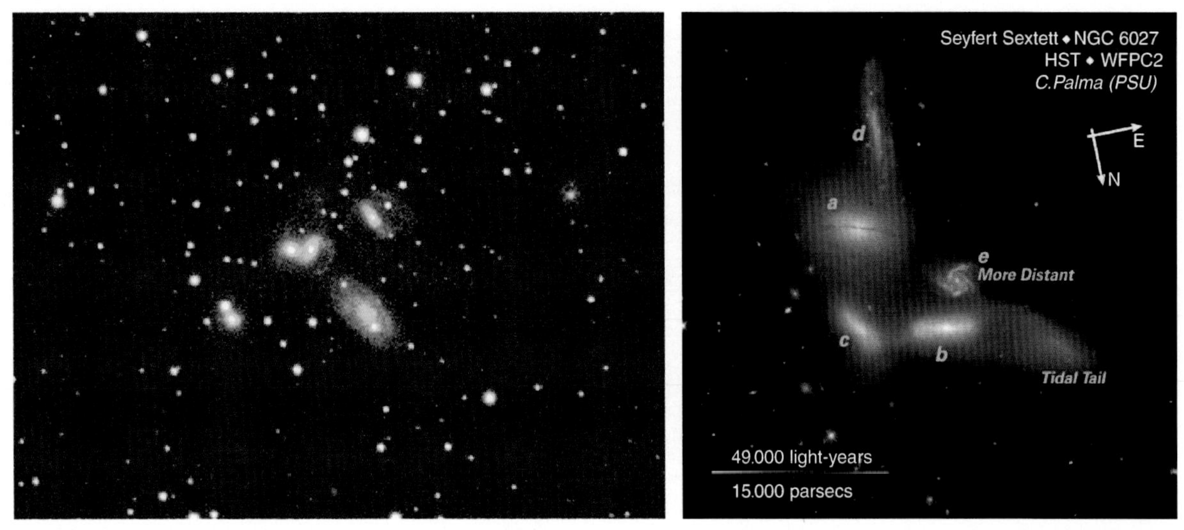

Fig. 6.10. Left: Stephan's Quintet, also known as Hickson Compact Group 92, is a very dense accumulation of galaxies with a diameter of about 80 kpc. Right: Seyfert's Sextet, an apparent accumulation of six galaxies located very close together on the sphere. Only four of the galaxies (a)–(d) are in fact galaxies belonging to the group; the spiral galaxy (e) is located at significantly higher distance. Another object originally classified as a galaxy is no galaxy but instead a tidal tail that was ejected in tidal interactions of galaxies in the group

space. Furthermore, about three quarters of HCGs with four or more member galaxies show extended X-ray emission, most likely coming from intragroup hot gas, providing additional evidence for the presence of a common gravitational potential well.

More recently, galaxy groups have been selected from spectroscopic surveys. For these, a three-dimensional overdensity criterion can be applied, which considerably reduces projection effects and which also allows the detection of groups in regions of larger mean projected galaxy number density. The velocity dispersion in groups is significantly smaller than that in clusters; typical values are $\sigma_v \sim 300\,\mathrm{km/s}$.

Compact groups have a lifetime which is much shorter than the age of the Universe. The dynamical timescale is $t_{\rm dyn} \sim R/\sigma_v \sim 0.02\,H_0^{-1}$, thus small compared to $t_0 \sim H_0^{-1}$. By dynamical friction (see Sect. 6.2.6), galaxies in groups lose kinetic (orbital) energy and move closer to the dynamical center where interactions and mergers with other group galaxies take place, as also seen by the high fraction of member galaxies with morphological signs of interactions. Since the lifetime of compact groups is shorter than the age of the Universe, they must have formed not too long ago. If we do not happen to live in a special epoch of cosmic history, such groups must therefore still be forming today. From dynamical studies, one finds that – as in clusters – the total mass of groups is significantly larger than the sum of the mass visible in galaxies; a typical mass-to-light ratio is $M/L \sim 50h$ (in Solar units), which is comparable to that of the Local Group.

As in clusters, the fraction of group members which are spirals is lower than the fraction of spirals among field (i.e., isolated) galaxies, and the relative abundance of spiral galaxies decreases with increasing σ_v of the group. Furthermore, galaxy groups are X-ray emitters, so they likewise contain hot intergalactic gas, albeit at lower temperatures and lower metallicities than clusters.

There are by now good indications that (compact) galaxy groups contain a diffuse optical light component, as is the case for galaxy clusters, and that the fraction of optical emission due to the diffuse component varies strongly between individual groups. Since the origin of the intragroup stellar component is most likely related to the history of galaxy interactions and tidal stripping by the group potential, the relative contribution of the intragroup light may contain valuable information about the evolution of groups.

6.2.9 The Morphology–Density Relation

As mentioned several times before, the mixture of galaxy types in clusters differs from that of isolated (field) galaxies. Whereas about 70% of the field galaxies are spirals, clusters are dominated by early-type galaxies, in particular in their inner regions. Furthermore, the fraction of spirals in a cluster depends on the distance to the center and increases for larger r. Obviously, the local density has an effect on the morphological mix of galaxies.

More generally, one may ask whether the mixture of the galaxy population depends on the local galaxy density. While earlier studies of this effect were frequently constrained to galaxies within and around clusters, new extensive redshift surveys like the 2dFGRS and the SDSS (see Sect. 8.1.2) allow us to systematically investigate this question with very large and carefully selected samples of galaxies. The morphological classification of such large samples is performed by automated software tools, which basically measure the light concentration in the galaxies. A comparison of galaxies classified this way with visual classifications shows very good agreement.

Results from the Sloan Digital Sky Survey. As an example of such an investigation, results from the Sloan Digital Sky Survey are shown in Fig. 6.11. The galaxies have been morphologically classified, based on SDSS photometry, and separated into four classes, corresponding to elliptical galaxies, S0 galaxies, and early (Sa) and late (Sc) types of spiral. In this analysis, only galaxies have been included for which the redshift was spectroscopically measured. Therefore, the spatial galaxy density can be estimated. However, one needs to take into account the fact that the measured redshift is a superposition of the cosmic expansion and the peculiar velocity of a galaxy. The peculiar velocity may have rather large values (~ 1000 km/s), in particular in clusters of galaxies. For this reason, for each galaxy in the sample the surface number density of galaxies which have a redshift within ± 1000 km/s of the target galaxy has been determined. The left panel in Fig. 6.11 shows the fraction of the different galaxy classes as a function of this local galaxy density. A very clear dependence, in particular of the fraction of late-type spirals, on the local density can be seen: in regions of higher galaxy density Sc spirals contribute less than 10% of the galaxies, whereas their fraction is about 30% in low-density re-

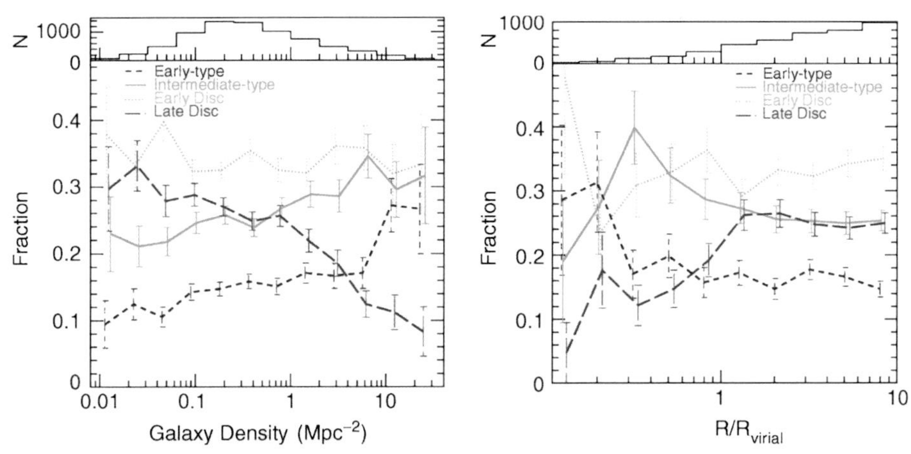

Fig. 6.11. The number fraction of galaxies of different morphologies is plotted as a function of the local galaxy density (left panel), and for galaxies in clusters as a function of the distance from the cluster center, scaled by the corresponding virial radius (right panel). Galaxies have been divided into four different classes. "Early-types" contain mainly ellipticals, "intermediates" are mainly S0 galaxies, "early and late disks" are predominantly Sa and Sc spirals, respectively. In both representations, a clear dependence of the galaxy mix on the density or on the distance from the cluster center, respectively, is visible. In the histograms at the top of each panel, the number of galaxies in the various bins is plotted

gions. Combined, the fraction of spirals decreases from $\sim 65\%$ in the field to about 35% in regions of high galaxy density. In contrast, the fraction of ellipticals and S0 galaxies increases towards higher densities, with the increase being strongest for ellipticals.

In the right-hand panel of Fig. 6.11, the mixture of galaxy morphologies is plotted as a function of the distance to the center of the nearest cluster, where the distance has been scaled by the virial radius of the corresponding cluster. As expected, a very strong dependence of the fraction of ellipticals and spirals on this distance is seen. Sc spirals contribute a mere 5% of galaxies in the vicinity of cluster centers, whereas the fraction of ellipticals and S0 galaxies strongly increases inwards.

The two diagrams in Fig. 6.11 are of course not mutually independent: a region of high galaxy density is very likely to be located in the vicinity of a cluster center, and the opposite is valid accordingly. Therefore, it is not immediately clear whether the mix of galaxy morphologies depends primarily on the respective density of the environment of the galaxies, or whether it is caused by morphological transformations in the inner regions of galaxy clusters.

The morphology–density relation is also seen in galaxy groups. The fraction of late-type galaxies decreases with increasing group mass. Furthermore, the fraction of early-type galaxies increases with decreasing distance from the group center, as is also the case in clusters.

Alternative Consideration: The Color–Density Relation. We pointed out in Sect. 3.7.2 that galaxies at fixed luminosity seem to have a bimodal color distribution (see Fig. 3.33). Using the same data set as that used for Fig. 3.33, the fraction of galaxies that are contained in the red population can be studied as a function of the local galaxy density. The result of this study is shown in the left-hand panel of Fig. 6.12, where the fraction of galaxies belonging to the red population is plotted against the local density of galaxies, measured in terms of the fifth-nearest neighboring galaxy within a redshift of ± 1000 km/s. The fraction of red galaxies increases towards higher local number density, and the relative increase is stronger for the less luminous galaxies. If we identify the red galaxies with the early-type galaxies in Fig. 6.11, these two results are in qualitative agreement. Surprisingly, the fraction of galaxies in the red sample seems to be a function of a combination of the local galaxy density and the luminosity of the galaxy, as is shown in the right-hand panel of Fig. 6.12.

Interpretation. A closer examination of Fig. 6.11 may provide a clue as to what physical processes are responsible for the dependence of the morphological mix on the local number density. We consider first the right-hand panel of Fig. 6.11. Three different regimes in radius can be identified: for $R \gtrsim R_{\rm vir}$, the fraction of the different galaxy types remains basically constant. In the intermediate regime, $0.3 \lesssim R/R_{\rm vir} \lesssim 1$, the fraction of S0 galaxies strongly increases inwards, whereas the fraction of late-type spirals decreases accordingly. This result is compatible with the interpretation that in the outer regions of galaxy clusters spirals lose gas (for instance, by their motion through the intergalactic medium), and these galaxies then transform into passive S0 galaxies. Below $R \lesssim 0.3 R_{\rm vir}$, the fraction of S0 galaxies decreases strongly, and the fraction of ellipticals increases substantially.

In fact, the ratio of the number densities of S0 galaxies and ellipticals, for $R \lesssim 0.3 R_{\rm vir}$, strongly decreases as R decreases. This may hint at a morphological transformation in which S0 galaxies are turned into ellipticals, probably by mergers. Such gas-free mergers, also called "dry mergers", may be the preferred explanation for the generation of elliptical galaxies. One of the nice properties of dry mergers is that such a merging process would not be accompanied by a burst of star formation, unlike the case of gas-rich collisions of galaxies. The existence of a population of newly-born stars in ellipticals would be difficult to reconcile with the generally old stellar population actually observed in these galaxies.

Considering now the dependence on local galaxy density (the left-hand panel of Fig. 6.11), a similar behavior of the morphological mix of galaxies is observed: there seem to exist two characteristic values for the galaxy density where the relative fractions of galaxy morphologies change noticeably. Interestingly, the relation between morphology and density seems to evolve only marginally between $z = 0.5$ and the local Universe.

One clue as to the origin of the morphological transformation of galaxies in clusters, as a function of distance from the cluster center, comes from the observation that the velocity dispersion of very bright cluster galaxies seems to be significantly smaller than that of

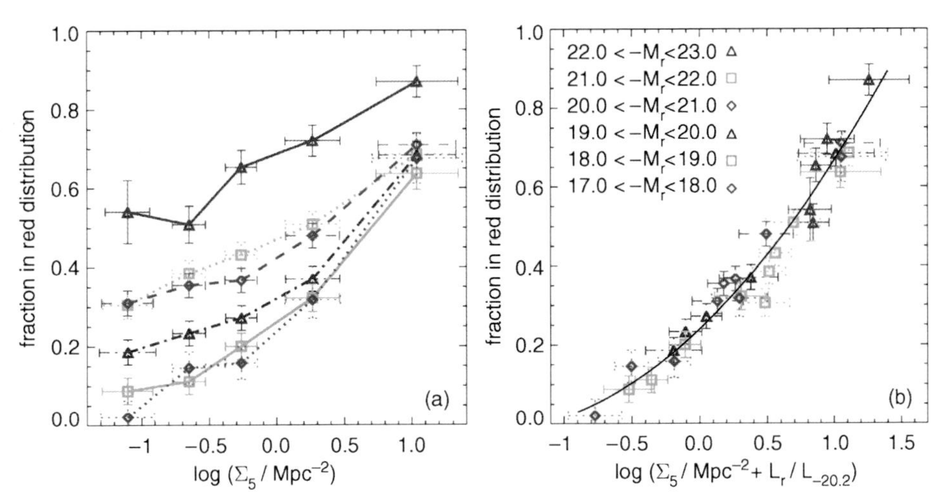

Fig. 6.12. Left: the fraction of galaxies in the red distribution (see Sect. 3.7.2) is shown as a function of Σ_5, an estimator of the local galaxy number density based on the projected distance of the fifth-nearest spectroscopically confirmed neighbor galaxy within $\pm 1000\,\mathrm{km/s}$. Different symbols correspond to different luminosity bins, as indicated. Right: the same red fraction is plotted against a combination of the local galaxy density Σ_5 and the luminosity of the galaxy

less luminous ones. Assuming that the mass-to-light ratio does not vary substantially among cluster members, this then indicates that the most massive galaxies have smaller velocity dispersions. One way to achieve this trend in the course of cluster evolution is by dynamical interactions between cluster galaxies. Such interactions tend to "thermalize" the velocity distribution of galaxies, so that the mean kinetic energy of galaxies tends to become similar. This then causes more massive galaxies to become slower on average. If this interpretation holds, then the morphology–density relation may be attributed to these dynamical interactions, rather than to the (so-called ram-pressure) stripping of the interstellar medium as the galaxies move through the intracluster medium.

E+A Galaxies. Galaxy clusters contain a class of galaxies which is defined in terms of spectral properties. These galaxies show strong Balmer line absorption in their spectra, characteristic of A stars, but no [OII] or Hα emission lines. The latter indicates that these galaxies are not undergoing strong star formation at present, whereas the former shows that there was an episode of star formation within the past ~ 1 Gyr, about as long ago as the main-sequence lifetime of A stars. These galaxies have been termed E+A galaxies since their spectra appears like a superposition of that of A-stars and that of otherwise normal elliptical galaxies. They are interpreted as being post-starburst galaxies. Since they were first seen in clusters, the interpretation of the origin of E+A galaxies was originally centered on the cluster environment – for example star-forming galaxies falling into a cluster and having their interstellar medium removed by tidal forces caused by the cluster potential well and/or stripping as the galaxies move through the intracluster medium. However, E+A galaxies were later also found in different environments, making the above interpretation largely obsolete. By investigating the spatial correlation of these galaxies with other galaxies shows that the phenomenon is not associated with the large-scale environment. An overdensity of neighboring galaxies can be seen only out to scales of ~ 100 kpc. If the sudden turn-off of the star-formation activity is indeed caused by an external perturbation, it is therefore likely that it is caused by the dynamical interaction of close neighboring galaxies. Indeed, about 30% of E+A galaxies are found to have morphological signatures of perturbations, such as tidal tails, supporting the interaction hypothesis.

In fact, the spiral galaxies in clusters seem to differ statistically from those of field spirals, in that the fraction of disk galaxies with absorption-line spectra, and thus no ongoing star formation, seems to be larger in clusters than in the field by a factor ~ 4, indicating that the cluster environment has a marked impact on the star-formation ability of these galaxies.

6.3 X-Ray Radiation from Clusters of Galaxies

One of the most important discoveries of the UHURU X-ray satellite, launched in 1970, was the detection of X-ray radiation from massive clusters of galaxies. With the later Einstein X-ray satellite and more recently ROSAT, X-ray emission was also detected from lower-mass clusters and groups. Three examples for the X-ray emission of galaxy clusters are displayed in Figs. 6.13–6.15. Figure 6.13 shows the Coma cluster of galaxies, observed with two different X-ray observatories. Although Coma was considered to be a fully relaxed cluster, distinct substructure is visible in its X-ray radiation. The cluster RXJ 1347−1145 (Fig. 6.14) is regarded as the most luminous cluster in the X-ray domain. A large mass estimate of this cluster also follows from the analysis of the gravitationally lensed arcs (see Sect. 6.5) that are visible in Fig. 6.14; the cover of this book shows a more recent image of this cluster, taken with the ACS camera on-board HST, where a large number of arcs can be readily detected. Finally, Fig. 6.15 shows a superposition of the X-ray emission and an optical image of the cluster MS 1054−03, which is situated at $z = 0.83$ and to which we will refer as an example frequently below.

6.3.1 General Properties of the X-Ray Radiation

Clusters of galaxies are the brightest extragalactic X-ray sources besides AGNs. Their characteristic luminosity is $L_{\rm X} \sim 10^{43}$ up to $\sim 10^{45}$ erg/s for the most massive clusters. This X-ray emission from clusters is spatially extended, so it does not originate in individual galaxies. The spatial region from which we can detect this radiation can have a size of 1 Mpc or even larger. Furthermore, the X-ray radiation from clusters does not vary on timescales over which it has been observed ($\lesssim 30$ yr). Variations would also not be expected if the radiation originates from an extended region.

Continuum Radiation. The spectral energy distribution of the X-rays leads to the conclusion that the emission process is optically thin thermal bremsstrahlung (free–free radiation) from a hot gas. This radiation is produced by the acceleration of elec-

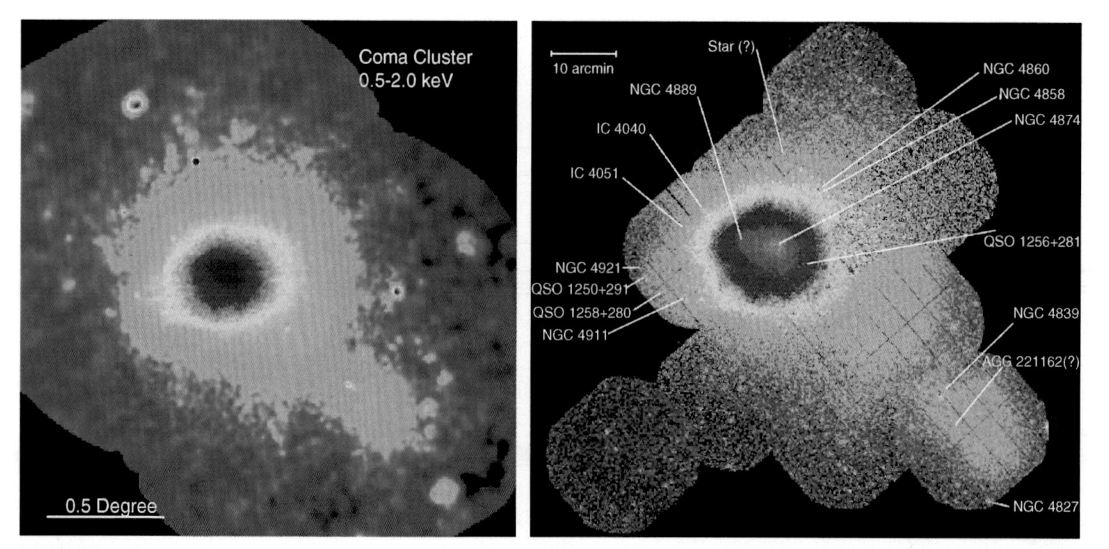

Fig. 6.13. X-ray images of the Coma cluster, taken with the ROSAT-PSPC (left) and XMM-EPIC (right). The image size in the left panel is 2.7° × 2.5°. A remarkable feature is the secondary maximum in the X-ray emission at the lower right of the cluster center which shows that even Coma, long considered to be a regular cluster, is not completely in an equilibrium state, but is dynamically evolving, presumably by the accretion of a galaxy group

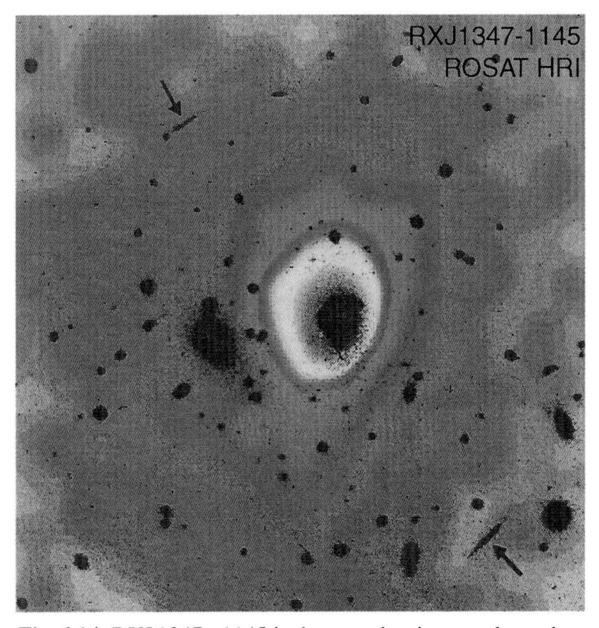

Fig. 6.14. RXJ 1347−1145 is the most luminous galaxy cluster in the X-ray domain. A color-coded ROSAT/HRI image of this cluster, which shows the distribution of the intergalactic gas, is superposed on an optical image of the cluster. The two arrows indicate giant arcs, images of background galaxies which are strongly distorted by the gravitational lens effect

trons in the Coulomb field of protons and atomic nuclei. Since an accelerated electrically charged particle emits radiation, such scattering processes between electrons and protons in an ionized gas yields emission of photons. From the spectral properties of this radiation, the gas temperature in galaxy clusters can be determined, which is, for clusters with mass between $\sim 10^{14} M_\odot$ and $\sim 10^{15} M_\odot$, in the range of 10^7–10^8 K, or 1–10 keV, respectively.

The emissivity of bremsstrahlung is described by

$$\epsilon_\nu^{\rm ff} = \frac{32\pi Z^2 e^6 n_{\rm e} n_{\rm i}}{3 m_{\rm e} c^3} \sqrt{\frac{2\pi}{3 k_{\rm B} T m_{\rm e}}}\, {\rm e}^{-h_{\rm P}\nu/k_{\rm B}T}\, g_{\rm ff}(T,\nu)\,, \tag{6.30}$$

where e denotes the elementary charge, $n_{\rm e}$ and $n_{\rm i}$ the number density of electrons and ions, respectively, Z the charge of the ions, and $m_{\rm e}$ the electron mass. The function $g_{\rm ff}$ is called Gaunt factor; it is a quantum mechanical correction factor of order 1, or, more precisely,

$$g_{\rm ff} \approx \frac{3}{\sqrt{\pi}} \ln\left(\frac{9 k_{\rm B} T}{4 h_{\rm P} \nu}\right) .$$

Hence, the spectrum described by (6.30) is flat for $h_{\rm P}\nu \ll k_{\rm B}T$, and exponentially decreasing for $h_{\rm P}\nu \gtrsim k_{\rm B}T$, as is displayed in Fig. 6.16.

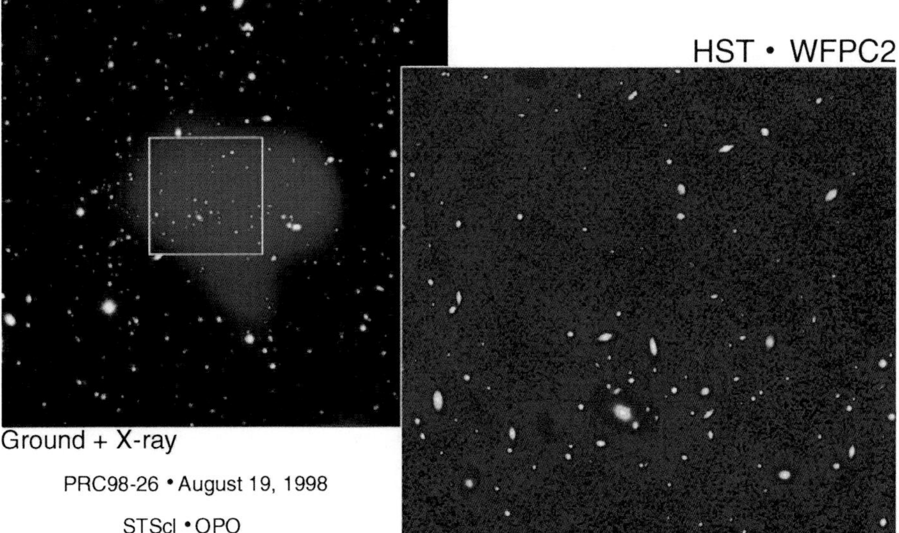

Fig. 6.15. The cluster of galaxies MS 1054−03 is, at $z = 0.83$, the highest-redshift cluster in the Einstein Medium Sensitivity Survey, which was compiled from observations with the Einstein satellite (see Sect. 6.3.5). On the right, an HST image of the cluster is shown, while on the left is an optical image, obtained with the 2.2-m telescope of the University of Hawaii, superposed (in blue) with the X-ray emission of the cluster measured with the ROSAT-HRI

The temperature of the gas in massive clusters is typically $T \sim 5 \times 10^7$ K, or $k_B T \sim 5$ keV – X-ray astronomers usually specify temperatures and frequencies in keV (see Appendix C). For a thermal plasma with Solar abundances, the total bremsstrahlung emission is

$$\epsilon^{\rm ff} = \int_0^\infty {\rm d}\nu\, \epsilon_\nu^{\rm ff} \approx 3.0 \times 10^{-27} \sqrt{\frac{T}{1\,{\rm K}}} \left(\frac{n_{\rm e}}{1\,{\rm cm}^{-3}}\right)^2 \,{\rm erg\,cm^{-3}\,s^{-1}} \,. \tag{6.31}$$

The energy resolution and the angular resolution of X-ray satellites prior to the Chandra and XMM-Newton observatories, which were both launched in 1999, did not permit detailed analyses of the spatial dependence of the gas temperature. Therefore, it is often assumed in modeling the X-ray emission from clusters that T is spatially constant. However, observations by these two recent satellites show that in many cases this assumption is not well justified because clear temperature gradients are observed.

Line Emission. The assumption that the X-ray emission originates from a hot, diffuse gas (intracluster medium, ICM) was confirmed by the discovery of line emission in the X-ray spectrum of clusters. The most prominent line in massive clusters is located at energies just below 7 keV: it is the Lyman-α line of 25-fold ionized iron (thus, of an iron nucleus with only a single electron). Slightly less ionized iron has a strong transition at somewhat lower energies of $E \sim 6.4$ keV. Later, other lines were also discovered in the X-ray spectrum of clusters. As a rule, the hotter the gas is, thus the more completely ionized it is, the weaker the line emission. The X-ray emission of clusters with relatively low temperatures, $k_B T \lesssim 2$ keV, is sometimes dominated by line emission from highly ionized atoms (C, N, O, Ne, Mg, Si, S, Ar, Ca; see Fig. 6.16). The emissivity of a thermal plasma with Solar abundance and temperatures in the range $10^5\,{\rm K} \lesssim T \lesssim 4 \times 10^7$ K can roughly be approximated by

$$\epsilon \approx 6.2 \times 10^{-19} \left(\frac{T}{1\,{\rm K}}\right)^{-0.6} \left(\frac{n_{\rm e}}{1\,{\rm cm}^{-3}}\right)^2 \,{\rm erg\,cm^{-3}\,s^{-1}} \,. \tag{6.32}$$

Equation (6.32) accounts for free–free emission as well as line emission. Compared to (6.31), one finds a different dependence on temperature: while the total emissivity for bremsstrahlung is $\propto T^{1/2}$, it increases again towards lower temperatures where the line emission becomes more important. It should be noted in particular that the emissivity depends quadratically on the density of the plasma, since both bremsstrahlung and the collisional excitation responsible for line emission are two-body processes. Thus in order to estimate the mass of the hot gas from its X-ray luminosity, the spatial distribution of the gas needs to be known. For example, if the gas in a cluster is locally inhomogeneous, the value of $\langle n_{\rm e}^2 \rangle$ which determines the X-ray emissivity may deviate significantly from $\langle n_{\rm e} \rangle^2$. As we will see later, clusters of galaxies satisfy a number of scaling relations, and one relation between the gas mass and the X-ray luminosity is found empirically, from which the gas mass can be estimated.

Morphology of the X-Ray Emission. From the morphology of their X-ray emission, one can roughly distinguish between regular and irregular clusters, as is also done in the classification of the galaxy distribution. In Fig. 6.17, X-ray surface brightness contours are superposed on optical images of four galaxy clusters or groups, respectively, covering a wide range of cluster mass and X-ray temperature. Regular clusters show a smooth brightness distribution, centered on the optical center of the cluster, and an outwardly decreasing surface brightness. Typically, regular clusters have a high X-ray luminosity $L_{\rm X}$ and high temperatures. In contrast, irregular clusters may have several brightness maxima, often centered on cluster galaxies or subgroups of cluster galaxies. Some of the irregular clusters show a high temperature as well, which is interpreted as a consequence of merger processes between clusters, in which the gas is heated by shock fronts. The trend emerges that in clusters with a larger fraction of spirals, $L_{\rm X}$ and T are lower. Irregular clusters also have a lower central galaxy density compared to regular clusters. Clusters of galaxies with a dominating central galaxy often show a strong central peak in X-ray emission. The X-ray emission often deviates from axial symmetry, so that the assumption of clusters being spherically symmetric is not well founded in these cases.

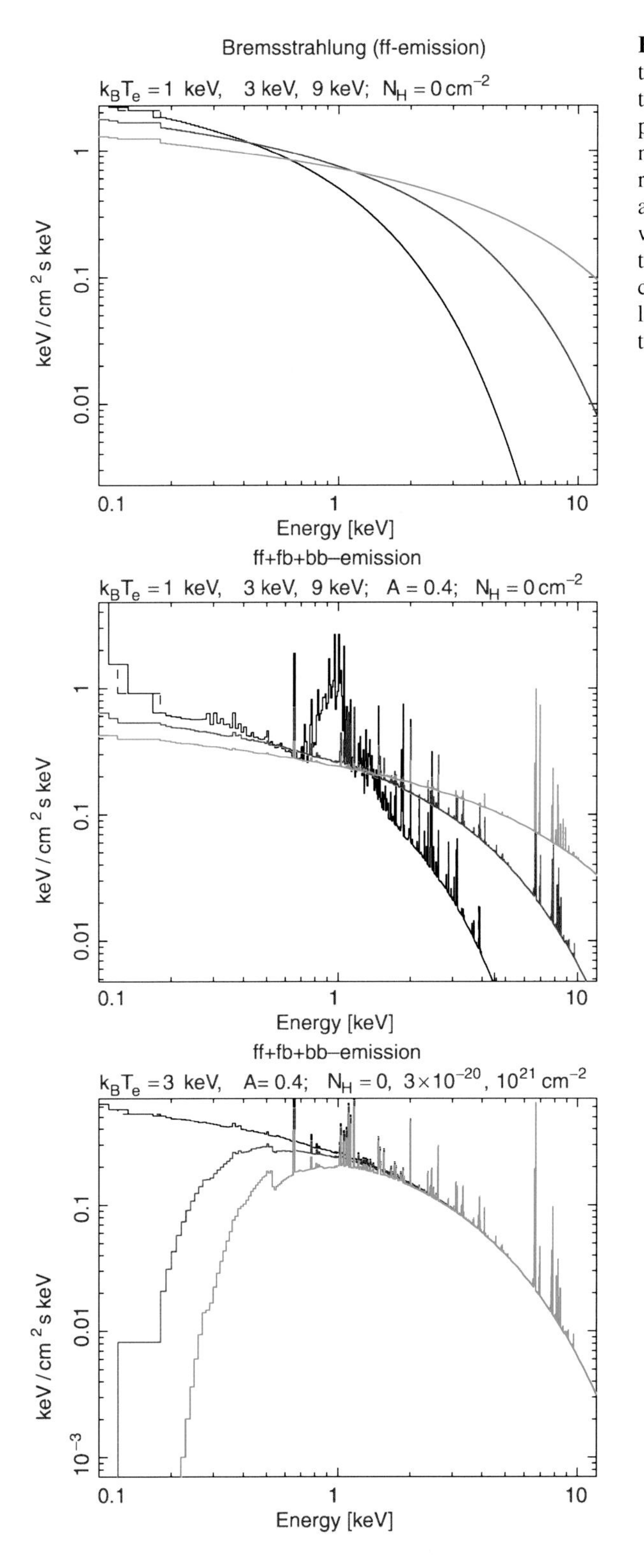

Fig. 6.16. X-ray emission of a hot plasma. In the top panel, the bremsstrahlung spectrum is shown, for three different gas temperatures; the radiation of hotter gas extends to higher photon energies, and above $E \sim k_B T$ the spectrum is exponentially cut off. In the central panel, atomic transitions and recombination radiation are also taken into account. These additional radiation mechanisms become more important towards smaller T, as can be seen from the $T = 1$ keV curve. In the bottom panel, photo-absorption is included, with different column densities in hydrogen and a metallicity of 0.4 in Solar units. This absorption produces a cut-off in the spectrum towards lower energies

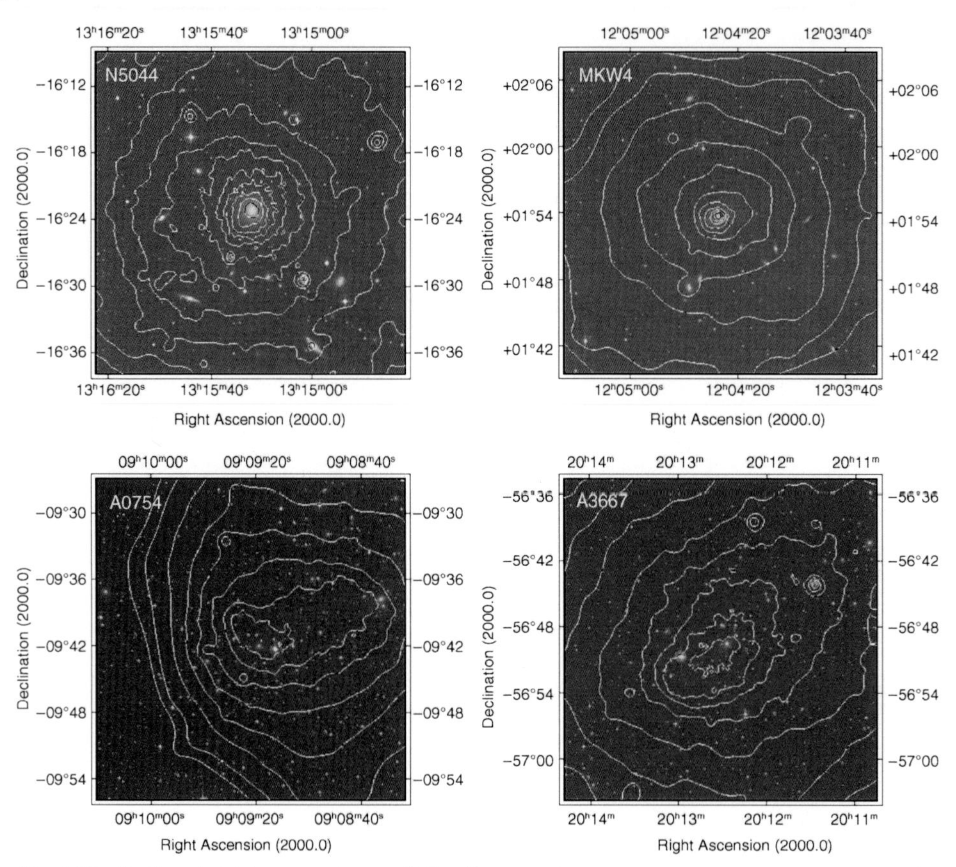

Fig. 6.17. Surface brightness contours of the X-ray emission for four different groups or clusters of galaxies. Upper left: the galaxy group NGC 5044, at redshift $z = 0.009$, with an X-ray temperature of $T \approx 1.07$ keV and a virial mass of $M_{200} \approx 0.32h^{-1} \times 10^{14} M_{\odot}$. Upper right: the group MKW4, at $z = 0.02$, with $T \approx 1.71$ keV and $M_{200} \approx 0.5h^{-1} \times 10^{14} M_{\odot}$. Lower left: the cluster of galaxies A 0754, at $z = 0.053$, with $T \approx 9.5$ keV and $M_{200} \approx 13.1h^{-1} \times 10^{14} M_{\odot}$. Lower right: the cluster of galaxies A 3667, at $z = 0.056$, with $T \approx 7.0$ keV and $M_{200} \approx 5.6h^{-1} \times 10^{14} M_{\odot}$. The X-ray data were obtained by ROSAT, and the optical images were taken from the Digitized Sky Survey. These clusters are part of the HIFLUGCS Survey, which we will discuss more thoroughly in Sect. 6.3.5

6.3.2 Models of the X-Ray Emission

Hydrostatic Assumption. To draw conclusions about the properties of the intergalactic (intracluster) medium from the observed X-ray radiation and about the distribution of mass in the cluster, the gas distribution needs to be modeled. For this, we first consider the speed of sound in the cluster gas,

$$c_{\mathrm{s}} \approx \sqrt{\frac{P}{\rho_{\mathrm{g}}}} = \sqrt{\frac{n k_{\mathrm{B}} T}{\rho_{\mathrm{g}}}} = \sqrt{\frac{k_{\mathrm{B}} T}{\mu\, m_{\mathrm{p}}}} \sim 1000\,\mathrm{km\,s^{-1}} ,$$

where P denotes the gas pressure, ρ_{g} the gas density, and n the number density of gas particles. Then, the *average molecular mass* is defined as the average mass of a gas particle in units of the proton mass,

$$\mu := \frac{\langle m \rangle}{m_{\mathrm{p}}} . \tag{6.33}$$

so that $\rho_{\mathrm{g}} = n \langle m \rangle = n \mu m_{\mathrm{p}}$. For a gas of fully ionized hydrogen, one gets $\mu = 1/2$ because in this case one has one proton and one electron per ~proton mass. Since the cluster gas also contains helium and heavier elements, one obtains $\mu \sim 0.63$. The sound-crossing time for the cluster is

$$t_{\mathrm{sc}} = \frac{2R_{\mathrm{A}}}{c_{\mathrm{s}}} \sim 7 \times 10^{8}\,\mathrm{yr} ,$$

and is thus, for a cluster with $T \sim 10^8$ K, significantly shorter than the lifetime of the cluster, which can be approximated roughly by the age of the Universe. Since the sound-crossing time defines the time-scale on which deviations from the pressure equilibrium are evened out, the gas can be in hydrostatic equilibrium. In this case, the equation

$$\nabla P = -\rho_g \nabla \Phi \tag{6.34}$$

applies, with Φ denoting the gravitational potential. Equation (6.34) describes how the gravitational force is balanced by the pressure force. In the spherically symmetric case in which all quantities depend only on the radius r, we obtain

$$\frac{1}{\rho_g}\frac{dP}{dr} = -\frac{d\Phi}{dr} = -\frac{GM(r)}{r^2}, \tag{6.35}$$

where $M(r)$ is the mass enclosed within radius r. Here, $M(r)$ is the total enclosed mass, i.e., not just the gas mass, because the potential Φ is determined by the total mass. By inserting $P = nk_BT = \rho_g k_B T/(\mu m_p)$ into (6.35), we obtain

$$M(r) = -\frac{k_B T r^2}{G\mu m_p}\left(\frac{d\ln\rho_g}{dr} + \frac{d\ln T}{dr}\right). \tag{6.36}$$

This equation is of central importance for the X-ray astronomy of galaxy clusters because it shows that we can derive the mass profile $M(r)$ from the radial profiles of ρ_g and T. Thus, if one can measure the density and temperature profiles, the mass of the cluster, and hence the total density, can be determined as a function of radius.

However, these measurements are not without difficulties. $\rho_g(r)$ and $T(r)$ need to be determined from the X-ray luminosity and the spectral temperature, using the bremsstrahlung emissivity (6.30). Obviously, they can be observed only in projection in the form of the surface brightness

$$I_\nu(R) = 2\int_R^\infty dr\, \frac{\epsilon_\nu(r)\, r}{\sqrt{r^2 - R^2}}, \tag{6.37}$$

from which the emissivity, and thus density and temperature, need to be derived by de-projection. Furthermore, the angular and energy resolution of X-ray telescopes prior to XMM-Newton and Chandra were not high enough to measure both $\rho_g(r)$ and $T(r)$ with sufficient accuracy, except for the nearest clusters. For this reason, the mass determination is often performed by employing additional, simplifying assumptions.

Isothermal Gas Distribution. From the radial profile of $I(R)$, $\epsilon(r)$ can be derived by inversion of (6.37). Since the spectral bremsstrahlung emissivity depends only weakly on T for $h_P\nu \ll k_BT$, due to (6.30), the radial profile of the gas density ρ_g can be derived from $\epsilon(r)$. The X-ray satellite ROSAT was sensitive to radiation of $0.1\,\mathrm{keV} \lesssim E \lesssim 2.4\,\mathrm{keV}$, so that the X-ray photons detected by it are typically from the regime where $h_P\nu \ll k_BT$.

Assuming that the gas temperature is spatially constant, $T(r) = T_g$, (6.36) simplifies, and the mass profile of the cluster can be determined from the density profile of the gas.

The β-Model. A commonly used method consists of fitting the X-ray data by a so-called β-model. This model is based on the assumption that the density profile of the total matter (dark and luminous) is described by an isothermal distribution, i.e., it is assumed that the temperature of the gas is independent of radius, and at the same time that the mass distribution in the cluster is described by the isothermal model that has been discussed in Sect. 6.2.4. With (6.8) and (6.11), we then obtain for the total density $\rho(r)$

$$\frac{d\ln\rho}{dr} = -\frac{1}{\sigma_v^2}\frac{GM}{r^2}. \tag{6.38}$$

On the other hand, in the isothermal case (6.36) reduces to

$$\frac{d\ln\rho_g}{dr} = -\frac{\mu m_p}{k_B T_g}\frac{GM}{r^2}. \tag{6.39}$$

The comparison of (6.38) and (6.39) then shows that $d\ln\rho_g/dr \propto d\ln\rho/dr$, or

$$\rho_g(r) \propto [\rho(r)]^\beta \quad \text{with} \quad \beta := \frac{\mu m_p \sigma_v^2}{k_B T_g} \tag{6.40}$$

must apply; thus the gas density follows the total density to some power. Here, the index β depends on the ratio of the dynamical temperature, measured by σ_v, and the gas temperature. Now, using the King approximation for an isothermal mass distribution – see (6.16) – as a model for the mass distribution, we obtain

$$\rho_g(r) = \rho_{g0} \left[1 + \left(\frac{r}{r_c} \right)^2 \right]^{-3\beta/2} , \qquad (6.41)$$

where ρ_{g0} is the central gas density. The brightness profile of the X-ray emission in this model is then, according to (6.37),

$$I(R) \propto \left[1 + \left(\frac{R}{r_c} \right)^2 \right]^{-3\beta+1/2} . \qquad (6.42)$$

The X-ray emission of many clusters is well described by this profile,[4] yielding values for r_c of 0.1 to $0.3h^{-1}$ Mpc and a value for the index $\beta = \beta_{fit} \approx 0.65$. Alternatively, β can be measured, with the definition given in (6.40), from the gas temperature T_g and the velocity dispersion of the galaxies σ_v, which yields typical values of $\beta = \beta_{spec} \approx 1$. Such a value would also be expected if the mass and gas distributions were both isothermal. In this case, they should have the same temperature, which was presumably determined by the formation of the cluster.

The fact that the two values for β determined above differ from each other (the so-called β-discrepancy) is as yet not well understood. The measured values for β_{fit} often depend on the angular range over which the brightness profile is fitted; the larger this range, the larger β_{fit} becomes, and thus the smaller the discrepancy. Furthermore, temperature measurements of clusters are often not very accurate because it is the emission-weighted temperature which is measured, which is, due to the quadratic dependence of the emissivity on ρ_g, dominated by the regions with the highest gas density. The fact that the innermost regions of clusters where the gas density is highest tend to have a temperature below the bulk temperature of the cluster may lead to an underestimation of "the" cluster temperature. In addition, the near independence of the spectral form of ϵ_ν^{ff} from T for $h_P\nu \ll k_B T$ renders the measurement of T difficult. Only with Chandra and XMM-Newton can the X-ray emission also be mapped at energies of up to $E \lesssim 10$ keV, which results in considerably improved temperature measurements.

Such investigations have revealed that the gas is not really isothermal. Typically, the temperature decreases towards the center and towards the edge, while it is rather constant over a larger range at intermediate radii. Many clusters are found, however, in which the temperature distribution is by no means radially symmetric, but shows distinct substructure. Finally, as another possible explanation for the β-discrepancy, it should be mentioned that the velocity distribution of those galaxies from which σ_v is measured may be anisotropic.

Besides all the uncertainty as to the validity of the β-model, we also need to mention that numerical simulations of galaxy clusters, which take dark matter and gas into account, have repeatedly come to the conclusion that the mass determination of clusters, utilizing the β-model, should achieve an accuracy of better than $\sim 20\%$, although different gas dynamical simulations have arrived at distinctly different results.

Dark Matter in Clusters from X-Ray Observations. Based on measurements of their X-ray emission, a mass estimate can be performed for galaxy clusters. It is found, in agreement with the dynamical method, that clusters contain much more mass than is visible in galaxies. The total mass of the intergalactic medium is clearly too low to account for the missing mass; its gas mass is only $\sim 15\%$ of the total mass of a cluster.

The mass of clusters of galaxies consists of $\sim 3\%$ contribution from stars in galaxies and $\sim 15\%$ from intergalactic gas, whereas the remaining $\sim 80\%$ consists of dark matter which therefore dominates the mass of the clusters.

6.3.3 Cooling Flows

In examining the intergalactic medium, we have assumed hydrostatic equilibrium, but we have disregarded the fact that the gas cools by its emission, thus it will lose internal energy. For this reason, once established, a hydrostatic equilibrium cannot be maintained over ar-

[4] We point out that the pair of equations (6.41) and (6.42) is valid independently of the validity of the assumptions from which (6.41) was obtained. If the observed X-ray emission is very well described by (6.42), the gas density profile (6.41) can be obtained from it, independently of the validity of the assumptions made before.

bitrarily long times. To decide whether this gas cooling is important for the dynamics of the system, the cooling time-scale needs to be considered. This cooling time turns out to be very long,

$$t_{\rm cool} := \frac{u}{\epsilon^{\rm ff}} \approx 8.5 \times 10^{10}\,{\rm yr} \left(\frac{n_{\rm e}}{10^{-3}\,{\rm cm}^{-3}}\right)^{-1} \left(\frac{T_{\rm g}}{10^8\,{\rm K}}\right)^{1/2}, \tag{6.43}$$

where $u = (3/2)nk_{\rm B}T_{\rm g}$ is the energy density of the gas and $n_{\rm e}$ the electron density. Hence, the cooling time is longer than the Hubble time nearly everywhere in the cluster, which allows a hydro*static* equilibrium to be established. In the centers of clusters, however, the density may be sufficiently large to yield $t_{\rm cool} \lesssim t_0 \sim H_0^{-1}$. Here, the gas can cool quite efficiently, by which its pressure decreases. This then implies that, at least close to the center, the hydrostatic equilibrium can no longer be maintained. To re-establish pressure equilibrium, gas needs to flow inwards and is thus compressed. Hence, an inward-directed mass flow should establish itself. The corresponding density increase will further accelerate the cooling process. Since the emissivity (6.32) of a relatively cool gas increases with decreasing temperature, this process should then very quickly lead to a strong compression and cooling of the gas in the centers of dense clusters. In parallel to this increase in density, the X-ray emission will strongly increase, because $\epsilon^{\rm ff} \propto n_{\rm e}^2$. As a result of this process, a radial density and temperature distribution should be established with a nearly unchanged pressure distribution. In Fig. 6.18, the cooler gas in the center of the Centaurus cluster is clearly visible.

These so-called *cooling flows* have indeed been observed in the centers of massive clusters, in the form of a sharp central peak in $I(R)$. However, we need to stress that, as yet, no inwards *flows* have been measured. Such a measurement would be very difficult, though, due to the small expected velocities. The amount of cooling gas can be considerable, with models predicting values of up to several $100 M_\odot$/yr. However, after spectroscopic observations by XMM-Newton became available, we have learned that these very high cooling rates implied by the models were significantly overestimated.

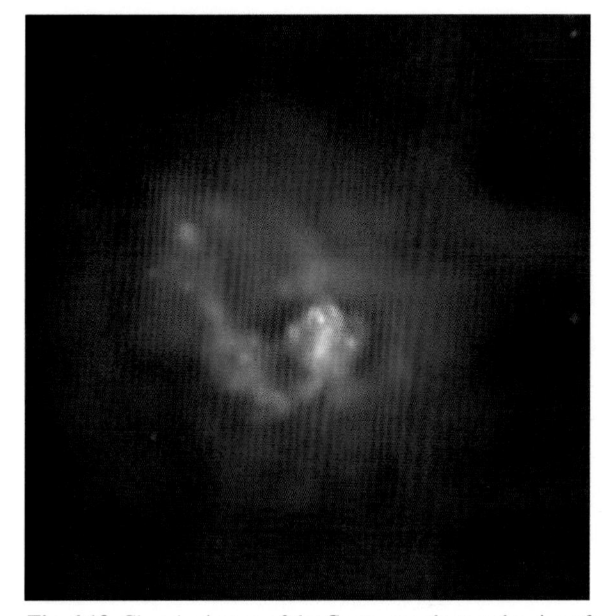

Fig. 6.18. Chandra image of the Centaurus cluster; the size of the field is $3' \times 3'$. Owing to the excellent angular resolution of the Chandra satellite, the complexity of the morphology in the X-ray emission of clusters can be analyzed. Colors indicate photon energies, from low to high in red, yellow, green, and blue. The relatively cool inner region might be the result of a cooling flow

The Fate of the Cooling Gas. The gas cooling in this way will accumulate in the center of the cluster, but despite the expected high mass of cold gas, no clear evidence has been found for it. In clusters harboring a cD galaxy, the cooled gas may, over a Hubble time, contribute a considerable fraction of the mass of this galaxy. Hence, the question arises whether cD galaxies may have formed by accretion in cooling flows. In this scenario, the gas would be transformed into stars in the cD galaxy. However, the star-formation rate in these central galaxies is much lower than the rate by which cluster gas cools, according to the "old" cooling flow models.

The sensitivity and spectral resolution achieved with XMM-Newton have strongly modified our view of cooling flows. In the standard model of cooling flows, the gas cools from the cluster temperature down to temperatures significantly below 1 keV. In this process many atomic lines are emitted, produced by various ionization stages, e.g., of iron, which change with decreasing temperature. Figure 6.19 (top panel) shows the expected spectrum of a cooling flow in which the gas cools down

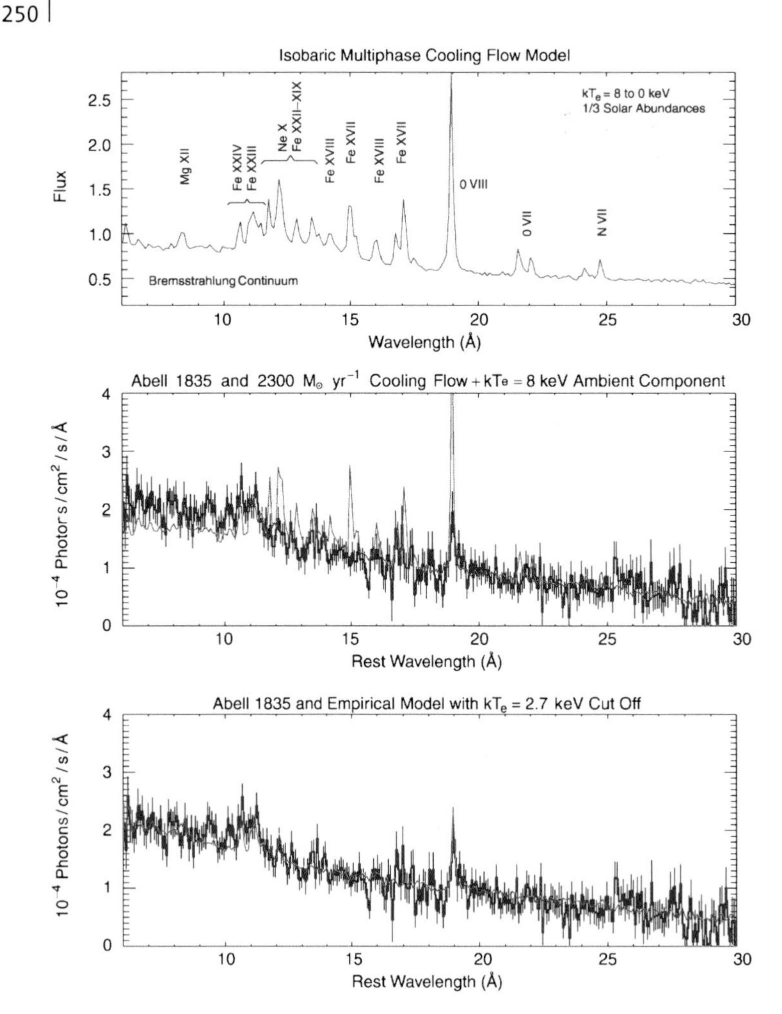

Fig. 6.19. In the top panel, a model spectrum of a cooling flow is shown, in which the gas cools down from 8 keV to $T_g = 0$. The strong lines of FeXVII can be seen. In the central panel, the spectrum of Abell 1835 is superposed on the model spectrum; clear discrepancies are visible, especially the absence of strong emission lines from FeXVII. If the gas is not allowed to cool down to temperatures below 3 keV (bottom panel), the agreement with observation improves visibly

from the cluster temperature of $T_g \approx 8$ keV to $T_g = 0$, where a chemical composition of 1/3 Solar abundance is assumed. In the central panel, this theoretical spectrum is compared with the spectrum of the cluster Abell 1835, where very distinct discrepancies become visible. In the bottom panel, the model was modified such that the gas cools down only to $T_g = 3$ keV; this model clearly matches the observed spectrum better.

Hence, cooler gas in the inner regions of clusters has now been directly detected spectroscopically. However, the temperature measurements from X-ray spectroscopy are significantly different from what one would expect. The above arguments imply that drastic cooling should take place in the gas, because the process of compression and cooling will accelerate for ever decreasing T_g. Therefore, one expects to find gas at all temperatures lower than the temperature of the cluster. But this seems not to be the case: whereas gas at $T_g \gtrsim 1$ keV is found, no gas seems to be present at even smaller temperatures, although the cooling flow models predict the existence of such gas. A minimum temperature seems to exist, below which the gas cannot cool, or the amount of gas that cools to $T_g = 0$ is considerably smaller than expected from the cooling flow model. This lower mass rate of gas that cools down completely would then also be compatible with the observed low star-formation rates in the central galaxies of clusters. In fact, a correlation between the cooling rate of gas as determined from XMM observations and the regions of star formation in clusters has been found.

What Prevents Massive Cooling Flows? One way to explain the clearly suppressed cooling rates in cooling flows is by noting that many clusters of galaxies harbor an active galaxy in their center, the activity of which, e.g., in form of (radio-)jets, may affect the ICM. For instance, energy could be transferred from the jet to the ICM, by which the ICM is heated. This heating might then prevent the temperature from dropping to arbitrarily small values. This hypothesis is supported by the fact that many clusters are known in which the ICM is clearly affected by the central AGN – see Fig. 6.20 for one of the first examples where this effect has been seen. Plasma from the jet seems to locally displace the X-ray emitting gas. By friction and mixing in the interface region between the jet and the ICM, the latter is certainly heated. It is unclear, though, whether this explanation is valid for every cluster, because not every cluster in which a very cool ICM is expected also contains an observed AGN. On the other hand, this is not necessarily an argument against the hypothesis of AGNs as heating sources, since AGNs often have a limited time of activity and may be switching on and off, depending on the accretion rate. Thus, the gas in a cluster may very well be heated by an AGN even if it is currently (at the time of observation) inactive. Another example of apparently underdense regions in the X-ray gas of a group is shown in Fig. 6.21.

The Bullet Cluster. Clusters of galaxies are indeed excellent laboratories for hydrodynamic and plasma-physical processes on large scales. In them, shock fronts, for instance in merging clusters, cooling fronts (which are also called "contact discontinuities" in hydrodynamics), and the propagation of sound waves can be observed. A particularly good example is the galaxy cluster 1E 0657−56 displayed in Fig. 6.22, the "bullet cluster". To the right of the cluster center, strong and relatively compact X-ray emission (the "bullet") is visible, while further to the right of it one sees an arc-shaped discontinuity in surface brightness. From the temperature distribution on both sides of the discontinuity one infers

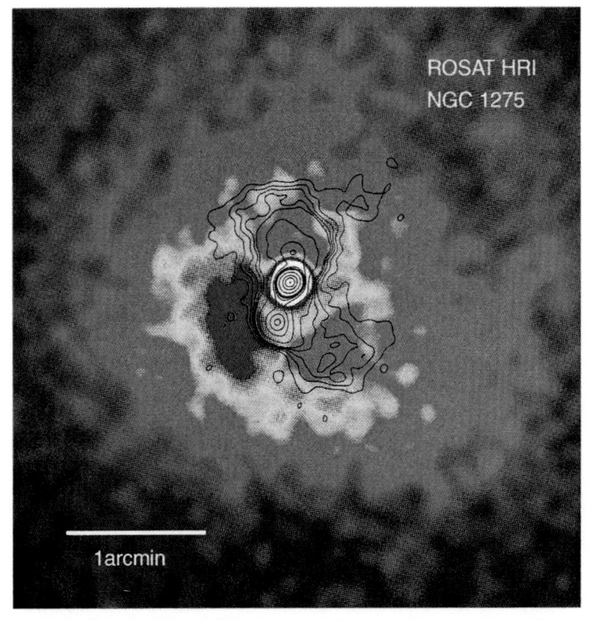

Fig. 6.20. A ROSAT-HRI-image of the central region of the Perseus cluster, with its central galaxy NGC 1275. The latter is the center of the emission in both the radio and the X-ray, here displayed as contours and color-coded, respectively. Clearly identifiable is the effect of the radio-jets on the X-ray emission – at the location of the radio lobes the X-ray emission is strongly suppressed

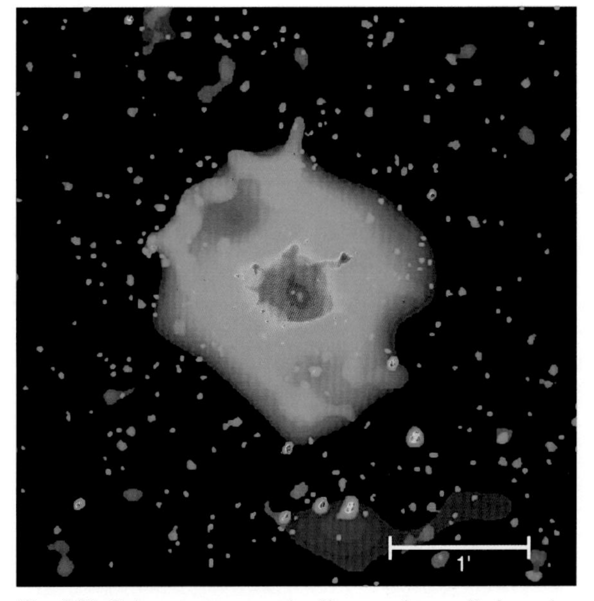

Fig. 6.21. Galaxy groups are also X-ray emitters, albeit weaker than clusters of galaxies. Moreover, the temperature of the ICM is lower than in clusters. This $4' \times 4'$ Chandra image shows HCG 62. Note the complexity of the X-ray emission and the two symmetrically aligned regions that seem to be virtually devoid of hot ICM – possibly holes blown free by jets from the central galaxy of this group (NGC 4761)

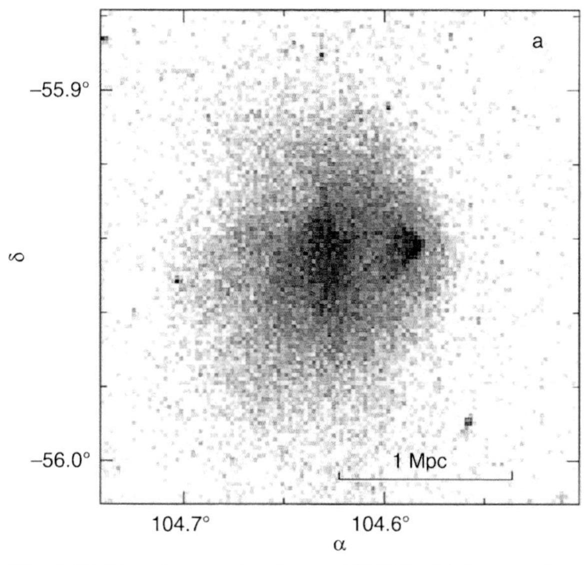

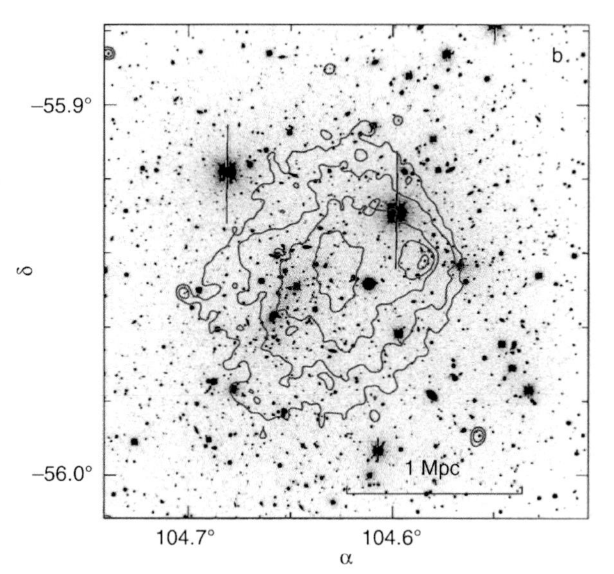

Fig. 6.22. The cluster of galaxies 1E 0657−56 is a perfect example of a merging cluster. On the left, a Chandra image of the cluster, while the right-hand image shows the superposition of the X-ray contours on an optical R-band image taken with the ESO NTT. The most remarkable feature in the X-ray map is the compact region to the right (westwards) of the cluster center (from which the cluster derives its name the "bullet cluster"), and the sharp transition in the surface brightness further to the right of it. An analysis of the brightness profile and of the X-ray temperature distribution shows that this must be a shock front moving at about 2.5 times the speed of sound, or $v \sim 3500\,\mathrm{km/s}$, through the gas. To the right of this shock front, a group of galaxies is visible

that it is a shock front. The strength of the shock implies that the "bullet" is moving at about $v \sim 3500\,\mathrm{km/s}$ (from left to right in the figure) through the intergalactic medium of the cluster. The interpretation of this observation is that we are witnessing the merger of two clusters, where one less massive cluster has passed, from left to right in Fig. 6.22, through a more massive one. The "bullet" in this picture is understood to be gas from the central region of the less massive cluster, which is still rather compact. This interpretation is impressively supported by the group of galaxies to the right of the shock front, which are probably the former member galaxies of the less massive cluster. As this cluster crosses through the more massive one, its galaxies and dark matter are moving collisionlessly, whereas the gas is decelerated by friction with the gas in the massive cluster: the galaxies and the dark matter are thus able to move faster through the cluster than the gas, which is lagging behind.

Indeed, a weak lensing analysis of this cluster (see Sect. 6.5.2) shows that the mass of the small cluster component is centered on the associated group of galaxies, not on the intracluster gas component causing the X-ray emission near the bullet. This result provides additional strong evidence for the interpretation given above, showing that galaxies and dark matter together behave as a collisionless gas as they cross the cluster, whereas the gas itself is held back by friction.

6.3.4 The Sunyaev–Zeldovich Effect

Electrons in the hot gas of the intracluster medium can scatter photons of the cosmic microwave background. The optical depth and thus the scattering probability for this Compton scattering is relatively low, but the effect is nevertheless observable and, in addition, is of great importance for the analysis of clusters, as we will now see.

A photon moving through a cluster of galaxies towards us will have a different direction after scattering and thus will not reach us. But since the cosmic background radiation is isotropic, for any CMB photon that is scattered out of the line-of-sight, another photon exists – statistically – that is scattered into it, so that the total number of photons reaching us is preserved. However, the energy of the photons changes slightly through scattering by the hot electrons, in a way that they have an (on average) higher frequency after scattering. Hence, by Compton scattering, energy is on average transferred from the electrons to the photons (see Fig. 6.23).

As a consequence, this scattering leads to a reduced number of photons at lower energies, relative to the Planck spectrum, and higher energy photons being added. This effect is called the *Sunyaev–Zeldovich effect* (SZ effect). It was predicted in 1970 and has now been observed in a large number of clusters.

> The CMB spectrum, measured in the direction of a galaxy cluster, deviates from a Planck spectrum; the degree of this deviation depends on the temperature of the cluster gas and on its density.

In the Rayleigh–Jeans domain of the CMB spectrum, thus at wavelengths larger than about 1 mm, photons are effectively removed by the SZ effect. For the change in specific intensity in the RJ part, one obtains

$$\frac{\Delta I_\nu^{\rm RJ}}{I_\nu^{\rm RJ}} = -2y \; , \tag{6.44}$$

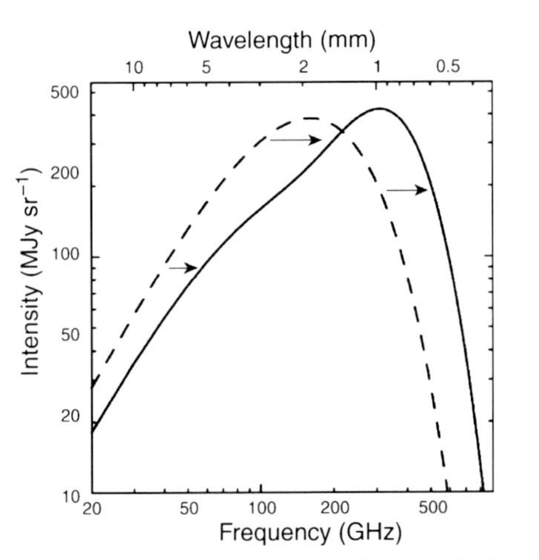

Fig. 6.23. The influence of the Sunyaev–Zeldovich effect on the cosmic background radiation. The dashed curve represents the Planck distribution of the unperturbed CMB spectrum, the solid curve shows the spectrum after the radiation has passed through a cloud of hot electrons. The magnitude of this effect, for clarity, has been very much exaggerated in this sketch

where

$$y = \int {\rm d}l \, \frac{k_{\rm B} T_{\rm g}}{m_{\rm e} c^2} \, \sigma_{\rm T} \, n_{\rm e} \quad \text{with} \quad \sigma_{\rm T} = \frac{8\pi}{3} \left(\frac{e^2}{m_{\rm e} c^2} \right)^2 \tag{6.45}$$

is the *Compton-y parameter* and $\sigma_{\rm T}$ the Thomson cross-section for electron scattering. Obviously, y is proportional to the optical depth with respect to Compton scattering, given as an integral over $n_{\rm e}\,\sigma_{\rm T}$ along the line-of-sight. Furthermore, y is proportional to the gas temperature, because that defines the average energy transfer per scattering event. Overall, y is proportional to the integral over the gas pressure $P = nk_{\rm B}T$ along the line-of-sight through the cluster.

Observations of the SZ effect provide another possibility for analyzing the gas in clusters. For instance, if one can spatially resolve the SZ effect, which is possible today with interferometric methods (see Fig. 6.24), one obtains information about the spatial density and temperature distribution. Here it is of crucial importance that the dependence on temperature and gas density is different from that in X-ray emission. Because of the quadratic dependence of the X-ray emissivity on $n_{\rm e}$, the X-ray luminosity depends not only on the total gas mass, but also on its spatial distribution. Small-scale clumps in the gas, for instance, would strongly affect the X-ray emission. In contrast, the SZ effect is linear in gas density and therefore considerably less sensitive with respect to inhomogeneities in the ICM.

The next generation of radio telescopes, which will operate in the mm-domain of the spectrum, will perform SZ surveys and search for clusters of galaxies by the SZ effect. The outcome of these surveys is expected to be a particularly useful sample of galaxy clusters because this selection criterion for clusters does not depend on the detailed gas distribution. Equations (6.44) and (6.45) also show that the SZ effect is independent of the cluster redshift, as long as the change in CMB temperature is spatially resolved. For this reason, it is expected that SZ surveys will identify a large number of high-redshift clusters.

Distance Determination. For a long time, the SZ effect was mainly considered a tool for measuring distances to clusters of galaxies, and from this the Hubble constant. We will now schematically show how the SZ effect,

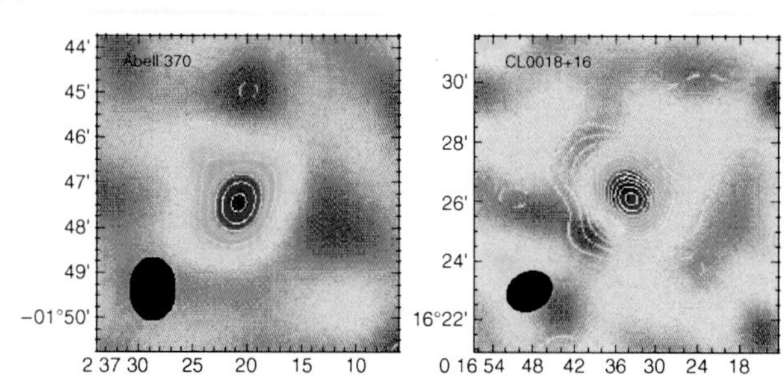

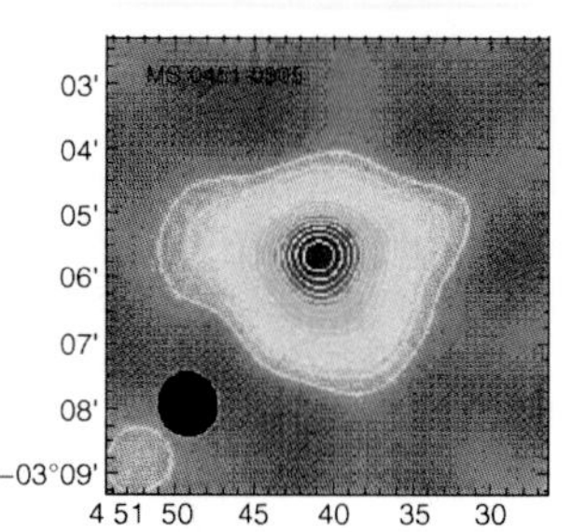

Fig. 6.24. Sunyaev–Zeldovich maps of three clusters of galaxies at $0.37 < z < 0.55$. Plotted is the temperature difference of the measured CMB relative to the average CMB temperature (or, at fixed frequency, the difference in radiation intensities). The black ellipse in each image specifies the instrument's beam size. For each of the clusters shown here, the spatial dependence of the SZ effect is clearly visible. Since the SZ effect is proportional to the electron density, the mass fraction of baryons in clusters can be measured if one additionally knows the total mass of the cluster from dynamical methods or from the X-ray temperature. The analysis of the clusters shown here yields for the mass fraction of the intergalactic gas $f_{\rm g} \approx 0.08\,h^{-1}$

in combination with the X-ray emission, allows us to determine the distance to a cluster. The change in the CMB intensity has the dependence

$$\frac{\left|\Delta I_\nu^{\rm RJ}\right|}{I_\nu^{\rm RJ}} \propto n_{\rm e}\, L\, T_{\rm g} ,$$

where L is the extent of the cluster along the line-of-sight. To obtain this relation, we replace the l-integration in (6.45) by a multiplication with L, which yields the correct functional dependence. On the other hand, the surface brightness of the X-ray radiation behaves as

$$I_{\rm X} \propto L n_{\rm e}^2 .$$

Combining these two relations, we are now able to eliminate $n_{\rm e}$. Since $T_{\rm g}$ is measurable by means of the X-ray spectrum, the dependence

$$\frac{\left|\Delta I_\nu^{\rm RJ}\right|}{I_\nu^{\rm RJ}} \propto \sqrt{L\, I_{\rm X}}$$

remains. Now assuming that the cluster is spherical, its extent L along the line-of-sight equals its transverse extent $R = \theta D_{\rm A}$, where θ denotes its angular extent and $D_{\rm A}$ the angular-diameter distance (4.45) to the cluster. With this assumption, we obtain

$$D_{\rm A} = \frac{R}{\theta} \sim \frac{L}{\theta} \propto \left(\frac{\Delta I_\nu^{\rm RJ}}{I_\nu^{\rm RJ}}\right)^2 \frac{1}{I_{\rm X}} . \qquad (6.46)$$

Hence, the angular-diameter distance can be determined from the measured SZ effect, the X-ray temperature of the ICM, and the surface brightness in the X-ray domain. Of course, this method is more complicated in practice than demonstrated here, but it is applied to the distance determination of clusters. In particular, the assumption of the same extent of the cluster along the line-of-sight as its transverse size is not well justified for any individual cluster, but one expects this assumption to be valid on average for a sample of clusters. Hence, the SZ effect is another method of distance determination, independent of the redshift of the cluster, and therefore suitable for determining the Hubble constant.

Discussion. The natural question arises whether this method, in view of the assumptions it is based on, can compete with the determination of the Hubble constant via the distance ladder and Cepheids. The same question also needs to be asked for the determination of H_0 by means of the time delay in gravitational lens systems, which we discussed in Sect. 3.8.4. In both cases, the answer to this question is the same: presumably neither of the two methods will provide a determination of the Hubble constant with an accuracy comparable to that achieved by the Hubble Key Project (Sect. 3.6.3) and from the angular fluctuations in the CMB (see Sect. 8.6). Nevertheless, both methods are of great value for cosmology: first, the distance ladder has quite a number of rungs. If only one of these contains an as yet undetected severe systematic error, it could affect the resulting value for H_0. Second, the Hubble Key Project measured the expansion rate in the local Universe, typically within $\sim 100\,$Mpc (the distance to the Coma cluster). As we will see later, the Universe contains inhomogeneities on these length-scales. Thus, it may well be that we live in a slightly overdense or underdense region of the Universe, where the Hubble constant deviates from the global value. In contrast to this, both the SZ effect and the lensing method measure the Hubble constant on truly cosmic scales, and both methods do so in only a single step – there is no distance ladder involved. For these reasons, these two methods are of great importance in additionally confirming our H_0 measurements. Another aspect adds to this, which must not be underestimated: even if the same or a similar value results from these measurements as the one from the Hubble Key Project, we still have learned an important fact, namely that the local Hubble constant agrees with the one measured on cosmological scales – this is one of the predictions of our cosmological model, which can thus be tested in an impressive way. Indeed, both methods have been applied to quite a number of lens systems and luminous clusters showing an SZ effect, respectively, and they yield values for H_0 which are slightly smaller than, but compatible within the error bars with the value of H_0 obtained from the Hubble Key Project.

6.3.5 X-Ray Catalogs of Clusters

Originally, clusters of galaxies were selected by overdensities of galaxies on the sphere using optical methods. As we have seen, projection effects may play a crucial role in this, in the form of coincidental overdensities in the projected galaxy distribution, which do not correspond to spatial overdensities. In addition, one has the superposition of foreground and background galaxies, which renders the selection more difficult the farther the clusters are away from us.

A more reliable way of selecting clusters is by their X-ray emission, since the hot X-ray gas signifies a deep potential well, thus a real three-dimensional overdensity of matter, so that projection effects become virtually negligible. The X-ray emission is $\propto n_e^2$, which again renders projection effects improbable. In addition, the X-ray emission, its temperature in particular, seems to be a very good measure for the cluster mass, as we will discuss further below. Whereas the selection of clusters is not based on their temperature, but on the X-ray luminosity, we shall see that L_X is also a good indicator for the mass of a cluster (see Sect. 6.4).

The first cosmologically interesting X-ray catalog of galaxy clusters was the EMSS (Extended Medium Sensitivity Survey) catalog. It was constructed from archival images taken by the *Einstein observatory* which were scrutinized for X-ray sources other than the primary target in the field-of-view of the respective observation. These were compiled and then further investigated using optical methods, i.e., photometry and spectroscopy. The EMSS catalog contains 835 sources, most of them AGNs, but it also contains 104 clusters of galaxies. Among these are six clusters at redshift ≥ 0.5; the most distant is MS1054−03 at $z = 0.83$ (see Fig. 6.15). Since the Einstein images all have different exposure times, the EMSS is not a strictly flux-limited catalog. But with the flux limit known for each exposure, the luminosity function of clusters can be derived from this.

The same method as was used to compile the EMSS was applied to ROSAT archival images by various groups, leading to several catalogs of X-ray-selected clusters. The selection criteria of these different groups, and therefore of the different catalogs, differ. Since ROSAT was more sensitive than the Einstein observatory, these catalogs contain a larger number of clusters, and also ones at higher redshift (Fig. 6.25). Furthermore, ROSAT performed a survey of the full sky, the ROSAT All Sky Survey (RASS). The RASS contains about 10^5 sources distributed over the whole sky. The identification of extended sources in the RASS (in contrast to non-extended sources – about five times more AGNs than clusters are expected) yielded a catalog of clusters as well which, owing to the relatively short exposure times in the RASS, contains the brightest clusters. The exposure time in the RASS is not uniform over the sky since the applied observing strategy led to particularly long exposures for the regions around the Northern and Southern ecliptic pole (see Fig. 6.26).

One of the cluster catalogs that were extracted from the RASS data is the HIFLUGCS catalog. It consists of the 63 X-ray-brightest clusters and is a strictly flux-limited survey, with $f_X(0.1–2.4\,\mathrm{keV}) \geq 2.0 \times 10^{-11}\,\mathrm{erg\,s^{-1}\,cm^{-1}}$; it excludes the Galactic plane, $|b| \geq 20°$, as well as other regions around the Magellanic clouds and the Virgo Cluster of galaxies in order to avoid large column densities of Galactic gas which lead to absorption, as well as Galactic and other nearby X-ray sources. The extended HIFLUGCS survey contains, in addition, several other

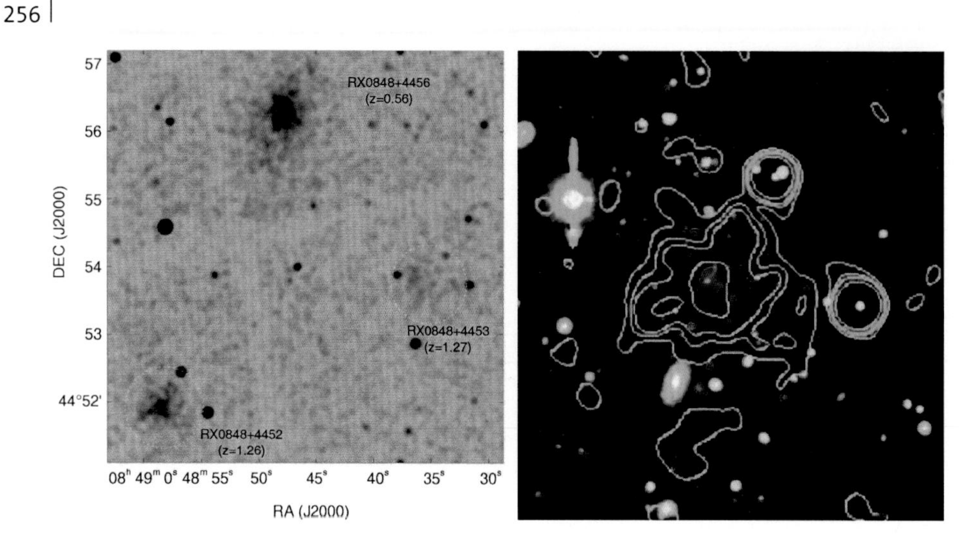

Fig. 6.25. Left: Chandra image of a $6' \times 6'$-field with two clusters of galaxies at high redshift. Right: a $2' \times 2'$-field centered on one of the clusters presented on the left (RX J0849+4452), in B, I, and K, overlaid with the X-ray brightness contours

clusters for which good measurements of the brightness profile and the X-ray temperature are available.

From the luminosity function of X-ray clusters, a mass function can be constructed, using the relation between L_X and the cluster mass that will be discussed in the following section. As we will explain in more detail in Sect. 8.2, this cluster mass function is an important probe for cosmological parameters.

6.4 Scaling Relations for Clusters of Galaxies

Our examination of galaxies revealed the existence of various scaling relations, for example the Tully–Fisher relation. These have proven to be very useful not only for the distance determination of galaxies, but also because any successful model of galaxy evolution needs be able to explain these empirical scaling relations. Therefore, it is of great interest to examine whether clusters of galaxies also fulfill any such scaling relation. As we will see, the X-ray properties of clusters play a central role in this.

6.4.1 Mass–Temperature Relation

It is expected that the larger the spatial extent, velocity dispersion of galaxies, temperature of the X-ray gas, and luminosity of a cluster are, the more massive it is. In fact, from theoretical considerations one can deduce the existence of relations between these parameters. The X-ray temperature T specifies the thermal energy per gas particle, which should be proportional to the binding energy for a cluster in virial equilibrium,

$$T \propto \frac{M}{r} .$$

Since this relation is based on the virial theorem, r should be chosen to be the radius within which the matter of the cluster is virialized. This value for r is called the *virial radius* $r_{\rm vir}$. From theoretical considerations of cluster formation (see Chap. 7), one finds that the virial radius is defined such that within a sphere of radius $r_{\rm vir}$, the average mass density of the cluster is about $\Delta_{\rm c} \approx 200$ times as high as the critical density $\rho_{\rm cr}$ of the Universe. The mass within $r_{\rm vir}$ is called the *virial mass* $M_{\rm vir}$ which is, according to this definition,

$$\boxed{M_{\rm vir} = \frac{4\pi}{3}\,\Delta_{\rm c}\,\rho_{\rm cr}\,r_{\rm vir}^3} . \quad (6.47)$$

Combining the two above relations, one obtains

$$\boxed{T \propto \frac{M_{\rm vir}}{r_{\rm vir}} \propto r_{\rm vir}^2 \propto M_{\rm vir}^{2/3}} . \quad (6.48)$$

This relation can now be tested on observations by using a sample of galaxy clusters with known temperature and with mass determined by the methods discussed in Sect. 6.3.2. An example of this is displayed in Fig. 6.27, in which the mass is plotted versus temperature for clusters from the extended HIFLUGCS sample. Since it is

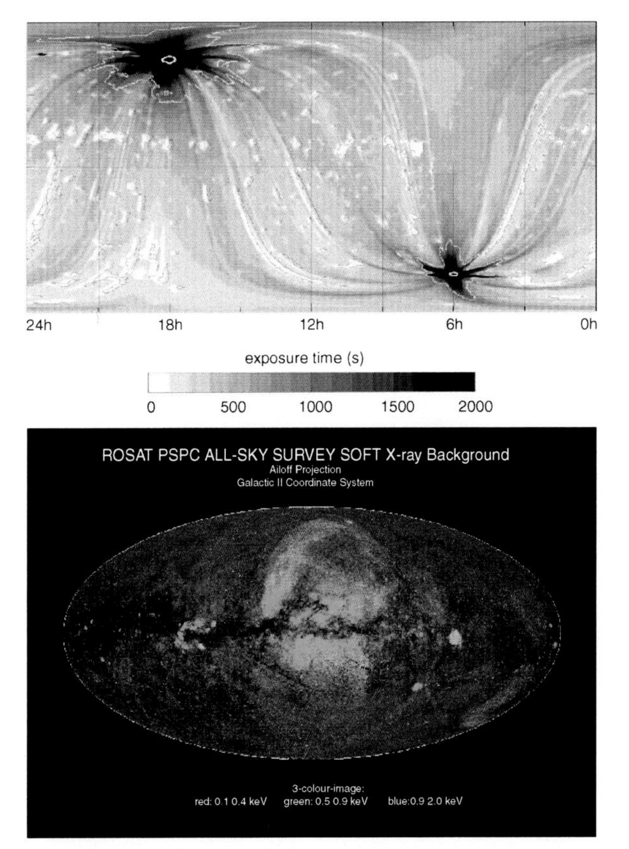

Fig. 6.26. The top panel shows the total exposure time in the ROSAT All Sky Survey as a function of sky position. Near the ecliptic poles the exposure time is longest, as a consequence of the applied observing strategy. Because of the "South Atlantic Anomaly" (a region of enhanced cosmic ray flux over the South Atlantic Ocean, off the coast of Brazil, caused by the shape of the Earth's magnetosphere), the exposure time is generally higher in the North than in the South. The X-ray sky, as observed in the RASS, is shown in the lower panel. The colors indicate the shape of the spectral energy distribution, where blue indicates sources with a harder spectrum

easier to determine the mass for small radii than the virial mass itself, the mass M_{500} within the radius r_{500}, the radius within which the average density is 500 times the critical density, has been plotted here. The measured values clearly show a very strong correlation, and best-fit straight lines describing power laws of the form $M = AT^{\alpha}$ are also shown in the figure. The exact values of the two fit parameters depend on the choice of the cluster sample; the right-hand panel of Fig. 6.27 shows in particular that galaxy groups (thus, "clusters" of low mass and temperature) are located below the power-law fit that is obtained from higher mass clusters. If one confines the sample to clusters with $M \geq 5 \times 10^{13} M_{\odot}$, the best fit is described by

$$M_{500} = 3.57 \times 10^{13} M_{\odot} \left(\frac{k_{\rm B} T}{1\,{\rm keV}} \right)^{1.58} , \tag{6.49}$$

with an uncertainty of slightly more than 10%. This relation is very similar to the one deduced from theoretical considerations, $M \propto T^{1.5}$. With only small variations in the parameters, the relation (6.49) is obtained both from a cluster sample in which the mass was determined based on an isothermal β-model, and from a cluster sample in which the measured radial temperature profile $T(r)$ was utilized in the mass determination (see Eq. 6.36). Constraining the sample to clusters with temperatures above 3 keV, one obtains a slope of 1.48 ± 0.1, in excellent agreement with theoretical expectations. Considerably steeper mass–temperature relations result from the inclusion of galaxy groups into the sample, from which we conclude that they do not follow the scaling argument sketched above in detail.

The X-ray temperature of galaxy clusters apparently provides a very precise measure for their virial mass, better than the velocity dispersion (see below).

With the current X-ray observatories, it will be possible to test these mass–temperature relations with even higher accuracy; the first preliminary results confirm the above result, and the improved accuracy of future observations will lead to an even smaller dispersion of the data points around the power law.

6.4.2 Mass–Velocity Dispersion Relation

The velocity dispersion of the galaxies in a cluster also can be related to the mass: from (6.25) we find

$$M_{\rm vir} = \frac{3 r_{\rm vir} \sigma_v^2}{G} . \tag{6.50}$$

Together with $T \propto \sigma_v^2$ and $T \propto r_{\rm vir}^2$, it then follows that

$$\boxed{M_{\rm vir} \propto \sigma_v^3} . \tag{6.51}$$

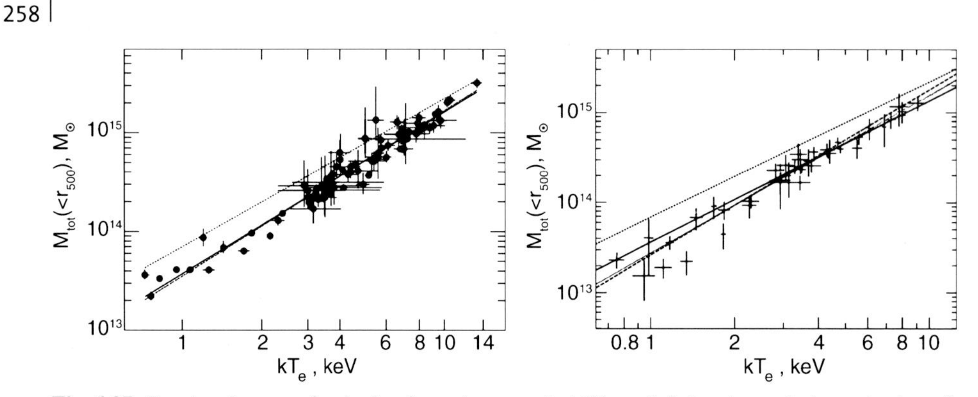

Fig. 6.27. For the clusters of galaxies from the extended HIFLUGCS sample (see Sect. 6.3.5), the mass within a mean overdensity of 500 is plotted as a function of X-ray temperature, where a dimensionless Hubble constant of $h = 0.5$ has been assumed. In the left-hand panel, the mass was determined by applying an isothermal β-model, while in the right-hand panel, the radial temperature profile $T(r)$ was used to determine the mass, by means of (6.36). Most of the temperature measurements are from observations by the ASCA satellite. The solid and dash-dotted curves in the left-hand panel show the best fit to the data, where for the latter only the clusters from the original HIFLUGCS sample were used. In the right-hand panel, the dotted line is a fit to all the data in the plot, while the solid line takes into account only clusters with a mass $\gtrsim 5 \times 10^{13} M_{\odot}$. In both panels, the upper dotted line shows the mass–temperature relation that was obtained from a simulation using simplified gas dynamics – the slope agrees with that found from the observations, but the amplitude is significantly too high

This relation can now be tested using clusters for which the mass has been determined using the X-ray method, and for which the velocity dispersion of the cluster galaxies has been measured. Alternatively, the relation $T \propto \sigma_v^2$ can be tested. One finds that these relations are essentially satisfied for the observed clusters. However, the relation between σ_v and M is not as tight as the M–T relation. Furthermore, numerous clusters exist which strongly deviate from this relation. These are clusters of galaxies that are not relaxed, as can be deduced from the velocity distribution of the cluster galaxies (which strongly deviates from a Maxwell distribution in these cases) or from a bimodal or even more complex galaxy distribution in the cluster. These outliers need to be identified, and removed, if one intends to apply the scaling relation between mass and velocity dispersion.

6.4.3 Mass–Luminosity Relation

The total X-ray luminosity that is emitted via bremsstrahlung is proportional to the squared gas density and the gas volume, hence it should behave as

$$L_X \propto \rho_g^2 \, T^{1/2} \, r_{vir}^3 \propto \rho_g^2 \, T^{1/2} \, M_{vir} \, . \tag{6.52}$$

Estimating the gas density through $\rho_g \sim M_g \, r_{vir}^{-3} = f_g \, M_{vir} \, r_{vir}^{-3}$, where $f_g = M_g / M_{vir}$ denotes the gas fraction with respect to the total mass of the cluster, and using (6.48), we obtain

$$L_X \propto f_g^2 \, M_{vir}^{4/3} \, . \tag{6.53}$$

This relation needs to be modified if the X-ray luminosity is measured within a fixed energy interval. Particularly for observations with ROSAT, which could only measure low-energy photons (below 2.4 keV), the received photons typically had $E_\gamma < k_B T$, so that the measured X-ray luminosity becomes independent of T. Hence, one expects a modified scaling relation between the X-ray luminosity measured by ROSAT $L_{<2.4\,keV}$ and the mass of the cluster,

$$L_{<2.4\,keV} \propto f_g^2 \, M_{vir} \, . \tag{6.54}$$

This scaling relation can also be tested empirically, as shown in Fig. 6.28, where the X-ray luminosity in the energy range of the ROSAT satellite is plotted against the virial mass. One can immediately see that clusters of galaxies indeed show a strong correlation between luminosity and mass, but with a clearly larger scatter than in the mass–temperature relation.[5] Therefore, the temper-

[5] It should be noted, though, that the determination of L_X and M are independent of each other, whereas in the mass determination the temperature is an explicit parameter so that the measurements of these two parameters are correlated.

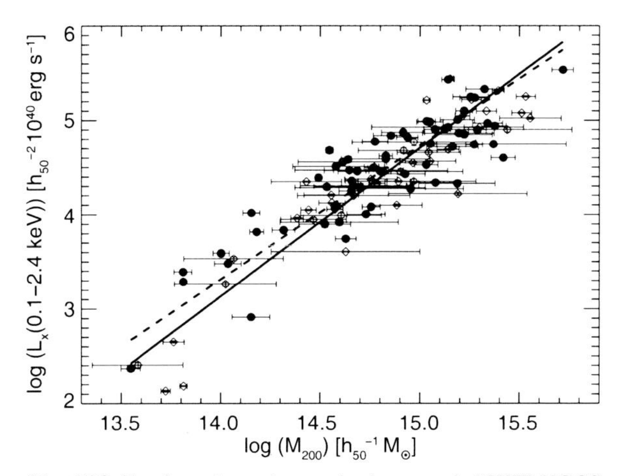

Fig. 6.28. For the galaxy clusters in the extended HIFLUGCS sample, the X-ray luminosity in the energy range of the ROSAT satellite is plotted versus the mass of the cluster. The solid points show the clusters of the HIFLUGCS sample proper. For the full sample and for the main HIFLUGCS sample, a best-fit power law is indicated by the solid line and dashed line, respectively

ature of the intergalactic gas is a better mass indicator than the X-ray luminosity or the velocity dispersion of the cluster galaxies.

However, determining the slope of the relation from the data approximately yields $L_{<2.4\,\mathrm{keV}} \propto M^{1.5}$, instead of the expected behavior ($L_{<2.4\,\mathrm{keV}} \propto M^{1.0}$). Obviously, the above scaling arguments are not valid with the assumption of a constant gas fraction. This discrepancy between theoretical expectations and observations has been found in several samples of galaxy clusters and is considered well established. An explanation is found in models where the intergalactic gas has not only been heated by gravitational infall into the potential well of the cluster. Other sources of heating may have been present or still are. For cooler, less massive clusters, this additional heating should have a larger effect than for the very massive ones, which could also explain the deviation of low-M clusters from the mass–luminosity relation of massive clusters visible in Fig. 6.28. As has already been argued in the discussion of cooling flows in Sect. 6.3.3, an AGN in the inner regions of the cluster may provide such a heating. The heating and additional kinetic energy provided by supernovae in cluster galaxies is also considered a potential source of additional heating of the intergalactic gas. It is obvious that solving this mystery will provide us with better insights into the formation and evolution of the gas component in clusters of galaxies.

Despite this discrepancy between the simple models and the observations, Fig. 6.28 shows a clear correlation between mass and luminosity, which can thus empirically be used after having been calibrated. Although the temperature is the preferred measure for a cluster's mass, one will in many cases resort to the relation between mass and X-ray luminosity because determining the luminosity (in a fixed energy range) is considerably simpler than measuring the temperature, for which significantly longer exposure times are required.

6.4.4 Near-Infrared Luminosity as Mass Indicator

Whereas the optical luminosity of galaxies depends not only on the mass of the stars but also on the star-formation history, the NIR light is much less dependent on the latter. As we have discussed before, the NIR luminosity is thus quite a reliable measure of the total mass in stars. For this reason, we would expect that the NIR luminosity of a cluster is very strongly correlated with its total stellar mass. Furthermore, if the latter is closely related to the total cluster mass, as would be the case if the stellar mass is a fixed fraction of the cluster's total mass, the NIR luminosity can be used to estimate the masses of clusters.

The Two-Micron All Sky Survey (2MASS) provides the first opportunity to perform such an analysis on a large sample of galaxy clusters. One selects clusters of galaxies for which masses have been determined by X-ray methods, and then measures the K-band luminosities of the galaxies within the cluster. Figure 6.29 presents the resulting mass–luminosity diagram within r_{500} for 93 galaxy clusters and groups, where the mass was derived from the clusters' X-ray temperatures (plotted on the top axis) by means of (6.49). A surprisingly close relation between these two parameters is seen, which can be described by a power law of the form

$$\frac{L_{500}}{10^{12} L_\odot} = 3.95 \left(\frac{M_{500}}{2 \times 10^{14} M_\odot} \right)^{0.69} , \qquad (6.55)$$

where a Hubble constant of $h = 0.7$ is assumed. The dispersion of individual clusters around this power law is about 32%, where at least part of this scatter originates in uncertainties in the mass determination – thus,

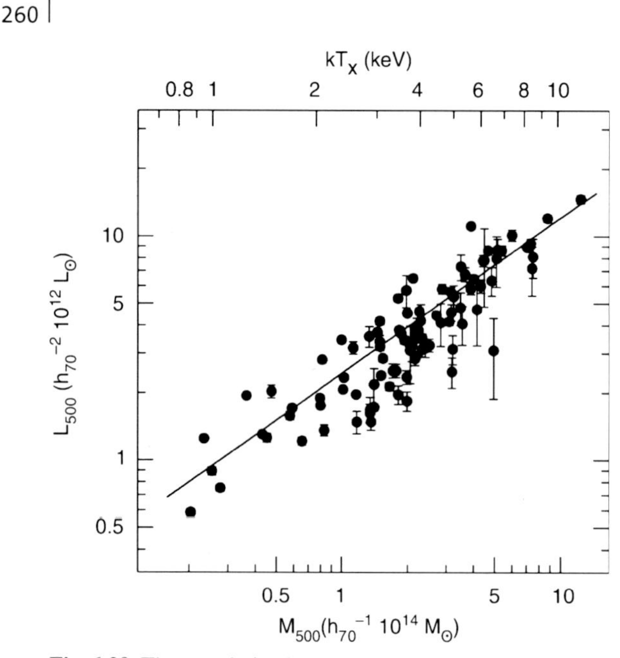

Fig. 6.29. The correlation between K-band luminosity and the mass of galaxy clusters, measured within the radius inside which the mean density is 500 times the critical density of the Universe. The cluster mass has been determined by the relation (6.49) between mass and temperature

Fig. 6.30. The cluster of galaxies A 370 at redshift $z = 0.375$ is one of the first two clusters in which giant luminous arcs were found in 1986. In this HST image, the arc is clearly visible; it is about 20″ long, tangentially oriented with respect to the center of the cluster which is located roughly halfway between the two bright cluster galaxies, and curved towards the center of the cluster. Only with HST images was it realized how thin these arcs are. In this image, several other lens effects are visible as well, for example a background galaxy that is imaged three-fold. The arc is the image of a galaxy at $z_s = 0.724$

the intrinsic scatter is even lower. This result is of great potential importance for future studies of galaxy clusters, and it renders the NIR luminosity a competitive method for the determination of cluster masses, which is of great interest in view of the next generation of NIR wide-field instruments (like VISTA on Paranal, for instance).

6.5 Clusters of Galaxies as Gravitational Lenses

6.5.1 Luminous Arcs

In 1986, two groups independently discovered unusually stretched, arc-shaped sources in two clusters of galaxies at high redshift (see Figs. 6.30 and 6.31). The nature of these sources was unknown at first; they were named *arcs*, or *giant luminous arcs*, which did not imply any interpretation originally. Different hypotheses for the origin of these arcs were formulated, like for instance emission by shock fronts in the ICM, originating from explosive events. All these scenarios were disproven when the spectroscopy of the arc in the cluster Abell 370 showed that the source is at a much higher redshift than the cluster itself. Thus, the arc is a background source, subject to the gravitational lens effect (see Sect. 3.8) of the cluster. By differential light deflection, the light beam of the source can be distorted in such a way that highly elongated arc-shaped images are produced.

The discovery that clusters of galaxies may act as strong gravitational lenses came as a surprise at that time. Based on the knowledge about the mass distribu-

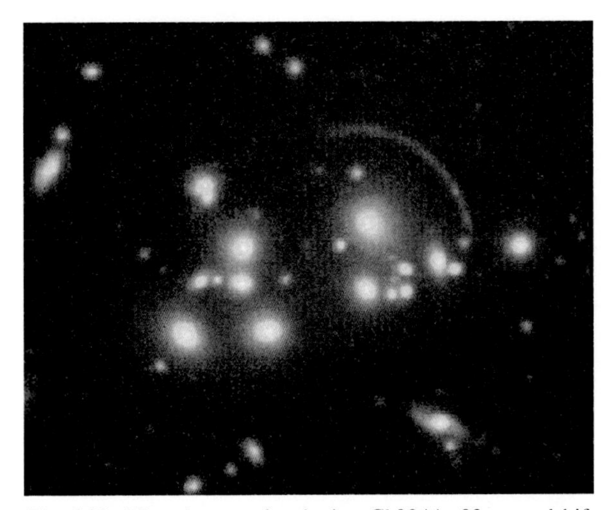

Fig. 6.31. The cluster of galaxies Cl 2244−02 at redshift $z = 0.33$ is the second cluster in which an arc was discovered. Spectroscopic analysis of this arc revealed the redshift of the corresponding source to be $z_s = 2.24$ – at the time of discovery in 1987, it was the first normal galaxy detected at a redshift > 2. This image was observed with the near-IR camera ISAAC at the VLT. Above the arc, one can see another strongly elongated source which is probably associated with a galaxy at very high redshift as well

tion of clusters, derived from X-ray observations before ROSAT, it was estimated that the central surface mass density of clusters is not sufficiently high for strong effects of gravitational light deflection to occur. This incorrect estimate of the central surface mass density in clusters originated from analyses utilizing the β-model which, as briefly discussed above, starts with some heavily simplifying assumptions.[6]

Hence, arcs are strongly distorted and highly magnified images of galaxies at high redshift. In some massive clusters several arcs were discovered and the unique angular resolution of the HST played a crucial role in such observations. Some of these arcs are so thin that their width is unresolved even by the HST, indicating an extreme length-to-width ratio. For many arcs, additional images of the same source were discovered, sometimes called "counter arcs". The identification of multiple images is performed either by optical spectroscopy (which is difficult in general, because one arc is highly magnified while the other images of the same source are considerably less strongly magnified and therefore much fainter in general, and also because spectroscopy of faint sources is very time-consuming), by multi-color photometry (all images of the same source should have the same color), or by common morphological properties.

[6] Another lesson that can be learned from the discovery of the arcs is one regarding the psychology of researchers. After the first observations of arcs were published, several astronomers took a second look at their own images of these two clusters and clearly detected the arcs in them. The reason why this phenomenon, which had been observed much earlier, was not published before can be explained by the fact that researchers were not completely sure about whether these sources were real. A certain tendency prevails in not recognizing phenomena that occur unexpectedly in data as readily as results which are expected. However, there are also those researchers who behave in exactly the opposite manner and even interpret phenomena expected from theory in some unusual way.

Lens Models. Once again, the simplest mass model for a galaxy cluster as a lens is the singular isothermal sphere (SIS). This lens model was discussed previously in Sect. 3.8.2. Its characteristic angular scale is specified by the Einstein radius (3.60), or

$$\theta_{\rm E} = 28.''8 \left(\frac{\sigma_v}{1000\,{\rm km/s}} \right)^2 \left(\frac{D_{\rm ds}}{D_{\rm s}} \right) . \tag{6.56}$$

Very high magnifications and distortions of images can occur only very close to the Einstein radius. This immediately yields an initial mass estimate of a cluster, by assuming that the Einstein radius is about the same as the angular separation of the arc from the center of the cluster. The projected mass within the Einstein radius can then be derived, using (3.66). Since clusters of galaxies are, in general, not spherically symmetric and may show significant substructure, so that the separation of the arc from the cluster's center may deviate significantly from the Einstein radius, this mass estimate is not very accurate in general; the uncertainty is estimated to be $\sim 30\%$. Models with asymmetric mass distributions predict a variety of possible morphologies for the arcs and the positions of multiple images, as is demonstrated in Fig. 6.32 for an elliptical lens. If several arcs are discovered in a cluster, or several images of the source of an arc, we can investigate detailed mass models for such a cluster. The accuracy of these models depends on the number and positions of the observed lensed images; e.g., on how many arcs and how many multiple image systems are available for modeling. The resulting mass models are not unambiguous, but they are robust. Clusters that contain many lensed images have very well-determined mass properties, for instance the mass and the mass

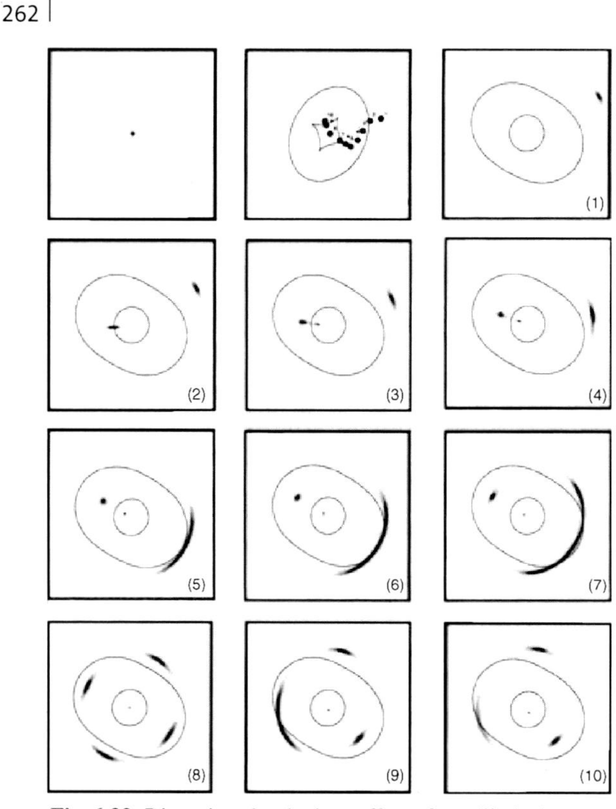

Fig. 6.32. Distortions by the lens effect of an elliptical potential, as a function of the source position. The first panel shows the source itself. The second panel displays ten positions of the source in the source plane (numbered from 1 to 10) relative to the center of the lens; the solid curves show the inner and outer caustics. The remaining panels (numbered from 1 to 10) show the inner and outer critical curves and the resulting images of the source

profile within the radii at which arcs are found, or the ellipticity of the mass distribution and its substructure.

Figure 6.33 shows two clusters of galaxies which contain several arcs. For a long time, A 2218 was the classic example of the existence of numerous arcs in a single galaxy cluster. Then after the installation of the ACS camera on-board HST in 2002, a spectacular image of the cluster A 1689 was obtained in which more than 100 arcs and multiple images were identified. Several sections of this image are shown in Fig. 6.34. For clusters of galaxies with such a rich inventory of lens phenomena, very detailed mass models can be constructed.

Such mass models have predictive power, allowing an iterative modeling process. An initial simple mass model is fitted to the most prominent lensed images in the observation, i.e., either giant arcs or clearly recognizable multiple images. In general, this model then predicts further images of the source producing the arc. Close to these predicted positions, these additional images are then searched for, utilizing the morphology of the light distribution and the color. If this initial model describes the overall mass distribution quite well, such images are found. The exact positions of the new images provide further constraints on the lens model which is then refined accordingly. Again, the new model will predict further multiple image systems, and so on. By this procedure, very detailed models can sometimes be obtained. Since the lens properties of a cluster depend on the distance or the redshift of the source, the redshift of lensed sources can be predicted from the identification of multiple image systems in clusters if a detailed mass model is available. These predictions can then be verified by spectroscopic analysis, and the success of this method gives us some confidence in the accuracy of the lens models.

Results. We can summarize the most important results of the examination of clusters using arcs and multiple images as follows: the mass of galaxy clusters is indeed much larger than the mass of their luminous matter. The lensing method yields a mass which is in very good agreement with mass estimates from the X-ray method or from dynamical methods. However, the core radius of clusters, i.e., the scale on which the mass profile flattens inwards, is significantly smaller than determined from X-ray observations. A typical value is $r_c \sim 30h^{-1}$ kpc, in contrast to $\sim 150h^{-1}$ kpc from the X-ray method. This difference leads to a discrepancy in the mass determination between the two methods on scales below $\sim 200h^{-1}$ kpc. We emphasize that, at least in principle, the mass determination based on arcs and multiple images is substantially more accurate because it does not require any assumptions about the symmetry of the mass distribution, about hydrostatic equilibrium of the X-ray gas, or about an isothermal temperature distribution. On the other hand, the lens effect measures the mass in cylinders because the lens equation contains only the projected mass distribution, whereas the X-ray method determines the mass inside spheres. The conversion between the two methods introduces uncertainties, in particular for clusters which deviate

Fig. 6.33. Top image: the cluster of galaxies A 2218 ($z_d = 0.175$) contains one of the most spectacular arc systems. The majority of the galaxies visible in the image are associated with the cluster and the redshifts of many of the strongly distorted arcs have now been measured. Bottom image: the cluster of galaxies Cl 0024+17 ($z = 0.39$) contains a rich system of arcs. The arcs appear bluish, stretched in a direction which is tangential to the cluster center. The three arcs to the left of the cluster center, and the arc to the right of it and closer to the center, are images of the same background galaxy which has a redshift of $z = 1.62$. Another image of the same source was found close to the cluster center. Also note the identical ("pretzel"-shaped) morphology of the images

significantly from spherical symmetry. Overestimating the core radius was the main reason why the discovery of the arcs was a surprise because clusters with core radii like the ones determined from the early X-ray measurements would in fact not act as strong gravitational lenses. Hence, the mere existence of arcs shows that the core radius must be small.

A closer analysis of galaxy clusters with cooling flows shows that, in these clusters, the mass profile estimated from X-ray observations is compatible with the observed arcs. Such clusters are considered dynamically relaxed, so that for them the assumption of a hydrostatic equilibrium is well justified. The X-ray analysis has to account explicitly for the existence of a cooling flow, though, and the accordingly modified X-ray emission profile is more sophisticated than the simple β-model. Clusters without cooling flows are distinctly more complex dynamically. Besides the discrepancy in mass determination, lensing and X-ray methods can lead to different estimates of the center of mass in such unrelaxed clusters, which may indicate that the gas has not had enough time since the last strong interaction or merging process to settle into an equilibrium state.

The mass distribution in clusters often shows significant substructure. Clusters of galaxies in which arcs are observed are often not relaxed. These clusters still undergo dynamical evolution – they are young systems with an age not much larger than $t_{\rm cross}$, or systems whose

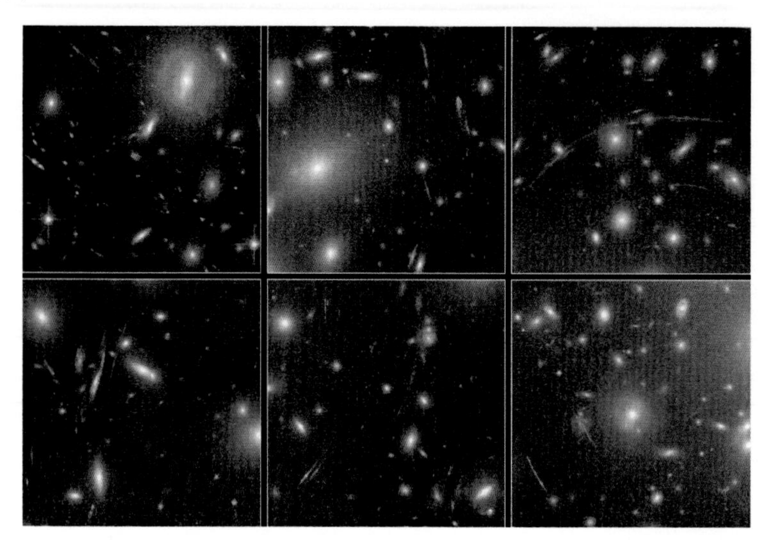

Fig. 6.34. The cluster of galaxies Abell 1689 has the richest system of arcs and multiple images found to date. In a deep ACS exposure of this cluster more than a hundred such lensed images were detected. Six sections of this ACS image are shown in which various arcs are visible, some with an extreme length-to-width ratio, indicating very high magnification factors

equilibrium was disturbed by a fairly recent merger process. For such clusters, the X-ray method is not well founded because the assumptions about symmetry and equilibrium are not satisfied. The distribution of arcs in the cluster A2218 (Fig. 6.33) clearly indicates a non-spherical mass distribution. Indeed, this cluster seems to consist of at least two massive components around which the arcs are curved, indicating that the cluster is currently undergoing a strong merging event. This is further supported by measurements of the temperature distribution of the intracluster gas, which shows a strong peak in the center, where the temperature is about a factor of 2 higher than in its surrounding region.

From lens models, we find that for clusters with a central cD galaxy, the orientation of the mass distribution follows that of the cD galaxy quite closely. We conclude from this result that the evolution of the cD galaxy must be closely linked to the evolution of the cluster, e.g., by accretion of a cooling flow onto the cD galaxy. Often, the shape of the mass distribution very well resembles the galaxy distribution and the X-ray emission.

> The investigation of galaxy clusters with the gravitational lens method provides a third, completely independent method of determining cluster masses. It confirms that the mass of galaxy clusters significantly exceeds that of the visible matter in stars and in the intracluster gas. We conclude from this result that clusters of galaxies are dominated by dark matter.

6.5.2 The Weak Gravitational Lens Effect

The Principle of the Weak Lensing Effect. In Sect. 3.8 we saw that gravitational light deflection does not only deflect light beams as a whole, but also that the size and shape of light beams are distorted by differential light deflection. This differential light deflection leads, e.g., to sources appearing brighter than they would be without the lens effect. The giant arcs discussed above are a very good example of these distortions and the corresponding magnifications.

If some background sources exist which are distorted in such an extreme way as to become visible as giant luminous arcs, then it appears plausible that many more background galaxies should exist which are less strongly distorted. Typically, these are located at larger angular separations from the cluster center, where the lens effect is weaker than at the location of the luminous arcs. Their distortion then is so weak that it cannot be identified in an individual galaxy image. The reason for this is that the intrinsic light distribution of galaxies is not circular; rather, the observed image shape is a superposition of the intrinsic shape and the gravitational lens distortion. The intrinsic ellipticity of galaxies is considerably larger than the shear, in general, and acts as a kind of noise in the measurement of the lensing effect. However, the distortion of adjacent galaxy images should be similar since the gravitational field their light beams are traversing is similar. By averaging over many such galaxy images, the distortion can then be measured (see Fig. 6.35) because no preferred direction exists in the intrinsic random orientation of galaxies. Since the results from the Hubble Deep Field (Fig. 1.27) became available, if not before, we have known that the sky is densely covered by small, faint galaxies. In deep optical images, one should therefore find a high number density of such galaxies located in the background of a galaxy cluster. Their measured shapes can be used for investigating the weak lensing effect of the cluster.

The distortion, obtained by averaging over image ellipticities, reflects the contribution of the tidal forces to the local gravitational field of the cluster. In this context, it is denoted as *shear*. It is given by the projection of the tidal contribution to the gravitational field along the line-of-sight. The shear results from the derivative of the deflection angle, where the deflection angle (3.47) depends linearly on the surface mass density of the lens. Hence, it is possible to reconstruct the surface mass density of galaxy clusters in a completely parameter-free way using the measured shear: it can be used to map the (dark) matter in a cluster.

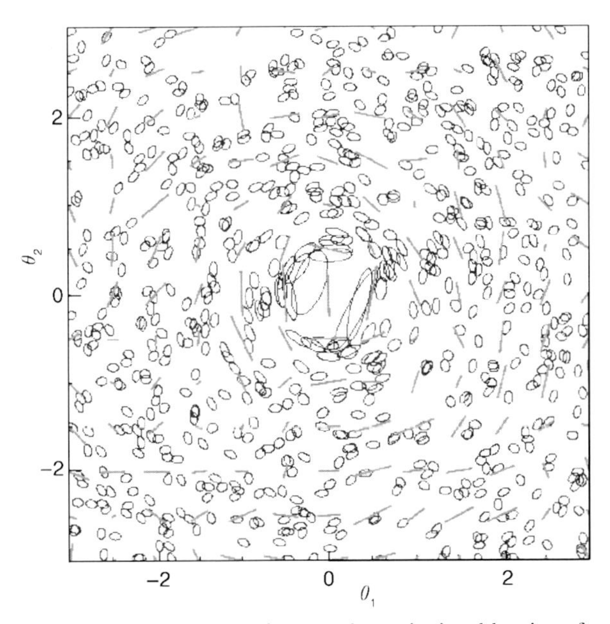

Fig. 6.35. The principle of the weak gravitational lensing effect is illustrated here with a simulation. Due to the tidal component of the gravitational field in a cluster, the shape of the images (ellipses) of background galaxies get distorted and, as for arcs, the galaxy images will be aligned, on average, tangentially to the cluster center. By local averaging over the ellipticities of galaxy images, a local estimate of the tidal gravitational field can be obtained (the direction of the sticks indicates the orientation of the tidal field, and their length is proportional to its strength). From this estimated tidal field, the projected mass distribution can then be reconstructed

Observations. Since shear measurements are based on averaging over image ellipticities of distant galaxies, this method of *weak gravitational lensing* requires optical images with as high a galaxy density as possible. This implies that the exposures need to be very deep to reach very faint magnitudes. But since very faint galaxies are also very distant and, as a consequence, have small angular extent, the observations need to be carried out under very good observing conditions, to be able to accurately measure the shape of galaxy images without them being smeared into circular images by atmospheric turbulence, i.e., the seeing. Typically, to apply this method images from 4-m class telescopes are used, with exposure times of one to three hours. This way, we reach a density of about 30 galaxies per square arcminute (thus, 10^5 per square degree) of which shapes are sufficiently well measurable. This corresponds to a limiting magnitude of about $R \sim 25$. The seeing during the exposure should not be larger than $\sim 0\farcs8$ to still be able to correct for seeing effects.

Systematic observations of the weak lensing effect only became feasible in recent years with the development of wide-field cameras.[7] This, together with the improvement of the dome seeing at many telescopes and the development of dedicated software for data analysis, rendered quantitative observational studies with weak lensing possible; the best telescopes at the best observatories regularly accomplish seeing below $1''$, and the dedicated software is specifically designed for measuring the shapes of extremely faint galaxy images and for correcting for the effects of seeing and anisotropy of the point-spread function.

[7] Prominent examples for such cameras are, for instance, the $\sim 12000 \times 8000$-pixel camera CFH12k, mounted on the Canada–France–Hawaii Telescope (CFHT), or the Wide-Field Imager (WFI), a $\sim (8000)^2$-pixel camera at the ESO/MPG 2.2-m telescope on La Silla. In 2003, the first square-degree camera was installed at the CFHT, Megacam, with $\sim (18000)^2$ pixels. Another square-degree camera, OmegaCAM, is due to start operations at ESO's newly built VLT Survey Telescope (VST) in 2007.

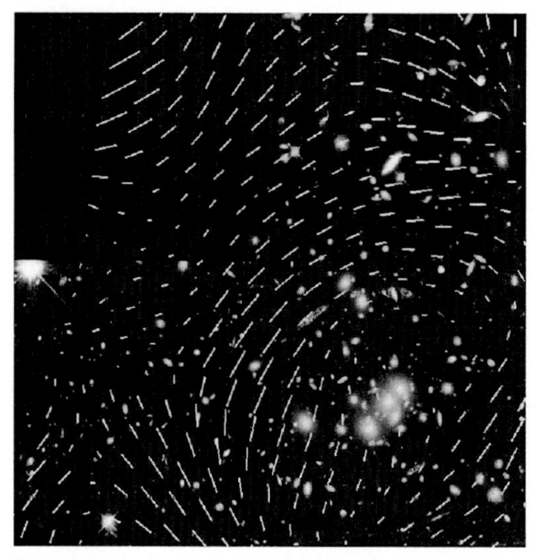

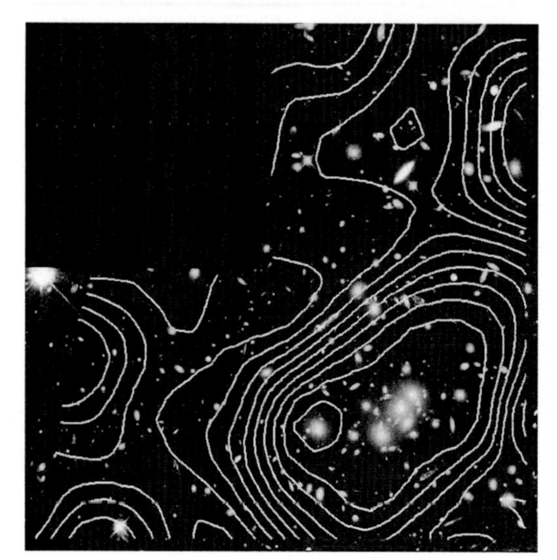

Fig. 6.36. Left: the tidal (or shear) field of the cluster Cl 0024+17 is indicated by sticks whose length and direction represent the strength and orientation of the tidal gravitational field. Right: the surface mass density is shown, reconstructed by means of the weak gravitational lens effect. The bright galaxies in the cluster are seen to follow the (dark) matter distribution; the orientation of the isodensity contours is the same as the orientation of the light in the center of the cluster

Mass Reconstruction of Galaxy Clusters. By means of this method, the reconstruction of the mass density of a large number of clusters became possible. The most important results of these investigations are as follows: the center of the mass distribution corresponds to the optical center of the cluster (see Fig. 6.36). If X-ray information is available, the mass distribution is, in general, found to be centered on the X-ray maximum. The shape of the mass distribution – e.g., its ellipticity and orientation – is in most cases very similar to the distribution of bright cluster galaxies. The comparison of the mass profile determined by this method and that determined from X-ray data agree well, typically within a factor of ~ 1.5 (see Fig. 6.37 for an example). Through the weak lensing effect, substructure in the mass distribution is also detected in some clusters (Fig. 6.38) which does not in all cases reflect the distribution of cluster galaxies. However, in general a good correspondence between light and mass exists (Fig. 6.39). From these lensing studies, we obtain a mass-to-light ratio for clusters that agrees with that found from X-ray analyses, about $M/L \sim 250h$ in Solar units. Clusters of galaxies that strongly deviate from this average value do exist, however. Two independent analyses for the cluster MS 1224+20 resulted in a mass-to-light ratio of

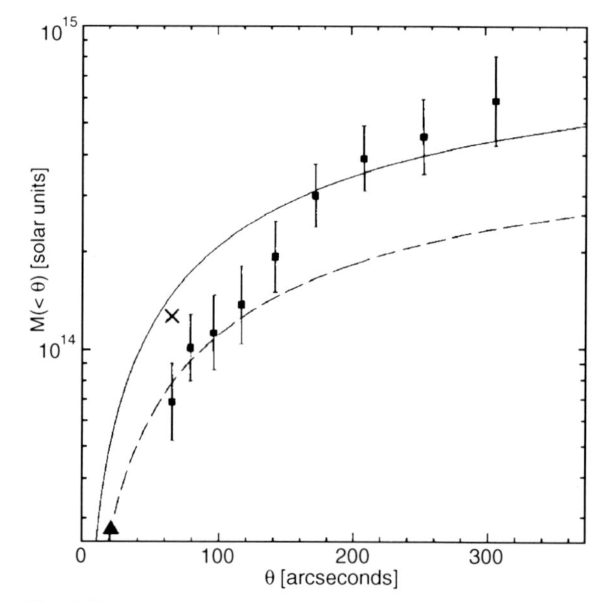

Fig. 6.37. Radial mass profile of the galaxy cluster Abell 2218. The data points with error bars are mass estimates from the weak lensing effect, the solid and dashed curves are isothermal sphere models assuming different velocity dispersions. The cross denotes the mass estimated from luminous arcs, and the triangle depicts the mass obtained from the central cD galaxy

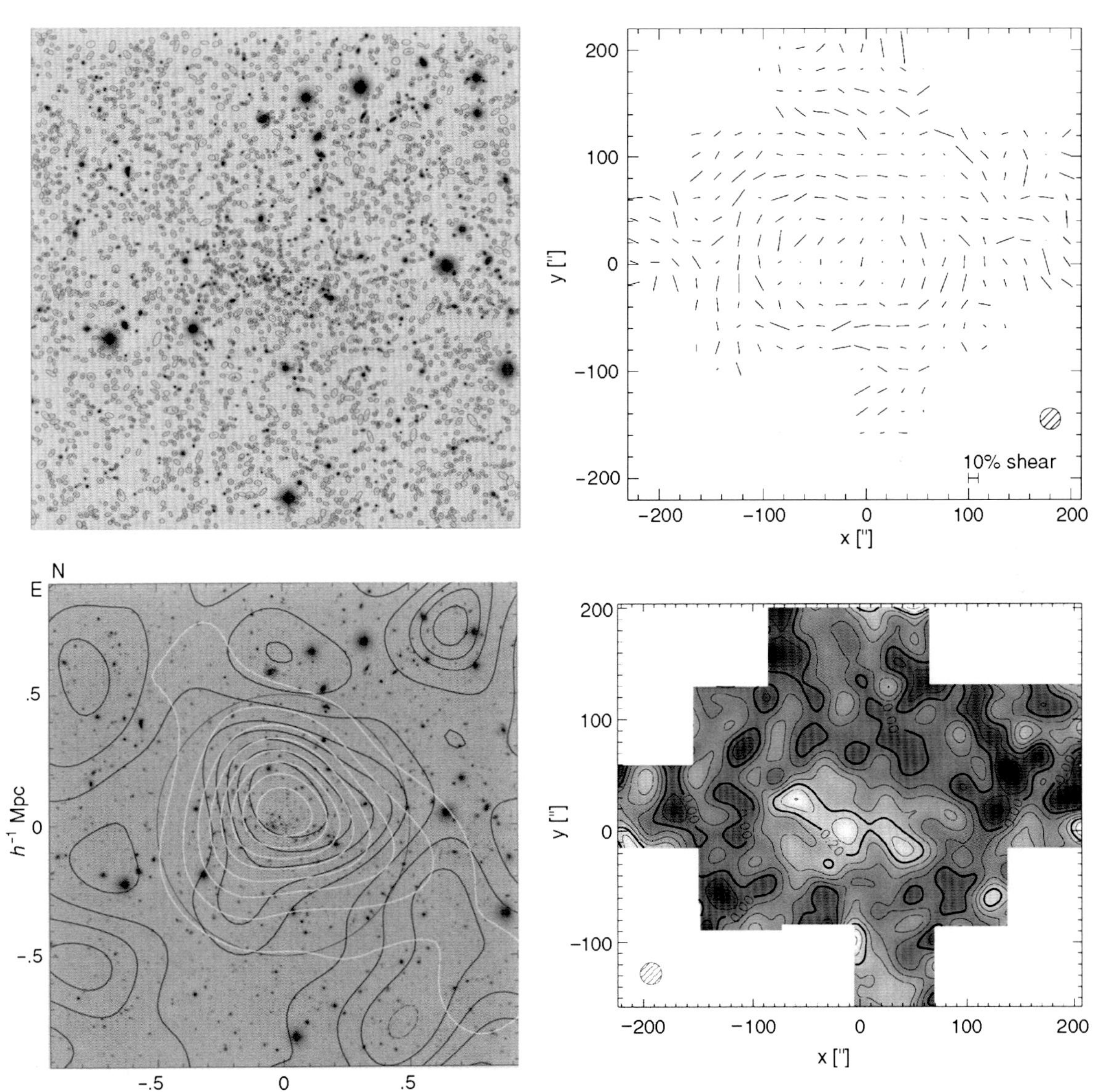

Fig. 6.38. Analysis of the cluster of galaxies MS 1054−03 by the weak lensing effect. In the upper left panel, a ground-based image is shown with a field size of 7.′5 × 7.′5. In this image, about 2400 faint objects are detected, the majority of which are galaxies at high redshift. From the measured ellipticities of the galaxies, the tidal field of the cluster can be reconstructed, and from this the projected mass distribution $\Sigma(\boldsymbol{\theta})$, presented in the lower left panel; the latter is indicated by the black contours, while the white contours represent the smoothed light distribution of the cluster galaxies. A mosaic of HST images allows the ellipticity measurement of a significantly larger number of galaxies, and with better accuracy. The tidal field resulting from these measurements is displayed in the upper right panel, with the reconstructed surface mass density shown in the lower right panel. One can clearly see that the cluster is strongly structured, with three density maxima which correspond to regions with bright cluster galaxies. This cluster seems to be currently in the process of formation through a merger of smaller entities

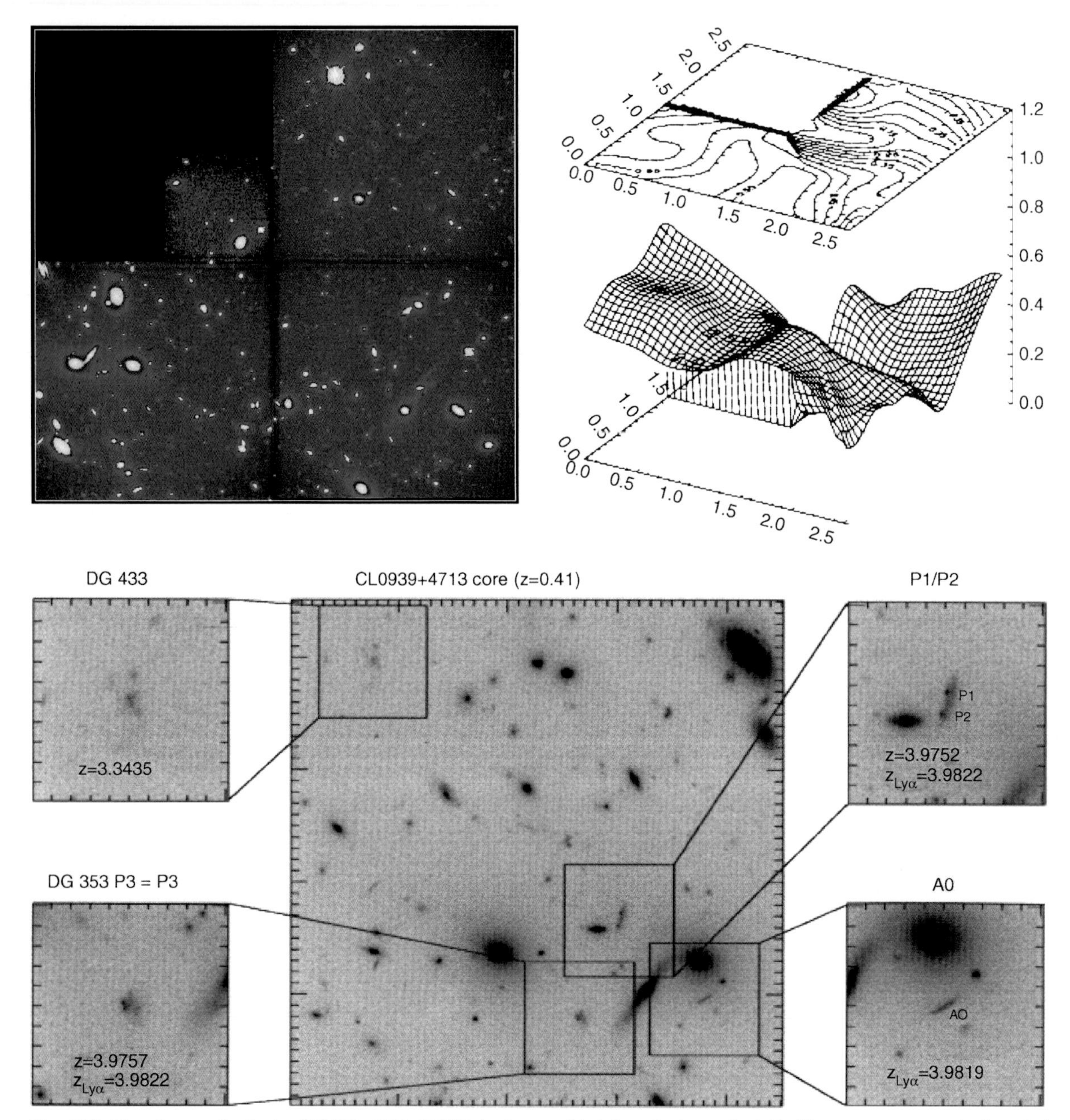

Fig. 6.39. The cluster of galaxies Cl0939+4713 (A851) is the cluster with the highest redshift in the Abell catalog. The HST image in the upper left panel was obtained shortly after the refurbishment of the HST in 1994; in this image, North is down, whereas it is up in the bottom images. The mass distribution of the cluster was reconstructed from this image and is shown in the upper right panel, both as the level surface and by the contours on top. We see that the distributions of bright galaxies and of (dark) matter are very similar: their respective centers are aligned, a secondary maximum exists in both the light and the matter distribution, as does the prominent minimum in which no bright galaxies are visible either. This cluster also shows strong lensing effects, which can be seen from the image at the bottom: a triple image system at $z \approx 3.98$ and an arc with $z = 3.98$ were confirmed spectroscopically

$M/L \approx 800h$ in Solar units, more than twice the value normally found in clusters.

The similarity of the mass and galaxy distributions is not necessarily expected because the lens effect measures the total mass distribution, and therefore mainly the dark matter in a cluster of galaxies. The similar distributions then imply that the galaxies in a cluster seem to basically follow the distribution of the dark matter, although there are some exceptions.

The Search for Clusters of Galaxies with Weak Gravitational Lensing. The weak lensing effect can not only be used to map the matter distribution of known clusters, but it can also be used to search for clusters. Mass concentrations generate a tangential shear field in their vicinity, which can specifically be searched for. The advantage of this method is that it detects clusters based solely on their mass properties, in contrast to all other methods which rely on the emission of electromagnetic radiation, whether in the form of optical light from cluster galaxies or as X-ray emission from a hot intracluster medium. In particular, if clusters with atypically low gas or galaxy content exist, they could be detected in this way.

With this method, quite a number of galaxy clusters have been detected already – see Fig. 6.40. Further candidates exist, in that from the shear signal a significant mass concentration is indicated but it cannot be identified with any concentration of galaxies on optical images. The clarification of the nature of these lens signals is of great importance: if in fact matter concentrations do exist which correspond to the mass of a cluster but which do not contain luminous galaxies, then our understanding of galaxy evolution needs to be revised. However, we cannot exclude the possibility that these statistically significant signals are statistical outliers, or result from projection effects – remember, lensing probes the line-of-sight integrated matter density. Together with the search for galaxy clusters by means of the SZ effect (Sect. 6.3.4), the weak lensing effect provides an interesting alternative for the detection of mass concentrations compared to traditionally methods.

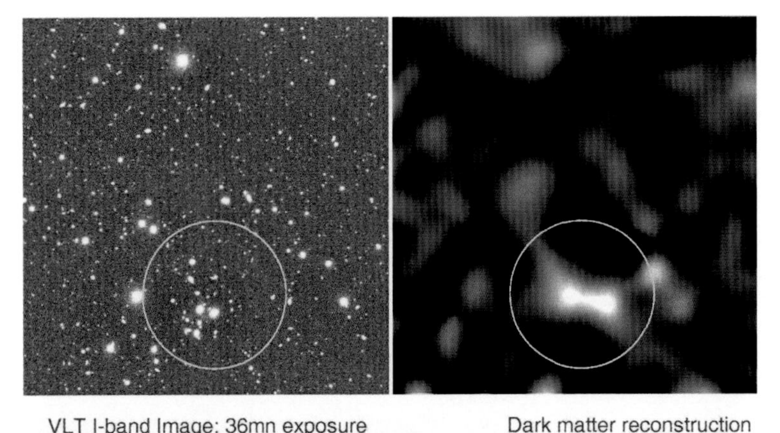

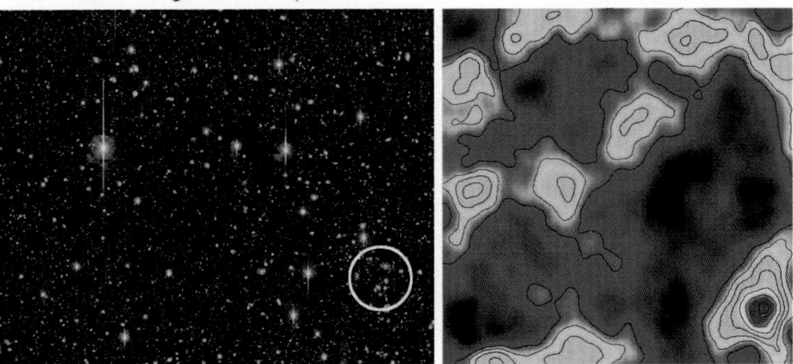

Fig. 6.40. Top left: a VLT/FORS1 image, taken as part of a survey of "empty fields". Top right: the mass reconstruction, as was obtained from the optical data by employing the weak lensing effect. Clearly visible is a peak in the mass distribution; the optical image shows a concentration of galaxies in this region. Hence, in this field a cluster of galaxies was detected for the first time by its lens properties. Bottom: as above, here a galaxy cluster was also detected through its lensing effect. On the left, an optical wide-field image is shown, obtained by the Big Throughput Camera, and the mass reconstruction is displayed on the right. The location of the peak in the latter coincides with a concentration of galaxies. Spectroscopic measurements yield that these form a cluster of galaxies at $z = 0.276$

6.6 Evolutionary Effects

Today, we are able to discover and analyze clusters of galaxies at redshifts $z \sim 1$ and higher; thus the question arises whether these clusters have the same properties as local clusters. At $z \sim 1$ the age of the Universe is only about half of that of the current Universe. One might therefore expect an evolution of cluster properties.

Luminosity Function. First, we shall consider the comoving number density of clusters as a function of redshift or, more precisely, the evolution of the luminosity function of clusters with z. As Fig. 6.41 demonstrates, such evolutionary effects are not very pronounced, and only at the highest luminosities or the most massive clusters, respectively, does an evolution become visible. This reveals itself by the fact that at high redshift, clusters of very high luminosity or very high mass are less abundant than they are today. The interpretation and the relevance of this fact will be discussed later (see Sect. 8.2.1).

Butcher–Oemler Effect. We saw in Chap. 3 that early-type galaxies are predominantly found in clusters and groups, whereas spirals are mostly field galaxies. For example, a massive cluster like Coma contains only 10% spirals, the other luminous galaxies are ellipticals or S0 galaxies (see also Sect. 6.2.9). Besides these morphological differences, the colors of galaxies are very useful for a characterization: early-type galaxies (ellipticals and S0 galaxies) have little ongoing star formation and therefore consist mainly of old, thus low-mass and cool stars. Hence they are red, whereas spirals feature active star formation and are therefore distinctly bluer. The fraction of blue galaxies in nearby clusters is very low.

Butcher and Oemler found that this changes if one examines clusters of galaxies at higher redshifts: these contain a larger fraction of blue galaxies, thus of spirals (see Fig. 6.42). This means that the mixture of galaxies changes over time. In clusters, spirals must become scarcer with increasing cosmic time, e.g., by transforming into early-type galaxies.

A possible and plausible explanation is that spirals lose their interstellar gas. Since they move through the intergalactic gas (which emits the X-ray radiation) at high velocities, the ISM in the galaxies may be torn away and mix with the ICM. This is plausible because the ICM also has a high metallicity. These metals can only originate in a stellar population, thus in the enriched material in the ISM of galaxies. Later, we will discuss some further evidence for transformations between galaxy types.

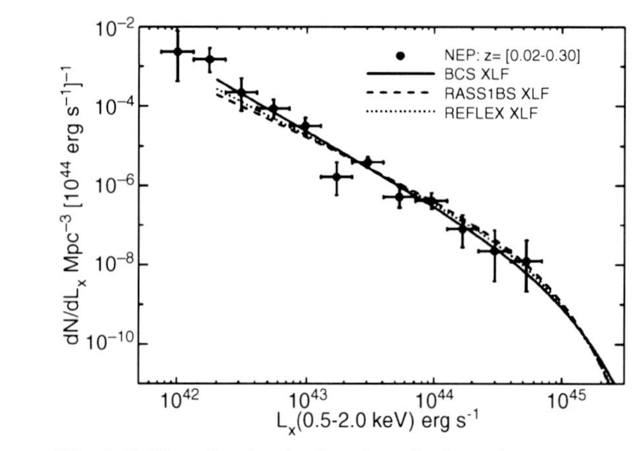

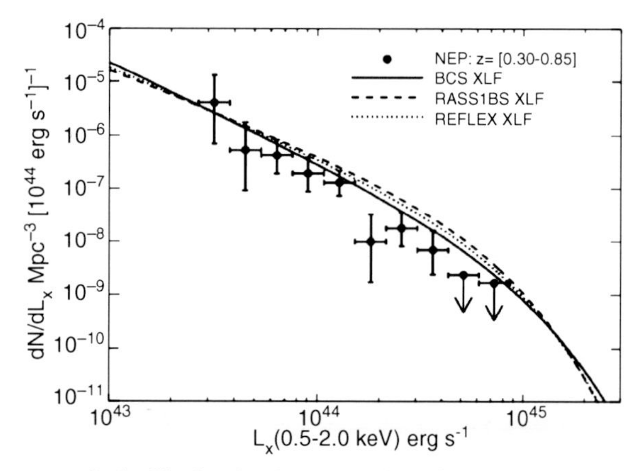

Fig. 6.41. X-ray luminosity function of galaxy clusters, as was obtained from a region around the North Ecliptic Pole (NEP), the region with the longest exposure time in the ROSAT All Sky Survey (see Fig. 6.26). Plotted is $\mathrm{d}N/\mathrm{d}L_{\mathrm{X}}$, the (comoving) number density per luminosity interval, for clusters with $0.02 \le z \le 0.3$ (left panel) and $0.3 \le z \le 0.85$ (right panel), respectively. The luminosity was derived from the flux in the photon energy range from 0.5 keV to 2 keV. The three different curves specify the local luminosity function of clusters as found in other cluster surveys at lower redshifts. We see that evolutionary effects in the luminosity function are relatively small and become visible only at high L_{X}

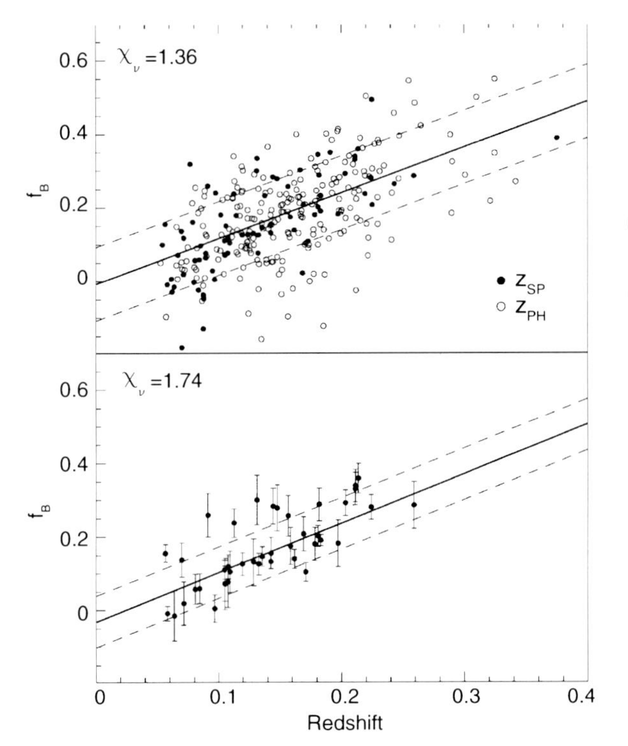

Fig. 6.42. Butcher–Oemler effect: in the upper panel, the fraction of blue galaxies f_b in a sample of 195 galaxy clusters is plotted as a function of cluster redshift, where open (filled) circles indicate photometric (spectroscopic) redshift data for the clusters. The lower panel shows a selection of clusters with spectroscopically determined redshifts and well-defined red cluster sequence. For the determination of f_b, foreground and background galaxies need to be statistically subtracted using control fields, which may also result in negative values for f_b. A clear increase in f_b with redshift is visible, and a line of regression yields $f_b = 1.34z - 0.03$

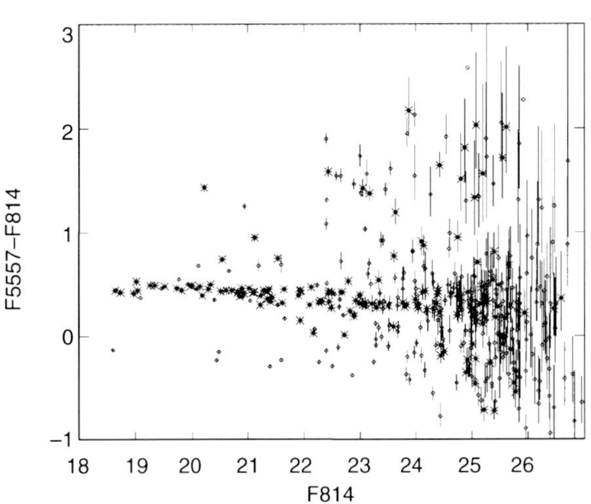

Fig. 6.43. Color–magnitude diagram of the cluster of galaxies Abell 2390, observed with the HST. Star symbols represent early-type galaxies, identified by their morphology, while diamonds denote other galaxies in the field. The red cluster sequence is clearly visible

Color–Magnitude Diagram. Plotting the color of cluster galaxies versus their magnitude, one finds a very well-defined, nearly horizontal sequence (Fig. 6.43). This red cluster sequence (RCS) is populated by the early-type galaxies in the cluster.

The scatter of early-type galaxies around this sequence is very small, which suggests that all early-type galaxies in a cluster have nearly the same color, only weakly depending on luminosity. Even more surprising is the fact that the color–magnitude diagrams of different clusters at the same redshift define a very similar red cluster sequence: cluster galaxies with the same redshift and luminosity have virtually the same color. Comparing the red sequences of clusters at different redshifts, one finds that the sequence of cluster galaxies is redder the higher the redshift is. In fact, the red cluster sequence is so precisely characterized that, from the color–magnitude diagram of a cluster alone, its redshift can be estimated, whereby a typical accuracy of $\Delta z \sim 0.1$ is achieved. The accuracy of this estimated redshift strongly depends on the choice of the color filters. Since the most prominent spectral feature of early-type galaxies is the 4000-Å break, the redshift is estimated best if this break is located right between two of the color bands used.

This well-defined red cluster sequence is of crucial importance for our understanding of the evolution of galaxies. We know from Sect. 3.9 that the composition of a stellar population depends on the mass spectrum at its birth (the initial mass function, IMF) and on its age: the older a population is, the redder it becomes. The fact that cluster galaxies at the same redshift all have roughly the same color indicates that their stellar populations have very similar ages. However, the only age that is singled out is the age of the Universe itself. In fact, the color of cluster galaxies is compatible with their stellar populations being roughly the same age as the Universe at that particular redshift. This also provides an explanation for why the red cluster sequence is

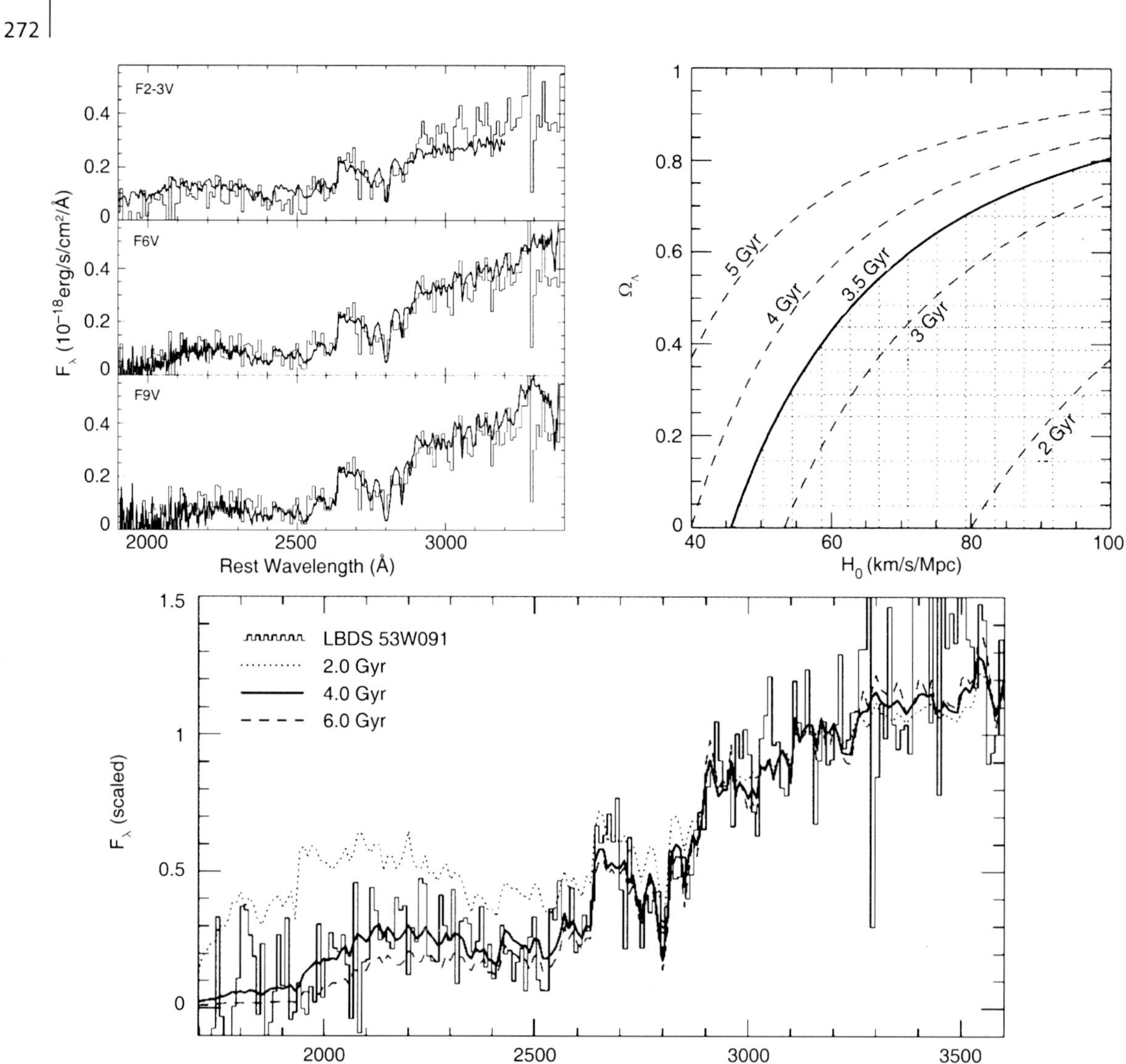

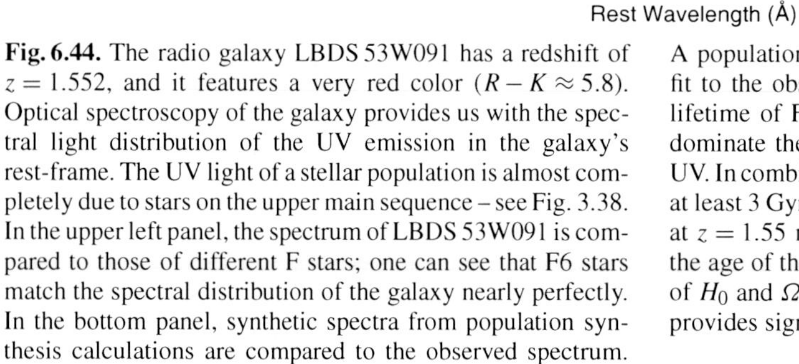

Fig. 6.44. The radio galaxy LBDS 53W091 has a redshift of $z = 1.552$, and it features a very red color ($R - K \approx 5.8$). Optical spectroscopy of the galaxy provides us with the spectral light distribution of the UV emission in the galaxy's rest-frame. The UV light of a stellar population is almost completely due to stars on the upper main sequence – see Fig. 3.38. In the upper left panel, the spectrum of LBDS 53W091 is compared to those of different F stars; one can see that F6 stars match the spectral distribution of the galaxy nearly perfectly. In the bottom panel, synthetic spectra from population synthesis calculations are compared to the observed spectrum. A population with an age of about 4 Gyr represents the best fit to the observed spectrum; this is also comparable to the lifetime of F6 stars: the most luminous (still existing) stars dominate the light distribution of a stellar population in the UV. In combination, this reveals that this galaxy at $z = 1.552$ is at least 3 Gyr old. Phrased differently, the age of the Universe at $z = 1.55$ must be at least 3 Gyr. In the upper right panel, the age of the Universe at $z = 1.55$ is displayed as a function of H_0 and Ω_Λ (for $\Omega_{\rm m} + \Omega_\Lambda = 1$). Hence, this single galaxy provides significant constraints on cosmological parameters

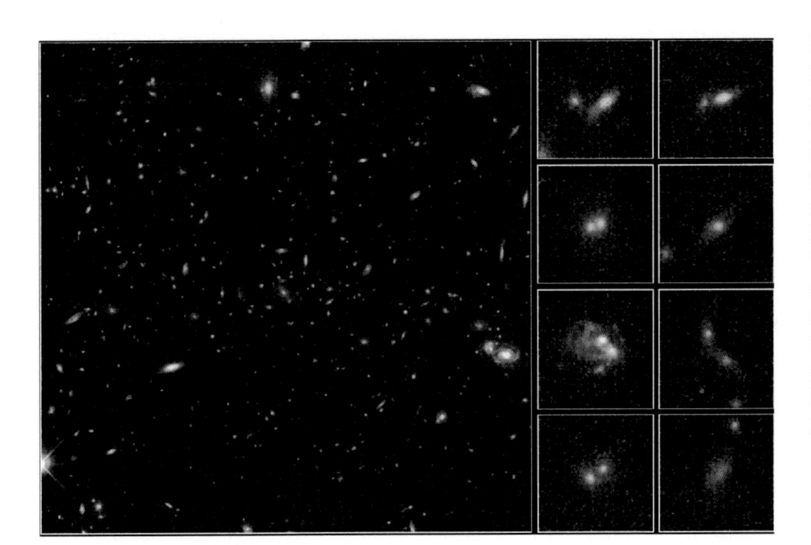

Fig. 6.45. The cluster of galaxies MS1054−03, observed with the HST, is the most distant cluster in the EMSS X-ray survey ($z = 0.83$). The reddish galaxies in the image on the left form a nearly linear structure. This cluster is far from being spherical, as we have also seen from its weak lensing results (Fig. 6.38) – it is not relaxed. The smaller images on the right show blow-ups of selected cluster fields where mergers of galaxies become visible: in this cluster, the merging of galaxies is directly observable. At least six of the nine merging pairs found in this cluster have been shown to be gravitationally bound systems

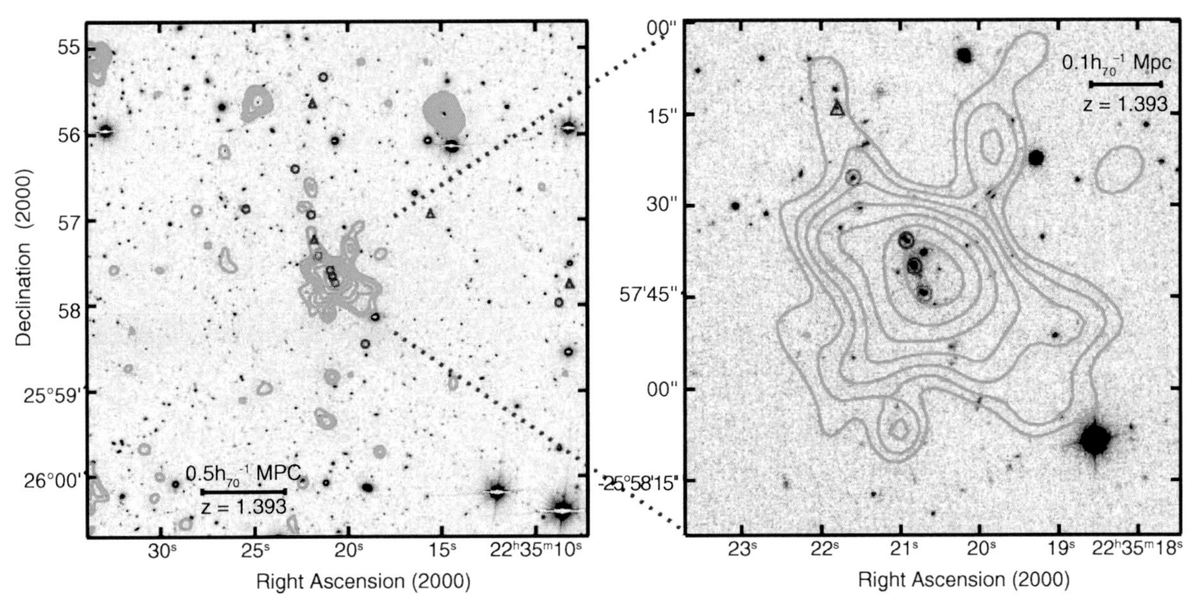

Fig. 6.46. The cluster of galaxies XMMU J2235.2−2557 was discovered in the field-of-view of an XMM-Newton image for which a different source was the original target. The image on the left shows the X-ray contours, superposed on an R-band image, while the image on the right shows the central section, here superposed on a K-band image. Galaxies in the field follow a red cluster sequence if the color is measured in $R-z$. The symbols denote galaxies at redshift $1.37 < z < 1.40$. The strong X-ray source to the upper right of the cluster center is a Seyfert galaxy at lower redshift. As of 2005, this cluster is the most distant X-ray selected cluster known, with a temperature of ~ 6 keV and a velocity dispersion of $\sigma \sim 750$ km/s

shifted towards bluer colors at higher redshifts – there, the age of the Universe was smaller, and thus the stellar population was younger. This effect is of particular importance at high redshifts. The fact that the color–magnitude diagram of early-type galaxies in clusters is not flat, in that more luminous galaxies are redder, follows from the dependence of galaxy colors on the metallicity of their stellar populations. The higher the

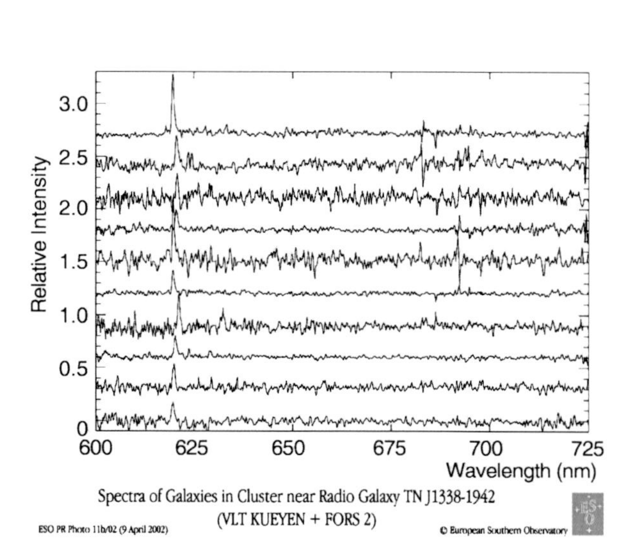

Fig. 6.47. The most distant known group of galaxies. The region around the radio galaxy TN J1338−1942 ($z = 4.1$) was scanned for galaxies at the same redshift; 20 such galaxies were found with the VLT, marked by circles in the left image. For 10 of these galaxies, the spectra are shown on the right; in all of them, the Lyα emission line is clearly visible. Hence, groups of galaxies were already formed in an early stage of the Universe

luminosity of a galaxy, and thus its stellar mass, the higher its metallicity.

Indeed, from the colors of cluster galaxies it is possible to derive very strict upper limits on their star formation in recent times. The color of cluster galaxies at high redshifts even provides interesting constraints on cosmological parameters – only those models are acceptable which have an age of the Universe, at the respective redshift, larger than the estimated age of the stellar population. One interesting example of this is presented in Fig. 6.44.

Therefore, we conclude from these observations that the stars in cluster galaxies formed at very early times in the Universe. But this does not necessarily mean that the galaxies themselves are also this old, because galaxies can be transformed into each other by merger processes (see Fig. 6.45). This changes the morphology of galaxies, but may leave the stellar populations largely unchanged.

Clusters of Galaxies at Very High Redshift. The search for clusters at high redshift is of great cosmological interest. As will be demonstrated in Sect. 7.5.2, the expected number density of clusters as a function of z strongly depends on the cosmological model. Hence, this search offers an opportunity to constrain cosmological parameters by the statistics of galaxy clusters.

The search for clusters in the optical (thus, by galaxy overdensities) becomes increasingly difficult at high z because of projection effects. Nevertheless, several groups have managed to detect clusters at $z \sim 0.8$ with this technique. In particular, the overdensity of galaxies in three-dimensional space can be analyzed if, besides the angular coordinates on the sphere, the galaxy colors are also taken into account. Because of the red cluster sequence, the overdensity is much more prominent in this space than in the sky projection alone.

Projection effects play a considerably smaller role in X-ray searches for clusters. Using sensitive X-ray satellites like ROSAT, some clusters with $z \sim 1.2$ have been found (see Fig. 6.25). The new X-ray satellites Chandra and XMM-Newton are even more sensitive. Therefore, one expects them to be able to find clusters at even higher redshifts; one example for a cluster at $z = 1.393$ is shown in Fig. 6.46. This example demonstrates combining deep X-ray images with observations in the optical and the NIR is an efficient method of compiling samples of distant clusters.

Through optical methods, it is also possible to identify galaxy concentrations at very high redshift. One approach is to assume that luminous AGNs at high redshift are found preferentially in regions of high overdensity, which is also expected from models of galaxy formation. With the redshift of the AGN known, the redshift at which one should search for an overdensity of galaxies near the AGN is defined. Those searches have proven to be quite successful; for instance, they are performed using narrow-band filter photometry, with the filter centered on the redshifted Lyα line, tuned to the redshift of the AGN. Candidates need to be verified spectroscopically afterwards. One example of a strong galaxy concentration at $z = 4.1$ is presented in Fig. 6.47. The identification of a strong spatial concentration of galaxies is not sufficient to have identified a cluster of galaxies though, because it is by no means clear whether one has found a gravitationally bound system of galaxies (and the corresponding dark matter). Rather, such galaxy concentrations are considered to be the predecessors of galaxy clusters which will only evolve into bound systems during later cosmological evolution.

7. Cosmology II: Inhomogeneities in the Universe

7.1 Introduction

In Chap. 4, we discussed homogeneous world models and introduced the standard model of cosmology. It is based on the cosmological principle, the assumption of a (spatially) homogeneous and isotropic Universe. Of course, the assumption of homogeneity is justified only on large scales because observations show us that our Universe is inhomogeneous on small scales – otherwise no galaxies or stars would exist.

The distribution of galaxies on the sky is not uniform or random (see Fig. 6.1), rather they form clusters and groups of galaxies. Also clusters of galaxies are not distributed uniformly, but their positions are correlated, grouped together in superclusters. The three-dimensional distribution of galaxies, obtained from redshift surveys, shows an interesting large-scale structure, as can be seen in Fig. 7.1 which shows the spatial distribution of galaxies in the two-degree Field Galaxy Redshift Survey (2dFGRS).

Even larger structures have been discovered. The Great Wall is a galaxy structure with an extent of $\sim 100h^{-1}$ Mpc, which was found in a redshift survey of galaxies (Fig. 7.2). Such surveys also led to the discovery of the so-called *voids*, nearly spherical regions which contain virtually no (bright) galaxies, and which have a diameter of typically $50h^{-1}$ Mpc. The discovery of these large-scale inhomogeneities raises the question of whether even larger structures might exist in the Universe, or more precisely: does a scale exist, averaged over which the Universe appears homogeneous? The existence of such a scale is a requirement for the homogeneous world models to provide a realistic description of the mean behavior of the Universe.

To date, no evidence of structures with linear dimension $\gtrsim 100h^{-1}$ Mpc have been found, as can also be seen from Fig. 7.1. Hence, the Universe seems to be basically homogeneous if averaged over scales of $R \sim 200h^{-1}$ Mpc. This "homogeneity scale" needs to be compared to the Hubble radius $R_{\rm H} \equiv c/H_0 \approx 3000$ h^{-1} Mpc. This implies $R \ll c/H_0$, so that after averaging, $[(c/H_0)/R]^3 \sim (15)^3 \sim 3000$ independent volume elements exist per Hubble volume. This justifies the approximation of a homogeneous world model when considering the mean history of the Universe.

On small scales, the Universe is inhomogeneous. Evidence for this is the galaxy distribution projected on the sky, the three-dimensional galaxy distribution determined by redshift surveys, and the existence of clusters of galaxies, superclusters, "Great Walls", and voids. In addition, the anisotropy of the cosmic microwave background (CMB), with relative fluctuations of $\Delta T/T \sim 10^{-5}$, indicates that the Universe already contained small inhomogeneities at redshift $z \sim 1000$, which we will discuss more thoroughly in Sect. 8.6.

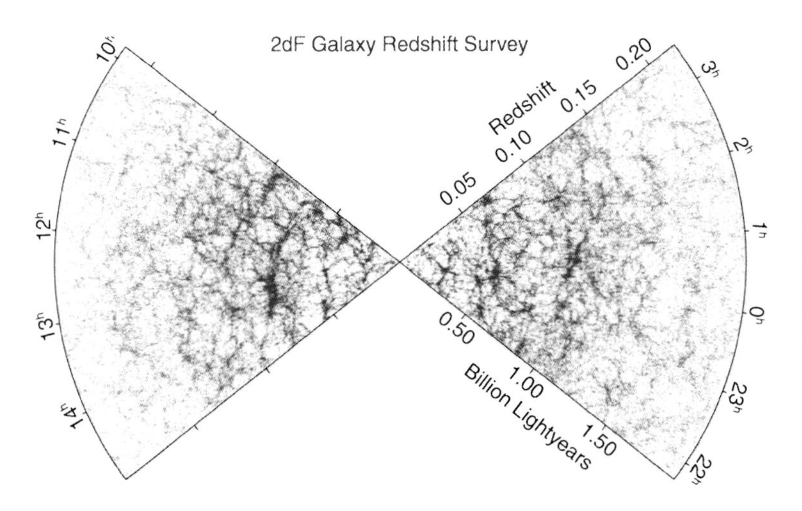

Fig. 7.1. The distribution of galaxies in the complete 2dF Galaxy Redshift Survey. In the radial direction, the escape velocity, or redshift, is plotted, and the polar angle is the right ascension. In this survey, more than 350 000 spectra were taken; plotted here is the distribution of more than 200 000 galaxies with reliable redshift measurements. The data from the complete survey are publicly available

Peter Schneider, Cosmology II: Inhomogeneities in the Universe.
In: Peter Schneider, Extragalactic Astronomy and Cosmology. pp. 277–307 (2006)
DOI: 10.1007/11614371_7

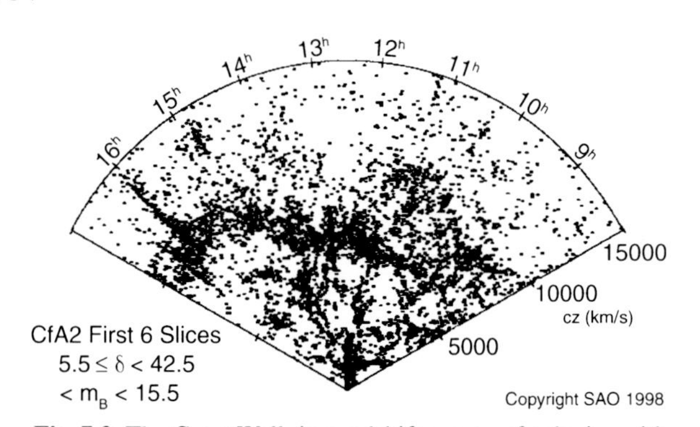

Fig. 7.2. The Great Wall: in a redshift survey of galaxies with radial velocities of $cz \leq 15\,000\,\mathrm{km/s}$, a galaxy structure was discovered which is located at a redshift of $cz \sim 6000\,\mathrm{km/s}$, extending in right ascension between $9^{\mathrm{h}} \leq \alpha \leq 16^{\mathrm{h}}$. Plotted are galaxies with declination $8.5^\circ \leq \delta \leq 42^\circ$

In this chapter, we will examine the evolution of such density inhomogeneities and their description.

7.2 Gravitational Instability

7.2.1 Overview

The smallness of the CMB anisotropy ($\Delta T/T \sim 10^{-5}$; see Sect. 8.6) suggests that the density inhomogeneities at redshift $z \sim 1000$ – this is the epoch where most of the CMB photons interacted with matter for the last time – must have had very small amplitudes. Today, the amplitudes of the density inhomogeneities are considerably larger; for example, a massive cluster of galaxies contains within a radius of $\sim 1.5h^{-1}$ Mpc more than 200 times more mass than an average sphere of this radius in the Universe. Thus, these are no longer small density fluctuations.

Obviously, the Universe became more inhomogeneous in the course of its evolution; as we will see, density perturbations grow over time. One defines the *relative density contrast*

$$\delta(\boldsymbol{r}, t) := \frac{\rho(\boldsymbol{r}, t) - \overline{\rho}(t)}{\overline{\rho}(t)} , \tag{7.1}$$

where $\overline{\rho}(t)$ denotes the mean cosmic matter density in the Universe at time t. From the definition of δ, one can immediately see that $\delta \geq -1$, because $\rho \geq 0$. The smallness of the CMB anisotropy suggests that at $z \sim 1000$, $|\delta| \ll 1$. The dynamics of the cosmic Hubble expansion is controlled by the gravitational field of the average matter density $\overline{\rho}(t)$, whereas the density fluctuations $\Delta\rho(\boldsymbol{r}, t) = \rho(\boldsymbol{r}, t) - \overline{\rho}(t)$ generate an additional gravitational field.

We shall here be interested only in very weak gravitational fields, for which the Newtonian description of gravity can be applied. Since the Poisson equation, which specifies the relation between matter density and the gravitational potential, is linear, the effects of the homogeneous matter distribution and of density fluctuations can be considered separately. The gravitational field of the total matter distribution is then the sum of the average matter distribution and that of the density fluctuations.

We consider a region in which $\Delta\rho > 0$, hence $\delta > 0$, so that the gravitational field in this region is stronger than the cosmic average. An overdense region produces a stronger gravitational field than that corresponding to the mean Hubble expansion. By this additional self-gravity, the overdense region will expand more slowly than the average Hubble expansion. Because of the delayed expansion, the density in this region will also decrease more slowly than in the cosmic mean, $\overline{\rho}(t) = (1+z)^3 \rho_0 = a^{-3}(t)\rho_0$, and hence the density contrast in this region will increase. As a consequence, the relative density will increase, which again produces an even stronger gravitational field, and so on. It is obvious that this situation is unstable. Of course, the argument also works the other way round: in an underdense region with $\delta < 0$, the gravitational field generated is weaker than in the cosmic mean, therefore the self-gravity is weaker than that which corresponds to the Hubble expansion. This implies that the expansion is decelerated less than in the cosmic mean, the underdense region expands faster than the Hubble expansion, and thus the local density will decrease more quickly than the mean density of the Universe. In this way, the density contrast decreases, i.e., δ becomes more negative over the course of time.

Density fluctuations grow over time due to their self-gravity; overdense regions increase their density contrast over the course of time, while underdense regions decrease their density contrast. In both cases, $|\delta|$ increases. Hence, this effect of *gravitational instability* leads to an increase of density fluctuations over the course of time. The evolution of structure in the Universe is described by the model of *gravitational instability*.

The evolution of structure in the Universe can be understood in the framework of this model. In this chapter we will describe structure formation quantitatively. This includes the analysis of the evolution of density perturbations over time, as well as a statistical description of such density fluctuations. We will then see that the evolution of inhomogeneities is directly observable, and that the Universe was less inhomogeneous at high redshift than it is today. Since the evolution of perturbations depends on the cosmological model, we need to examine whether this evolution can be used to obtain an estimate of cosmological parameters. In Chap. 8, we will give an affirmative answer to this question. Finally, we will briefly discuss the origin of density fluctuations.

7.2.2 Linear Perturbation Theory

We first will examine the growth of density perturbations. For this discussion, we will concentrate on length-scales that are substantially smaller than the Hubble radius. On these scales, structure growth can be described in the framework of the Newtonian theory of gravity. The effects of spacetime curvature and thus of General Relativity need to be accounted for only for density perturbations on length-scales comparable to, or larger than the Hubble radius. In addition, we assume for simplicity that the matter in the Universe consists only of dust (i.e., pressure-free matter), with density $\rho(\boldsymbol{r}, t)$. The dust will be described in the *fluid approximation*, where the velocity field of this fluid shall be denoted by $\boldsymbol{v}(\boldsymbol{r}, t)$.[1]

[1] Strictly speaking, the cosmic dust cannot be described as a fluid because the matter is assumed to be collisionless. This means that no interactions occur between the particles, except for gravitation. Two flows of such dust can thus penetrate each other. This situation can be compared to that of a fluid whose molecules are interacting by collisions. Through these collisions, the velocity distribution of the molecules will, at each position, assume an approximate Maxwell distribution, with a well-defined average velocity that corresponds to the flow velocity at this point. Such an unambiguous velocity does not exist for dust in general. However, at early times, when deviations from the Hubble flow are still very small, no multiple flows are expected, so that in this case, the velocity field is unambiguously defined.

Equations of Motion. The behavior of this fluid is described by the continuity equation

$$\frac{\partial \rho}{\partial t} + \nabla \cdot (\rho\, \boldsymbol{v}) = 0\,, \tag{7.2}$$

which expresses the fact that matter is conserved: the density decreases if the fluid has a diverging velocity field (thus, if particles are moving away from each other). In contrast, a converging velocity field will lead to an increase in density. Furthermore, the Euler equation applies,

$$\frac{\partial \boldsymbol{v}}{\partial t} + (\boldsymbol{v} \cdot \nabla)\, \boldsymbol{v} = -\frac{\nabla P}{\rho} - \nabla \Phi\,, \tag{7.3}$$

which describes the conservation of momentum and the behavior of the fluid under the influence of forces. The left-hand side of (7.3) is the time derivative of the velocity as would be measured by an observer moving with the flow, because $\partial \boldsymbol{v}/\partial t$ is the derivative at a fixed point in space, whereas the total left-hand side of (7.3) is the time derivative of the velocity measured along the flow lines. The latter is affected by the pressure gradient and the gravitational field Φ, the latter satisfying the Poisson equation

$$\nabla^2 \Phi = 4\pi G \rho\,. \tag{7.4}$$

Since we are only considering dust, the pressure vanishes, $P = 0$. These three equations for the description of a self-gravitating fluid can in general not be solved analytically. However, we will show that a special, cosmologically relevant exact solution can be found, and that by linearization of the system of equations approximate solutions can be constructed for $|\delta| \ll 1$.

Hubble Expansion. The special exact solution is the flow that we have already encountered in Chap. 4: the homogeneous expanding cosmos. By substituting into the above equations it is immediately shown that

$$\boldsymbol{v}(\boldsymbol{r}, t) = H(t)\boldsymbol{r}$$

is a solution of the equations if ρ is homogeneous and satisfies (4.11), and if the Friedmann equation (4.13) for the scale factor applies.

As long as the density contrast $|\delta| \ll 1$, the deviations of the velocity field from the Hubble expansion will be small. We expect that in this case, physically relevant solution of the above equations are those which deviate only slightly from the homogeneous case.

It is convenient to consider the problem in comoving coordinates; hence we define, as in (4.4),

$$\boldsymbol{r} = a(t)\,\boldsymbol{x}\,.$$

In a homogeneous cosmos, $\boldsymbol{x}$ is a constant for every matter particle, and its spatial position $\boldsymbol{r}$ changes only due to the Hubble expansion. Likewise, the velocity field is written in the form

$$\boldsymbol{v}(\boldsymbol{r}, t) = \frac{\dot{a}}{a}\boldsymbol{r} + \boldsymbol{u}\left(\frac{\boldsymbol{r}}{a}, t\right)\,, \tag{7.5}$$

where $\boldsymbol{u}(\boldsymbol{x}, t)$ is a function of the comoving coordinate $\boldsymbol{x}$. In (7.5), the first term represents the homogeneous Hubble expansion, whereas the second term describes the deviations from this homogeneous expansion. For this reason, $\boldsymbol{u}$ is called the *peculiar velocity*.

Transforming the Fluid Equations to Comoving Coordinates. We will now show how the above equations read in comoving coordinates. For this, we first note that the partial derivative $\partial/\partial t$ in (7.2) means a time derivative at fixed $\boldsymbol{r}$. If the equations are to be written in comoving coordinates, this partial time derivative needs to be transformed into one where $\boldsymbol{x}$ is kept fixed. For example,

$$\begin{aligned}\left(\frac{\partial}{\partial t}\right)_r \rho(\boldsymbol{r}, t) &= \left(\frac{\partial}{\partial t}\right)_r \rho_x\left(\frac{\boldsymbol{r}}{a}, t\right) \\ &= \left(\frac{\partial}{\partial t}\right)_x \rho_x(\boldsymbol{x}, t) - \frac{\dot{a}}{a}\boldsymbol{x}\cdot\nabla_x \rho_x(\boldsymbol{x}, t)\,,\end{aligned} \tag{7.6}$$

where ∇_x is the gradient with respect to comoving coordinates, and where we define the function $\rho_x(\boldsymbol{x}, t) \equiv \rho(a\boldsymbol{x}, t)$. Note that $\rho_x(\boldsymbol{x}, t)$ and $\rho(\boldsymbol{x}, t)$ both describe the same *physical* density field, but that ρ and ρ_x are different *mathematical* functions of their arguments. After these transformations, (7.2) becomes

$$\frac{\partial \rho}{\partial t} + \frac{3\dot{a}}{a}\rho + \frac{1}{a}\nabla\cdot(\rho\boldsymbol{u}) = 0\,, \tag{7.7}$$

where from now on all spatial derivatives are to be considered with respect to $\boldsymbol{x}$. For notational simplicity we from now on set $\rho \equiv \rho_x$ and $\delta \equiv \delta(\boldsymbol{x}, t)$, and note that the partial time derivative is to be understood to mean at fixed $\boldsymbol{x}$. Writing $\rho = \overline{\rho}(1+\delta)$ and using $\overline{\rho} \propto a^{-3}$, (7.7) reads in comoving coordinates

$$\frac{\partial \delta}{\partial t} + \frac{1}{a}\nabla\cdot[(1+\delta)\,\boldsymbol{u}] = 0\,. \tag{7.8}$$

Accordingly, the gravitational potential Φ is written as

$$\Phi(\boldsymbol{r}, t) = \frac{2\pi}{3}G\overline{\rho}(t)|\boldsymbol{r}|^2 + \phi(\boldsymbol{x}, t)\,; \tag{7.9}$$

the first term is the Newtonian potential for a homogeneous density field, and ϕ satisfies the Poisson equation for the density inhomogeneities,

$$\begin{aligned}\nabla^2\phi(\boldsymbol{x}, t) &= 4\pi G a^2(t)\overline{\rho}(t)\delta(\boldsymbol{x}, t) \\ &= \frac{3H_0^2\Omega_{\rm m}}{2a(t)}\,\delta(\boldsymbol{x}, t)\,,\end{aligned} \tag{7.10}$$

where in the last step we used $\overline{\rho} \propto a^{-3}$ and the definition of the density parameter $\Omega_{\rm m}$. Then, the Euler equation (7.3) becomes

$$\frac{\partial \boldsymbol{u}}{\partial t} + \frac{\boldsymbol{u}\cdot\nabla}{a}\boldsymbol{u} + \frac{\dot{a}}{a}\boldsymbol{u} = -\frac{1}{\overline{\rho}\,a}\nabla P - \frac{1}{a}\nabla\phi\,, \tag{7.11}$$

where (4.13) has been utilized.

Linearization. In the homogeneous case, $\delta \equiv 0$, $\boldsymbol{u} \equiv 0$, $\phi \equiv 0$, $\rho = \overline{\rho}$, and (7.7) then implies $\dot{\overline{\rho}} + 3H\overline{\rho} = 0$, which also follows immediately from (4.17) in the case of $P = 0$. Now we will look for approximate solutions of the above set of equations which describe only small deviations from this homogeneous solution. For this reason, in these equations we only consider first-order terms in the small parameters δ and $\boldsymbol{u}$, i.e., we disregard terms that contain $\boldsymbol{u}\delta$ or are quadratic in the velocity $\boldsymbol{u}$. After this linearization, we can eliminate the peculiar velocity $\boldsymbol{u}$ and the gravitational potential ϕ from the equations[2] and then obtain a second-order differential equation for the density contrast δ,

$$\frac{\partial^2 \delta}{\partial t^2} + \frac{2\dot{a}}{a}\frac{\partial \delta}{\partial t} = 4\pi G\overline{\rho}\delta\,. \tag{7.12}$$

It is remarkable that neither does this equation contain derivatives with respect to spatial coordinates, nor do the coefficients in the equation depend on $\boldsymbol{x}$. Therefore, (7.12) has solutions of the form

$$\delta(\boldsymbol{x}, t) = D(t)\,\tilde{\delta}(\boldsymbol{x})\,,$$

[2] For this, the linearized form of (7.8), $\partial\delta/\partial t + a^{-1}\nabla\cdot\boldsymbol{u} = 0$, is differentiated with respect to time and combined with the divergence of the linearized form of equation (7.11) for the pressure-free case, $\partial\boldsymbol{u}/\partial t + H\boldsymbol{u} = -a^{-1}\nabla\phi$. Finally, the Laplacian of ϕ is replaced by the Poisson equation (7.10).

i.e., the spatial and temporal dependences factorize in these solutions. Here, $\tilde{\delta}(\boldsymbol{x})$ is an arbitrary function of the spatial coordinate, and $D(t)$ satisfies the equation

$$\ddot{D} + \frac{2\dot{a}}{a}\dot{D} - 4\pi G\overline{\rho}(t)\, D = 0 \,. \tag{7.13}$$

The Growth Factor. The differential equation (7.13) has two linearly independent solutions. One can show that one of them increases with time, whereas the other decreases. If, at some early time, both functional dependences were present, the increasing solution will dominate at later times, whereas the solution decreasing with t will become irrelevant. Therefore, we will consider only the increasing solution, which is denoted by $D_+(t)$, and normalize it such that $D_+(t_0) = 1$. Then, the density contrast becomes

$$\delta(\boldsymbol{x}, t) = D_+(t)\, \delta_0(\boldsymbol{x}) \,. \tag{7.14}$$

This mathematical consideration allows us to draw immediately a number of conclusions. First, the solution (7.14) indicates that in linear perturbation theory *the spatial shape of the density fluctuations is frozen in comoving coordinates*, only their amplitude increases. The *growth factor* $D_+(t)$ of the amplitude follows a simple differential equation that is easily solvable for any cosmological model. In fact, one can show that for arbitrary values of the density parameter in matter and vacuum energy, the growth factor has the form

$$D_+(a) \propto \frac{H(a)}{H_0} \int_0^a \frac{\mathrm{d}a'}{\left[\Omega_\mathrm{m}/a' + \Omega_\Lambda a'^2 - (\Omega_\mathrm{m} + \Omega_\Lambda - 1)\right]^{3/2}} \,,$$

where the factor of proportionality is determined from the condition $D_+(t_0) = 1$.

In accordance with $D_+(t_0) = 1$, $\delta_0(\boldsymbol{x})$ would be the distribution of density fluctuations today if the evolution was indeed linear until the present epoch. Therefore, $\delta_0(\boldsymbol{x})$ is denoted as the *linearly extrapolated density fluctuation field*. However, the linear approximation breaks down if $|\delta|$ is no longer $\ll 1$. In this case, the terms that have been neglected in the above derivations are no longer small and have to be included. The problem then becomes *considerably* more difficult and defies analytical treatment. Instead one needs, in general, to rely on numerical procedures for analyzing the growth of density perturbations. Furthermore, it shall be noted once again that, for large density perturbations, the fluid approximation is no longer valid, and that up to now we have assumed the Universe to be matter dominated. At early times, i.e., for $z \gtrsim z_\mathrm{eq}$ (see Eq. 4.54), this assumption becomes invalid, so that the above equations need to be modified for these early epochs.

Example: Einstein–de Sitter Model. In the special case of a universe with $\Omega_\mathrm{m} = 1$, $\Omega_\Lambda = 0$, (7.13) can be solved explicitly. In this case, $a(t) = (t/t_0)^{2/3}$, so that

$$\left(\frac{\dot{a}}{a}\right) = \frac{2}{3t} \,, \quad \text{and } \overline{\rho}(t) = a^{-3}\rho_\mathrm{cr} = \frac{3H_0^2}{8\pi G}\left(\frac{t}{t_0}\right)^{-2} \,;$$

furthermore, in this model $t_0\, H_0 = 2/3$, so that (7.13) reduces to

$$\ddot{D} + \frac{4}{3t}\dot{D} - \frac{2}{3t^2}D = 0 \,. \tag{7.15}$$

This equation is easily solved by making the ansatz $D \propto t^q$; this ansatz is suggested because (7.15) is equidimensional in t, i.e., each term has the dimension $D/(\mathrm{time})^2$. Inserting into (7.15) yields a quadratic equation for q,

$$q(q-1) + \frac{4}{3}q - \frac{2}{3} = 0 \,,$$

with solutions $q = 2/3$ and $q = -1$. The latter corresponds to fluctuations decreasing with time and will be disregarded in the following. So, for the Einstein–de Sitter model, the increasing solution

$$D_+(t) = \left(\frac{t}{t_0}\right)^{2/3} = a(t) \,, \tag{7.16}$$

is found, i.e., in this case the growth factor equals the scale factor. For different cosmological parameters this is not the case, but the qualitative behavior is quite similar, which is demonstrated in Fig. 7.3 for three models. In particular, fluctuations were able to grow by a factor ~ 1000 from the epoch of recombination at $z \sim 1000$, from which the CMB photons originate, to the present day.

Evidence for Dark Matter on Cosmic Scales. At the present epoch, $\delta \gg 1$ certainly on scales of clusters of galaxies (~ 2 Mpc), and $\delta \sim 1$ on scales of superclusters

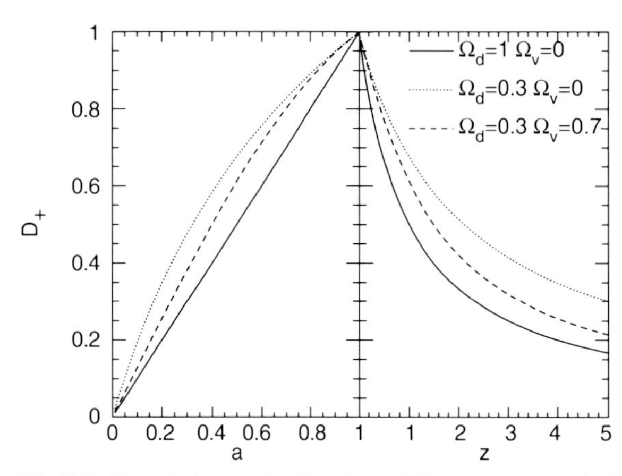

Fig. 7.3. Growth factor D_+ for three different cosmological models, as a function of the scale factor a (left panel) and of redshift (right panel). It is clearly visible how quickly D_+ decreases with increasing redshift in the EdS model, in comparison to the models of lower density

($\sim$ 10 Mpc). Hence, because of the law of linear structure growth (7.14) and the behavior of $D_+(t)$ shown in Fig. 7.3, we would expect $\delta \gtrsim 10^{-3}$ at $z = 1000$ for these structures to be able to grow to non-linear structures at the current epoch. For this reason, we should also expect CMB fluctuations to be of comparable magnitude, $\Delta T/T \gtrsim 10^{-3}$. The observed fluctuation amplitude is $\Delta T/T \sim 10^{-5}$, however. The corresponding density fluctuations therefore cannot have grown sufficiently strongly up to today to form non-linear structures.

This contradiction can be resolved by the dominance of dark matter. Since photons interact with baryonic matter only, the CMB anisotropies basically provide (at least on angular scales below $\sim 1°$) information on the density contrast of *baryons*. Dark matter may have had a higher density contrast at recombination and may have formed potential wells, into which the baryons then "fall" after recombination.

7.3 Description of Density Fluctuations

We will now examine the question of how to describe an inhomogeneous universe quantitatively, i.e., how to quantify the structures it contains. This task sounds easier at first sight than it is in reality. One has to realize that the aim of such a theoretical description cannot be to describe the complete function $\delta(\boldsymbol{x}, t)$ for a particular universe. No model of the Universe will be able to describe, for instance, the matter distribution in the vicinity of the Milky Way in detail. No model based on the laws of physics alone will be able to predict that at a distance of $\sim$ 800 kpc from the Galaxy a second massive spiral galaxy is located, because this specific feature of our local Universe depends on the specific initial conditions of the matter distribution in the early Universe. We can at best hope to predict the statistical properties of the mass distribution, such as, for example, the average number density of clusters of galaxies above a given mass, or the probability of a massive galaxy being found within 800 kpc of another one. Likewise, numerical sim ulations of the Universe (see below) cannot reproduce our Universe; instead, they are at best able to generate cosmological models that have the same statistical properties as our Universe.

It is quite obvious that a very large number of statistical properties exist for the density field, all of which we can examine and which we hope can be explained quantitatively by the correct model of structure formation in the Universe. To make any progress at all, the statistical properties need to be sorted or classified. How can the statistical properties of a density field best be described?

Two universes are considered equivalent if their density fields δ have the same statistical properties. One may then imagine considering a large (statistical) ensemble of universes whose density fields all have the same statistical properties, but for which the individual functions $\delta(\boldsymbol{x})$ are all different. This statistical ensemble is called a *random field*, and any individual distribution with the respective statistical properties is called a *realization of the random field*.

An example may clarify these concepts. We consider the waves on the surface of a large lake. The statistical properties of these waves – such as how many of them there are with a certain wavelength, and how their amplitudes are distributed – depend on the shape of the lake, its depth, and the strength and direction of the wind blowing over its surface. If we assume that the wind properties are not changing with time, the statistical properties of the water surface are constant over time. Of course, this does not mean that the amplitude of the surface as a function of position is constant.

Rather, it means that two photographs of the surface that are taken at different times are statistically indistinguishable: the distribution of the wave amplitudes will be the same, and there is no way of deciding which of the snapshots was taken first. Knowing the surface topography and the wind properties sufficiently well, one is able to compute the distribution of the wave amplitudes, but there is no way to predict the amplitude of the surface of the lake as a function of position at a particular time. Each snapshot of the lake is a realization of the random field, which in turn is characterized by the statistical properties of the waves.

7.3.1 Correlation Functions

Galaxies are not randomly distributed in space, but rather they gather in groups, clusters, or even larger structures. Phrased differently, this means that the probability of finding a galaxy at location $\boldsymbol{x}$ is not independent of whether there is a galaxy in the vicinity of $\boldsymbol{x}$. It is more probable to find a galaxy in the vicinity of another one than at an arbitrary location. This phenomenon is described such that one considers two points $\boldsymbol{x}$ and $\boldsymbol{y}$, and two volume elements $\mathrm{d}V$ around these points. If $\overline{n}$ is the average number density of galaxies, the probability of finding a galaxy in the volume element $\mathrm{d}V$ around $\boldsymbol{x}$ is then

$$P_1 = \overline{n}\,\mathrm{d}V\,,$$

independent of $\boldsymbol{x}$ if we assume that the Universe is statistically homogeneous. We choose $\mathrm{d}V$ such that $P_1 \ll 1$, so that the probability of finding two or more galaxies in this volume element is negligible.

The probability of finding a galaxy in the volume element $\mathrm{d}V$ at location $\boldsymbol{x}$ and at the same time finding a galaxy in the volume element $\mathrm{d}V$ at location $\boldsymbol{y}$ is then

$$P_2 = (\overline{n}\,\mathrm{d}V)^2\left[1+\xi_{\mathrm{g}}(\boldsymbol{x},\boldsymbol{y})\right]\,. \tag{7.17}$$

If the distribution of galaxies was uncorrelated, the probability P_2 would simply be the product of the probabilities of finding a galaxy at each of the locations $\boldsymbol{x}$ and $\boldsymbol{y}$ in a volume element $\mathrm{d}V$, so $P_2 = P_1^2$. But since the distribution is correlated, the relation does not apply in this simple form; rather, it needs to be modified, as was done in (7.17). Equation (7.17) defines the *two-point correlation function* (or simply "correlation function") of galaxies $\xi_{\mathrm{g}}(\boldsymbol{x},\boldsymbol{y})$.

By analogy to this, the correlation function for the total matter density can be defined as

$$\begin{aligned}\langle\rho(\boldsymbol{x})\,\rho(\boldsymbol{y})\rangle &= \overline{\rho}^2\,\langle[1+\delta(\boldsymbol{x})]\,[1+\delta(\boldsymbol{y})]\rangle\\ &= \overline{\rho}^2\,(1+\langle\delta(\boldsymbol{x})\,\delta(\boldsymbol{y})\rangle)\\ &=: \overline{\rho}^2\,[1+\xi(\boldsymbol{x},\boldsymbol{y})]\,,\end{aligned} \tag{7.18}$$

because the mean (or expectation) value $\langle\delta(\boldsymbol{x})\rangle = 0$ for all locations $\boldsymbol{x}$.

In the above equations, angular brackets denote averaging over an ensemble of distributions that all have identical statistical properties. In our example of the lake, the correlation function of the wave amplitudes at positions $\boldsymbol{x}$ and $\boldsymbol{y}$, for instance, would be determined by taking a large number of snapshots of its surface and then averaging the product of the amplitudes at these two locations over all these realizations.

Since the Universe is considered statistically homogeneous, ξ can only depend on the difference $\boldsymbol{x}-\boldsymbol{y}$ and not on $\boldsymbol{x}$ and $\boldsymbol{y}$ individually. Furthermore, ξ can only depend on the separation $r = |\boldsymbol{x}-\boldsymbol{y}|$, and not on the direction of the separation vector $\boldsymbol{x}-\boldsymbol{y}$ because of the assumed statistical isotropy of the Universe. Therefore, $\xi = \xi(r)$ is simply a function of the separation between two points.

For a homogeneous random field, the ensemble average can be replaced by spatial averaging, i.e., the correlation function can be determined by averaging over the density products for a large number of pairs of points with given separation r. The equivalence of ensemble average and spatial average is called the ergodicity of the random field. Only by this can the correlation function (and all other statistical properties) in our Universe be measured at all, because we are able to observe only a single – namely our – realization of the hypothetical ensemble. From the measured correlations between galaxy positions, as determined from spectroscopic redshift surveys of galaxies (see Sect. 8.1.2), one finds the approximate relation

$$\xi_{\mathrm{g}}(r) = \left(\frac{r}{r_0}\right)^{-\gamma}\,, \tag{7.19}$$

for galaxies of luminosity $\sim L^*$ (see Fig. 7.4), where $r_0 \simeq 5h^{-1}\,\mathrm{Mpc}$ denotes the correlation length, and

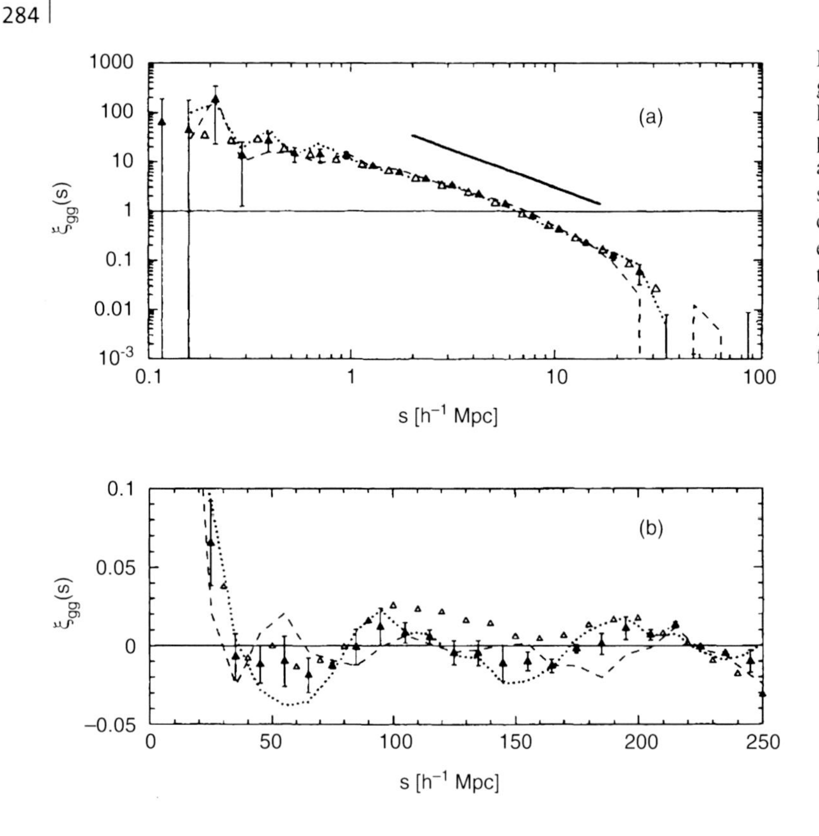

Fig. 7.4. The correlation function ξ_g of galaxies, as it was determined from the Las Campanas Redshift Survey. In the top panel, ξ_g is shown for small and intermediate separations, whereas the bottom panel shows it for large separations. Dashed and dotted lines indicate the northern and southern part, respectively, of the survey, and the solid triangles denote the correlation function obtained from combining both. A power law with slope $\gamma = 1.52$ is plotted for comparison (bold solid curve)

where the slope is about $\gamma \simeq 1.8$. This relation is approximately valid over a range of separations $2h^{-1}\,\text{Mpc} \lesssim r \lesssim 30h^{-1}\,\text{Mpc}$.

Hence, the correlation function provides a means to characterize the structure of the matter distribution in the Universe. Besides this two-point correlation function, correlations of higher order may also be defined, leading to general n-point correlation functions. These are more difficult to determine from observation, though. It can be shown that the statistical properties of a random field are fully specified by the set of all n-point correlations.

7.3.2 The Power Spectrum

An alternative (and equivalent) description of the statistical properties of a random field, and thus of the structure of the Universe, is the *power spectrum* $P(k)$. Roughly speaking, the power spectrum $P(k)$ describes the level of structure as a function of the length-scale $L \simeq 2\pi/k$; the larger $P(k)$, the larger the amplitude of the fluctuations on a length-scale $2\pi/k$. Here, k is a *wave number*. Phrased differently, the density fluctuations are decomposed into a sum of plane waves of the form $\delta(\boldsymbol{x}) = \sum a_{\boldsymbol{k}} \cos(\boldsymbol{x} \cdot \boldsymbol{k})$, with a wave vector $\boldsymbol{k}$ and an amplitude $a_{\boldsymbol{k}}$. The power spectrum $P(k)$ then describes the distribution of amplitudes with equal $k = |\boldsymbol{k}|$. Technically speaking, this is a Fourier decomposition. Referring back to the example of waves on the surface of a lake, one finds that a characteristic wavelength L_c exists, which depends, among other factors, on the wind speed. In this case, the power spectrum will have a prominent maximum at $k = 2\pi/L_c$.

The power spectrum $P(k)$ and the correlation function are related through a Fourier transform; formally, one has[3]

$$P(k) = 2\pi \int_0^\infty \mathrm{d}r\, r^2\, \frac{\sin kr}{kr}\, \xi(r)\,, \tag{7.20}$$

[3]This may not look like a "standard" Fourier transform on first sight. However, the relation between $P(k)$ and $\xi(r)$ is given by a three-dimensional Fourier transform. Since the correlation function depends only on the separation $r = |\boldsymbol{r}|$, the two integrals over the angular coordinates can be performed explicitly, leading to the form of (7.20).

i.e., the integral over the correlation function with a weight factor depending on $k \sim 2\pi/L$. This relation can also be inverted, and thus $\xi(r)$ can be computed from $P(k)$.

In general, knowing the power spectrum is not sufficient to unambiguously describe the statistical properties of any random field – in the same way as the correlation function $\xi(r)$ only provides an incomplete characterization. However, random fields do exist, so-called *Gaussian random fields*, which are uniquely characterized by $P(k)$. Such Gaussian random fields play an important role in cosmology because it is assumed that at very early epochs, the density field obeyed Gaussian statistics.

7.4 Evolution of Density Fluctuations

$P(k)$ and $\xi(r)$ both depend on cosmological time or redshift because the density field in the Universe evolves over time. Therefore, the dependence on t is explicitly written $P(k,t)$ and $\xi(r,t)$. Note that $P(k,t)$ is linearly related to $\xi(r,t)$, according to (7.20), and ξ in turn depends quadratically on the density contrast δ. If we interpret $\boldsymbol{x}$ as a *comoving* separation vector, from (7.14) we then know the time dependence of the density fluctuations, $\delta(\boldsymbol{x},t) = D_+(t)\delta_0(\boldsymbol{x})$. Thus, within the scope of the validity of (7.14),

$$\xi(x,t) = D_+^2(t)\,\xi(x,t_0)\,, \tag{7.21}$$

and accordingly

$$P(k,t) = D_+^2(t)\,P(k,t_0) =: D_+^2(t)\,P_0(k)\,, \tag{7.22}$$

where k is a *comoving wave number*. We shall stress once again that these relations are valid only in the framework of Newtonian, linear perturbation theory in the matter dominated era of the Universe, to which we had restricted ourselves in Sect. 7.2.2. Equation (7.22) states that the knowledge of $P_0(k)$ is sufficient to obtain the power spectrum $P(k,t)$ at any time, again within the framework of linear perturbation theory.

7.4.1 The Initial Power Spectrum

The Harrison–Zeldovich Spectrum. Initially it may seem as if $P_0(k)$ is a function that can be chosen arbitrarily, but one objective of cosmology is to calculate this power spectrum and to compare it to observations. More than thirty years ago, arguments were already developed to specify the functional form of the initial power spectrum.

At early times, the expansion of the Universe follows a power law, $a(t) \propto t^{1/2}$ in the radiation-dominated era. At that time, no natural length-scale existed in the Universe to which one might compare a wavelength. The only mathematical function that depends on a length but does not contain any characteristic scale is a power law;[4] hence for very early times one should expect

$$P(k) \propto k^{n_s}\,. \tag{7.23}$$

Many years ago, Harrison, Zeldovich, Peebles and others argued, based on scaling relations, that it should be $n_s = 1$. For this reason, the spectrum (7.23) with $n_s = 1$ is called *Harrison–Zeldovich spectrum*. With such a spectrum, we may choose a time t_i after the inflationary epoch and write

$$P(k,t_i) = D_+^2(t_i)\,A\,k^{n_s}\,, \tag{7.24}$$

where A is a normalization constant that cannot be determined from theory but has to be fixed by observations. Assuming the validity of (7.22),

$$P_0(k) = A\,k^{n_s}$$

would then apply.

The Transfer Function. This relation above needs to be modified for several reasons. In linear perturbation theory, which led to $\delta(\boldsymbol{x},t) = D_+(t)\,\delta_0(\boldsymbol{x})$, we assumed the validity of Newtonian dynamics, considered only the matter-dominated epoch of the Universe, and disregarded any pressure terms. The evolution of perturbations in the radiation-dominated cosmos proceeds differently though, also depending on the scale of the perturbations in comparison to the length of the horizon, so that a correction term of the form

$$P_0(k) = A\,k^{n_s}\,T^2(k) \tag{7.25}$$

[4] You can convince yourself of this by trying to find another type of function of a scale that does not involve a characteristic length; e.g., $\sin x$ does not work if x is a length, since the sine of a length is not defined; one thus needs something like $\sin(x/x_0)$, hence introducing a length-scale. The same arguments apply to other functions, such as the logarithm, the exponential, etc. Also note that the sum of two power laws, e.g., $Ax^\alpha + Bx^\beta$ defines a characteristic scale, namely that value of x where the two terms become equal.

needs to be introduced. $T(k)$ is called the *transfer function*; it can be computed for any cosmological model if the matter content of the Universe is specified. In particular, $T(k)$ depends on the nature of dark matter. One distinguishes between *cold dark matter* (CDM) and *hot dark matter* (HDM). These two kinds of dark matter differ in the thermal velocities of their constituents at time t_{eq}, when radiation and matter had equal density. The particles of CDM were non-relativistic at this time, whereas those of HDM had velocities of order c. If dark matter consists of weakly interacting elementary particles, the difference between CDM and HDM depends on the mass m of the particles. Assuming that the "temperature" of the dark matter particles is close to the temperature of the Universe, then a particle mass m satisfying the relation

$$mc^2 \gg k_B T(t_{eq}) \simeq k_B \times 2.73\,\mathrm{K}\,(1+z_{eq}) = k_B \times 2.73\,\mathrm{K} \times 23\,900\,\Omega_m h^2 \sim 6\Omega_m h^2\,\mathrm{eV}$$

indicates CDM, whereas HDM is characterized by the opposite inequality, i.e., $mc^2 \ll k_B T(t_{eq})$; for instance, neutrinos belong to HDM. The important distinction between HDM and CDM follows from the considerations below.

If density fluctuations become too large on a certain scale, linear perturbation theory breaks down and (7.25) is no longer valid. Then the true current power spectrum $P(k, t_0)$ will deviate from $P_0(k)$. Nevertheless, in this case it is still useful to examine $P_0(k)$ – it is then called the *linearly extrapolated power spectrum*.

7.4.2 Growth of Density Perturbations

Within the framework of linear Newtonian perturbation theory in the "cosmic fluid", $\delta(\boldsymbol{x}, t) = D_+(t)\,\delta_0(\boldsymbol{x})$ applies. Modifications to this behavior are necessary for several reasons:

- If dark matter consists of relativistic particles, these are not gravitationally bound in the potential well of a density concentration. In this case, they are able to move freely and to escape from the potential well, which in the end leads to its dissolution if these particles dominate the matter overdensity. From this, it follows immediately that for HDM small-scale density perturbations cannot form. For CDM this effect of *free-steaming* does not occur.
- At redshifts $z \gtrsim z_{eq}$, radiation dominates the density of the Universe. Since the expansion law $a(t)$ is then distinctly different from that in the matter-dominated phase, the growth rate for density fluctuations will also change.
- As discussed in Sect. 4.5.2, a horizon exists with comoving scale $r_{H,com}(t)$. Physical interactions can take place only on scales smaller than $r_{H,com}(t)$. For fluctuations of length-scales $L \sim 2\pi/k \gtrsim r_{H,com}(t)$, Newtonian perturbation theory will cease to be valid, and one needs to apply linear perturbation theory in the framework of the General Relativity.

CDM and HDM. The first of the above points immediately implies that a clear difference must exist between HDM and CDM models as regards structure formation and evolution. In HDM models, small-scale fluctuations are washed out by free-streaming of relativistic particles, i.e., the power spectrum is completely suppressed for large k, which is expressed by the transfer function $T(k)$ decreasing exponentially for large k. In the context of such a theory, very large structures will form first, and galaxies can form only later by fragmentation of large structures. However, this formation scenario is in clear contradiction with observations. For example, we observe galaxies and QSOs at $z \sim 6$ so that small-scale structure is already present at times when the Universe had less than 10% of its current age. In addition, the observed correlation function of galaxies, both in the local Universe (see Fig. 7.4) and at higher redshift, is incompatible with cosmological models in which the dark matter is composed mainly of HDM.

Hot dark matter leads to structure formation that does not agree with observation. Therefore we can exclude HDM as the dominant constituent of dark matter. For this reason, it is now commonly assumed that the dark matter is "cold". The achievements of the CDM scenario in the comparison between model predictions and observations fully justify this assumption.

We shall elaborate on the last statement in quite some detail in Chap. 8.

In linear perturbation theory, fluctuations grow on all scales, or for all wave numbers, independent of each other. This applies not only in the Newtonian case, but also remains valid in the framework of General Relativity as long as the fluctuation amplitudes are small. Therefore, the behavior on any (comoving) length-scale can be investigated independently of the other scales. At very early times, perturbations with a comoving scale L are larger than the (comoving) horizon, and only for $z < z_{\rm enter}(L)$ does the horizon become larger than the considered scale L. Here, $z_{\rm enter}(L)$ is defined as the redshift at which the (comoving) horizon equals the (comoving) length-scale L,

$$r_{\rm H,com}(z_{\rm enter}(L)) = L \; . \tag{7.26}$$

It is common to say that at $z_{\rm enter}(L)$ the perturbation under consideration "enters the horizon", whereas actually the process is the opposite – the horizon outgrows the perturbation. Relativistic perturbation theory shows that density fluctuations of scale L grow as long as $L > r_{\rm H,com}$, namely $\propto a^2$ if radiation dominates (thus, if $z > z_{\rm eq}$), or $\propto a$ if matter dominates (thus, if $z < z_{\rm eq}$). Free-streaming particles or pressure gradients cannot impede the growth on scales larger than the horizon length because, according to the definition of the horizon, physical interactions – which pressure or free-steaming particles would be – cannot extend to scales larger than the horizon size.

Qualitative Behavior of the Transfer Function. The behavior of the growth of a density perturbation on a scale L for $z < z_{\rm enter}(L)$ depends on $z_{\rm enter}$ itself. If a perturbation enters the horizon in the radiation-dominated phase, $z_{\rm eq} \lesssim z_{\rm enter}(L)$, the fluctuation cannot grow during the epoch $z_{\rm eq} \lesssim z \lesssim z_{\rm enter}(L)$. In this period, the energy density in the Universe is dominated by radiation, and the resulting expansion rate prevents an efficient perturbation growth. At later epochs, when $z \lesssim z_{\rm eq}$, the growth of density perturbation continues. If $z_{\rm enter}(L) \lesssim z_{\rm eq}$, thus if the perturbation enters the horizon during the matter-dominated epoch of the Universe, these perturbations will grow as described in Sect. 7.2.2, with $\delta \propto D_+(t)$. This implies that a length-scale L_0 is singled out, namely the one for which

$$z_{\rm eq} = z_{\rm enter}(L_0) \; , \tag{7.27}$$

or

$$L_0 = r_{\rm H,com}(z_{\rm eq}) = \frac{c}{\sqrt{2}H_0} \frac{1}{\sqrt{(1+z_{\rm eq})\Omega_{\rm m}}} \simeq \frac{c}{\sqrt{2}H_0} \frac{1}{\sqrt{23\,900\,\Omega_{\rm m}h}} \simeq 12\left(\Omega_{\rm m}h^2\right)^{-1}\,{\rm Mpc}\; , \tag{7.28}$$

where the expression for $r_{\rm H,com}(z)$ generalizes (4.69), and where (4.54) has been used for $z_{\rm eq}$.

Density fluctuations with $L > L_0$ enter the horizon after matter started to dominate the energy density of the Universe; hence their growth is not impeded by a phase of radiation-dominance. In contrast, density fluctuations with $L < L_0$ enter the horizon at a time when radiation dominates. These then cannot grow further as long as $z > z_{\rm eq}$, and only in the matter-dominated epoch will their amplitudes proceed to grow again. Their relative growth up to the present time has therefore grown by a smaller factor than that of fluctuations with $L > L_0$ (see Fig. 7.5). The quantitative consideration of these effects allows us to compute the transfer function. In general, this needs to be done numerically, but very

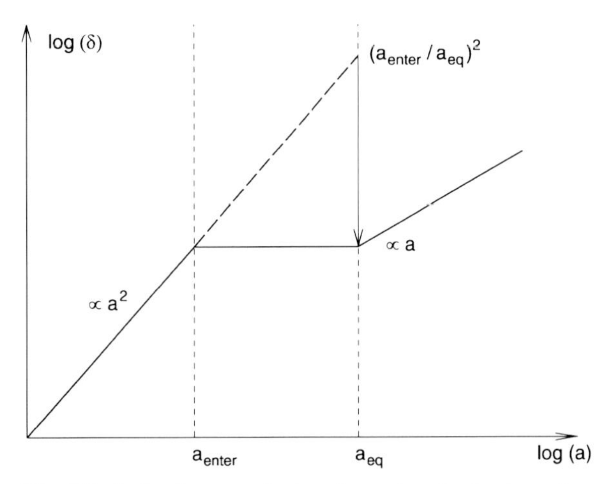

Fig. 7.5. A density perturbation that enters the horizon during the radiation-dominated epoch of the Universe ceases to grow until matter starts to dominate the energy content of the Universe. In comparison to a perturbation that enters the horizon later, during the matter-dominated epoch, the amplitude of the smaller perturbation is suppressed by a factor $(a_{\rm eq}/a_{\rm enter})^2$, which explains the qualitative behavior (7.29) of the transfer function

good approximations exist. Two limiting cases are easily treated analytically,

$$T(k) \approx 1 \text{ for } k \ll 1/L_0 ,$$
$$T(k) \approx (kL_0)^{-2} \text{ for } k \gg 1/L_0 ; \qquad (7.29)$$

but the important point is:

> In the framework of the CDM model, the transfer function can be computed, and thus, by means of (7.19), also the power spectrum of the density fluctuations as a function of length-scale and redshift. The amplitude of the power spectrum has to be obtained from observations.

The Shape Parameter. The transfer function depends on the combination kL_0, which is the inverse of the ratio of the length-scale under consideration ($\sim 2\pi/k$) and the horizon scale at the epoch of equality, and thus on $k(\Omega_m h^2)^{-1}$. Since distances determined from redshift are measured in units of h^{-1} Mpc, the shape of the transfer function, and thus also that of the power spectrum, depends on $\Gamma = \Omega_m h$. Γ is called the *shape parameter* of the power spectrum. It is sometimes used as a free parameter instead of being identified with $\Omega_m h$. A detailed analysis shows that Γ depends also on Ω_b, but since $\Omega_b \lesssim 0.05$ is small, according to primordial nucleosynthesis (see Sect. 4.4.4), this effect is relatively small and often neglected.

If the galaxy distribution follows the distribution of dark matter, the former can be used to determine the correlation function or the power spectrum. Both from the distribution of galaxies projected onto the sphere (angular correlation function) and from its three-dimensional distribution (which is determined from redshift surveys), values in the range $\Gamma \sim 0.15$–0.25 are found. From $T(k) \approx 1$ for $kL_0 \ll 1$, and with (7.24), we find that $P(k) \propto k$ for $kL_0 \ll 1$. This behavior is compatible with the CMB anisotropy measurements by COBE on large scales, as we will discuss in detail in Chap. 8.

In Fig. 7.6, the power spectrum is plotted for several cosmological models that have different density parameter, shape parameter, and normalization of the power spectrum. The thin curves show $P(k)$ as derived from linear perturbation theory, and the bold curves display the power spectrum with non-linear structure evolution taken into account. The power spectra displayed all have a characteristic wave number at which the slope of $P(k)$ changes. It is specified by $\sim 2\pi/L_0$, with the characteristic length L_0 being defined in (7.28).

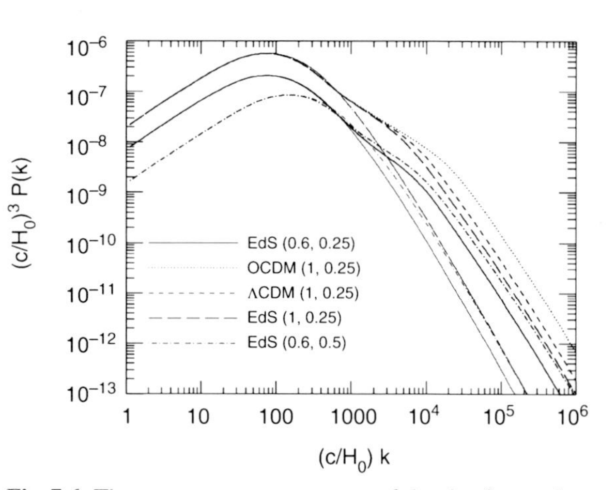

Fig. 7.6. The current power spectrum of density fluctuations for CDM models. The wave number k is given in units of H_0/c, and $(H_0/c)^3$ $P(k)$ is dimensionless. The various curves have different cosmological parameters: EdS: $\Omega_m = 1$, $\Omega_\Lambda = 0$; OCDM: $\Omega_m = 0.3$, $\Omega_\Lambda = 0$; ΛCDM: $\Omega_m = 0.3$, $\Omega_\Lambda = 0.7$. The values in parentheses specify (σ_8, Γ), where σ_8 is the normalization of the power spectrum (which will be discussed below), and where Γ is the shape parameter. The thin curves correspond to the power spectrum $P_0(k)$ linearly extrapolated to the present day, and the bold curves take the non-linear evolution into account

Besides pure CDM and HDM models (the latter being excluded by observation), one can consider models which are dominated by CDM, but which have a (small) contribution by HDM; these are called mixed dark matter (MDM) models. Such a contribution has indeed now become part of the standard model, due to the detected finite rest mass of neutrinos which implies $0 < \Omega_\nu \ll 1$. With this contribution, $T(k)$ is changed in such a way that small scales (i.e., large k) are slightly suppressed in the power spectrum. We will see later that by observing the power spectrum we can constrain the rest mass of neutrinos very well, and cosmological observations provide, in fact, by far the most stringent mass limits for neutrinos.

Density Distribution of Baryons. The evolution of density fluctuations of baryons differs from that of dark matter. The reason for this is essentially the interaction of baryons with photons: although matter dominates the Universe for $z < z_{eq}$, the density of baryons remains smaller than that of radiation for a long time, until after recombination begins. Since photons and baryons interact with each other by photon scattering on free electrons, which again are tightly coupled electromagnetically to protons and helium nuclei, and since radiation cannot fall into the potential wells of dark matter, baryons are hindered from doing so as well. Hence, the baryons are subject to radiation pressure. For this reason, the density distribution of baryons is initially much smoother than that of dark matter. Only after recombination does the interaction of baryons with photons cease to exist, and the baryons can fall into the potential wells of dark matter, i.e., some time later the distribution of baryons will closely resemble that of the dark matter.

The linear theory of the evolution of density fluctuations will break down at the latest when $|\delta| \sim 1$; the above equations for the power spectrum $P(k, t)$ are therefore valid only if the respective fluctuations are small. However, very accurate fitting formulae now exist for $P(k, t)$ which are also valid in the non-linear regime. For some cosmological models, the non-linear power spectrum is displayed in Fig. 7.6.

7.5 Non-Linear Structure Evolution

Linear perturbation theory has a limited range of applicability; in particular, the evolution of structures like clusters of galaxies cannot be treated within the framework of linear perturbation theory. One might imagine that one can evolve the system of equations (7.2)–(7.4) to higher orders in the small variables δ and $|\boldsymbol{u}|$, and so consider a non-linear perturbation theory. In fact, a quite extensive literature exists on this topic in which such calculations have been performed. It is worth mentioning, though, that while this higher-order perturbation theory indeed allows us to follow density fluctuations to slightly larger values of $|\delta|$, the achievements of this theory do not, in general, justify the large mathematical effort. In addition, the fluid approximation is no longer valid if gravitationally bound systems form because, as mentioned earlier, multiple steams of matter will occur in this case.

However, for some interesting limiting cases, analytical descriptions exist which are able to represent the non-linear evolution of the mass distribution in the Universe. We shall now investigate a special and very important case of such a non-linear model. In general, studying the non-linear structure evolution requires the use of numerical methods. Therefore, we will also discuss some aspects of such numerical simulations.

7.5.1 Model of Spherical Collapse

We consider a spherical region in an expanding Universe, with its density $\rho(t)$ enhanced compared to the mean cosmic density $\overline{\rho}(t)$,

$$\rho(t) = [1 + \delta(t)]\, \overline{\rho}(t) , \tag{7.30}$$

where we use the density contrast δ as defined in (7.1). For reasons of simplicity we assume that the density within the sphere is homogeneous although, as we will later see, this is not really a restriction. The density perturbation is assumed to be small for small t, so that it will grow linearly at first, $\delta(t) \propto D_+(t)$, as long as $\delta \ll 1$. If we consider a time t_i which is sufficiently early such that $\delta(t_i) \ll 1$, then $\delta(t_i) = \delta_0\, D_+(t_i)$, where δ_0 is the density contrast linearly extrapolated to the present day. It should be mentioned once again that $\delta_0 \neq \delta(t_0)$, because the latter is determined by the non-linear evolution.

Let R_{com} be the initial *comoving* radius of the overdense sphere; as long as $\delta \ll 1$, the *comoving* radius will change only marginally. The mass within this sphere is

$$M = \frac{4\pi}{3}\, R_{com}^3 \rho_0\, (1 + \delta_i) \approx \frac{4\pi}{3}\, R_{com}^3 \rho_0 , \tag{7.31}$$

because the physical radius is $R = a R_{com}$, and $\overline{\rho} = \rho_0 / a^3$. This means that a unique relation exists between the initial *comoving* radius and the mass of this sphere, independent of the choice of t_i and δ_0, if only we choose $\delta(t_i) = \delta_0\, D_+(t_i) \ll 1$.

Due to the enhanced gravitational force, the sphere will expand slightly more slowly than the Universe as a whole, which again will lead to an increase in its density contrast. This then decelerates the expansion rate even further, relative to the cosmic expansion rate. Indeed, the equations of motion for the radius of the sphere are identical to the Friedmann equations for the cosmic expansion, only with the sphere having an effective Ω_m

different from that of the mean Universe. If the initial density is sufficiently large, the expansion of the sphere will come to a halt, i.e., $R(t)$ will reach a maximum; after this, the sphere will recollapse.

If $t_{\max}$ is the time of maximum expansion, then the sphere will, theoretically, collapse to a single point at time $t_{\rm coll} = 2t_{\max}$. The relation $t_{\rm coll} = 2t_{\max}$ follows from the time reversal symmetry of the equation of motion: the time to the maximum expansion is equal to the time from that point back to complete collapse.[5] The question of whether the expansion of the sphere will come to a halt depends on the density contrast $\delta(t_{\rm i})$ or δ_0 – compare the discussion of the expansion of the Universe in Sect. 4.3.1 – and on the model for the cosmic background.

Special Case: The Einstein–de Sitter Model. In the special case of $\Omega_{\rm m} = 1$ and $\Omega_\Lambda = 0$, this behavior can easily be quantified analytically; we thus treat this case separately. In this cosmological model, any sphere with $\delta_0 > 0$ is a "closed universe" and will therefore recollapse at some time. For the collapse to take place before t_1, $\delta(t_{\rm i})$ or δ_0 needs to exceed a threshold value. For instance, for a collapse at $t_{\rm coll} \le t_0$, a linearly extrapolated overdensity of

$$\delta_0 \ge \delta_{\rm c} = \frac{3}{20}(12\pi)^{2/3} \simeq 1.69 \qquad (7.32)$$

is required. More generally, one finds that $\delta_0 \ge \delta_{\rm c}\,(1+z)$ is needed for the collapse to occur before redshift z.

Violent Relaxation and Virial Equilibrium. Of course, the sphere will not really collapse to a single point. This would only be the case if the sphere was perfectly homogeneous and if the particles in the sphere moved along perfectly radial orbits. In reality, small-scale density and gravitational fluctuations will exist within such a sphere. These then lead to deviations of the particles' tracks from perfectly radial orbits, an effect that is more important the higher the density of the sphere becomes. The particles will scatter on these fluctuations in the gravitational field and will virialize; this process of *violent relaxation* has already been described in Sect. 6.2.6 and occurs on short time-scales – roughly the dynamical time-scale, i.e., the time it takes the particles to fully cross the sphere. In this case, the virialization is essentially complete at $t_{\rm coll}$. After that, the sphere will be in virial equilibrium, and its average density will be[6]

$$\langle\rho\rangle = (1+\delta_{\rm vir})\,\overline{\rho}(t_{\rm coll})\,,$$

$$\text{where} \quad (1+\delta_{\rm vir}) \simeq 178\,\Omega_{\rm m}^{-0.6}\,. \qquad (7.33)$$

This relation forms the basis for the statement that the virialized region, e.g., of a cluster, is a sphere with an average density ~ 200 times the critical density $\rho_{\rm cr}$ of the Universe at the epoch of collapse. Another conclusion from this consideration is that a massive galaxy cluster with a virial radius of $1.5\,h^{-1}$Mpc must have formed from the collapse of a region that originally had a comoving radius of about six times this size, roughly $10\,h^{-1}$Mpc. Such a virialized mass concentration of dark matter is called a *dark matter halo*.

Up to now, we have considered the collapse of a homogeneous sphere. From the above arguments one can easily convince oneself that the model is still valid if the sphere has a radial density gradient, e.g., if the density decreases outwards. In this case, the initial density contrast will also decrease as a function of radius. The inner regions of such a sphere will then collapse faster than the outer ones; a halo of lower mass will form first, and only later, when the outer regions have also collapsed, will a halo with higher mass form. From this it follows that halos of low initial mass will grow in mass by further accretion of matter.

The spherical collapse model is a simple model for the non-linear evolution of a density perturbation in the Universe. Despite being simplistic, it represents the fundamental principles of gravitational collapse and yields approximate relations, e.g., for the collapse time and mean density inside the virialized region, as they are found from numerical simulations.

[5] For the same reason that it takes a stone thrown up into the air the same time to reach its peak altitude as to fall back to the ground from there

[6] This result is obtained from conservation of energy and from the virial theorem. The total energy $E_{\rm tot}$ of the sphere is a constant. At the time of maximum expansion, it is given solely by the gravitational binding energy of the system since then the expansion velocity, and thus the kinetic energy, vanishes. On the other hand, the virial theorem implies that in virial equilibrium $E_{\rm kin} = -E_{\rm pot}/2$, and by combining this with the conservation of energy $E_{\rm tot} = E_{\rm kin} + E_{\rm pot}$ one is then able to compute $E_{\rm pot}$ in equilibrium and hence the radius and density of the collapsed sphere.

7.5.2 Number Density of Dark Matter Halos

Press–Schechter Model. The model of spherical collapse allows us to approximately compute the number density of dark matter halos as a function of their mass and redshift; this model is called the *Press–Schechter model*.

We consider a field of density fluctuations $\delta_0(\boldsymbol{x})$, featuring fluctuations on all scales according to the power spectrum $P_0(k)$. Assume that we smooth this field with a *comoving* smoothing length R, by convolving it with a filter function of this scale. In our example of the waves on a lake, we could examine a picture of its surface taken through a pane of milk-glass, by which all the contours on small scales would be blurred. Then, let $\delta_R(\boldsymbol{x})$ be the smoothed density field, linearly extrapolated to the present day. This field does not contain any fluctuations on scales $\lesssim R$, because these have been smoothed out. Each maximum in $\delta_R(\boldsymbol{x})$ corresponds to a peak with characteristic scale $\gtrsim R$ and, according to (7.31), each of these maxima corresponds to a mass peak of mass $M \sim (4\pi R^3/3)\rho_0$. If the amplitude δ_R of the density peak is sufficiently large, a sphere of (comoving) radius R around the peak will decouple from the linear growth of density fluctuations and will begin to grow non-linearly. Its expansion will come to a halt, and then it will recollapse. This process is similar to that in the spherical collapse model and can be described approximately by this model. The density contrast required for the collapse, $\delta_R \geq \delta_{\rm min}$, can be computed for any cosmological model and for any redshift.

If the statistical properties of $\delta_0(\boldsymbol{x})$ are Gaussian – which is expected for a variety of reasons – the statistical properties of the fluctuation field δ_0 are completely defined by the power spectrum $P(k)$. Then the number density of density maxima with $\delta_R \geq \delta_{\rm min}$ can be computed, and hence the (comoving) number density $n(M, z)$ of relaxed dark matter halos in the Universe as a function of mass M and redshift z can be determined.

The Mass Spectrum. The most important results of the Press–Schechter model are easily explained (see Fig. 7.7). The number density of halos of mass M depends of course on the amplitude of the density fluctuation δ_0 – i.e., on the normalization of the power spectrum $P_0(k)$. Hence, the normalization of $P_0(k)$

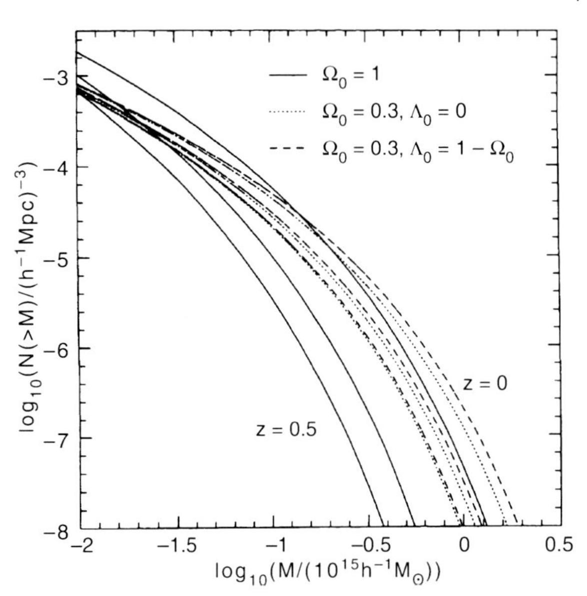

Fig. 7.7. Number density of dark matter halos with mass $> M$, computed from the Press–Schechter model. The comoving number density is shown for three different redshifts, $z = 0$ (upper curves), $z = 0.33$, and $z = 0.5$ (lower curves), for three different cosmological models: an Einstein–de Sitter model (solid lines), a low-density open model with $\Omega_{\rm m} = 0.3$ and $\Omega_\Lambda = 0$ (dotted lines), and a flat universe of low density with $\Omega_{\rm m} = 1 - \Omega_\Lambda = 0.3$ (dashed lines). The normalization of the density fluctuation field has been chosen such that the number density of halos with $M > 10^{14} h^{-1} M_\odot$ at $z = 0$ in all models agrees with the local number density of galaxy clusters. Note the dramatic redshift evolution in the EdS model

can be determined by comparing the prediction of the Press–Schechter model with the observed number density of galaxy clusters, as we will discuss further in Sect. 8.2.1 below. The corresponding result is called the "cluster-normalized power spectrum".

Furthermore, we find that $n(M, z)$ is a decreasing function of halo mass M. This follows immediately from the previous argument, since a larger M requires a larger smoothing length R, together with the fact that the number density of mass peaks of a given amplitude $\delta_{\rm min}$ decreases with increasing smoothing length. For large M, $n(M, z)$ decreases exponentially because sufficiently high peaks become very rare for large smoothing lengths. Therefore, *very* few clusters with mass $\gtrsim 2 \times 10^{15} M_\odot$ exist. From Fig. 7.7, we can see that the number density of clusters with $M \gtrsim 10^{15} M_\odot$ today is about $10^{-7}\,{\rm Mpc}^{-3}$, so the average separation between two such clusters is larger than 100 Mpc, which is compatible with the observation that the most nearby massive cluster (Coma) is about 90 Mpc away from us.

The density contrast $\delta_{\rm min}$ required for a collapse before redshift z is a function of z, as we have seen above. In particular, for the Einstein–de Sitter model we have $\delta_{\rm min} \simeq 1.69(1+z)$. In general, $\delta_{\rm min} = \delta_{\rm c}/D_+(z)$, where $\delta_{\rm c}$ and $D_+(z)$ each depend on the cosmological model. This means that the redshift dependence of $\delta_{\rm min}$ depends on the cosmological model and is basically described by the growth factor $D_+(z)$. Since $D_+(z)$ is, at fixed z (we recall that, by definition, $D_+(0)=1$), larger for smaller $\Omega_{\rm m}$ (see Fig. 7.3), the ratio of the number density of halos at redshift z to the one in the current Universe, $n(M,z)/n(M,0)$, is larger the smaller $\Omega_{\rm m}$ is. For cluster masses ($M \sim 10^{15} M_\odot$), the evolution of this ratio in the Einstein–de Sitter model is dramatic, whereas it is less strong in open and in flat, Λ-dominated universes (see Fig. 7.7).

By comparing the number density of galaxy clusters at high redshift with the current abundance, we can

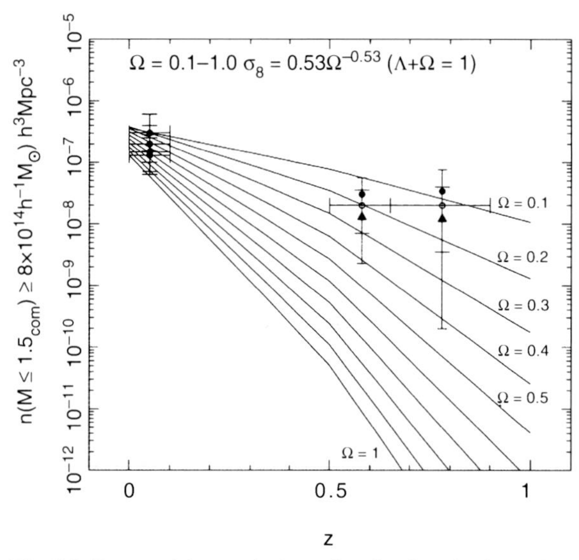

Fig. 7.8. Expected (comoving) number density of galaxy clusters with mass $> 8 \times 10^{14} h^{-1} M_\odot$ within a (comoving) radius of $R < 1.5h^{-1}$ Mpc, for flat cosmological models and different values of the density parameter $\Omega_{\rm m}$. The normalization of the power spectrum in the models has been chosen such that the current cluster number density is approximately reproduced. The points with error bars show results from observations of galaxy clusters at different redshift – although the error bars at high redshift are very large, a high-density universe seems to be excluded

thus obtain constraints on $\Omega_{\rm m}$, and in some sense also on Ω_Λ. Even a few very massive clusters at $z \gtrsim 0.5$ are sufficient to exclude the Einstein–de Sitter model by this argument. As a matter of fact, the existence of the cluster MS1054−03 (Fig. 6.15) alone, the mass of which was determined by dynamical methods, from its X-ray emission, and by the lens effect, is already nearly sufficient to falsify the Einstein–de Sitter model (see Fig. 7.8). However, at least one problem exists in the application of this method, namely making a sufficiently accurate mass determination for distant clusters and, in addition, determining whether they are relaxed and thus accounted for in the Press–Schechter model. Also the completeness of the local cluster sample is a potential problem.

A Special Case. To get a more specific impression of the Press–Schechter mass spectrum, we consider the special case where the power spectrum $P_0(k)$ is described by a power law, $P_0(k) \propto k^n$. From Fig. 7.6, we can see that this provides quite a good description over a large range of k if one concentrates on scales either clearly above or far below the maximum of P_0. The length-scale at which P_0 has its maximum is specified roughly by (7.28). As we can also see from Fig. 7.6, the nonlinear evolution that the Press–Schechter model refers to is relevant only for scales considerably smaller than this maximum, rendering the power law a valid approximation, with $n \sim -1.5$. In this case, the mass function can be written in closed form,

$$n(M,z) = \frac{\rho_{\rm cr}\Omega_{\rm m}}{\sqrt{\pi}} \frac{\gamma}{M^2} \left(\frac{M}{M^*(z)}\right)^{\gamma/2} \times \exp\left[-\left(\frac{M}{M^*(z)}\right)^{\gamma}\right] , \qquad (7.34)$$

where $\gamma = 1 + n/3 \sim 0.5$, and where $M^*(z)$ is the z-dependent mass-scale above which the mass spectrum is exponentially cut off. For masses considerably smaller than $M^*(z)$, the Press–Schechter mass spectrum is basically a power law in M. The characteristic mass-scale $M^*(z)$ depends on the normalization of the power spectrum and on the growth factor,

$$M^*(z) = M_0^* \left[D_+(z)\right]^{2/\gamma} = M_0^* (1+z)^{-2/\gamma} , \quad (7.35)$$

where the final expression applies to an Einstein–de Sitter universe. Hence, the characteristic mass-scale grows over time, and it describes the mass-scale on which the mass distribution in the Universe is just becoming non-linear for a particular redshift. This mass-scale at the current epoch, M_0^*, depends on the normalization of the power spectrum; it approximately separates groups from clusters of galaxies, and explains the fact that clusters are (exponentially) less abundant than groups.

Furthermore, the Press–Schechter model describes a very general property of structure formation in a CDM model, namely that low-mass structures – like galaxy-mass dark matter halos – form at early times, whereas large mass accumulations evolve only later. The explanation for this is found in the shape of the power spectrum $P(k)$ as described in (7.25) together with the asymptotic form (7.29) of the transfer function $T(k)$. A model like this is also called a *hierarchical structure formation* or a "bottom-up" scenario. In such a model, small structures that form early later merge to form large structures.

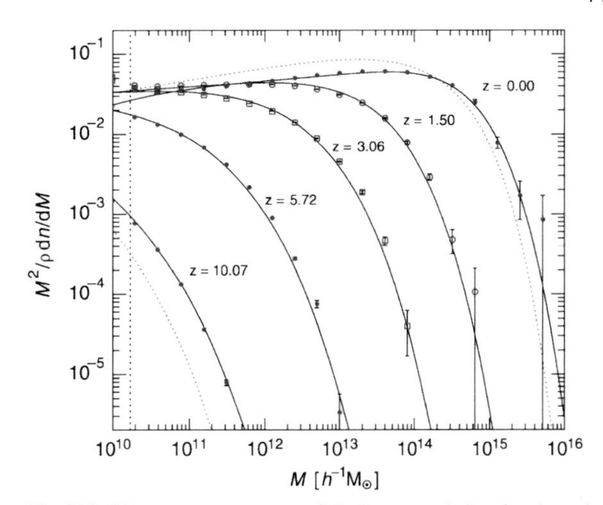

Fig. 7.9. The mass spectrum of dark matter halos is plotted for five different redshifts (data points with error bars), as determined in the Millennium simulation (which we will discuss more extensively below – see Fig. 7.12). The solid curves describe an approximation for the mass spectrum, which has been obtained from *different* simulations, and which obviously provides an excellent description of the simulation results. For $z = 0$ and $z = 10$, the prediction of the Press–Schechter model is indicated by the dotted curves, underestimating the abundance of very massive halos and overestimating the density of lower-mass halos. The vertical dotted line indicates the lowest halo mass which can still be resolved in these simulations

Comparison with Numerical Simulations. The Press–Schechter model is a very simple model, based on assumptions that are not really justified in detail. Nevertheless, its predictions are in astounding agreement with the number density of halos determined from simulations, and this model, published in 1974, has for nearly 25 years predicted the halo density with an accuracy that was difficult to achieve in numerical simulations. Only since the mid-1990s have the precision and statistics of numerical simulations of structure formation reached a level on which significant discrepancies with the Press–Schechter model become clearly noticeable. However, the analytical description has also been improved; instead of a spherical collapse, the more realistic ellipsoidal collapse has been investigated, by which the number density of halos is modified relative to the Press–Schechter model. This advanced model is found to be in very good agreement with the numerical results, as demonstrated in Fig. 7.9, so that today we have a good description of $n(M, z)$ that very accurately resembles the results from numerical simulations.

7.5.3 Numerical Simulations of Structure Formation

Analytical considerations – such as, for instance, linear perturbation theory or the spherical collapse model – are only capable of describing limiting cases of structure formation. In general, gravitational dynamics is too complicated to be analytically examined in detail. For this reason, experiments to simulate structure formation by means of numerical methods have been performed for some time already. The results of these simulations, when compared to observations, have contributed very substantially to establishing the standard model of cosmology, because only through them did it become possible to quantitatively distinguish the predictions of this model from those of other models. Of course, the enormous development in computer hardware rendered corresponding progress in simulations possible; in addition, the continuous improvement of numerical algorithms has allowed steadily improved spatial resolution of the simulations.

Since the Universe is dominated by dark matter, it is often sufficient to compute the behavior of this dark matter and thus to consider solely gravitational interactions. Only in recent years has computing power increased to a level where hydrodynamic processes can also approximately be taken into account, so that the baryonic component of the Universe can be traced as well. In addition, radiative transfer can be included in such simulations, hence the influence of radiation on the heating and cooling of the baryonic component can also be examined.

The Principle of Simulations. Representative Dark Matter Particles. We will now give a brief description of the principle of such simulations, where we confine ourselves to dark matter. Of course, no individual particles of dark matter are traced in the simulations: since it presumably consists of elementary particles, which therefore have a high number density, one would only be able to simulate an extremely small, microscopic section of the Universe. Rather, one examines the behavior of dark matter in the expanding Universe by representing its particles by bodies of mass M, and by then assuming that these "macroscopic particles" behave like the dark matter particles in a volume $V = M/\rho$. Effectively, this corresponds to the assumption that dark matter consists of particles of mass M. Since this assumption cannot be valid in detail, we will later need to modify the resulting equations.

Choice of Simulation Volume. The next point one needs to realize right from the start is that one cannot simulate the full spatial volume of the Universe (which may be infinite) but only a representative section of it. Typically, a *comoving* cube with side length L is chosen. For this section to be representative, the linear extent L should be larger than the largest observed structures in the Universe. Otherwise, the effects of the large-scale structure would be neglected. For example, hardly any structure is found in the Universe on scales $\gtrsim 200\,h^{-1}$ Mpc, so that $L = 200\,h^{-1}$ Mpc is a reasonable value for the comoving size of the cube. Since the numerical effort scales with the number of grid points at which the gravitational force is computed, and which is limited by the computer's speed and memory, the choice of L also immediately implies the length-scale of the numerical resolution. Furthermore, the total mass within the numerical volume is $\propto \Omega_{\rm m} L^3$, so that for a given maximum number of particles, the minimum mass that can be resolved in the simulation is also known.

Periodic Boundary Conditions. Since particles close to the boundaries of the cube also feel gravitational forces from matter outside the cube, one cannot simply assume the region outside the cube to be empty. We need to make assumptions about the matter distribution outside the numerical volume. Since one assumes that the Universe is essentially homogeneous on scales $> L$, the cube is extended periodically – for instance, a particle leaving the cube at its upper boundary will immediately re-enter the cube from the lower side. The mass distribution (and with it also the force field) is periodic in these simulations, with a period of L. This assumption of periodicity has an effect on the results for the mass distribution on scales comparable to L; the quantitative analysis of the results from these simulations should therefore be confined to scales $\lesssim L/2$.

Softening Length. With the above assumptions, the equation of motion for all particles can now be set up. The force on the i-th particle is

$$\boldsymbol{F}_i = \sum_{j \neq i} \frac{M^2\,(\boldsymbol{r}_j - \boldsymbol{r}_i)}{|\boldsymbol{r}_j - \boldsymbol{r}_i|^3}\,, \tag{7.36}$$

thus the sum of forces exerted by all the other particles, where these are periodically extended. This aspect may appear at first sight more difficult than it actually is, as we will explain below. In particular, the force law (7.36) also describes strong collisions of particles, e.g., where a particle changes its velocity direction by $\sim 90°$ in a collision if it comes close enough to another particle. Of course, this effect is a consequence of replacing the dark matter constituents by macroscopic "particles" of mass M. As we have seen in Sect. 3.2.4, the typical relaxation time-scale for a system is $\propto N/\ln N$, and since the mass in the numerical volume is defined by L, one has $N \propto 1/M$. Reducing the particles' mass and increasing N accordingly, the abundance of strong collisions would decrease, but computer power and memory is then a limiting factor. Thus to correct for the artefact of strong collisions, the force law is modified for small separations such that strong collisions no longer occur. The length-scale below which the force equation is modified ("softened") and deviates from $\propto 1/r^2$ is called *softening length* and is chosen to be about the mean separation of two particles of mass M – the smaller M, the smaller the softening length. This then also defines a limit for the spatial resolution in the simulation: scales below or comparable to the softening length are not resolved, and the behavior on these small scales is affected by numerical artefacts.

Computation of the Force Field. The computation of the force acting on individual particles by summation, as in (7.36), is not feasible in practice, as can be seen as follows. Assume the simulation to trace 10^8 particles, then in total 10^{16} terms need to be calculated using (7.36) – for each time step. Even on the most powerful computers this is not feasible today. To handle this problem, one evaluates the force in an approximate way. One first notes that the force experienced by the i-th particle, exerted by the j-th particle, is not very sensitive to small variations in the separation vector $\boldsymbol{r}_i - \boldsymbol{r}_j$, as long as these variations are much smaller than the separation itself. Except for the nearest particles, the force on the i-th particle can then be computed by introducing a grid into the cube and shifting the particles in the simulation to the closest grid point.[7] With this, a discrete mass distribution on a regular grid is obtained. The force field of this mass distribution can then be computed by means of a Fast Fourier Transform (FFT), a fast and very efficient algorithm. However, the introduction of the grid establishes a lower limit to the spatial force resolution; this is often chosen such that it agrees with the softening length. Because the size of the grid cells also defines the spatial resolution of the force field, it is chosen to be roughly the mean separation between two particles, so that the number of grid points is typically of the same order as the number of particles. This is called the PM (particle–mesh) method. To achieve better spatial resolution, the interaction of closely neighboring particles is considered separately. Of course, this force component first needs to be removed from the force field as computed by FFT. This kind of calculation of the force is called the $\mathrm{P}^3\mathrm{M}$ (particle–particle particle–mesh) method.

Initial Conditions and Evolution. The initial conditions for the simulation are set at very high redshift. The particles are then distributed such that the power spectrum of the resulting mass distribution resembles a Gaussian random field with the theoretical (linear) power spectrum $P(k, z)$ of the cosmological model. The equations of motion for the particles with the force field described above are then integrated in time. The choice of the time step is a critical issue in this integration, as can be seen from the fact that the force on particles with relatively close neighbors will change more quickly than that on rather isolated particles. Hence, the time step is either chosen such that it is short enough for the former particles – which requires substantial computation time – or the time step is varied for different particles individually, which is clearly the more efficient strategy. For different times in the evolution, the particle positions and velocities are stored; these results are then available for subsequent analysis.

Examples of Simulations. The size of the simulations, measured by the number of particles considered, has increased enormously in recent years with the corresponding increase in computing capacities and the development of efficient algorithms. In modern simulations, 512^3 or even more particles are traced. One example of such a simulation is presented in Fig. 7.10, where the structure evolution was computed for four different cosmological models. The parameters for these simulations and the initial conditions (i.e., the initial realization of the random field) were chosen such that the resulting density distributions for the current epoch (at $z = 0$) are as similar as possible; by this, the dependence of the redshift evolution of the density field on the cosmological parameters can be recognized clearly. Comparing simulations like these with observations has contributed substantially to our realizing that the matter density in our Universe is considerably smaller than the critical density.

Massive clusters of galaxies have a very low number density, which can be seen from the fact that the massive cluster closest to us (Coma) is about 90 Mpc away. This is directly related to the exponential decrease of the abundance of dark matter halos with mass, as described by the Press–Schechter model (see Sect. 7.5.2). In simulations such as that shown in Fig. 7.10, the simulated volume is still too small to derive statistically meaningful results on such sparse mass concentrations. This difficulty has been one of the reasons for simulating considerably larger volumes. The Hubble Volume Simulations (see Fig. 7.11) use a cube with a side length of $3000h^{-1}$ Mpc, not much less than the currently visible Universe. This simulation is particularly well-suited

[7] In practice, the mass of a particle is distributed to all 8 neighboring grid points, with the relative proportion of the mass depending on the distance of the particle to each of these grid points.

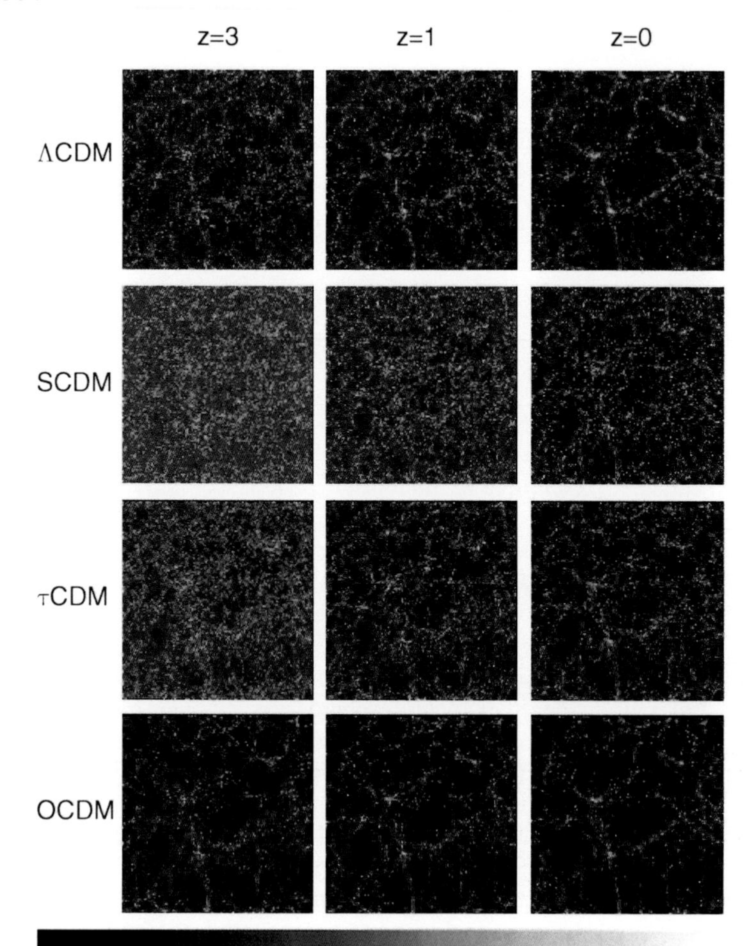

Fig. 7.10. Simulations of the dark matter distribution in the Universe for four different cosmological models: $\Omega_m = 0.3$, $\Omega_\Lambda = 0.7$ (ΛCDM), $\Omega_m = 1.0$, $\Omega_\Lambda = 0.0$ (SCDM and τCDM), and $\Omega_m = 0.3$, $\Omega_\Lambda = 0.$ (OCDM). The two Einstein–de Sitter models differ in their shape parameter Γ which specifies the shape of the power spectrum $P(k)$. For each of the models, the mass distribution is presented for three different redshifts, $z = 3$, $z = 1$, and today, $z = 0$. Whereas the current mass distribution is quite similar in all four models (the model parameters were chosen as such), they clearly differ at high redshift. We can see, for instance, that significantly less structure has formed at high redshift in the SCDM model compared to the other models. From the analysis of the matter distribution at high redshift, one can therefore distinguish between the different models. In these simulations by the VIRGO Consortium, 256^3 particles were traced; the side length of the simulated volume is $\sim 240h^{-1}$ Mpc

to studying the statistical properties of very massive structures, like, e.g., the distribution of galaxy clusters. On the other hand, this large volume, together with the limited total number of particles that can be followed, means that the mass and spatial resolution of this simulation are insufficient for studying galaxies.

To date (2006), by far the largest simulation is the Millennium simulation, carried out for a cosmological model with $\Omega_m{=}0.25$, $\Omega_\Lambda{=}0.75$, a power spectrum normalization of $\sigma_8{=}0.9$, and a Hubble constant of $h{=}0.73$. A cube of side length $500h^{-1}$ Mpc was considered, in which $(2160)^3 \approx 10^{10}$ particles with a mass of $8.6 \times 10^8 h^{-1} M_\odot$ each were traced. With this choice of parameters, one can spatially resolve the halos of galaxies. At the same time, the volume is large enough for the simulation to contain a large number of massive

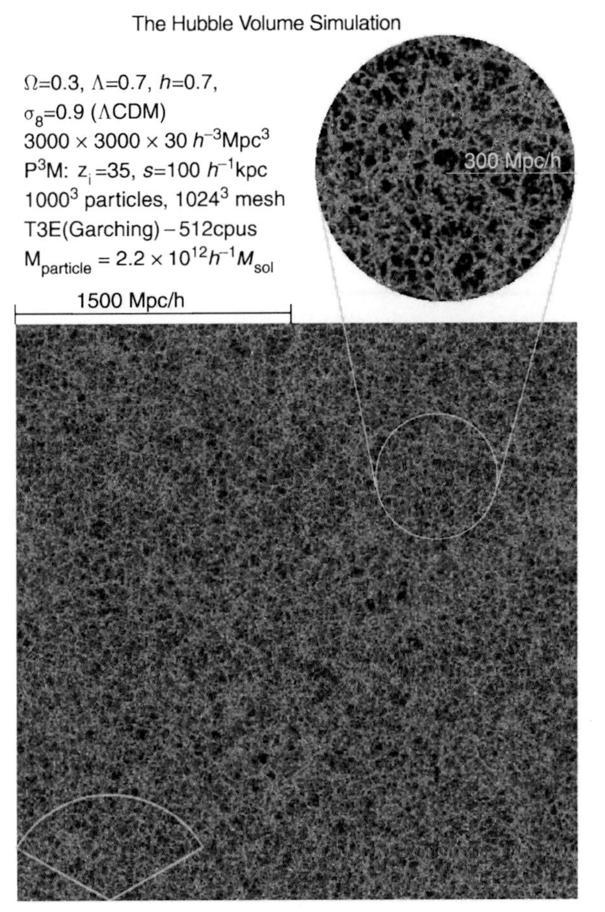

Fig. 7.11. The Hubble Volume Simulations: simulated is a box of volume $(3000h^{-1}\,\text{Mpc})^3$, containing 10^9 particles, where a ΛCDM model with $\Omega_\text{m} = 0.3$ and $\Omega_\Lambda = 0.7$ was chosen. Displayed is the projection of the density distribution of a $30h^{-1}$ Mpc thick slice of the cube. Simulations like this can be used to analyze the statistical properties of the mass distribution in the Universe on large scales. The sector in the lower left corner represents roughly the size of the CfA redshift survey (see Fig. 7.2)

clusters whose evolutionary history can be followed. The spatial resolution of the simulation is $\sim 5h^{-1}$ kpc, yielding a linear dynamic range of $\sim 10^5$. The resulting mass distribution at $z{=}0$ is displayed in Fig. 7.12 in slices of $15h^{-1}$ Mpc thickness each, where the linear scale changes by a factor of four from one slice to the next. The images zoom in to a region around a massive cluster that becomes visible with its rich substructure in the uppermost slice, as well as filaments of the matter distribution, at the intersections of which massive halos form. The mass distribution in the Millennium simulation is of great interest for numerous different investigations. We will discuss some of its results further in Chap. 9.

Analysis of Numerical Results. The analysis of the numerical results is nearly as intricate as the simulation itself because the positions and velocities of $\sim 10^9$ particles alone do not provide any new insights. The output of the simulation needs to be analyzed with respect to specific questions. Obviously, the (non-linear) power spectrum $P(k, z)$ of the matter distribution can be computed from the spatial distribution of particles; the corresponding results have led to the construction of the analytic fit formulae presented in Fig. 7.6. Furthermore, one can search for voids in the resulting particle distribution, which can then be compared to the observed abundance and typical size of voids.

One of the main applications is the search for collapsed mass concentrations (i.e., dark matter halos), and their number density can be compared to predictions from the Press–Schechter model and to observations. From this, it has been found that the Press–Schechter mass function represents the basic aspects of the mass spectrum astonishingly well, but even more accurate formulae for the mass spectrum of halos have been constructed from the simulations (see Fig. 7.9). However, the identification of a halo and the determination of its mass from the positions and velocities of the particles is by no means trivial, and various methods for this are applied. For instance, we can concentrate on spatial overdensities of particles and define a halo as a spherical region, within which the average density is just 200 times the critical density – this definition of a halo is suggested by the spherical collapse model. Alternatively, those particles which are gravitationally bound, as can be obtained from the particle velocities, can be assigned to a halo.

The direct link between the results from dark matter simulations and the observed properties of the Universe requires an understanding of the relation between dark matter and luminous matter. Dark matter halos in simulations cannot be compared to the observed galaxy distribution without further assumptions, e.g., on the mass-to-light ratio. We will return to these aspects later.

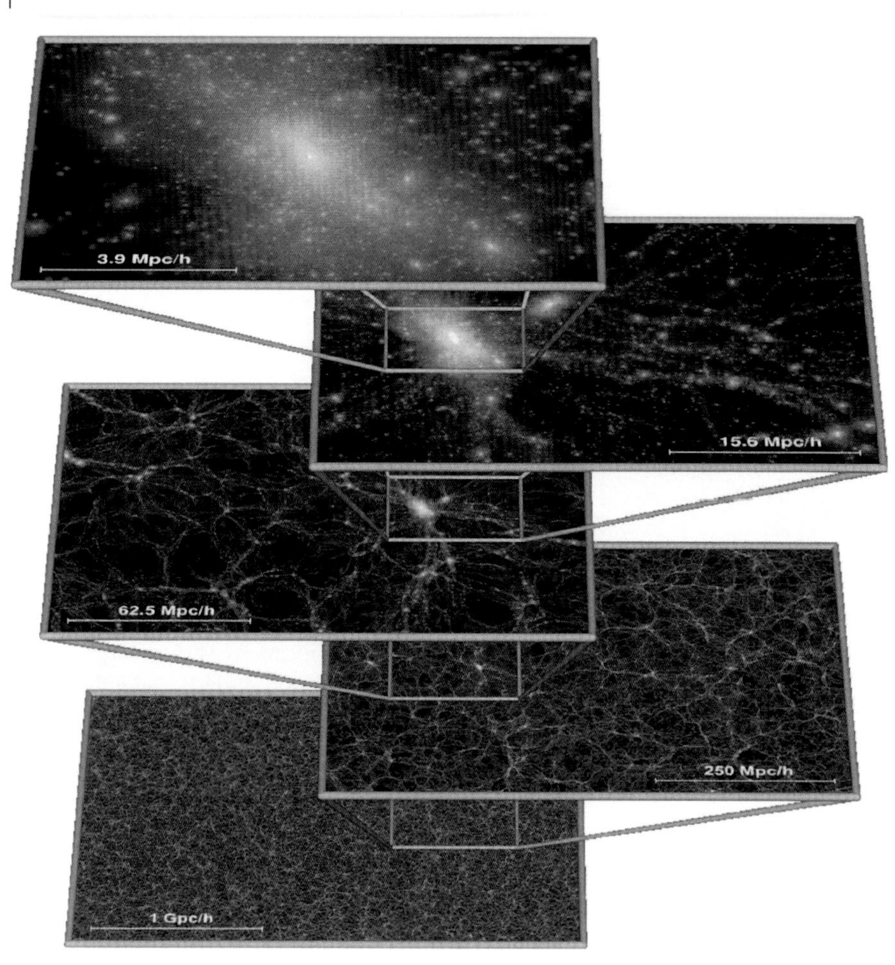

Fig. 7.12. Distribution of matter in slices of thickness $15h^{-1}$ Mpc each, computed in the Millennium simulation. This simulation took about a month, running on 512 CPU processors. The output of the simulation, i.e., the position and velocities of all 10^{10} particles at 64 times steps, has a data volume of ~ 27 TB. The region shown in the two lower slices is larger than the simulated box which has a sidelength of $500h^{-1}$ Mpc; nevertheless, the matter distribution shows no periodicity in the figure as the slice was cut at a skewed angle to the box axes

7.5.4 Profile of Dark Matter Halos

As already mentioned above, dark matter halos can be identified in mass distributions generated by numerical simulations. Besides the abundance of halos as a function of their mass and redshift, their radial mass profile can also be analyzed if individual halos are represented by a sufficient number of dark matter particles. The ability to obtain halo mass profiles depends on the mass resolution of a simulation. A surprising result has been obtained from these studies, namely that halos seem to show a universal density profile. We will briefly discuss this result in the following.

If we define a halo as described above, i.e., as a spherical region within which the average density is ~ 200 times the critical density at the respective redshift, the mass M of the halo is related to its (virial) radius r_{200} by

$$M = \frac{4\pi}{3} r_{200}^3 \, 200 \, \rho_{\rm cr}(z) \; .$$

Since the critical density at redshift z is specified by $\rho_{\rm cr}(z) = 3H^2(z)/(8\pi G)$, we can write this as

$$M = \frac{100 r_{200}^3 H^2(z)}{G} \; , \tag{7.37}$$

so that at each redshift, a unique relation exists between the halo mass and its radius. We can also define the virial velocity V_{200} of a halo as the circular velocity at the virial radius,

$$V_{200}^2 = \frac{GM}{r_{200}} \; . \tag{7.38}$$

Combining (7.37) and (7.38), we can express the halo mass and virial radius as a function of the virial velocity,

$$M = \frac{V_{200}^3}{10 G H(z)} , \qquad r_{200} = \frac{V_{200}}{10 H(z)} . \tag{7.39}$$

Since the Hubble function $H(z)$ increases with redshift, the virial radius at fixed virial velocity decreases with redshift. From (7.37) we also see that r_{200} decreases with redshift at fixed halo mass. Hence, halos at a given mass (or given virial velocity) are more compact at higher redshift than they are today.

The NFW Profile. The *density profile of halos* averaged over spherical shells seems to have a universal functional form, which was first reported by Julio Navarro, Carlos Frenk & Simon White in a series of articles in the mid-1990s. This *NFW-profile* is described by

$$\boxed{\rho(r) = \frac{\rho_s}{(r/r_s)(1 + r/r_s)^2}} , \tag{7.40}$$

where ρ_s is the amplitude of the density profile, and r_s specifies a characteristic radius. For $r \ll r_s$ we find $\rho \propto r^{-1}$, whereas for $r \gg r_s$, the profile follows $\rho \propto r^{-3}$. Therefore, r_s is the radius at which the slope of the density profile changes (see Fig. 7.13). ρ_s can be expressed in terms of r_s, since, according to the definition of r_{200},

$$\overline{\rho} = 200 \rho_{cr}(z) = \frac{3}{4\pi r_{200}^3} \int_0^{r_{200}} 4\pi r^2 \, dr \, \rho(r)$$

$$= 3\rho_s \int_0^1 \frac{dx \, x^2}{c x (1 + cx)^2} , \tag{7.41}$$

where in the last step the integration variable was changed to $x = r/r_{200}$, and the *concentration index*

$$\boxed{c := \frac{r_{200}}{r_s}} \tag{7.42}$$

was defined. The larger the value of c, the more strongly the mass is concentrated towards the inner regions.

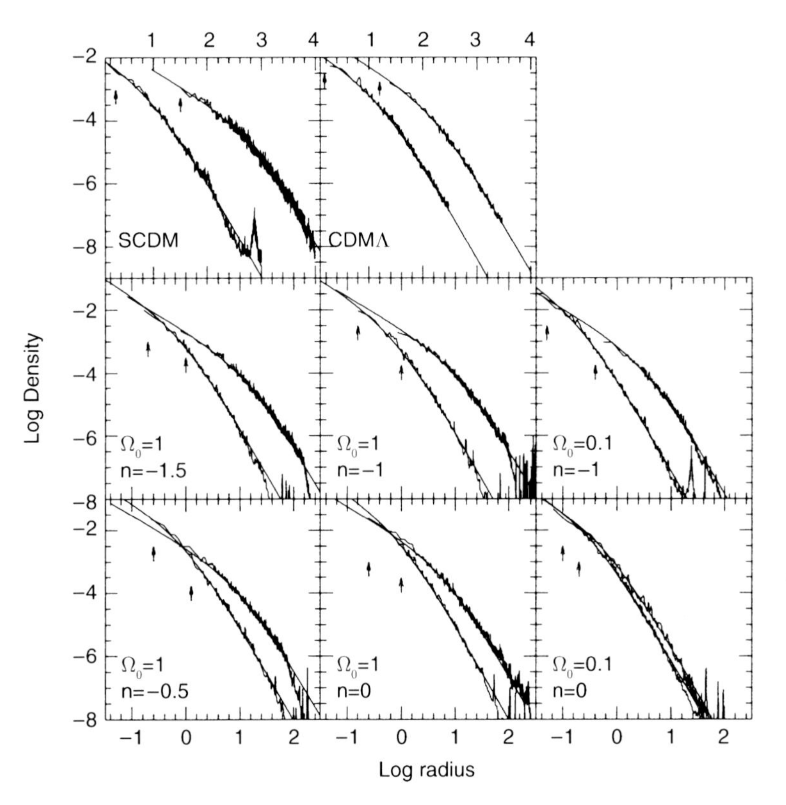

Fig. 7.13. For eight different cosmological simulations, the density profile is shown for the most massive and the least massive halo, each as a function of the radius, together with the best fitting density profile (7.40). The cosmological models represent an EdS model (here denoted by SCDM), a ΛCDM model, and different models with power spectra that are assumed to be power laws locally, $P(k) \propto k^n$. The arrows indicate the softening length in the gravitational force for the respective halos; thus, the major part of the profiles is numerically well resolved

Equation (7.41) implies that ρ_s can be expressed in terms of $\rho_{cr}(z)$ and c, and performing the integration in (7.41) yields

$$\rho_s = \frac{200}{3}\rho_{cr}(z)\,\frac{c^3}{\ln(1+c)-c/(1+c)}\;.$$

Since M is determined by r_{200}, the NFW profile is parametrized by r_{200} (or by the mass of the halo) and by the concentration c that describes the shape of the distribution. Simulations show that the concentration index c is strongly correlated with the mass and the redshift of the halo; one finds approximately

$$c \propto \left(\frac{M}{M^*}\right)^{-1/9}(1+z)^{-1}\;,$$

where M^* is the non-linear mass scale already mentioned in Sect. 7.5.2. This result can also be obtained from analytical scaling arguments, under the assumption of the existence of a universal density profile. In Fig. 7.14, the density profile of dark halos is plotted as a function of the scaled radius r/r_{200}, where the similarity in the profile shapes for the different simulations becomes clearly visible, as well as the dependence of the concentration index on the halo mass. The range over which the density distribution of numerically simulated halos is described by the profile (7.40) is limited by the virial radius r_{200}, whereas in the central region of halos the numerical resolution of the simulations is too low to test (7.40) for very small r. The latter comment concerns the inner $\sim 1\%$ of the halo mass.

Generalization. No good analytical argument has yet been found for the existence of such a universal density profile, in particular not for the specific functional form of the NFW profile. As a matter of fact, other numerical simulations find a slightly different density profile that can be expressed as

$$\rho \propto \frac{1}{(r/r_s)^\alpha\,(1+r/r_s)^{3-\alpha}}\;,$$

with $\alpha \sim 1.5$, whereas the NFW profile is characterized by $\alpha = 1$. The reason for the difference between different simulations has not conclusively been established, but probably the density profile in the innermost region

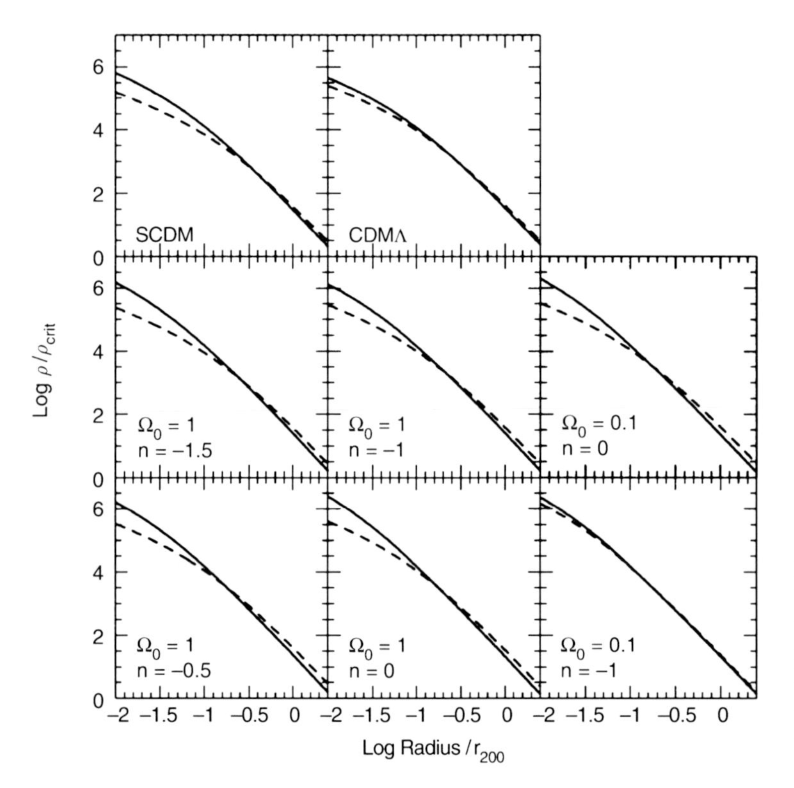

Fig. 7.14. The density profiles from Fig. 7.13, but now the density is scaled by the critical density, and the radius scaled by r_{200}. Solid (dashed) curves correspond to halos of low (high) mass – thus, halos of low mass are relatively denser close to the center, and they have a higher concentration index c

(which is difficult to resolve numerically) is more complicated than a power law. For large radii, the different research groups agree on the shape of the profile $\propto r^{-3}$.

Comparison with Observations. The comparison of these theoretical profiles with an observed density distribution is by no means simple because the density profile of dark matter is of course not directly observable. For instance, in normal spiral galaxies, $\rho(r)$ is dominated by baryonic matter at small radii. For example, in the Milky Way, roughly half of the matter within R_0 consists of stars and gas, so that only little information is provided on ρ_{DM} in the central region. In general, it is assumed that galaxies with very low surface brightness (LSBs) are dominated by dark matter well into the center. The rotation curves of LSB galaxies are apparently *not* in agreement with the expectations from the NFW model (Fig. 7.15); in particular, they provide no evidence of a cusp in the central density distribution ($\rho \to \infty$ for $r \to 0$).

Part of this discrepancy may perhaps be explained by the finite angular resolution of the 21-cm line measurements of the rotation curves; however, the discrepancy remains if higher-resolution rotation curves are measured using optical long-slit and integral-field spectroscopy. As an additional point, the kinematics of these galaxies may be more complicated, and in some cases their dynamical center is difficult to determine. The orbits of stars and gas in these galaxies may show a more complex behavior than expected from a smooth density profile. The mass distribution in the (inner parts of a) dark matter halo is neither smooth nor axially symmetric, and stars and gas do not move on circular orbits in a thin plane of symmetry. Instead, simulations show that the pressure support of the gas, together with non-circular motions and projection effects systematically underestimate the rotational velocity in the center of dark matter halos, thereby creating the impression of a constant density core. Nevertheless, the observed rotation curves of LSB galaxies may prove to be a major problem for the CDM model – hence, this potential discrepancy must be resolved.

An additional complication is the fact that not only is baryonic matter present in the inner regions of galaxies

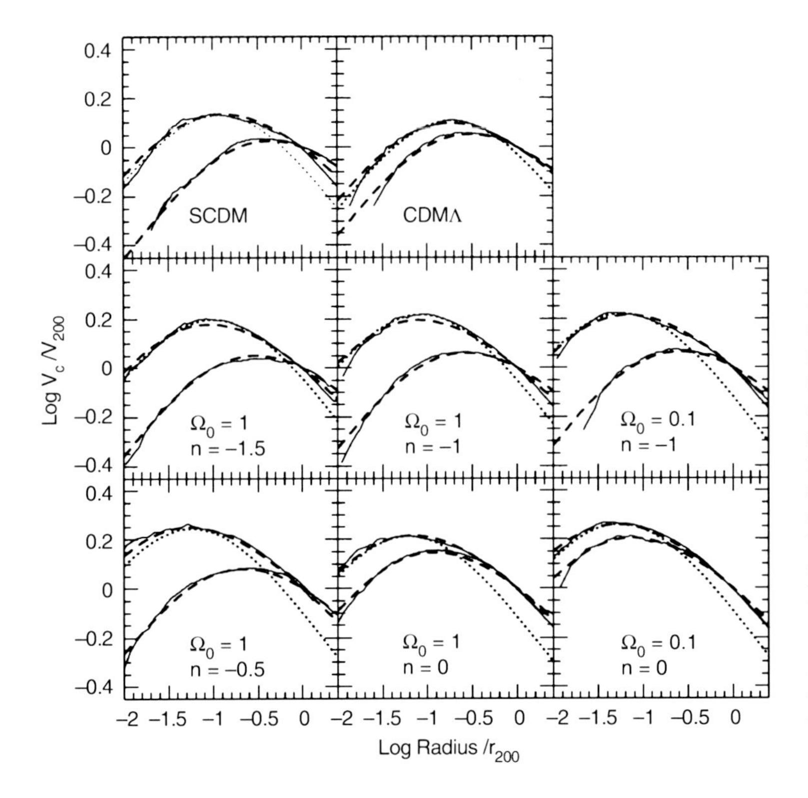

Fig. 7.15. The rotation curves in the NFW density profiles from Fig. 7.13, in units of the rotational velocity at r_{200}. All curves initially increase, reach a maximum, and then decrease again; over a fairly wide range in radius, the rotation curves are approximately flat. The solid curves are taken directly from the simulation, while dashed curves indicate the rotation curves expected from the NFW profile. The dotted curve in each panel presents a fit to the low-mass halo data with the so-called Hernquist profile, a mass distribution frequently used in modeling – it fits the rotation curve very well in the inner part of the halo, but fails beyond $\sim 0.1 R_{200}$. In these scaled units, halos of low mass have a relatively higher maximum rotational velocity

(and clusters), thus contributing to the density, but also these baryons have modified the density profile of dark matter halos in the course of cosmic evolution. Baryons are dissipative, they can cool, form a disk, and accrete inwards. The change in the resulting density distribution of baryons by dissipative processes cause a change of the gravitational potential over time, to which dark matter also reacts. The dark matter profile in real galaxies is thus modified compared to pure dark matter simulations.

Despite these difficulties, it has been found that the X-ray data of many clusters are compatible with an NFW profile. Analyses based on the weak lensing effect also show that an NFW mass profile provides a very good description for shear data. In Fig. 7.16 it is shown that the radial profile of the galaxy density in clusters on average follows an NFW profile, where the mean concentration index is $c \approx 3$, i.e., smaller than expected for the *mass* profile of clusters. One interpretation of this result is that the galaxy distribution in clusters is less strongly concentrated than the density of dark matter.

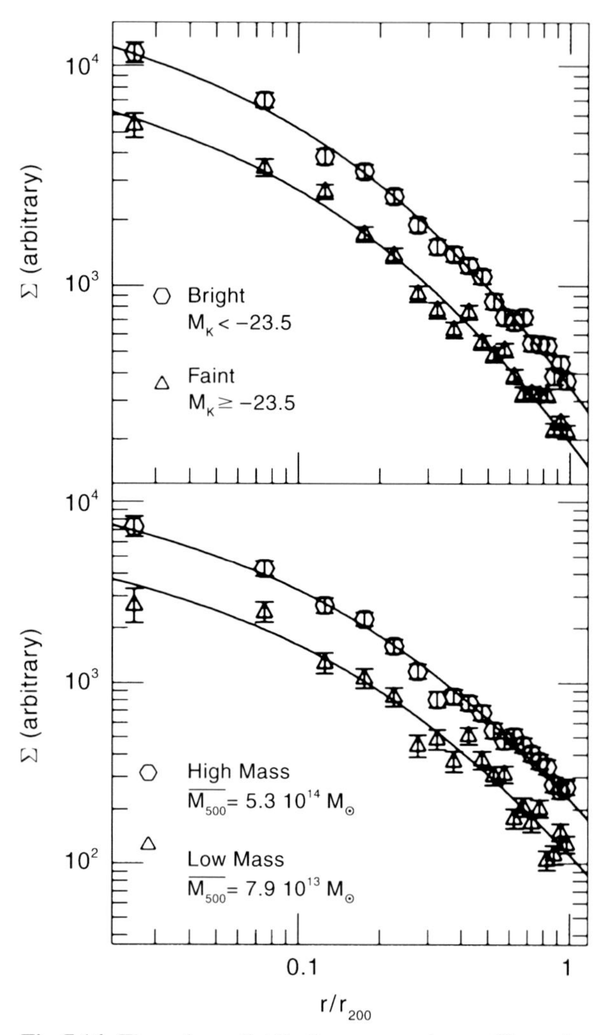

Fig. 7.16. The galaxy distribution averaged over 93 nearby clusters of galaxies, as a function of the projected distance to the cluster center. Galaxies have been selected in the NIR, and cluster masses, and thus r_{200}, have been determined from X-ray data. Plotted is the projected number density of cluster galaxies, averaged over the various clusters, versus the scaled radius r/r_{200}. In the top panel the galaxy sample is split into luminous and less luminous galaxies, while in the bottom panel the cluster sample is split according to the cluster mass. The solid curves show a fit of the projected NFW profile, which turns out to be an excellent description in all cases. The concentration index is, with $c \approx 3$, roughly the same in all cases, and smaller than expected for the mass profile of clusters

7.5.5 The Substructure Problem

As we will discuss in detail in the next chapter, the CDM model of cosmology has proven to be enormously successful in describing and predicting cosmological observations. Because this model has achieved this success and is therefore considered the standard model, results that apparently do not fit into the standard model are of particular interest. The rotation curves of LSB galaxies mentioned above are one such result. Either one finds a good reason for this apparent discrepancy between observation and the predictions of the CDM model or, otherwise, results of this kind indicate the necessity to introduce extensions to the CDM model. In the former case, the model would have overcome another hurdle in demonstrating its validity and would be confirmed even further, whereas in the latter case, new insights would be gained into the physics of cosmology.

Sub-Halos of Galaxies and Clusters of Galaxies. Besides the rotation curves of LSB galaxies, there is another observation that does not seem to fit into the picture of the CDM model at first sight. Numerical simulations of structure formation show that a halo of mass M contains numerous halos of much lower mass, so-called sub-halos. For instance, a halo with the mass of a galaxy cluster contains hundreds or even thousands of halos with masses that are orders of magnitude lower.

Indeed, this can be expected because clusters of galaxies contain substructure, visible in the form of the cluster galaxies. In the upper part of Fig. 7.17, the simulation of a cluster and its substructure is displayed. Indeed, this mass distribution looks just like the mass distribution expected in a cluster of galaxies, with the main cluster halo and its distribution of member galaxies. The lower part of Fig. 7.17 shows the simulation of a halo with mass $\sim 2 \times 10^{12} M_{\odot}$, which corresponds to a massive galaxy. As one can easily see, its mass distribution shows a large number of sub-halos as well. In fact, the two mass distributions are nearly indistinguishable, except for their scaling in the total mass.[8] The presence of substructure over a very wide range in mass is a direct consequence of hierarchical structure formation, in which objects of higher mass each contain smaller structures that have been formed earlier in the cosmic evolution.

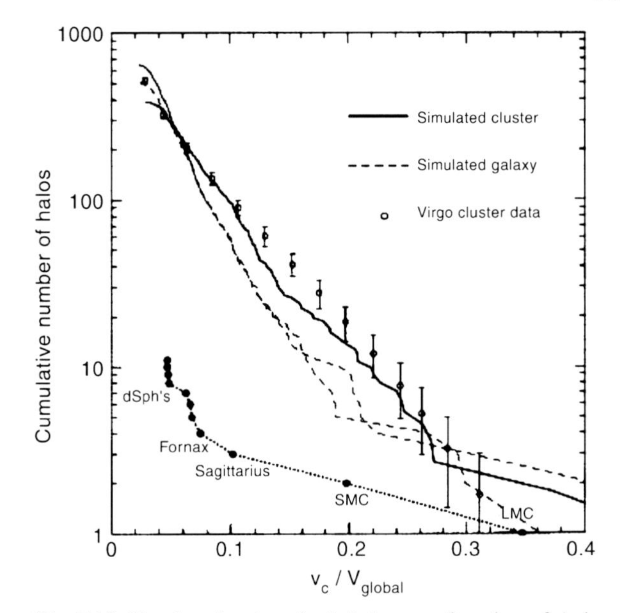

Fig. 7.18. Number density of sub-halos as a function of their mass. The mass is expressed by the corresponding Keplerian rotational velocity v_c, measured in units of the corresponding rotational velocity of the main halo. The curves show this number density of sub-halos with rotational velocity $\geq v_c$ for a halo of either cluster mass or galaxy mass. The observed numbers of sub-halos (i.e., of galaxies) in the Virgo Cluster are plotted as open circles with error bars, and the number of satellite galaxies of the Milky Way as filled circles. One can see that the simulations describe the abundance of cluster galaxies quite well, but around the Galaxy significantly fewer satellite galaxies exist than predicted by a CDM model

Fig. 7.17. Density distribution of two simulated dark matter halos. In the top image, the halo has a virial mass of $5 \times 10^{14} M_{\odot}$, corresponding to a cluster of galaxies. The halo in the bottom image has a mass of $2 \times 10^{12} M_{\odot}$, representing a massive galaxy. In both cases, the presence of substructure in the mass distribution can be seen. It can be identified with individual cluster galaxies in the case of the galaxy cluster. The substructure in a galaxy can not be identified easily with any observable source population; one may expect that these are satellite galaxies, but observations show that these are considerably less abundant than the substructure seen here. Apart from the length-scale (and thus also the mass-scale), both halos appear very similar from a qualitative point of view

Whereas this substructure in clusters is easily identified with the cluster member galaxies, the question arises as to what the sub-halos in galaxies can possibly correspond to. These show a broad mass spectrum, as displayed in Fig. 7.18. Some of these sub-halos are rec-

[8] The reason for this is found in the property of the power spectrum of density fluctuations that has been discussed in Sect. 7.5.2, namely that $P(k)$ can be approximated by a power law over a wide range in k. Such a power law features no characteristic scale. For this reason, the properties of halos of high and low mass are scale-invariant, as is clearly visible in Fig. 7.17.

ognized in our Milky Way, namely the known satellite galaxies like, e.g., the Magellanic Clouds. In a similar way, the satellite galaxies of the Andromeda galaxy may also be identified with sub-halos. However, as we have seen in Sect. 6.1, fewer than 40 members of the Local Group are known – whereas the numerical simulations predict hundreds of satellite galaxies for the Galaxy. This apparent deficit in the number of observed sub-halos is considered to be another potential problem of CDM models.

However, one always needs to remember that the simulations only predict the distribution of mass, and not that of light (which is accessible to observation). One possibility of resolving this apparent discrepancy centers on the interpretation that these sub-halos do in fact exist, but that most of them do not, or only weakly, emit radiation. What appears as a cheap excuse at first sight is indeed already part of the models of the formation and evolution of galaxies. As will be discussed in Sect. 9.6.3 in more detail, it is difficult to form a considerable stellar population in halos of masses below $\sim 10^9 M_\odot$. Most halos below this mass threshold will therefore be hardly detectable because of their low luminosity. In this picture, sub-halos in galaxies would in fact be present, as predicted by the CDM models, but most of these would be "dark".

Evidence for the Presence of CDM Substructure in Galaxies. A direct indication of the presence of substructure in the mass distribution of galaxies indeed exists, which originates from gravitational lens systems. As we have seen in Sect. 3.8, the image configuration of multiple quasars can be described by simple mass models for the gravitational lens. Concentrating on those systems with four images of a source, for which the position of the lens is also observed (e.g., with the HST), a simple mass model for the lens has fewer free parameters than the coordinates of the observed quasar images that need to be fitted. Despite of this, it is possible, with very few exceptions, to describe the angular positions of the images with such a model very accurately. This result is not trivial, because for some lens systems which were observed using VLBI techniques, the image positions are known with a precision of better than 10^{-4} arcsec, with an image separation of the order of $1''$. This result demonstrates that the mass distribution of lens galaxies is, on scales of the image separation, quite well described by simple mass models.

Besides the image positions, such lens models also predict the magnifications μ of the individual images. Therefore, the ratio of the magnifications of two images should agree with the flux ratio of these images of the background source. The surprising result from the analysis of lens systems is that, although the image positions of (nearly) all quadruply imaged systems are very precisely reproduced by a simple mass model, in not a single one of these systems does the mass model reproduce the flux ratios of the images!

Perhaps the simplest explanation for these results is that the simple mass models used for the lens are not correct and other kinds of lens models should be used. However, this explanation can be excluded for many of the observed systems. Some of these systems contain

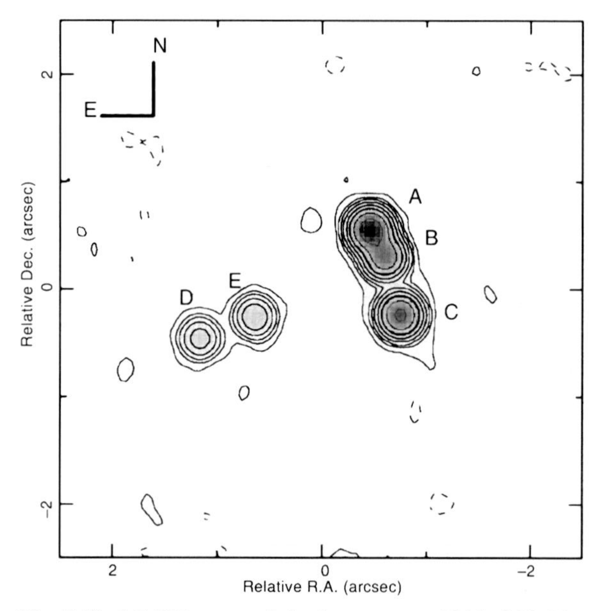

Fig. 7.19. 8.5-GHz map of the lens system 2045+265. The source at $z_s = 1.28$ is imaged four-fold (components A-D) by a lens galaxy at $z_s = 0.867$, while component E represents emission from the lens, as is evident from its different radio spectrum. From the general properties of the gravitational lens mapping, one can show that any "smooth" mass model of the lens predicts the flux of B to be roughly the same as the sum of the fluxes of components A and C. Obviously, this rule is strongly violated in this lens system, because B is weaker than both A and C. This result can only be explained by small-scale structure in the mass distribution of the lens galaxy

two or three images of the source that are positioned very closely together, for which one therefore knows that they are located close to a critical curve. In such a case, the magnification can be estimated quite well analytically; in particular, it no longer depends on the exact form of the lens model employed. Hence, the existence of such "universal properties" of the lens mapping excludes the existence of *simple* (i.e., "smooth") mass models capable of describing the observed flux ratios. One example of this is presented in Fig. 7.19.

The natural explanation for these flux discrepancies is the fact that a lensing galaxy does not only have a smooth large-scale mass profile, but that there is also small-scale substructure in its density. In the case of spiral galaxies, this may be the spiral arms, which can be seen as a small-scale perturbation in an otherwise smooth mass profile. However, most lens galaxies are ellipticals. The sub-halos that are predicted by the CDM model may then represent the substructure in their mass distribution. For a further discussion of this model, we first should mention that a small-scale perturbation of the mass profile only slightly changes the deflection angle caused by the lens, whereas the magnification μ may be modified much more strongly. As a matter of fact, by means of simulations, it has been demonstrated that lens galaxies containing sub-halos of about the same abundance as postulated by the CDM model give rise to a statistical distribution of discrepancies in the flux ratios which is very similar to that found in the observed lens systems. Furthermore, these simulations show that, on average, a particular image of the source is clearly demagnified compared to the predictions by simple, smooth lens models, again in agreement with the observational results. And finally, in the case that a relatively massive sub-halo is located close to one of the images, the image position should also be slightly shifted, compared to the smooth mass model. This effect was in fact directly detected in two lens systems: in these cases, a sub-halo exists in the lens galaxy which is massive enough to form stars, and which therefore can be observed. Its effect on the magnification and the image position can then be inferred from the lens model (see Fig. 7.20).

For these reasons, it is probable that galaxies contain sub-halos, as predicted by the CDM model, but most sub-halos, in particular those with low mass, contain

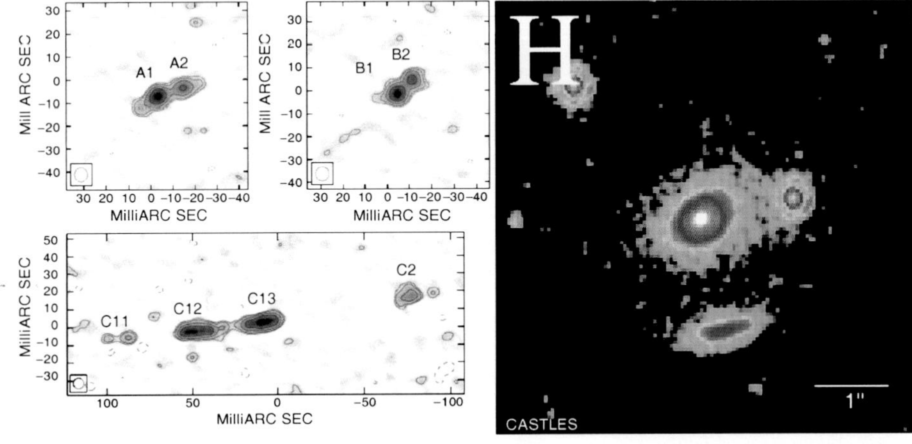

Fig. 7.20. On the right, an H-band image of the lens system MG 2016+112 is shown, consisting of a lens galaxy in the center and four images of the background source, the two southernmost of which are nearly merged in this image. On the left, VLBI maps of these components are presented; the radio source consists of a compact core and a jet component, clearly visible in images A and B. The VLBI map of component C reveals that is it in fact a double image of the source, in which the core and jet components each are visible twice. Any smooth mass model for the lens galaxy predicts that the separation C12–C11 should roughly be the same as that between C13–C2, which obviously contradicts the observation. In this case, the substructure in the mass distribution is even visible: if one includes the weak emission south of component C, which is visible in the image on the right, into the lens model as a mass component, the separation of the components in image C can be well modeled

only few stars and are therefore not visible. One consequence of this explanation is that the low-mass satellite galaxies that are seen in our Local Group should be dominated by dark matter. Given the faintness and low surface brightness of these galaxies, obtaining kinematical information for them is very difficult and requires large telescopes for spectroscopy of individual stars in these objects. The results of such investigations indicate that the dwarf galaxies in the Local Group are indeed dark matter dominated, with a mass-to-light ratio of ~ 100 in Solar units. However, this conclusion is based on the assumption that the stars in these systems are in dynamical equilibrium, an assumption which is difficult to test.

7.6 Peculiar Velocities

As mentioned on several occasions before, cosmic sources do not exactly follow the Hubble expansion, but have an additional peculiar velocity. Deviations from the Hubble flow are caused by local gravitational fields, and such fields are in turn generated by local density fluctuations. These inevitably lead to an acceleration, which affects the matter and generates peculiar velocities. In numerical simulations, the peculiar velocities of individual particles are followed in the computations automatically. In this brief section, we will investigate the large-scale peculiar velocities as they are derived from linear perturbation theory.

Since the spatial dependence of the density contrast δ is constant in time, $\delta(\boldsymbol{x},t)=\delta_0(\boldsymbol{x})\,D_+(t)$ (see Eq. 7.14), the acceleration vector $\boldsymbol{g}$ has a constant direction in the framework of linear perturbation theory. Hence, one obtains the peculiar velocity in the form

$$\boldsymbol{u}(\boldsymbol{x}) \sim \int \mathrm{d}t\, \boldsymbol{g}(\boldsymbol{x},t)\,,$$

i.e., parallel to $\boldsymbol{g}(\boldsymbol{x})$. Quantitatively, we obtain for today, thus for $t=t_0$, a relation between the velocity and acceleration field:

$$\boldsymbol{u}(\boldsymbol{x}) = \frac{2}{3H_0\Omega_\mathrm{m}}\, f(\Omega_\mathrm{m})\,\boldsymbol{g}(\boldsymbol{x})\,, \tag{7.43}$$

where we defined the function

$$f(\Omega_\mathrm{m}) := \frac{a(t)}{D_+(t)}\,\frac{\mathrm{d}D_+}{\mathrm{d}a} = \frac{\mathrm{d}\log D_+}{\mathrm{d}\log a}\,. \tag{7.44}$$

For $t=t_0$, the function $f(\Omega_\mathrm{m})$ can be expressed by a very simple and very accurate approximation, $f(\Omega_\mathrm{m})\approx\Omega_\mathrm{m}^{0.6}$. This was first discovered for the case where $\Omega_\Lambda=0$, but it was later found that a cosmological constant has only a marginal effect on this relation. Introducing corrections arising from Ω_Λ, one obtains the slightly more accurate approximation

$$f \approx \Omega_\mathrm{m}^{0.6} + \frac{\Omega_\Lambda}{70}\left(1+\frac{\Omega_\mathrm{m}}{2}\right)\,. \tag{7.45}$$

From the smallness of the last term, one can see that the correction for Λ is marginal indeed, because of which one sets $f=\Omega_\mathrm{m}^{0.6}$ in most cases.

On the other hand, $\boldsymbol{g}(\boldsymbol{x})$ is the gradient of the gravitational potential, $\boldsymbol{g}\propto-\nabla\phi$. This implies that $\boldsymbol{u}(\boldsymbol{x})$ is a gradient field, i.e., a scalar function $\psi(\boldsymbol{x})$ exists such that $\boldsymbol{u}=\nabla\psi$, where the gradient is taken with respect to the comoving spatial coordinate $\boldsymbol{x}$. Therefore, $\nabla\cdot\boldsymbol{g}\propto-\nabla^2\phi\propto-\delta$, so that also $\nabla\cdot\boldsymbol{u}\propto-\delta$; here, the Poisson equation (7.10) has been utilized. Taken together, these results yield for today

$$\nabla\cdot\boldsymbol{u}(\boldsymbol{x}) = -H_0\,\Omega_\mathrm{m}^{0.6}\,\delta_0(\boldsymbol{x})\,. \tag{7.46}$$

We would like to derive this result in somewhat more detail and begin with the linearized form of Eq. (7.8),

$$\frac{\partial\delta}{\partial t} + \frac{1}{a}\,\nabla\cdot\boldsymbol{u} = 0\,, \tag{7.47}$$

where the gradient is, here and in the following, always taken with respect to comoving coordinates. The fact that $\delta(\boldsymbol{x},t)$ factorizes (see Eq. 7.14) immediately yields

$$\frac{\partial\delta}{\partial t} = \frac{\dot D_+}{D_+}\,\delta\,.$$

Combining this equation with (7.47) and, as above, defining $\boldsymbol{u}=\nabla\psi$ leads to

$$\begin{aligned}\nabla^2\psi = \nabla\cdot\boldsymbol{u} &= -a\,\frac{\dot D_+}{D_+}\delta = -a\,\dot a\,\frac{1}{D_+}\,\frac{\mathrm{d}D_+}{\mathrm{d}a}\,\delta\\ &= -a\,H(a)\,f(\Omega_\mathrm{m})\,\delta \approx -a\,H(a)\,\Omega_\mathrm{m}^{0.6}\delta\,,\end{aligned} \tag{7.48}$$

where we used the previously defined function $f(\Omega_\mathrm{m})$. This Poisson equation for ψ can be solved, and by computing the gradient the peculiar velocity field can be calculated,

$$u(x, t) = \frac{\Omega_{\rm m}^{0.6}}{4\pi} \, a \, H(a) \int {\rm d}^3 y \, \delta(y, t) \, \frac{y - x}{|y - x|^3} \; . \tag{7.49}$$

Equation (7.49) shows that the velocity field can be derived from the density field. If the density field in the Universe were observable, one would obtain a direct prediction for the corresponding velocity field from the above relations. This depends on the matter density $\Omega_{\rm m}$, so that from a comparison with the observed velocity field, one could estimate the value for $\Omega_{\rm m}$. We will come back to this in Sect. 8.1.6.

7.7 Origin of the Density Fluctuations

We have seen in Sect. 4.5.3 that the horizon and the flatness problem in the normal Friedmann–Lemaître evolution of the Universe can be solved by postulating an early phase of very rapid – exponential – expansion of the cosmos. In this inflationary phase of the Universe, any initial curvature of space is smoothed away by the tremendous expansion. Furthermore, the exponential expansion enables the complete currently visible Universe to have been in causal contact prior to the inflationary phase. These two aspects of the inflationary model are so attractive that today most cosmologists consider inflation as part of the standard model, even if the physics of inflation is as yet not understood in detail.

The inflationary model has another property that is considered very promising. Through the huge expansion of the Universe, microscopic scales are blown up to macroscopic dimensions. The large-scale structure in the current Universe corresponds to microscopic scales prior to and during the inflationary phase. From quantum mechanics, we know that the matter distribution cannot be fully homogeneous, but it is subject to quantum fluctuations, expressed, e.g., by Heisenberg's uncertainty relation. By inflation, these small quantum fluctuations are expanded to large-scale density fluctuations. For this reason, the inflationary model also provides a natural explanation for the presence of initial density fluctuations.

In fact, one can study these effects quantitatively and attempt to calculate the initial power spectrum of these fluctuations. The result of such investigations will depend slightly on the details of the inflationary model they are based on. However, these models agree in their prediction that the initial power spectrum should have a form very similar to the Harrison–Zeldovich fluctuation spectrum, except that the spectral index $n_{\rm s}$ of the primordial power spectrum should be slightly smaller than the Harrison–Zeldovich value of $n_{\rm s} = 1$. Thus, the model of inflation can be directly tested by measuring the power spectrum and, as we shall see in Chap. 8, the power-law slope $n_{\rm s}$ indeed seems to be slightly flatter that unity, as expected from inflation.

The various inflationary models also differ in their predictions of the relative strength of the fluctuations of spacetime, which should be present after inflation. Such fluctuations are not directly linked to density fluctuations, but they are a consequence of General Relativity, according to which spacetime itself is also a dynamical parameter. One consequence of this is the existence of gravitational waves. Although no gravitational waves have been directly detected until now, the analysis of the double pulsar PSR J1915+1606 proves the existence of such waves.[9] Primordial gravitational waves provide an opportunity to empirically distinguish between the various models of inflation. These gravitational waves leave a "footprint" in the polarization of the cosmic microwave background that is measurable in principle. A satellite mission to perform these measurements is currently being discussed.

[9] The double pulsar PSR J1915+1606 was discovered in 1974. From the orbital motion of the pulsar and its companion star, gravitational waves are emitted, according to General Relativity. Through this, the system loses kinetic (orbital) energy, so that the size of the orbit decreases over time. Since pulsars represent excellent clocks, and we can measure time with extremely high precision, this change in the orbital motion can be observed with very high accuracy and compared with predictions from General Relativity. The fantastic agreement of theory and observation is considered a definite proof of the existence of gravitational waves. For the discovery of the double pulsar and the detailed analysis of this system, Russell Hulse and Joseph Taylor were awarded the Nobel Prize in Physics in 1993. In 2003, a double neutron star binary was discovered where pulsed radiation from both components can be observed. This fact, together with the small orbital period of 2.4 h implying a small separation of the two stars, makes this an even better laboratory for studying strong-field gravity.

8. Cosmology III: The Cosmological Parameters

In Chaps. 4 and 7, we described the fundamental aspects of the standard model of cosmology. Together with the knowledge of galaxies, clusters of galaxies, and AGNs that we have gained in the other chapters, we are now ready to discuss the determination of the various cosmological parameters. In the course of this discussion, we will describe a number of methods, each of which is in itself useful for estimating cosmological parameters, and we will present the corresponding results from these methods. The most important aspect of this chapter is that we now have more than one independent estimate for each cosmological parameter, so that the determination of these parameters is highly redundant. This very aspect is considerably more important than the precise values of the parameters themselves, because it provides a test for the consistency of the cosmological model.

We will give an example in order to make this point clear. In Sect. 4.4.4, we discussed how the cosmic baryon density can be determined from primordial nucleosynthesis and the observed ratio of deuterium to hydrogen in the Universe. Thus, this determination is based on the correctness of our picture of the thermal history of the early Universe, and on the validity of the laws of nuclear physics shortly after the Big Bang. As we will see later, the baryon density can also be derived from the angular fluctuations in the cosmic background radiation, for which the structure formation in a CDM model, discussed in the previous chapter, is needed as a foundation. If our standard model of cosmology was inconsistent, there would be no reason for these two values of the baryon density to agree – as they do in a remarkable way. Therefore, in addition to obtaining a more precise value of Ω_b from this comparison than from each of the individual methods alone, the agreement is also a strong indication of the validity of the standard model.

We will begin in Sect. 8.1 with the observation of the large-scale distribution of matter, the large-scale structure (LSS). It is impossible to observe the large-scale structure of the matter distribution itself; rather, only the spatial distribution of visible galaxies can be measured. Assuming that the galaxy distribution follows, at least approximately (which we will specify later), that of the dark matter, the power spectrum of the density fluctuations can be estimated from that of the galaxies. As we discussed in the previous chapter, the power spectrum in turn depends on the cosmological parameters. In Sect. 8.2, we will summarize some aspects of clusters of galaxies which are relevant for the determination of the cosmological parameters.

In Sect. 8.3, Type Ia supernovae will be used as cosmological tools, and we will discuss their Hubble diagram. Since SN Ia are considered to be standard candles, their Hubble diagram provides information on the density parameters Ω_m and Ω_Λ. These observations provided the first clear indication, around 1998, that the cosmological constant differs from zero. We will then analyze the lensing effect of the LSS in Sect. 8.4, by means of which information about the statistical properties of the LSS of matter is obtained directly, without the necessity for any assumptions on the relation between matter and galaxies. As a matter of fact, this galaxy-mass relation can be directly inferred from the lens effect. In Sect. 8.5, we will turn to the properties of the intergalactic medium and, in particular, we will introduce the Lyman-α forest in QSO spectra as a cosmological probe.

Finally, we will discuss the anisotropy of the cosmic microwave background in Sect. 8.6. Through observations of the cosmic microwave background and their analysis, a vast amount of very accurate information about the cosmological parameters are obtained. In particular, we will report on the recent and exciting results concerning CMB anisotropies, and will combine these findings with the results obtained by other methods. This combination yields a set of parameters for the cosmological model which is able to describe nearly all observations of cosmological relevance in a self-consistent manner, and which today defines the standard model of cosmology.

8.1 Redshift Surveys of Galaxies

8.1.1 Introduction

The inhomogeneous large-scale distribution of matter that was described in Chap. 7 is not observable directly

Peter Schneider, Cosmology III: The Cosmological Parameters.
In: Peter Schneider, Extragalactic Astronomy and Cosmology. pp. 309–354 (2006)
DOI: 10.1007/11614371_8 © Springer-Verlag Berlin Heidelberg 2006

because it consists predominantly of dark matter. If it is assumed that the distribution of galaxies traces the underlying distribution of dark matter fairly, the properties of the LSS of matter could be studied by observing the galaxy distribution in the Universe. Quite a few good reasons exist for this assumption not to be completely implausible. For instance, we observe a high galaxy density in clusters of galaxies, and with the methods discussed in Chap. 6, we are able to verify that clusters indeed represent strong mass concentrations. Qualitatively, this assumption therefore seems to be justified. We will later modify it slightly.

In any case, the distribution of galaxies on the sphere appears inhomogeneous and features large-scale structure. Since galaxies have evolved from the general cosmic density field, they should contain information about the latter. It is consequently of great interest to examine and quantify the properties of the galaxy distribution.

In principle, two possible ways exist to accomplish this study of the galaxy distribution. With photometric sky surveys, the two-dimensional distribution of galaxies on the sphere can be mapped. To also determine the third spatial coordinate, it is necessary to measure the redshift of the galaxies using spectroscopy, deriving the distance from the Hubble law (1.6). It is obvious that we can learn considerably more about the statistical properties of the galaxy distribution from their three-dimensional distribution; hence, redshift surveys are of particular interest.

The graphical representation of the spatial galaxy positions is accomplished with so-called wedge diagrams. They represent a sector of a circle, with the Milky Way at its center. The radial coordinate is proportional to z (or cz – by this, the distance is measured in km/s), and the polar angle of the diagram represents an angular coordinate in the sky (e.g., right ascension), where an interval in the second angular coordinate is selected in which the galaxies are located. An example for such a wedge diagram is shown in Fig. 8.1.

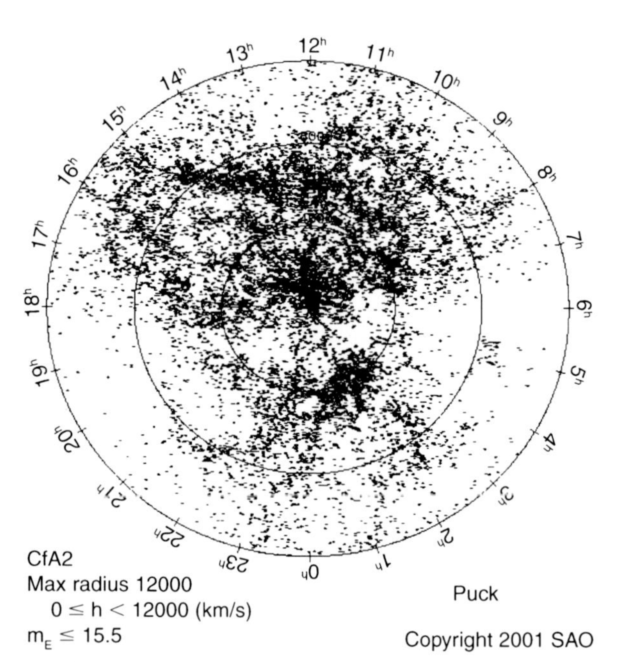

Fig. 8.1. The CfA redshift survey, in equatorial coordinates. Along its radial axis, this wedge diagram shows the escape velocity cz up to 12 000 km/s, and the polar angle specifies the right ascension of a galaxy. The Great Wall extends from $9^{\rm h}$ to $15^{\rm h}$. The overdensity at $1^{\rm h}$ and $cz = 4000$ km/s is the Pisces–Perseus supercluster

8.1.2 Redshift Surveys

Performing redshift surveys is a very time-consuming task compared to making photometric sky maps, because recording a spectrum requires much more observing time than the mere determination of the apparent magnitude of a source. Hence, the history of redshift surveys, like that of many other fields in astronomy, is driven by the development of telescopes and instruments. The introduction of CCDs in astronomy in the early 1980s provided a substantial increase in sensitivity and accuracy of optical detectors, and enabled us to carry out redshift surveys of galaxies in the nearby Universe containing several thousand galaxies (see Fig. 8.1). Using a single slit in the spectrograph implied that in each observation the spectra of only one or very few galaxies could be recorded simultaneously. The situation changed with the introduction of spectrographs with high multiplexity which were designed specifically to perform redshift surveys. With them, the spectra of many objects (up to a thousand) in the field-of-view of the instrument can be observed simultaneously.

The Strategy of Redshift Surveys. Such a survey is basically defined by two criteria. The first is its geometry:

a region of the sky is chosen in which the survey is performed. Second, those objects in this region need to be selected for which spectra should be obtained. In most cases, for practical reasons the objects are selected according to their brightness, i.e., spectra are taken of all galaxies above a certain brightness threshold. The latter defines the number density of galaxies in the survey, as well as the required exposure time. To apply the second criterion, a photometric catalog of sources is required as a starting point. The criteria may be refined further in some cases. For instance, a minimum angular extent of objects may be chosen to avoid the inclusion of stars. The spectrograph may set constraints on the selection of objects; e.g., a multi-object spectrograph is often unable to observe two sources that are too close together on the sky.

Examples of Redshift Surveys. In the 1980s, the *Center for Astrophysics (CfA) Survey* was carried out which measured the redshifts of more than 14 000 galaxies in the local Universe (Fig. 8.1). The largest distances of these galaxies correspond to about $cz \sim 15\,000$ km/s. One of the most spectacular results from this survey was the discovery of the "Great Wall", a huge structure in the galaxy distribution (see also Fig. 7.2).

In the *Las Campanas Redshift Survey (LCRS)*, carried out in the first half of the 1990s, the redshifts of more than 26 000 galaxies were measured. They are located in six narrow strips of 80° length and 1.5° width each. With distances of up to $\sim 60\,000$ km/s, this survey is considerably deeper than the CfA Redshift Survey. The distribution of galaxies is displayed in Fig. 8.2, from which we can recognize the typical bubble or honeycomb structure. Galaxies are distributed along filaments, which are surrounding large regions in which virtually no galaxies exist – the aforementioned voids. The galaxy distribution shows a structure which is qualitatively very similar to the dark matter distribution generated in numerical simulations (see, e.g., Fig. 7.12). In addition, we see from the galaxy distribution that no structures exist with scales comparable to the extent of the survey. Thus, the LCRS has probed a scale larger than that where significant structures of the mass distribution are found. The survey volume of the LCRS therefore covers a representative section of the Universe.

A different kind of redshift survey became possible through the sky survey carried out with the IRAS satellite (see Sect. 1.3.2). In these redshift surveys of IRAS galaxies, the selection of objects for which spectra were obtained was based on the 60 μm flux measured by IRAS in its (near) all-sky survey. Various redshift surveys are based on this selection, differing in the flux limit applied; for example the 2 Jy survey (hence, $S_{60\,\mu\mathrm{m}} \geq 2$ Jy), or the 1.2 Jy survey. The QDOT and PSCz surveys both have a limiting flux of $S_{60\,\mu\mathrm{m}} \geq 0.6$ Jy, where QDOT observed spectra for one out of six randomly chosen galaxies from the IRAS sample, while PSCz is virtually complete and contains $\sim 15\,500$ redshifts. One of the advantages of the IRAS

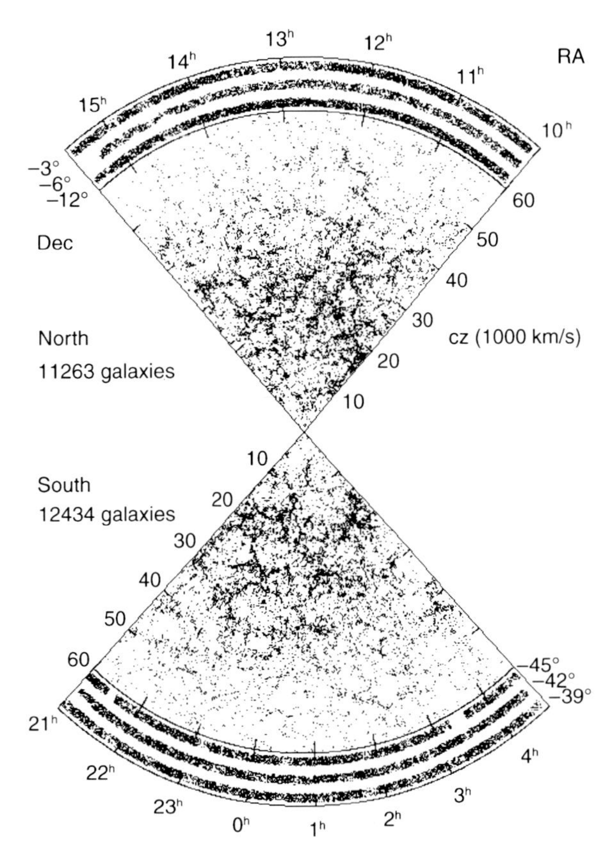

Fig. 8.2. The Las Campanas Redshift Survey consists of three fields each at the North and South Galactic Pole. Each of these fields is a strip 1.5° wide and 80° long. Overall, the survey contains about 26 000 galaxies, and the median of their redshift is about 0.1. The six strips show the distribution of galaxies on the sphere, and the wedge diagram indicates, for galaxies with measured redshift, the right ascension versus distance from the Milky Way, measured in units of 1000 km s^{-1}

surveys is that the FIR flux is nearly unaffected by Galactic absorption, an effect that needs to be corrected for when galaxies are selected from optical photometry. Furthermore, the PSCz is an "all-sky" survey, containing the galaxy distribution in a sphere around us, so that we obtain a complete picture of the local galaxy distribution. However, one needs to be aware of the fact that in selecting galaxies via their FIR emission one is thus selecting a particular type of galaxy, predominantly those which have a high dust content and active star formation which heats the dust.

The *Canada–France Redshift Survey (CFRS)* obtained spectroscopy of faint galaxies with $17.5 \le I \le 22.5$, with a median redshift of about 0.5. The resulting catalog contains 948 objects, 591 of which are galaxies. This survey was performed by a multi-object spectrograph at the CFHT (see Sect. 1.3.3) which was able to take the spectra of up to 100 objects simultaneously. For the first time, due to its faint limiting magnitude it enabled us to study the evolution of (optically-selected) galaxies, for example by means of their luminosity function and their star-formation rate, and to investigate the redshift dependence of the galaxy correlation function – and thus to see the evolution of the large-scale structure.

Currently, two large spectroscopic surveys with faint limiting magnitudes are being carried out. Both of them use high multiplex spectrographs mounted on 10-m class telescopes: the VIMOS instrument on the VLT and the DEIMOS instrument on Keck. The target of both surveys, the VIMOS VLT Deep Survey (VVDS) and the DEEP2 survey, is to obtain spectra of several tens of thousands of galaxies with $z \sim 1$, thus extending the CFRS by more than an order of magnitude in sample size and by ~ 1.5 magnitudes in depth.

The 2dF Survey and the Sloan Digital Sky Survey. The scientific results from the first redshift surveys motivated the production of considerably more extended surveys. By averaging over substantially larger volumes in the Universe, it was expected that the statistics on the galaxy distribution could be significantly improved. In addition, the analysis of the galaxy distribution at higher redshift would also enable a measurement of the evolution in the galaxy distribution. Two very extensive redshift surveys were performed with these main objectives in mind: the *two-degree Field Galaxy Redshift Survey (2dFGRS)* and the *Sloan Digital Sky Survey (SDSS)*.

The 2dFGRS was carried out using a spectrograph specially designed for this project, which was mounted at the 4-m Anglo Australian Telescope. Using optical fibers to transmit the light of the observed objects from the focal plane to the spectrograph, up to 400 spectra could be observed simultaneously over a usable field with a diameter of 2°. The positioning of the individual fibers on the location of the pre-selected objects was performed by a robot. The redshift survey covered two large connected regions in the sky, of $75° \times 15°$ and $75° \times 7.5°$, plus 100 additional, randomly distributed fields. This survey geometry was chosen so as to yield the optimal cosmological information about the galaxy distribution, that is, the most precise measurement of the correlation function at relevant scales. The photometric input catalog was the APM galaxy catalog which had been compiled from digitized photographic plates. The limiting magnitude of the galaxies for which spectra were obtained is approximately $B \lesssim 19.5$, where this value is corrected for Galactic extinction. The 2dFGRS has been completed, and it contains redshifts for more than 230 000 galaxies (see Fig. 7.1). The spectra and redshifts are publicly available. The scientific yield from this large data set is already very impressive, as we will show further below.

For the SDSS, a dedicated telescope was built, equipped with two instruments. The first is a camera with 30 CCDs which has scanned nearly a quarter of the sky in five photometric bands, generating by far the largest photometric sky survey with CCDs. The amount of data collected in this survey is enormous, and its storage and reduction required a tremendous effort. For this photometric part of the Sloan Survey, a new photometric system was developed, with its five filters (u, g, r, i, z) chosen such that their transmission curves overlap as little as possible (see Appendix A.4). The selection of targets for spectroscopy was carried out using this photometric information. As in the 2dF Survey, the multi-object spectrograph used optical fibers, and in this case these had to be manually installed in holes that had been punched into a metal plate. With about 640 simultaneously observed spectra, the strategy was similar to that for the 2dFGRS. The aim of the spectroscopic survey was to obtain about a million galaxy spectra. The data products of the SDSS have been made publicly available at regular intervals, and currently (2006) about half of the survey has been published. For the

SDSS, the scientific yield has also been very high already, and not just based on the redshift survey. In fact, the photometric data has been used in a large variety of other Galactic and extragalactic projects.

Both the 2dF survey and the SDSS also recorded, besides the spectra of galaxies, those of QSOs which were selected based on their optical colors; this yielded by far the most extensive QSO surveys.

8.1.3 Determination of the Power Spectrum

We will now return to the question of whether the distribution of (dark) matter in the Universe can be derived from the observed distribution of galaxies. If galaxies trace the distribution of dark matter fairly, the power spectrum of dark matter can be determined from the galaxy distribution. However, since the formation and evolution of galaxies is as yet not understood sufficiently well to allow us to quantitatively predict the relation between galaxies and dark matter (at least not without introducing a number of model assumptions), this assumption is not justified a priori. For instance, it may be that there is a threshold in the local density of dark matter, below which the formation of galaxies does not occur or is at least strongly suppressed.

The connection between dark matter and galaxies is parametrized by the so-called *linear bias factor b*. It is defined by

$$\delta_g := \frac{\Delta n}{\overline{n}} = b \frac{\Delta \rho}{\overline{\rho}} = b\,\delta \,, \tag{8.1}$$

where $\overline{n}$ is the average density of the galaxy population considered, and $\Delta n = n - \overline{n}$ is the deviation of the local number density of galaxies from their average density. Hence, the bias factor is the ratio of the relative overdensities of galaxies to dark matter. Such a linear relation is not strictly justified from theory. However, it is a plausible ansatz on scales where the density field is linear. In principle, the bias factor b may depend on the galaxy type, on redshift, and on the length-scale that is considered.

The definition given in (8.1) must be understood in a statistical sense. In a volume V, we expect on average $\overline{N} = \overline{n}\,V$ galaxies, whereas the observed number of galaxies is $N = n\,V$. Hence,

$$\left(\frac{\Delta n}{\overline{n}}\right)_V = \frac{n - \overline{n}}{\overline{n}} = \frac{N - \overline{N}}{\overline{N}} = b\,\delta_V \,,$$

where δ_V is the density contrast of matter, averaged over the volume V. Under the assumption of linear biasing, we can then infer the statistical properties of matter from those of the galaxy distribution. A physical model for biasing is sketched in Fig. 8.3.

Normalization of the Power Spectrum. In Sect. 7.4.2, we demonstrated that the power spectrum of the density fluctuations can be predicted in the framework of a CDM model, except for its normalization which has to be measured empirically. A convenient way for its parametrization is through the parameter σ_8. This parameter is motivated by the following observation.

Analyzing spheres of radius $R = 8h^{-1}$ Mpc in the local Universe, it is found that optically-selected galaxies have, on this scale, a fluctuation amplitude of about 1,

$$\sigma_{8,g}^2 := \left\langle \left(\frac{\Delta n}{\overline{n}}\right)^2 \right\rangle_8 \approx 1 \,, \tag{8.2}$$

where the averaging is performed over different spheres of identical radius $R = 8h^{-1}$ Mpc. Accordingly, we define the dispersion of the matter density contrast, averaged over spheres of radius $R = 8h^{-1}$ Mpc as

$$\sigma_8^2 = \left\langle \delta^2 \right\rangle_8 \,. \tag{8.3}$$

Using the definition of the bias factor (8.1), we then obtain

$$\sigma_8 = \frac{\sigma_{8,g}}{b} \approx \frac{1}{b} \,. \tag{8.4}$$

Because of this simple relation, it has become common practice to use σ_8 as a parameter for the normalization of the power spectrum.[1] If $b = 1$, thus if galaxies trace the matter distribution fairly, then one has $\sigma_8 \approx 1$. If b is not too different from unity, we see that the density fluctuations on a scale of $\sim 8h^{-1}$ Mpc are becoming non-linear at the present epoch, in the sense that $\delta \sim 1$. On larger scales, the evolution of the density contrast can approximately be described by linear perturbation theory.

[1] More precisely, one considers σ_8 the normalization of the power spectrum linearly extrapolated to the present day, $P_0(k)$, so that the relation (8.4) needs to be modified slightly.

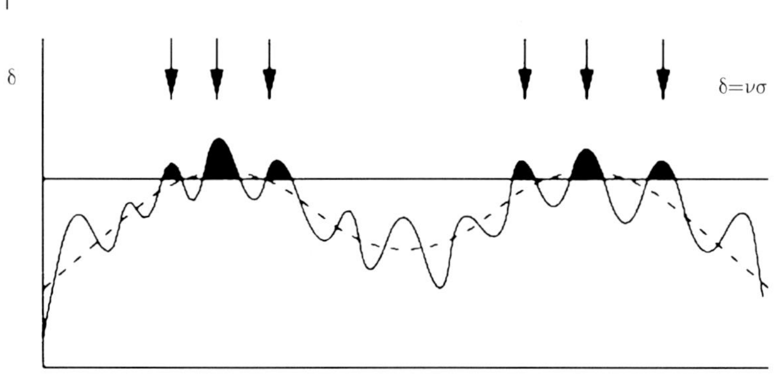

Fig. 8.3. The sketch represents a particular model of biasing. Let the one-dimensional density profile of matter be specified by the solid curve, which results from a superposition of a large-scale (represented by the dashed curve) and a small-scale fluctuation. Assuming that galaxies can form only at locations where the density field exceeds a certain threshold – plotted as a straight line – the galaxies in this density profile will be localized at the positions indicated by the arrows. Obviously, the locations of the galaxies are highly correlated; they only form near the peaks of the large-scale fluctuation. In this picture, the correlation of galaxies on small scales is much stronger than the correlation of the underlying density field

Shape of the Power Spectrum. If one assumes that b does not depend on the length-scale considered, the *shape* of the dark matter power spectrum can be determined from the power spectrum $P_g(k)$ of the galaxies, whereas its amplitude depends on b. As we have seen in Sect. 7.4.2, the shape of $P(k)$ is described by the shape parameter $\Gamma = h\,\Omega_m$ in the framework of CDM models.

The comparison of the shape of the power spectrum of galaxies with that of CDM models yields $\Gamma \sim 0.25$ (see Fig. 8.4). Since $\Gamma = h\Omega_m$, this result indicates a Universe of low density (unless h is unreasonably low).

In the 2dFGRS, the power spectrum of galaxies was measured with a much higher accuracy than had previously been possible. Since a constant b can be expected, at best, in the linear domain, i.e., on scales above $\sim 10h^{-1}$ Mpc, only such linear scales are used in the comparison with the power spectra from CDM models. As the density parameter Ω_m seems to be relatively small, the baryonic density plays a noticeable role in the transfer function (see Eq. 7.25) which depends on Ω_b as well as on Γ. The measurement accuracy of the galaxy distribution in the 2dFGRS is high enough to be sensitive to this dependence. In Fig. 8.5, the measured power spectrum of galaxies from the 2dFGRS is shown, together with predictions from CDM models for different shape parameters Γ. Two families of model curves are drawn: one where the baryon density is set to zero, and the other for the value of Ω_b which results from the analysis of primordial nucleosynthesis (see Sect. 4.4.4).

Considering models in which Ω_b is a free parameter, there are two domains in parameter space for which good fits of the power spectrum of the galaxy distribution are obtained (see Fig. 8.5). One of the two domains is characterized by a very high baryon fraction of the matter density, and by a very large value for $\Omega_m h$. These parameter values are incompatible with virtually every other determination of cosmological parameters. On the other hand, a good fit to the shape of the power spectrum is obtained by

$$\Gamma = \Omega_m h = 0.18 \pm 0.02\,, \quad \Omega_b/\Omega_m = 0.17 \pm 0.06\,. \tag{8.5}$$

As is seen in Fig. 8.5, and as we will show further below, these values for the parameters are in very good agreement with those obtained from other cosmological observations.

Comparing the power spectra of two different types of galaxies, we should observe that they are proportional to each other and to the power spectrum of dark matter. Their amplitudes, however, may differ if their bias factors are different. The comparison of red and blue galaxies in the 2dFGRS shows that these indeed have a very similar shape, supporting the assumption of a linear biasing on large scales. However, the bias factor

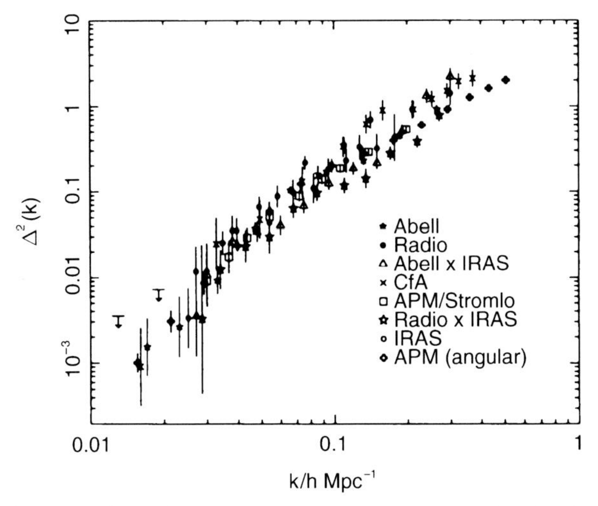

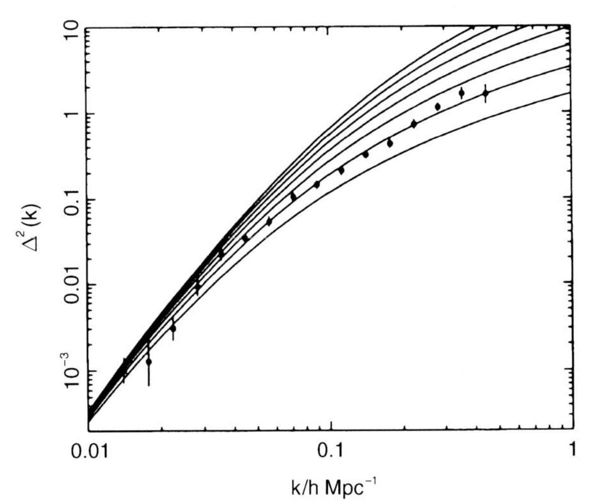

Fig. 8.4. Left: the power spectrum of galaxies is displayed, as determined from different galaxy surveys, where $\Delta^2(k) \propto k^3\,P(k)$ is a dimensionless description of the power spectrum. Right: model spectra for $\Delta(k)$ are plotted, where Γ varies from 0.5 (uppermost curve) to 0.2; the data from the various surveys have been suitably averaged. We see that a value of $\Gamma \sim 0.25$ for the shape parameter fits the observations quite well

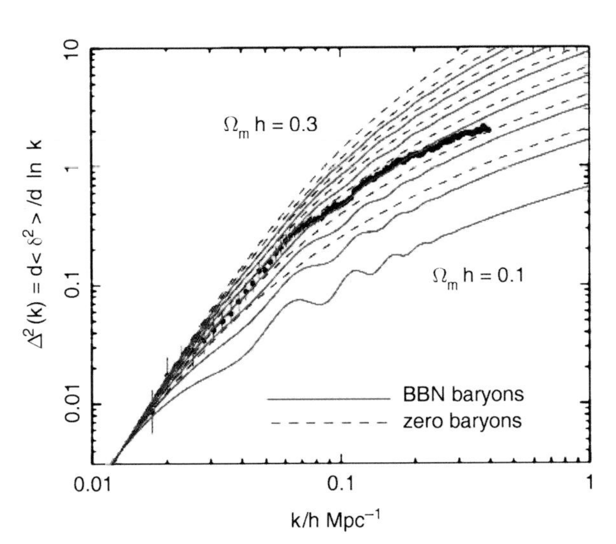

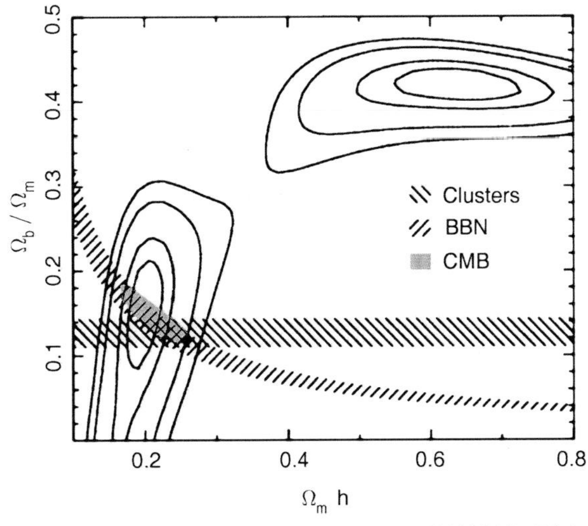

Fig. 8.5. Left: power spectrum of the galaxy distribution as measured in the 2dFGRS (points with error bars), here represented as $\Delta^2(k) \propto k^3\,P(k)$. The curves show power spectra from CDM models with different shape parameter $\Gamma = \Omega_{\rm m}h$, and two values of $\Omega_{\rm b}$: one as obtained from primordial nucleosynthesis (BBN, solid curves), and the other for models without baryons (dashed curves). The Hubble constant $h = 0.7$ and the slope $n_{\rm s} = 1$ of the primordial power spectrum were assumed. A very good fit to the observational data is obtained for $\Gamma \approx 0.2$ (from Peacock, 2003, astro-ph/0309240). Right: confidence contours in the $\Omega_{\rm m}h$–$\Omega_{\rm b}/\Omega_{\rm m}$-plane. Two regions in parameter space are seen to provide good fits to the data. The upper region is incompatible with many other cosmological data. In contrast, the lower left domain in parameter space is in remarkable agreement with measurements from BBN (see Sect. 4.4.4), with the baryon fraction in clusters of galaxies (see Sect. 8.2.3), and with CMB anisotropy measurements (see Sect. 8.6)

for red galaxies is larger by about a factor 1.4 than for blue galaxies. This result is not completely unexpected, because red galaxies are located preferentially in clusters of galaxies, whereas blue galaxies are rarely found in massive clusters. Hence, red galaxies seem to follow the density concentrations of the dark matter much more closely than blue galaxies.

8.1.4 Effect of Peculiar Velocities

Redshift Space. The relative velocities of galaxies in the Universe are not only due to the Hubble expansion but, in addition, galaxies have peculiar velocities. The peculiar velocity of the Milky Way is measurable from the CMB dipole (see Fig. 1.17). Owing to these peculiar velocities, the observed redshift of a source is the superposition of the cosmic expansion velocity and its peculiar velocity v along the line-of-sight,

$$c z = H_0 D + v \, . \tag{8.6}$$

The measurement of the other two spatial coordinates (the angular position on the sky) is not affected by the peculiar velocity. The peculiar velocity therefore causes a distortion of galaxy positions in wedge diagrams, yielding a shift in the radial direction relative to their true positions. Since, in general, only the redshift is measurable and not the true distance D, the observed three-dimensional position of a source is specified by the angular coordinates and the redshift distance

$$s = \frac{c z}{H_0} = D + \frac{v}{H_0} \, . \tag{8.7}$$

The space that is spanned by these three coordinates is called *redshift space*. In particular, we expect that the correlation function of galaxies is not isotropic in redshift space.

Galaxy Distribution in Redshift Space. The best known example of this effect is the "Fingers of God". To understand their origin, we consider galaxies in a cluster. They are situated in a small region in space, all at roughly the same distance D and within a small solid angle on the sphere. However, due to the high velocity dispersion, the galaxies span a broad range in s, which is easily identified in a wedge diagram as a highly stretched structure pointing towards us, as can be seen in Fig. 7.2.

Mass concentrations of smaller density have the opposite effect: galaxies that are closer to us than the center of this overdensity move towards the concentration, hence away from us. Therefore their redshift distance s is larger than their true distance D. Conversely, the peculiar velocity of galaxies behind the mass concentration is pointing towards us, so their s is smaller than their true distance. If we now consider galaxies that are located on a spherical shell around this mass concentration, this sphere in physical space becomes an oblate ellipsoid with symmetry axis along the line-of-sight in redshift space. This effect is illustrated in Fig. 8.6.

Hence, the distortion between physical space and redshift space is caused by peculiar velocities which manifest themselves in the transformation (8.7) of the radial coordinate in space (thus, the one along the line-of-sight). Due to this effect, the correlation function of galaxies is not isotropic in redshift space. The reason for this is the relation between the density field and the corresponding peculiar velocity field. Specializing (7.49) to the current epoch and using the relation (8.1) between the density fields of matter and of galaxies, we obtain

$$\boldsymbol{u}(\boldsymbol{x}) = \beta \, \frac{H_0}{4\pi} \int \mathrm{d}^3 y \, \delta_{\mathrm{g}}(\boldsymbol{y}) \, \frac{\boldsymbol{y} - \boldsymbol{x}}{|\boldsymbol{y} - \boldsymbol{x}|^3} \, , \tag{8.8}$$

where we defined the parameter

$$\beta := \frac{\Omega_{\mathrm{m}}^{0.6}}{b} \, . \tag{8.9}$$

This relation between the density field of galaxies and the peculiar velocity is valid in the framework of linear perturbation theory under the assumption of linear biasing. The anisotropy of the correlation function is now caused by this correlation between $\boldsymbol{u}(\boldsymbol{x})$ and $\delta_{\mathrm{g}}(\boldsymbol{x})$, and the degree of anisotropy depends on the parameter β. Since the correlation function is anisotropic, this likewise applies to the power spectrum.

Cosmological Constraints. Indeed, the anisotropy of the correlation function can be measured, as is shown in Fig. 8.7 for the 2dFGRS (where in this figure, the usual

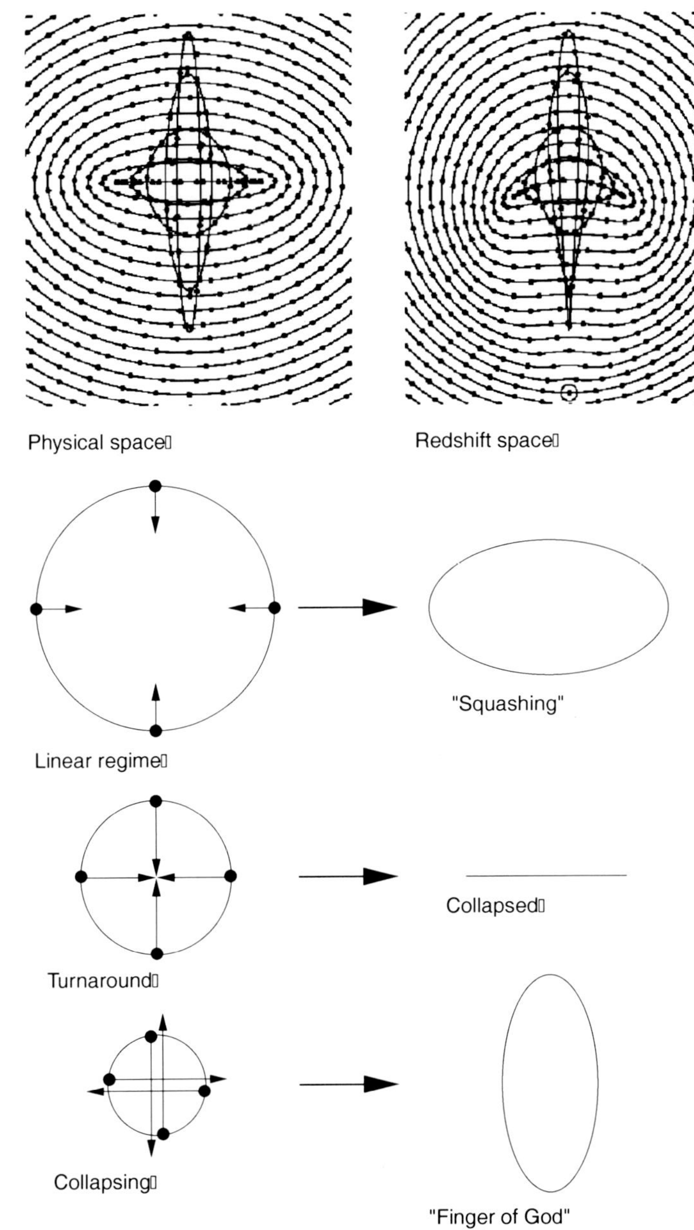

Fig. 8.6. The influence of peculiar velocities on the location of galaxies in redshift space. The upper left panel shows the positions of galaxies (points) in redshift space, which are in reality located on spherical shells. Galaxies connected by curves have the same separation from the center of a spherically-symmetric overdensity (such as a galaxy cluster) in real space. The explanation for the distortion in redshift space is given in the lower panel. On large scales, galaxies are falling into the cluster, so that galaxies closer to us have a peculiar velocity directed away from us. Thus, in redshift space they appear to be more distant than they in fact are. The inner virialized region of the cluster generates a "Finger of God", shown by the highly elongated ellipses in redshift space directed toward the observer. Here, galaxies from a small spatial region are spread out in redshift space due to the large velocity dispersion yielding large radial patterns in corresponding wedge diagrams. In the upper right panel, the same effect is shown for the case where the cluster is situated close to us (small circle in lower center)

convention of denoting the transverse separation as σ and that along the line-of-sight in redshift space as π is followed). Clearly visible is the oblateness of the curves of equal correlation strength along the line-of-sight for separations $\gtrsim 10h^{-1}$ Mpc, for which the density field is still linear, whereas for smaller separation the finger-of-god effect emerges. This oblateness at large separations depends directly on β, due to (8.8), so that β can be determined from this anisotropy.[2] However, one needs to take into account the fact that galaxies are not strictly

[2] In fact, one can decompose the correlation function into multipole components, such as the monopole (which is the isotropic part of the correlation function), quadrupole (describing the oblateness), etc. The ratio of the quadrupole and monopole components in the linear regime is independent of the underlying power spectrum, and depends only on β.

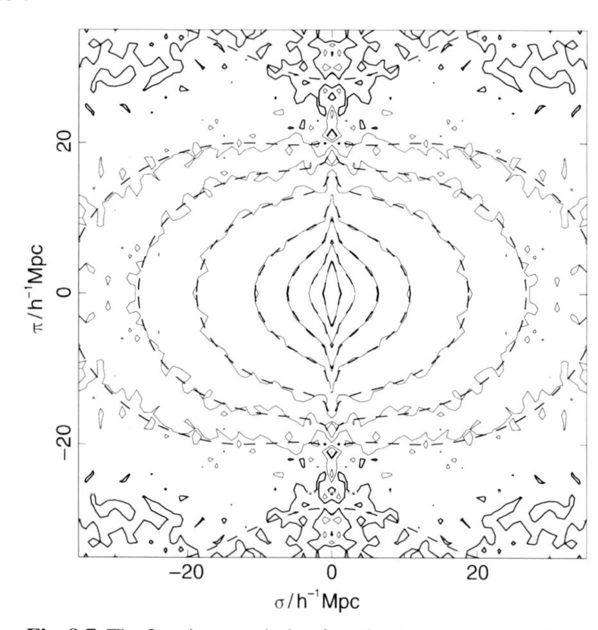

Fig. 8.7. The 2-point correlation function ξ_g, as measured from the 2dFGRS, plotted as a function of the transverse separation σ and the radial separation π in redshift space. Solid contours connect values of constant ξ_g. The dashed curves show the same correlation function, determined from a cosmological simulation that accounts for small-scale velocities. The oblateness of the distribution for large separation and the Fingers of God are clearly visible

following the cosmic velocity field. Due to small-scale gravitational interactions they have a velocity dispersion σ_p around the velocity field as predicted by linear theory. A quantitative interpretation of the anisotropy of the correlation function needs to account for this effect, which causes an additional smearing of galaxy positions in redshift space along the line-of-sight. Therefore, the derived value of β is related to σ_p. It is possible to determine both quantities simultaneously, by comparing the observed correlation function with models for different values of β and σ_p. From this analysis, confidence regions in the β-σ_p-plane are obtained, which feature a distinct minimum in the corresponding χ^2 function and by which both parameters can be estimated simultaneously. For the best estimate of these values, the 2dFGRS yielded

$$\beta = 0.51 \pm 0.05\,; \quad \sigma_p \approx 520\,\text{km/s}\,. \tag{8.10}$$

8.1.5 Angular Correlations of Galaxies

Measuring the correlation function or the power spectrum is not only possible with extensive redshift surveys of galaxies, which have become available only relatively recently. In fact, the correlation properties of galaxies can also be determined from their angular positions on the sphere. The three-dimensional correlation of galaxies in space implies that their angular positions are likewise correlated. These angular correlations are easily visible in the projection of bright galaxies onto the sphere (see Fig. 6.2).

The angular correlation function $w(\theta)$ is defined in analogy with the three-dimensional correlation function $\xi(r)$ (see Sect. 7.3.1). Considering two solid angle element $d\omega$ at $\boldsymbol{\theta}_1$ and $\boldsymbol{\theta}_2$, the probability of finding a galaxy at $\boldsymbol{\theta}_1$ is $P_1 = \overline{n}\,d\omega$, where $\overline{n}$ denotes the average density of galaxies on the sphere (with well-defined properties like, for instance, a minimum magnitude limit). The probability of finding a galaxy near $\boldsymbol{\theta}_1$ and another one near $\boldsymbol{\theta}_2$ is then

$$P_2 = (\overline{n}\,d\omega)^2\,\left[1 + w(|\boldsymbol{\theta}_1 - \boldsymbol{\theta}_2|)\right]\,, \tag{8.11}$$

where we utilize the statistical homogeneity and isotropy of the galaxy distribution, by which the correlation function w depends only on the absolute angular separation. The angular correlation function $w(\theta)$ is of course very closely related to the three-dimensional correlation function ξ_g of galaxies. Furthermore, $w(\theta)$ depends on the redshift distribution of the galaxies considered; the broader this distribution is, the fewer pairs of galaxies are found at a given angular separation which are also located close to each other in three-dimensional space, and hence are correlated. This means that the broader the redshift distribution of galaxies, the smaller the expected angular correlation.

The relation between $w(\theta)$ and $\xi_g(r)$ is given by the *Limber equation*, which can, in its simplest form, be written as

$$w(\theta) = \int dz\; p^2(z) \int d(\Delta z) \times \xi_g\left(\sqrt{[D_A(z)\theta]^2 + \left(\frac{dD}{dz}\right)^2 (\Delta z)^2}\right)\,, \tag{8.12}$$

where $D_A(z)$ is the angular diameter distance (4.45), $p(z)$ describes the redshift distribution of galaxies, and $\mathrm{d}D$ specifies the physical distance interval corresponding to a redshift interval $\mathrm{d}z$,

$$\mathrm{d}D = -c\,\mathrm{d}t = -\frac{c\,\mathrm{d}a}{a\,H} \quad \Rightarrow \quad \frac{\mathrm{d}D}{\mathrm{d}z} = \frac{c}{(1+z)\,H(z)} .$$

Long before extensive redshift surveys were performed, the correlation $w(\theta)$ had been measured. Since it is linearly related to ξ_g, and since ξ_g in turn is related to the power spectrum of the matter fluctuations and to the bias factor, the measured angular correlation function could be compared to cosmological models. For some time, such analyses have hinted at a small value for the shape parameter $\Gamma = \Omega_m h$ of about 1/4 (see Fig. 8.4), which is incompatible with an Einstein–de Sitter model. Figure 8.8 shows $w(\theta)$ for four magnitude intervals measured from the SDSS. We see that $w(\theta)$ follows a power law over a wide angular range, which we would also expect from (8.12) and from the fact that $\xi(r)$ follows a power law.[3] In addition, the figure shows that $w(\theta)$ becomes smaller the fainter the galaxies are, because fainter galaxies have a higher redshift on average and they define a broader redshift distribution.

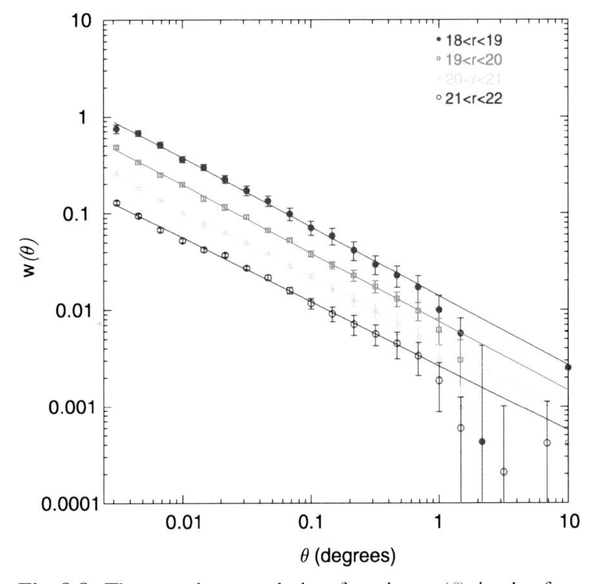

Fig. 8.8. The angular correlation function $w(\theta)$ in the four magnitude intervals $18 < r^* < 19$, $19 < r^* < 20$, $20 < r^* < 21$, and $21 < r^* < 22$, as measured from the first photometric data of the SDSS, together with a power law fit to the data in the angular range $1' \le \theta \le 30'$; the slope in all cases is very close to $\theta^{-0.7}$

8.1.6 Cosmic Peculiar Velocities

The relation (8.8) between the density field of galaxies and the peculiar velocity can also be used in a different context. To see this, we assume that the distance of galaxies can be determined independently of their redshift. In the relatively local Universe this is possible by using secondary distance measures (such as, e.g., the scaling relations for galaxies that were discussed in Sect. 3.4). With the distance known, we are then able to determine the radial component of the peculiar velocity by means of the redshift,

$$v = cz - H_0\,D .$$

To measure values of v of order $\sim 500\,\mathrm{km/s}$, D needs to be determined with a relative accuracy of

$$\frac{v}{cz} .$$

[3] It is an easy exercise to show that a power law $\xi(r) \propto r^{-\gamma}$ implies an angular correlation function $w(\theta) \propto \theta^{-(\gamma-1)}$.

With the distance measurements being accurate to about 10%, the distance to which this method can be applied is limited to $cz/H_0 \sim 100\,\mathrm{Mpc}$, corresponding to an expansion velocity $cz \sim 6000\,\mathrm{km/s}$. Thus, the peculiar velocity field can be determined only relatively locally. In order to measure D, one typically uses the Tully–Fisher relation for spirals, and the fundamental plane or the D_n–σ relation for ellipticals. In most cases, these measurements are carried out for groups of galaxies which then all have roughly the same distance; in this way, the measurement accuracy of their common (or average) distance is improved.

Equation (8.8) now allows us to predict the peculiar velocity field from the measured density field of galaxies, which can then be compared with the measured peculiar velocities – where this relation depends on β (see Eq. 8.9). Therefore, we can estimate β from this comparison. The inverse of this method is also possible: to derive the density distribution from the peculiar velocity field, and then to compare this with the observed

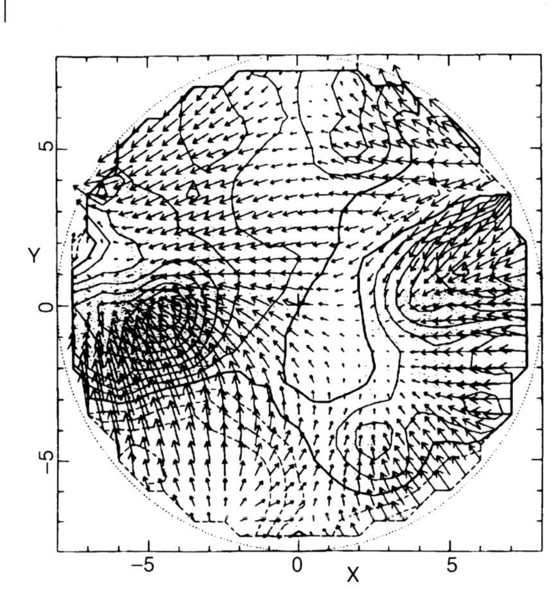

Potent Mass Density

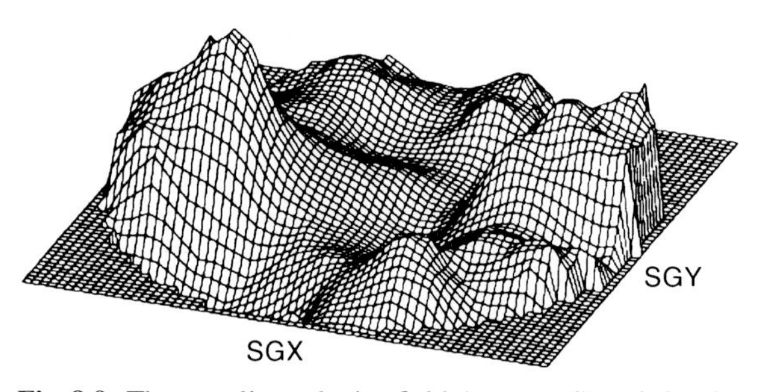

Fig. 8.9. The peculiar velocity field (top panel) and the derived density field (bottom panel) in our neighborhood. The distances here are specified as expansion velocities in units of 1000 km/s. The mass concentration on the left, towards which the velocity vectors are pointing, is the Great Attractor (see text), and on the right is the Pisces–Perseus supercluster. By comparing this reconstructed mass distribution with the distribution of galaxies, β can be determined. Early analyses of this kind resulted in relatively large values of β, whereas more recent results show $\beta \sim 0.5$. Since the bias factor may be different for the various types of galaxies, β may depend on the type of observed galaxies as well. For instance, one finds that IRAS galaxies have a lower β, thus also a lower b, than galaxies which have been selected optically

galaxy distribution.[4] Such a comparison is displayed in Fig. 8.9.

Measurements of the peculiar velocity field in the mid-1980s led to the conclusion that an unseen mass concentration, i.e., one that could not, at that time, be identified with a large concentration of galaxies, was having a significant effect on the local velocity field. This mass concentration (which was termed the "Great Attractor") was located roughly in the direction of the Galactic Center, which is the reason why it was not directly observable.

X-ray cluster samples are much less affected by Galactic absorption than optically selected clusters, and therefore provide a much clearer view of the mass distribution surrounding the Local Group, including the Zone of Avoidance. In recent years, based on such X-ray selected clusters, the simple picture of the Great Attractor has been modified. In fact, at the proposed distance to the Great Attractor of ~ 80 Mpc, the matter density seems to be considerably smaller than originally thought. However, behind the Great Attractor there seems to be a significant overdensity of clusters at larger distances (see Fig. 8.10).

The Velocity Dipole of the Galaxy Distribution. A related aspect of these studies is the question of whether the observed peculiar velocity of the Local Group, as determined from the dipole anisotropy of the CMB, can be traced back to the matter distribution around us. We would expect to find a related dipole in the matter distribution which caused an acceleration of the Local Group to the observed value of the peculiar velocity of 627 ± 22 km/s towards the direction $\ell = 273° \pm 3$, $b = 29° \pm 3°$ in Galactic coordinates (this value is obtained from the direct measurement of the dipole velocity in the rest-frame of the Sun, to which the motion of the Sun relative to the Local Group rest-frame is added).

In principle, this question can be answered from photometric galaxy surveys alone. We found a relation (8.8) between the fractional galaxy overdensity δ_g and the pe-

[4]At first sight this seems to be impossible, since only the radial component of the peculiar velocity can be measured – proper motions of galaxies are far too small to be observable. However, in the linear regime we can assume the velocity field to be a gradient field, $\boldsymbol{u} = \nabla\psi$; see Sect. 7.6. The velocity potential ψ can be obtained by integrating the peculiar velocity, $\psi(\boldsymbol{x}) - \psi(\boldsymbol{0}) = \int_0^x \mathrm{d}\boldsymbol{l} \cdot \boldsymbol{u}$, where the integral is taken over a curve connecting the observer at $\boldsymbol{0}$ to a point $\boldsymbol{x}$. Choosing radial curves, only the radial component of the peculiar velocity enters the integral. Therefore, this component is sufficient, in principle, to construct the velocity potential ψ, and therefore the three-dimensional velocity field.

Fig. 8.10. An optical image taken in the direction of the Great Attractor. This image has a side length of half a degree and was observed by the WFI at the ESO/MPG 2.2-m telescope on La Silla. The direction of this pointing is only $\sim 7°$ away from the Galactic disk. For this reason, the stellar density in the image is extremely high (about 200 000 stars can be found in this image) and, due to extinction in the disk of the Milky Way, much fewer faint galaxies at high redshift are found in this image than in comparable images at high Galactic latitude. Nevertheless, a large number of galaxies are visible (greenish), belonging to a huge cluster of galaxies (ACO 3627, at a distance of about 80 Mpc), which is presumably the main contributor to the Great Attractor

culiar velocity $\boldsymbol{u}(\boldsymbol{x})$ which we can specialize to the point of origin $\boldsymbol{x} = \boldsymbol{0}$. This relation is based on the assumption of linear biasing. A galaxy at distance D contributes to the peculiar velocity by an amount $\propto m/D^2$, where m is its mass. If we assume that the mass-to-light ratio of galaxies are all the same, then $m \propto L$ and the contribution of this galaxy to $\boldsymbol{u}$ is $\propto L/D^2 \propto S$. Hence, under these simplifying assumptions the contribution of a galaxy to the peculiar velocity depends only on its observed flux.

To apply this simple idea to real data, we need an all-sky map of the galaxy distribution. This is difficult to obtain, due to the presence of extinction towards the Galactic plane. However, if the galaxy distribution is mapped at infrared wavelengths, these effects are minimized. It is therefore not surprising that most of the studies on the dipole distribution of galaxies concentrate on infrared surveys. The IRAS source catalog still provides one of the major catalogs for such an analysis. More recently, the Two-Micron All Sky Survey (2MASS) catalog provided an all-sky map in the near-IR which can be used as well. The NIR also has the advantage that the luminosity at these wavelengths traces the mass of the stellar population of a galaxy quite well, in contrast to shorter wavelength for which the mass-to-light ratio among galaxies varies much more. The results of these studies is that the dipole of the galaxy distribution lies within $\sim 20°$ of the CMB dipole. This is quite a satisfactory result, if we consider the number of assumptions that are made in this method. The amplitude of the expected velocity depends on the factor $\beta = \Omega_{\rm m}^{0.6}/b$. Thus, by comparing the predicted velocity from the galaxy distribution with the observed dipole of the CMB this factor can be determined, yielding $\beta = 0.49 \pm 0.04$.

Supplementing the photometric surveys with redshifts allows the determination of the distance out to which the galaxy distribution has a marked effect on the Local Group velocity, by adding up the contributions of galaxies within a maximum distance from the Local Group. Although the detailed results from different groups vary slightly, the characteristic distance turns out to be $\sim 150h^{-1}$ Mpc, i.e., larger than the distance to the putative Great Attractor. In fact, earlier results suggested a considerably smaller distance, which was one of the reasons for postulating the presence of the Great Attractor.

8.2 Cosmological Parameters from Clusters of Galaxies

Being the most massive and largest gravitationally bound and relaxed objects in the Universe, clusters of galaxies are of special value for cosmology. In this section, we will explain various methods by which cosmological parameters have been derived from observations of galaxy clusters.

8.2.1 Number Density

In Sect. 7.5.2, we demonstrated that it is possible, for a given cosmological model, to calculate the number density of halos as a function of mass and redshift. This finding suggests that we should now compare the observed number density of galaxy clusters with these theoretical results and draw conclusions from this. We saw in Chap. 6 that the selection of clusters by their X-ray emission is currently viewed as the most reliable method of finding clusters. Hence, we use the X-ray cluster catalogs described in Sect. 6.3.5 for a comparison of the halo number density with model predictions.

In order to perform this comparison, the masses of clusters need to be determined. We discussed various methods of cluster mass determination in Chap. 6. Since a very detailed mass determination is possible only for individual clusters, but not for a large sample (which is required for a statistical comparison), one usually applies the scaling relations discussed in Sect. 6.4. In particular, the relation (6.52) between X-ray temperature, X-ray luminosity, and virial mass plays a central role. The scaling relations are then calibrated on clusters for which detailed mass estimates have been performed.

The comparison of the number density of observed clusters to the halo density in cosmological models can be performed either in the local Universe or as a function of redshift. In the former case, one obtains the normalization of the power spectrum from this comparison, hence σ_8, for a given matter density parameter Ω_m. More precisely, the number density of halos depends on the combination $\sigma_8 \Omega_m^{0.5}$, where the exact value of the exponent of Ω_m depends on the mass range of the halos that are considered. The analysis of cluster catalogs like the ones compiled from the ROSAT All Sky Survey (RASS) yields a value of about

$$\sigma_8 \Omega_m^{0.5} \approx 0.5 , \tag{8.13}$$

where the uncertainty in this value mainly comes from the calibration of the scaling relations.

The degeneracy between Ω_m and σ_8 can be broken by considering the redshift evolution of the number density of clusters. As we have seen in Sect. 7.2.2, the growth factor D_+ of the density perturbations depends on the cosmological parameters. For a low-density universe, the growth factor D_+ at high redshift is considerably larger than in an Einstein–de Sitter universe (see Fig. 7.3). Hence, the expected number density of clusters at high redshift is considerably smaller in an EdS model than in one of low density, for a fixed local number density of clusters. Indeed, in an EdS universe virtually no clusters of high mass are expected at $z \gtrsim 0.5$ (see Fig. 7.7), whereas the evolution of the halo number density is significantly weaker for cosmological models with small Ω_m. The fact that very massive clusters have been discovered at redshift $z > 0.5$ is therefore incompatible with a cosmological model of high matter density.

8.2.2 Mass-to-Light Ratio

On average, the mass-to-light ratio of cosmic objects seems to be an increasing function of their mass. In Chap. 3 we saw that M/L is smaller for spirals than for ellipticals, and furthermore that for ellipticals M/L increases with mass. In Chap. 6, we argued that galaxy groups like the Local Group have $M/L \sim 100h$, and that for galaxy clusters M/L is several hundreds, where all these values are quoted in Solar units. We conclude from this sequence that M/L increases with the length- or mass-scale of objects. Going to even larger scales – superclusters, for instance – M/L seems not to increase any further, rather it seems to approach a saturation value (see Fig. 8.11).

Thus, if we assume the M/L ratio of clusters to be characteristic of the average M/L ratio in the Universe, the average density of the Universe ρ_0 can be calculated from the measured luminosity density $\mathcal{L}$ and the M/L ratio for clusters,

$$\rho_0 = \left\langle \frac{M}{L} \right\rangle \mathcal{L} .$$

Here, L and $\mathcal{L}$ refer to a fixed frequency interval, e.g., to radiation in the B-band; $\mathcal{L}$ can be measured, for instance, by determining the local luminosity function of galaxies, yielding

$$\Omega_m = \frac{\langle M/L \rangle_B}{1200\,h} . \tag{8.14}$$

Since several methods of determining cluster masses now exist (see Chap. 6), and since L is measurable as well, (8.14) can be applied to clusters in order to estimate Ω_m. Typically, this results in $\Omega_m \sim 0.2$, a value for Ω_m which is slightly smaller than that obtained by other methods. However, this method is presumably less reli-

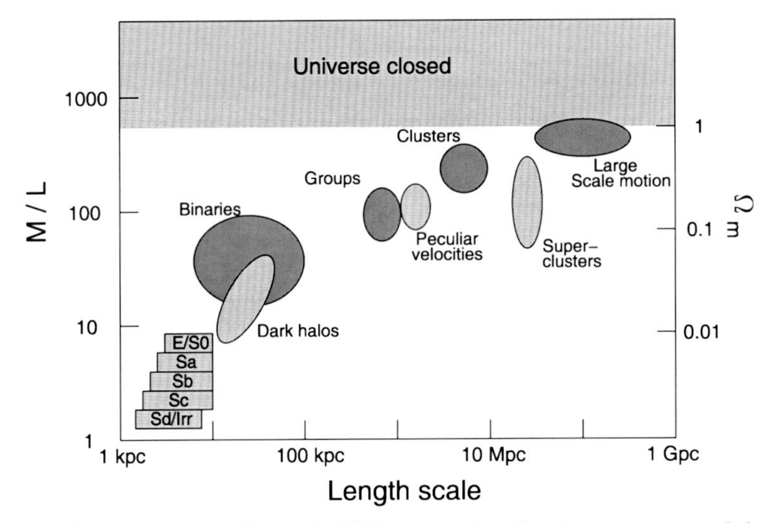

Fig. 8.11. The mass-to-light ratio M/L seems to be a function of the length- or mass-scale of cosmic objects. The luminous region in spirals has $M/L \sim 3$ (all values in Solar units of $M_\odot/L_\odot$), whereas that of ellipticals has $M/L \sim 10$. However, galaxies have a dark matter halo, so that the true mass of galaxies, and thus their M/L, is much larger than that which is measured in their visible region. Masses can also be estimated from the dynamics of galaxy pairs, typically yielding $M/L \sim 50$ for galaxies, including their dark halo. Galaxy groups and clusters have an even higher M/L ratio, hence they are particularly strongly dominated by dark matter, reaching $M/L \sim 250$. If the M/L ratio in clusters corresponds to the average M/L in the Universe, it is possible to determine the matter density in the Universe from the luminosity density, and to obtain a value of $\Omega_m \sim 0.2$. Only some early investigations of large-scale peculiar motions in the Universe have indicated even higher M/L, but these values seem not to be confirmed by more recent measurements

able than the other ones described in this section: $\mathcal{L}$ is not easily determined (e.g., the normalization of the Schechter luminosity function has been revised considerably in recent years, and its accuracy is not better than $\sim 20\%$), and the M/L ratio in clusters is not necessarily representative. For instance, the evolution of galaxies in a cluster is different from that of a "mean galaxy".

8.2.3 Baryon Content

As discussed in Chap. 6, clusters of galaxies largely consist of dark matter. Only about 15% of their mass is baryonic, the major part of which is contributed by hot intergalactic gas, visible through its X-ray emission. Within the accuracy of the measurements, the baryon content of clusters does not seem to vary between different clusters, rather it seems to have a uniform value. The existence of a universal baryon fraction is to be expected, since it is difficult to imagine how for structures as large as clusters the mixture of baryons and dark matter would strongly differ from the cosmic average. A massive cluster with a current virial radius of ~ 1.5 Mpc has formed by the gravitational contraction of a comoving volume with a linear extent of about ~ 10 Mpc. Effects like feedback from supernova explosions or other outflow phenomena, that are occurring in galaxies and which may reduce their baryon mass, are not effective in galaxy clusters due to their size.

Assuming the baryon fraction f_b in clusters to be representative of the Universe, the density parameter of the Universe can be determined, because the cosmic baryon density is presumed to be known from primordial nucleosynthesis (Sect. 4.4.4). This yields

$$\Omega_m \approx \frac{\Omega_m}{\Omega_b}\,\Omega_b \approx \frac{\Omega_b}{f_b} \approx 0.3\,. \tag{8.15}$$

8.2.4 The LSS of Clusters of Galaxies

Under the assumption that the galaxy distribution follows that of dark matter, the galaxy distribution enables us to draw conclusions about the statistical properties of the dark matter distribution, e.g., its power spectrum. At least on large scales, where structure evolution still proceeds almost linearly today, this assumption seems to be justified if an additional bias factor is allowed for. Hence, it is obvious to also examine the large-scale distribution of galaxy clusters, which should follow the distribution of dark matter on linear scales as well, although probably with a different bias factor.

The ROSAT All Sky Survey (see Sect. 6.3.5) allowed the compilation of a homogeneous sample of galaxy clusters with which the analysis of the large-scale distribution of clusters became possible for the first time. Figure 8.12 shows that the power spectrum of clusters has the same shape as that of galaxies, however with a considerably larger normalization. The ratio of the two power spectra displayed in this figure is based on different bias factors for galaxies and clusters, $b_{\rm clusters} \approx 2.6 b_{\rm g}$. For this reason the power spectrum

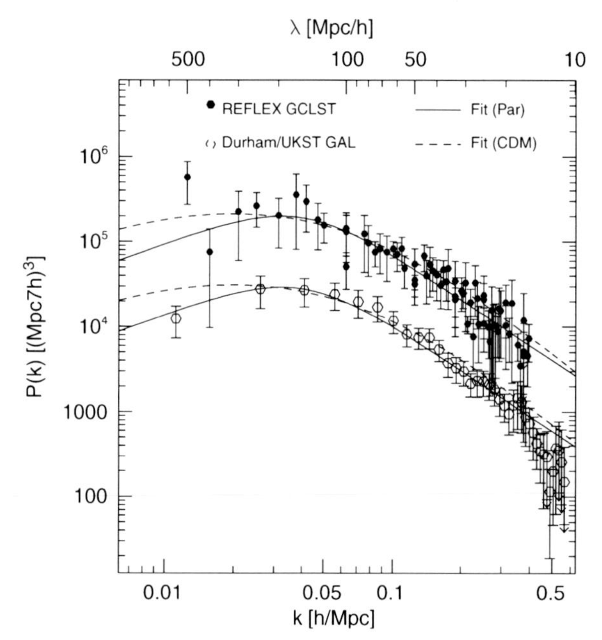

Fig. 8.12. The power spectrum of galaxies (open symbols) and of galaxy clusters from the REFLEX survey (filled symbols). The two power spectra have basically the same shape, but they differ by a multiplicative factor. This factor specifies the square of the ratio of the bias factors of optically selected galaxies and of X-ray clusters, respectively. Particularly on large scales, mapping the power spectrum from clusters is of substantial importance

for clusters has an amplitude that is larger by a factor of about $(2.6)^2$ than that for galaxies. Since clusters of galaxies are much less abundant than galaxies, the density maxima of the dark matter corresponding to the former need to have a higher threshold than those of galaxies, which will, in the biasing model illustrated in Fig. 8.3, result in stronger correlations.

The analysis of the power spectrum by means of clusters is interesting, particularly on large scales, yielding an additional data point for the shape parameter $\Gamma = \Omega_{\rm m} h$. Together with the cluster abundance, their correlation properties yield values of $\Omega_{\rm m} \approx 0.34$ and $\sigma_8 \approx 0.71$.

8.3 High-Redshift Supernovae and the Cosmological Constant

8.3.1 Are SN Ia Standard Candles?

As mentioned in Sect. 2.3.2, Type Ia supernovae are supposed to be the result of explosion processes of white dwarfs which cross a critical mass threshold by accretion of additional matter. This threshold should be identical for all SNe Ia, making it at least plausible that they all have the same luminosity. If this were the case, they would be ideal for standard candles: owing to their high luminosity, they can be detected and examined even at large distances.

However, it turns out that SNe Ia are not really standard candles, since their maximum luminosity varies from object to object with a dispersion of about 0.4 mag in the blue band light. This is visible in the top panel of Fig. 8.13. If SNe Ia were standard candles, the data points would all be located on a straight line, as described by the Hubble law. Clearly, deviations from the Hubble law can be seen, which are significantly larger than the photometric measurement errors.

It turns out that there is a strong correlation between the luminosity and the shape of the light curve of SNe Ia. Those of higher maximum luminosity show a slower decline in the light curve, as measured from its maximum. Furthermore, the observed flux is possibly affected by extinction in the host galaxy, in addition to the extinction in the Milky Way. With the resulting reddening of the spectral distribution, this effect can be

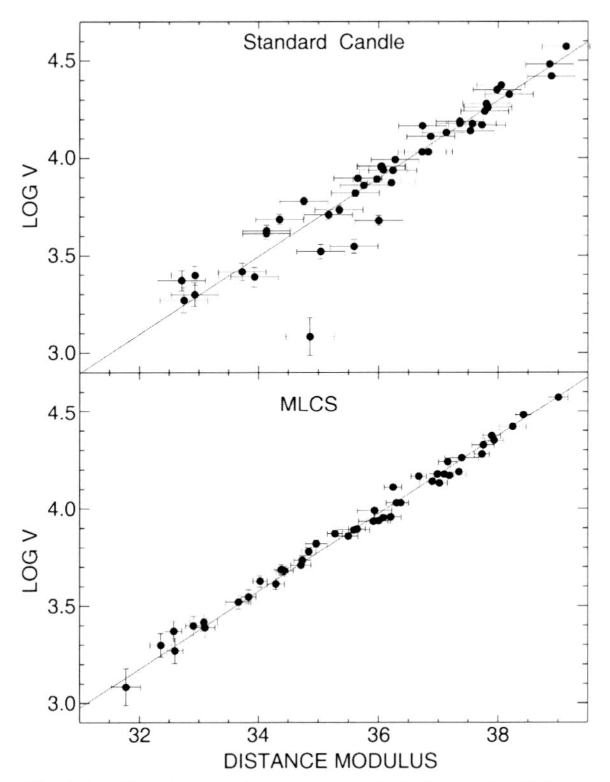

Fig. 8.13. The Hubble diagram for relatively nearby SNe Ia. Plotted is the measured expansion velocity cz as a function of the distance modulus for the individual supernovae. In the top panel, it is assumed that all sources have the same luminosity. If this was correct, all data points should be aligned along the straight line, as follows from the Hubble law. Obviously, the scatter is significant. In the bottom panel, the luminosities have been corrected by means of the so-called MLCS method in which the shape of the light curve and the colors of the SN are used to "standardize" the luminosity (see text for more explanations). By this the deviations from the Hubble law become dramatically smaller – the dispersion is reduced from 0.42 mag to 0.15 mag

derived from the observed colors of the SN. The combined analysis of these effects provides a possibility for deducing an empirical correction to the maximum luminosity from the observed light curves in several filters, accounting both for the relation of the width of the curve to the observed luminosity and for the extinction. This correction was calibrated on a sample of SNe Ia for which the distance to the host galaxies is very accurately known. With this correction applied, the SNe Ia follow the Hubble law much more closely, as can be seen in the bottom panel of Fig. 8.13. A scatter of only $\sigma = 0.15$ mag around the Hubble relation remains. Figure 8.14 demonstrates the effect of this correction on the light curves of several SNe Ia which initially appear to have very different maximum luminosities and widths. After correction they become nearly identical. The left panel of Fig. 8.14 suggests that the light curves of SN Ia can basically be described by a one-parameter family of functions, and that this parameter can be deduced from the shape, in particular the width, of the light curves.

With this correction, SNe Ia become standardized candles, i.e., by observing the light curves in several bands their "corrected" maximum luminosity can be determined. Since the observed flux of a source depends on its luminosity and its luminosity distance $D_{\rm L}$, and the latter also depends, besides redshift, on the cosmological model, SNe Ia can be used for the determination of cosmological parameters by measuring the luminosity distance as a function of redshift. To apply this method, it is necessary to detect and observe SNe Ia at appreciable redshifts, where deviations from the linear Hubble law become visible.

8.3.2 Observing SNe Ia at High Redshifts

An efficient strategy for the discovery of supernovae at large distances has been developed, and two large international teams have performed extensive searches for SNe Ia at high redshifts in recent years. Two photometric images of the same field, observed about four weeks apart, are compared and searched for sources which are not visible in the image taken first but which are seen in the later one. Of these candidates, spectra are then immediately taken to verify the nature of the source as a SN Ia and to determine its redshift. Subsequently, these sources become subject to extensive photometric monitoring in order to obtain precise light curves with a time coverage (sampling rate) as complete as possible. For this observation strategy to be feasible, the availability of observing time for both spectroscopy and subsequent photometry needs to be secured well before the search for candidates begins. Hence, this kind of survey requires a very well-planned strategy and coordination involving several telescopes. Since SNe Ia at high redshift are very faint, the new 8-m

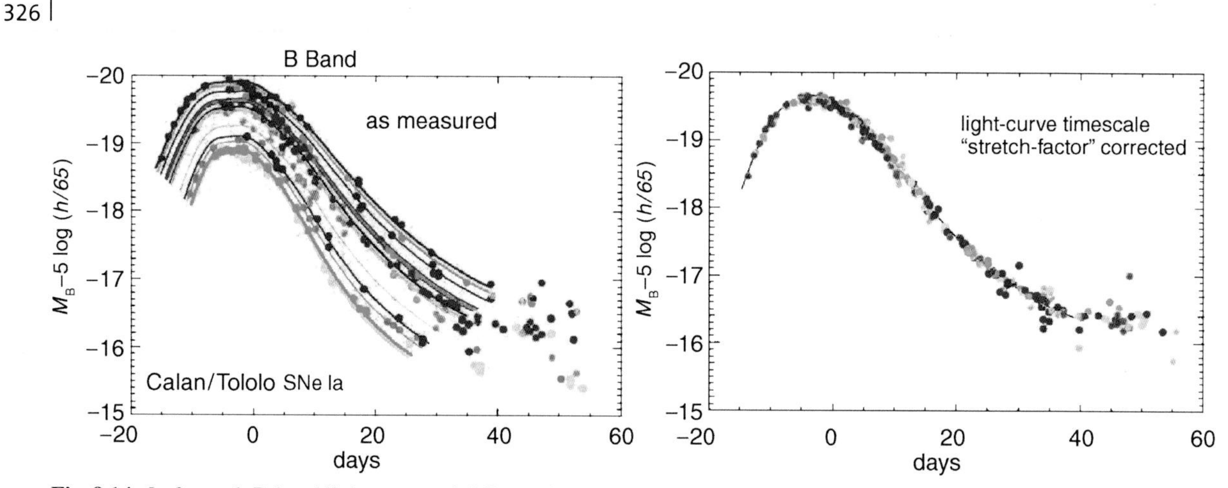

Fig. 8.14. Left panel: B-band light curves of different SNe Ia. One sees that the shape of the light curves and the maximum luminosity of the SNe Ia differ substantially among the sample. A transformation was found empirically with a single parameter described by the width of the light curve. By means of this transformation, the different light curves can all be made congruent, as displayed in the right panel

class telescopes need to be used for the spectroscopic observations.

Both teams were very successful in detecting distant SNe Ia. In their first large campaigns, the results of which were published in 1998, they detected and analyzed sources out to redshifts of $z \lesssim 0.8$. Since then, further SNe Ia have been found, some with redshifts ≥ 1. Substantial advances have also been made by observing with the HST. Among other achievements, the HST detected a SN Ia at redshift $z = 1.7$. Of special relevance is that the conclusions of both teams are in extraordinary agreement. Since they use slightly different methods in the correction of the maximum luminosity, this agreement serves as a significant test of the systematic uncertainties intrinsic to this method.

8.3.3 Results

As a first result, we mention that the width of the light curve is larger for SNe Ia at higher redshift than it is for local objects. This is expected because, due to redshift, the *observed* width evolves by a factor $(1+z)$. This dependence has been convincingly confirmed, showing in a direct way the transformation of the intrinsic to the observed time interval as a function of redshift.

Plotting the observed magnitudes in a Hubble diagram, one can look for the set of cosmological parameters which best describes the dependence of observed magnitudes $m_{\rm obs}$ on redshift, as is illustrated in Fig. 8.15.

Comparing the maximum magnitude of the measured SNe Ia, or their distance modulus respectively, with that which would be expected for an empty universe ($\Omega_{\rm m} = 0 = \Omega_\Lambda$), one obtains a truly surprising

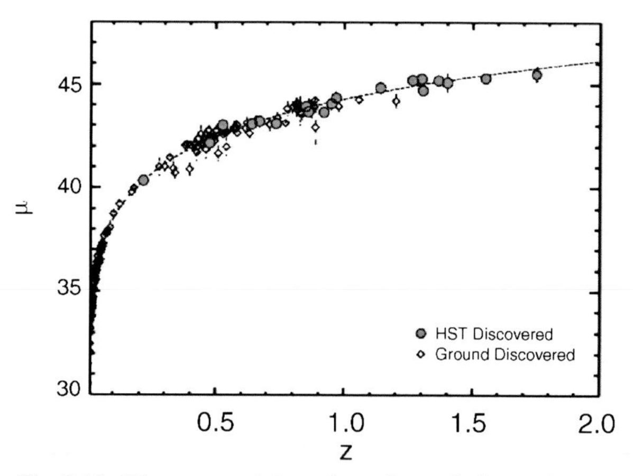

Fig. 8.15. Distance modulus of nearby and distant SNe Ia, determined from the corrected maximum flux of the source. Diamond symbols represent supernovae that were detected from the ground, circles those that were found by the HST. Particularly remarkable is the small scatter of the data points around the curve that corresponds to a cosmological model with $\Omega_{\rm m} = 0.29$, $\Omega_\Lambda = 0.71$

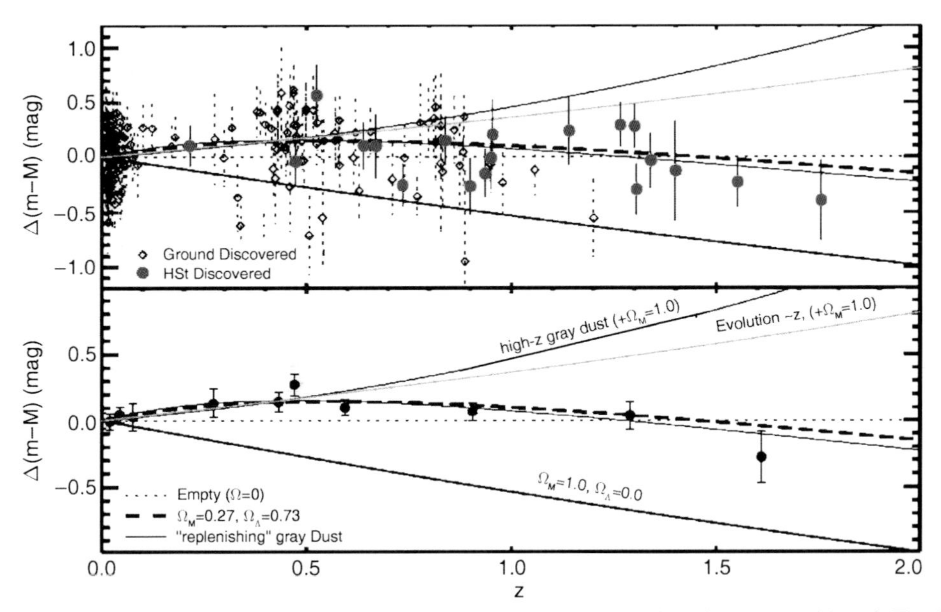

Fig. 8.16. Difference between the maximum brightness of SNe Ia and that expected in an empty universe ($\Omega_m = 0 = \Omega_\Lambda$). Diamond symbols represent events that were detected from the ground, circles the ones discovered by the HST. In the top panel, the individual SNe Ia are presented, whereas in the bottom panel they are averaged in redshift bins. An empty universe would correspond to the dotted straight line, $\Delta(m - M) = 0$. The dashed curve corresponds to a cosmological model with $\Omega_m = 0.27$, $\Omega_\Lambda = 0.73$. Furthermore, model curves for universes with constant acceleration are drawn; these models, which are not well-motivated from physics, and models including "gray dust" (in which the extinction is assumed to be independent of the wavelength), can be excluded

result (see Fig. 8.16). Considering at first only the supernovae with $z \lesssim 1$, one finds that these are fainter than predicted even for an empty universe. It should be mentioned that, according to (4.13), such an empty universe would expand at constant rate, $\ddot{a} = 0$. The luminosity distance in such a universe is therefore larger than in any other universe with a vanishing cosmological constant. The luminosity distance can only be increased by assuming that the Universe expanded *more slowly* in the past than it does today, hence that *the expansion has accelerated over time.* From (4.19) it follows that such an accelerated expansion is possible only if $\Omega_\Lambda > 0$. This result, first published in 1998, meant a turnaround in our physical world view because, until then, we were convinced that the cosmological constant was zero.

More recently, this result has been confirmed by ever more detailed investigations. In particular, the sample of SNe Ia was enlarged and (by employing the HST) extended to higher redshifts. From this, it was shown that for $z \gtrsim 1$ the trend is reversed and SN Ia become brighter than they would be in an empty universe (see Fig. 8.16). At these high redshifts the matter density dominates the Universe, proceeding as $(1+z)^3$ in contrast to the constant vacuum energy.

The corresponding constraints on the density parameters Ω_m and Ω_Λ are plotted in Fig. 8.17, in comparison to those that were obtained in 1998. As becomes clear from the confidence contours, the SN Ia data are not compatible with a universe without a cosmological constant. An Einstein–de Sitter model is definitely excluded, but also a model with $\Omega_m = 0.3$ (a value derived from galaxy redshift surveys) and $\Omega_\Lambda = 0$ is incompatible with these data.

We conclude from these results that a non-vanishing dark energy component exists in the Universe, causing an accelerated expansion through its negative pressure. The simplest form of this dark energy is the vacuum energy or the cosmological constant.

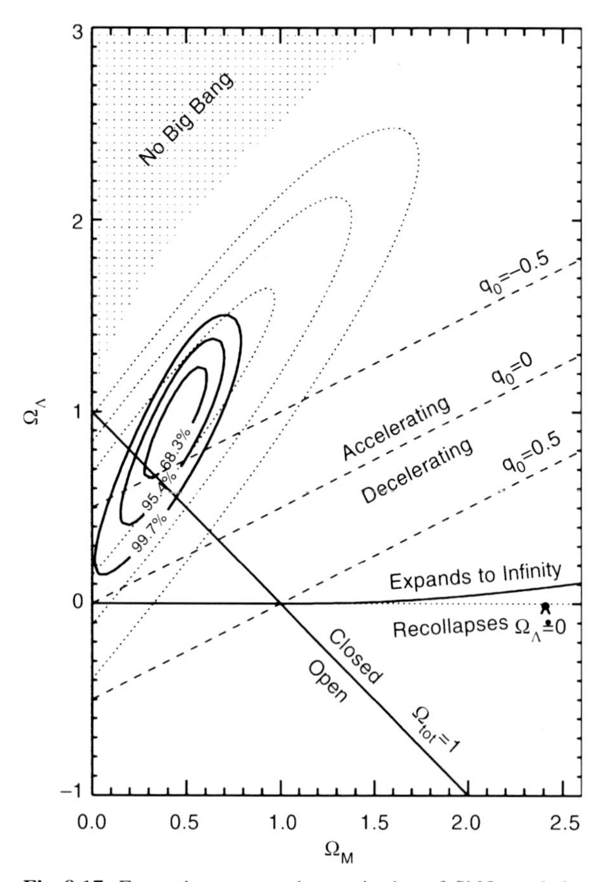

Fig. 8.17. From the measured magnitudes of SN Ia and the correspondingly implied values for the luminosity distances, confidence regions in the Ω_m–Ω_Λ plane are plotted here. The solid contours result from the 157 SNe Ia that are also plotted in Fig. 8.16, whereas the dotted contours represent the results from 1998. Dashed lines represent cosmological models with the same deceleration parameter q_0

Other forms of dark energy, such as that with a modified equation of state $P = w\,\rho\,c^2$, with $-1/3 > w > -1$, where $w = -1$ corresponds to a cosmological constant, are currently the subject of intense discussion. Constraints on w will, in the foreseeable future, only be possible through astronomical observations, and they will help us to shed some light on the physical nature of dark energy.

8.3.4 Discussion

The discovery of the Hubble diagram of SNe Ia being incompatible with a universe having a vanishing vacuum energy came as a surprise. It was the first evidence of the existence of dark energy. The cosmological constant, first introduced by Einstein, then later discarded again, seems to indeed have a non-vanishing value.

This far-reaching conclusion, with its consequences for fundamental physics, obviously needs to be critically examined. Which options do we have to explain the observations without demanding an accelerated expansion of the Universe?

Evolutionary Effects. The above analysis is based on the implicit assumption that, on average, SNe Ia all have the same maximum (corrected) luminosity, independent of their redshift. As for other kinds of sources for which a Hubble diagram can be constructed and from which cosmological parameters can be derived, the major difficulty lies in distinguishing the effects of spacetime curvature from evolutionary effects. A z-dependent evolution of SNe Ia, in such a way that they become less luminous with increasing redshift, could have a similar effect on a Hubble diagram as would an accelerated expansion.

At first sight, such an evolution seems improbable since, according to our current understanding, the explosion of a white dwarf close to the Chandrasekhar mass limit is responsible for these events, and this mass threshold solely depends on fundamental physical constants. On the other hand, the exact mass at which the explosion will be triggered may well depend on the chemical composition of the white dwarf, and this in turn may depend on redshift. Although it is presumably impossible to prove that such evolutionary effects are not involved or that their effect is at least smaller than cosmological effects, one can search for differences between SNe Ia at low and at high z. For instance, it has been impressively demonstrated that the spectra of high-redshift SNe Ia are very similar to those of nearby ones. Hence, no evidence for evolutionary effects has been found from these spectral studies. Furthermore, the time until the maximum is reached is independent of z, if one accounts for the time dilation $(1+z)$.

Extinction. The correction to the luminosity for extinction in the host galaxy and in the Milky Way is determined from reddening. The relation between extinction and reddening depends on the properties of the dust – if these evolve with z the correction may become systematically wrong. To test for this possibility one can separately investigate SNe Ia that occur in early-type galaxies, in which only little dust exists, and compare these to events in spiral galaxies. In this test, no systematic differences are found, neither in events at high redshift nor in nearby SNe Ia.

One possibility that has been discussed is the existence of "gray dust": dust that causes an absorption independent of wavelength. In such a case extinction would not reveal itself by reddening. However, this hypothesis lacks any theoretical explanation for the physical nature of the dust particles. In addition, the observation of SNe Ia at $z \gtrsim 1$ shows that the evolution of their magnitude at maximum is compatible with a Λ-universe. In contrast, in a scenario involving "gray dust", a monotonic decrease of the brightness with redshift would be expected, relative to an empty universe.

Although it cannot be completely ruled out that the results from SN Ia investigations are affected by systematic effects that mimic a cosmological effect, all tests that have been performed for such systematics have been negative. For this reason, the results are a very strong indication of a universe with finite vacuum energy density. The confirmation of this conclusion by the CMB anisotropies (see Fig. 8.6) is indeed impressive.

8.4 Cosmic Shear

On traversing the inhomogeneous matter distribution in the Universe, light beams are deflected and distorted, where the distortion is caused by the tidal gravitational field of the inhomogeneously distributed matter. As was already discussed in the context of the reconstruction of the matter distribution in galaxy clusters (see Sect. 6.5.2), by measuring the shapes of images of distant galaxies this tidal field can be mapped. From probing the tidal field, conclusions can be drawn about

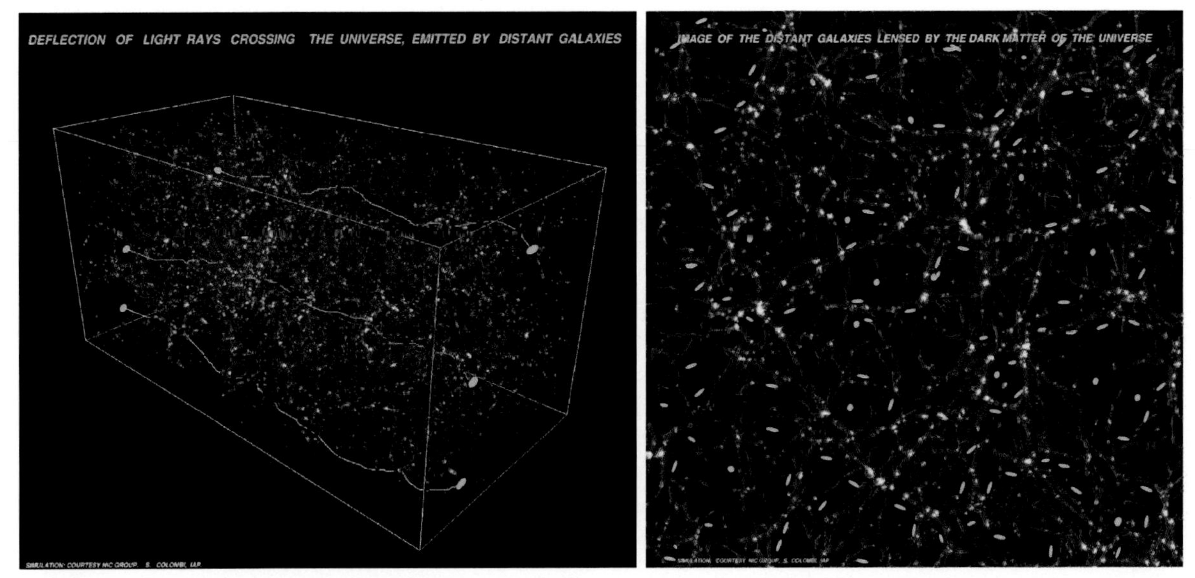

Fig. 8.18. As light beams propagate through the Universe they are affected by the inhomogeneous matter distribution; they are deflected, and the shape and size of their cross-section changes. This effect is displayed schematically here – light beams from sources at the far side of the cube are propagating through the large-scale distribution of matter in the Universe, and we observe the distorted images of the sources. In particular, the image of a circular source is elliptical to a first approximation. Since the distribution of matter is highly structured on large scales, the image distortion caused by light deflection is coherent: the distortion of two neighboring light beams is very similar, so that the observed ellipticities of neighboring galaxies are correlated. From a statistical analysis of the shapes of galaxy images, conclusions about the statistical properties of the matter distribution in the Universe can be drawn. Hence, the ellipticities of images of distant sources are closely related to the (projected) matter distribution, as displayed schematically in the right panel

the matter distribution. This effect, called *cosmic shear*, is sketched in Fig. 8.18. In contrast to the case of a galaxy cluster in which the tidal field is rather strong, the large-scale distribution of matter causes a very much weaker tidal field: a typical value for this shear is about 1% on angular scales of a few arcminutes, meaning that the image of an intrinsically circular source attains an axis ratio of 0.99:1.

The shear field results from the projection of the three-dimensional tidal field along the line-of-sight. Hence, we are able to obtain information about the statistical properties of the density inhomogeneities in the Universe, by a statistical analysis of the image shapes of distant galaxies. For instance, the two-point correlation function of the image ellipticities can be measured. This is linked to the power spectrum $P(k)$ of the matter distribution. Thus, by comparing measurements of cosmic shear with cosmological models we obtain constraints on the cosmological parameters, without the need to make any assumptions about the relation between luminous matter (galaxies) and dark matter.

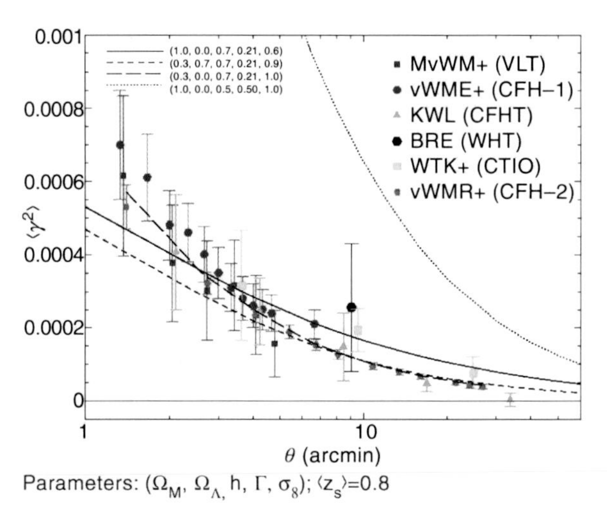

Fig. 8.19. Early measurements of cosmic shear. Plotted is the shear dispersion, measured from the ellipticities of faint and small galaxy images on deep CCD exposures, as a function of angular scale. Data from different teams are represented by different symbols. For instance, MvWM+ resulted from a VLT project, vWMR+ from a large survey (VIRMOS-Descartes) at the CFHT. For this latter project, the images of about 450 000 galaxies have been analyzed; the corresponding error bars from this survey are significantly smaller than those of the earlier surveys. The curves indicate cosmic shear predictions in different cosmological models, where the curves are labeled by the cosmological parameters $\Omega_{\rm m}$, Ω_Λ, h, Γ and σ_8

Since the size of the effect is expected to be very small, systematic effects like the anisotropy of the point-spread function or distortions in the telescope optics need to be understood very well, and they need to be corrected for in the measurements. In principle, the problems are the same as in the mass reconstruction of galaxy clusters with the weak lensing effect (Sect. 6.5.2), but they are substantially more difficult to deal with since the measurable signal is considerably smaller.

In March 2000, four research groups published, quasi-simultaneously, the first measurements of cosmic shear, and in the fall of 2000 another measurement was obtained from VLT observations. Since then, several teams worldwide have successfully performed measurements of cosmic shear, for which a large number of different telescopes have been used, including the HST. The development of wide-field cameras and of special software for data analysis are mainly responsible for these achievements. Some of the early results are compiled in Fig. 8.19.

By comparison of these measurement results with theoretical models, constraints on cosmological parameters are obtained; one example of this is presented in Fig. 8.20. Currently, a major source of uncertainty in this cosmological interpretation is our insufficient knowledge of the redshift distribution of the faint galaxies that are used for the measurements. In the coming years this uncertainty will be greatly reduced, as extensive redshift surveys of faint galaxies will be conducted with the next generation of multi-object spectrographs at 10-m class telescopes.

The most significant result that has been obtained from cosmic shear so far is a derivation of a combination of the matter density $\Omega_{\rm m}$ and the normalization σ_8 of the power spectrum of density fluctuations, which can also be seen in Fig. 8.20. The near-degeneracy of these two parameters has roughly the same functional form as for the number density of galaxy clusters, since with both methods we probe the matter distribution on similar physical length-scales. For an assumed value of $\Omega_{\rm m} = 0.3$, σ_8 can thus be constrained. Although the values obtained by different groups differ slightly, they are compatible with $\sigma_8 \approx 0.8$ within the range of uncertainty.

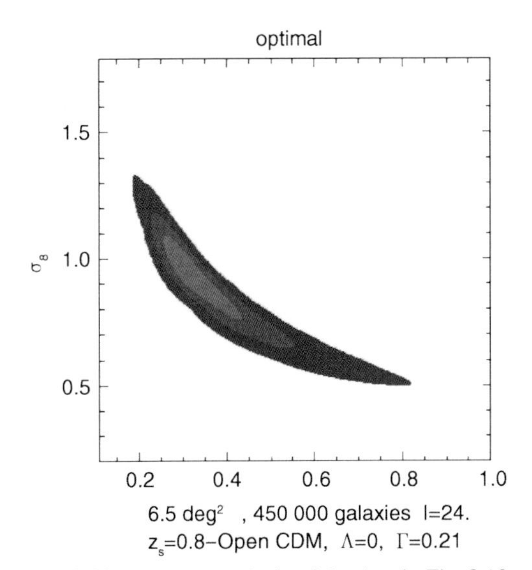

Fig. 8.20. From the analysis of the data in Fig. 8.19 and comparison with model predictions, constraints on cosmological parameters can be derived. Here, confidence contours in the Ω_m–σ_8 parameter plane are shown, where $\Omega_\Lambda = 0$ was assumed. The effect of Ω_Λ on the prediction of the shear is relatively small. The data suggest a Universe of low density. Presently the largest uncertainty in the quantitative analysis of shear data is the insufficiently known redshift distribution of the faint galaxies

8.5 Origin of the Lyman-α Forest

We have seen in Sect. 5.6.3 that in the spectrum of any QSO a large number of absorption lines at wavelengths shorter than the Lyα emission line of the QSO are found. The major fraction of these absorption lines originate from the Lyα transition of neutral hydrogen located along the line-of-sight to the source. Since the absorption is found in the form of a line spectrum, the absorbing hydrogen cannot be distributed homogeneously. A homogeneous intergalactic medium containing neutral hydrogen would be visible in continuum absorption. In this section, we will first examine this continuum absorption. We will then summarize some observational results on the Lyα forest and explain why studying this provides us with valuable information about the cosmological parameters.

8.5.1 The Homogeneous Intergalactic Medium

We first ask whether part of the baryons in the Universe may be contained in a homogeneous intergalactic medium. This question can be answered by means of the *Gunn–Peterson test*. Neutral hydrogen absorbs photons at a rest wavelength of $\lambda = \lambda_{\mathrm{Ly}\alpha} = 1216\,\text{Å}$. Photons from a QSO at redshift z_{QSO} attain this wavelength $\lambda_{\mathrm{Ly}\alpha}$ somewhere along the line-of-sight between us and the QSO, if they are emitted by the QSO at $\lambda_{\mathrm{Ly}\alpha}\left(1+z_{\mathrm{QSO}}\right)^{-1} < \lambda < \lambda_{\mathrm{Ly}\alpha}$. However, if the wavelength at emission is $> \lambda_{\mathrm{Ly}\alpha}$, the radiation can nowhere on its way to us be absorbed by neutral hydrogen. Hence, a jump in the observed continuum radiation should occur between the red and the blue side of the Lyα emission line of the QSO: this is the Gunn–Peterson effect. The optical depth for absorption is, for models with $\Omega_\Lambda = 0$, given by

$$\tau = 4.14 \times 10^{10}\,h^{-1}\,\frac{n_{\mathrm{HI}}(z)/\mathrm{cm}^{-3}}{(1+z)\sqrt{1+\Omega_{\mathrm{m}} z}}\,, \qquad (8.16)$$

where $n_{\mathrm{HI}}(z)$ is the density of neutral hydrogen at the absorption redshift z, with $(1+z) = \lambda/\lambda_{\mathrm{Ly}\alpha} < (1+z_{\mathrm{QSO}})$.

Such a jump in the continuum radiation of QSOs across their Lyα emission line, with an amplitude $S(\text{blue})/S(\text{red}) = \mathrm{e}^{-\tau}$, has not been observed for QSOs at $z \lesssim 5$. Tight limits for the optical depth were obtained by detailed spectroscopic observation, yielding $\tau < 0.05$ for $z \lesssim 3$ and $\tau < 0.1$ for $z \sim 5$. At even higher redshift observations become increasingly difficult, because the Lyα forest then becomes so dense that hardly any continuum radiation is visible between the individual absorption lines (see, e.g., Fig. 5.40 for a QSO at $z_{\mathrm{QSO}} = 3.62$). From the upper limit for the optical depth, one obtains bounds for the density of neutral hydrogen,

$$n_{\mathrm{HI}}(\text{comoving}) \lesssim 2 \times 10^{-13}\,h\,\mathrm{cm}^{-3}$$
$$\text{or} \quad \Omega_{\mathrm{HI}} \lesssim 2 \times 10^{-8}\,h^{-1}\,.$$

From this we conclude that hardly any homogeneously distributed baryonic matter exists in the intergalactic medium, or that hydrogen in the intergalactic medium is virtually fully ionized. However, from primordial nucleosynthesis we know the average density of hydrogen – it is much higher than the above limits – so that hydrogen must be present in essentially fully ionized form. We will discuss in Sect. 9.4 how this reionization of the intergalactic medium presumably happened.

In recent years, QSOs at redshifts > 6 have been discovered, not least by careful color selection in data from the Sloan Digital Sky Survey (see Sect. 8.1.2). The spectrum of one of these QSOs is displayed in Fig. 5.42. For this QSO, we can see that virtually no radiation bluewards of the Lyα emission line is detected. After this discovery, it was speculated whether the redshift had been identified at which the Universe was reionized. The situation is more complicated, though. First, the Lyα forest is so dense at these redshifts that lines blend together, making it very difficult to draw conclusions about a homogeneous absorption. Second, in spectra of QSOs at even higher redshift, radiation bluewards of the Lyα emission line has been found. As we will soon see, the reionization of the Universe probably took place at a redshift significantly higher than $z \sim 6$.

8.5.2 Phenomenology of the Lyman-α Forest

Neutral hydrogen in the IGM *is* being observed in the Lyα forest. For the observation of this Lyα forest, spectra of QSOs with high spectral resolution are required because the typical width of the lines is very small, corresponding to a velocity dispersion of $\sim 20\,\mathrm{km/s}$. To obtain spectra of high resolution and of good signal-to-noise ratio, very bright QSOs are selected. In this field, enormous progress has been made since the emergence of 10 m-class telescopes.

As mentioned before, the line density in the Lyα forest is a strong function of the absorption redshift. The number density of Lyα absorption lines with equivalent width (in the rest-frame of the absorber) $W \geq 0.32$ Å at $z \gtrsim 2$ is found to follow

$$\frac{\mathrm{d}N}{\mathrm{d}z} \sim k(1+z)^{\gamma} , \tag{8.17}$$

with $\gamma \sim 2.5$ and $k \sim 4$, which implies a strong redshift evolution. At lower redshift, where the Lyα forest is located in the UV part of the spectrum and therefore is considerably more difficult to observe (only by UV-sensitive satellites like the IUE, FUSE, and the HST), the evolution is slower and the number density deviates from the power law given above.

From the line strength and width, the HI column density N_{HI} of a line can be measured. The number density of lines as a function of N_{HI} is

$$\frac{\mathrm{d}N}{\mathrm{d}N_{\mathrm{HI}}} \propto N_{\mathrm{HI}}^{-\beta} , \tag{8.18}$$

with $\beta \sim 1.6$. This power law approximately describes the distribution over a wide range of column densities, $10^{12}\,\mathrm{cm}^{-2} \lesssim N_{\mathrm{HI}} \lesssim 10^{22}\,\mathrm{cm}^{-2}$, including Ly-limit systems and damped Lyα systems.

The temperature of the absorbing gas can be estimated from the line width as well, by identifying the width with the thermal line broadening. As typical values, one obtains $\sim 10^4$ K to 2×10^4 K which, however, are somewhat model-dependent.

The Proximity Effect. The statistical properties of the Lyα forest depend only on the redshift of the absorption lines, and not on the redshift of the QSO in the spectrum of which they are measured. This is as expected if the absorption is not physically linked to the QSO, and this observational fact is one of the most important indicators for an intergalactic origin of the absorption.

However, there is one effect in the statistics of Lyα absorption lines which is directly linked to the QSO. One finds that the number density of Lyα absorption lines at those redshifts which are only slightly smaller than the emission line redshift of the QSO itself, is lower than the mean absorption line density at this redshift (averaged over many different QSO lines-of-sight). This effect indicates that the QSO has some effect on the absorption lines, if only in its immediate vicinity; for this reason, it is named the *proximity effect*. An explanation of this effect follows directly from considering the ionization stages of hydrogen. The gas is ionized by energetic photons which originate from hot stars and AGNs and which form an ionizing background. On the other hand, ionized hydrogen can recombine. The degree of ionization results from the equilibrium between these two processes.

The number of photoionizations of hydrogen atoms per volume element and unit time is proportional to the density of neutral hydrogen atoms and given by

$$\dot{n}_{\mathrm{ion}} = \Gamma_{\mathrm{HI}}\, n_{\mathrm{HI}} , \tag{8.19}$$

where Γ_{HI}, the photoionization rate, is proportional to the density of ionizing photons. The corresponding number of recombinations per volume and time is proportional to the density of free protons and electrons,

$$\dot{n}_{\rm rec} = \alpha \, n_{\rm p} \, n_{\rm e} \,, \tag{8.20}$$

where the recombination coefficient α depends on the gas temperature. The Gunn–Peterson test tells us that the intergalactic medium is essentially fully ionized, and thus $n_{\rm HI} \ll n_{\rm p} = n_{\rm e} \approx n_{\rm b}$ (we disregard the contribution of helium in this consideration). We then obtain for the density of neutral hydrogen in an equilibrium of ionization and recombination

$$n_{\rm HI} = \frac{\alpha}{\Gamma_{\rm HI}} \, n_{\rm p}^2 \,. \tag{8.21}$$

This results shows that $n_{\rm HI}$ is inversely proportional to the number density of ionizing photons. However, the intergalactic medium in the vicinity of the QSO does not only experience the ionizing background radiation field but, in addition, the energetic radiation from the QSO itself. Therefore, the degree of ionization of hydrogen in the immediate vicinity of the QSO is higher, and consequently less Lyα absorption can take place there.

Since the contribution of the QSO to the ionizing radiation depends on the distance of the gas from the QSO ($\propto r^{-2}$), and since the spectrum and ionizing flux of the QSO is observable, examining the proximity effect provides an estimate of the intensity of the ionizing background radiation as a function of redshift. This value can then be compared to the total ionizing radiation which is emitted by QSOs and young stellar populations at the respective redshift. This comparison, in which the luminosity function of AGNs and the star-formation rate in the Universe are taken into account, yields good agreement, thus confirming our model for the proximity effect.

8.5.3 Models of the Lyman-α Forest

Since the discovery of the Lyα forest, various models have been developed in order to explain its nature. Since about the mid-1990s, one model has been established that is directly linked to the evolution of large-scale structure in the Universe.

The "Old" Model of the Lyman-α Forest. Prior to this time, models were designed in which the Lyα forest was caused by quasi-static hydrogen clouds. These clouds (Lyα clouds) were postulated and were initially seen as a natural picture given the discrete nature of the absorption lines. From the statistics of the number density of lines, the cloud properties (such as radius and density) could then be constrained. If the line width represented a thermal velocity distribution of the atoms, the temperature and, together with the radius, also the mass of the clouds could be derived (e.g., by utilizing the density profile of an isothermal sphere). The conclusion from these arguments was that such clouds would evaporate immediately unless they were gravitationally bound in a dark matter halo (mini-halo model), or confined by the pressure of a hot intergalactic medium.[5]

The New Picture of the Lyman-α Forest. For about a decade now, a new paradigm has existed for the nature of the Lyα forest. Its establishment became possible through advances in hydrodynamic cosmological simulations.

We discussed structure formation in Chap. 7, where we concentrated mainly on dark matter. After recombination at $z \sim 1100$ when the Universe became neutral and therefore the baryonic matter no longer experienced pressure by the photons, baryons were, just like dark matter, only subject to gravitational forces. Hence the behavior of baryons and dark matter became very similar up to the time when baryons began to experience significant pressure forces by heating (e.g., due to photoionization) and compression. The spatial distribution of baryons in the intergalactic medium thus followed that of dark matter, as is also confirmed by numerical simulations. In these simulations, the intensity of ionizing radiation is accounted for – it is estimated, e.g., from the proximity effect. Figure 8.21 shows the column density distribution of neutral hydrogen which results from such a simulation. It shows a structure similar to the distribution of dark matter, however with a higher density contrast due to the quadratic dependence of the HI density on the baryon density – see (8.21).

From the distribution of neutral gas simulated this way, synthetic absorption line spectra can then be computed. For these, the temperature of the gas and its peculiar velocity are used, the latter resulting from the simulation as well. Such a synthetic spectrum is

[5] The latter assumption was excluded at last by the COBE measurements of the CMB spectrum, because such a hot intergalactic medium would cause deviations of the CMB spectrum from its Planck shape, by Compton scattering of the CMB photons.

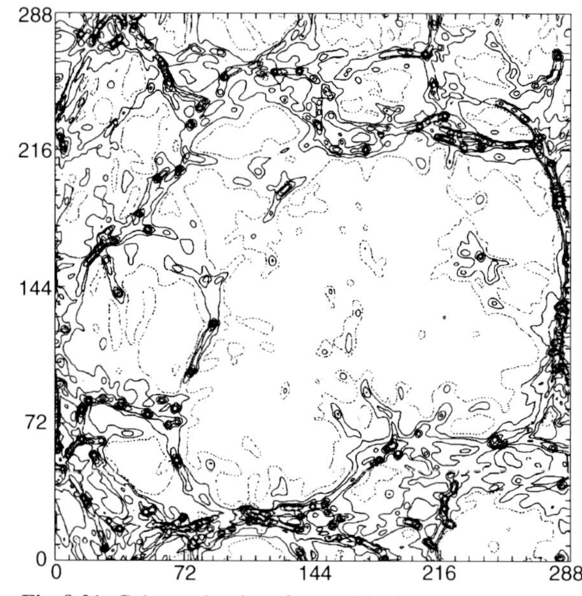

Fig. 8.21. Column density of neutral hydrogen, computed in a joint simulation of dark matter and gas. The size of the cube displayed here is $10h^{-1}$ Mpc (comoving). By computing the Lyα absorption of photons crossing a simulated cube like this, simulated spectra of the Lyα forest are obtained, which can then be compared statistically with observed spectra

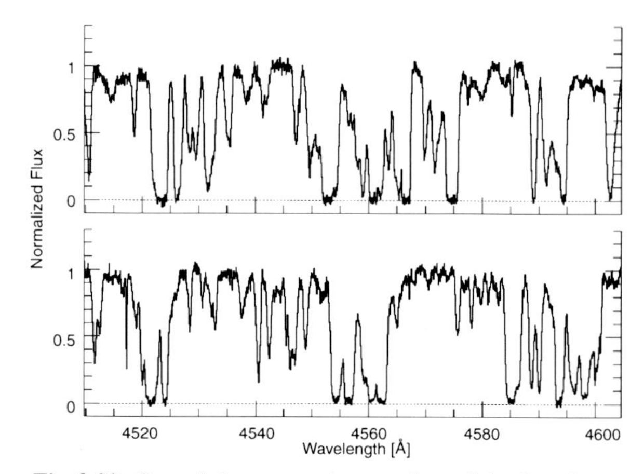

Fig. 8.22. One of the spectra is a section of the Lyα forest towards the QSO 1422+231 (see also Fig. 5.40), the other is a simulated spectrum; both are statistically so similar that it is impossible to distinguish them – which one is which?

displayed in Fig. 8.22, together with a measured Lyα spectrum. These two spectra are, from a statistical point of view, virtually identical, i.e., their density of lines, the width and optical depth distributions, and their correlation properties are equal. For this reason, the evolution of cosmic structure provides a natural explanation for the Lyα forest, without the necessity of additional free parameters or assumptions. In this model, the evolution of $\mathrm{d}N/\mathrm{d}z$ is driven mainly by the Hubble expansion and the resulting change in the degree of ionization in the intergalactic medium.

Besides the correlation properties of the Lyα lines in an individual QSO spectrum, we can also consider the correlation between absorption line spectra of QSOs which have a small angular separation on the sky. In this case, the corresponding light rays are close together, probing neighboring spatial regions of the intergalactic medium. If the neutral hydrogen is correlated on scales larger than the transverse separation of the two lines-of-sight towards the QSOs, correlated Lyα absorption lines should be observable in the two spectra. As a matter of fact, it is found that the absorption line spectra of QSOs show correlations, provided that the angular separation is sufficiently small. The correlation lengths derived from these studies are $\gtrsim 100\,h^{-1}$ kpc, in agreement with the results from numerical simulations. In particular, the lines-of-sight corresponding to different images of multiple-imaged QSOs in gravitational lens systems are very close together, so that the correlation of the absorption lines in these spectra can be very well verified.

Where are the Baryons Located? As another result of these investigations it is found that at $2 \lesssim z \lesssim 4$ the majority ($\sim 85\%$) of baryonic matter is contained in the Lyα forest, mainly in systems with column densities of $10^{14}\,\mathrm{cm}^{-2} \lesssim N_{\mathrm{HI}} \lesssim 3\times10^{15}\,\mathrm{cm}^{-2}$. Thus, at these high redshifts we observe nearly the full inventory of baryons. At lower redshift, this is no longer the case. Indeed, only a fraction of the baryons can be observed in the local Universe, for instance in stars or in the intergalactic gas in clusters of galaxies. From theoretical arguments, we expect that the majority of baryons today should be found in the form of intergalactic gas, for example in galaxy groups and large-scale filaments that are seen in simulations of structure formation. This gas is expected to have a temperature between $\sim 10^5$ K and $\sim 10^7$ K and is therefore very difficult to detect; it is called the warm-hot intergalactic medium. At these temperatures, the gas is essentially fully ionized so that

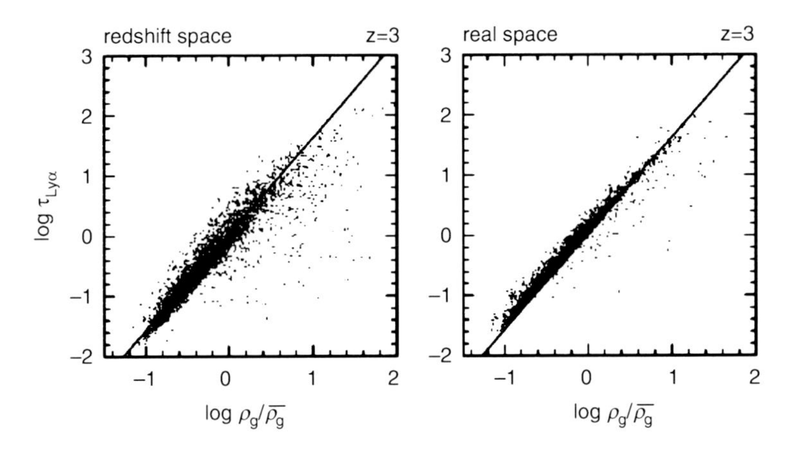

Fig. 8.23. Optical depth for Lyα absorption versus gas density, obtained from a cosmological simulation. Each data point represents a line-of-sight through a gas distribution like the one presented in Fig. 8.21. For the panel on the right, peculiar motion of the gas was neglected; in this case, the points follow the relation (8.22) very accurately. With the peculiar motions and thermal line broadening taken into account (left panel), the points also follow this relation on average

it cannot be detected in absorption line spectra. However, the temperature and density are too low to expect significant X-ray emission from this gas.[6]

8.5.4 The Lyα Forest as Cosmological Tool

The aforementioned simulations of the Lyα forest predict that most of the lines originate in regions of the intergalactic medium where the gas density is $\rho_g \lesssim 10\overline{\rho}_g$. Hence, the density of the absorbing gas is relatively low, compared, e.g., to the average gas density in a galaxy. The temperature of the gas causing the absorption is about $\sim 10^4$ K. At these densities and temperatures, pressure forces are small compared to gravitational forces, so that the gas follows the density distribution of dark matter very closely. From the absorption line statistics, it is therefore possible to derive the statistical properties of the dark matter distribution. More precisely, the two-point correlation function of the Lyα lines reflects the spectrum of density fluctuations in the Universe, and hence it can be used to measure the power spectrum $P(k)$.

[6] Although hydrogen is not detectable in this intergalactic medium due to its complete ionization, lines from metal ions at a high ionization stage can be observed in UV absorption line spectra, for instance the lines of OVI, the five times ionized oxygen. To derive a baryon density from observations of these lines, assumptions about the temperature of the gas and about its metallicity are required. The latest results, which have mainly been obtained using the UV satellite FUSE, are compatible with the idea that today the major fraction of baryons is contained in this warm-hot intergalactic medium.

We will consider some aspects of this method in more detail. The temperature of the intergalactic gas is not homogeneous because gas heats up by compression. Thus at a fixed redshift dense gas is hotter than the average baryon temperature T_0. As long as the compression proceeds adiabatically, T basically depends on the density, $T = T_0(\rho_g/\overline{\rho}_g)^\alpha$, where T_0 and the exponent α depend on the ionization history and on the spectrum of the ionizing photons. Typical values are $4000\,\mathrm{K} \lesssim T_0 \lesssim 10\,000\,\mathrm{K}$ and $0.3 \lesssim \alpha \lesssim 0.6$. The density of neutral hydrogen is specified by (8.21), $n_{\rm HI} \propto \rho_g^2 T^{-0.7}/\Gamma_{\rm HI}$, where the temperature dependence of the recombination rate has been taken into account. Since the temperature depends on the density, one obtains for the optical depth of Lyα absorption

$$\tau = A\left(\frac{\rho_g}{\overline{\rho}_g}\right)^\beta , \tag{8.22}$$

where $\beta = 2 - 0.7\alpha \approx 1.6$, with the prefactor depending on the observed redshift, the ionization rate $\Gamma_{\rm HI}$, and the average temperature T_0.

In Fig. 8.23, the distribution of optical depth and gas density at redshift $z = 3$ is plotted, obtained from a hydrodynamical simulation. As is seen from the right-hand panel, the distribution follows the relation (8.22) very closely, which means that a major fraction of the gas has not been heated by shock fronts, but rather by adiabatic compression. Even with peculiar motion of the gas and thermal broadening taken into account, as is the case in the panel on the left, the average distribution still follows the analytical relation very closely.

From the observed distribution of τ, it is thus possible to draw conclusions about the distribution of the gas overdensity $\rho_g/\overline{\rho}_g$. As argued above, the latter is basically the same as the corresponding overdensity of dark matter. From an absorption line spectrum, $\tau(\lambda)$ can be determined (wavelength-)pixel by pixel, where λ corresponds, according to $\lambda = (1+z)\,1216\,\text{Å}$, to a distance along the line-of-sight, at least if peculiar velocities are disregarded. From $\tau(\lambda)$, the overdensity as a function of this distance follows with (8.22), and thus a one-dimensional cut through the density fluctuations is obtained. The correlation properties of this density are determined by the power spectrum of the matter distribution, which can be measured in this way.

This probe of the density fluctuations is applied at redshifts $2 \lesssim z \lesssim 4$, where, on the one hand, the Lyα forest is in the optical region of the observed spectrum, and on the other hand, the forest is not too dense for this analysis to be feasible. This technique therefore probes the large-scale structure at significantly earlier epochs than is the case for the other cosmological probes described earlier. At such earlier epochs the density fluctuations are linear down to smaller scales than they are today. For this reason, the Lyα forest method yields invaluable information about the power spectrum on smaller scales than can be probed with, say, galaxy redshift surveys. We shall come back to the use of this method in combination with the CMB anisotropies in Sect. 8.7.

8.6 Angular Fluctuations of the Cosmic Microwave Background

The cosmic microwave background consists of photons that last interacted with matter at $z \sim 1000$. Since the Universe must have already been inhomogeneous at this time, in order for the structures present in the Universe today to be able to form, it is expected that these spatial inhomogeneities are reflected in a (small) anisotropy of the CMB: the angular distribution of the CMB temperature reflects the matter inhomogeneities at the redshift of decoupling of radiation and matter.

Since the discovery of the CMB in 1965, such anisotropies have been searched for. Under the assumption that the matter in the Universe only consists of baryons, the expectation was that we would find relative fluctuations in the CMB temperature of $\Delta T/T \sim 10^{-3}$ on scales of a few arcminutes. This expectation is based on the theory of gravitational instability for structure growth: to account for the density fluctuations observed today, one needs relative density fluctuations at $z \sim 1000$ of order 10^{-3}. Despite increasingly more sensitive observations, these fluctuations were not detected. The upper limits resulting from these searches for anisotropies provided one of the arguments that, in the mid-1980s, caused the idea of the existence of dark matter on cosmic scales to increasingly enter the minds of cosmologists. As we will see soon, in a Universe which is dominated by dark matter the expected CMB fluctuations on small angular scales are considerably smaller than in a purely baryonic Universe. Only with the COBE satellite were temperature fluctuations in the CMB finally observed in 1992 (Fig. 1.17). Over the last few years, sensitive and significant measurements of the CMB anisotropy have also been carried out using balloons and ground-based telescopes.

We will first describe the physics of CMB anisotropies, before turning to the observational results and their interpretation. As we will see, the CMB anisotropies depend on nearly all cosmological parameters, such as Ω_m, Ω_b, Ω_Λ, Ω_{HDM}, H_0, the normalization σ_8, the primordial slope n_s, and the shape parameter Γ of the power spectrum. Therefore, from an accurate mapping of the angular distribution of the CMB and by comparison with theoretical expectations, all these parameters can, in principle, be determined.

8.6.1 Origin of the Anisotropy: Overview

The CMB anisotropies reflect the conditions in the Universe at the epoch of recombination, thus at $z \sim 1000$. Temperature fluctuations originating at this time are called *primary anisotropies*. Later, as the CMB photons propagate through the Universe, they may experience a number of distortions along their way which, again, may change their temperature distribution on the sky. These effects then lead to *secondary anisotropies*.

The most basic mechanisms causing primary anisotropies are the following:

- Inhomogeneities in the gravitational potential cause photons which originate in regions of higher density to climb out of a potential well. As a result of this, they loose energy and are redshifted (gravitational redshift). This effect is partly compensated for by the fact that, besides the gravitational redshift, a gravitational time delay also occurs: a photon that originates in an overdense region will be scattered at a slightly earlier time, and thus at a slightly higher temperature of the Universe, compared to a photon from a region of average density. Both effects always occur side by side. They are combined under the term *Sachs–Wolfe effect*. Its separation into two processes is necessary only in a simplified description; a general relativistic treatment of the Sachs–Wolfe effect jointly yields both processes.

- We have seen that density fluctuations are always related to peculiar velocities of matter. Hence, the electrons that scatter the CMB photons for the last time do not follow the pure Hubble expansion but have an additional velocity that is closely linked to the density fluctuations (compare Sect. 7.6). This results in a Doppler effect: if photons are scattered by gas receding from us with a speed larger than that corresponding to the Hubble expansion, these photons experience an additional redshift which reduces the temperature measured in that direction.
- In regions of a higher dark matter density, the baryon density is also enhanced. On scales larger than the horizon scale at recombination (see Sect. 4.5.2), the distribution of baryons follows that of the dark matter. On smaller scales, the pressure of the baryon–photon fluid is effective because, prior to recombination, these two components had been closely coupled by Thomson scattering. Baryons are adiabatically compressed and thus get hotter in regions of higher baryon density, hence their temperature – and with it the temperature of the photons coupled to them – is also larger.
- The coupling of baryons and photons is not perfect since, owing to the finite mean free path of photons, the two components are decoupled on small spatial scales. This implies that on small length-scales, the temperature fluctuations can be smeared out by the diffusion of photons. This process is known as *Silk damping*, and it implies that on angular scales below about $\sim 5'$, only very small primary fluctuations exist.

Obviously, the first three of these effects are closely coupled to each other. In particular, on scales $> r_{\rm H,com}(z_{\rm rec})$ the first two effects can partially compensate each other. Although the energy density of matter is, at recombination, higher than that of the radiation (see Eq. 4.54), the energy density in the baryon–photon fluid is dominated by radiation, so that it is considered a relativistic fluid. Its speed of sound is thus $c_{\rm s} \approx \sqrt{P/\rho} \approx c/\sqrt{3}$. The high pressure of this fluid causes oscillations to occur. The gravitational potential of the dark matter is the driving force, and pressure the restoring force. These oscillations, which can only occur on scales below the sound horizon at recombination, then lead to adiabatic compression and peculiar velocities of the baryons, hence to anisotropies in the background radiation.

Secondary anisotropies result, among other things, from the following effects:

- Thomson scattering of CMB photons. Since the Universe is currently transparent for optical photons (since we are able to observe objects at $z > 6$), it must have been reionized between $z \sim 1000$ and $z \sim 6$, presumably by radiation from the very first generation of stars and/or by the first QSOs. After this reionization, free electrons are available again, which may then scatter the CMB photons. Since Thomson scattering is essentially isotropic, the direction of a photon after scattering is nearly independent of its incoming direction. This means that scattered photons no longer carry information about the CMB temperature fluctuations. Hence, the scattered photons form an isotropic radiation component whose temperature is the average CMB temperature. The main effect resulting from this scattering is a reduction of the measured temperature anisotropies, by the fraction of photons which experience such scattering.
- Photons propagating towards us are traversing a Universe in which structure formation takes place. Due to this evolution of the large-scale structure, the gravitational potential is changing over time. If it was time-independent, photons would enter and leave a potential well with their frequency being unaffected, compared to photons that are propagating in a homogeneous Universe: the blueshift they experience when falling into a potential well is exactly balanced by the redshift they attain when climbing out. However, this "conservation" of photon energy

no longer occurs if the potential is varying with time. One can show that for an Einstein–de Sitter model, the peculiar gravitational potential ϕ (7.10) is constant over time,[7] and hence, the propagation in the evolving Universe yields no net frequency shift. For other cosmological models this effect does occur; it is called the *integrated Sachs–Wolfe effect*.

- The gravitational deflection of CMB photons, caused by the gravitational field of the cosmic density fluctuations, leads to a change in the photon direction. This means that two lines-of-sight separated by an angle θ at the observer have a physical separation at recombination which may be different from $D_A(z_{rec})\theta$, due to the gravitational light deflection. Because of this, the correlation function of the temperature fluctuations is slightly smeared out. This effect is relevant on small angular scales.
- The Sunyaev–Zeldovich effect, which we discussed in Sect. 6.3.4 in the context of galaxy clusters, also affects the temperature distribution of the CMB. Some of the photons propagating along lines-of-sight passing through clusters of galaxies are scattered by the hot gas in these clusters, yielding a temperature change in these directions. We recall that in the direction of clusters the measured intensity of the CMB radiation is reduced at low frequencies, whereas it is increased at high frequencies. Hence, the SZ effect can be identified in the CMB data if measurements are conducted over a sufficiently large frequency range.

8.6.2 Description of the Cosmic Microwave Background Anisotropy

Correlation Function and Power Spectrum. In order to characterize the statistical properties of the angular distribution of the CMB temperature, the two-point correlation function of the temperature on the sphere can be employed, in the same way as it is used for describing the density fluctuations or the angular correlation function of galaxies. To do this, the relative temperature fluctuations $\mathcal{T}(\boldsymbol{n}) = [T(\boldsymbol{n}) - T_0]/T_0$ are defined, where $\boldsymbol{n}$ is a unit vector describing the direction on the sphere, and T_0 is the average temperature of the CMB. The correlation function of the temperature fluctuations is then defined as

$$C(\theta) = \langle \mathcal{T}(\boldsymbol{n})\, \mathcal{T}(\boldsymbol{n}') \rangle \,, \qquad (8.23)$$

where the average extends over all pairs of directions $\boldsymbol{n}$ and $\boldsymbol{n}'$ with angular separation θ. As for the description of the density fluctuations in the Universe, for the CMB it is also common to consider the power spectrum of the temperature fluctuations, instead of the correlation function.

We recall (see Sect. 7.3.2) that the power spectrum $P(k)$ of the density fluctuations is defined as the Fourier transform of the correlation function. However, exactly the same definition cannot be applied to the CMB. The difference here is that the density fluctuations $\delta(\boldsymbol{x})$ are defined on a flat space (approximately, at the relevant length-scales). In this space, the individual Fourier modes (plane waves) are orthogonal, which enables a decomposition of the field $\delta(\boldsymbol{x})$ into Fourier modes in an unambiguous way. In contrast to this, the temperature fluctuations $\mathcal{T}$ are defined on the sphere. The analog to the Fourier modes in a flat space are spherical harmonics on the sphere, a complete orthogonal set of functions into which $\mathcal{T}(\boldsymbol{n})$ can be expanded.[8] On small angular scales, where a sphere can be considered locally flat, spherical harmonics approximately behave like plane waves. The power spectrum of temperature fluctuations, in most cases written as $\ell(\ell+1)C_\ell$, then describes the amplitude of the fluctuations on an angular scale $\theta \sim \pi/\ell = 180°/\ell$. $\ell = 1$ describes the dipole anisotropy, $\ell = 2$ the quadrupole anisotropy, and so on.

Line-of-Sight Projection. The CMB temperature fluctuations on the sphere result from projection, i.e., the integration along the line-of-sight of the three-dimensional temperature fluctuations which have been discussed above. This integration also needs to account for the secondary effects, those in the propagation of photons from $z \sim 1000$ to us. Overall, this is a relatively complicated task that, moreover, requires the explicit consideration of some aspects of General Relativity. The necessity for this can clearly be seen by

[7] This is seen with (7.10) due to the dependence $\overline{\rho} \propto a^{-3}$ and $\delta \propto D_+ = a$ for an EdS model.

[8] Spherical harmonics are encountered in many problems in mathematical physics, for instance in the quantum mechanical treatment of the hydrogen atom or, more generally, in all spherically symmetric problems in physics.

considering the fact that two directions which are separated by more than $\sim 1°$ have a spatial separation at recombination which is larger than the horizon size at that time – so spacetime curvature explicitly plays a role. Fortunately, the physical phenomena that need to be accounted for are (nearly) all of a linear nature. This means that, although the corresponding system of coupled equations is complicated, it can nevertheless easily be solved, since the solution of a system of linear equations is not a difficult mathematical problem. Generally accessible software packages exist (i.e., CMBFAST), which compute the power spectrum C_ℓ for any combination of cosmological parameters.

8.6.3 The Fluctuation Spectrum

Horizon Scale. To explain the basic features of CMB fluctuations, we first point out that a characteristic length-scale exists at $z_{\rm rec}$, namely the horizon length. It is specified by (4.71). For cosmological models with $\Omega_\Lambda = 0$, the horizon spans an angle of – see (4.72) –

$$\theta_{\rm H,rec} \approx 1.8° \sqrt{\Omega_{\rm m}} .$$

This angle is modified for models with a cosmological constant; if the Universe is flat ($\Omega_{\rm m} + \Omega_\Lambda = 1$), one finds

$$\theta_{\rm H,rec} \approx 1.8° , \tag{8.24}$$

with a very weak dependence on the matter density, about $\propto \Omega_{\rm m}^{-0.1}$. As we will demonstrate in the following, this angular scale of the horizon is directly observable.

Fluctuations on Large Scales. On scales $\gg \theta_{\rm H,rec}$ the Sachs–Wolfe effect dominates, since oscillations in the baryon–photon fluid can occur only on scales below the horizon length. For this reason, the CMB angular spectrum directly reflects the fluctuation spectrum $P(k)$ of matter. In particular, for a Harrison–Zeldovich spectrum, $P(k) \propto k$ one expects that

$$\ell(\ell+1)C_\ell \approx \text{const} \quad \text{for} \quad \ell \ll \frac{180°}{\theta_{\rm H,rec}} \simeq 100 ,$$

and the amplitude of the fluctuations immediately yields the amplitude of $P(k)$. This flat behavior of the fluctuation spectrum for $n_{\rm s} = 1$ is modified by the integrated Sachs–Wolfe effect.

Sound Horizon and Acoustic Peaks. On angular scales $< \theta_{\rm H,rec}$, fluctuations are observed that were inside the horizon prior to recombination, hence physical effects may act on these scales. As already mentioned, the fluid of baryons and photons is dominated by the energy density of the photons. Their pressure prevents the baryons from falling into the potential wells of dark matter. Instead, this fluid oscillates. Since the energy density is dominated by photons, i.e., by relativistic particles, this fluid is relativistic and its sound speed is $c_{\rm s} \approx c/\sqrt{3}$. Therefore, the maximum wavelength at which a wave may establish a full oscillation prior to recombination is

$$\lambda_{\rm max} \simeq t_{\rm rec}\, c_{\rm s} = r_{\rm H}(t_{\rm rec})/\sqrt{3} . \tag{8.25}$$

This length-scale is called the *sound horizon*. It corresponds to an angular scale of $\theta_1 \approx \theta_{\rm H,rec}/\sqrt{3} \sim 1°$, or $\ell_1 \sim 200$ for a flat cosmological model with $\Omega_{\rm m} + \Omega_\Lambda = 1$. By the Doppler effect and by adiabatic compression, these oscillations generate temperature fluctuations that should be visible in the temperature fluctuation spectrum C_ℓ. Hence, $\ell(\ell+1)C_\ell$ should have a maximum at $\ell_1 \sim 200$; additional maxima are expected at integer multiples of ℓ_1. These maxima in the angular fluctuation spectrum are termed *acoustic peaks* (or Doppler peaks); their ℓ-values and their amplitudes are the most important cosmological means of diagnostics on the CMB anisotropies.

Silk Damping. Since recombination is not instantaneous but extends over a finite range in redshift, CMB photons are last scattered within a shell of finite thickness. Considering a length-scale that is much smaller than the thickness of this shell, several maxima and minima of T are located within this shell along a line-of-sight. For this reason, the temperature fluctuations on these small scales are averaged out in the integration along the line-of-sight. The thickness of the recombination shell is roughly equal to the diffusion length of the photons, therefore this effect is relevant on the same length-scales as the aforementioned Silk damping. This means that on scales $\lesssim 5'$ ($\ell \gtrsim 2500$), one expects a damping of the anisotropy spectrum and, as a consequence, only very small (primary) temperature fluctuations on such small scales.

Model Dependence of the Fluctuation Spectrum. Figure 8.24 shows the power spectra of CMB fluc-

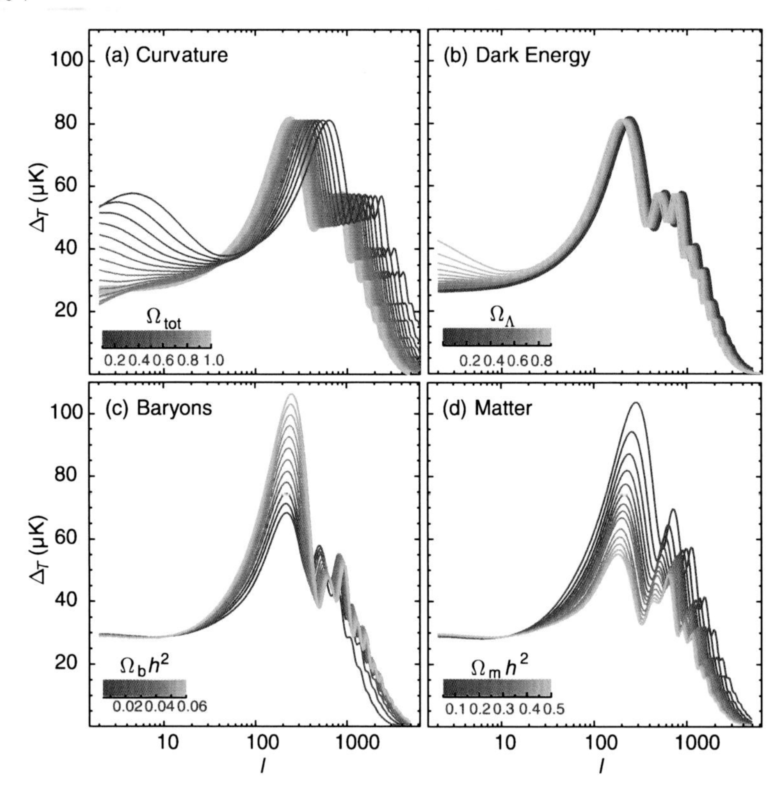

Fig. 8.24. Dependence of the CMB fluctuation spectrum on cosmological parameters. Plotted is the square root of the power per logarithmic interval in ℓ, $\Delta_T = \sqrt{\ell(\ell+1)C_\ell/(2\pi)}\,T_0$. These power spectra were obtained from an accurate calculation, taking into account all the processes previously discussed in the framework of perturbation theory in General Relativity. In all cases, the reference model is defined by $\Omega_m + \Omega_\Lambda = 1$, $\Omega_\Lambda = 0.65$, $\Omega_b h^2 = 0.02$, $\Omega_m h^2 = 0.147$, and a slope in the primordial density fluctuation spectrum of $n_s = 1$, corresponding to the Harrison–Zeldovich spectrum. In each of the four panels, one of these parameters is varied, and the other three remain fixed. The various dependences are discussed in detail in the main text

tuations where, starting from some reference model, individual cosmological parameters are varied. First we note that the spectrum is basically characterized by three distinct regions in ℓ (or in the angular scale). For $\ell \lesssim 100$, $\ell(\ell+1)C_\ell$ is a relatively flat function if – as in the figure – a Harrison–Zeldovich spectrum is assumed. In the range $\ell \gtrsim 100$, local maxima and minima can be seen that originate from the acoustic oscillations. For $\ell \gtrsim 2000$, the amplitude of the power spectrum is strongly decreasing due to Silk damping.

Figure 8.24(a) shows the dependence of the power spectrum on the curvature of the Universe, thus on $\Omega_{tot} = \Omega_m + \Omega_\Lambda$. We see that the curvature has two fundamental effects on the spectrum: first, the locations of the minima and maxima of the Doppler peaks are shifted, and second, the spectral shape at $\ell \lesssim 100$ depends strongly on Ω_{tot}. The latter is a consequence of the integrated Sachs–Wolfe effect because the more the world model is curved, the stronger the time variations of the gravitational potential ϕ. The shift in the acoustic peaks is essentially a consequence of the change in the geometry of the Universe: the size of the sound horizon depends only weakly on the curvature, but the angular diameter distance $D_A(z_{rec})$ is a very sensitive function of this curvature, so that the angular scale that corresponds to the sound horizon changes accordingly.

The dependence on the cosmological constant for flat models is displayed in Fig. 8.24(b). Here one can see that the effect of Ω_Λ on the locations of the acoustic peaks is comparatively small, so that these basically depend on the curvature of the Universe. The most important influence of Ω_Λ is seen for small ℓ. For $\Omega_\Lambda = 0$, the integrated Sachs–Wolfe effect vanishes and the power spectrum is flat (for $n_s = 1$), whereas larger Ω_Λ always produce a strong integrated Sachs–Wolfe effect.

The influence of the baryon density is presented in Fig. 8.24(c). An increase in the baryon density causes the amplitude of the first Doppler peak to rise, whereas that of the second peak decreases. In general, the amplitudes of the odd-numbered Doppler peaks increase, and those of the even-numbered peaks decrease with increasing $\Omega_b h^2$. Furthermore, the damping of fluctuations sets in at smaller ℓ (hence, larger angular scales)

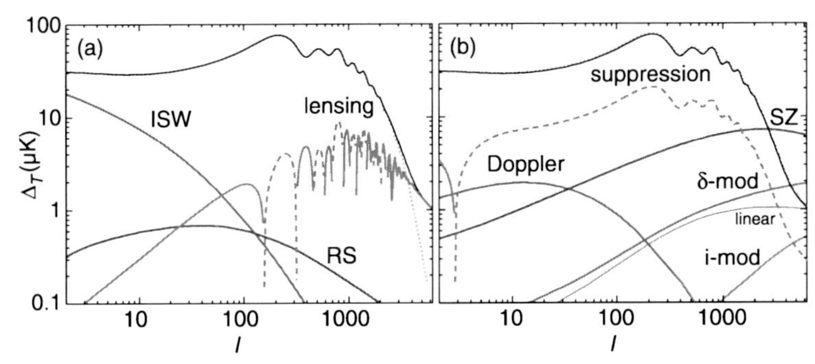

Fig. 8.25. The uppermost curve in each of the two panels shows the spectrum of primary temperature fluctuations for the same reference model as used in Fig. 8.24, whereas the other curves represent the effect of secondary anisotropies. On large angular scales (small ℓ), the integrated Sachs–Wolfe effect dominates, whereas the effects of gravitational light deflection (lensing) and of the Sunyaev–Zeldovich effect (SZ) dominate at large ℓ. On intermediate angular scales, the scattering of photons by free electrons which are present in the intergalactic gas after reionization (curve labeled "suppression") is the most efficient secondary process. Other secondary effects which are included in these plots are considerably smaller than the ones mentioned above and are thus of little interest here

if Ω_b is reduced, since in this case the mean free path of photons increases, and so the fluctuations are smeared out over larger scales. Finally, Fig. 8.24(d) demonstrates the dependence of the temperature fluctuations on the density parameter $\Omega_m h^2$. Changes in this parameter result in both a shift in the locations of the Doppler peaks and in changes of their amplitudes.

From this discussion, it becomes obvious that the CMB temperature fluctuations can provide an enormous amount of information about the cosmological parameters. Thus, from an accurate measurement of the fluctuation spectrum, very tight constraints on these parameters can be obtained.

Secondary Anisotropies. In Fig. 8.25, the secondary effects in the CMB anisotropies are displayed and compared to the reference model used above. Besides the already extensively discussed integrated Sachs–Wolfe effect, the influence of free electrons after reionization of the Universe has to be mentioned in particular. Scattering of CMB photons on these electrons essentially reduces the fluctuation amplitude on all scales, by a factor $e^{-\tau}$, where τ is the optical depth with respect to Thomson scattering. The latter depends on the reionization redshift, since the earlier the Universe was reionized, the larger τ is. Also visible in Fig. 8.25 is the fact that, on small angular scales, gravitational light deflection and the Sunyaev–Zeldovich effect become dominant. The identification of the latter is possible by its characteristic frequency dependence, whereas distinguishing the lens effect from other sources of anisotropies is not directly possible.

8.6.4 Observations of the Cosmic Microwave Background Anisotropy

To understand why so much time lies between the discovery of the CMB in 1965 and the first measurement of CMB fluctuations in 1992, we note that these fluctuations have a relative amplitude of $\sim 2 \times 10^{-5}$. The smallness of this effect means that in order to observe it very high precisions is required. The main difficulty with ground-based measurements is emission by the atmosphere. To avoid this, or at least to minimize it, satellite experiments or balloon-based observations are strongly preferred. Hence, it is not surprising that the COBE satellite was the first to detect CMB fluctuations.[9] Besides mapping the temperature distribution on the sphere (see Fig. 1.17) at an angular resolution of $\sim 7°$, COBE also found that the CMB is the most perfect blackbody that has ever been measured. The power spectrum for $\ell \lesssim 20$ measured by COBE was almost flat, and therefore compatible with the Harrison–Zeldovich spectrum.

[9] With the exception of the dipole anisotropy, caused by the peculiar velocity of the Sun, which has an amplitude of $\sim 10^{-3}$; this was identified earlier

Galactic Foreground. The measured temperature distribution of the microwave radiation is a superposition of the CMB and of emission from Galactic (and extragalactic) sources. In the vicinity of the Galactic disk, this foreground emission dominates, which is clearly visible in Fig. 1.17, whereas it seems to be considerably weaker at higher Galactic latitudes. However, due to its different spectral behavior, the foreground emission can be identified and subtracted. We note that the Galactic foreground basically consists of three components: synchrotron radiation from relativistic electrons in the Galaxy, thermal radiation by dust, and bremsstrahlung from hot gas. The synchrotron component defines a spectrum of about $I_\nu \propto \nu^{-0.8}$, whereas the dust is much warmer than 3 K and thus shows a spectral distribution of about $I_\nu \propto \nu^{3.5}$ in the spectral range of interest for CMB measurements. Bremsstrahlung has a flat spectrum in the relevant spectral region, $I_\nu \approx$ const. This can be compared to the spectrum of the CMB, which has a form $I_\nu \propto \nu^2$ in the Rayleigh–Jeans region.

There are two ways to extract the foreground emission from the measured intensity distribution. First, by observing at several frequencies the spectrum of the microwave radiation can be examined at any position, and the three aforementioned foreground components can be identified by their spectral signature and subtracted. As a second option, external datasets may be taken into account. At larger wavelengths, the synchrotron radiation is significantly more intense and dominates. From a sky map at radio frequencies, the distribution of synchrotron radiation can be obtained and its intensity at the frequencies used in the CMB measurements can be extrapolated. In a similar way, the infrared emission from dust, as measured, e.g., by the IRAS satellite (see Fig. 2.11), can be used to estimate the dust emission of the Galaxy in the microwave domain. Finally, one expects that gas that is emitting bremsstrahlung also shows strong Balmer emission of hydrogen, so that the bremsstrahlung pattern can be predicted from an Hα map of the sky. Both options, the determination of the foregrounds from multifrequency data in the CMB experiment and the inclusion of external data, are utilized in order to obtain a map of the CMB which is as free from foreground emission as possible – which indeed seems to have been accomplished in the bottom panel of Fig. 1.17.

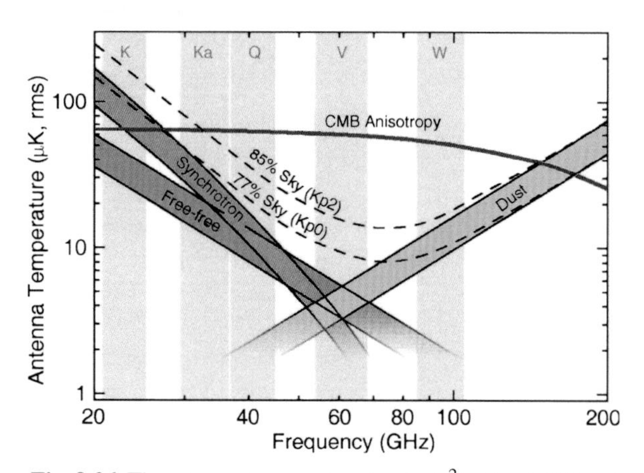

Fig. 8.26. The antenna temperature ($\propto I_\nu\, \nu^{-2}$) of the CMB and of the three foreground components discussed in the text, as a function of frequency. The five frequency bands of WMAP are marked. The dashed curves specify the average antenna temperature of the foreground radiation in the 77% and 85% of the sky, respectively, in which the CMB analysis was conducted. We see that the three high-frequency channels are not dominated by foreground emission

The optimal frequency for measuring the CMB anisotropies is where the foreground emission has a minimum; this is the case at about 70 GHz (see Fig. 8.26). Unfortunately, this frequency lies in a spectral region that is difficult to access from the ground.

From COBE to WMAP. In the years after the COBE mission, different experiments performed measurements of the anisotropy from the ground, focusing mainly on smaller angular scales. In around 1997, evidence was accumulating for the presence of the first Doppler peak, but the error bars of individual experimental results were too large at that time to clearly localize this peak. The breakthrough was then achieved in March 2000, when two groups published their CMB anisotropy results: BOOMERANG and MAXIMA. Both are balloon-based experiments, each observing a large region of the sky at different frequencies. In Fig. 8.27, the maps from the BOOMERANG experiment are presented. Both experiments have unambiguously measured the first Doppler peak, localizing it at $\ell \approx 200$. From this, it was concluded that we live in a nearly flat Universe – the quantitative analysis of the data yielded $\Omega_{\rm m} + \Omega_\Lambda \approx 1 \pm 0.1$. Furthermore, clear

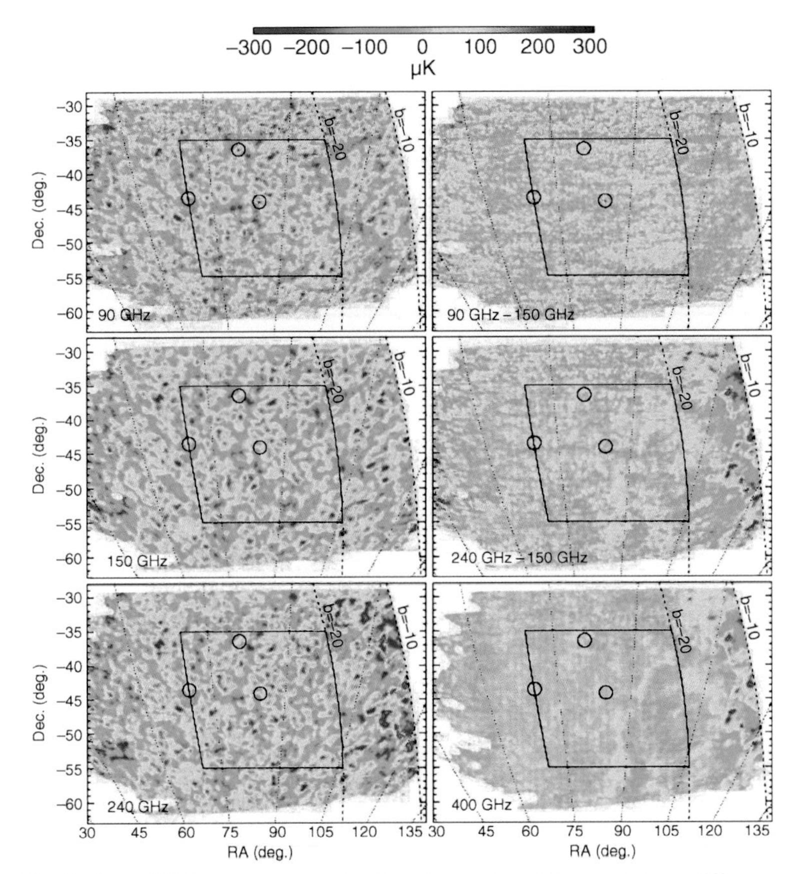

Fig. 8.27. In 2000, two groups published the results of their CMB observations, BOOMERANG and MAXIMA. This figure shows the BOOMERANG data. On the left, the temperature distributions at 90 GHz, 150 GHz, and 240 GHz are displayed, while the lower right panel shows that at 400 GHz. The three small circles in each panel denote the location of known strong point sources. The two upper panels on the right show the differences of temperature maps obtained at two different frequencies, e.g., the temperature map obtained with the 90-GHz data minus that obtained from the 150-GHz data. These difference maps feature considerably smaller fluctuations than the individual maps. This is compatible with the idea that the major fraction of the radiation originates in the CMB and not, e.g., in Galactic radiation which has a different spectral distribution and would thus be more prominent in the difference maps. Only the region within the dashed rectangle was used in the original analysis of the temperature fluctuations, in order to avoid boundary effects. The fluctuation spectrum computed from the difference maps is compatible with pure noise

indications of the presence of the second Doppler peak were found.

In April 2001, refined CMB anisotropy measurements from three experiments were released, BOOMERANG, MAXIMA, and DASI. For the former two, the observational data were the same as in the year before, but improved analysis methods were applied; in particular, a better instrumental calibration was obtained. The resulting temperature fluctuation spectrum is presented in Fig. 8.28, demonstrating that it was now possible to determine the locations of the first three Doppler peaks.

The status of measurements of the CMB anisotropy as of the end of 2002 is shown in Fig. 8.29. In the left-hand panel, the results of numerous experiments are plotted individually. The panel on the right shows the weighted mean of these experiments. Although it might not be suspected at first sight, the results of all experiments shown on the left are compatible with each other. With that we mean that the individual measure-

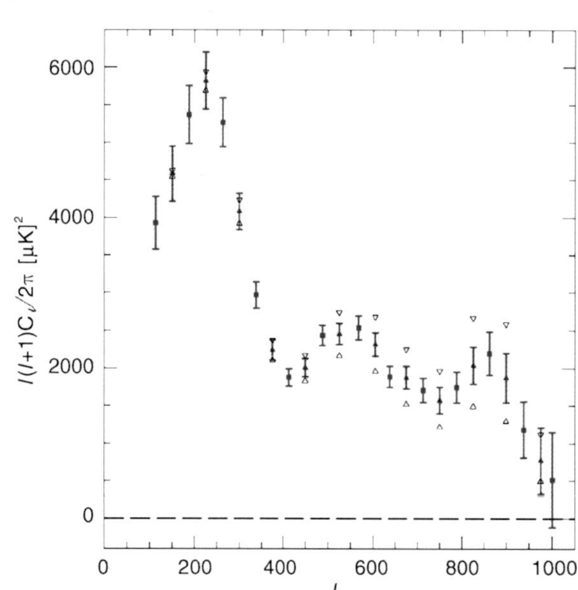

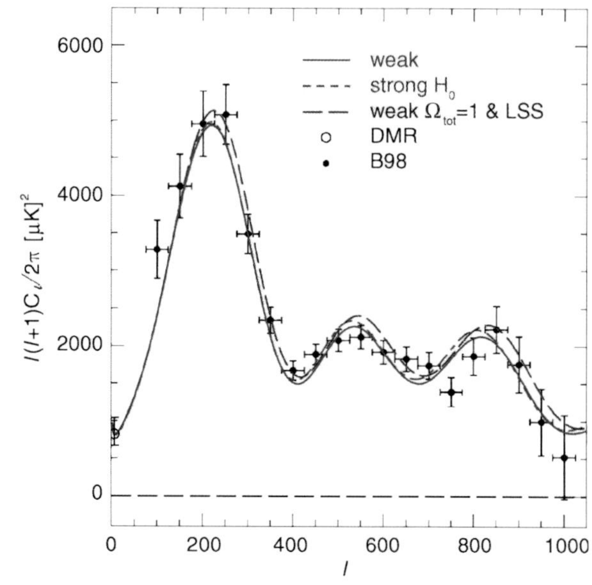

Fig. 8.28. Power spectrum of the CMB angular fluctuations, measured with the BOOMERANG experiment. These results were published in 2002, based on the same data as the previously released results, but using an improved analysis. Plotted are the coefficients $\ell(\ell+1)C_\ell/(2\pi)$ as a function of wave number or the multipole order $\ell \sim 180°/\theta$, respectively. The first three peaks can clearly be distinguished; they originate from oscillations in the photon–baryon fluid at the time of recombination. In the panel on the right, the fluctuation spectra of several cosmological models which provide good fits to the CMB data are plotted. The model denoted "weak" (solid curve) uses the constraints $0.45 \le h \le 0.90$, $t_0 > 10$ Gyr, and it has $\Omega_\Lambda = 0.51$, $\Omega_m = 0.51$, $\Omega_b h^2 = 0.022$, $h = 0.56$, and accordingly $t_0 = 15.2$ Gyr. The short-dashed curve ("strong H_0") uses a stronger constraint $h = 0.71 \pm 0.08$, and yields $\Omega_\Lambda = 0.62$, $\Omega_m = 0.40$, $\Omega_b h^2 = 0.022$, $h = 0.65$, and accordingly $t_0 = 13.7$ Gyr

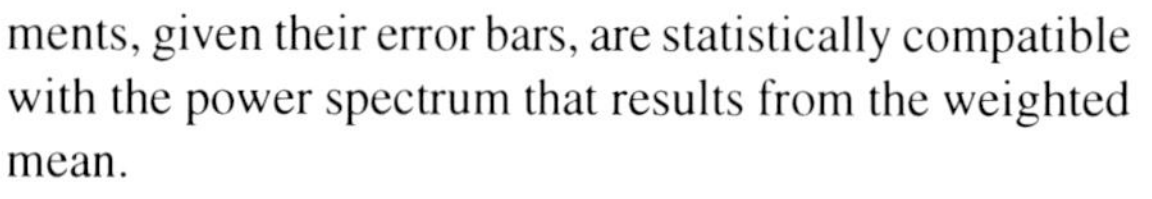

ments, given their error bars, are statistically compatible with the power spectrum that results from the weighted mean.

With the optimally averaged power spectrum, we can now determine the cosmological model which best describes these data. Under the assumption of a flat model, we obtain $\Omega_\Lambda = 0.71 \pm 0.11$ and a baryon density of $\Omega_b h^2 = 0.023 \pm 0.003$, in excellent agreement with the value obtained from primordial nucleosynthesis (see Eq. 4.62). Furthermore, the spectral index of the primordial density fluctuations is constrained to $n_s = 0.99 \pm 0.06$, which is very close to the Harrison–Zeldovich value of 1. In addition, the Hubble constant is estimated to be $h = 0.71 \pm 0.13$, again in extraordinarily good agreement with the value obtained from local investigations using the distance ladder, which is a completely independent measurement. These agreements are truly impressive if one recalls the assumptions our cosmological model is based upon.

Baryonic Oscillations in the Galaxy Distribution. As an aside, though a very interesting one, it should be mentioned here that the baryonic oscillations which are responsible for generating the acoustic peaks in the CMB anisotropy spectrum have now also been observed in the large-scale distribution of galaxies. To understand how this can be the case, we consider what happens after recombination. Imagine that recombination happened instantaneously; then right at that moment there are density fluctuations in the dark matter component as well as the acoustic oscillations in the baryons. The photons can stream freely, due to the absence of free electrons, and the sudden drop of pressure in the baryon component reduces the sound speed from $c/\sqrt{3}$ essentially to zero.

We said before that the baryons can then fall into the potential wells of the dark matter. However, since the cosmic baryon density is only about six times smaller than that of the dark matter, the baryonic density fluctuations at recombination are not completely negligible

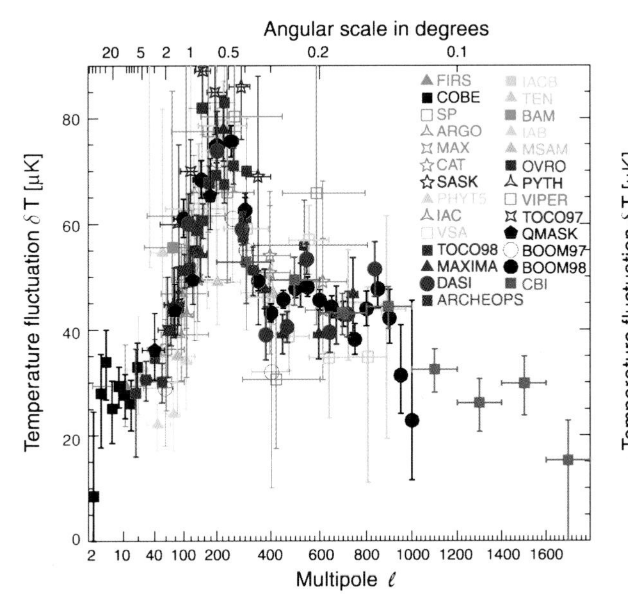

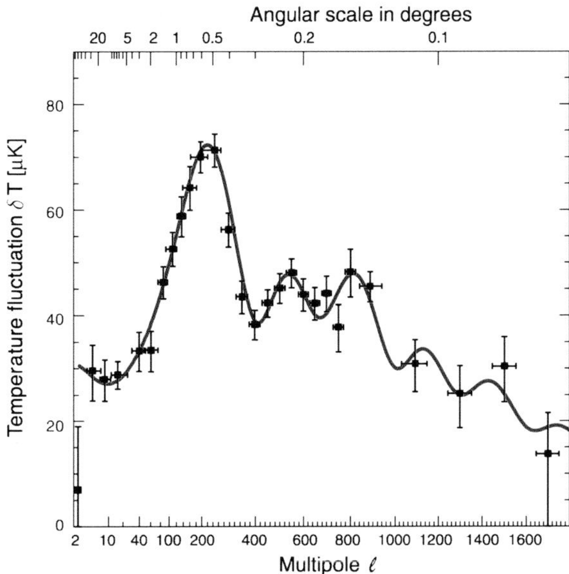

Fig. 8.29. This figure summarizes the status of the CMB anisotropy measurements as of the end of 2002. Left: the results from a large number of individual experiments are shown. Right: the "best" spectrum of the fluctuations is plotted, obtained by a weighted mean of the individual results where the corresponding error bars have been taken into account for the weighting. The red curve shows the fluctuation spectrum of the best-fitting cosmological model

compared to those of the dark matter. Therefore, they form their own potential wells, and part of the dark matter will fall into them. After some time, baryons and dark matter have about the same spatial distribution, which is described by the linear evolution of the density field, where the initial condition is a superposition of the dark matter fluctuations at recombination plus that of the baryonic oscillations. Whereas the corresponding density contrast of the latter is small compared to that of the dark matter, it has the unique feature that it carries a well-defined length-scale, namely the sound horizon at recombination. The matter correlation function in the local Universe should therefore contain a feature at just this length-scale. If galaxies trace the underlying matter distribution, this length-scale should then be visible in the galaxy correlation function.

In 2005, this feature in the galaxy correlation function was indeed observed, using the 2dFGRS and the SDSS redshift surveys. In these correlations, we thus see the same features as displayed by the acoustic oscillations in the CMB, but at much lower redshift. The reason why this discovery is of great importance is seen from the fact that the baryonic oscillations define a specific length-scale. The ratio of this length-scale to the observed angular scale yields the angular diameter distance to the redshift specified by the sample of galaxies considered. Thus, we have a "standard rod" in the Universe by which we can directly measure the distance–redshift relation, which in turn depends on the cosmological parameters, yielding a new and very valuable probe for cosmology. The fact that the baryonic oscillations cause a feature in the galaxy correlation function at a separation of ~ 100 Mpc immediately implies that very large redshift surveys are needed to measure this effect, encompassing very large volumes of the Universe.

8.6.5 WMAP: Precision Measurements of the Cosmic Microwave Background Anisotropy

In June 2001, the Wilkinson Microwave Anisotropy Probe satellite was launched, named in honor of David Wilkinson, one of the pioneers of CMB research. WMAP is, after COBE, only the second experiment to obtain an all-sky map in the microwave regime. Compared to COBE, WMAP observes over a wider

Fig. 8.30. Comparison of the CMB anisotropy measurements by COBE (top) and WMAP (bottom), after subtraction of the dipole originating from the motion of the Sun relative to the CMB rest-frame. The enormously improved angular resolution of WMAP is easily seen. Although these maps were recorded at different frequencies, the similarity in the temperature distribution is clearly visible and could be confirmed quantitatively. From this, the COBE results have, for the first time, been confirmed independently

frequency range, using five (instead of three) frequencies; it has a much improved angular resolution (which is frequency-dependent; about 20′, compared to $\sim 7°$ for COBE), and in addition, WMAP is able to measure the polarization of the CMB. Results from the first year of observation with WMAP were published in 2003. These excellent results confirmed our cosmological world model in such a way that we are now justified in calling it *the* standard model of cosmology. The most important results from WMAP will be discussed in the following.

Comparison to COBE. Since WMAP is the first satellite after COBE to map the full sky in the relevant frequency range, its first year results allowed the first verification of the COBE measurements. In Fig. 8.30, sky maps by COBE and by WMAP are displayed. The dramatically improved angular resolution of the WMAP map is obvious. In addition, it can clearly be seen that both maps are very similar if one compares them at a common angular resolution. This comparison can be performed quantitatively by "blurring" the WMAP map to the COBE resolution using a smoothing algorithm. Since WMAP is not observing at exactly the same frequencies as COBE, it is necessary to interpolate between two frequencies in the WMAP maps to match the frequency of the COBE map. The comparison then shows that, when accounting for the noise, the two maps are completely identical, with the exception of a single location in the Galactic disk. This can be explained, e.g., by a deviation of the spectral behavior of this source from the 2.73 K blackbody spectrum that was implicitly assumed for the aforementioned interpolation between two WMAP frequencies. The confirmation of the COBE measurements is indeed highly impressive.

Cosmic Variance. Before we continue discussing the WMAP results we need to explain the concept of cosmic variance. The angular fluctuation spectrum of CMB anisotropies is quantified by the multipole coefficients C_ℓ. For instance, C_1 describes the strength of the dipole. The dipole has three components; these can be described, for example, by an amplitude and two angles which specify a direction on the sphere. Accordingly, the quadrupole has five independent components, and in general, C_ℓ is defined by $(2\ell+1)$ independent components.

Cosmological models of the CMB anisotropies predict the *expectation value* of the amplitude of the individual components C_ℓ. In order to compare measurements of the CMB with these models one needs to understand that we will never measure the expectation value, but instead we measure only the mean value of the components contributing to the C_ℓ on *our* microwave sky. Since the quadrupole has only five independent components, the expected statistical deviation of the average from the expectation value is $C_2/\sqrt{5}$. In general, the statistical deviation of the average of C_ℓ from the expectation value is

$$\Delta C_\ell = \frac{C_\ell}{\sqrt{2\ell+1}} . \tag{8.26}$$

In contrast to many other situations, in which the statistical uncertainties can be reduced by analyzing a larger sample, this is not possible in the case of the CMB: there is only one microwave sky that we can observe. Hence, we cannot compile a sample of microwave maps, but instead depend on the one map of our sky. Observers at another location in the Universe will see a different CMB sky, and thus will measure different values C_ℓ, since their CMB sky corresponds to a different realization of the random field which is specified by the power spectrum $P(k)$ of the density fluctuations. This means that (8.26) is a fundamental limit to the statistical accuracy, which cannot be overcome by any improvements in instrumentation. This effect is called *cosmic variance*. The precision of the WMAP measurements is, for all $\ell \lesssim 350$, better than the cosmic variance (8.26). Therefore, the fluctuation spectrum for $\ell \lesssim 350$ measured by WMAP is "definite", i.e., further improvements of the accuracy in this angular range will not provide additional cosmological information (however, in future measurements one may test for potential systematic effects).

The Fluctuation Spectrum. Since WMAP observes at five different frequencies, the Galactic foreground radiation can, in principle, be separated from the CMB due to the different spectral behavior. Alternatively, external datasets may be utilized for this, as described in Sect. 8.6.4. This second method is preferred because, by using multifrequency data in the foreground subtraction, the noise properties of the resulting CMB map would get very complicated. The sky regions in which the foreground emission is particularly strong – mainly in the Galactic disk – are disregarded in the determination of C_ℓ. Furthermore, known point sources are also masked in the map.

The resulting fluctuation spectrum is presented in Fig. 8.31. In this figure, instead of plotting the individual C_ℓ, the fluctuation amplitudes have been averaged in ℓ-bins. The solid curve indicates the expected fluctuation spectrum in a ΛCDM-Universe whose parameters are quantitatively discussed further below. The gray region surrounding the model spectrum specifies the width of the cosmic variance, according to (8.26) and modified with respect to the applied binning.

The first conclusion is that the measured fluctuation spectrum agrees with the model extraordinarily well. Virtually no statistically significant deviations of the data points from the model are found. Smaller deviations which are visible are expected to occur as statistical outliers. The agreement of the data with the model is in fact spectacular: despite its enormous potential for new discoveries, WMAP "only" confirmed what had already been concluded from earlier measurements. Hence, the results from WMAP confirmed the cosmological model in an impressive way and, at the same time, considerably improved the accuracy of the parameter values.

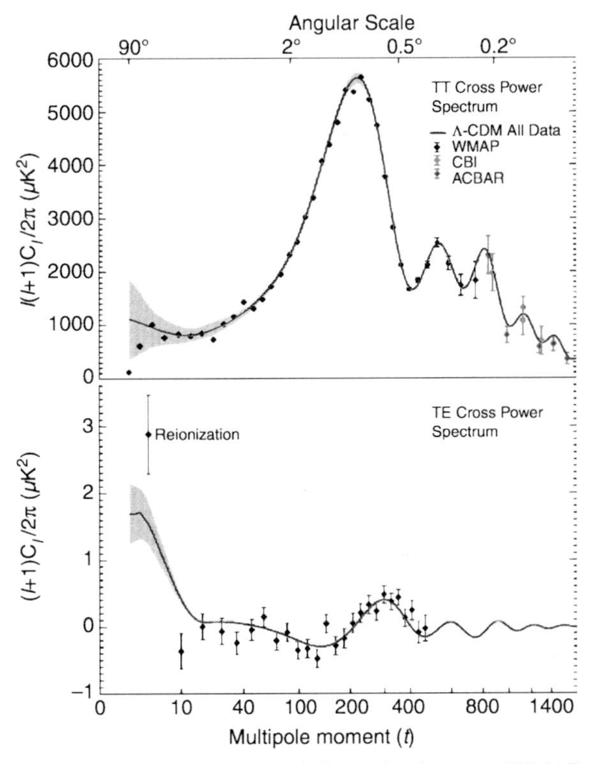

Fig. 8.31. As the central result from the first-year WMAP measurements, the top panel shows the fluctuation spectrum of the CMB temperature (TT), whereas the bottom panel displays the power spectrum of the correlation between the temperature distribution and polarization amplitude (TE). Besides the data points from WMAP, which are plotted here in ℓ-bins, the results from two other CMB experiments (CBI and ACBAR) are also plotted, at larger ℓ. The curve in each panel shows the best-fitting ΛCDM model, and the gray region surrounding it indicates the cosmic variance. The large amplitude of the point in the TE spectrum at small ℓ indicates an unexpectedly high polarization on large angular scales, which suggests an early reionization of the Universe

The only data point which deviates substantially from the model is that of the quadrupole, $\ell = 2$. In the COBE measurements, the amplitude of the quadrupole was also smaller than expected, as can be seen in Fig. 8.29. If one assigns physical significance to this deviation, this discrepancy may provide the key to possible extensions of the standard model of cosmology. Indeed, shortly after publication of the WMAP results, a number of papers were published in which an explanation for the low quadrupole amplitude was sought. Another kind of explanation may be found in the fact that for the analysis of the angular spectrum about 20% of the sky was disregarded, mainly the Galactic disk. The foreground emission is concentrated towards the disk, and we cannot rule out the possibility that it has a measurable impact on those regions of the sphere that have not been disregarded. This influence would affect the spectrum mainly at low ℓ. Anomalies in the orientation of the low-order multipoles have in fact been found in the data. Currently, there is probably no reason to assume that the low quadrupole amplitude is of cosmological relevance. If, on the other hand, future analysis of the data can rule out a substantial foreground contribution from the Galaxy (or even from the Solar System), the low quadrupole amplitude may be a smoking gun for modifications of the standard model.

Polarization of the CMB. The cosmic background radiation is blackbody radiation and should therefore be unpolarized. Nevertheless, polarization measurements of the CMB have been conducted which revealed a finite polarization. This effect shall be explained in the following.

The scattering of photons on free electrons not only changes the direction of the photons, but also produces a linear polarization of the scattered radiation. The direction of this polarization is perpendicular to the plane spanned by the incoming and the scattered photons. Consider now a region of space with free electrons. Photons from this direction have either propagated from the epoch of recombination to us without experiencing any scattering, or they have been scattered into our direction by the free electrons. Through this scattering, the radiation is, in principle, polarized. Roughly speaking, photons that have entered this region from the "right" or the "left" are polarized in north–south direction, and photons infalling from "above" or "below" show a polarization in east-west direction after scattering. If the CMB, as seen from the scattering electrons, was isotropic, an equal number of photons would enter from right and left as from above and below, so that the net polarization would vanish. However, the scattering electrons see a slightly anisotropic CMB sky, in much the same way as we observe it; therefore, the net-polarization will not completely vanish.

This picture implies that the CMB radiation may be polarized. The degree of polarization depends on the probability of a CMB photon having been scattered since recombination, thus on the optical depth with respect to Thomson scattering. Since the optical depth depends on the redshift at which the Universe was reionized, this redshift can be estimated from the degree of polarization.

In the lower part of Fig. 8.31, the power spectrum of the correlation between the temperature distribution and the polarization is plotted. One finds a surprisingly large value of this cross-power for small ℓ. This measurement is probably the most unexpected discovery in the WMAP data from the first year of observation, because it requires a very early reionization of the Universe, $z_{\rm ion} \sim 15$, hence much earlier than derived from, e.g., the spectra of QSOs at $z \gtrsim 6$.

The Future of CMB Measurements. Before discussing the cosmological parameters that result from the WMAP data, we will briefly outline the prospects of CMB measurements in the years after 2005. On the one hand, WMAP will continue to carry out measurements for several years, improving the accuracy of the measurements and, in particular, testing the results from the first year. The power spectrum of the polarization itself, to data (February 2006) has not yet been published, so that we can expect new insights (or another confirmation of the standard model) from that as well, in particular regarding the reionization redshift. As for COBE, the WMAP data will also be a rich source of research for many years.

Balloon and ground-based observations will conduct CMB measurements on small angular scales and so extend the results from WMAP towards larger ℓ. For example, the experiments DASI, CBI, and BOOMERANG have measured polarization fluctuations of the CMB, as well as temperature-polarization cross-correlations. As their measurements extend to smaller angular scales than WMAP, the damping tail in the angular power spectrum was detected by these experiments. In particular, the results from the 2003 flight of BOOMERANG have confirmed the large scattering optical depth found by WMAP and thus the high redshift of reionization.

The progress in such measurements has already been enormous in recent years, and outstanding results can be expected for the near future. Finally, the ESA satellite Planck is due to be launched in 2008, observing in a frequency range between 30 and 850 GHz and at an angular resolution of about 5′. Like WMAP, Planck will also map the full sky. Besides measuring the CMB, this mission will produce very interesting astrophysical results; it is expected, for instance, that the Planck satellite will discover about 10^4 clusters of galaxies by the Sunyaev–Zeldovich effect.

8.7 Cosmological Parameters

For a long time, the determination of the cosmological parameters has been one of the prime challenges in cosmology, and numerous different methods were developed and applied to determine H_0, Ω_m, Ω_Λ, etc. Until a few years ago, these different methods yielded results with relatively large error margins, some of which did not even overlap. In recent years, the situation has fundamentally changed, as already discussed in the previous sections. The measurements by WMAP form the current highlight in the determination of the cosmological parameters, and thus we begin this section with a presentation of these results.

8.7.1 Cosmological Parameters with WMAP

The precise measurement of the first Doppler peak, together with consideration of the full angular power spectrum, provides very tight constraints on the deviation of the cosmological model from a model with vanishing curvature. In Fig. 8.32, the confidence regions in the Ω_m–Ω_Λ plane are given, determined either from the CMB data alone (WMAP, combined with measurements at small angular scales, which in the following is called WMAPext), or by combining these with SNe Ia data and/or the value of H_0 as determined from the Hubble Key Project (see Sect. 3.6.2). As in earlier CMB measurements presented in Sect. 8.6.4, the results from WMAP also show that the deviation of $\Omega_m + \Omega_\Lambda$ from unity is very small.

For this reason, we consider the other cosmological parameters under the assumption of a flat Universe, $\Omega_m + \Omega_\Lambda = 1$. The WMAP team analyzed a six-dimensional cosmological parameter space, spanned by the amplitude A of the density fluctuations (this amplitude is directly linked to σ_8, but it is measured on considerably larger scales than $8\,h^{-1}$ Mpc because the CMB data probe the power spectrum on such large scales), by the slope n_s of the spectrum of primordial fluctuations (with $n_s = 1$ for a Harrison–Zeldovich spectrum), by the optical depth τ with respect to Thomson scattering after the reionization of the Universe, by the scaled Hubble constant h, and by the density parameters $\Omega_m h^2$ and $\Omega_b h^2$. Table 8.1 lists the best-fit values for these parameters, where four different combinations of data were used: WMAP alone, WMAP in combination with measurements of the CMB fluctuations on smaller scales (WMAPext), WMAPext in combination with the results from the 2dFGRS, and finally the combination of WMAPext, the 2dFGRS, and Lyα forest results.

Considering first the CMB results alone, we find that the value of n_s is very close to unity, hence the primordial fluctuation spectrum must have a slope very similar to, but slightly smaller than the Harrison–Zeldovich spectrum – in agreement with the predictions from inflationary models (see Sect. 7.7). The value of the Hubble constant, $h = 0.73 \pm 0.05$, is in excellent agreement with that determined from the Hubble Key Project. The derived baryon density $\Omega_b h^2$ is also in outstanding agreement with the value obtained from primordial nucleosynthesis. Combining the values for $\Omega_m h^2$ and h stated in the table, we obtain a value for $\Gamma = \Omega_m h$ which is in very good agreement with that found from the galaxy distribution in the 2dFGRS (see Eq. 8.5). We should recall once again that we are dealing with completely independent methods for the determination of these parameters.

The measurement of the integrated Sachs–Wolfe effect in the fluctuation spectrum is a verification of the value for Ω_Λ being different from zero, fully independent of the supernovae results. As a matter of fact, the

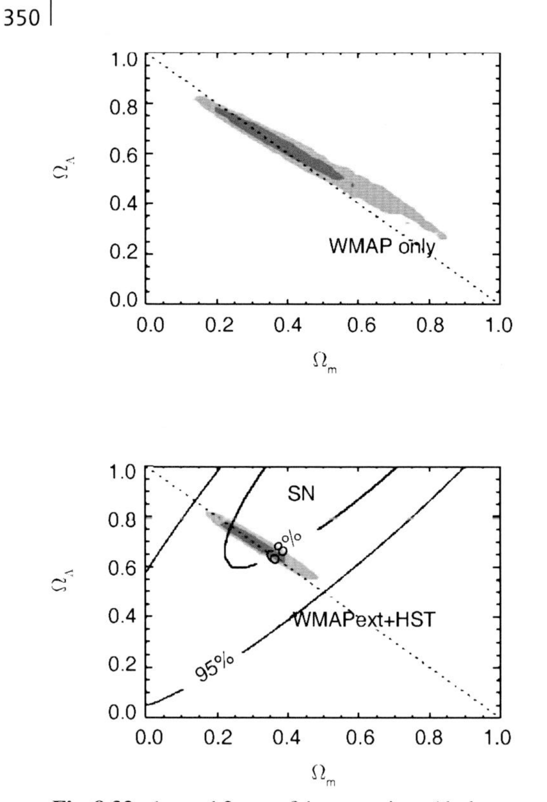

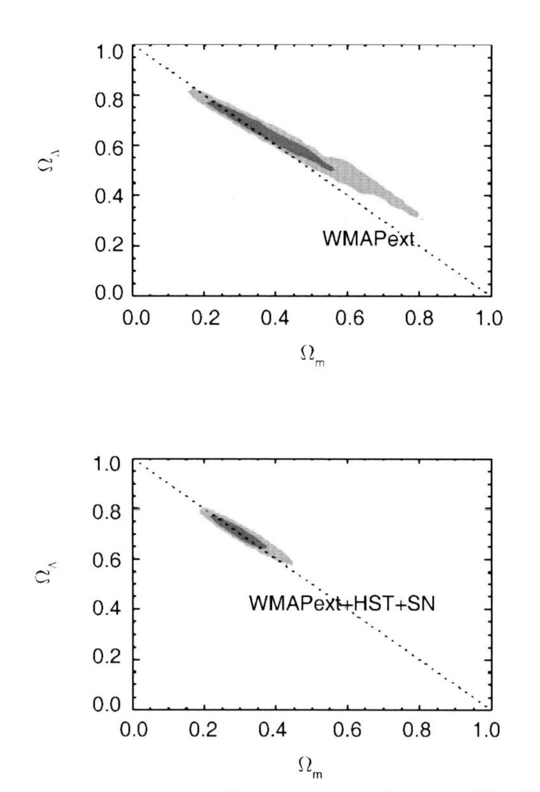

Fig. 8.32. 1σ and 2σ confidence regions (dark gray and light gray areas, respectively) in the Ω_m–Ω_Λ plane. In the upper left panel, only the WMAP data were used. In the upper right panel, the WMAP data were combined with CMB measurements on smaller angular scales (WMAPext). In the lower left panel, the WMAPext data were combined with the determination of the Hubble constant from the HST Key Project, and the confidence region which is obtained from SN Ia measurements is included only for comparison. In the lower right panel, the SN Ia data are included in addition. The dashed line indicates models of vanishing curvature, $\Omega_m + \Omega_\Lambda = 1$

Table 8.1. The six basic parameters determined from the WMAP data, where a flat cosmological model ($\Omega_m + \Omega_\Lambda = 1$) is assumed. A is the amplitude and n_s the slope of the primordial power spectrum, and τ is the optical depth with respect to Thomson scattering after reionization. $\chi^2_{\rm eff}/\nu$ is a statistical measure for the agreement of the best-fit model and the data. The different columns list the best-fit parameters obtained by using the WMAP data alone, WMAP in combination with CMB measurements on small angular scales (WMAPext), WMAPext with additional inclusion of the power spectrum from the 2dFGRS (WMAPext+2dFGRS), and finally the additional inclusion of the power spectrum from the Lyα forest (WMAPext+2dFGRS+Lyα) (from Spergel et al., 2003, ApJS, **148**, 175)

	WMAP	WMAPext	WMAPext+2dFGRS	WMAPext+2dFGRS+Lyα
A	0.9 ± 0.1	0.8 ± 0.1	0.8 ± 0.1	$0.75^{+0.08}_{-0.07}$
n	0.99 ± 0.04	0.97 ± 0.03	0.97 ± 0.03	0.96 ± 0.02
τ	$0.166^{+0.076}_{-0.071}$	$0.143^{+0.071}_{-0.062}$	$0.148^{+0.073}_{-0.071}$	$0.117^{+0.057}_{-0.053}$
h	0.72 ± 0.05	0.73 ± 0.05	0.73 ± 0.03	0.72 ± 0.03
$\Omega_m h^2$	0.14 ± 0.02	0.13 ± 0.01	0.134 ± 0.006	0.133 ± 0.006
$\Omega_b h^2$	0.024 ± 0.001	0.023 ± 0.001	0.023 ± 0.001	0.0226 ± 0.0008
$\chi^2_{\rm eff}/\nu$	1429/1341	1440/1352	1468/1381	...

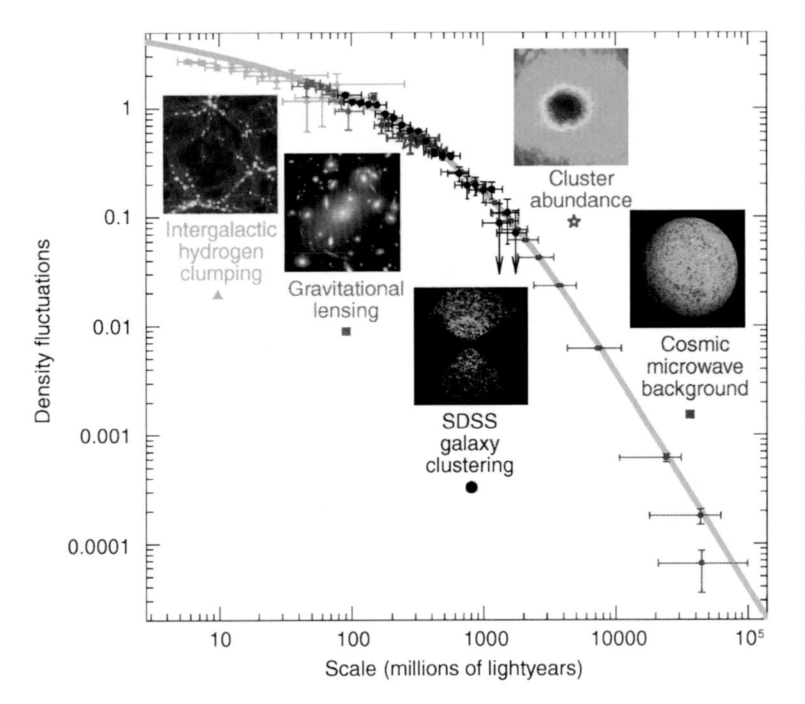

Fig. 8.33. The power spectrum of density fluctuations in the Universe, as determined by different methods. Here, $\Delta^2(k) \propto k^3 P(k)$ is plotted. Note that small length-scales (or large k, respectively) are towards the left in the plot. Going from large to small scales, the results presented here are obtained from CMB temperature fluctuations, from the abundance of galaxy clusters, from the large-scale distribution of galaxies, from cosmic shear, and from the statistical properties of the Lyα forest. One can see that the power spectrum of a ΛCDM model is able to describe all these data over many orders of magnitude in scale

physical origin of this effect can be proven directly because the integrated Sachs–Wolfe effect is produced at relatively low redshifts (where the influence of a cosmological constant is noticeable), as a result of the time evolution of the gravitational potential. Therefore, it should be directly correlated with the large-scale matter overdensities which are observable in the distribution of galaxies and clusters of galaxies, assuming a bias model. For example, one can correlate the CMB temperature map with luminous elliptical galaxies, as they are observed photometrically in the Sloan Digital Sky Survey over very large regions on the sky (see Sect. 9.1.2). The significant correlation signal found in this analysis yields very strong evidence for the temperature fluctuations having originated from the Sachs–Wolfe effect on large angular scales, hence providing a direct proof of $\Omega_\Lambda \neq 0$.

A big surprise in the WMAP results is the large value of τ, which is derived in particular from the TE power spectrum. This value for τ implies that the Universe was reionized at a fairly high redshift of $z \sim 15$.

The combination of CMB results with those from the large-scale distribution of galaxies and the statistics of the Lyα forest allows us to measure the power spectrum at smaller length-scales, as shown in Fig. 8.33. This combination therefore provides stronger constraints on the cosmological parameters.

We can see from Table 8.1 that with this combination the error margins of some parameters can indeed be reduced, compared to considering the WMAP data alone; in particular, this is the case for $\Omega_{\rm m} h^2$. It is important to note that by combining the different data sets, the values of the parameters change only within the range of their error bars as determined from the CMB data, which means that the different datasets are compatible with each other (and with the flat ΛCDM model). With these primary parameters known, further parameters may now be derived; these are listed in Table 8.2.

The combined data yield, as a best value for the total density of the Universe,

$$\Omega_{\rm m} + \Omega_\Lambda = 1.02 \pm 0.02 \; , \tag{8.27}$$

in outstanding agreement with the prediction from the inflationary model. Furthermore, the fraction of hot dark matter can be constrained, for which the small-scale observations (here from the Lyα forest) are of particular importance, since HDM reduces the power on small

Table 8.2. Cosmological parameters, as derived from the CMB data (WMAP) and the combination of these with the data from 2dFGRS and the Lyα forest. Here, $z_{\rm ion}$ is the redshift of the reionization of the Universe (where it is assumed that the reionization was homogeneous, instantaneous and complete), $z_{\rm rec}$ is the redshift of the recombination (this is the redshift at which the z-distribution of the last scattering of CMB photons has its maximum), $z_{\rm eq}$ is the redshift where matter and radiation had the same energy density, $n_{\rm b}$ is the number density of baryons today, and η is the number density ratio of baryons to photons (from Spergel et al., 2003, ApJS, **148**, 175)

	WMAP	WMAPext+2dFGRS+Lyα
h	0.72 ± 0.05	$0.71^{+0.04}_{-0.03}$
σ_8	0.9 ± 0.1	0.84 ± 0.04
$\sigma_8\Omega_{\rm m}^{0.6}$	0.44 ± 0.10	$0.38^{+0.04}_{-0.05}$
$\Omega_{\rm b}$	0.047 ± 0.006	0.044 ± 0.004
$\Omega_{\rm m}$	0.29 ± 0.07	0.27 ± 0.04
t_0	13.4 ± 0.3 Gyr	13.7 ± 0.2 Gyr
$z_{\rm ion}$	17 ± 5	17 ± 4
$z_{\rm rec}$	1088^{+1}_{-2}	1089 ± 1
$z_{\rm eq}$	3454^{+385}_{-392}	3233^{+194}_{-210}
$n_{\rm b}$	$(2.7\pm0.1)\times10^{-7}\,{\rm cm}^{-3}$	$(2.5\pm0.1)\times10^{-7}\,{\rm cm}^{-3}$
η	$\left(6.5^{+0.4}_{-0.3}\right)\times10^{-10}$	$\left(6.1^{+0.3}_{-0.2}\right)\times10^{-10}$

scales. We obtain

$$\Omega_\nu h^2 < 0.0076\,, \tag{8.28}$$

with a 2σ significance, which implies a strict upper limit for the neutrino mass of $m_\nu < 0.23$ eV, where (4.63) was used. This limit is significantly tighter than that which is currently achievable in laboratory measurements. Besides the Lyα forest, cosmic shear can also be utilized for measurements on small scales, as is demonstrated in Fig. 8.34.

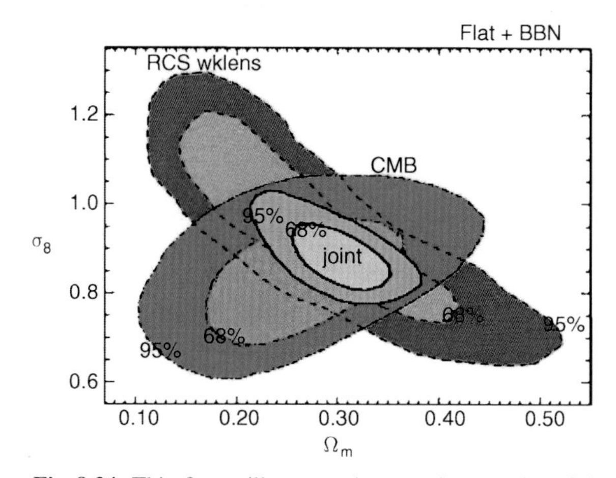

Fig. 8.34. This figure illustrates the complementarity of the CMB data with those from cosmic shear. The individual confidence regions of both methods (blue for the CMB, orange for cosmic shear) in the $\Omega_{\rm m}$–σ_8 plane are nearly orthogonal, so that a combination of both methods leads to a significantly smaller region (green) of allowed parameter pairs

8.7.2 Cosmic Harmony

With the exception of the high optical depth τ, the WMAP results to data have not brought big surprises. However, this fact in itself is surprising: given the high sensitivity and angular resolution of this satellite, it could well have been possible that the measured fluctuation spectrum showed discrepancies with respect to our standard model. Remarkably, this does not seem to be the case.

Hence we are in a situation in which the basic cosmological parameters are not only known with an accuracy that had been unimaginable only a few years ago, but also each of these individual values has been measured by more than one independent method, confirming the self-consistency of the model in an impressive manner.

- **Hubble Constant.** H_0 has been determined with the Hubble Key Project, by means of the distance ladder, particularly using Cepheids. The resulting value is in outstanding agreement with that derived from CMB anisotropies. Other estimates of H_0 yield comparable values. Although the determination of H_0 by means of the time delay measurement for galaxy-scale gravitational lenses, and by means of the SZ effect typically yield somewhat smaller values, these are still compatible with the values from the Hubble Key Project and the CMB measurements within the range of the expected statistical errors and systematic effects which are difficult to control.
- **Contribution of Baryons to the Total Density.** The ratio $\Omega_{\rm b}/\Omega_{\rm m}$ has been determined from the baryon fraction in clusters of galaxies, from redshift surveys, and from the CMB fluctuations, all yielding $\Omega_{\rm b}/\Omega_{\rm m}\approx0.15$.
- **Baryon Density.** The value for $\Omega_{\rm b}h^2$ determined from primordial nucleosynthesis combined with

measurements of the deuterium abundance in Lyα systems has also impressively been confirmed by the WMAP results.

- **Matter Density.** Assuming the value of H_0 to be known, Ω_m has been determined from the distribution of galaxies in redshift surveys, from the CMB, and from the evolution of the number density of galaxy clusters.
- **Vacuum Energy.** The very tight limits on the curvature of the Universe obtained from the CMB measurements, and the implied tight limits on the deviation of $\Omega_m + \Omega_\Lambda$ from unity, allows us to determine Ω_Λ from the measurement of Ω_m and the integrated Sachs–Wolfe effect. These values are in excellent agreement with the SN Ia measurements, as shown in Fig. 8.35.
- **Normalization of the Power Spectrum.** Since the CMB fluctuations measure the power spectrum at large length-scales, the normalization obtained from these measurements can be compared to the value of σ_8 only if the shape of $P(k)$ is very well known. However, with the shape of $P(k)$ being tightly constrained within the framework of CDM models by the accurate determination of the other parameters, the CMB measurements yield a value for σ_8 that is in very good agreement with that obtained from the abundance of galaxy clusters and from cosmic shear (see Fig. 8.34). Furthermore, these values are compatible with those derived from the peculiar velocity field of galaxies. However, the uncertainties in σ_8 for the individual methods are about $\pm 10\%$ each, so that σ_8, at the present time, is perhaps the least accurately determined cosmological parameter.
- **Age of the Universe.** The age of the Universe derived from the WMAP data, $t_0 \approx 13.4 \times 10^9$ yr, is compatible with the age of globular clusters and of the oldest white dwarfs in our Galaxy.

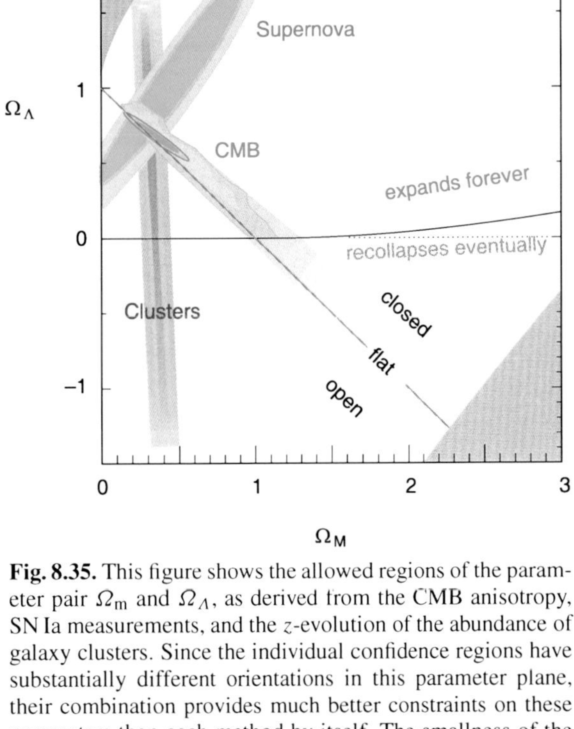

Fig. 8.35. This figure shows the allowed regions of the parameter pair Ω_m and Ω_Λ, as derived from the CMB anisotropy, SN Ia measurements, and the z-evolution of the abundance of galaxy clusters. Since the individual confidence regions have substantially different orientations in this parameter plane, their combination provides much better constraints on these parameters than each method by itself. The smallness of the individual confidence regions and the fact that they are overlapping is an impressive demonstration of the self-consistency of our cosmological model

The observational results described in this chapter opened an era of precision cosmology. On the one hand, the accuracy of the individual cosmological parameters will doubtlessly be improved in the coming years by new observational results; on the other hand, the interest of cosmology will increasingly shift towards observations of the early Universe. Studies of the evolution of cosmic structure, of the formation of galaxies and clusters, and of the history of the reionization of the Universe will increasingly become the focus of cosmological research.

Another central objective of future cosmological research will remain the investigation of dark matter and of dark energy. For the foreseeable future, the latter in particular will be accessible only through astronomical observations. Due to the enormous importance of a non-vanishing dark energy density for fundamental physics, studying its properties will be at the center of interest of more than just astrophysicists. It is expected that a successful theory describing dark energy will necessitate a significant breakthrough in our general understanding of fundamental physics.

The search for the constituents of dark matter will keep physicists busy in the coming years. Experiments at future particle accelerators (e.g., the LHC at CERN) and the direct search, in underground laboratories, for particles which may account for candidates of dark matter, are promising. In any case, dark matter (if it indeed consists of elementary particles) will open up a new field in particle physics. For these reasons, the interests of cosmology and particle physics are increasingly converging – in particular since the Universe is the largest and cheapest laboratory for particle physics.

9. The Universe at High Redshift

In the previous chapter we explained by what means the cosmological parameters may be determined, and what progress has been achieved in recent years. This might have given the impression that, with the determination of the values for $\Omega_{\rm m}$, Ω_Λ, etc., cosmology is nearing its conclusion. As a matter of fact, for several decades cosmologists have considered the determination of the density parameter and the expansion rate of the Universe their prime task, and now this goal has seemingly largely been achieved. However, from this point on, the future evolution of the field of cosmology will probably proceed in two directions. First, we will try to uncover the nature of dark energy and to gain new insights into fundamental physics along the way. Second, astrophysical cosmology is much more than the mere determination of a few parameters. We want to understand how the Universe evolved from a very primitive initial state into what we are observing around us today – galaxies of different morphologies, the large-scale structure of their distribution, clusters of galaxies, and active galaxies. We seek to study the formation of stars and of metals, and also the processes that reionized the intergalactic medium.

The boundary conditions for studying these processes are now very well defined. A few years ago, the cosmological parameters in models of galaxy evolution, for instance, could vary freely because they had not been determined sufficiently well at that time. Today, a successful model needs to come up with predictions compatible with observations, but using the parameters of the standard model. There is little freedom left in designing such models. In other words, the stage on which the formation and evolution of objects and structure takes place is prepared, and now the cosmic play can begin.

Progress in recent years, with developments in instrumentation having played a vital role, has allowed us to examine the Universe at very high redshift. An obvious indication of this progress is the increasingly high maximum redshift of sources that can be observed; as an example, Fig. 9.1 presents the spectrum of a QSO at redshift $z = 6.43$. Today, we know quite a few galaxies at redshift $z > 6$, i.e., we observe these objects at a time when the Universe had less than 10% of its current age. Besides larger telescopes, which enabled these deep images of the Universe, gaining access to new wavelength domains is of particular importance for our studies of the distant Universe. This can be seen, for example, from the fact that the optical radiation of a source at redshift $z \sim 1$ is shifted into the NIR. Because of this, near-infrared astronomy is about as important for galaxies at $z \gtrsim 1$ as optical astronomy is for the local Universe. Furthermore, the development of submillimeter astronomy has provided us with a view of sources that are

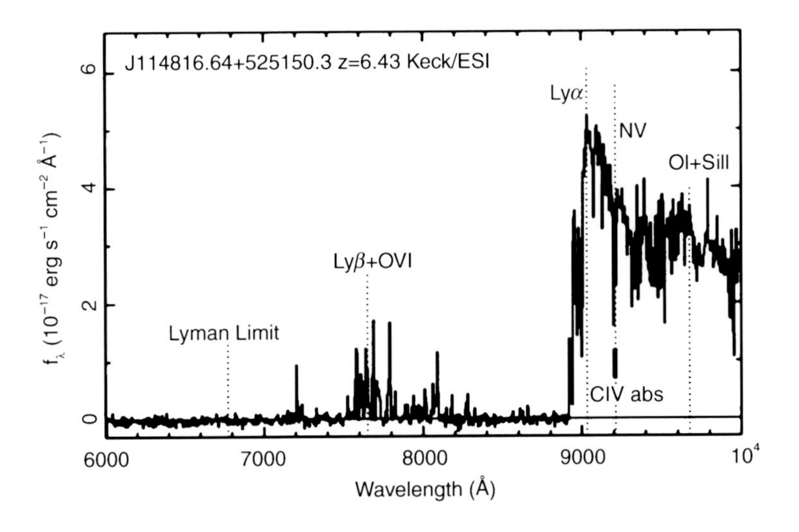

Fig. 9.1. Spectrum of a QSO at the high redshift of $z = 6.43$. Like many other QSOs at very high redshift, this source was discovered with the Sloan Digital Sky Survey. The spectrum was obtained with the Keck telescope. The redshifted Lyα line is clearly visible, its blue side "eaten" away by intergalactic absorption. Almost all radiation bluewards of the Lyα line is absorbed, with only the emission from the Lyβ line still getting through. For $\lambda \leq 7200$ Å the spectral flux is compatible with zero; intergalactic absorption is too strong here

Peter Schneider, The Universe at High Redshift.
In: Peter Schneider, Extragalactic Astronomy and Cosmology. pp. 355–405 (2006)
DOI: 10.1007/11614371_9

nearly completely hidden to the optical eye because of strong dust absorption.

In this chapter, we will attempt to provide an impression of astronomy of the distant Universe, and shed light on some interesting aspects that are of particular importance for our understanding of the evolution of the Universe. This field of research is currently developing very rapidly, so we will simply address some of the main topics in this field today. We begin in Sect. 9.1 with a discussion of methods to specifically search for high-redshift galaxies, and we will then focus on a method by which galaxy redshifts can be determined solely from photometric information in several bands (thus, from the color of these objects). This method can be applied to deep sky images observed by HST, and we will present some of the results of these HST surveys. Finally, we will emphasize the importance of gravitational lenses as "natural telescopes", which, due to their magnification, provide us with a deeper view into the Universe.

Gaining access to new wavelength domains paves the way for the discovery of new kinds of sources; in Sect. 9.2 we will present galaxy populations which have been identified by submillimeter and NIR observations, and whose relation to the other known types of galaxies is yet to be uncovered. In Sect. 9.3 we will show that, besides the CMB, background radiation also exists at other wavelengths, but whose nature is considerably different from that of the CMB. The question of when and by what processes the Universe was reionized will be discussed in Sect. 9.4. Then, in Sect. 9.5, we will focus on the history of cosmic star formation, and show that at redshift $z \gtrsim 1$ the Universe was much more active than it is today – in fact, most of the stars which are observed in the Universe today were already formed in the first half of cosmic history. This empirical discovery is one of the aspects that one attempts to explain in the framework of models of galaxy formation and evolution. In Sect. 9.6 we will highlight some aspects of these models and their link to observations. Finally, we will discuss the sources of gamma-ray bursts. These are explosive events which, for a very short time, appear brighter than all other sources of gamma rays on the sky put together. For about 25 years the nature of these sources was totally unknown; even their distance estimates were spread over at least seven orders of magnitude. Only since 1997 has it been known that these sources are of extragalactic origin.

9.1 Galaxies at High Redshift

In this section we will first consider the question of how distant galaxies can be found, and how to identify them as such. The properties of these high-redshift galaxies can then be compared with those of galaxies in the local Universe, which were described in Chap. 3. The question then arises as to whether galaxies at high z, and thus in the early Universe, look like local galaxies, or whether their properties are completely different. One might, for instance, expect that the mass and luminosity of galaxies are evolving with redshift. Examining the galaxy population as a function of redshift, one can trace the history of global cosmic star formation and analyze when most of the stars visible today have formed, and how the density of galaxies changes as a function of redshift. We will investigate some of these questions in this and the following sections.

9.1.1 Lyman-Break Galaxies (LBGs)

How to Find High-Redshift Galaxies? Until about 1995 only a few galaxies with $z > 1$ had been known; most of them were radio galaxies discovered by optical identification of radio sources. The most distant normal galaxy with $z > 2$ then was the source of the giant luminous arc in the galaxy cluster Cl 2244−02 (see Fig. 6.31). Very distant galaxies are faint, and so the question arises of how galaxies at high z can be detected at all.

The most obvious answer to this question may perhaps be by spectroscopy of a sample of faint galaxies. This method is not feasible though, since galaxies with $R \lesssim 22$ have redshifts $z \lesssim 0.5$, and spectra of galaxies with $R > 22$ are only observable with 4-m telescopes and with a very large investment of observing time. Also, the problem of finding a needle in a haystack arises: most galaxies with $R \lesssim 24.5$ have redshifts $z \lesssim 2$ (a fact that was not known before 1995), so how can we detect the small fraction of galaxies with larger redshifts?

Narrow-Band Photometry. A more systematic method that has been applied is narrow-band photometry. Since hydrogen is the most abundant element in the Universe, one expects that some fraction of galaxies feature a Lyα emission line (as do all QSOs). By comparing two sky images, one taken with a narrow-band filter centered on a wavelength λ, the other with a broader filter also centered roughly on λ, this line emission can be searched for specifically. If a galaxy at $z \approx \lambda/(1216\,\text{Å}) - 1$ has a strong Lyα emission line, it should be particularly bright in the narrow-band image in comparison to the broad-band image, relative to other sources. This search for Lyα emission line galaxies had been almost without success until the mid-1990s. Among other reasons, one did not know what to expect, e.g., how faint galaxies at $z \sim 3$ are and how strong their Lyα line would be.

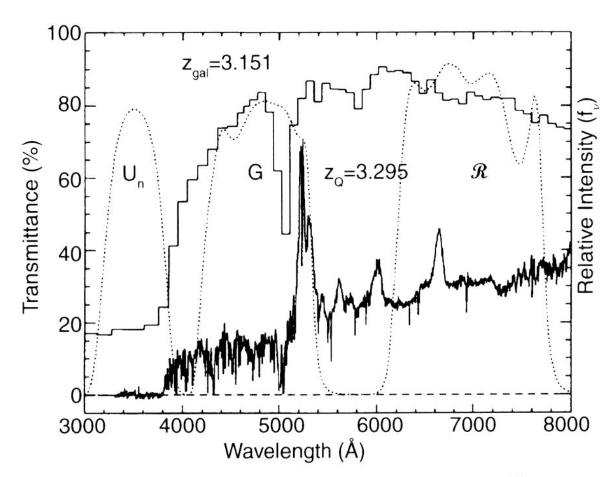

Fig. 9.2. Principle of the Lyman-break method. The histogram shows the synthetic spectrum of a galaxy at $z = 3.15$, generated by models of population synthesis; the spectrum belongs to a QSO at slightly higher redshift. Clearly, the decline of the spectrum at $\lambda \leq 912(1+z)$Å is noticeable. The three dotted curves are the transmission curves of three broad-band filters, chosen such that one of them (U_n) blocks all photons with wavelengths above the Lyman-break. The color of this galaxy would then be blue in $G - \mathcal{R}$, and very red in $U_n - G$

The Lyman-Break Method. The breakthrough was obtained with a method that became known as the *Lyman-break method*. Since hydrogen is so abundant and its ionization cross-section so large, one can expect that photons with $\lambda < 912$ Å are very heavily absorbed by neutral hydrogen in its ground state. Therefore, photons with $\lambda < 912$ Å have a low probability of escaping from a galaxy without being absorbed.

Intergalactic absorption also contributes. In Sect. 5.6.3 we saw that each QSO spectrum features a Lyα forest and Lyman-limit absorption. The intergalactic gas absorbs a large fraction of photons emitted by a high-redshift source at $\lambda < 1216$ Å, and virtually all photons with a rest-frame wavelength $\lambda \lesssim 912$ Å. As also discussed in Sect. 8.5.2, the strength of this absorption increases with increasing redshift. Combining these facts, we conclude that spectra of high-redshift galaxies should display a distinct feature – a "break" at $\lambda = 1216$ Å. Furthermore, radiation with $\lambda \lesssim 912$ Å should be strongly suppressed by intergalactic absorption, as well as by absorption in the interstellar medium of the galaxies themselves, so that only a very small fraction of these ionizing photons will reach us.

From this, a strategy for the detection of galaxies at $z \gtrsim 3$ emerges. We consider three broad-band filters with central wavelengths $\lambda_1 < \lambda_2 < \lambda_3$, where their spectral ranges are chosen to not (or only marginally) overlap. If $\lambda_1 \lesssim (1+z)\,912\,\text{Å} \lesssim \lambda_2$, a galaxy containing young stars should appear relatively blue as measured with the filters λ_2 and λ_3, and be virtually invisible in the λ_1-filter: because of the absorption, it will drop out of the λ_1-filter (see Fig. 9.2). For this reason, galaxies that have been detected in this way are called *Lyman-break galaxies* (LBG) or *drop-outs*. An example of this is displayed in Fig. 9.3.

Large Samples of LBGs. The method was first applied systematically in 1996, using the filters specified in Fig. 9.2. As can be read from Fig. 9.4, the expected location of a galaxy at $z \sim 3$ in a color–color diagram with this set of filters is nearly independent of the type and star-formation history of the galaxy. Hence, sources in the relevant region of the color–color diagram are very good candidates for being galaxies at $z \sim 3$. The redshift needs to be verified spectroscopically, but the crucial point is that the color selection of candidates yields a very high success rate per observed spectrum, and thus spectroscopic observing time at the telescope is spent very efficiently in confirming the redshift of distant galaxies. With the commissioning of the Keck telescope (and later also of other telescopes of the 10-m class), spectroscopy of galaxies with $B \lesssim 25$ became possible (see Fig. 9.5). Employing this method, more than 1000 galaxies with $2.5 \lesssim z \lesssim 3.5$ have been detected and spectroscopically verified to date.

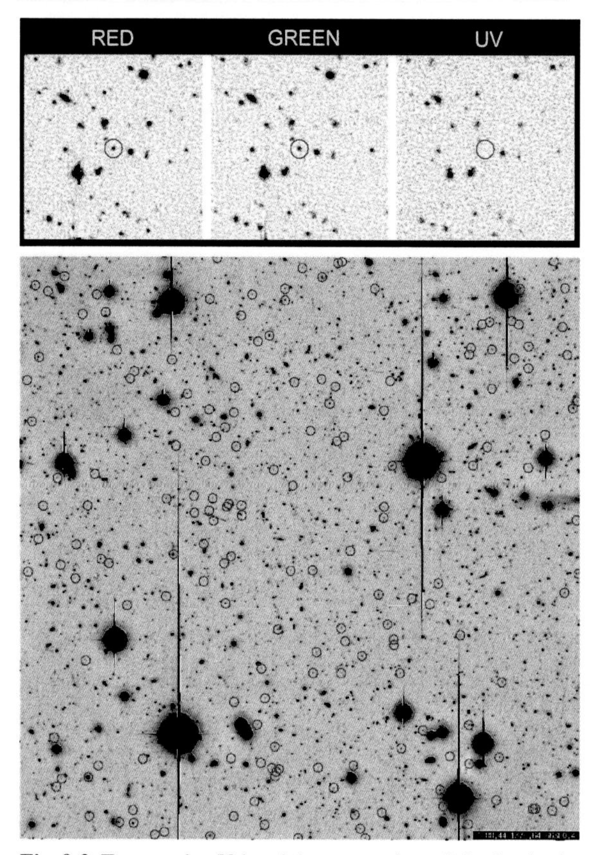

Fig. 9.3. Top panel: a U-band drop-out galaxy. It is clearly detected in the two redder filters, but vanishes almost completely in the U-filter. Bottom panel: in a single CCD frame, a large number of candidate Lyman-break galaxies are found. They are marked with circles here; their density is about 1 per square arcminute

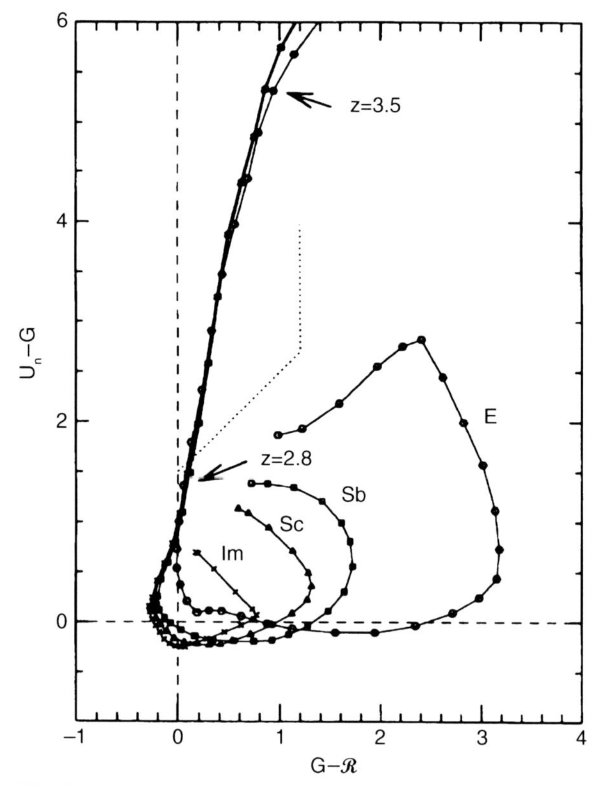

Fig. 9.4. Evolutionary tracks of galaxies in the $(G-\mathcal{R})$ – (U_n-G) color–color diagram, for different types of galaxies, as obtained from population synthesis models. All evolutionary tracks start at $z=0$, and the symbols along the curves mark intervals of $\Delta z = 0.1$. The colors of the various galaxy types are very different at lower redshift, but for $z \geq 2.7$, the evolutionary tracks for the different types nearly coincide – a consequence of the Lyα absorption in the intergalactic medium. Hence, a color selection of galaxies in the region between the dotted and dashed curves should select galaxies with $z \geq 3$. Indeed, this selection of candidates has proven to be very successful; more than 1000 galaxies with $z \sim 3$ have been spectroscopically verified

From the spectra shown in Fig. 9.5, it also becomes apparent that not all galaxies which fulfill the selection criteria also show a Lyα emission line, which provides one possible explanation for the lack of success in earlier searches for high-redshift galaxies using narrow-band filters. The spectra of the high-redshift galaxies which were found by this method are very similar to those of starburst galaxies at low redshift. Obviously, the galaxies selected in this way feature active star formation. Due to the chosen selection criterion, such sources are, of course, preferentially selected, since star formation produces a blue spectrum at (rest-frame) wavelengths above 1216 Å; in addition the luminosity of galaxies in the UV range strongly depends on the star-formation rate.

The Correlation Function of LBGs. For a large variety of objects, and over a broad range of separations, the correlation function of objects can be described by the power law (7.19), with a slope of typically $\gamma \sim 1.7$. However, the amplitude of this correlation function varies between different classes of objects. For exam-

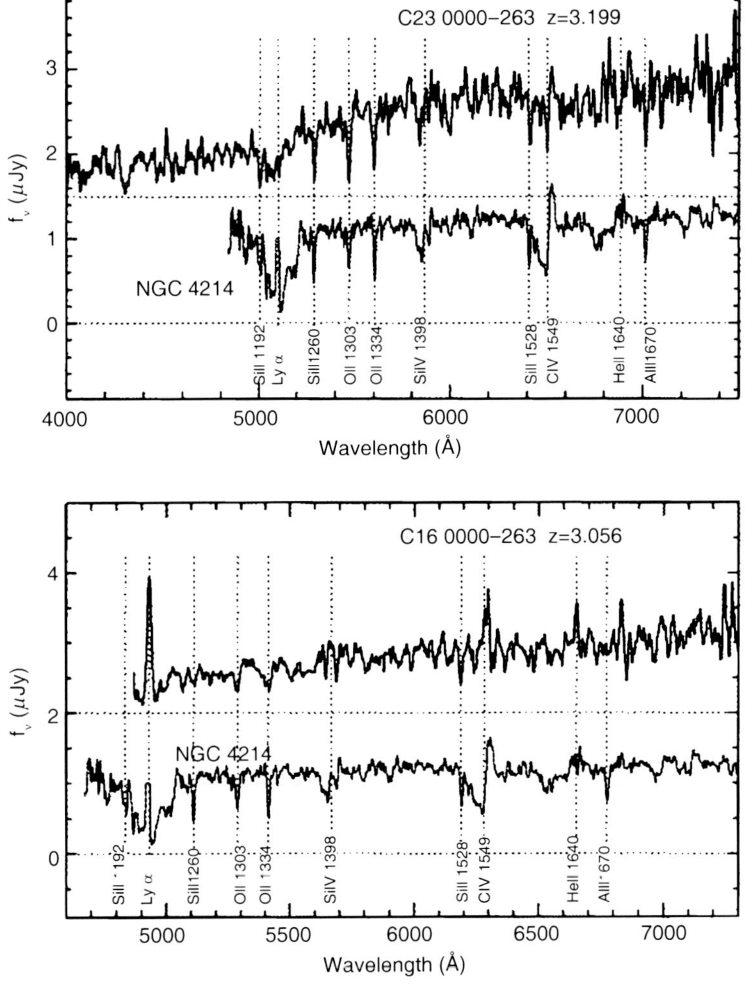

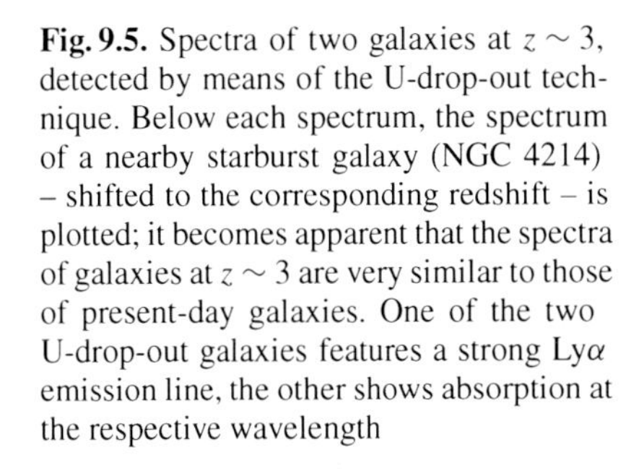

Fig. 9.5. Spectra of two galaxies at $z \sim 3$, detected by means of the U-drop-out technique. Below each spectrum, the spectrum of a nearby starburst galaxy (NGC 4214) – shifted to the corresponding redshift – is plotted; it becomes apparent that the spectra of galaxies at $z \sim 3$ are very similar to those of present-day galaxies. One of the two U-drop-out galaxies features a strong Lyα emission line, the other shows absorption at the respective wavelength

ple, we saw in Sect. 8.2.4 that the power spectrum of galaxy clusters is larger by about a factor 7 than that of galaxies (see Fig. 8.12); the same ratio holds of course for the corresponding correlation functions. As we argued there, the strength of the correlation depends on the mass of objects; in the simple picture of biasing shown in Fig. 8.3, the correlation of objects is larger the rarer they are. High-mass peaks exceeding the density threshold needed for gravitational collapse have a lower mean density than low-mass peaks, so they are therefore expected to be more biased (see Sect. 8.1.3) and thus more strongly correlated.

In fact, these qualitative arguments can be substantiated with numerical simulations, as well as with quantitative analytical estimates. From the LSS simulations described in Sect. 7.5.3, one can identify dark matter halos (employing, e.g., the overdensity criterion which follows from the spherical collapse model in Sect. 7.5.1) and compare their correlation function with that of the overall dark matter distribution. From such analyses it is concluded that massive halos are more strongly clustered than the dark matter itself, whereas low-mass halos are less correlated. The division between these two cases occurs roughly at the mass scale $M_*(z)$ – see (7.34) – which describes the non-linear mass scale at a given epoch. Therefore, measuring the correlation function of objects and comparing it with the correlation of dark matter at the corresponding redshift, the characteristic mass of the halos in which these objects reside can be determined, as well as their bias.

The Halo Mass of LBGs. If we consider the spatial distribution of LBGs, we find a large correlation amplitude. The (comoving) correlation length of LBGs at redshifts $1.5 \lesssim z \lesssim 3.5$ is $r_0 \sim 4.2h^{-1}$ Mpc, i.e., not very different from the correlation length of L_*-galaxies in the present Universe. Since the bias factor of present-day galaxies is about unity, implying that they are clustered in a similar way to the dark matter distribution, this result then implies that the bias of LBGs at high redshift must be considerably larger than unity. This conclusion is based on the fact that the dark matter correlation at high redshifts was smaller than today by the factor $D_+^2(z)$. Thus we conclude that LBGs are rare objects and thus correspond to high-mass dark matter halos. Comparing the observed correlation length r_0 with numerical simulations, the characteristic halo mass of LBGs can be determined, yielding $\sim 3 \times 10^{11} M_\odot$ at redshifts $z \sim 3$, and $\sim 10^{12} M_\odot$ at $z \sim 2$. Furthermore, the correlation length is observed to increase with the luminosity of the LBG, indicating that more luminous galaxies are hosted by more massive halos, which are more strongly biased than less massive ones. If these results are combined with the observed correlation functions of galaxies in the local Universe and at $z \sim 1$, and with the help of numerical simulations, then this indicates that a typical high-redshift LBG will evolve into an elliptical galaxy by today.

Proto-Clusters. Furthermore, the clustering of LBGs shows that the large-scale galaxy distribution was already in place at high redshifts. In some fields the observed overdensity in angular position and galaxy redshift is so large that one presumably observes galaxies which will later assemble into a galaxy cluster – hence, we observe some kind of proto-cluster. We have already shown such a proto-cluster in Fig. 6.47. Galaxies in such a proto-cluster environment seem to have about twice the stellar mass of those LBGs outside such structures, and the age of their stellar population appears older by a factor of two. This result indicates that the stellar evolution of galaxies in dense environments proceeds faster than in low-density regions, in accordance with expectations from structure formation. It also reveals a dependence of galaxy properties on the environment, which we have seen before manifested in the morphology–density relation (see Sect. 6.2.9). Proto-clusters of galaxies have also been detected at higher redshifts up to $z \sim 6$, using narrow-band imaging searches for Lyman-α emission galaxies.

Whereas the clustering of LBGs is well described by the power law (7.19) over a large range of scales, the correlation function exhibits a significant deviation from this power law on very small scales: the angular correlation function exceeds the power law at $\Delta\theta \lesssim 7''$, corresponding to physical length-scales of ~ 200 kpc. It thus seems that this scale marks a transition in the distribution of galaxies. To get an idea of the physical nature of this transition, we note that this length scale is about the virial radius of a dark matter halo with $M \sim 3 \times 10^{11} M_\odot$, i.e., the mass of halos which host the LBGs. On scales below this virial radius, the correlation function thus no longer describes the correlation between two distinct dark matter halos. An interpretation of this fact is provided in terms of merging: when two galaxies and their dark matter halos merge, the resulting dark matter halo hosts both galaxies, with the more massive one close to the center and the other one as "satellite galaxy". The correlation function on scales below the virial radius thus indicates the clustering of galaxies within the same halo, whereas on larger scales, where it follows the power-law behavior, it indicates the correlation between different halos.

Winds of Star-Forming Galaxies. The inferred high star-formation rates of LBGs implies an accordingly high rate of supernova explosions. These release part of their energy in the form of kinetic energy to the interstellar medium in these galaxies. This process will have two consequences. First, the ISM in these galaxies will be heated locally, which slows down (or prevents) further star formation in these regions. This thus provides a feedback effect for star formation which prevents all the gas in a galaxy from turning into stars on a very short time-scale, and is essential for understanding the formation and evolution of galaxies, as we shall see in Sect. 9.6. Second, if the amount of energy transferred from the SNe to the ISM is large enough, a galactic wind may be launched which drives part of the ISM out of the galaxy into its halo. Evidence for such galactic winds has been found in nearby galaxies, for example from neutral hydrogen observations of edge-on spirals which show an extended gas distribution outside the disk. Furthermore, the X-ray corona of spirals (see Fig. 3.18) is most likely linked to a galactic wind in these systems.

Indeed, there is now clear evidence for the presence of massive winds from LBGs. The spectra of LBGs often show strong absorption lines, e.g., of CIV, which are blueshifted relative to the velocity of the emission lines in the galaxy. An example of this effect can be seen in the spectra of Fig. 9.5, where in the upper panel the emission line of CIV is accompanied by an absorption to the short-wavelength side of the emission line. Such absorption can be produced by a wind moving out from the star-forming regions of the galaxy, so that its redshift is smaller than that of the emission regions. Characteristic velocities are $\sim 200\,\mathrm{km/s}$. In one case where the spectral investigation has been performed in most detail (the LBG cB58; see Fig. 9.13), the outflow velocity is $\sim 255\,\mathrm{km/s}$, and the outflowing mass rate exceeds the star-formation rate. Whereas these observations clearly show the presence of outflowing gas, it remains undetermined whether this is a fairly local phenomenon, restricted to the star-formation sites, or whether it affects the ISM of the whole galaxy.

Connection to QSO Absorption Lines. A slightly more indirect argument for the presence of strong winds from LBGs comes from correlating the absorption lines in background QSO spectra with the position of LBGs. These studies have shown that whenever the sightline of a QSO passes within $\sim 40\,\mathrm{kpc}$ of an LBG, very strong CIV absorption lines (with column density exceeding $10^{14}\,\mathrm{cm}^{-2}$) are produced, and that the corresponding absorbing material spans a velocity range of $\Delta v \gtrsim 250\,\mathrm{km/s}$; for about half of the cases, strong CIV absorption is produced for impact parameters within 80 kpc. This frequency of occurrence implies that about 1/3 of all CIV metal absorption lines with $N \gtrsim 10^{14}\,\mathrm{cm}^{-2}$ in QSO spectra are due to gas within $\sim 80\,\mathrm{kpc}$ from those LBGs which are bright enough to be included in current surveys. It is plausible that the remaining 2/3 are due to fainter LBGs.

The association of CIV absorption line systems with LBGs by itself does not prove the existence of winds in such galaxies; in fact, the absorbing material may be gas orbiting in the halo in which the corresponding LBG is embedded. In this case, no outflow phenomenon would be implied. However, in that case one might wonder where the large amount of metals implied by the QSO absorption lines is coming from. They could have been produced by an earlier epoch of star formation, but in that case the enriched material must have been expelled from its production site in order to be located in the outer part of $z \sim 3$ halos. It appears more likely that the production of metals in QSO absorption systems is directly related to the ongoing star formation in the LBGs. We shall see in Sect. 9.2.5 that clear evidence for superwinds has been discovered in one massive star-forming galaxy at $z \sim 3$.

Finally, we mention another piece of evidence for the presence of superwinds in star-forming galaxies. There are indications that the density of absorption lines in the Lyα forest is reduced when the sightline to the QSO passes near a foreground LBG. This may well be explained by a wind driven out from the LBG, pushing neutral gas away and thus leaving a gap in the Lyα forest. The characteristic size of the corresponding "bubbles" is $\sim 0.5\,\mathrm{Mpc}$ for luminous LBGs.

Lyman-Break Galaxies at Low Redshifts. One might ask whether galaxies similar to the LBGs at $z \sim 3$ exist in the current Universe. Until recently this question was difficult to investigate, since it requires imaging of lower redshift galaxies at ultraviolet wavelengths. With the launch of GALEX an appropriate observatory became available with which to observe galaxies with rest-frame UV luminosities similar to those of LBGs. UV-selected galaxies show a strong inverse correlation between the stellar mass and the surface brightness in the UV. Lower-mass galaxies are more compact than those of higher stellar mass. On the basis of this correlation we can consider the population of large and compact UV-selected galaxies separately. The larger ones show a star-formation rate of a few $M_\odot/\mathrm{yr}$; at this rate, their stellar mass content can be built up on a time-scale comparable to the Hubble time, i.e., the age of the Universe. These galaxies are typically late-type spiral galaxies, and they show a metallicity similar to our Galaxy. In contrast, the compact galaxies have a lower stellar mass and about the same star-formation rate, which allows them to generate their stellar population much faster, in about 1 Gyr. Their metallicity is smaller by about a factor of 2. These properties of these compact UV-selected galaxies are quite similar to those of the LBGs seen at higher redshifts, and hence, they may be closely related to the LBG population.

Lyman-Break Galaxies at High Redshift. By variation of the filter set, drop-outs can also be discovered at larger wavelengths, thus at accordingly higher redshifts. The object selection at higher z implies an increasingly dominant role of the Lyα forest whose density is a strongly increasing function of redshift (see Sect. 8.5.2). This method has been routinely applied up to $z \sim 4.5$, yielding so-called B-drop-outs. Galaxies at considerably higher redshifts are difficult to access from the ground with this method. One reason for this is that galaxies become increasingly faint with redshift, rendering observations substantially more problematic. Furthermore, one needs to use increasingly redder filter sets. At such large wavelengths the night sky gets significantly brighter, which further hampers the detection of very faint objects. For detecting a galaxy at redshift, say, $z = 5.5$ with this method, the Lyα line, now at $\lambda \approx 7900$ Å, is located right in the I-band, so that for an efficient application of the drop-out technique only the I- and z-band filters or NIR-filters are viable, and with those filters the brightness of the night sky is very problematic (see Fig. 9.6 for an example of a drop-out galaxy at very high redshift). Furthermore, candidate very high-redshift galaxies detected as drop-outs are very difficult to verify spectroscopically due to their very low flux. In spite of this, we will see later that the drop-out method has achieved spectacular results even at redshifts considerably higher than $z \sim 4$, where the HST played a central role. But the new generation of 10-m class telescopes, equipped with instruments sensitive in the appropriate wavelength regimes, can also reveal a population of high-redshift drop-out candidates. In particular, the Subaru telescope, which carries a wide-field camera, has produced a deep field survey with several broad-band filters and two narrow-band filters situated at wavelengths of 8840 Å and 9840Å, respectively, which is ideally suited to selecting $z \sim 6$ LBGs. A survey conducted with Subaru has detected about 12 LBG candidates at this redshift. Calculating the spatial number density of these objects indicates that luminous star-forming galaxies were rarer by an order of magnitude at $z \sim 6$ than at $z \sim 3$.

9.1.2 Photometric Redshift

The Lyman-break technique is a special case of a method for estimating the redshift of galaxies (and QSOs) by multicolor photometry. This technique can be employed due to the spectral break at $\lambda = 912$ Å and $\lambda = 1216$ Å, respectively. Spectra of galaxies also show other characteristic features. As was discussed in detail in Sect. 3.9, the broad-band energy distribution is basically a superposition of stellar radiation. A stellar population of age $\gtrsim 10^8$ yr features a 4000-Å

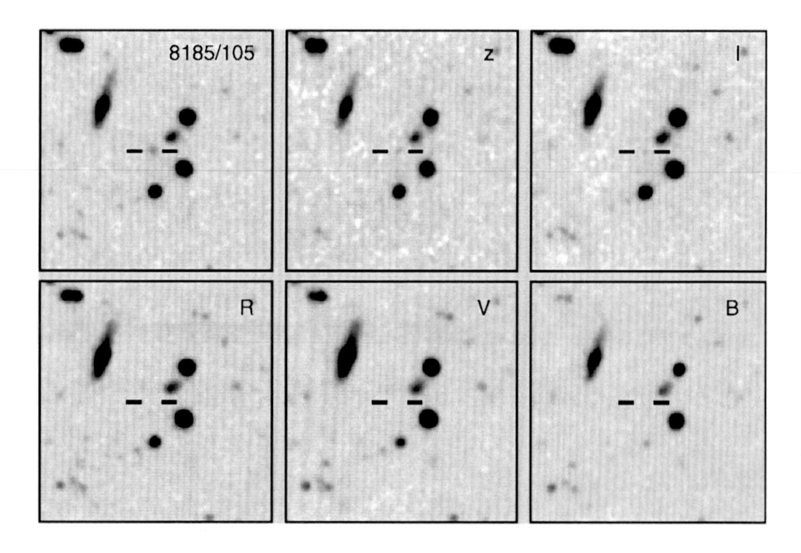

Fig. 9.6. A galaxy at $z = 5.74$, which is visible in the narrow-band filter (upper left panel) and in the I- and z-band (located between the two horizontal dashes), but which does not show any flux in the three filters at shorter wavelength

break because, due to a sudden change in the opacity at this wavelength, the spectra of most stars show such a break at about 4000 Å (see Fig. 3.47). Hence, the radiation from a stellar population at $\lambda < 4000$ Å is less intense than at $\lambda > 4000$ Å; this is the case particularly for early-type galaxies (see Fig. 3.50).

If we assume that the star-formation histories of galaxies are not too diversified, galaxies will not be located at an arbitrary location in a multidimensional color diagram; rather, they should be concentrated in certain regions. In this context the 4000-Å break and the Lyα-break play a central role, as is illustrated in Fig. 9.7. Once these characteristic domains in color space where (most of) the galaxies are situated are identified, the redshift of galaxies can be estimated solely from their observed colors, since they are functions of the redshift. The corresponding estimate is called the *photometric redshift*.

More precisely, a number of standard spectra of galaxies (so-called templates) are used, which are either selected from observed galaxies or computed by population synthesis models. Each of these template spectra can then be redshifted in wavelength, from which a K-correction (see Sect. 5.6.1) results. For each template spectrum and any redshift, the expected galaxy colors are determined by integrating the spectral energy distribution, multiplied by the transmission functions of the applied filters, over wavelength (see Eq. A.25). This set of colors can then be compared with the observed colors of galaxies, and the set best resembling the observation is taken as an estimate for not only the redshift but also the galaxy type.

The advantage of this method is that multicolor photometry is much less time-consuming than spectroscopy of individual galaxies. In addition, this method can be extended to much fainter magnitudes than are achievable for spectroscopic redshifts. The disadvantage of the method becomes obvious when an insufficient number of colors are available, since then the photometric redshift estimates can yield a completely wrong z. One example for the occurrence of extremely wrong redshift estimates is provided by a break in the spectral energy distribution. Depending of whether this break is identified as the Lyman-break or the 4000-Å break, the resulting redshift estimates will be very different. To break the corresponding degeneracy, a sufficiently large number of filters must be available to probe the spectral energy distribution over a wide range in wavelengths. As a general rule, the more photometric bands that are available and the smaller the uncertainties in the measured magnitudes, the more accurate the estimated redshift. Normally, data from four or five photometric bands are required to obtain useful redshift estimates. In particular, the reliability of the photometric redshift benefits from data over a large wavelength range, so that a combination of several optical and NIR filters is desirable.

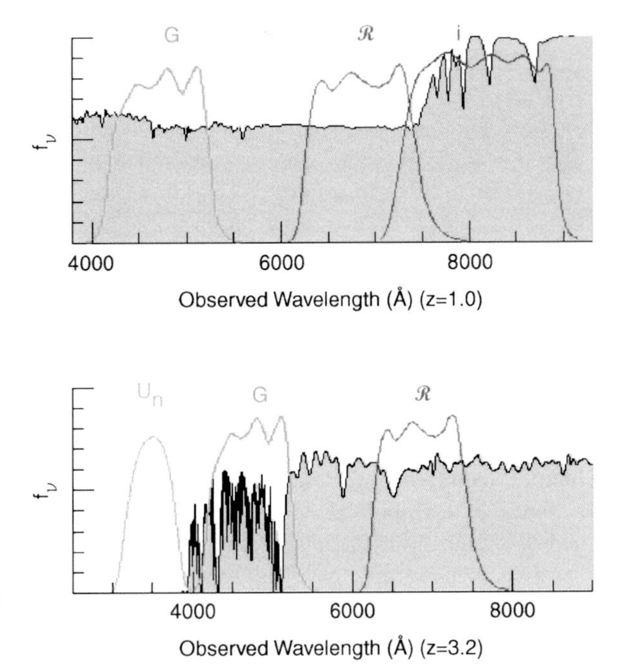

Fig. 9.7. The bottom panel illustrates again the principle of the drop-out method, for a galaxy at $z \sim 3.2$. Whereas the Lyman-α forest absorbs part of the spectral flux between (rest-frame wavelength) 912 Å and 1216 Å, the flux below 912 Å vanishes almost completely. By using different combinations of filters (top panel), an efficient selection of galaxies at other redshifts is also possible. The example shows a galaxy at $z = 1$ where the 4000-Å break is utilized, which occurs in stellar populations after several 10^7 yr (see Fig. 3.47) and which is considered to be one of the most important features for the method of photometric redshift

The successful application of this method also depends on the type of the galaxies. As we have seen in Sect. 6.6, early-type galaxies form a relatively well-defined color–magnitude sequence at any redshift, due to their old stellar populations (manifested in clusters of galaxies in form of the red cluster sequence), so that the redshift of this type of galaxy can be estimated very accurately from multicolor information. However, this is only the case if the 4000-Å break is located in between two of the applied filters. For $z \gtrsim 1$ this is no longer the case in the optical range of the spectrum. Other types of galaxies show larger variations in their spectral energy distribution, depending, e.g., on the star-formation history.

Photometric redshifts are particularly useful for statistical purposes, for instance in situations in which the exact redshift of each individual galaxy in a sample is of little relevance. However, by using a sufficient number of filters a redshift accuracy of $\Delta z \sim 0.03(1+z)$ is achievable, as demonstrated in Fig. 9.8 by a comparison of photometric redshifts with redshifts determined spectroscopically for galaxies in the field of the HDF-North.

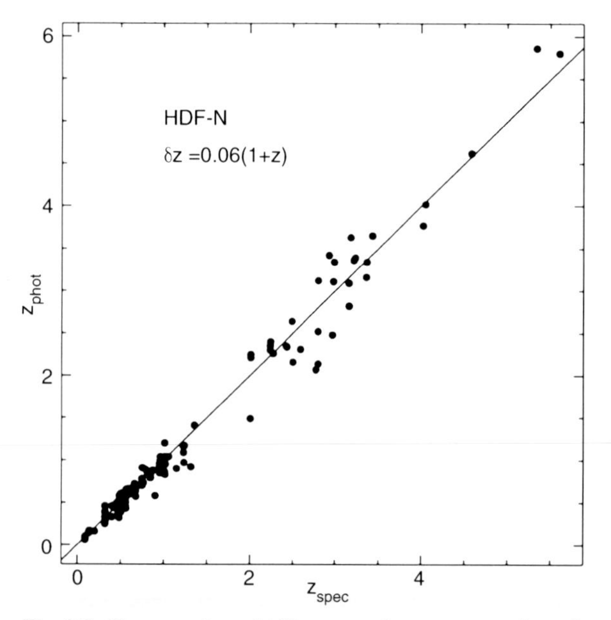

Fig. 9.8. Photometric redshift versus the spectroscopic redshift for galaxies in the HDF-North. Photometric data in four optical and two NIR bands have been used here. We see how accurate photometric redshifts can be – their quality depends on the photometric accuracy in the individual filters, the number of filters used, the redshift and the type of the galaxy, and also on details of the applied analysis method

9.1.3 Hubble Deep Field(s)

The HDF-N. In 1995, an unprecedented observing program was conducted with the HST. A deep image in four filters (U_{300}, B_{450}, V_{606}, and I_{814}) was observed with the Wide Field/Planetary Camera 2 (WFPC2) on-board HST, covering a field of $\sim 5.3\,\mathrm{arcmin}^2$, with a total exposure time of about 10 days. This resulted in the deepest sky image of that time, displayed in Fig. 9.9. The observed field was carefully selected such that it did not contain any bright sources. Furthermore, the position of the field was chosen such that the HST was able to continually point into this direction, a criterion excluding all but two relatively small regions on the sky, due to the low HST orbit around the Earth. Another special feature of this program was that the data became public immediately after reduction, less than a month after the final exposures had been taken. Astronomers worldwide immediately had the opportunity

Fig. 9.9. The Hubble Deep Field (North), at its time by far the deepest image of the sky. In December 1995, the HST was pointed to this field for about 10 days, and observations were conducted in four different filters. The raw and reduced data were made publicly available worldwide as early as Jan. 15, 1996. In this image, which spans about 5 square arcminutes, about 3000 galaxies are visible, extending over a wide range in redshift

to scientifically exploit these data and to compare them with data at other frequency ranges or to perform their own follow-up observations. Such a rapid and wide release was uncommon at that time, but is now seen more frequently. Rarely has a single data set inspired and motivated a large community of astronomers as much as the *Hubble Deep Field* (North) – HDF(N) – did.

Follow-up observations of the HDF(N) – have been made in nearly all accessible wavelength ranges, so that it is the best-observed region of the extragalactic sky. The field contains ~ 3000 galaxies, six X-ray sources, 16 radio sources, and fewer than 20 stars. For more than 150 galaxies in this field, redshifts have been determined spectroscopically, and about 30 have been found at $z > 2$. Never before could galaxy counts be conducted to magnitudes as faint as it became possible in the HDF-N (see Fig. 9.10); several hundred galaxies per square arcminute could be photometrically analyzed in this field.

Detailed spectroscopic follow-up observations were conducted by several groups, through which the HDF became, among other things, a calibration field for photometric redshifts (see, for instance, Fig. 9.8). Most galaxies in the HDF are far too weak to be analyzed spectroscopically, so that one often has to rely on photometric redshifts.

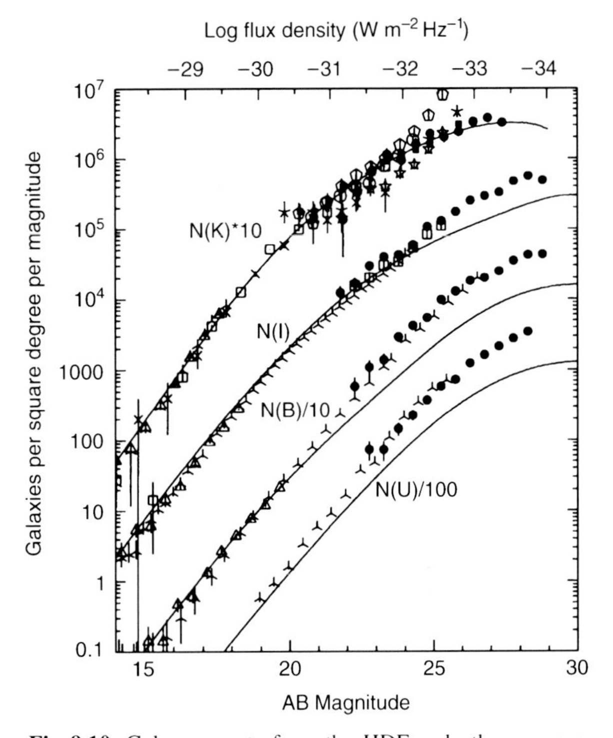

Fig. 9.10. Galaxy counts from the HDF and other surveys. Solid symbols are from the HDF, open symbols from various ground-based observations. The curves represent predictions from models in which the spectral energy distribution of the galaxies does not evolve – the counts lie significantly above these so-called non-evolution models: clearly, the galaxy population must be evolving. Note that the counts in the different color filters are shifted by a factor 10 each, simply for display purposes

HDF-S and the Hubble Ultra-Deep Field. Later, in 1998, a second HDF was observed, this time in the southern sky. In contrast to the HDF-N, which had been chosen to be as empty as possible, the HDF-S contains a QSO. Its absorption line spectrum can be compared with the galaxies found in the HDF-S, by which one hopes to obtain information on the relation between QSO absorption lines and galaxies. In addition to the WFPC2 camera, the HDF-S was simultaneously observed with the cameras STIS ($51'' \times 51''$ field-of-view, where the CLEAR "filter" was used, which has a very broad spectral sensitivity; in total, STIS is considerably more sensitive than WFPC2) and NICMOS (a NIR camera with a maximum field-of-view of $51'' \times 51''$) which had both been installed in the meantime. Nevertheless, the impact of the HDF-S was smaller than that of the HDF-N; one reason for this may be that the requirement of the presence of a QSO, combined with the need for a field in the continuous viewing zone of HST, led to a field close to several very bright Galactic stars. This circumstance makes photometric observations from the ground very difficult, e.g., due to stray light.

In 2002, an additional camera was installed on-board HST. The *Advanced Camera for Surveys* (ACS) has, with its side length of $3\overset{\prime}{.}4$, a field-of-view about twice as large as WFPC2, and with half the pixel size ($0\overset{\prime\prime}{.}05$) it better matches the diffraction-limited angular resolution of HST. Therefore, ACS is a substantially more powerful camera than WFPC2 and is, in particular, best suited for surveys. With the *Hubble Ultra-Deep Field* (HUDF), the currently, and presumably for quite some years to come, deepest image of the sky was observed and published in 2004 (see Fig. 9.11). The HUDF is, in all filters, deeper by about one magnitude than the HDF. The depth of the ACS images in combination with the

Fig. 9.11. The Hubble Ultra-Deep Field, a field of $\sim 3\rlap{.}'4 \times 3\rlap{.}'4$ observed by the ACS camera. The limiting magnitude up to which sources are detected in this image is about one magnitude fainter than in the HDF. More than 10 000 galaxies are visible in the image, many of them at redshifts $z \geq 5$

relatively red filters that are available provides us with an opportunity to identify drop-out candidates at redshift $z \sim 6$; several such candidates have already been verified spectroscopically.

One of the immediate results from the HDF was the finding that the morphology of faint galaxies is quite different from those in the nearby Universe. Locally, most luminous galaxies fit into the morphological Hubble sequence of galaxies. This ceases to be the case for high-redshift galaxies. In fact, galaxies at $z \sim 2$ are much more compact than local luminous galaxies, they show irregular light distributions and do not resemble any of the Hubble sequence morphologies. By redshifts $z \sim 1$, the Hubble sequence seems to have been partly established.

Further Deep-Field Projects with HST: GOODS, GEMS, COSMOS. The great scientific harvest from the deep HST images, particularly in combination with data from other telescopes and the readiness to make such data available to the scientific community for multifrequency analyses, provided the motivation for additional HST surveys. The GOODS (Great Observatories Origins Deep Surveys) project is a joint observational campaign of several observatories, centering on two fields of $\sim 16' \times 10'$ size each that have been observed by the ACS camera at several epochs between 2003 and 2005. One of these two regions contains the HDF-N, the other a field that became known as the Chandra Deep Field South (CDF-S). The Chandra satellite observed both GOODS fields with a total exposure time of $\sim 1 \times 10^6$ s and $\sim 2 \times 10^6$ s, respectively. Also, the Spitzer observatory took long exposures of these two fields. In addition, several ground-based observatories are involved in this survey, for instance by contributing an ultra-deep wide field image ($\sim 30' \times 30'$) centered on the CDF-S. The data themselves and the data products (like object catalogs, color information, etc.) are all publicly available and have already led to a large number of scientific results. Even larger surveys using

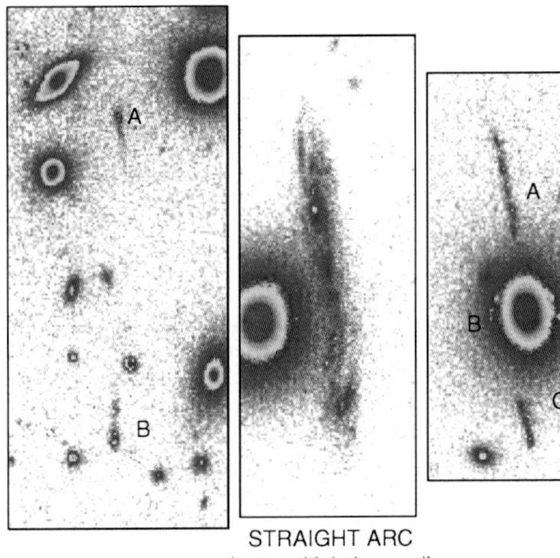

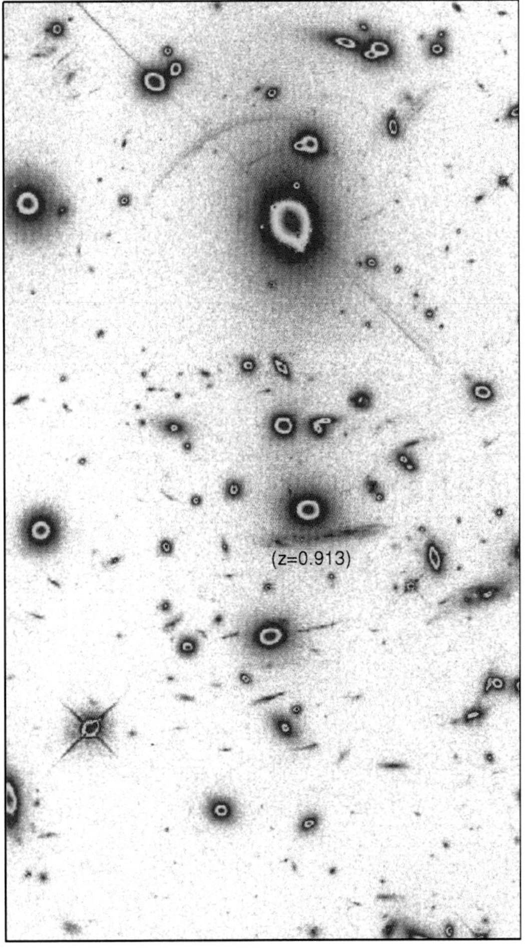

the HST (GEMS, a field of $30' \times 30'$ centered on the CDF-S, and the 2 deg^2 COSMOS survey) will further improve the statistics of the results obtained from the HUDF and GOODS.

The multiwavelength approach by GOODS yields an unprecedented view of the high-redshift Universe. Although these studies and scientific analyses are ongoing (at the time of writing), quite a large number of very high-redshift ($z \gtrsim 5$) galaxies have already now been discovered and studied: a sample of more than 500 I-band drop-outs has been obtained from deep ACS/HST images. Lyman-break galaxies at $z \sim 6$ seem to have stellar populations with masses and lifetimes comparable to those at $z \sim 3$. This implies that at a time when the Universe was 1 Gyr old, a stellar population with mass $\sim 3 \times 10^{10} M_\odot$ and age of a few hundred million years (as indicated by the observed 4000-Angstrom break) was already in place. This, together with the apparently high metallicity of these sources, is thus another indication of how quickly the early Universe has evolved. The $z \sim 6$ galaxies are very compact, with half-light radii of ~ 1 kpc, and thus differ substantially from the galaxy population known in the lower-redshift Universe.

9.1.4 Natural Telescopes

Galaxies at high redshift are faint and therefore difficult to observe spectroscopically. For this reason, the brightest galaxies are preferentially selected (for detailed examination), i.e., basically those which are the most luminous at a particular z – resulting in undesired, but hardly avoidable selection effects. For example, those Lyman-break galaxies at $z \sim 3$ for which the red-

Fig. 9.12. A particularly interesting and efficient way to examine galaxies at high redshift is provided by the strong lensing effect in clusters of galaxies. Since a gravitational lens can magnify the light of background galaxies (by magnification of the solid angle), one can expect to detect apparently brighter galaxies at high redshift in the background of clusters. Here, an HST image of the cluster Abell 2390 is shown, in which several lens systems are visible. On the left, the central region of the cluster is shown. Three systems with a strong lens effect in this cluster are presented in the blow-ups at top. In the center, the so-called "straight arc" is visible which has a redshift of about 0.91. On the right and left, two multiply imaged systems are displayed, the images indicated by letters; the two sources associated with these images have redshifts of $z = 4.04$ and 4.05, respectively

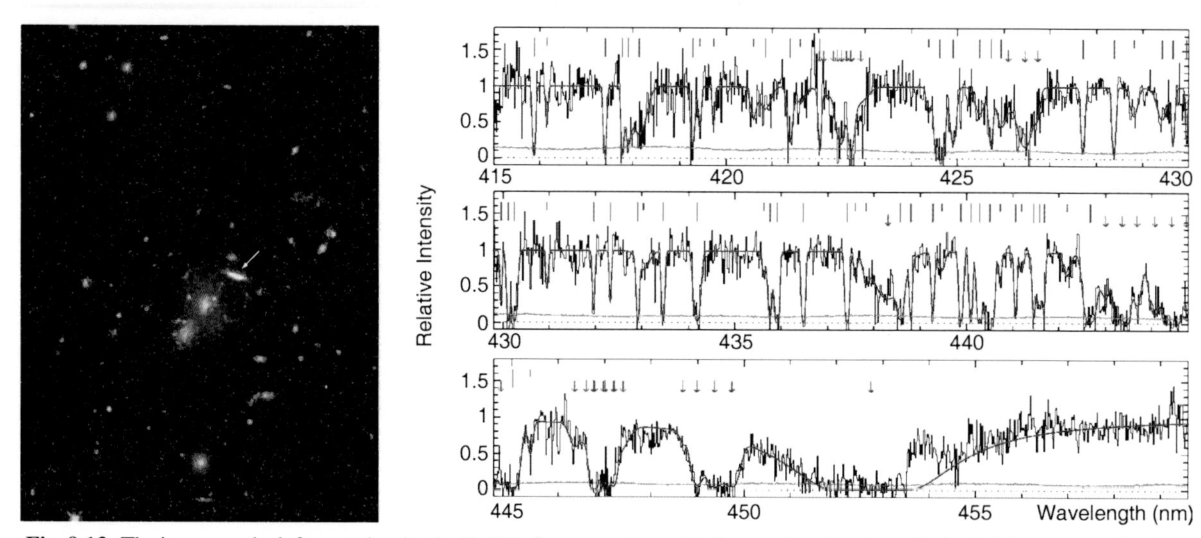

Fig. 9.13. The image on the left was taken by the Hubble Space Telescope. It shows the cluster of galaxies MS1512+36, which has a redshift of $z = 0.37$. To the right, and slightly above the central cluster galaxy, an extended and apparently very blue object is seen, marked by an arrow. This source is not physically associated with the cluster but is a background galaxy at a redshift of $z = 2.72$. With this HST image it was proved that this galaxy is strongly lensed by the cluster and, by means of this, magnified by a factor of ~ 30. Due to the magnification, this Lyman-break galaxy is the brightest normal galaxy at redshift $z \sim 3$, a fact that can be profitably used for a detailed spectroscopic analysis. On the right, a small section from a high-resolution VLT spectrum of this galaxy is shown. The Lyα transition of the galaxy is located at $\lambda = 4530$ Å, visible as a broad absorption line. Absorption lines at shorter wavelengths originate from the Lyα-forest along the line-of-sight (indicated by short vertical lines) or by metal lines from the galaxy itself (indicated by arrows)

shift is verified spectroscopically are among the most luminous of their kind. The sensitivity of our telescopes is insufficient in most cases to spectroscopically analyze a rather more typical galaxy at $z \sim 3$.

The magnification by gravitational lenses can substantially alter the apparent magnitude of sources; gravitational lenses can then act as natural (and inexpensive!) telescopes. Examples are the arcs in clusters of galaxies: many of them have a very high redshift, are magnified by a factor $\gtrsim 5$, and hence are brighter by about ~ 1.5 mag than they would be without the lens effect (see Fig. 9.12). It should be mentioned that a factor of 5 in magnification corresponds to a factor 25 in the exposure time required for spectroscopy.[1]

An extreme example of this effect is represented by the galaxy cB58 at $z = 2.72$, which is displayed in Fig. 9.13. It was discovered in the background of a galaxy cluster and is magnified by a factor ~ 30. Hence, it appears brighter by more than three magnitudes than a typical Lyman-break galaxy. For this reason, the most detailed spectra of all galaxies at $z \sim 3$ have been taken of this particular source.

One can argue that there is a high probability that the flux of the apparently most luminous sources from a particular source population is magnified by lensing. The apparently most luminous IRAS galaxy, F10214+47, is magnified by a factor ~ 50 by the lens effect of a foreground galaxy (where the exact value of the magnification depends on the wavelength, since the intrinsic structure and size of the source is wavelength-dependent – hence the magnification is differential). Other examples are the QSOs B1422+231 and APM 08279+5255, which are among the brightest quasars despite their high redshifts. In both cases, multiple images of the QSOs were discovered, verifying the action of the lens effect. Their magnification, and therefore their brightness, renders these sources preferred objects for QSO absorption line spectroscopy (see Fig. 5.40). The Lyman-break galaxy cB58 men-

[1] This factor of 25 makes the difference between an observation that is feasible and one that is not. Whereas the proposal for a spectroscopic observation of 3 hours exposure time at an 8-m telescope may be successful, a similar proposal of 75 hours would be hopelessly doomed to failure.

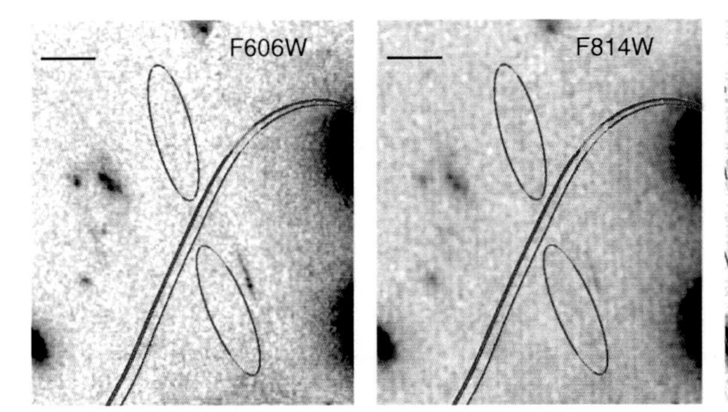

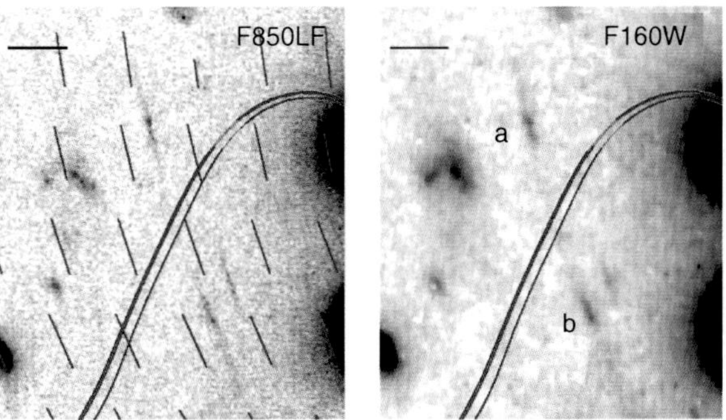

Fig. 9.14. A section of the galaxy cluster Abell 2218 ($z = 0.175$), observed with the HST in four different filters. This region was selected because the magnification by the gravitational lens effect for sources at high redshift is expected to be very large here. This fact has been established by a detailed mass model of this cluster which could be constructed from the geometrical constraints provided by the numerous arcs and multiple images (Fig. 6.33). The red lines denote the critical curves of this lens for source redshifts of $z = 5$, 6.5, and 7. A double image of an extended source is clearly visible in the NIR image (on the right); this double image was not detected at shorter wavelengths – the expected position is marked by two ellipses in the two images on the left. The direction of the local shear, i.e., of the expected image distortion, is plotted in the second image from the right; the observed elongation of the two images a and b is compatible with the shear field from the lens model. Together with the photometry of these two images, a redshift between $z = 6.8$ and $z = 7$ is derived for the source of this double image

tioned previously is another example. One important result of such investigations of high-redshift sources should be mentioned here. These galaxies and QSOs have a high metal abundance, from which we conclude that star formation must have already set in during a very early phase of the Universe. We will later return to this point.

The magnification effect is also utilized deliberately, by searching for highly redshifted sources in fields around clusters of galaxies. For a massive cluster, one knows that distant sources located behind the cluster center are substantially magnified. It is therefore not surprising that some of the most distant galaxies known have been detected in systematic searches for drop-out galaxies near the centers of massive clusters. One example of this is shown in Fig. 9.14, where a galaxy at $z \sim 7$ is doubly imaged by the cluster Abell 2218 (see Fig. 6.33), and by means of this it is magnified by a factor ~ 25.

9.2 New Types of Galaxies

The Lyman-break galaxies discussed above are not the only galaxies that are expected to exist at high redshifts. We have argued that LBGs are galaxies with active star formation. Moreover, the UV radiation from their newly-born hot stars must be able to escape from the galaxies. From observations in the local Universe we know, however, that a large fraction of star formation is hidden from our direct view, since the star-formation region is enveloped by dust. The latter is heated by absorbing the UV radiation, and re-emits this energy in the form of thermal radiation in the FIR domain of the spectrum. At high redshifts such galaxies would certainly not be detected by the Lyman-break method.

Instrumental developments opened up new wavelength regimes which yield access to other types of galaxies. Two of these will be described in more detail here: EROs (Extremely Red Objects) and submillimeter (sub-mm) sources, the latter often being called SCUBA galaxies because they were first observed in large numbers by the SCUBA camera. But before we discuss these objects we will first investigate starburst galaxies in the relatively local Universe.

9.2.1 Starburst Galaxies

One class of galaxies, the so-called *starburst galaxies*, is characterized by a strongly enhanced star-formation rate, compared to normal galaxies. Whereas our Milky

Way is forming stars with a rate of $\sim 3 M_\odot/\mathrm{yr}$, the star-formation rate in starburst galaxies can be larger by a factor of more than a hundred. Dust heated by hot stars radiates in the FIR, rendering starbursts very strong FIR emitters. Many of them were discovered by the IRAS satellite ("IRAS galaxies"); they are also called ULIRGs (ultra-luminous infrared galaxies).

The reason for this strongly enhanced star formation is presumably the interaction with other galaxies or the result of merger processes, an impressive example of which is the merging galaxy pair known as the "Antennae" (see Fig. 9.15). In this system, stars and star clusters are currently being produced in very large numbers. The images show a large number of star clusters with a characteristic mass of $10^5 M_\odot$, some of which are spatially resolved by HST. Furthermore, particularly luminous individual stars (supergiants) are also observed. The ages of the stars and star clusters span a wide range and depend on the position within the galaxies. For instance, the age of the predominant population is about 5–10 Myr, with a tendency for the youngest stars to be located in the vicinity of strong dust absorption. However, stellar populations with an age of 100 and 500 Myr, respectively, have also been discovered; the latter presumably originates from the time of the first encounter of these two galaxies, which then led to the ejection of the tidal tails. This seems to be a common phenomenon; for example, in the starburst galaxy Arp 220 (see Fig. 1.12) one also finds star clusters of a young population with age $\lesssim 10^7$ yr, as well as older ones with age $\sim 3 \times 10^8$ yr. It thus seems that during the merging process several massive bursts of star-cluster formation are triggered.

It was shown by the ISO satellite that the most active regions of star formation are not visible on optical images, since they are completely enshrouded by

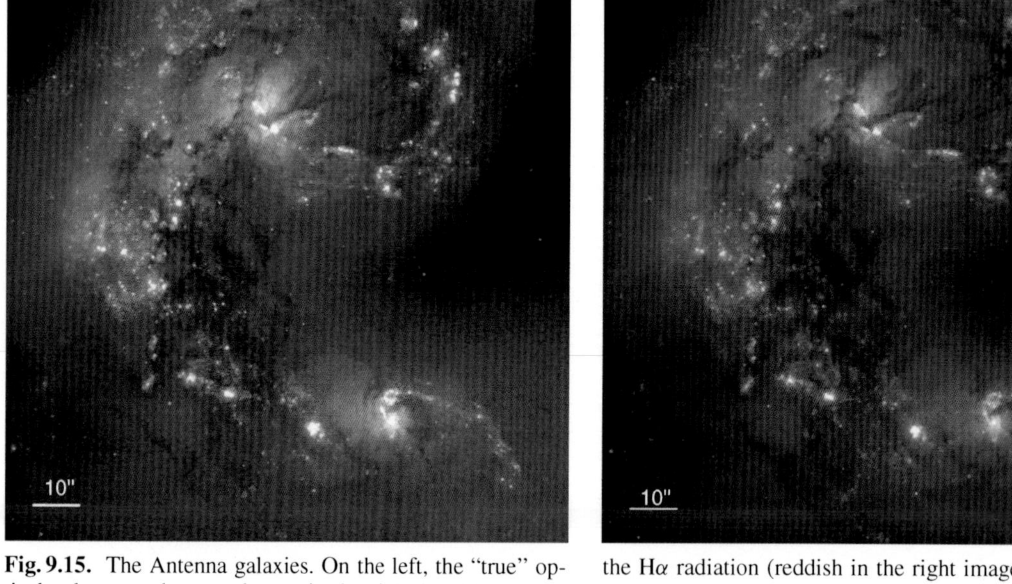

Fig. 9.15. The Antenna galaxies. On the left, the "true" optical colors are shown, whereas in the right-hand image the reddish color shows Hα emission. This pair of merging galaxies (also see Fig. 1.13 and Fig. 3.4 for other examples of merging galaxies) is forming an enormous number of young stars. Both the UV emission (bluish in the left image) and the Hα radiation (reddish in the right image) are considered indicators of star formation. The individual knots of bright emission are not single stars but star clusters with typically $10^5 M_\odot$; however, it is also possible to resolve individual stars (red and blue supergiants) in these galaxies

dust. A map at 15 μm shows the hot dust heated by young stars (see Fig. 9.16), where this IR emission is clearly anticorrelated with the optical radiation. Obviously, a complete picture of star formation in such galaxies can only be obtained from a combination of optical and IR images.

Combining deep optical and NIR photometry with MIR imaging from the Spitzer telescope, star-forming galaxies at high redshifts can be detected even if they contain an appreciable amount of dust (and thus may fail to satisfy the LBG selection criteria). These studies find that the comoving number density of ULIRGs with $L_{\rm IR} \gtrsim 10^{12} L_\odot$ at $z \sim 2$ is about three orders of magnitude larger than the local ULIRG density. These results seem to imply that the high-mass tail of the local galaxy population with $M \gtrsim 10^{11} M_\odot$ was largely in place at redshift $z \sim 1.5$ and evolves passively from there on. We shall come back to this aspect below.

Observations with the Chandra satellite have shown that starburst galaxies contain a rich population of very luminous compact X-ray sources (Ultra-luminous Compact X-ray Sources, or ULXs; see Fig. 9.17). Similar sources, though with lower luminosity, are also detected in the Milky Way, where these are binary systems with one component being a compact star (white dwarf, neutron star, or black hole). The X-ray emission is caused by accretion of matter (which we discussed in Sect. 5.3.2) from the companion star onto the compact component.

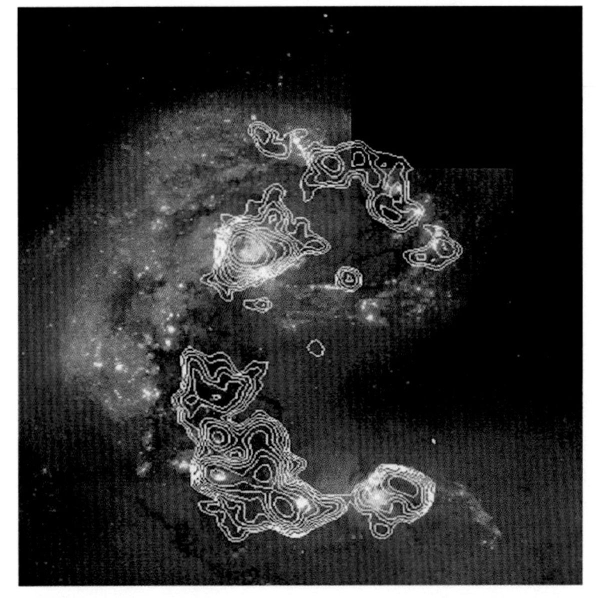

Fig. 9.16. The Antenna galaxies: superposed on the optical HST image are contours of infrared emission at 15 μm, measured by ISO. The strongest IR emission originates in optically dark regions. A large fraction of the star formation in this galaxy pair (and in other galaxies?) is not visible on optical images because it is hidden by dust absorption

Some of the ULXs in starbursts are so luminous, however, that the required mass of the compact star by far exceeds $1 M_\odot$ if the Eddington luminosity is assumed as an upper limit for the luminosity (see Eq. 5.23). Hence, one concludes that either the emission of these sources is highly anisotropic, hence beamed towards us, or that the sources are black holes with masses of up to $\sim 200 M_\odot$. In the latter case, we may just be witnessing the formation of supermassive black holes in these starbursts.

This latter interpretation is also supported by the fact that the ULXs are concentrated towards the center of the galaxies – hence, these BHs may spiral into the galaxy's center by dynamical friction, and there merge to a SMBH. This is one of the possible scenarios for the formation of SMBHs in the cores of galaxies, a subject to which we will return in Sect. 9.6.3.

9.2.2 Extremely Red Objects (EROs)

As mentioned several times previously, the population of galaxies detected in a survey depends on the selection criteria. Thus, using the Lyman-break method, it is mainly those galaxies at high redshift which feature active star formation and therefore have a blue spectral distribution at wavelengths longwards of Lyα that are discovered. The development of NIR detectors enabled the search for galaxies at longer wavelengths. Of particular interest here are surveys of galaxies in the K-band, the longest wavelength window that is reasonably accessible from the ground (with the exception of the radio domain).

The NIR waveband is of particular interest because the luminosity of galaxies at these wavelengths is not dominated by young stars. As we have seen in Fig. 3.48, the luminosity in the K-band depends only weakly on the age of the stellar population, so that it provides a reliable measure of the total stellar mass of a galaxy.

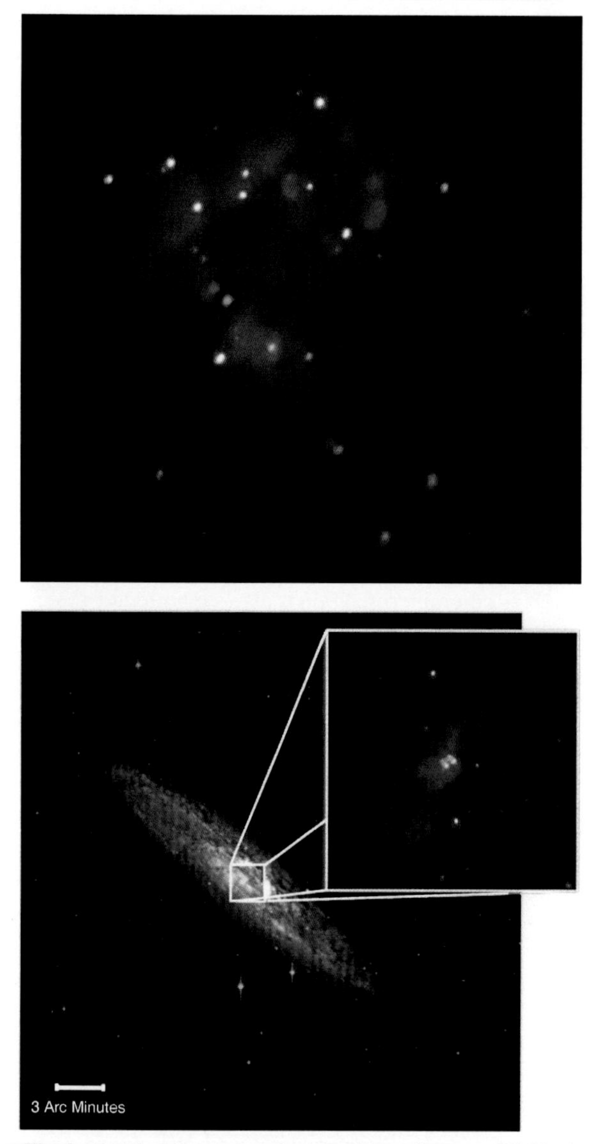

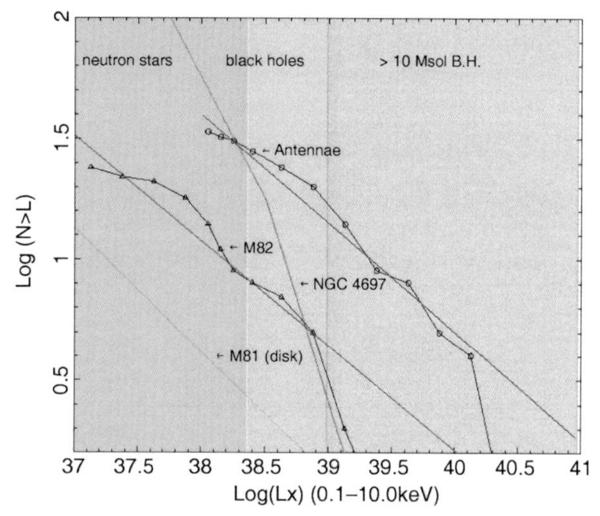

Fig. 9.17. Ultra-luminous Compact X-ray Sources (ULXs) in starburst galaxies. Upper left: the discrete X-ray sources in the Antenna galaxies; the size of the image is $4' \times 4'$. Lower left: optical (image) and (inlaid) Chandra image of the starburst galaxy NGC 253. Four of the ULXs are located within one kiloparsec from the center of the galaxy. The X-ray image is $2\rlap{.}'2 \times 2\rlap{.}'2$. Upper right: $5' \times 5'$ Chandra image of the starburst galaxy M82; the diffuse radiation (red) is emitted by gas at $T \sim 10^6$ K which is heated by the starburst and flows out from the central region of the galaxy. It is supposed that M82 had a collision with its companion M81 (see Fig. 6.7) within the last 10^8 yr, by which the starburst has been triggered. Lower right: the luminosity function of the ULXs in some starburst galaxies

Characteristics of EROs. Examining galaxies with a low K-band flux, one finds either galaxies with low stellar mass at low redshifts, or galaxies at high redshift with high optical (due to redshift) luminosity. But since the luminosity function of galaxies is relatively flat for $L \lesssim L_*$, one expects the latter to dominate the surveys, due to the larger volume at higher z. In fact, K-band surveys detect galaxies with a broad redshift distribution.

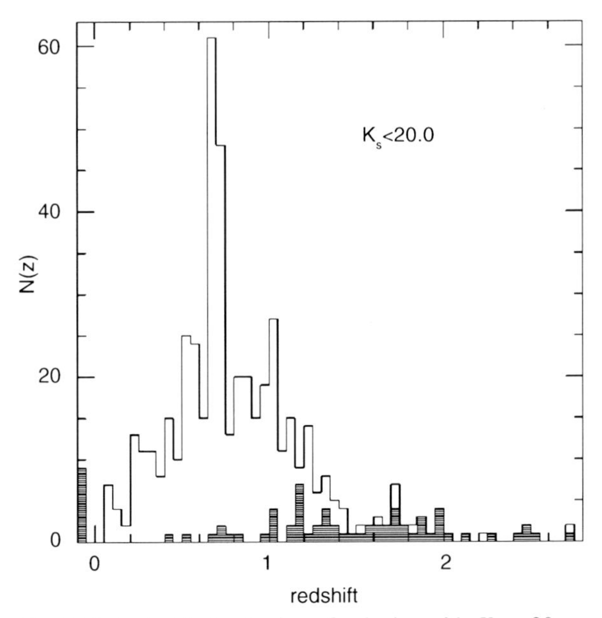

Fig. 9.18. Redshift distribution of galaxies with $K_s < 20$, as measured in the K20 survey. The shaded histogram represents galaxies for which the redshift was determined solely by photometric methods. The bin at $z < 0$ contains those 9 galaxies for which it has not been possible to determine z. The peak at $z \sim 0.7$ is produced by two clusters of galaxies in the fields of the K20 survey

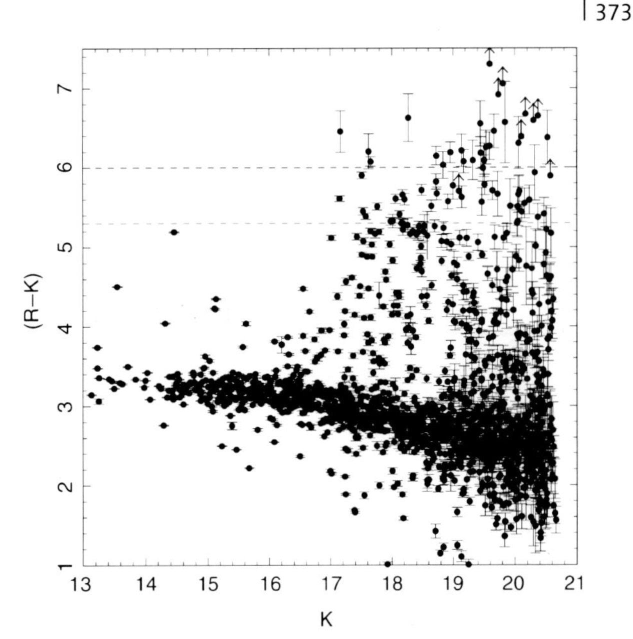

Fig. 9.19. Color–magnitude diagram, i.e., $R-K$ as a function of K, for sources in ten fields around clusters of galaxies. We see that for faint magnitudes (roughly $K \geq 19$), a population of sources with a very red color (about $R-K \geq 5.3$) turns up. These objects are called EROs

In Fig. 9.18 the z-distribution of galaxies in the K20 survey is shown. In this survey, objects with $K_s < 20$ have been selected in two fields with a combined area of 52 arcmin^2, where K_s is a filter at a wavelength slightly shorter than the classic K-band filter. After excluding stars and type 1 AGNs, 489 galaxies were found, 480 of which have had their redshifts determined. The median redshift in this survey is $z \approx 0.8$.

Considering galaxies in a $(R-K)$ vs. K color–magnitude diagram (Fig. 9.19), one can identify a population of particularly red galaxies, thus those with a large $R-K$. These objects have been named *Extremely Red Objects* (*EROs*); about 10% of the galaxies in K-selected surveys at faint magnitudes are EROs, typically defined by $R-K > 5$. Spectroscopic analysis of these galaxies poses a big challenge because an object with $K = 20$ and $R-K > 5$ necessarily has $R > 25$, i.e., it is extremely faint in the optical domain of the spectrum. With the advent of 10-m class telescopes, spectroscopy of these objects has become possible in recent years.

The Nature of EROs: Passive Ellipticals Versus Dusty Starbursts. From these spectroscopic results, it was found that the class of EROs contains rather different kinds of sources. To understand this point we will first consider the possible explanations for a galaxy with such a red spectral distribution. As a first option, the object may be an old elliptical galaxy with the 4000-Å break being redshifted to the red side of the R-band filter, i.e., typically an elliptical galaxy at $z \gtrsim 1.0$. For these galaxies to be sufficiently red to satisfy the selection criterion for EROs, they need to already contain an old stellar population by this redshift, which implies a very high redshift for the star formation in these objects; it is estimated from population synthesis models that their formation redshift must be $z_{\rm form} \gtrsim 2.5$. A second possible explanation for large $R-K$ is reddening by dust. Such EROs may be galaxies with active star formation where the optical light is strongly attenuated by dust extinction. If these galaxies are located at a redshift of $z \sim 1$, the measured R-band flux corresponds to a rest-frame emission in the UV region of the spectrum where extinction is very efficient.

Spectroscopic analysis reveals that both types of EROs are roughly equally abundant. Hence, about half of the EROs are elliptical galaxies that already have, at $z \sim 1$, a luminosity similar to that of today's ellipticals, and are at that epoch already dominated by an old stellar population. The other half are galaxies with active star formation which do not show a 4000-Å break but which feature the emission line of [OII] at $\lambda = 3727$ Å, a clear sign of star formation. Further analysis of EROs by means of very deep radio observations confirms the large fraction of galaxies with high star-formation rates. Utilizing the close relation of radio emissivity and FIR luminosity, we find a considerable fraction of EROs to be ULIRGs at $z \sim 1$.

Spatial Correlations. EROs are very strongly correlated in space. The interpretation of this strong correlation may be different for the passive ellipticals and for those with active star formation. In the former case the correlation is compatible with a picture in which these EROs are contained in clusters of galaxies or in overdense regions that will collapse to a cluster in the future. The correlation of the EROs featuring active star formation can probably not be explained by cluster membership, but the origin of the correlation may be the same as for the correlation of the LBGs.

The number density of passive EROs, thus of old ellipticals, is surprisingly large compared with expectations from the model of hierarchical structure formation that we will discuss in Sect. 9.6.

9.2.3 Submillimeter Sources: A View Through Thick Dust

FIR emission from hot dust is one of the best indicators of star formation. However, observations in this waveband are only possible from space, such as was done with the IRAS and ISO satellites. Dust emission has its maximum at about 100 μm, which is not observable from the ground. At longer wavelengths there are spectral windows where observations through the Earth's atmosphere are possible, for instance at 450 μm and 850 μm in the submillimeter waveband. However, the observing conditions at these wavelengths are extremely dependent on the amount of water vapor in the atmosphere, so that the observing sites must by dry and at high elevations. In the submillimeter (sub-mm) range, the long wavelength domain of thermal dust radiation can be observed, which is illustrated in Fig. 9.20.

Since about 1998 sub-mm astronomy has experienced an enormous boom, with two instruments having been put into operation: the Submillimeter Common User Bolometer Array (SCUBA), operating at 450 μm and 850 μm, with a field-of-view of 5 arcmin^2, and the Max-Planck Millimeter Bolometer (MAMBO), operating at 1300 μm. Both are bolometer arrays which initially had 37 bolometers each, but which since then have been upgraded to a considerably larger number of bolometers. Figure 9.21 shows a $20' \times 17'$ MAMBO image of a field in the region of the COSMOS survey.

The Negative K-Correction of Submillimeter Sources. The emission of dust at these wavelenghs is described by a Rayleigh–Jeans spectrum, modified by an emissivity function that depends on the dust properties (chemical composition, distribution of dust grain sizes); typically, one finds

$$S_\nu \propto \nu^{2+\beta} \quad \text{with} \quad \beta \sim 1 \ldots 2 \, .$$

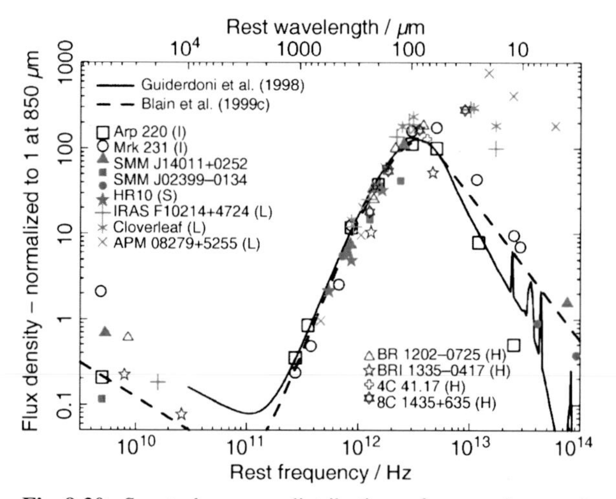

Fig. 9.20. Spectral energy distribution of some dusty galaxies with known redshift z (symbols), together with two model spectra (curves). Four types of galaxy are distinguished: (I) IRAS galaxies at low z; (S) luminous sub-mm galaxies; (L) distant sources that are magnified by the gravitational lens effect and multiply imaged; (H) AGNs. Only a few sources among the lens systems (presumably due to differential magnification) and the AGNs deviate significantly from the model spectra

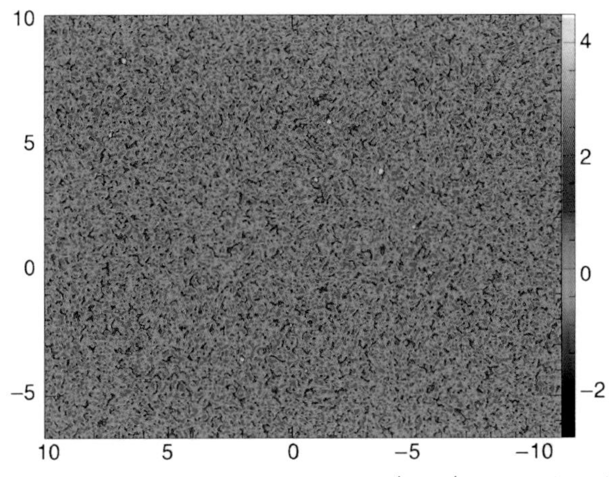

Fig. 9.21. The image shows a field of $20' \times 17'$ in the region of the COSMOS survey, observed by the 117-channel MAMBO instrument at the IRAM 30-m telescope on Pico Veleta. Coded in color is the signal-to-noise ratio of the map, where the noise level is about 0.9 mJy per $11''$ beam. About a dozen sources with $S/N \geq 4$ are visible

This steep spectrum for frequencies below the peak of the thermal dust emission at $\lambda \sim 100\,\mu m$ implies a very strong negative K-correction (see Sect. 5.6.1) for wavelengths in the sub-mm domain: at a fixed observed wavelength, the rest-frame wavelength becomes increasingly smaller for sources at higher redshift, and there the emissivity is larger. As Fig. 9.22 demonstrates, this spectral behavior causes the effect that the flux in the sub-mm range does not necessarily decrease with redshift. For $z \lesssim 1$, the $1/D^2$-dependence of the flux dominates, so that up to $z \sim 1$ sources at fixed luminosity get fainter with increasing z. However, between $z \sim 1$ and $z \sim z_{\rm flat}$ the sub-mm flux as a function of redshift remains nearly constant or even increases with z, where $z_{\rm flat}$ depends on the dust temperature $T_{\rm d}$; for $T_{\rm d} \sim 40$ K and $\lambda \sim 850\,\mu m$ one finds $z_{\rm flat} \sim 8$. We therefore have the quite amazing situation that sources appear brighter when they are moved to larger distances. This is caused by the very negative K-correction which more than compensates for the $1/D^2$-decrease of the flux. Only for $z > z_{\rm flat}$ does the flux begin to rapidly decrease with redshift, since then, due to redshift, the correspond-

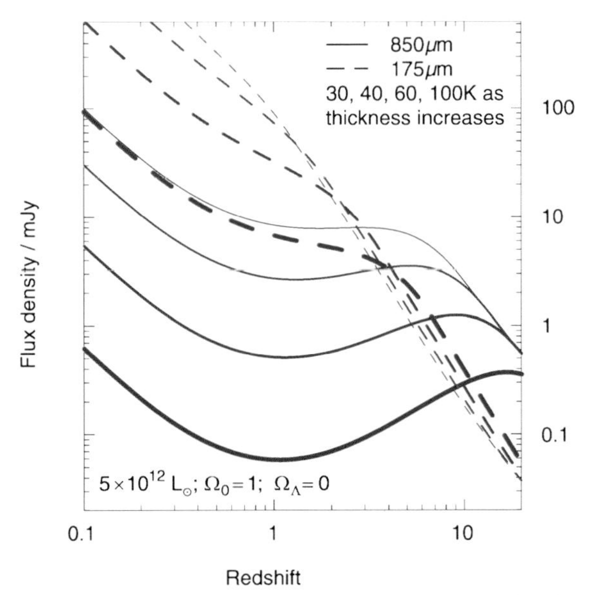

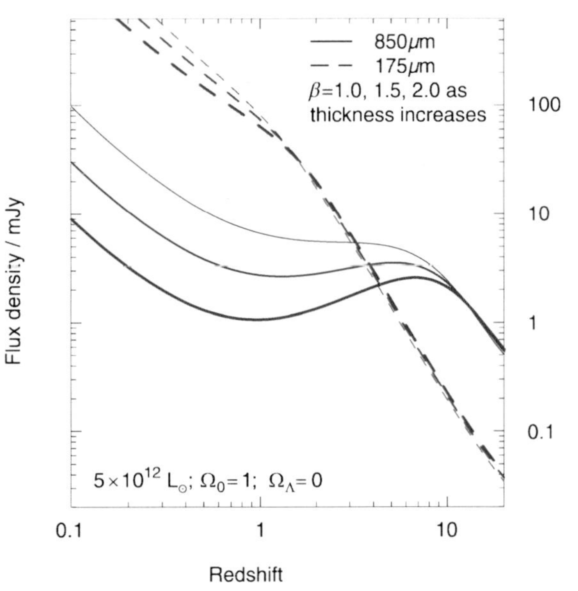

Fig. 9.22. Predicted flux from dusty galaxies as a function of redshift. The bolometric luminosity of these galaxies is kept constant. The solid red and the blue dashed curves show the flux at $\lambda = 850\,\mu m$ and $\lambda = 175\,\mu m$, respectively. On the right, the index β of the dust emissivity is varied, and the temperature of the dust $T_{\rm d} = 38$ K is kept fixed. On the left, $\beta = 1.5$ is fixed and the temperature is varied. It is remarkable how flat these curves are over a very wide range in redshift, in particular at $850\,\mu m$; this is due to the very strong negative K-correction which derives from the spectral behavior of thermal dust emission, shown in Fig. 9.20

ing rest-frame frequency is shifted to the far side of the maximum of the dust spectrum (see Fig. 9.20). Hence, a sample of galaxies that is flux-limited in the sub-mm domain should have a very broad z-distribution. The dust temperature is about $T_d \sim 20$ K for low-redshift spirals, and $T_d \sim 40$ K is a typical value for galaxies at higher redshift featuring active star formation. The higher T_d, the smaller the sub-mm flux at fixed bolometric luminosity.

Counts of sub-mm sources at high Galactic latitudes have yielded a far higher number density than was predicted by galaxy evolution models. For the density of sources as a function of limiting flux S, at wavelength $\lambda = 850\,\mu$m, one obtains

$$N(> S) \simeq 7.9 \times 10^3 \left(\frac{S}{1\,\mathrm{mJy}}\right)^{-1.1} \mathrm{deg}^{-2}\,. \qquad (9.1)$$

The Identification of SCUBA Sources. At first, the optical identification of these sources turned out to be extremely difficult: due to the relatively low angular resolution of SCUBA and MAMBO the positions of sources could only be determined with an accuracy of $\sim 15''$. A large number of faint galaxies can be identified on deep optical images within an error circle of this radius. Furthermore, Fig. 9.22 suggests that these sources have a relatively high redshift, thus they should be very faint in the optical. An additional problem is reddening and extinction by the same dust that is the source of the sub-mm emission.

The identification of SCUBA sources was finally accomplished by means of their radio emission, since about half of the sources selected at sub-mm wavelengths can be identified in very deep radio observations at 1.4 GHz. Since the radio sky is far less crowded than the optical one, and since the VLA achieves an angular resolution of $\sim 1''$ at $\lambda = 20$ cm, the optical identification of the corresponding radio source becomes relatively easy. One example of this identification process is shown in Fig. 9.23. With the accurate radio position of a sub-mm source, the optical identification can then be performed. In most cases, they are very faint optical sources indeed, so that spectroscopic analysis is difficult and very time-consuming. Another method for estimating the redshift results from the spectral energy distribution shown in Fig. 9.20. Since this spectrum seems to be nearly universal, i.e., not varying much among different sources, some kind of photometric redshift can be estimated from the ratio of the fluxes at 1.4 GHz and 850 μm, yielding quite accurate values in many cases.

Until 2004, redshifts had been measured for about 100 sub-mm sources with a median of roughly $\bar{z} \sim 2.5$. In some of these sources an AGN component, which heats the dust, was identified, but in general newly born stars seem to be the prime source of the energetic pho-

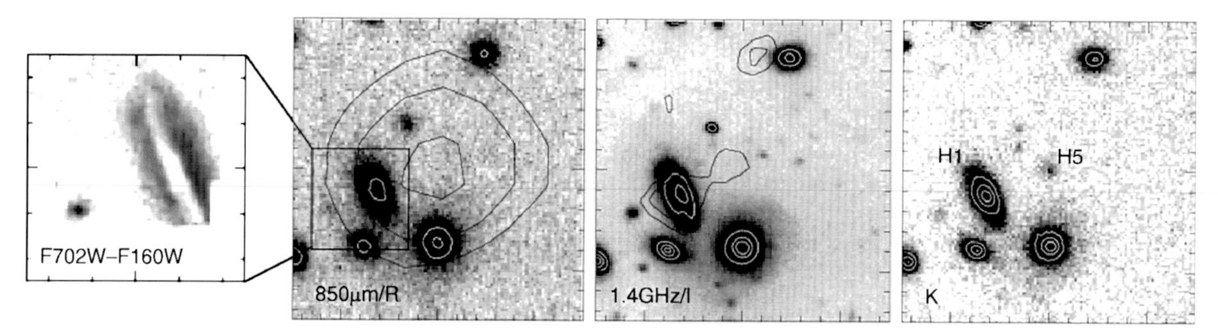

Fig. 9.23. The sub-mm galaxy SMM J09429+4658. The three images on the right have a side length of 30″ each, centered on the center of the error box of the 850 μm observation. The smaller image on left is the difference of two HST images in red and infrared filters, showing the dust disk in the spiral galaxy H1. The second image from the left displays an R-band image, superposed with the contours of the SCUBA 850 μm emission. The second image from the right is an I-band image, superposed with the contours of radio emission at 1.4 GHz, and the right-most panel shows a K-band image. The radio contours show emission from the galaxy H1 ($z = 0.33$), but also weaker emission right at the center of the sub-mm map. In the K-band, a NIR source (H5) is found exactly at this position. It remains unclear which of these two sources is the sub-mm source, but the ratio of sub-mm to 1.4-GHz emission would be atypical if H1 is identified with the sub-mm source

tons which heat the dust. The optical morphology and the number density of the sub-mm sources suggest that we are witnessing the formation of elliptical galaxies in these sub-mm sources.

Additional support for this idea is provided by the fact that the sub-mm galaxies are typically brighter and redder than (rest-frame) UV-selected galaxies at redshifts $z \sim 2.5$. This indicates that the stellar masses in sub-mm galaxies are higher than those of LBGs.

A joint investigation of $z \sim 2$ sub-mm galaxies at X-ray, optical and MIR wavelengths yields that these sources are not only forming stars at a high rate, but that they also already contain a substantial stellar population with $M \sim 10^{11} M_\odot$, roughly an order of magnitude more massive than LBGs at similar redshifts. The large AGN fraction among sub-mm galaxies indicates that the growth of the stellar population is accompanied by accretion and thus the growth of supermassive black holes in these objects. Nevertheless, the relatively faint X-ray emission from these galaxies suggests that either their SMBHs have a mass well below the local relation between $M_\bullet$ and stellar properties of (spheroidal) galaxies, or that they accrete at well below the Eddington rate. Furthermore, the typical ratio of X-ray to sub-mm luminosity of these sources is about one order of magnitude smaller than in typical AGNs, which seems to imply that the total luminosity of these sources is dominated by the star-formation activity, rather than by accretion power. This conclusion is supported by the fact that the optical counterparts of sub-mm sources show strong signs of merging and interactions, together with their larger size compared to optically-selected galaxies at the same redshifts. This latter point shows that the emission comes from an extended region, as expected from star formation in mergers, rather than AGN activity.

9.2.4 Damped Lyman-Alpha Systems

In our discussion of QSO absorption lines in Sect. 5.6.3, we mentioned that the Lyα lines are broadly classed into three categories: the Lyα forest, Lyman-limit systems, and damped Lyα systems, which are separated by a column density of $N_{\rm HI} \sim 10^{17}\,{\rm cm}^{-2}$ and $N_{\rm HI} \sim 2 \times 10^{20}\,{\rm cm}^{-2}$, respectively. The origin of the Lyα forest, as discussed in some detail in Sect. 8.5, is diffuse highly ionized gas with fairly small density contrast. In comparison, the large column density of damped Lyα systems (DLAs) strongly suggests that hydrogen is mostly neutral in these systems. The reason for this is self-shielding: for column densities of $N_{\rm HI} \gtrsim 2 \times 10^{20}\,{\rm cm}^{-2}$ the background of ionizing photons is unable to penetrate deep into the corresponding hydrogen "cloud", so that only its surface is highly ionized. Interestingly enough, this column density is about the same as that observed in 21-cm hydrogen emission at the optical radius of nearby spiral galaxies.

DLAs can be observed at all redshifts $z \lesssim 5$. For $z > 5$ the Lyα forest becomes so dense that these damped absorption lines are very difficult to identify. For $z \lesssim 1.6$ the Lyα transition cannot be observed from the ground; since the apertures of optical/UV telescopes in space are considerably smaller than those on the ground, observing low-redshift DLAs is substantially more complicated than that of higher z.

The Neutral Hydrogen Mass Contained in DLAs. The column density distribution of Lyα forest lines is a power law, given by (8.18). The relatively flat slope of $\beta \sim 1.6$ indicates that most of the neutral hydrogen is contained in systems of high column density. This can be seen as follows: the total column density of neutral hydrogen above some minimum column density $N_{\rm min}$ is

$$N_{\rm HI,tot}(N_{\rm max}) \propto \int_{N_{\rm min}}^{N_{\rm max}} {\rm d}N_{\rm HI}\, N_{\rm HI} \frac{{\rm d}N}{{\rm d}N_{\rm HI}} \propto \int_{N_{\rm min}}^{N_{\rm max}} {\rm d}N_{\rm HI}\, H_{\rm HI}^{1-\beta} = \frac{N_{\rm max}^{2-\beta} - N_{\rm min}^{2-\beta}}{2-\beta} , \tag{9.2}$$

and is, for $\beta < 2$, dominated by the highest column density systems. In fact, unless the distribution of column densities steepens for very high $N_{\rm HI}$, the integral diverges. From the extended statistics now available for DLAs, it is known that ${\rm d}N/{\rm d}N_{\rm HI}$ attains a break at column densities above $N_{\rm HI} \gtrsim 10^{21}\,{\rm cm}^{-2}$, rendering the above integral finite. Nevertheless, this consideration implies that most of the neutral hydrogen in the Universe visible in QSO absorption lines is contained in DLAs. From the observed distribution of DLAs as a function of column density and redshift, the density parameter $\Omega_{\rm HI}$ in neutral hydrogen as a function of redshift can be inferred. Apparently, $\Omega_{\rm HI} \sim 10^{-3}$ over the whole redshift interval $0 < z < 5$, with perhaps a small redshift dependence. Compared to the current density of stars, this neutral hydrogen density is smaller only by a factor ~ 3. Therefore, the hydrogen contained in DLAs is an important reservoir for star formation, and DLAs may represent condensations of gas that turn into "normal" galaxies once star-formation sets in. Since DLAs have low metallicities, typically $1/10$ of the Solar abundance, it is quite plausible that they have not yet experienced much star formation.

The Nature of DLAs. This interpretation is supported by the kinematical properties of DLAs. Whereas the fact that the Lyα line is damped implies that its observed shape is essentially independent of the Doppler velocity of the gas, velocity information can nevertheless be obtained from metal lines. Every DLA is associated with metal absorption line systems, covering low- and high-ionization species (such as SiII and CIV, respectively) which can be observed by choosing the appropriate wavelength coverage of the spectrum. The profiles of these metal lines are usually split up into several components. Interpreted as ionized "clouds", the velocity range Δv thus obtained can be used as an indicator of the characteristic velocities of the DLA. The values of Δv cover a wide range, with a median of ~ 90 km/s for the low-ionization lines and ~ 190 km/s for the high-ionization transitions. The observed distribution is largely compatible with the interpretation that DLAs are rotating disks with a characteristic rotational velocity of $v_{\rm c} \sim 200$ km/s, once random orientations and impact parameters of the line-of-sight to the QSO are taken into account.

Search for Emission from DLAs. If this interpretation is correct, then we might expect that the DLAs can also be observed as galaxies in emission. This, however, is exceedingly difficult for the high-redshift DLAs. Noting that they are discovered as absorption lines in the spectrum of QSOs, we face the difficulty of imaging a high-redshift galaxy very close to the line-of-sight to a bright QSO (to quote characteristic numbers, the typical QSO used for absorption-line spectroscopy has $B \sim 18$, whereas an L_*-galaxy at $z \sim 3$ has $B \sim 24.5$). Due to the size of the point-spread function this is nearly hopeless from the ground. But even with the resolution of HST, it is a difficult undertaking. Another possibility is to look for the Lyα emission line at the absorption redshift, located right in the wavelength range where the DLA fully blocks the QSO light. However, as we discussed for LBGs above, not all galaxies show Lyα in emission, and it is not too surprising that these searches have largely failed. To data, only three DLA have been detected in emission, with two of them seen only through the Lyα emission line at the trough of the damped absorption line, but with no observable continuum radiation. This latter fact indicates that the blue light from DLAs is considerably fainter than that from a typical LBG at $z \sim 3$, consistent with the interpretation that DLAs are not strong star-forming objects. One of these three DLAs, however, is observed to be considerably brighter and seems to share some characteristics of LBGs, including a high star-formation rate. In addition, two DLAs have been detected by [OIII] emission lines. Overall, then, the nature of high-redshift DLAs is still unclear, due to the small number of direct identifications.

For DLAs at low redshifts the observational situation is different, in that a fair fraction of them have counterparts seen in emission. Whereas the interpretation of the data is still not unambiguous, it seems that the low-redshift population of DLAs may be composed of normal galaxies.

The spatial abundance of DLAs is largely unknown. The observed frequency of DLAs in QSO spectra is the product of the spatial abundance and the absorption cross-section of the absorbers. This product can be compared with the corresponding quantity of local galaxies: the detailed mapping of nearby galaxies in the 21-cm line shows that their abundance and gaseous cross-section are compatible with the frequency of DLAs for $z \lesssim 1.5$, and falls short by a factor ~ 2 for the higher-redshifts DLAs.

9.2.5 Lyman-Alpha Blobs

The search for high-redshift galaxies with narrow-band imaging, where the filter is centered on the redshifted Lyα emission line, has revealed a class of objects which are termed "Lyman-α blobs". These are luminous and very extended sources of Lyα emission; their characteristic flux in the Lyα line is $\sim 10^{44}$ erg/s, and their typical size is ~ 100 kpc. Some of these sources show no detectable continuum emission in any broad-band optical filter.

The nature of these high-redshifts objects is currently unknown. Suggested explanations are wide-ranging, including a hidden QSO, strong star formation and associated superwinds, as well as "cold accretion", where gas is accreted onto a dark matter halo and hydrogen is collisionally excited in the gas of temperature $\sim 10^4$ K, yielding the observed Lyα emission. It even seems plausible that the Lyman-α blobs encompass a range of different phenomena, and that all three modes of powering the line emission indeed occur.

Two of these Lyα blobs were discovered by narrow-band imaging of the aforementioned proto-cluster of LBGs at $z = 3.09$. Both of them are sub-mm sources and therefore star-forming objects; the more powerful one has a sub-mm flux suggesting a star-formation rate of $\sim 1000 M_\odot/\text{yr}$. Spatially resolved spectroscopy extending over the full ~ 100 kpc size of one of the Lyα blobs shows that across the whole region there is an absorption line centered on the Lyα emission line. The optical depth of the absorption line suggests an HI column density of $\sim 10^{19}\,\text{cm}^{-2}$, and its centroid is blueshifted relative to the underlying emission line by ~ 250 km/s. The spatial extent of the blueshifted absorption shows that the outflowing material is a global phenomenon in this object – a true superwind, most likely driven by energetic star formation and subsequent supernova explosions in these objects.

9.3 Background Radiation at Smaller Wavelengths

The cosmic microwave background (CMB) is a remnant of the early hot phase of the Universe, namely thermal radiation from the time before recombination. As we extensively discussed in Sect. 8.6, the CMB contains a great deal of information about our Universe. Therefore, one might ask whether background radiation also exists in other wavebands, which then might be of similar value for cosmology. The neutrino background that should be present as a relic from the early epochs of the Universe, in the form of a thermal distribution of all three neutrino families with $T \approx 1.9$ K (see Sect. 4.4.2), is likely to remain undiscovered for quite some time due to the very small cross-section of these low-energy neutrinos.

Indeed, apparently isotropic radiation has been found in wavelength domains other than the microwave regime (Fig. 9.24). Following the terminology of the CMB, these are called background radiation as well. However, the name should not imply that it is a background radiation of cosmological origin, in the same sense as the CMB. From the thermal cosmic history (see Sect. 4.4), no optical or X-ray radiation is expected from the early phases of the Universe. Hence, for a long time it was unknown what the origin of these different background radiations may be.

At first, the early X-ray satellites discovered a background in the X-ray regime (cosmic X-ray background, CXB). Later, the COBE satellite detected an apparently isotropic radiation component in the FIR, the cosmic infrared background (CIB).

In the present context, we simply denote the flux in a specific frequency domain, averaged over sky position at high Galactic latitudes, as background radiation. Thus, when talking about an optical background here,

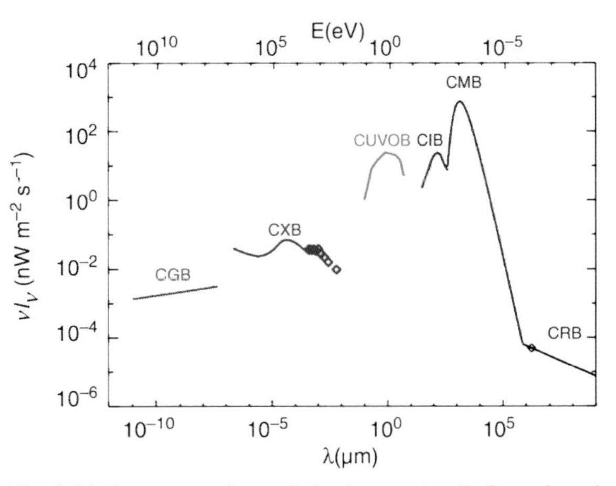

Fig. 9.24. Spectrum of cosmic background radiation, plotted as νI_ν versus wavelength. Besides the CMB, background radiation exists in the radio domain (cosmic radio background, CRB), in the infrared (CIB), in the optical/UV (CUVOB), in the X-ray (CXB), and at gamma-ray energies (CGB). With the exception of the CMB, probably all of these backgrounds can be understood as a superposition of the emission from discrete sources. Furthermore, this figure shows that the energy density in the CMB exceeds that of other radiation components, as was assumed when we considered the radiation density in the Universe in Chap. 4

we refer to the sum of the radiation of all galaxies and AGNs per solid angle. The interpretation of such a background radiation depends on the sensitivity and the angular resolution of the telescopes used. Imagine, for instance, observing the sky with an optical camera that has an angular resolution of only one arcminute. A relatively isotropic radiation would then be visible at most positions in the sky, featuring only some very bright or very large sources. On improving the angular resolution, more and more individual sources would become visible – culminating in the observations of the Ultra-Deep Fields – and the background could then be identified as the sum of the emission of individual sources. In analogy to this thought-experiment, one may wonder whether the CXB or the CIB can likewise be understood as a superposition of radiation from discrete sources.

9.3.1 The IR Background

Observations of background radiation in the infrared are very difficult to accomplish. First, it is problematic to measure absolute fluxes due to the thermal emission of the detector. In addition, the emission by interplanetary dust (and by the interstellar medium in our Milky Way) is much more intense than the infrared flux from extragalactic sources. For these reasons, the absolute level of the infrared background has been determined only with relatively large uncertainties, as displayed in Fig. 9.25.

The ISO satellite was able to resolve about 10% of the CIB at $\lambda = 175\,\mu$m into discrete sources. Also in the sub-mm range (at about 850 μm) almost all of the CIB seems to originate from discrete sources which consist mainly of dust-rich star-formation regions (see Sect. 9.2.3, where the source population in the sub-mm domain was discussed).

In any case, no indication has yet been found that the origin of the CIB is different from the emission by a population of discrete sources, in particular of high-redshift starburst galaxies. Further resolving the background radiation into discrete sources will become possible by future FIR satellites such as, for instance, Herschel.

9.3.2 The X-Ray Background

In the 1970s, the first X-ray satellites discovered not only a number of extragalactic X-ray sources (such as AGNs and clusters of galaxies), but also an apparently isotropic radiation component, the CXB. Its spectrum is a very hard (i.e., flat) power law, cut off at an energy above ~ 40 keV, which can roughly be described by

$$I_\nu \propto E^{-0.3} \exp\left(-\frac{E}{E_0}\right) , \tag{9.3}$$

with $E_0 \sim 40$ keV. Initially, the origin of this radiation was unknown, since its spectral shape was different from the spectra of sources that were known at that time. For example, it was not possible to obtain this spectrum by a superposition of the spectra of know AGNs.

ROSAT, with its substantially improved angular resolution compared to earlier satellites (such as the *Einstein* observatory), conducted source counts at much lower fluxes, based on some very deep images. From this, it was shown that at least 80% of the CXB in the energy range between 0.5 keV and 2 keV is emitted by discrete sources, of which the majority are AGNs. Hence it is natural to assume that the total CXB at these low X-ray energies originates from discrete sources, and observations by XMM-Newton seem to confirm this.

However, the X-ray spectrum of normal AGNs is different from (9.3), namely it is considerably steeper (about $S_\nu \propto \nu^{-0.7}$). Therefore, if these AGNs contribute the major part of the CXB at low energies, the CXB at higher energies cannot possibly be produced by the same AGNs. Subtracting the spectral energy of

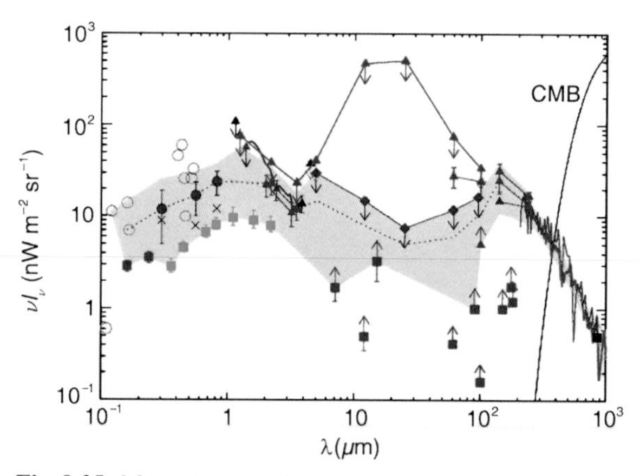

Fig. 9.25. Measurement of, and limits to, the CIB. Squares denote lower limits derived from the integration of observed source counts, diamonds are upper limits from flux measurements, and other symbols show absolute flux measurements. The shaded yellow range indicates the current observational limits to the CIB

the AGNs found by ROSAT from the CXB spectrum (9.3), one obtains an even harder spectrum, resembling very closely that of thermal bremsstrahlung. Therefore, it was supposed for a long time that the CXB is, at higher energies, produced by a hot intergalactic gas at temperatures of $k_B T \sim 30$ keV.

This model was excluded, however, by the precise measurement of the thermal spectrum of the CMB by COBE, showing that the CMB has a perfect blackbody spectrum. If a postulated hot intergalactic gas were able to produce the CXB, it would cause significant deviations of the CMB from the Planck spectrum, namely by the inverse Compton effect (the same effect that causes the SZ effect in clusters of galaxies – see Sect. 6.3.4). Thus, the COBE results clearly ruled out this possibility.

By now, the nature of the CXB at higher energies has also essentially been determined (see Fig. 9.26), mainly through very deep observations with the Chandra satellite. An example of a very deep observation, the Chandra Deep Field South, is shown in Fig. 9.27. From source counts performed in such fields, about 75% of the CXB in the energy range of $2\,\mathrm{keV} \le E \le 10\,\mathrm{keV}$ could be resolved into discrete sources. Again, most of these sources are AGNs, but typically with a significantly harder (i.e., flatter) spectrum than the AGNs that are producing the low-energy CXB. Such a flat X-ray spectrum can be produced by photoelectric absorption of an intrinsically steep power-law spectrum, where photons closer to the ionization energy are more efficiently absorbed than those at higher energy. According to the

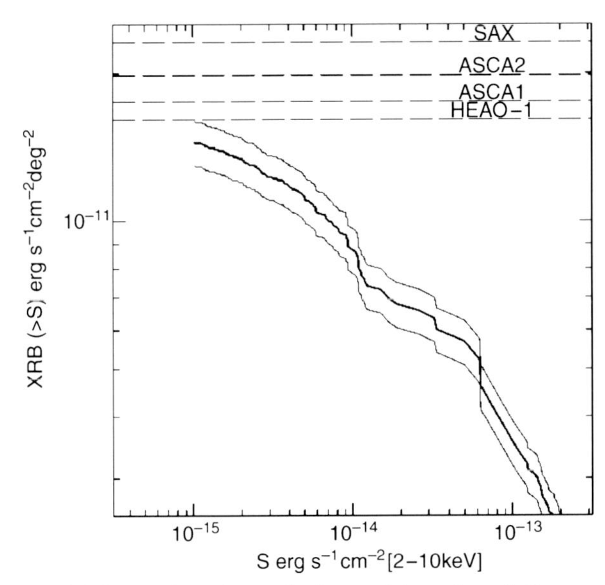

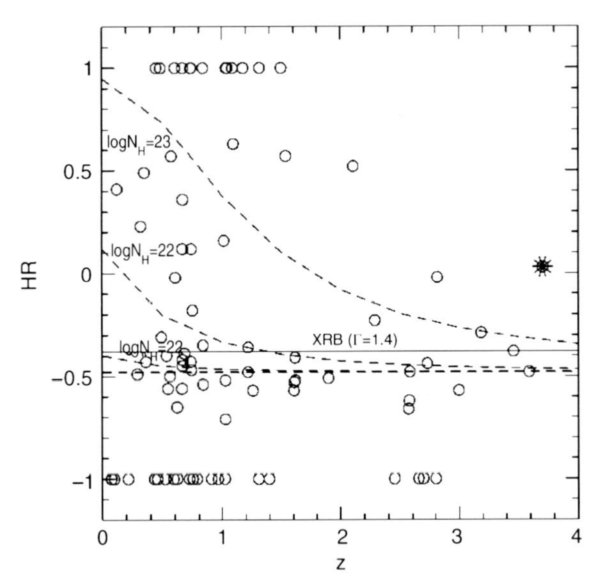

Fig. 9.26. In the left panel, the total intensity of discrete sources with an individual flux $> S$ in the energy range $2\,\mathrm{keV} \le E \le 10\,\mathrm{keV}$ is plotted (thick curve), together with the uncertainty range (between the two thin curves). Most of the data are from a 3×10^5 s exposure of the Chandra Deep Field. The dashed lines show different measurements of the CXB flux in this energy range; depending on which of these values is the correct one, between 60% and 90% of the CXB in the Chandra Deep Field at this energy is resolved into discrete sources. In the right panel, the hardness ratio HR – specifying the ratio of photons in the energy range $2\,\mathrm{keV} \le E \le 10\,\mathrm{keV}$ to those in $0.5\,\mathrm{keV} \le E \le 2\,\mathrm{keV}$, $\mathrm{HR} = (S_{>2\,\mathrm{keV}} - S_{<2\,\mathrm{keV}})/(S_{>2\,\mathrm{keV}} + S_{<2\,\mathrm{keV}})$ – is plotted as a function of redshift, for 84 sources in the Chandra Deep Field with measured redshifts. This plot indicates that the HR decreases with redshift; this trend is expected if the X-ray spectrum of the AGNs is affected by intrinsic absorption. The dashed curves show the expected value of HR for a source with an intrinsic power-law spectrum $I_\nu \propto \nu^{-0.7}$, which is observed through an absorbing layer with a hydrogen column density of N_H, by which these curves are labeled. Since low-energy photons are more strongly absorbed by the photoelectric effect than high-energy ones, the absorption causes the spectrum to become harder, thus flatter, at relatively low X-ray energies. This implies an increase of the HR (also see the bottom panel of Fig. 6.16). This effect is smaller for higher redshift sources, since the photon energy at emission is then larger by a factor of $(1+z)$

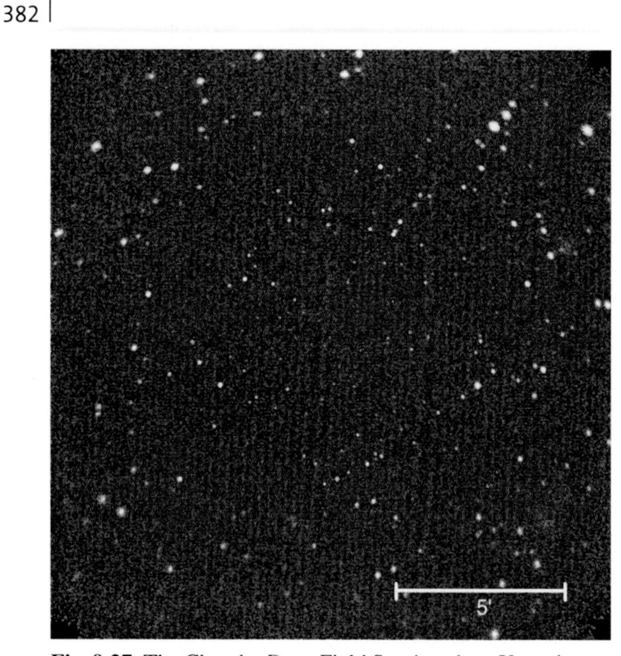

Fig. 9.27. The Chandra Deep Field South, a deep X-ray image of a $16' \times 16'$ field with an exposure time of 10^6 s – one of the deepest X-ray images ever obtained. Most of the sources visible in this field are AGNs, but galaxies, groups, and clusters are also detected. The photon energy is color-coded, from lower to higher energies in red, yellow, and blue. One of the sources in this field is a very distant QSO of Type 2. A radial variation of the PSF in the field is visible by the increasing size of individual sources towards the edges

classification scheme of AGNs discussed in Sect. 5.5, these are Type 2 AGNs, thus Seyfert 2 galaxies and QSOs with strong intrinsic self-absorption. We should recall that Type 2 QSOs have only been detected by Chandra – hence, it is no coincidence that the same satellite has also been able to resolve the high-energy CXB.

9.4 Reionization of the Universe

After recombination at $z \sim 1100$, the intergalactic gas became neutral, with a residual ionization of only $\sim 10^{-4}$. Had the Universe remained neutral we would not be able to receive any photons that were emitted bluewards of the Lyα line of a source, because the absorption cross-section for Lyα photons is too large (see Eq. 8.16). Since such photons are observed from QSOs, as can be seen for instance in the spectra of the $z > 5.7$ QSOs in Fig. 9.28, and since an appreciable fraction of homogeneously distributed neutral gas in the intergalactic medium can be excluded for $z \lesssim 5$, from the tight upper bounds on the strength of the Gunn–Peterson effect (Sect. 8.5.1) the Universe must have been reionized between the recombination epoch and the redshift $z \sim 6.5$ of the most distant known sources. From the WMAP results (see Sect. 8.7.1) one concludes that reionization must have taken place at very high redshift, $z \sim 15$.

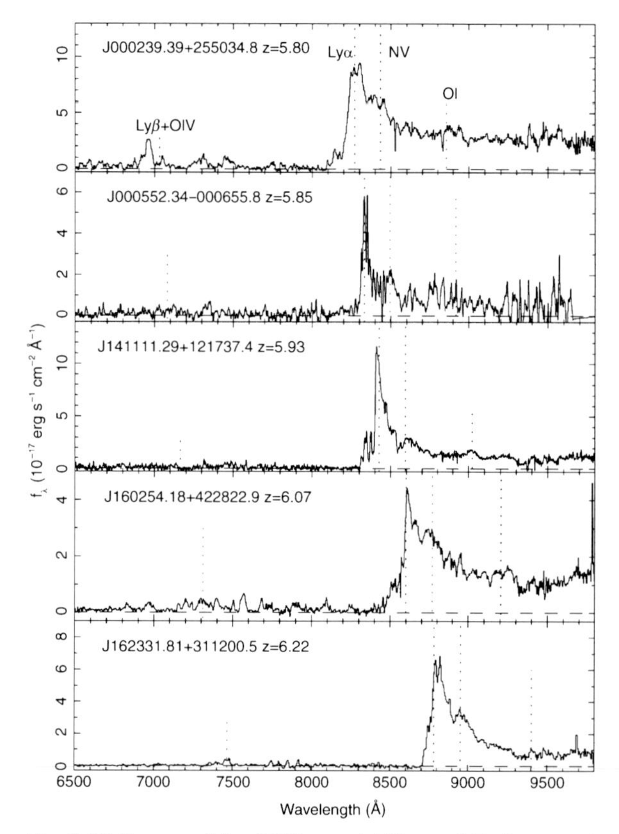

Fig. 9.28. Spectra of five QSOs at redshifts $z > 5.7$, discovered in multicolor data from the Sloan Digital Sky Survey. The positions of the most important emission lines are marked. Particularly remarkable is the complete lack of flux bluewards of the Lyα emission line in some of the QSOs, indicating a strong Gunn–Peterson effect. However, this absorption is not complete in all QSOs, which points at strong variations in the density of neutral hydrogen in the intergalactic medium at these high redshifts. Either the hydrogen density varies strongly for different lines-of-sight, or the degree of ionization is very inhomogeneous

This raises the question of how this reionization occurred, in particular which process was responsible for it. The latter question is easy to answer – reionization must have happened by photoionization. Collisional ionization can be ruled out because for it to be efficient the IGM would need to be very hot, a scenario which can be excluded due to the perfect Planck spectrum of the CMB – the argument here is the same as above, where we excluded the idea of a hot IGM as the source of the CXB. Hence, the next question is where the energetic photons that caused the photoionization of the IGM come from.

Two kinds of sources may account for them – hot stars or AGNs. Currently, it is not unambiguously clear which of these is the predominant source of energetic photons causing reionization since our current understanding of the formation of supermassive black holes is still insufficient. However, it is currently thought that the main source of photoionization photons is the first generation of hot stars.

9.4.1 The First Stars

Following on from the above arguments, understanding reionization is thus directly linked to studying the first generation of stars. In the present Universe star formation occurs in galaxies; thus, one needs to examine when the first galaxies could have formed. From the theory of structure formation, the mass spectrum of dark matter halos at a given redshift can be computed by means of, e.g., the Press–Schechter model (see Sect. 7.5.2). Two conditions need to be fulfilled for stars to form in these halos. First, gas needs to be able to fall into the dark halos. Since the gas has a finite temperature, pressure forces may impede the infall into the potential well. Second, this gas also needs to be able to cool, condensing into clouds in which stars can then be formed. We will now examine these two conditions.

The Jeans Mass. By means of a simple argument, we can estimate under which conditions pressure forces are unable to prevent the infall of gas into a potential well. To do this, we consider a slightly overdense spherical region of radius R whose density is only a little larger than the mean cosmic matter density $\overline{\rho}$. If this sphere is homogeneously filled with baryons, the gravitational binding energy of the gas is about

$$|E_{\rm grav}| \sim \frac{G M M_{\rm b}}{R} ,$$

where M and $M_{\rm b}$ denote the total mass and the baryonic mass of the sphere, respectively. The thermal energy of the gas can be computed from the kinetic energy per particle, multiplied by the number of particles in the gas, or

$$E_{\rm th} \sim c_{\rm s}^2 M_{\rm b} .$$

Here,

$$c_{\rm s} \approx \sqrt{\frac{k_{\rm B} T_{\rm b}}{\mu m_{\rm p}}}$$

is the speed of sound in the gas, which is about the average speed of the gas particles, and $\mu m_{\rm p}$ denotes the average particle mass in the gas. For the gas to be bound in the gravitational field, its gravitational binding energy needs to be larger than its thermal energy, $|E_{\rm grav}| > E_{\rm th}$, which yields the condition $GM > c_{\rm s}^2 R$. Since we have assumed an only slightly overdense region, the relation $M \sim \overline{\rho} R^3$ between mass and radius of the sphere applies. From the two latter equations, the radius can be eliminated, yielding the condition

$$M > \left(\frac{c_{\rm s}^2}{G}\right)^{3/2} \frac{1}{\sqrt{\overline{\rho}}} . \tag{9.4}$$

Thus, as a result of our simple argument we find that the mass of the halo needs to exceed a certain threshold for gas to be able to fall in. A more accurate treatment yields the condition

$$M > M_{\rm J} \equiv \frac{\pi^{5/2}}{6} \left(\frac{c_{\rm s}^2}{G}\right)^{3/2} \frac{1}{\sqrt{\overline{\rho}}} . \tag{9.5}$$

In the final step we defined the *Jeans mass* $M_{\rm J}$, which describes the minimum mass of a halo required for the gravitational infall of gas. The Jeans mass depends on the temperature of the gas, expressed through the sound speed $c_{\rm s}$, and on the mean cosmic matter density $\overline{\rho}$. The latter can easily be expressed as a function of redshift, $\overline{\rho}(z) = \overline{\rho}_0 (1+z)^3$.

The baryon temperature has a more complicated dependence on redshift. For sufficiently high redshifts, the small fraction of free electrons that remain after recombination – the gas has a degree of ionization of $\sim 10^{-4}$ –

provide a thermal coupling of the baryons to the cosmic background radiation, by means of Compton scattering. This is the case for redshifts $z \gtrsim z_t$, where

$$z_t \approx 140 \left(\frac{\Omega_b h^2}{0.022} \right)^{2/5} ;$$

hence, $T_b(z) \approx T(z) = T_0(1+z)$ for $z \gtrsim z_t$. For smaller redshifts, the density of photons gets too small to maintain this coupling, and baryons start to adiabatically cool down by the expansion, so that for $z \lesssim z_t$ we obtain approximately $T_b \propto \rho_b^{2/3} \propto (1+z)^2$.

From this temperature dependence, the Jeans mass can then be calculated as a function of redshift. For $z_t \lesssim z \lesssim 1000$, M_J is independent of z because $c_s \propto T^{1/2} \propto (1+z)^{1/2}$ and $\overline{\rho} \propto (1+z)^3$, and its value is

$$M_J = 1.35 \times 10^5 \left(\frac{\Omega_m h^2}{0.15} \right)^{-1/2} M_\odot , \tag{9.6}$$

whereas for $z \lesssim z_t$ we obtain, with $T_b \simeq 1.7 \times 10^{-2}$ $(1+z)^2$ K,

$$M_J = 5.7 \times 10^3 \left(\frac{\Omega_m h^2}{0.15} \right)^{-1/2} \times \left(\frac{\Omega_b h^2}{0.022} \right)^{-3/5} \left(\frac{1+z}{10} \right)^{3/2} M_\odot . \tag{9.7}$$

Cooling of the Gas. The Jeans criterion is a necessary condition for the formation of proto-galaxies. In order to form stars, the gas in the halos needs to be able to cool further. Here, we are dealing with the particular situation of the first galaxies, whose gas is metal-free, so metal lines cannot contribute to the cooling. This means that cooling can only happen via hydrogen and helium. Since the first excited state of hydrogen has a high energy (that of the Lyα transition, thus $E \sim 10.2$ eV), this cooling is efficient only above $T \gtrsim 2 \times 10^4$ K. However, the halos which form at high redshift have low mass, so that their virial temperature is considerably below this energy scale. Therefore, atomic hydrogen is a very inefficient coolant for these first halos, insufficient to initiate the formation of stars. Furthermore, helium is of no help in this context, since its excitation temperature is even higher than that of hydrogen. The problem resulting from these arguments has become even worse given the WMAP discovery of a higher reionization redshift than previously estimated.

Only in recent years has it been discovered that molecular hydrogen represents an extremely important component in cooling processes. Despite its very small transition probability, H_2 dominates the cooling rate of primordial gas at temperatures below $T \sim 10^4$ K – see Fig. 9.29 – where the precise value of this temperature depends on the abundance of H_2.

By means of H_2, the gas can cool in halos with a temperature exceeding about $T_{vir} \gtrsim 3000$ K, corresponding to a mass of $M \gtrsim 10^4 M_\odot$; the exact values depend on the redshift. In these halos, stars may then be able to form. However, these stars will certainly be different from those known to us, because they do not contain any metals. Therefore, the opacity of the stellar plasma is much lower. Such stars, which at the same mass presumably have a much higher temperature and luminosity (and thus a shorter lifetime), are called *Population III stars*. Due to their high temperature they are much more efficient sources of ionizing photons than stars with "normal" metallicity.

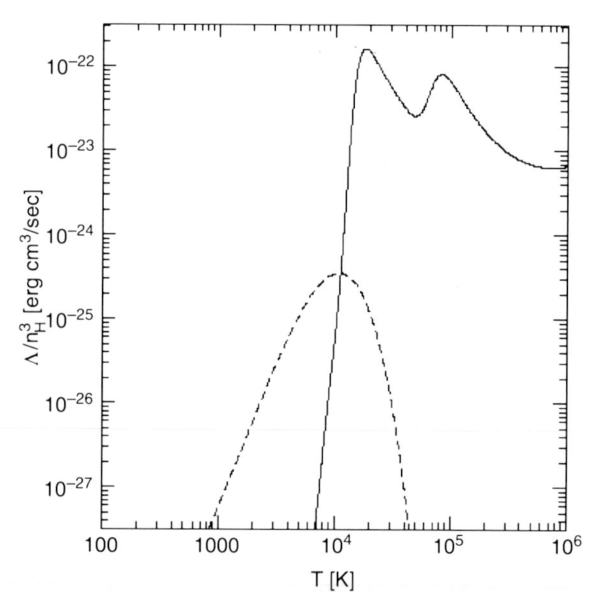

Fig. 9.29. Cooling rate as a function of the temperature for a gas consisting of atomic and molecular hydrogen (with 0.1% abundance) and of helium. The solid curve describes the cooling by atomic gas, the dashed curve that by molecular hydrogen; thus, the latter is extremely important at temperatures below $\sim 10^4$ K. At considerably lower temperatures the gas cannot cool, hence no star formation will take place

9.4.2 The Reionization Process

Dissociation of Molecular Hydrogen. The energetic photons from these Population III stars are now capable of ionizing hydrogen in their vicinity. More important still is another effect: the binding energy of H_2 is only 11.26 eV. Since the Universe is transparent for photons with energies below 13.6 eV, photons with $11.26\,\text{eV} \leq E_\gamma \leq 13.6\,\text{eV}$ can propagate very long distances and dissociate molecular hydrogen. This means that as soon as the first stars have formed in a region of the Universe, molecular hydrogen in their vicinities will be destroyed and further star formation will then be prevented.[2]

Metal Enrichment of the Intergalactic Medium. Soon after Population III stars have formed, they will explode as supernovae. Through this process, the metals produced by them are ejected into the intergalactic medium, by which the initial metal enrichment of the IGM occurs. The kinetic energy transferred by SNe to the gas within the halo can exceed its binding energy, so that the baryons of the halo can be blown away and further star formation is prevented. Whether this effect may indeed lead to gas-free halos, or whether the released energy can instead be radiated away, depends on the geometry of the star-formation regions. In any case, it can be assumed that in those halos where the first generation of stars was born, further star formation was considerably suppressed, particularly since all molecular hydrogen was destroyed.

We can assume that the metals produced in these first SN explosions are, at least partially, ejected from the halos into the intergalactic medium, thus enriching the latter. The existence of metal formation in the very early Universe is concluded from the fact that even sources at very high redshift (like QSOs at $z \sim 6$) have a metallicity of about one tenth the Solar value. Furthermore, the Lyα forest also contains gas with non-vanishing metallicity. Since the Lyα forest is produced by the intergalactic medium, this therefore must have been enriched.

The Final Step to Reionization. For gas to cool in halos without molecular hydrogen, their virial temperature needs to exceed about 10^4 K (see Fig. 9.29). Halos of this mass form with appreciable abundance at redshifts of $z \sim 10$, as follows, e.g., from the Press–Schechter model (see Sect. 7.5.2). In these halos, efficient star formation can then take place; the first proto-galaxies will form. These will then ionize the surrounding IGM in the form of HII regions, as sketched in Fig. 9.30. The corresponding HII regions will expand because increasingly more photons are produced. If the halo density is sufficiently high, these HII regions will start to overlap. Once this occurs, the IGM is ionized, and reionization is completed.

We therefore conclude that reionization is a two-stage process. In a first phase, Population III stars form through cooling of gas by molecular hydrogen, which is then destroyed by these very stars. Only in a later epoch and in more massive halos is cooling provided by atomic hydrogen, then leading to reionization.

A Luminous J-band Drop-Out? The aforementioned fact that, even at redshifts as large as $z \sim 6$, massive galaxies with a fairly old stellar population were already in place shows that there was an epoch of intense star formation at even earlier times. This clearly suggests that these galaxies must have played an important role in the reionization process of the Universe. In fact, in the HUDF an object was found that appears to be a J-band drop-out, with no radiation seen at wavelengths shorter than the J-band. Spectroscopy of this source revealed the presence of a strong 4000-Å break, not only giving further support to its high redshift of $z \sim 6.5$, but also indicating that the source contains a post-starburst stellar population. The bolometric luminosity of this source is estimated to be $\sim 10^{12} L_\odot$, and using a Salpeter initial mass function (see Eq. 3.67), a stellar mass of $\sim 6 \times 10^{11} M_\odot$ is obtained. The strong 4000-Angstrom break indicates that the spectral energy distribution is dominated by A0 stars of masses $\lesssim 3 M_\odot$. This provides a clear indication of an old age of the population of $\gtrsim 300$ Myr, implying that the stars must have formed at redshifts $z > 9$, but possibly at even higher redshift.

We can estimate the comoving volume that this galaxy was able to reionize, based on its high luminosity. With all uncertainties entering such an estimate (such as the escape fraction of ionizing photons from the galaxy), it is concluded that this galaxy could ionize a volume of $\sim 10^5\,\text{Mpc}^3$. This needs to be compared to the high-redshift volume within which such a source would have

[2] To destroy all the H_2 in the Universe one needs less than 1% of the photon flux that is required for the reionization.

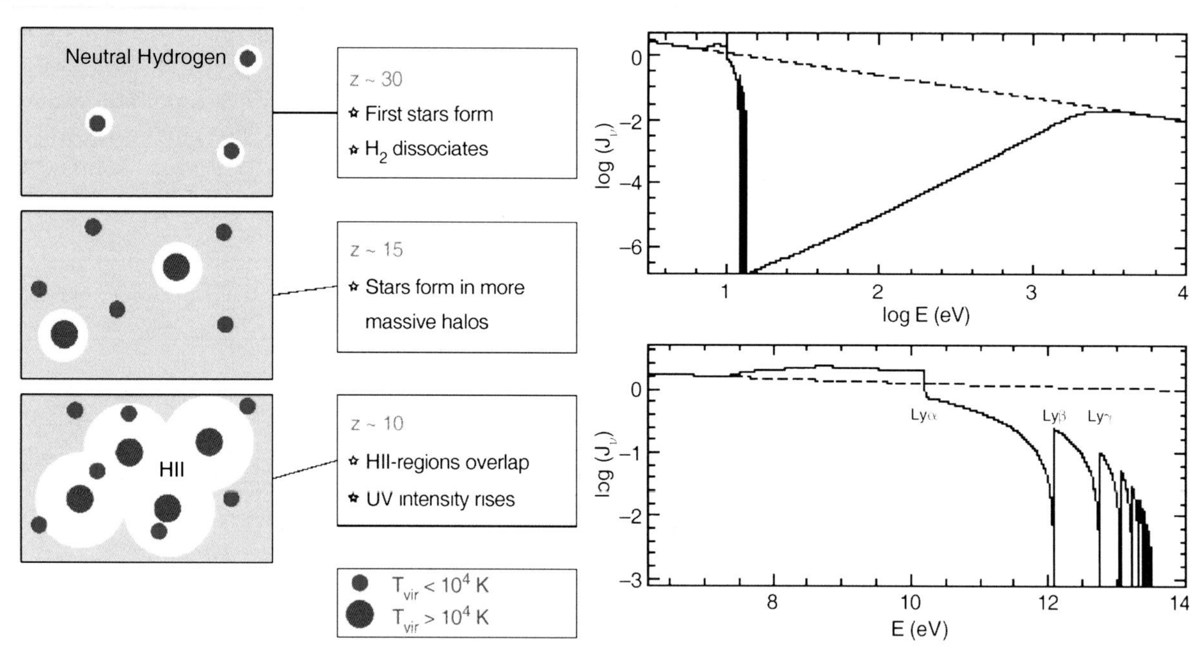

Fig. 9.30. Left: a sketch of the geometry of reionization is shown: initially, relatively low-mass halos collapse, a first generation of stars ionizes and heats the gas in these halos. By heating, the temperature increases so strongly (to about $T \sim 10^4$ K) that gas can escape from the potential wells; these halos may never again form stars efficiently. Only when more massive halos have collapsed will continuous star formation set in. Ionizing photons from this first generation of hot stars produce HII regions around their halos, which is the onset of reionization. The regions in which hydrogen is ionized will grow until they start to overlap; at that time, the flux of ionizing photons will strongly increase. Right: the average spectrum of photons at the beginning of the reionization epoch is shown; here, it has been assumed that the flux from the radiation source follows a power law (dashed curve). Photons with an energy higher than that of the Lyα transition are strongly suppressed because they are efficiently absorbed. The spectrum near the Lyman limit shows features which are produced by the combination of breaks corresponding to the various Lyman lines, and the redshifting of the photons

been detected in the HUDF. The result is that these two volumes are quite comparable. Hence, it seems that this galaxy was capable of reionizing "its" volume of the Universe. Again we should warn that this is a preliminary conclusion, based on a single object; nevertheless, it indicates that we might be seeing direct evidence for early reionization, in accordance with the results from WMAP.

Prospects for Observing Reionization Directly. We note that only a small fraction of the baryons needs to burn in hot stars to ionize all hydrogen, as we can easily estimate: by fusing four H-nuclei (protons) to He, an energy of about 7 MeV per nucleon is released. However, only 13.6 eV per hydrogen atom is required for ionization.

Furthermore, we point out again that the very dense Lyα forest seen towards QSOs at high redshift is no unambiguous sign for approaching the redshift of reionization, because a very small fraction of neutral atoms (about 1%) is already sufficient to produce a large optical depth for Lyα photons. Direct observation of reionization will probably be quite difficult; an illustration of this is sketched in Fig. 9.31.

We have confined our discussion to the ionization of hydrogen, and disregarded helium. The ionization energy of helium is higher than that of hydrogen, so that its ionization will be completed later. From the statistical analysis of the Lyα forest and from the analysis of helium absorption lines in high-redshift QSOs, a reionization redshift of $z \sim 3.2$ for helium is obtained.

With the upcoming Next Generation Space Telescope (which has more recently been named the James Webb

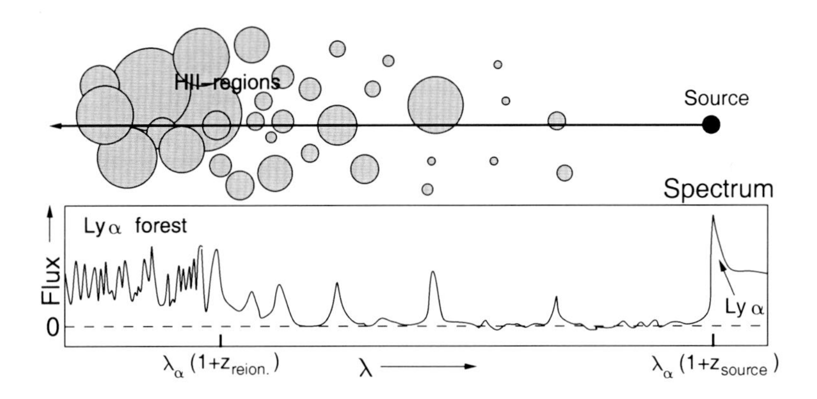

Fig. 9.31. Sketch of a potential observation of reionization: light from a very distant QSO propagates through a partially ionized Universe; at locations where it passes through HII regions, radiation will get through – flux will be visible at the corresponding wavelengths. When the HII regions start to overlap, the normal Lyα forest will be produced

Space Telescope, JWST), one hopes to discover the first light sources in the Universe; this space telescope, with a diameter of 6.5 m, will be optimized for operation at wavelengths between 1 and 5 μm.

9.5 The Cosmic Star-Formation History

The scenario for reionization as described above should, at least for the main part, be close to reality, with its details still being subject to intense discussion. In particular, the star-formation rate of the Universe as a function of redshift can be computed only by making relatively strong model assumptions, because too many physical processes are involved; we will elaborate on this in the next section. However, observations of galaxies at very high redshifts have also been accomplished in recent years, through which it has become possible to empirically trace the star-formation rate up to large redshifts.

9.5.1 Indicators of Star Formation

We define the star-formation rate (SFR) as the mass of the stars that are formed per year, typically given in units of $M_\odot/\mathrm{yr}$. For our Milky Way, we find a SFR of $\sim 3 M_\odot/\mathrm{yr}$. Since the signatures for star formation are obtained only from massive stars, their formation rate needs to be extrapolated to lower masses to obtain the full SFR, by assuming an IMF (initial mass function; see Sect. 3.9.4). Typically, a Salpeter-IMF is chosen between $0.1 M_\odot \le M \le 100 M_\odot$. We will start by listing the most important indicators of star formation:

- Emission in the far infrared (FIR). This is radiation emitted by warm dust which is heated by hot young stars. For the relation of FIR luminosity to the SFR, observation yields the approximate relation
$$\frac{\mathrm{SFR_{FIR}}}{M_\odot/\mathrm{yr}} \sim \frac{L_{\rm FIR}}{5.8\times 10^9 L_\odot} \; .$$
- Radio emission by galaxies. A very tight correlation exists between the radio luminosity of galaxies and their luminosity in the FIR, over many orders of magnitude of the corresponding luminosities. Since $L_{\rm FIR}$ is a good indicator of the star-formation rate, this should apply for radiation in the radio as well (where we need to disregard the radio emission from a potential AGN component). The radio emission of normal galaxies originates mainly from supernova remnants (SNRs). Since SNRs appear shortly after the beginning of star formation, caused by core-collapse supernovae at the end of the life of massive stars in a stellar population, radiation from SNRs is a nearly instantaneous indicator of the SFR. Once again from observations, one obtains
$$\frac{\mathrm{SFR_{1.4\,GHz}}}{M_\odot/\mathrm{yr}} \sim \frac{L_{1.4\,\rm GHz}}{8.4\times 10^{27}\,\mathrm{erg\,s^{-1}\,Hz^{-1}}} \; .$$
- Hα emission. This line emission comes mainly from the HII regions that form around young hot stars. As an estimate of the SFR, one uses
$$\frac{\mathrm{SFR_{H\alpha}}}{M_\odot/\mathrm{yr}} \sim \frac{L_{\rm H\alpha}}{1.3\times 10^{41}\,\mathrm{erg\,s^{-1}}} \; .$$
- UV radiation. This is only emitted by hot young stars, thus indicating the SFR in the most recent past, with
$$\frac{\mathrm{SFR_{UV}}}{M_\odot/\mathrm{yr}} \sim \frac{L_{\rm UV}}{7.2\times 10^{27}\,\mathrm{erg\,s^{-1}\,Hz^{-1}}} \; .$$

Applied to individual galaxies, each of these estimates is quite uncertain, which can be seen by comparing the resulting estimates from the various methods (see Fig. 9.32). For instance, Hα and UV photons are readily absorbed by dust in the interstellar medium of the galaxy or in the star-formation regions themselves. Therefore, the relations above should be corrected for this self-absorption, which is possible when the redding can be obtained from multicolor data. It is also expected that the larger the dust absorption, the stronger the FIR luminosity will be, causing deviations from the linear relation $\mathrm{SFR_{FIR}} \propto \mathrm{SFR_{UV}}$. After the appropriate corrections, the values for the SFR derived from the various indicators are quite similar, but still have a relatively large scatter.

There are also a number of other indicators of star formation. The fine-structure line of singly ionized carbon at $\lambda = 157.7\,\mu\mathrm{m}$ is of particular importance as it is one of the brightest emission lines in galaxies, which can account for a fraction of a percent of their total luminosity. The emission is produced in regions which are subject to UV radiation from hot stars, and thus associated with star-formation activity. Due to its wavelength, this line is difficult to observe and has, until recently, been detected only in star-forming regions in our Galaxy and in other local galaxies. However, recently this

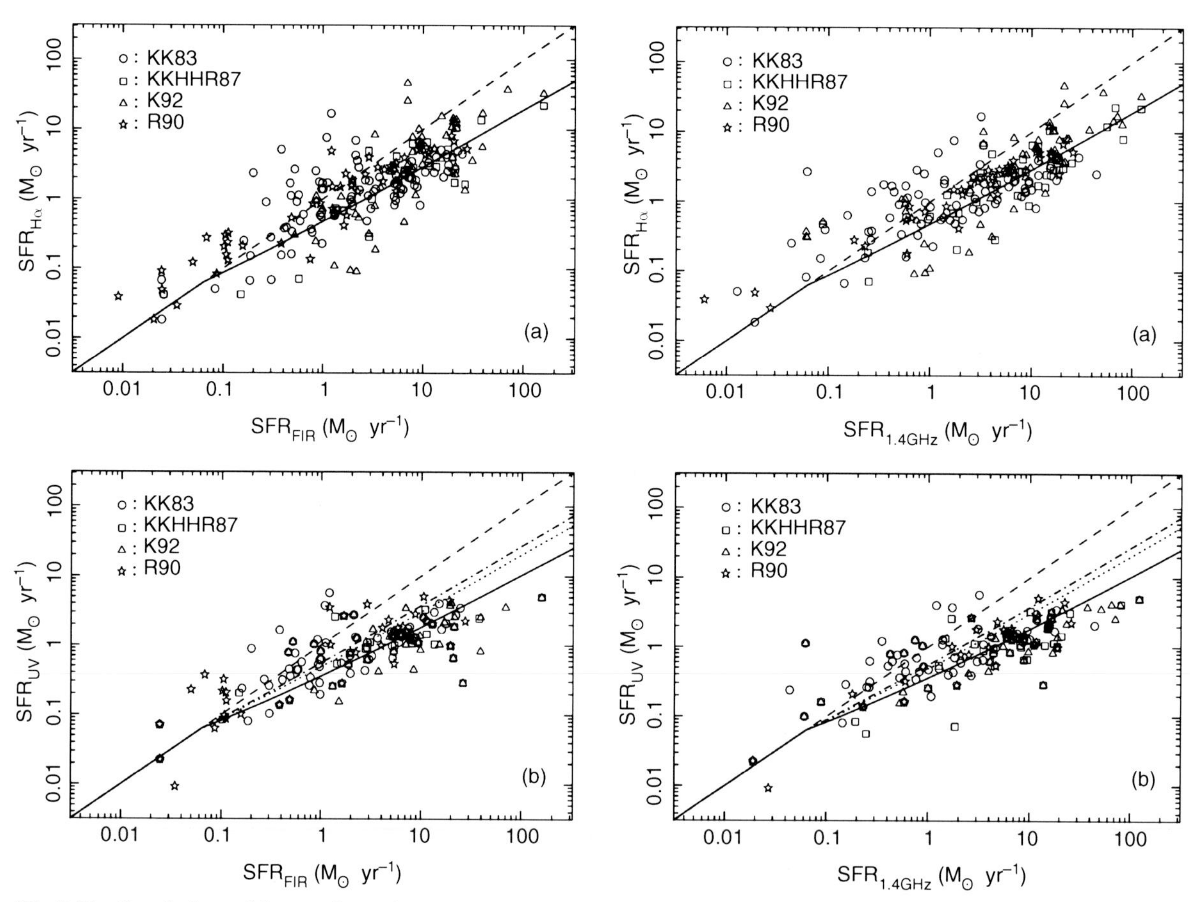

Fig. 9.32. Correlations of the star-formation rates in a sample of galaxies, as derived from observation in different wavebands. In all four diagrams, the dashed line marks the identity relation $\mathrm{SFR}_i = \mathrm{SFR}_2$; as is clearly seen, using the Hα luminosity and UV radiation as star-formation indicators seems to underestimate the SFR. Since radiation may be absorbed by dust at these wavelengths, and also since the amount of warm dust probably depends on the SFR itself, this effect can be corrected for, as shown by the solid curves in the four panels

line was detected from the most distant QSO known, at $z = 6.42$, where it is redshifted into the submillimeter part of the spectrum. This not only suggests that the host galaxy of the QSO undergoes an intense burst of star formation, but also that the material in this host galaxy is already significantly enriched with metals.

9.5.2 Redshift Dependence of the Star Formation: The Madau Diagram

The density of star formation, $\rho_{\rm SFR}$, is defined as the mass of newly formed stars per year per unit (comoving) volume, typically measured in $M_\odot\,{\rm yr}^{-1}\,{\rm Mpc}^{-3}$. Therefore, $\rho_{\rm SFR}$ as a function of redshift specifies how many stars have formed at any time. By means of the star-formation density we can examine the question, for instance, of whether the formation of stars began only at relatively low redshifts, or whether the conditions in the early Universe were such that stars formed efficiently even at very early times.

Investigations of the SFR in galaxies, by means of the above indicators, and source counts of such star-forming galaxies, allow us to determine $\rho_{\rm SFR}$. The plot of these results (Fig. 9.33) is called a "Madau diagram". In about 1996, Piero Madau and his colleagues accomplished, for the first time, a determination of the SFR at high redshifts from Lyman-break galaxies, where the intrinsic extinction was neglected in these first estimates. Correcting for this extinction (for which the progress in submillimeter astronomy has been extremely important, as we saw in Sect. 9.2.3), a nearly constant $\rho_{\rm SFR}$ is found for $z \gtrsim 1$, together with a decline by about a factor of 10 from $z \sim 1$ to the present time. These results have more recently been confirmed by investigations with the Spitzer satellite, observing a large sample of galaxies at FIR wavelengths. Whereas the star-formation rate density at low redshifts is dominated by galaxies which are not very prominent at FIR wavelength, this changes drastically for redshifts $z \gtrsim 0.7$, above which most of the star-formation activity is hidden from the optical view by dust. From this we conclude that most stars in our neighborhood were already formed at high redshift: star formation at earlier epochs was considerably more active than it is today.

Although the redshift-integrated star-formation rate and the mass density of stars determined from galaxy surveys, as displayed in Fig. 9.34, slightly deviate from each other, the degree of agreement is quite satisfactory if one recalls the assumptions that are involved in the determination of the two quantities: besides the uncertainties discussed above in the determination of the star-formation rate, we need to mention in particular the shape of the IMF of the newly formed stars for the determination of the stellar mass density. In fact, Fig. 9.34 shows that we have observed the formation of essentially the complete current stellar density.

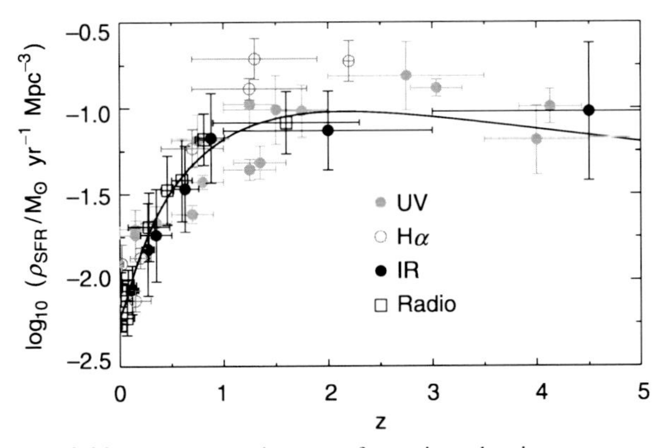

Fig. 9.33. The comoving star-formation density $\rho_{\rm SFR}$ as a function of redshift, where the different symbols denote different indicators used for the determination of the star-formation rate. This plot, known as the "Madau diagram", shows the history of star formation in the Universe. Clearly visible is the decline for $z < 1$; towards higher redshifts, ρ^* seems to remain nearly constant. The curve is an empirical fit to the data

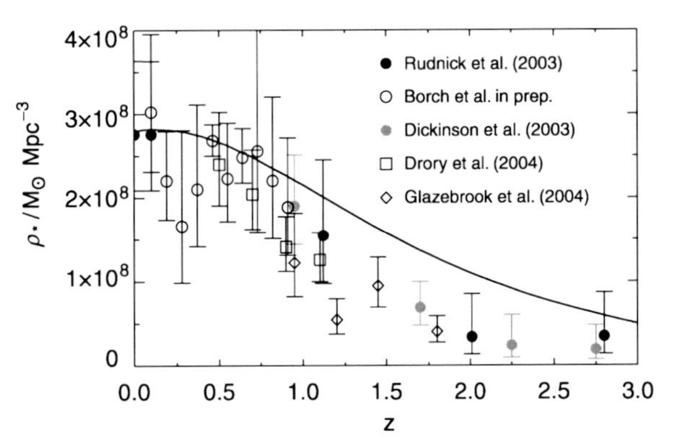

Fig. 9.34. Redshift evolution of the mass density in stars, as measured from various galaxy surveys. The solid curve specifies the integrated star-formation density from Fig. 9.33

A simple argument supports the idea that a significant fraction of cosmic star formation occurs in sources which are hidden from our view. From Fig. 9.25 we conclude that the energy density of the FIR background radiation is of the same order as that of the optical/UV background. The latter originates from star-forming regions, from which energetic optical/UV photons can escape and which can be observed as starbursts. In contrast to that, the CIB comes from the dusty regions in sub-mm sources, heated by hot and thus newly formed stars. The comparable energy density in these two radiation fields then indicates that both modes of star formation – with and without strong dust obscuration – are about equally abundant. In fact, more than half of cosmic star formation seems to hide in such dust regions.

The relative proportion of star formation in dusty regions seems to be a strong function of redshift. Whereas in the local Universe very luminous infrared galaxies are rare, their abundance increases rapidly with redshift, at least out to $z \sim 1$. The total comoving infrared luminosity density evolves as $\propto (1+z)^4$ out to $z \sim 1$, whereas the ultraviolet luminosity density increases more slowly, $\propto (1+z)^{2.5}$. Together, this then indicates that dusty and hidden star formation was even more important at higher redshifts than it is today, and it dominates the star-formation activity beyond $z \sim 0.7$.

Whereas most of the star formation in the local Universe occurs in spiral and irregular galaxies at a modest rate (so-called quiescent star formation), the star-formation activity at higher redshifts was dominated by bursts of star formation, as evidenced in the sub-mm galaxies and in LBGs. At a redshift $z \sim 1$, the latter has apparently ceased to dominate, yielding the strong decline of the star-formation rate density from then until today. This behavior may be expected if bursts of star formation are associated with the merging of galaxies; the merger rate declines strongly with time in models of the Universe dominated by a cosmological constant. This transition may also be responsible for the onset of the Hubble sequence of galaxy morphologies after $z \sim 1$.

The derivation of the star-formation rate as a function of redshift is largely drawn from galaxy surveys which are based on color selection, such as LBGs, EROs and sub-mm galaxies. The possibility cannot be excluded that additional populations of galaxies which are luminous but do not satisfy any of these photometric selection criteria are present at high redshift. Such galaxies can be searched for by spectroscopic surveys, extending to very faint magnitude limits. This opportunity now arises as several of the 10-m class telescopes are now equipped with high multiplex spectrographs which can thus take spectra of many objects at the same time. One of them is VIMOS at the VLT, another is DEIMOS on Keck. With both instruments, extensive spectroscopic surveys are being carried out on flux-limited samples of galaxies. Among the first results of these surveys is the finding that there are indeed more bright galaxies at redshift $z \sim 3$ than previously found, by about a factor of 2, leading to a corresponding correction of the star-formation rate at high redshifts. In a color–color diagram, these galaxies are preferentially located just outside the selection box for LBGs (see Fig. 9.4). Given that this selection box was chosen such as to yield a high reliability of the selected candidates, it is not very surprising that a non-negligible fraction of galaxies lying outside, but near to it are galaxies at high redshift with similar properties.

9.6 Galaxy Formation and Evolution

The extensive results from observations of galaxies at high redshift which were presented earlier might suggest that the formation and evolution of galaxies is quite well understood today. We are able to examine galaxies at redshifts up to $z \sim 6$ and therefore observe galaxies at nearly all epochs of cosmic evolution. This seems to imply that we can study the evolution of galaxies directly. However, this is true only to a certain degree. Although we have now found a large number of galaxies at nearly every redshift, the relation between galaxies at different redshifts is not easily understood. We cannot suppose that galaxies seen at different redshifts represent various subsequent stages of evolution of the same kind of galaxy. The main reason for this difficulty is that different selection criteria need to be applied to find galaxies at different redshifts.

We shall explain this point with an example. Actively star-forming galaxies with $z \gtrsim 2.5$ are efficiently detected by applying the Lyman-break criterion, but only those which do not experience much reddening by dust. Actively star-forming galaxies at $z \sim 1$ are discovered as EROs if they are sufficiently reddened by dust. The relation between these two galaxy populations depends, of course, on how large the fraction of galaxies is whose star-formation regions are enshrouded by dense dust. To determine this fraction, one would need to find Lyman-break galaxies at $z \sim 1$, or EROs at $z \sim 3$. Both observations are virtually impossible today, however. For the former this is because the Lyman break is then located in the UV domain of the spectrum and we have no sufficiently sensitive UV observatory available. For the latter it is because the rest wavelength corresponding to the observed R-band is so small that virtually no optical radiation from such objects would be visible, rendering spectroscopy of these objects impossible. In addition to this, there is the problem that galaxies with $1.3 \lesssim z \lesssim 2.5$ are difficult to discover because, for objects at those redshifts, hardly any spectroscopic indicators are visible in the optical range of the spectrum – both the 4000-Å break and the $\lambda = 3727$ Å line of [OII] are redshifted into the NIR, as are the Balmer lines of hydrogen, and the Lyman lines of hydrogen are located in the UV part of the spectrum. For these reasons, this range in redshift is also called the "redshift desert".[3] Thus, it is difficult to trace the individual galaxy populations as they evolve into each other at the different redshifts. Do the LBGs at $z \sim 3$ possibly represent an early stage of today's ellipticals (and the passive EROs at $z \sim 1$), or are they an early stage of spiral galaxies?

The difficulties just mentioned are the reasons why our understanding of the evolution of the galaxy population is only possible within the framework of models, with the help of which the different observational facts are being interpreted. We will discuss some aspects of such models in this section.

Another challenge for galaxy evolution models are the observed scaling relations of galaxy properties. We expect that a successful theory of galaxy evolution can predict the Tully–Fisher relation for spiral galaxies, the fundamental plane for ellipticals, as well as the tight correlation between galaxy properties and the central black hole mass. This latter point also implies that the evolution of AGNs and galaxies must be considered in parallel, since the growth of black hole mass with time is expected to occur via accretion, i.e., during phases of activity in the corresponding galaxies. The hierarchical model of structure formation implies that high-mass galaxies form by the merging of smaller ones; if the aforementioned scaling relations apply at high redshifts (and there are indications for this to be true, although with redshift-dependent pre-factors that reflect the evolution of the stellar population in galaxies), then the merging process must preserve the scaling laws, at least on average.

[3]Spectroscopy in the NIR is possible in principle, but the high level of night-sky brightness and, in particular, the large number of atmospheric transition lines renders spectroscopic observations in the NIR much more time consuming than optical spectroscopy.

9.6.1 Expectations from Structure Formation

Cosmological N-body simulations predict the evolution of the dark matter distribution as a function of redshift, in particular the formation of halos and their merger processes. At the beginning of the evolution, gas follows the dark matter. However, as soon as the gas becomes dense enough, physical effects like heating by dissipation, friction, and cooling start to play a prominent role. Since dark matter is not susceptible to these processes, the behavior and the spatial distribution of the two components begins to differ.

In a CDM model, halos of lower mass form first; only later can more massive halos form. This "bottom-up" scenario of structure formation follows from the shape of the power spectrum of density fluctuations, which itself is defined by the nature of dark matter – namely cold dark matter. The formation of halos of increasingly higher mass then happens by the merging of lower-mass halos. Such merging processes are directly observable; the Antennae (see Fig. 9.15) are only one very prominent example. Merging should be particularly frequent in regions where the galaxy density is high, in clusters of galaxies for instance. As shown in Fig. 6.45 for one cluster, a large number of such merging processes are detected in galaxy clusters at high redshift.

If the gas in a halo can cool efficiently, stars may form, as we previously discussed in the context of reionization. Since cooling is a two-body process, i.e., the cooling rate per volume element is $\propto \rho^2$, only dense gas can cool efficiently. One expects that the gas, having a finite amount of angular momentum like the dark matter halo itself, will initially accumulate in a disk, as a consequence of its own dissipation. The gas in the disk then reaches densities at which efficient star formation can set in. In this way, the formation of disk galaxies, thus of spirals, can be understood qualitatively.

However, the formation of disk galaxies by dissipational collapse of gas inside relaxed dark matter halos is not without problems. The hierarchical nature of structure growth implies that due to subsequent merging events, the disks can be significantly perturbed or even destroyed. Furthermore, the disks can lose angular momentum in the course of galaxy collisions. It is likely that an understanding of the formation of disk galaxies requires additional ingredients; for example, disks may form as a result of gas-rich mergers, where the resulting angular momentum of the baryons is sufficient to form a rotating and flat structure through dissipation.

9.6.2 Formation of Elliptical Galaxies

Properties of Ellipticals. Whereas the formation of disk galaxies can be explained qualitatively in a relatively straightforward way, the question of the formation of ellipticals is considerably more difficult to answer. Stars in ellipticals feature a very high velocity dispersion, indicating that the gas out of which they have formed cannot have kinematically cooled down beforehand into a disk by dissipation. On the other hand, it is hard to comprehend how star formation may proceed without gas compression induced by dissipation and cooling.

In Sect. 3.4.3 we saw that the properties of ellipticals are very well described by the fundamental plane. It is also found that the evolution of the fundamental plane with redshift can almost completely be explained by passive evolution of the stellar population in ellipticals. In the same way, we stated in Sect. 6.6 that the ellipticals in a cluster follow a very well-defined color–magnitude relation (the red cluster sequence), which suggests that the stellar populations of ellipticals at a given redshift all have a similar age. By comparing the colors of stellar populations in ellipticals with models of population synthesis, an old age for the stars in ellipticals is obtained, as shown in Fig. 3.49.

Monolithic Collapse. A simple model is capable of coherently describing these observational facts, namely monolithic collapse. According to this description, the gas in a halo is nearly instantaneously transformed into stars. In this process, most of the gas is consumed, so that no further generations of stars can form later. For all ellipticals with the same redshift to have nearly identical colors, this formation must have taken place at relatively high redshift, $z \gtrsim 2$, so that the ellipticals are all of essentially the same age. This scenario thus requires the formation of stars to happen quickly enough, before the gas can accumulate in a disk. The process of star formation remains unexplained in this picture, however.

Minor Mergers. We rather expect, according to the model of hierarchical structure formation, that massive galaxies form by the mergers of smaller entities. Consider what may happen in the merging of two halos or two galaxies, respectively. Obviously, the outcome of a merger depends on several parameters, like the relative velocity, the impact parameter, the angular momenta, and particularly the mass ratio of the two merging halos. If a smaller galaxy merges with a massive one, the properties of the dominating galaxy are expected to change only marginally: the dark halo gains slightly more mass from the companion, the stars of which are simply added to the stellar population of the massive galaxy. Such a "minor merger" is currently taking place in the Milky Way, where the Sagittarius dwarf galaxy is being torn apart by the tidal field of the Galaxy, and its stars are being incorporated into the Milky Way as an additional population. This population has, by itself, a relatively small velocity dispersion, forming a cold stream of stars that can also be identified as such by its kinematic properties. The large-scale structure of the Galaxy is nearly unaffected by a minor merger like this.

Major Mergers and Morphological Transformations of Galaxies. The situation is different in a merger process where both partners have a comparable mass. In such "major mergers" the galaxies will change completely. The disks will be destroyed, i.e., the disk population attains a high velocity dispersion and can transform into a spheroidal component. Furthermore, the gas orbits are perturbed, which may trigger massive starbursts like, e.g., in the Antenna galaxies. By means of this perturbation of gas orbits, the SMBH in

Fig. 9.35. The galaxy Centaurus A. The optical image is displayed in grayscales, the contours show the radio emission, and in red, an infrared image is presented, taken by the ISO satellite. The ISO map indicates the distribution of dust, which is apparently that of a barred spiral. It seems that this elliptical galaxy features a spiral that is stabilized by the gravitational field of the elliptical. Presumably, this galaxy was formed in a merger process; this may also be the reason for the AGN activity

the centers of the galaxies can be fed, initiating AGN activity, as is presumably seen in the galaxy Centaurus A shown in Fig. 9.35. Due to the violence of the interaction, part of the matter is ejected from the galaxies. These stars and the respective gas are observable as tidal tails in optical images or by the 21-cm emission of neutral hydrogen. From these arguments, which are also confirmed by numerical simulations, one expects that in a "major merger" an elliptical galaxy may form. In the violent interaction, the gas is either ejected, or heated so strongly that any further star formation is suppressed.

This scenario for the formation of ellipticals is expected from models of structure formation. Thus far it has been quite successful. For instance, it provides a straightforward explanation for the Butcher–Oemler effect (see Sect. 6.6), which states that clusters of galaxies at higher redshift contain a larger fraction of blue galaxies. Because of the particularly frequent mergers in clusters, due to the high galaxy density, such blue galaxies are transformed more and more into early-type galaxies. However, galaxies in clusters may also lose their gas in their motion through the hot intergalactic medium, by which the gas is ripped out due to the so-called ram pressure. The fact that the fraction of ellipticals in a cluster remains rather constant as a function of redshift, whereas the abundance of S0 galaxies increases with decreasing z, indicates the importance of the latter process as an explanation of the Butcher–Oemler effect. In this case, the gas of the disk is stripped, no further star formation takes place, and the spiral galaxy is changed into a disk galaxy without any current star formation – hence, into a galaxy that features the basic properties of S0 galaxies (see Fig. 9.36). On the other hand, we have seen in Sect. 3.2.5 that many ellipticals show signs of complex evolution which can be interpreted as the consequence of such mergers. Therefore, it is quite possible that the formation of ellipticals in galaxy groups happens by violent merger processes, and that these then contribute to the cluster populations by the merging of groups into clusters.

This model also has its problems, though. One of these is that merger processes of galaxies are also observed to occur at lower redshifts. Ellipticals formed in these mergers would be relatively young, which is hardly compatible with the above finding of a consistently old age of ellipticals. However, ellipticals are predominantly located in galaxy clusters whose members are already galaxies with a low gas content. In the merging process of such galaxies, the outcome will be an elliptical, but no starburst will be induced by merging because of the lack of gas – such mergers are sometimes called "dry mergers". In this context, we need to mention that the phrase "age of ellipticals" refers to the age of their stellar populations – the stars in the ellipticals are old, but not necessarily the galaxies themselves.

The importance of dry mergers was recognized more recently for a number of reasons. First, wide-field imaging with HST, using mosaics of single fields, have shown a large number of pairs of spheroidal galaxies at $z \lesssim 0.7$ which show signs of interactions. A dramatic example of this is also seen in Fig. 6.45, where several gravitationally bound pairs of early-type galaxies are seen in the outskirts of a cluster at $z = 0.83$. These pairs will merge on a time-scale of $\lesssim 1$ Gyr. Second, numer-

Fig. 9.36. An HST image of NGC 4650A, a polar-ring galaxy. Spectroscopy shows that the inner disk-like part of the galaxy rotates around its minor axis. This part of the galaxy is surrounded by a ring of stars and gas which is intersected by the polar axis of the disk and which rotates as well. Hence, the inner disk and the polar ring have angular momentum vectors that are pretty much perpendicular to each other; such a configuration cannot form from the "collapse" of the baryons in a dark matter halo. Instead, the most probable explanation for the formation of such special galaxies is a huge collision of two galaxies in the past. Originally the disk may have been the disk of the more massive of the two collision partners, whereas the less massive galaxy has been torn apart and its material has been forced into a polar orbit around the more massive galaxy. New stars have then formed in the disk, visible here in the bluish knots of bright emission. Since the polar ring is deep inside the halo of the other galaxy, the halo mass distribution can be mapped out to large radii using the kinematics of the ring

ical simulations indicate that gas-free mergers preserve the fundamental plane, in the sense that the merging of two ellipticals that live on the fundamental plane will lead to merger remnant that lies on there as well. Third, dry mergers may provide the explanation for the structural differences between high-luminosity and low-luminosity ellipticals. The former ones tend to have boxy isophotes and very little rotation, whereas the latter tend to have more disky isophotes and substantial rotational support (see Fig. 3.9). In this model, the disky ellipticals form from mergers of gas-rich disk galaxies; the merger remnants will then contain a population of young stars that were formed in the process of merging, and depending on the relative orientation of the angular momentum vectors of the two galaxies, the merged galaxy may have a substantial contribution of rotational support. In contrast, no new stars are formed in the merger of two early-type galaxies, preserving the age of the stellar population. Since early-type galaxies are on average more massive that late-type ones, the remnants of dry mergers are expected to be more massive on average than those of gas-rich mergers.[4]

Redshift Evolution of Ellipticals. The luminosity function of luminous red galaxies up to $z \sim 1$ is not much different from their local luminosity function, whereas that of blue galaxies shows a very strong evolution. We deduce from this that the formation of ellipticals is already concluded at a very early time. As also shown by the Madau diagram (Fig. 9.33), most of the star formation takes place at high redshifts, whereas the local Universe is rather quiet in comparison. In a Universe with low density, the evolution of the growth factor $D_+(z)$ over time (see Fig. 7.3) also implies that cosmic evolution will slow down considerably after $z \sim 1$, hence the majority of mergers happen at higher redshifts. In fact, the fraction of irregular and peculiar galaxies increases with decreasing magnitude, hence with increasing mean redshift.

Indeed, the evolution of massive ellipticals seems to have occurred even earlier than $z \sim 1$. Until recently, the radio galaxy presented in Fig. 6.44 was the most distant known elliptical galaxy, with a redshift of $z = 1.55$. In the K20 survey (see Sect. 9.2.2), four EROs were recently discovered with redshifts $1.6 \le z \le 1.9$. The spectroscopic verification of these objects is extremely challenging because they are very faint at optical wavelengths ($R \lesssim 24$), and furthermore, at these redshifts no distinct spectral signatures are visible in the optical. The redshifts of these four EROs were determined by a correlation of their spectra with the spectrum of the aforementioned galaxy LBDS 53W091 (Fig. 6.44). The spectra of these galaxies can, in the framework of population synthesis (Sect. 3.9), be explained by a stellar population with an age of about one to two billion years, and they are similar to the spectra of EROs at $z \sim 1$, with the exact age of the population depending on the assumed metallicity. HST images of these objects strongly indicate that their morphology also identifies them as early-type galaxies. The stellar mass in these galaxies,

[4] The fact that spectacular images of merging galaxies show mainly gas-rich mergers (such as in Fig. 9.15 or Fig. 1.13) can be attributed to selection effects. On the one hand, gas-rich mergers lead to massive star formation, yielding an increased luminosity of the systems, whereas dry mergers basically preserve the luminosity. On the other hand, gas-rich mergers can be recognized as such for a longer period of time than dry ones, owing to the clearly visible tidal tails traced by luminous newly formed stars.

derived from the NIR magnitude, is $M_* \gtrsim 10^{11} M_\odot$, so they are comparable to elliptical galaxies in the local Universe. Although this is only a small sample of objects, it is possible to estimate from them the density of massive early-type galaxies at these high redshifts. This yields a density which is comparable to that of massive star-forming galaxies at similar redshift. This means that at $z \sim 2$, not only were a large fraction of the stars which are visible today formed (Fig. 9.34), but also that a comparable fraction of the stars were already present at that time in the form of an old stellar population. The cosmic mass density of stars in these early-type galaxies is about 10% of the current density in systems with stellar masses above $\sim 10^{11} M_\odot$. The early appearance of massive ellipticals at such high redshifts, with number densities as observed, are difficult to explain by hierarchical models of galaxy evolution, which we will discuss next.

9.6.3 Semi-Analytic Models

One can try to understand the above qualitative arguments in greater detail and quantitatively. Note that this is not possible by means of a cosmological hydrodynamic simulation: the physical processes that determine the formation of stars in galaxies occur on very small length-scales, whereas the evolution of structures, which defines, e.g., the merger rate, happens on cosmological scales. Hence it is impossible to treat both scales together in a single simulation. Furthermore, the physical laws determining the behavior of gas (hydrodynamical processes such as shock fronts and friction; radiation processes) are too complicated to be modeled in a detailed simulation, except in those which are confined to a single galaxy. In addition, many of the gas processes are not understood sufficiently well to compute their effects from basic physical laws. Star formation is just one example of this, although arguably the most important one.

To make progress, we can parametrize the functional behavior of those processes which we are unable to describe with a quantitative physical model. To give one example, the star-formation rate in a galactic disk is expected (and observed) to depend on the local surface mass density $\Sigma_{\rm g}$ of gas in the disk. Therefore, the star-formation rate is parametrized in the form $\dot{M}_* = A\Sigma_{\rm g}^{\beta}$, and the parameters A and β adjusted by comparison of the model predictions with observations. Such *semi-analytic models* of galaxy formation and evolution have in recent years contributed substantially to our understanding and interpretation of observations. We will discuss some of the properties of these models in the following.

Merger Trees. In the CDM model, massive halos are formed by the merging of halos of lower mass. An extension to the Press–Schechter theory (see Sect. 7.5.2) allows us to compute the statistical properties of these merger processes of halos. By means of these, it is then possible to generate a statistical ensemble of merger histories for any halo of mass M today. Each individual halo is then represented by a merger tree (see Fig. 9.37). Alternatively, such merger trees can also be extracted from numerical simulations of structure formation, by following the mass assemble history of individual halos. The statistical properties of halos of mass M at redshift z are then obtained by analyzing the ensemble of merger trees. Each individual merger tree specifies

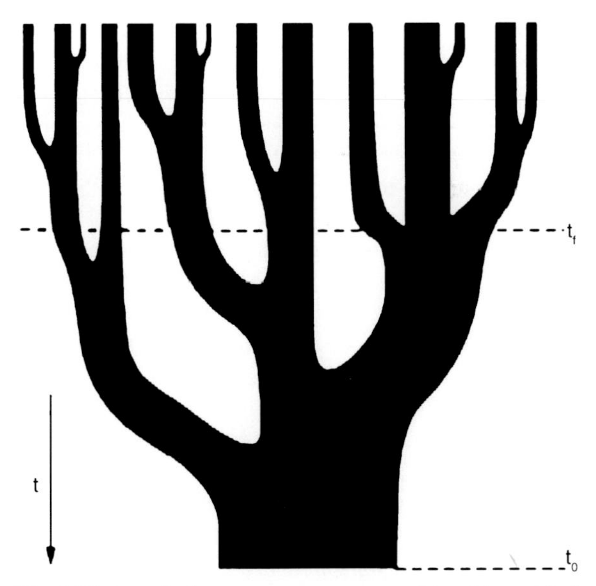

Fig. 9.37. A typical merger tree, as expected in a hierarchical CDM model of structure formation. The time axis runs from top to bottom. A massive halo at the present time t_0 has formed by mergers of numerous halos of lower mass, as indicated in the figure. One defines the time of halo formation as the time $t_{\rm f}$ at which one of the sub-halos had reached half the mass of the current halo

the merger processes that have led to the formation of a particular halo.

Cooling Processes and Star Formation. In a halo which does not undergo any merger process at a given time, gas can cool, where the cooling rate is determined by the chemical composition and the density of the gas. Besides atomic radiation, bremsstrahlung (free–free radiation) is also relevant for cooling, in particular at higher temperatures. If the density is sufficiently high and cooling is efficient, gas can be transformed into stars. Star formation is parametrized by a factor of proportionality between the star-formation rate and the rate at which gas cools. The newly formed stars are associated with a "disk component".

Feedback. Shortly after the formation of stars, the more massive of them will explode in the form of supernovae. This will re-heat the gas, since the radiation from the SN explosions and, in particular, the kinetic energy of the expanding shell, transfers energy to the gas. By this heating process, the amount of gas that can efficiently cool is reduced; this reduction increases with the star-formation rate. This leads to a self-regulation of star formation, which prevents all the gas in a halo from being transformed into stars. This kind of self-regulation by the feedback from supernovae (and, to some extent, also by the winds from the most massive stars) is also the reason why the star formation in our Milky Way is moderate, instead of all the gas in the disk being involved in the formation of stars.

Suppression of Low-Mass Galaxies. Besides heating by the feedback process described above, the gas in a halo can also be heated by intergalactic UV radiation which is produced by AGNs and starbursts, and which is responsible for maintaining the high ionization level in the intergalactic medium (see Sect. 8.5). This radiation has two effects on the gas: first, the gas is heated by photoionization due to the energetic photons, and second, the degree of ionization in the halo gas is increased. Both effects act in the same direction, by impeding an efficient cooling of the gas and hence the formation of stars. For halos of larger mass, intergalactic radiation is of fairly little importance because the corresponding heating rate is substantially smaller than that occurring by the dissipation of the gas. For low-mass halos, however, this effect is important. In this case, gas heating can be strong enough for the generated pressure to prevent the infall of gas into the gravitational potential of the dark halo. For this reason, one expects that halos of lower mass have a lower baryon fraction than that of the cosmic mixture, $f_\mathrm{b} = \Omega_\mathrm{b}/\Omega_\mathrm{m}$. The actual value of the baryon fraction depends on the details of the merger history of a halo. Quantitative studies yield an average baryon mass of

$$\overline{M}_\mathrm{b} = \frac{f_\mathrm{b}\, M}{\left[1+(2^{1/3}-1)M_\mathrm{C}/M\right]^3}\;, \tag{9.8}$$

where $M_\mathrm{C} \sim 10^9 M_\odot$ is a characteristic mass, defined such that for a halo with mass M_C, $\overline{M}_\mathrm{b}/M = f_\mathrm{b}/2$. For halos of mass smaller than M_C, the baryon fraction is suppressed, whereas for halo masses $\gg M_\mathrm{C}$, the baryon fraction corresponds to the cosmic average.

The low baryon fraction in low-mass halos is also expected because of the different shape of the halo mass spectrum in the Press–Schechter model, compared to the shape of the luminosity function of galaxies. The former is roughly $\propto M^{-2}$ for masses below M^*, whereas the galaxy luminosity function behaves like $\propto L^{-1}$. This different functional form is obviously not compatible with a constant mass-to-light ratio.

Another process for the suppression of baryons in low-mass halos is feedback; the transfer of kinetic energy from SN explosions to the gas can eject part of the gas from the potential well of the halo, and this effect becomes more efficient the smaller the binding energy of the gas is, i.e., the less massive the halo is. The suppression of the formation of low-mass galaxies by the effects mentioned here is a possible explanation for the apparent problem of CDM sub-structure in halos of galaxies discussed in Sect. 7.5.5. In this model, CDM sub-halos would be present, but they would be unable to have experienced an efficient star-formation history – hence, they would be dark.

Whereas the abundance of dark matter sub-halos in galaxies no longer presents a problem for CDM models of structure formation, the spatial distribution of satellite galaxies around the Milky Way requires more explanation. As we mentioned in Sect. 6.1.1, the 11 satellites of the Galaxy seem to form a planar distribution. Such a distribution would be extremely unlikely if the satellite population was drawn from a near-isotropic probability distribution. Therefore, the planar satellite distribution has been considered as a further potential problem for CDM-like models. However, using semi-analytic models of galaxy formation, combined with simulations of the large-scale structure, a different picture emerges. Since galaxies preferentially form in filaments of the large-scale structure, the accretion of smaller mass halos onto a high-mass halo occurs predominantly in the direction of the filament. The most massive sub-halos therefore tend to form a planar distribution, not unlike the one seen in the Milky Way's satellite distribution. The anisotropy of the distribution of massive satellites may also serve to explain the Holmberg effect.

Major Mergers. In the framework of semi-analytic models, a spheroidal stellar population may form in a "major merger", which may be defined in terms of the mass ratio of the merging halos (e.g., larger than 1:3) – the disk populations of the two merging galaxies are dynamically heated to commonly form an elliptical galaxy. The gas in the two components is heated by shocks to the virial temperature of the resulting halo, which suppresses future star formation.

Minor Mergers. If the masses of the two components in a merger are very different, the gas of the smaller component will basically be accreted onto the more massive halo, where it can cool again and form new stars. By this process, a new disk population may form. In this model, a spiral galaxy is created by forming a bulge in a "major merger" at earlier times, with the disk of stars and gas being formed later in minor mergers and by the accretion of gas. Hence the bulge of a spiral is, in this picture, nothing but a small elliptical galaxy, which is also suggested by the very similar characteristics of bulges and ellipticals, including the fact that both types of object seem to follow the same relation between black hole mass and velocity dispersion, as explained in Sect. 3.5.3.

The merging process of the two components does not occur instantaneously, but since the smaller galaxy will have, in general, a finite orbital angular momentum, it will first enter into an orbit around the more massive component. One example of this is the Sagittarius dwarf galaxy, but also the Magellanic Clouds will, in a distant future, merge with the Milky Way in this way. By dynamical friction, the satellite galaxy then loses its orbital

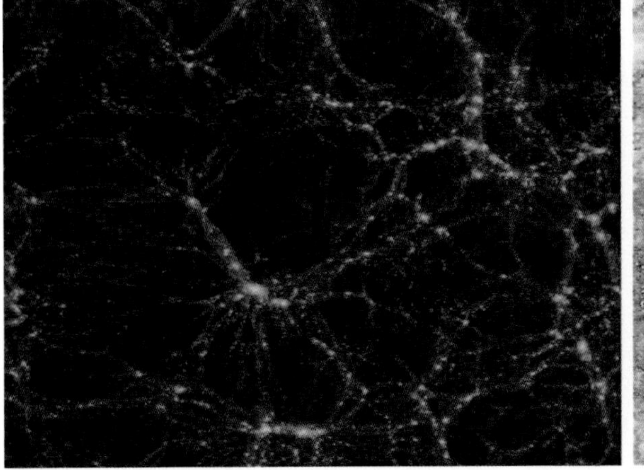

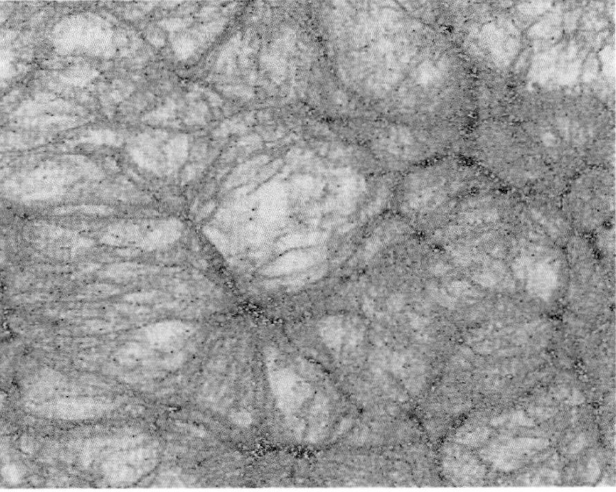

Fig. 9.38. On the left, the distribution of dark matter resulting from an N-body simulation is shown. The dark matter halos identified in this mass distribution were then modeled as the location of galaxy formation – the formation of halos and their merger history can be followed explicitly in the simulations. Semi-analytic models describe the processes which are most important for the gas and the formation of stars in halos, from which a model for the distribution of galaxies results. In the panel on the right, the resulting distribution of model galaxies is represented by colored dots, where the color indicates the spectral energy distribution of the respective galaxy: blue indicates galaxies with active star formation, red are galaxies which are presently not forming any new stars. The latter are particularly abundant in clusters of galaxies – in agreement with observations

energy, and the tidal component of the gravitational field removes stars, gas, and dark matter away from it; the Magellanic Stream (see Fig. 6.6) is presumably the result of such a process. Only after several orbits – the number of which depends on the initial conditions and on the mass ratio – will the satellite galaxy finally merge with the larger one.

Results from Semi-Analytic Models. The free parameters in semi-analytic models – such as the star-formation efficiency or the fraction of energy from SNe that is transferred into the gas – are fixed by comparison with some key observational results. For example, one requires that the models reproduce the correct normalization of the Tully–Fisher relation and that the number counts of galaxies match those observed. Although these models are too simplistic to trace the processes of galaxy evolution in detail, they are highly successful in describing the basic aspects of the galaxy population, and they are continually being refined. For instance, this model predicts that galaxies in clusters basically consist of old stellar populations, because here the merger processes were already concluded quite early in cosmic history. Therefore, at later times gas was no longer available for the formation of stars. Figure 9.38 shows the outcome of such a model in which the merger history of the individual halos has been taken straight from the numerical N-body simulation, hence the spatial locations of the individual galaxies are also described by these simulations.

By comparison of the results from such semi-analytic models with the observed properties of galaxies and their spatial distribution, the models can be increasingly refined. In this way, we obtain more realistic descriptions of those processes which are included in the models in a parametrized form. This comparison is of central importance for achieving further progress in our understanding of the complex processes that are occurring in galaxy evolution, which can neither be studied in detail by observation, nor be described by more fundamental simulations.

As a result of such models, the correlation function of galaxies as it is obtained from the Millennium simulation (see Sect. 7.5.3) is presented in Fig. 9.39, in comparison to the correlation function observed in the 2dFGRS. The agreement between the model and the observations is quite impressive; both show a nearly perfect power law. In particular, the correlation function of galaxies distinctly deviates from the correlation function of dark matter on small scales, implying a scale-dependent bias factor. The question arises as to which processes in the evolution of galaxies may produce such a perfect power law: why does the bias factor behave just such that ξ_g attains this simple shape. The answer is found by analyzing luminous and less luminous galaxies separately, or galaxies with and without active star formation – for each of these subpopulations of galaxies, ξ_g is *not* a power law. For this reason, the simple shape of the correlation function shown in Fig. 9.39 is probably a mere coincidence ("cosmic conspiracy").

Another result from such models is presented in Fig. 9.40, also from the Millennium simulation (see Sect. 7.5.3). Here, one of the most massive dark matter halos in the simulation box at redshift $z = 6.2$ is shown, together with the mass distribution in this spatial region at redshift $z = 0$. In both cases, besides the distribution of dark matter, the galaxy distribution is also

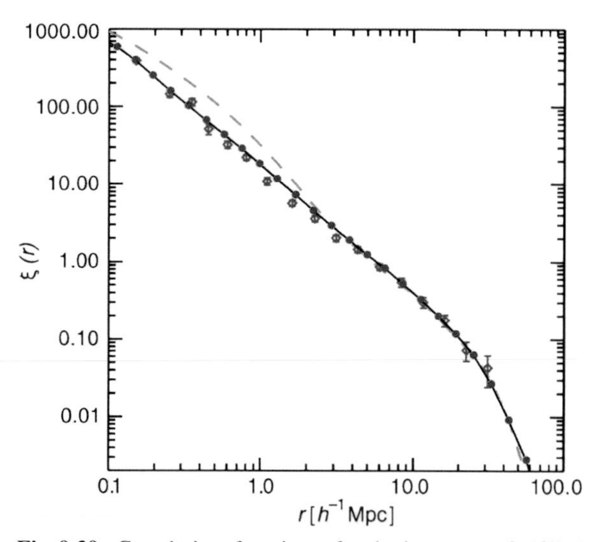

Fig. 9.39. Correlation function of galaxies at $z = 0$ (filled circles connected by the solid curve), computed from the Millennium simulation in combination with semi-analytic models of galaxy evolution. This is compared to the observed galaxy correlation function as derived from the 2dFGRS (diamonds with error bars). The dashed curve shows the correlation function of matter

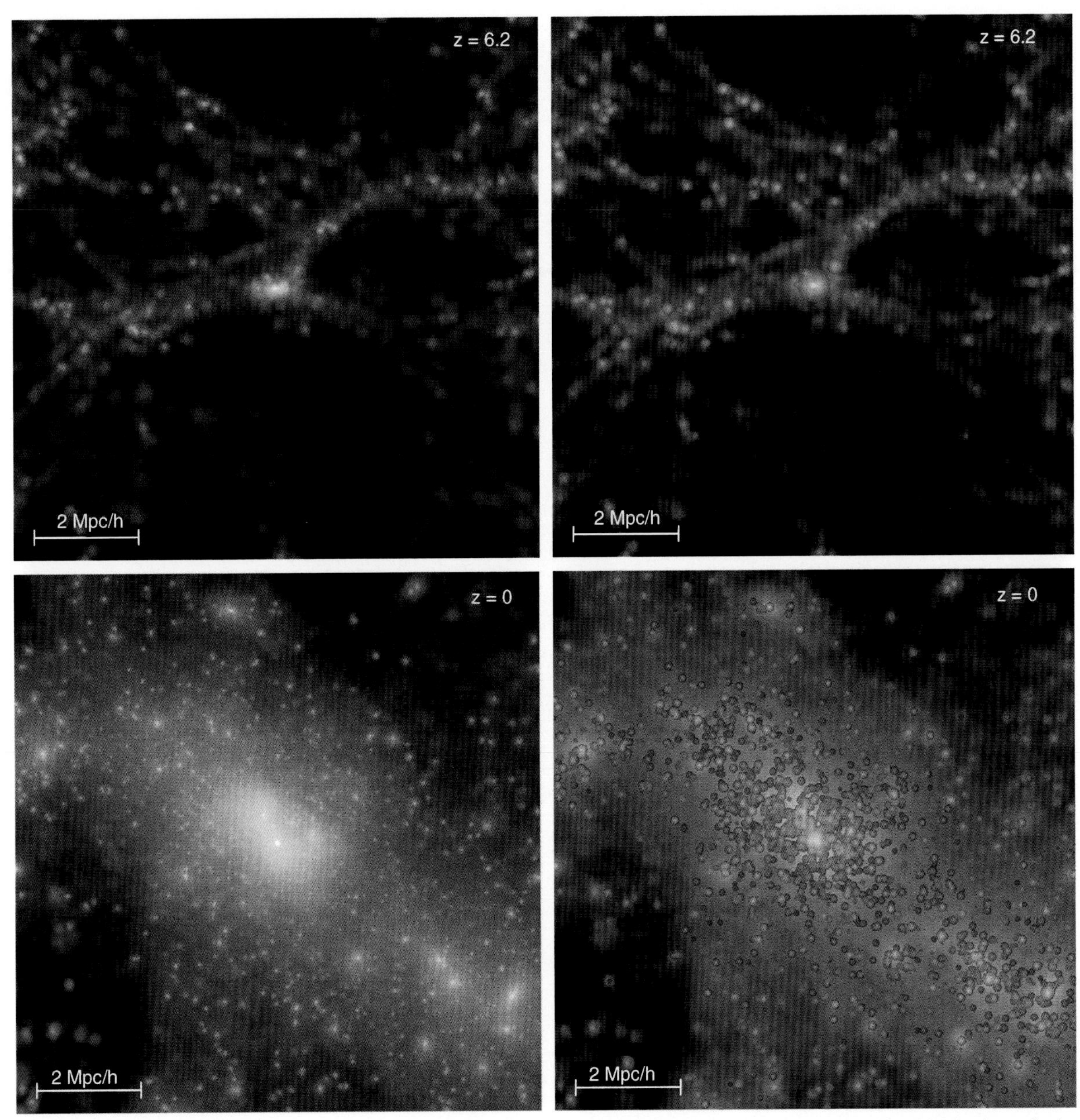

Fig. 9.40. In the top panels, one of the most massive halos at $z = 6.2$ from the Millennium simulation (see Fig. 7.12) is shown, whereas in the bottom panels, the corresponding distribution in this spatial region at $z = 0$ is shown. Thus, this early massive halo is now located in the center of a very massive galaxy cluster. In the panels on the left, the mass distribution is displayed, with the corresponding distribution of galaxies as determined from a semi-analytic model, which is shown in the right-hand panels. Galaxies at $z = 6.2$ are all blue since their stellar population must be young, whereas at $z = 0$, most galaxies contain an old stellar population, here indicated by the red color. Each of the panels shows the projected distribution in a cube with a comoving side length of $10h^{-1}$ Mpc

displayed, obtained from semi-analytic models. Massive halos which have formed early in cosmic history are currently found predominantly in the centers of very massive galaxy clusters. Assuming that the luminous QSOs at $z \sim 6$ are harbored in the most massive halos of this epoch, we can deduce that these may today be identified as the central galaxies in clusters. This may provide an explanation as to why so many central, dominating cluster galaxies show AGN activity, though with a smaller luminosity due to small accretion rates.

Abundance and Evolution of Supermassive Black Holes. Within the framework of these models, predictions are also made about the statistical evolution of SMBHs in the cores of galaxies. When two galaxies merge, their SMBHs will also coalesce after some time, where an accurate estimate for this time-scale is difficult to obtain. Owing to the high initial orbital angular momentum, the two SMBHs are, at the beginning of a merger, on an orbit with rather large mutual separation. By dynamical friction (see Sect. 6.2.6), caused by the matter distribution in the newly formed galaxy, the pair of SMBHs will lose orbital energy after the merger of the galaxies, and the two black holes will approach each other. Since this process takes a relatively long time, and since a massive galaxy will, besides a few major mergers, undergo numerous minor mergers, it is conceivable that many of the black holes that were originally the nuclei of low-mass satellite galaxies are today still on orbits at relatively large distances from the center of galaxies.

In this model, phases in the evolution of galaxies exist in which two SMBHs are located close to the center. Indeed, there are a number of indications that such galaxies are actually observed. For instance, galaxies with two active nuclei have been found. Also a class of radio sources exists with an X-shaped morphology (instead of the usual bipolar radio structure), which can be interpreted as pairs of active SMBHs. Another signature of a binary system of SMBHs would be a periodicity in the emission, reflecting the orbital period. In some AGNs, periodic variations in the brightness have in fact been detected, the blazar OJ 287 being the best known example, with a period of 11.86 years.

When two SMBHs merge, the initially wide orbit shrinks, in the final stages due to the emission of gravitational waves. This will cause the orbits to become more circular, as well as a decrease of the separation of the two black holes. According to the theory of black holes, there is a closest separation at which an orbit still is stable. Once the separation has shrunk to that size, the merging occurs, accompanied by a burst of gravitational wave emission. If the two SMBHs have the same mass, each of them will emit the same amount of gravitational wave energy, but in opposite directions, so that the net amount of momentum carried away by the gravitational waves is zero. However, if the masses are not equal, this cancellation no longer occurs, and the waves carry away a net linear momentum. According to momentum conservation, this will yield a recoil to the merged SMBH, and it will therefore move out of the galactic nucleus. Depending on the recoil velocity, it may return to the center in a few dynamical time-scales. However, if the recoil velocity is larger than the escape velocity from the galaxy, it may actually escape from the gravitational potential and become an intergalactic black hole. The importance of this effect is not quantitatively known, since the amplitude of the recoil velocity as determined from theoretical models is uncertain. It is zero for equal masses, and very small if one of the black hole masses is much smaller than the other. The recoil velocity attains a maximum value when the mass ratio of the two SMBHs is about $1/3$.

The mass increase of SMBHs in the course of cosmic history then has two different origins, first the merging with other low-mass SMBHs as a consequence of merger events, and second the accretion of gas that leads to the activity of SMBHs. Hierarchical models of galaxy evolution with central SMBHs are able to both reproduce, under certain assumptions, the correlation (see Sect. 3.5.3) between the SMBH mass and the properties of the spheroidal stellar component, and to successfully model the integrated AGN luminosity and the redshift-dependent luminosity function of AGNs.

In the course of the merger of two SMBHs, an intense emission of gravitational waves will occur in the final phase. The space project LISA, which is planned to be launched sometime after 2013, is capable of observing the emission of gravitational waves from such merger processes, essentially throughout the visible Universe. Hence, it will become possible to directly trace the merger history of galaxies.

9.6.4 Cosmic Downsizing

The hierarchical model of structure formation predicts that smaller-mass objects are formed first, with more massive systems forming later in the cosmic evolution. As discussed before, there is ample evidence for this to be the case; e.g., galaxies are in place early in the cosmic history, whereas clusters are abundant only at redshifts $z \lesssim 1$. However, looking more closely into the issue, apparent contradictions are discovered. For example, the most massive galaxies in the local Universe, the massive ellipticals, contain the oldest population of stars, although their formation should have occurred later than those of less massive galaxies. In turn, most of the star formation in the local Universe seems to be associated with low- or intermediate-mass galaxies, whereas the most massive ones are passively evolving. Now turning to high redshift: for $z \sim 3$, the bulk of star formation seems to occur in the LBGs, which, according to their clustering properties (see Sect. 9.1.1), are associated with high-mass halos. The study of passively evolving EROs indicates that massive old galaxies were in place as early as $z \sim 2$, hence they must have formed very early in the cosmic history. The phenomenon that massive galaxies form their stars in the high-redshift Universe, whereas most of the current star formation occurs in galaxies of lower mass, has been termed "downsizing".

This downsizing can be studied in more detail using redshift surveys of galaxies. The observed line width of the galaxies yields a measure of the characteristic velocity and thus the mass of the galaxies (and their halos). Such studies have been carried out in the local Universe, showing that local galaxies have a bimodal distribution in color (see Sect. 3.7.2), which in turn is related to a bimodal distribution in the specific star-formation rate. Extending such studies to higher redshifts, by spectroscopic surveys at fainter magnitudes, we can study whether this bimodal distribution changes over time. In fact, such studies reveal that the characteristic mass separating the star-forming galaxies from the passive ones evolves with redshift, such that this dividing mass increases with z. For example, this characteristic mass decreased by a factor of ~ 5 between $z = 1.4$ and $z = 0.4$. Hence, the mass scale above which most galaxies are passively evolving decreases over time, restricting star formation to increasingly lower-mass galaxies.

Studies of the fundamental plane for field ellipticals at higher redshift also point to a similar conclusion. Whereas the massive ellipticals at $z \sim 0.7$ lie on the fundamental plane of local galaxies when passive evolution of their stellar population is taken into account, lower-mass ellipticals at these redshifts have a smaller mass-to-light ratio, indicating a younger stellar population. Also here, the more massive galaxies seem to be older than less massive ones.

Another problem with which models of galaxy formation are faced is the absence of very massive galaxies today. The luminosity function of galaxies is described reasonably well by a Schechter luminosity function, i.e., there is a luminosity scale L_* above which the number density of galaxies decreases exponentially. Assuming plausible mass-to-light ratios, this limiting luminosity translates into a halo mass which is considerably lower than the mass scale $M_*(z = 0)$ above which the abundance of dark matter halos is exponentially cut-off. In fact, the shape of the mass spectrum of dark matter halos is quite different from that of the stellar mass (or luminosity) spectrum of galaxies. Why, then, is there some kind of maximum luminosity (or stellar mass) for galaxies? It has been suggested that the value of L_* is related to the ability of gas in a dark matter halo to cool; if the mass is too high, the corresponding virial temperature of the gas is large and the gas density low, so that the cooling times are too large to make gas cooling, and thus star formation, efficient. With a relatively high cosmic baryon density of $\Omega_b = 0.045$, however, this argument fails to provide a valid quantitative explanation.

The clue to the solution of these problems may come from the absence of cooling flows in galaxy clusters. As we saw in Sect. 6.3.3, the gas density in the inner regions of clusters is large enough for the gas to cool in much less than a Hubble time. However, in spite of this fact, the gas seems to be unable to cool, for otherwise the cool gas would be observable by means of intense line radiation. This situation resembles that of the massive galaxies: if they were already in place at high redshifts, why has additional gas in their halos (visible, e.g., through its X-ray emission, and expected in structure formation models to accrete onto the host halo) not cooled and formed stars? The solution for this problem in galaxy clusters was the hypothesis that

AGN activity in their central galaxy puts out enough energy to heat the gas and prevent it from cooling to low temperatures. Direct and indirect evidence for the validity of this hypothesis exists, as explained in Sect. 6.3.3.

A similar mechanism may occur in galaxies as well. We have learned that galaxies host a supermassive black hole whose mass scales with the velocity dispersion of the spheroidal stellar component and thus, in elliptical galaxies, with the mass of the dark matter halo. If gas accretes onto these halos, it may cool and, on the one hand, form stars; on the other hand, this process will lead to accretion of gas onto the central black hole and make it active again. This activity can then heat the gas and thus prevent further star formation. If the time needed to cool the gas and form stars is shorter than the free-fall time to the center of the galaxy, stars can form before the AGN activity is switched on. In the opposite case, star formation is prevented. A quantitative analysis of these two time-scales shows that they are about equal for a halo of mass $\sim 2 \times 10^{11} h^{-1} M_\odot$, about the right mass-scale for explaining the cut-off luminosity L_* in the Schechter function.

Accounting for the AGN feedback explicitly in semi-analytic models of galaxy evolution provides a good match to the observed galaxy luminosity function in the current Universe. Furthermore, such models closely reproduce the observed evolution of the stellar mass function, which provides a framework for understanding the "downsizing" phenomenon. In fact, since the mass of the central black hole was accumulated by accretion, and since the total energy output that can be generated in the course of growing a black hole to a mass of $\sim 10^8 M_\odot$ is very large, it should not be too surprising that this nuclear activity has a profound impact on the galaxy hosting the SMBH. The fact that the hosts of luminous QSOs show no signs of strong star formation may be another indication that the AGN luminosity prevents efficient star formation in its local environment.

9.7 Gamma-Ray Bursts

Discovery and Phenomenology. In 1968, surveillance satellites for the monitoring of nuclear test ban treaties discovered γ-flashes similar to those that are observed in nuclear explosions. However, these satellites found that the flashes were not directed from Earth but from the opposite direction – hence, these γ-flashes must be a phenomenon of cosmic origin. Since the satellite missions were classified, the results were not published until 1973. The sources were named *gamma-ray bursts* (GRB).

The flashes are of very different duration, from a few milliseconds up to ~ 100 s, and they differ strongly in their respective light curves (see Fig. 9.41). They are observed in an energy range from ~ 100 keV up to several MeV, sometimes to even higher energies.

The nature of GRBs had been completely unclear initially, because the positional accuracy of the bursts as determined by the satellites was far too large to allow an identification of any corresponding optical source. The angular resolution of these γ-detectors was many degrees (for some, a 2π solid angle). A more precise position was determined from the time of arrival of the bursts at the location of several satellites, but the error box was still too large to search for counterparts of the source in other spectral ranges.

Early Models. The model favored for a long time included accretion phenomena on neutron stars in our Galaxy. If their distance was $D \sim 100$ pc, the corresponding luminosity would be about $L \sim 10^{38}$ erg/s, thus about the Eddington luminosity of a neutron star. Furthermore, indications of absorption lines in GRBs at about 40 keV and 80 keV were found, which were interpreted as cyclotron absorption corresponding to a magnetic field of $\sim 10^{12}$ Gauss – again, a characteristic value for the magnetic field of neutron stars. Hence, most researchers before the early 1990s thought that GRBs occur in our immediate Galactic neighborhood.

The Extragalactic Origin of GRBs. A fundamental breakthrough was then achieved with the BATSE experiment on-board the Compton Gamma Ray Observatory, which detected GRBs at a rate of about one per day over a period of nine years. The statistics of these GRBs shows that GRBs are isotropically distributed on the sky (see Fig. 9.42), and that the flux distribution $N(>F)$ clearly deviates, at low fluxes, from the $F^{-1.5}$-law. These two results meant an end to those models that had linked GRBs to neutron stars in our Milky Way, which becomes clear from the following argument.

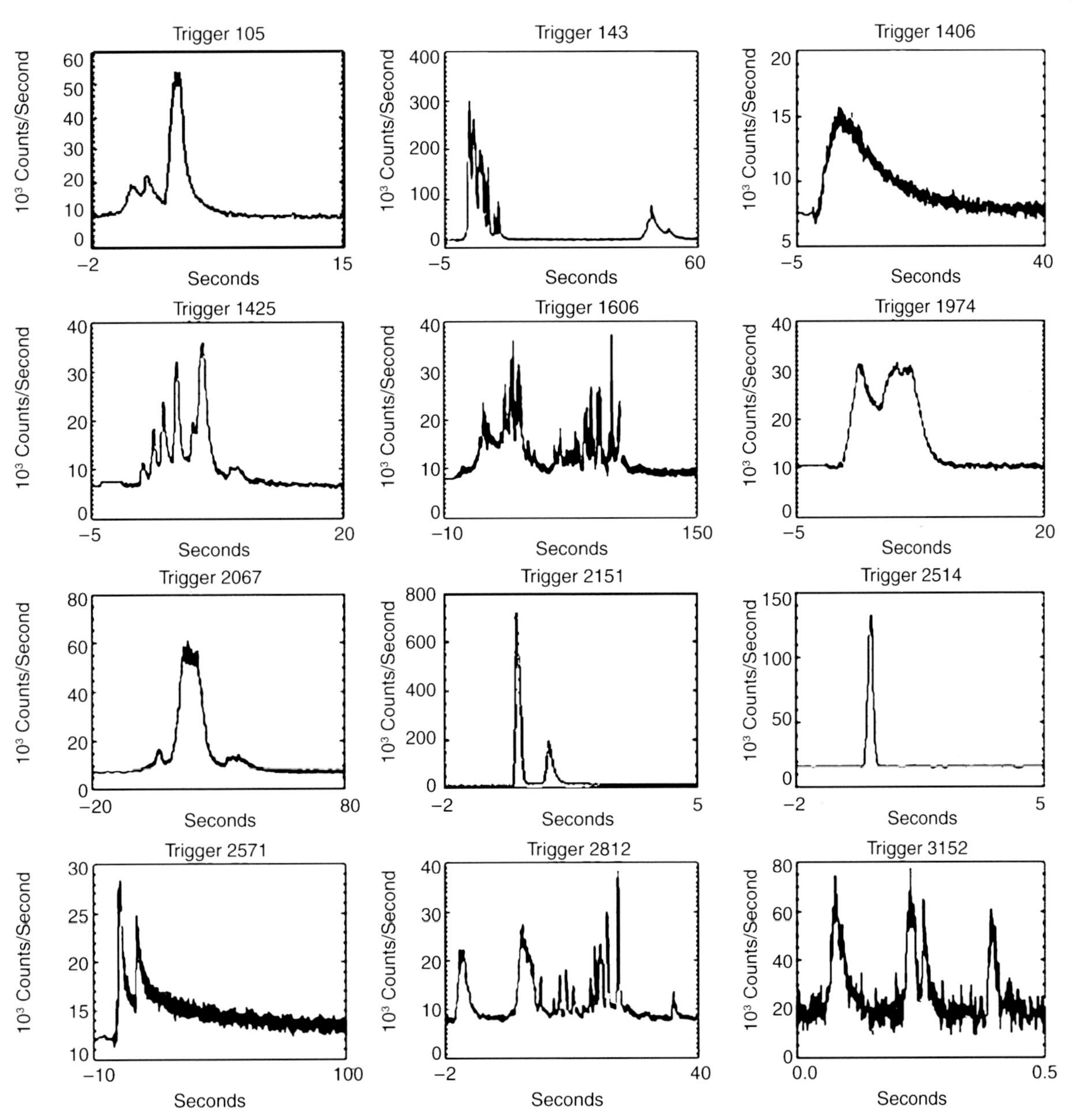

Fig. 9.41. Gamma-ray light curves of various gamma-ray bursts; the different time-scales on the x-axis should be particularly noted. All these light curves appear to be very dissimilar

Neutron stars are concentrated towards the disk of the Galaxy, hence the distribution of GRBs should feature a clear anisotropy – except for the case that the typical distance of the sources is very small ($\lesssim 100$ pc), much smaller than the scale-height of the disk. In the latter case, the distribution might possibly be isotropic, but the flux distribution would necessarily have to follow the law $N(>F) \propto F^{-3/2}$, as expected for a homogeneous distribution of sources, which was discussed in Sect. 4.1.2. Because this is clearly not the case, a dif-

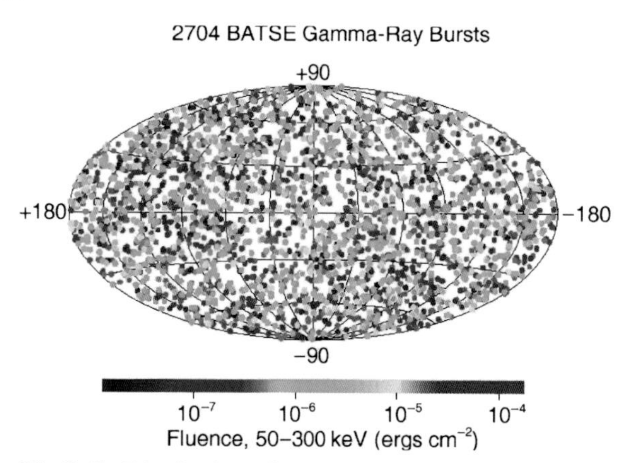

Fig. 9.42. Distribution of gamma-ray bursts on the sphere as observed by BATSE, an instrument on-board the CGRO-satellite, during the about nine year mission; in total, 2704 GRBs are displayed. The color of the symbols represents the observed strength (fluence, or energy per unit area) of the bursts. One can see that the distribution on the sky is isotropic to a high degree

ferent distribution of sources is required, hence also a different kind of source.

The only way to obtain an isotropic distribution for sources which are typically more distant than the disk scale-height is to assume sources at distances considerably larger than the distance to the Virgo Cluster, hence $D \gg 20\,\mathrm{Mpc}$; otherwise, one would observe an overdensity in this direction. In addition, the deviation from the $N(>F) \propto F^{-3/2}$-law means that we observe sources up to the edge of the distribution (or, more precisely, that the curvature of spacetime, or the cosmic evolution of the source population, induces deviations from the Newtonian counts), so that the typical distance of GRBs should correspond to a relatively high redshift. This implies that the total energy in a burst has to be $E \sim 10^{51}$ to 10^{54} erg. This energy corresponds to the rest mass Mc^2 of a star. The major part of this energy is emitted within ~ 1 s, so that GRBs are, during this short time-span, more luminous than all other γ-sources in the Universe put together.

Identification and Afterglows. In February 1997, the first identification of a GRB in another wavelength band was accomplished by the X-ray satellite Beppo-SAX. Within a few hours of the burst, Beppo-SAX observed the field within the GRB error box and discovered a transient source, by which the positional accuracy was increased to a few arcminutes. In optical observations of this field, a transient source was then detected as well, very accurately defining the position of this GRB. The optical source was identified with a faint galaxy. Optical spectroscopy of the source revealed the presence of absorption features at redshift $z = 0.835$; hence, this GRB must have a redshift equal or larger than this. For the first time, the extragalactic nature of GRBs was established. In fast progression, other GRBs could be identified with a transient optical source, and some of them show transient radiation also at other wavelengths, from the radio band up to X-rays. The lower-energy radiation of a GRB after the actual burst in gamma-rays is called an *afterglow*.

GRBs can be broadly classified into short- and long-duration bursts, with a division at a duration of $t_{\rm burst} \sim 2$ s. The spectral index of the short-duration bursts is considerably harder at γ-ray energies that that of long-duration bursts. Until 2005, only afterglows from long-duration bursts had been discovered. Long-duration bursts occur in galaxies at high redshift, typically $z \sim 1$ or higher, with the highest-redshift burst identified to date having a redshift of $z = 6.3$. In one case, an optical burst was discovered about 30 seconds after the GRB, with the fantastic brightness of $V \sim 9$ mag, at a redshift of $z = 1.6$. For a short period of time, this source was apparently more luminous than any quasar in the Universe. Thus, during or shortly after the burst at high energies, GRBs are also very bright in the optical.

Fireball Model. Whereas the distance of the sources, and therefore also their luminosity, was then known, the question of the nature of GRBs still remained unanswered. One model of GRBs quite accurately describes their emission characteristics, including the afterglow. In this fireball model, the radiation is released in the relativistic outflow of electron–positron pairs with a Lorentz factor of $\gamma \geq 100$. However, different hypotheses exist as to how this fireball is produced, hence what the physical origin of a GRB might be. One of these states that a GRB is caused by the merger of two neutron stars, or a neutron star and a black hole. In this case, the emission will probably be highly anisotropic, so that estimates of the luminosity from the observed flux, based on the assumption of isotropic emission, may not be correct.

Hypernovae. At the present time, the sources of long-duration GRBs have pretty much been identified. With the accurate location obtained from observations of their afterglows, it was found that they occur in star-forming galaxies, not in passive early-type galaxies. This finding is similar to that of core-collapse supernovae which are also found only in galaxies with star-formation activity. It was therefore speculated that the origin of GRBs is closely linked to star formation. The discovery of a coincidence of several GRBs with supernova explosions suggests that long-duration GRBs are extraordinarily energetic explosions of stars, so-called hypernovae. Even if the emission is highly anisotropic, as expected from the fireball model, the corresponding energy released by the hypernovae is very large.

For the purpose of identifying GRBs, the SWIFT satellite was launched in November 2004. This satellite is equipped with three instruments: a wide-field gamma-ray telescope to discover the GRBs, an X-ray telescope, and a UV-optical telescope. Within a few seconds of the discovery of a GRB, the satellite targets the location of the burst, so that it can be observed by the latter two telescopes, obtaining an accurate position. This information is then immediately transmitted to the ground, where other telescopes can follow the afterglow emission. SWIFT is expected to discover about 100 GRBs per year and to obtain significantly improved statistics of their afterglow light curves and redshifts. Already in its first year of operation, a large number of GRBs were found by SWIFT, including the one at $z = 6.3$.

Counterparts of Short-Duration GRBs. SWIFT has allowed the identification of four short-duration GRBs in 2005. In contrast to the long-duration bursts, some of these seem to be associated with elliptical galaxies; this essentially precludes any association with supernova explosions. In fact, for one of these short burst, very sensitive limits on the optical brightness explicitly rules out any contribution from a supernova explosion. Furthermore, the host galaxies of short bursts are at substantially lower redshift, $z \sim 0.2$. Given that both kinds of GRBs have about the same observed flux (or energy), this implies that short-duration bursts are less energetic than long-duration ones, by approximately two orders of magnitude. All of these facts clearly indicate that short- and long-duration GRBs are due to different populations of sources. The lower energies of short bursts and their occurrence in early-type galaxies with old stellar populations are consistent with them being due to the merging of compact objects, either two neutron stars, or a neutron star and a black hole.

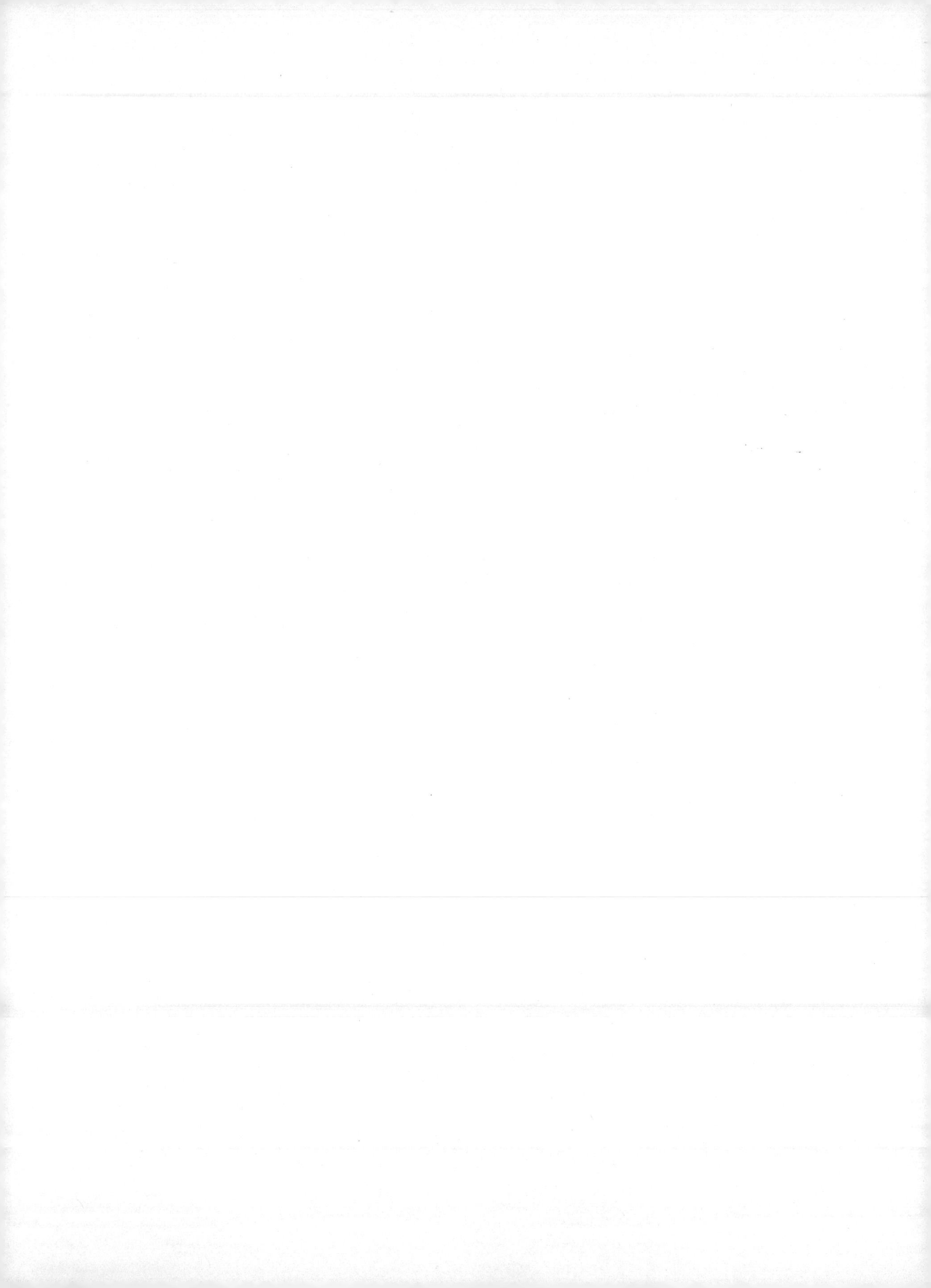

10. Outlook

In the final chapter of this book, we will dare to give an outlook for the fields of extragalactic astronomy and cosmology for the next few years from the perspective of 2006.

Progress in (extragalactic) astronomy is achieved through information obtained from increasingly improving instruments and by refining our theoretical understanding of astrophysical processes, which in turn is driven by observational results. It is easy to foresee that the evolution of instrumental capabilities will continue rapidly in the near future, enabling us to perform better and more detailed studies of cosmic sources. A few examples illustrating this statement will be given here. The size of optical wide-field cameras had reached a value of $\sim (20\,000)^2$ pixels by 2003 with the installment of Megacam at the CFHT. This multi-chip camera allows the mapping of one square degree of the sky in a single exposure and, with a pixel size of $0\farcs2$, it is well matched to the excellent seeing conditions typically met on Mauna Kea (see Fig. 10.1). Additional instruments with similar characteristics have been recently finished or are about to be commissioned. One of them is OmegaCAM, a square-degree camera at the newly-built VLT Survey Telescope on Paranal. Furthermore, the development of NIR detectors is rapid, and soon wide-field cameras in the NIR regime will be considerably larger than current ones. For instance, in 2007 the new 4-meter telescope VISTA will go into operation on Paranal, which will be equipped initially with a single instrument, a wide-field NIR camera. The combination of deep and wide optical and NIR images will no doubt lead to great strides in astronomy. For example, in the field of galaxy surveys, accurate photometric redshifts will become available. The same holds true for weak gravitational lensing or the search for very rare objects, for which surveying large regions of the sky is obviously necessary.

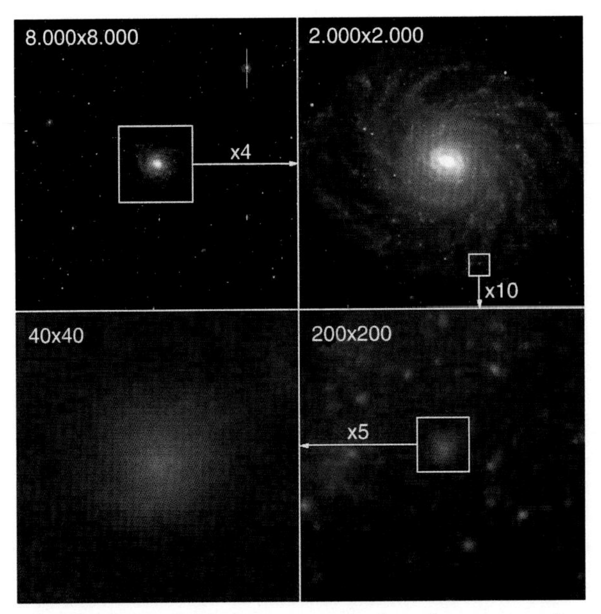

Fig. 10.1. Wide-field cameras, attached to telescopes on sites with excellent atmospheric conditions, can obtain detailed images of a large number of objects simultaneously. This is illustrated here with the CFH12K camera at the CFHT. Numbers in each panel, which show subsequent enlargements, denote the number of pixels displayed, where the pixel size is $0\farcs2$

Within only a decade, the total collecting area of large optical telescopes has increased by a large factor, as is illustrated in Fig. 10.2. At the present time, about 10 telescopes of the 10-meter class are in operation, the first of which, Keck I, was put into operation in 1993. In addition, the development of adaptive optics will allow us to obtain diffraction-limited angular resolution from ground-based observations (see Fig. 10.3).

In another step to improve angular resolution, optical and NIR interferometry will increasingly be employed. For example, the two Keck telescopes (Fig. 1.28) are mounted such that they can be used for interferometry. The four unit telescopes of the VLT can be combined, either with each other or with additional (auxiliary) smaller telescopes, to act as an interferometer (see Fig. 1.31). The auxiliary telescopes can be placed at different locations, thus yielding different baselines and thereby increasing the coverage in angular resolution. Finally, the Large Binocular Telescope (LBT), which consists of two 8.4-meter telescopes mounted on the same platform, was developed and constructed for the specific purpose of optical and NIR interferometry and had first light in October 2005. Once in operation, expected to occur by the end of 2006, this

Peter Schneider, Outlook.
In: Peter Schneider, Extragalactic Astronomy and Cosmology. pp. 407–414 (2006)
DOI: 10.1007/11614371_10

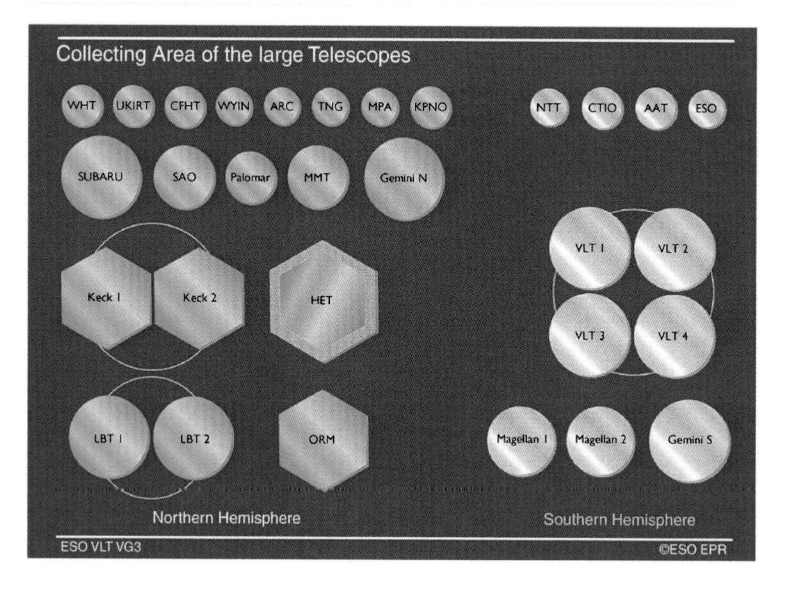

Fig. 10.2. The collecting area of large optical telescopes is displayed. Those in the Northern hemisphere are shown on the left, whereas southern telescopes are shown on the right. The joint collecting area of these telescopes has been increased by a large factor over the past decade: only the telescopes shown in the upper row plus the 5-meter Palomar telescope and the 6-meter SAO were in operation before 1993. If, in addition, the parallel development of detectors is considered, it is easy to understand why observational astronomy is making such rapid progress

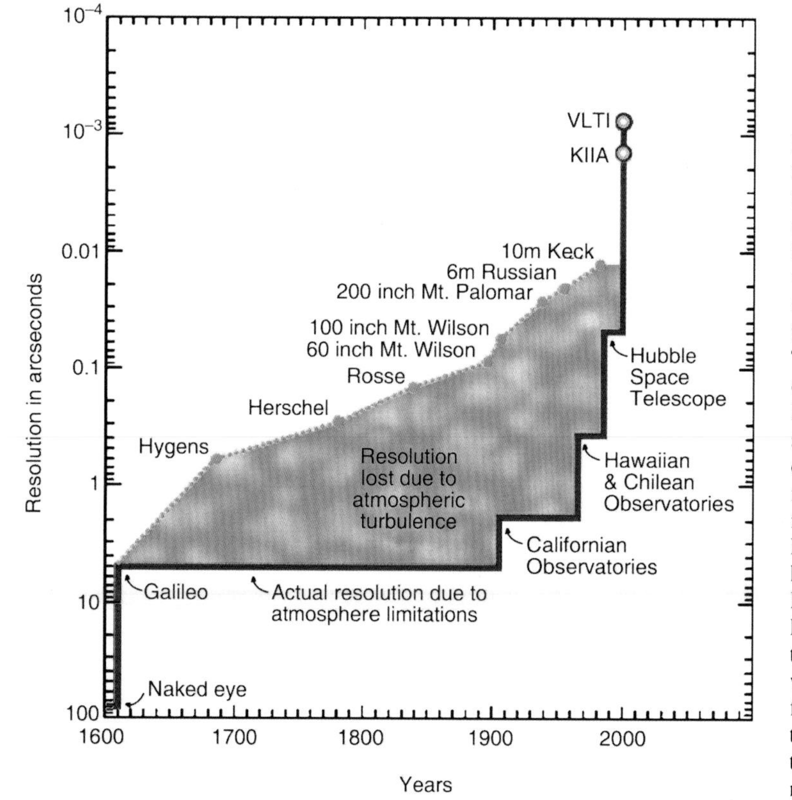

Fig. 10.3. This figure illustrates the evolution of angular resolution as a function of time. The upper dotted curve describes the angular resolution that would be achieved in the case of diffraction-limited imaging, which depends, at fixed wavelength, only on the aperture of the telescope. Some historically important telescopes are indicated. The lower curve shows the angular resolution actually achieved. This is mainly limited by atmospheric turbulence, i.e., seeing, and thus is largely independent of the size of the telescope. Instead, it mainly depends on the quality of the atmospheric conditions at the observatories. For instance, we can clearly recognize how the opening of the observatories on Mount Palomar, and later on Mauna Kea, La Silla and Paranal have lead to breakthroughs in resolution. A further large step was achieved with HST, which is unaffected by atmospheric turbulence and is therefore diffraction limited. Adaptive optics and interferometry will characterize the next essential improvements

telescope will start a new era in high-resolution optical astronomy.

The Hubble Space Telescope has turned out to be the most successful astronomical observatory of all time (although it certainly was also the most expensive one). This success can be explained predominantly by its angular resolution, which is enormously superior compared to ground-based telescopes, and by the significantly reduced night-sky brightness, in particular at longer wavelengths. The importance of HST for extragalactic astronomy is not least based on the characteristics of galaxies at high redshifts. Before the launch of HST, it was not known that such objects are small and therefore have, at a given flux, a high surface brightness. This demonstrates the advantage of the high resolution that is achieved with HST. Several service missions to the observatory led to the installment of new and more powerful instruments which have continuously improved the capacity of HST. At present, the future of HST is very uncertain. After the fatal disaster of the Space Shuttle Challenger, NASA initially canceled the next planned servicing mission; this mission is vital for HST since its gyroscopes need to be replaced. In addition, this servicing mission was scheduled to bring two new powerful instruments on-board, further increasing the scientific capabilities of HST. At present (2006) it is unclear whether this servicing mission will be launched, thereby prolonging the lifetime of HST for several more years and bridging the time until JWST will be launched (see below).

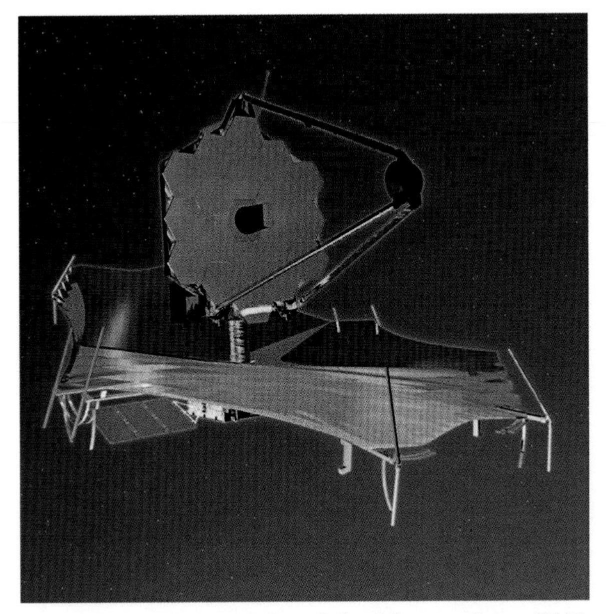

Fig. 10.4. Artist's impression of the 6.5-meter James Webb Space Telescope. Like the Keck telescopes, the mirror is segmented and protected against Solar radiation by a giant heat shield. Keeping the mirror and the instruments permanently in the shadow will permit a passive cooling at a temperature of ~ 35 K. This will be ideal for conducting observations at NIR wavelengths, with unprecedented sensitivity

Fortunately, the successor of HST is already at an intensive stage of planning and is currently scheduled to be launched in 2013. This Next Generation Space Telescope (which was named James Webb Space Telescope – JWST) will have a mirror of 6.5-meters diameter and therefore will be substantially more sensitive than HST. Furthermore, JWST will be optimized for observations in the NIR (1 to 5 μm) and thus be able, in particular, to observe sources at high redshifts whose stellar light is redshifted into the NIR regime of the spectrum. The Spitzer Space Telescope already operates at NIR and MIR wavelength. Despite the fact that Spitzer carries only a 60 cm mirror, it is far more sensitive and efficient in this wavelength regime than previous satellites.

We hope that JWST will be able to observe the first galaxies and the first AGN, i.e., those sources responsible for reionizing the Universe. Besides a NIR camera, JWST will carry the first multi-object spectrograph in space, which is optimized for spectroscopic studies of high-redshift galaxy samples and whose sensitivity will exceed that of all previous instruments by a huge factor. Furthermore, JWST will carry a MIR instrument which is being developed for imaging and spectroscopy in the wavelength range $5\,\mu\text{m} \le \lambda \le 28\,\mu\text{m}$.

A new kind of observatory is planned for X-ray astronomy where the focal length will be so large as to require two spacecraft. One of them will carry the mirror system, whereas the other will host the instruments. Operating such a telescope will require that the separation between the telescope and the focal plane be kept constant with very high precision. This poses a technological challenge for formation flight; formation flights also need to be mastered for future IR interferometers in space. The Next Generation X-ray Telescope will be capable of observing galaxy clusters to the highest redshifts and to extend the studies of AGNs to much lower luminosities than is currently possible. In particular we hope to study gas physics in the close vicinity of the event horizon of black holes.

Far-infrared astronomy will receive its next boost in 2008, when the Herschel satellite will be launched by ESA. Its 3.5 meter mirror will provide a far better sensitivity in this wavelength regime than previous FIR telescopes. Herschel will be launched together with the Planck satellite, which will yield a far more detailed image of the microwave sky than even WMAP. While mainly targeted at measurements of the CMB anisotropy, with better angular resolution and far better wavelength coverage than WMAP, Planck will not only be a very important mission for cosmology; its sky survey at many frequencies will also benefit many other fields of astronomy. The discovery of galaxy clusters by means of the Sunyaev–Zeldovich effect should be mentioned as just one example.

There will also be revolutionary developments in radio astronomy. New mm and sub-mm telescopes, such as the recently commissioned APEX, will provide much more detailed maps of the dust emission from star-forming regions than before. APEX will conduct a Sunyaev–Zeldovich survey for galaxy clusters and therefore follow a new strategy for selecting clusters. In a way, this provides a connection to the future Planck mission. In particular, we expect a large number of clusters at high redshift which are of special value when using clusters as cosmological probes. Towards the end of this decade, ALMA (Atacama Large Millimeter Array, Fig. 10.5), a 64 antenna interferometer operating at mm and sub-mm wavelengths, will start to make its first observations. Its enormously increased angular resolution and sensitivity will allow us to study, among other issues, the dust emission and molecules of very high redshift galaxies and QSOs. Furthermore, future telescopes constructed in the Antarctic would provide further opportunities for infrared and sub-mm astronomy owing to the extremely dry atmosphere.

At even longer wavelengths, a technological revolution will take place. Currently being developed are concepts for radio telescopes whose radio lobes will be digitally generated on computers. Such digital radio interferometers not only allow a much improved sensitivity and angular resolution, but they also enable us to observe many different sources in vastly different sky regions simultaneously. LOFAR will be the prototype of such an instrument and will operate at frequencies below about 200 MHz. In the more distant future, the Square Kilometer Array (SKA) will be a much larger observatory – its name is derived from its effective collecting area. SKA will provide a giant boost to astronomy; for the first time ever, the achievable number density of sources on the radio sky will be comparable to or even larger than that in the optical. The limits of such instruments are no longer bound by the properties of the individual antennas, but rather by the capacity of the computers which analyze the data. To exploit the full capacity of these digital radio interferometers, a giant evolution in the hardware and software of such supercomputers will be required.

Fig. 10.5. Artist's impression of the Atacama Large Millimeter Array (ALMA) which is currently being built on the Llano de Chajnantor in Chile, a plateau at 5000 meters altitude (this is also the site of APEX). The 64 antennas will have a diameter of 12 meters each. They will be operated in an interferometric mode, and they will start a totally new era in (sub-)mm astronomy, owing to the large collecting area and the excellent atmospheric conditions at this site

New windows to the Universe will be opened. The first gravitational wave antennas are already in place, and their next generation will probably be able to discover the signals from relatively nearby supernova explosions. With LISA, mergers of supermassive black holes will become detectable throughout the visible Universe, as we mentioned before. Giant neutrino detectors will open the field of neutrino astronomy and will be able, for example, to observe processes in the innermost parts of AGNs. Observatories for cosmic rays are being built. The Pierre-Auger observatory in Argentina is one such example that has been in operation since 2004; in particular, it will study the highest-energy cosmic rays.

Parallel to these developments in telescopes and instruments, theory is progressing steadily. The continuously increasing capacity of computers available for numerical simulations is only one aspect, albeit an important one. New approaches for modeling, triggered by new observational results, are of equal importance. The close connection between theory, modeling, and observations will become increasingly important since the complexity of data requires an advanced level of modeling and simulations for their quantitative interpretation.

The huge amount of data obtained with current and future instruments is useful not only for the observers taking the data, but also for others in the astronomical community. Realizing this fact, many observatories have set up archives from which data can be retrieved. Space observatories pioneered such data archives, and a great deal of science results from the use of archival data. Examples here are the use of the HST deep fields by a large number of researchers, or the analysis of serendipitous sources in X-ray images which led to the EMSS (see Sect. 6.3.5). Together with the fact that an understanding of astronomical sources usually requires data taken over a broad range of frequencies, there is a strong motivation for the creation of *virtual observatories*: infrastructures which connect archives containing astronomical data from a large variety of instruments and which can be accessed electronically. In order for such virtual observatories to be most useful, the data structures and interfaces of the various archives need to become mutually compatible. Intensive activities in creating such virtual observatories are ongoing; they will doubtlessly play in increasingly important role in the future.

One of the major challenges for the next few years will certainly be the investigation of the very distant Universe, studying the evolution of cosmic objects and structures at very high redshift up to the epoch of reionization. To relate the resulting insights of the distant Universe to those obtained more locally and thus to obtain a consistent view about our cosmos, major theoretical investigations will be required as well as extensive observations across the whole redshift range, using the broadest wavelength range possible. Furthermore, the new astrometry satellite GAIA will offer us the unique opportunity to study cosmology in our Milky Way. With GAIA, the aforementioned stellar streams, which were created in the past by the tidal disruption of satellite galaxies during their merging with the Milky Way, can be verified. New insights gained with GAIA will certainly also improve our understanding of other galaxies.

The second major challenge for the near future is the fundamental physics on which our cosmological model is based. From observations of galaxies and galaxy clusters, and also from our determinations of the cosmological parameters, we have verified the presence of dark matter. Since there seem to be no plausible astrophysical explanations for its nature, dark matter most likely consists of new kinds of elementary particles. Two different strategies to find these particles are currently being followed. First, experiments aim at directly detecting these particles, which should also be present in the immediate vicinity of the Earth. These experiments are located in deep underground laboratories, thus shielded from cosmic rays. Several such experiments, which are an enormous technical challenge due to the sensitivity they are required to achieve, are currently running. They will obtain increasingly tighter constraints on the properties of WIMPS with respect to their mass and interaction cross-section. Such constraint will, however, depend on the mass model of the dark matter in our Galaxy. As a second approach, the Large Hadron Collider at CERN will start operating in 2007 and should establish a new energy range for elementary particle physics. In particular, we hope to find evidence for or against the validity of the supersymmetric model for particle physics, as an extension of the current standard model. Indeed, we might expect the detection of the lightest supersymmetric particle, the neutralino, which would be an excellent candidate for the dark matter particle.

Whereas at least plausible ideas exist about the nature of dark matter which can be experimentally tested in the coming years, the presence of a non-vanishing density of dark energy, as evidenced from cosmology, presents an even larger mystery for fundamental physics. Though from quantum physics we might expect a vacuum energy density to exist, its estimated energy density is tremendously larger than the cosmic dark energy density. The interpretation that dark energy is a quantum mechanical vacuum energy therefore seems highly implausible. As astrophysical cosmologists, we could take the view that vacuum energy is nothing more than a cosmological constant, as originally introduced by Einstein; this would then be an additional fundamental constant in the laws of nature. From a physical point of view, it would be much more satisfactory if the nature of dark energy could be derived from the laws of fundamental physics. The huge discrepancy between the density of dark energy and the simple estimate of the vacuum energy density clearly indicates that we are currently far from a physical understanding of dark energy. To achieve this understanding, we might well assume that a new theory must be developed which unifies quantum physics and gravity – in a manner similar to the way other 'fundamental' interactions (like electromagnetism and the weak force) have been unified within the standard model of particle physics. Deriving such a theory of quantum gravity turns out to be enormously problematic despite intensive research over several decades. However, the density of dark energy is so incredibly small that its effects can only be recognized on the largest length-scales, implying the necessity of further astronomical and cosmological experiments. Only astronomical techniques are able to probe the properties of dark energy empirically.

To investigate the nature of dark energy, two different approaches are currently seen as the most promising: studying the Hubble diagram of type Ia supernovae, and cosmic shear. To increase the sensitivity of both methods substantially, a satellite mission is currently being planned which will allow a precision application of these methods by conducting wide-field multicolor photometry from space. This will yield accurate lightcurves of SN Ia, as well as accurate shape measurements of very faint galaxies which are needed for cosmic shear studies. Furthermore, there are several planned ground-based projects to build telescopes, or instruments for existing telescopes, which predominantly aim at applying these two cosmological probes. One of them is the Large Synoptic Survey Telescope, an 8-meter telescope with a 7 square degree field camera.

Although inflation is currently part of the standard model of cosmology, the physical processes occurring during the inflationary phase have not been understood up to now. The fact that different field-theoretical models of inflation yield very similar cosmological consequences is an asset for cosmologists: from their point-of-view, the details of inflation are not immediately relevant, as long as a phase of exponential expansion occurred. But the same fact indicates the size of the problem faced in studying the process of inflation, since different physical models yield rather similar outcomes with regard to cosmological observables. Perhaps the most promising probe of inflation is the polarization of the cosmic microwave background, since it allows us to study whether, and with what amplitude, gravitational waves were generated during inflation. Predictions of the ratio of gravity wave energy to that of density fluctuations are different in different physical models of inflation. After the Planck satellite has been put in orbit, a mission which is able to measure the CMB polarization with sufficient accuracy to test inflation will probably be considered.

Another cosmological observation poses an additional challenge to fundamental physics. We observe baryonic matter in our Universe, but we see no signs of appreciable amounts of antimatter. If certain regions in the Universe consisted of antimatter, there would be observable radiation from matter-antimatter annihilation at the interface between the different regions. The question therefore arises, what processes caused an excess of matter over antimatter in the early Universe? We can easily quantify this asymmetry – at very early times, the abundance of protons, antiprotons and photons were all quite similar, but after proton-antiproton annihilation at $T \sim 1$ GeV, a fraction of $\sim 10^{-10}$ – the current baryon-to-photo ratio – is left over. This slight asymmetry of the abundance of protons and neutrons over their antiparticles in the early Universe, often called baryogenesis, has not been explained in the framework of the standard model of particle physics. Furthermore, we would like to understand why the densities of baryons and dark matter are essentially the same, differing by a mere factor of ~ 6.

The aforementioned issues are arguably the best examples of the increasingly tight connection between cosmology and fundamental physics. Progress in either field can only be achieved by the close collaboration between theoretical and experimental particle physics and astronomy.

Finally, and perhaps too late in the opinion of some readers, we should note again that this book has assumed throughout that the physical laws, as we know them today, can be used to interpret cosmic phenomena. We have no real proof that this assumption is correct, but the successes of this approach justify this assumption in hindsight. If this assumption had been grossly violated, there would be no reason why the values of the cosmological parameters, estimated with vastly different methods and thus employing very different physical processes, mutually agree. The price we pay for the acceptance of the standard model of cosmology, which results from this approach, is high though: the standard model implies that we accept the existence and even dominance of dark matter and dark energy in the Universe.

Not every cosmologist is willing to pay this price. For instance, M. Milgrom introduced the hypothesis that the flat rotation curves of spiral galaxies are not due to the existence of dark matter. Instead, they could arise from the possibility that the Newtonian law of gravity ceases to be valid on scales of 10 kpc – on such large scales, and the correspondingly small accelerations, the law of gravity has not been tested. Milgrom's *Modified Newtonian Dynamics (MOND)* is therefore a logically possible alternative to the postulate of dark matter on scales of galaxies. Indeed, MOND offers an explanation for the Tully–Fisher relation of spiral galaxies.

There are, however, several reasons why only a few astrophysicists follow this approach. MOND has an additional free parameter which is fixed by matching the observed rotation curves of spiral galaxies with the model, without invoking dark matter. Once this parameter is fixed, MOND cannot explain the dynamics of galaxies in clusters without needing additional matter – dark matter. Thus, the theory has just enough freedom to fix a problem on one length- (or mass-)scale, but apparently fails on different scales. We can circumvent the problem again by postulating warm dark matter, which would be able to fall into the potential wells of clusters, but not into the shallower ones of galaxies, thereby replacing one kind of dark matter (CDM) with another.

In fact, the consequences of accepting MOND would be far reaching: if the law of gravity deviates from the Newtonian law, the validity of General Relativity would be questioned, since it contains the Newtonian force law as a limiting case of weak gravitational fields. General Relativity, however, forms the basis of our world models. Rejecting it as the correct description of gravity, we would lose the physical basis of our cosmological model – and thus the impressive quantitative agreement of results from vastly different observations that we described in Chap. 8. The acceptance of MOND therefore demands an even higher price than the existence of dark matter, but it is an interesting challenge to falsify MOND empirically.

This example shows that the modification of one aspect of our standard model has the consequence that the whole model is threatened: due to the large internal consistency of the standard model, modifying one aspect has a serious impact on all others. This does not mean that there cannot be other cosmological models which can provide as consistent an explanation of the relevant observational facts as our standard model does. However, an alternative explanation of a single aspect cannot be considered in isolation, but must be seen in its relation to the others. Of course, this poses a true challenge to the promoters of alternative models: whereas the overwhelming majority of cosmologists are working hard to verify and to refine the standard model and to construct the full picture of cosmic evolution, the group of researchers working on alternative models is small[1] and thus hardly able to put together a convincing and consistent model of cosmology. This fact finds its justification in the successes of the standard model, and in the agreement of observations with the predictions of this model.

We have, however, just uncovered an important sociological aspect of the scientific enterprise: there is

[1] However, there has been a fairly recent increase in research activity on MOND. This was triggered mainly by the fact that after many years of research, a theory called TeVeS (for Tensor-Vector-Scalar field) was invented, containing General Relativity, MOND and Newton's law in the respective limits – though at the cost of introducing three new arbitrary functions.

a tendency to 'jump on the bandwagon'. This results in the vast majority of research going into one (even if the most promising) direction – and this includes scientific staff, research grants, observing time etc. The consequence is that new and unconventional ideas have a hard time getting heard. Hopefully (and in the view of this author, very likely), the bandwagon is heading in the right direction. There are historical examples to the contrary, though – we now know that Rome is not at the center of the cosmos, nor the Earth, nor the Sun, nor the Milky Way, despite long epochs when the vast majority of scientists were convinced of the veracity of these ideas.

Appendix

A. The Electromagnetic Radiation Field

In this appendix, we will briefly review the most important properties of a radiation field. We thereby assume that the reader has encountered these quantities already in a different context.

A.1 Parameters of the Radiation Field

The electromagnetic radiation field is described by the *specific intensity* I_ν, which is defined as follows. Consider a surface element of area $\mathrm{d}A$. The radiation energy which passes through this area per time interval $\mathrm{d}t$ from within a solid angle element $\mathrm{d}\omega$ around a direction described by the unit vector $\boldsymbol{n}$, with frequency in the range between ν and $\nu+\mathrm{d}\nu$, is

$$\mathrm{d}E = I_\nu \, \mathrm{d}A \, \cos\theta \, \mathrm{d}t \, \mathrm{d}\omega \, \mathrm{d}\nu \, , \tag{A.1}$$

where θ describes the angle between the direction $\boldsymbol{n}$ of the light and the normal vector of the surface element. Then, $\mathrm{d}A \, \cos\theta$ is the area projected in the direction of the infalling light. The specific intensity depends on the considered position (and, in time-dependent radiation fields, on time), the direction $\boldsymbol{n}$, and the frequency ν. With the definition (A.1), the dimension of I_ν is energy per unit area, time, solid angle, and frequency, and it is typically measured in units of $\mathrm{erg\,cm^{-2}\,s^{-1}\,ster^{-1}\,Hz^{-1}}$. The specific intensity of a cosmic source describes its surface brightness.

The *specific net flux* F_ν passing through an area element is obtained by integrating the specific intensity over all solid angles,

$$F_\nu = \int \mathrm{d}\omega \, I_\nu \, \cos\theta \, . \tag{A.2}$$

The flux that we receive from a cosmic source is defined in exactly the same way, except that cosmic sources usually subtend a very small solid angle on the sky. In calculating the flux we receive from them, we may therefore drop the factor $\cos\theta$ in (A.2); in this context, the specific flux is also denoted as S_ν. However, in this Appendix (and only here!), the notation S_ν will be reserved for another quantity. The flux is measured in units of $\mathrm{erg\,cm^{-2}\,s^{-1}\,Hz^{-1}}$. If the radiation field is isotropic, F_ν vanishes. In this case, the same amount of radiation passes through the surface element in both directions.

The *mean specific intensity* J_ν is defined as the average of I_ν over all angles,

$$J_\nu = \frac{1}{4\pi} \int \mathrm{d}\omega \, I_\nu \, , \tag{A.3}$$

so that, for an isotropic radiation field, $I_\nu = J_\nu$. The *specific energy density* u_ν is related to J_ν according to

$$u_\nu = \frac{4\pi}{c} J_\nu \tag{A.4}$$

where u_ν is the energy of the radiation field per volume element and frequency interval, thus measured in $\mathrm{erg\,cm^{-3}\,Hz^{-1}}$. The total energy density of the radiation is obtained by integrating u_ν over frequency. In the same way, the intensity of the radiation is obtained by integrating the specific intensity I_ν over ν.

A.2 Radiative Transfer

The specific intensity of radiation in the direction of propagation between source and observer is constant, as long as no emission or absorption processes are occurring. If s measures the length along a line-of-sight, the above statement can be formulated as

$$\frac{\mathrm{d}I_\nu}{\mathrm{d}s} = 0 \, . \tag{A.5}$$

An immediate consequence of this equation is that the surface brightness of a source is independent of its distance. The observed flux of a source depends on its distance, because the solid angle, under which the source is observed, decreases with the square of the distance, $F_\nu \propto D^{-2}$ (see Eq. A.2). However, for light propagating through a medium, emission and absorption (or scattering of light) occurring along the path over which the light travels may change the specific intensity. These effects are described by the *equation of radiative transfer*

$$\frac{\mathrm{d}I_\nu}{\mathrm{d}s} = -\kappa_\nu \, I_\nu + j_\nu \, . \tag{A.6}$$

The first term describes the absorption of radiation and states that the radiation absorbed within a length interval $\mathrm{d}s$ is proportional to the incident radiation.

Peter Schneider, The Electromagnetic Radiation Field.
In: Peter Schneider, Extragalactic Astronomy and Cosmology. pp. 417–423 (2006)
DOI: 10.1007/11614371_A

The factor of proportionality is the *absorption coefficient* κ_ν, which has the unit of cm^{-1}. The *emission coefficient* j_ν describes the energy that is added to the radiation field by emission processes, having a unit of $\mathrm{erg\,cm^{-3}\,s^{-1}\,Hz^{-1}\,ster^{-1}}$; hence, it is the radiation energy emitted per volume element, time interval, frequency interval, and solid angle. Both, κ_ν and j_ν depend on the nature and state (such as temperature, chemical composition) of the medium through which light propagates.

The absorption and emission coefficients both account for true absorption and emission processes, as well as the scattering of radiation. Indeed, the scattering of a photon can be considered as an absorption that is immediately followed by an emission of a photon.

The *optical depth* τ_ν along a line-of-sight is defined as the integral over the absorption coefficient,

$$\tau_\nu(s) = \int_{s_0}^{s} \mathrm{d}s'\, \kappa_\nu(s') \,, \tag{A.7}$$

where s_0 denotes a reference point on the sightline from which the optical depth is measured. Dividing (A.6) by κ_ν and using the relation $\mathrm{d}\tau_\nu = \kappa_\nu\, \mathrm{d}s$ in order to introduce the optical depth as a new variable along the light ray, the equation of radiative transfer can be written as

$$\frac{\mathrm{d}I_\nu}{\mathrm{d}\tau_\nu} = -I_\nu + S_\nu \,, \tag{A.8}$$

where the source function

$$S_\nu = \frac{j_\nu}{\kappa_\nu} \tag{A.9}$$

is defined as the ratio of the emission and absorption coefficients. In this form, the equation of radiative transport can be formally solved; as can easily be tested by substitution, the solution is

$$I_\nu(\tau_\nu) = I_\nu(0)\, \exp(-\tau_\nu) + \int_0^{\tau_\nu} \mathrm{d}\tau'_\nu\, \exp\left(\tau'_\nu - \tau_\nu\right)\, S_\nu(\tau'_\nu) \,. \tag{A.10}$$

This equation has a simple interpretation. If $I_\nu(0)$ is the incident intensity, it will have decreased by absorption to a value $I_\nu(0)\exp(-\tau_\nu)$ after an optical depth of τ_ν. On the other hand, energy is added to the radiation field by emission, accounted for by the τ'-integral. Only a fraction $\exp\left(\tau'_\nu - \tau_\nu\right)$ of this additional energy emitted at τ' reaches the point τ, the rest is absorbed.

In the context of (A.10), we call this a *formal* solution for the equation of radiative transport. The reason for this is based on the fact that both the absorption and the emission coefficient depend on the physical state of the matter through which radiation propagates, and in many situations this state depends on the radiation field itself. For instance, κ_ν and j_ν depend on the temperature of the matter, which in turn depends, by heating and cooling processes, on the radiation field to which it is exposed. Hence, one needs to solve a coupled system of equations in general: on the one hand the equation of radiative transport, and on the other hand the equation of state for matter. In many situations, very complex problems arise from this, but we will not consider them further in the context of this book.

A.3 Blackbody Radiation

For matter in thermal equilibrium, the source function S_ν is solely a function of the matter temperature,

$$S_\nu = B_\nu(T) \,, \quad \text{or} \quad j_\nu = B_\nu(T)\, \kappa_\nu \,, \tag{A.11}$$

independent of the composition of the medium (Kirchhoff's law). We will now consider radiation propagating through matter in thermal equilibrium at constant temperature T. Since in this case $S_\nu = B_\nu(T)$ is constant, the solution (A.10) can be written in the form

$$\begin{aligned} I_\nu(\tau_\nu) &= I_\nu(0)\, \exp(-\tau_\nu) + B_\nu(T) \int_0^{\tau_\nu} \mathrm{d}\tau'_\nu\, \exp\left(\tau'_\nu - \tau_\nu\right) \\ &= I_\nu(0)\, \exp(-\tau_\nu) + B_\nu(T)\left[1 - \exp(-\tau_\nu)\right] \,. \end{aligned} \tag{A.12}$$

From this it follows that $I_\nu = B_\nu(T)$ is valid for sufficiently large optical depth τ_ν. The radiation propagating through matter which is in thermal equilibrium is described by the function $B_\nu(T)$ if the optical depth is sufficiently large, independent of the composition of the matter. A specific case of this situation can be illustrated by imagining the radiation field inside a box whose opaque walls are kept at a constant temperature T. Due to the opaqueness of the walls, their optical depth is infinite, hence the radiation field within the box is given by $I_\nu = B_\nu(T)$. This is also valid if the volume is filled with matter, as long as the latter is in thermal equilibrium at temperature T. For these reasons, this kind of radiation field is also called blackbody radiation.

The function $B_\nu(T)$ was first obtained in 1900 by Max Planck, and in his honor, it was named the *Planck function*; it reads

$$B_\nu(T) = \frac{2h_P\nu^3}{c^2}\frac{1}{e^{h_P\nu/k_BT}-1}\,, \tag{A.13}$$

where $h_P = 6.625\times10^{-27}$ erg s is the *Planck constant* and $k_B = 1.38\times10^{-16}$ erg K^{-1} is the Boltzmann constant. The shape of the spectrum can be derived from statistical physics. *Blackbody radiation* is defined by $I_\nu = B_\nu(T)$, and *thermal radiation* by $S_\nu = B_\nu(T)$. For large optical depths, thermal radiation converges to blackbody radiation.

The Planck function has its maximum at

$$\frac{h_P\nu_{max}}{k_BT} \approx 2.82\,, \tag{A.14}$$

i.e., the frequency of the maximum is proportional to the temperature. This property is called *Wien's law*. This law can also be written in more convenient units,

$$\nu_{max} = 5.88\times10^{10}\,\text{Hz}\,\frac{T}{1\,\text{K}}\,. \tag{A.15}$$

The Planck function can also be formulated depending on wavelength $\lambda = c/\nu$, such that $B_\lambda(T)\,d\lambda = B_\nu(T)\,d\nu$,

$$B_\lambda(T) = \frac{2h_Pc^2/\lambda^5}{\exp(h_Pc/\lambda k_BT)-1}\,. \tag{A.16}$$

Two limiting cases of the Planck function are of particular interest. For low frequencies, $h_P\nu \ll k_BT$, one can apply the expansion of the exponential function for small arguments in (A.13). The leading-order term in this expansion then yields

$$B_\nu(T) \approx B_\nu^{RJ}(T) = \frac{2\nu^2}{c^2}k_BT\,, \tag{A.17}$$

which is called the *Rayleigh–Jeans approximation* of the Planck function. We point out that the Rayleigh–Jeans equation does not contain the Planck constant, and this law had been known even before Planck derived his

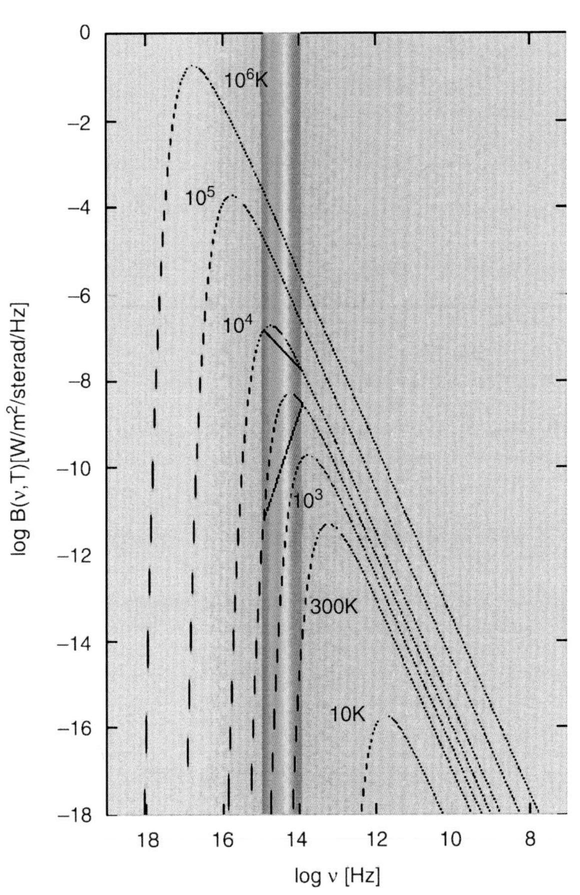

Fig. A.1. The Planck function (A.13) for different temperatures T. The plot shows $B_\nu(T)$ as a function of frequency ν, where high frequencies are plotted towards the left (thus large wavelengths towards the right). The exponentially decreasing Wien part of the spectrum is visible on the left, the Rayleigh–Jeans part on the right. The *shape* of the spectrum in the Rayleigh–Jeans part is independent of the temperature, which is determining the amplitude however

exact equation. In the other limiting case of very high frequencies, $h_P\nu \gg k_BT$, the exponential factor in the denominator in (A.13) becomes very much larger than unity, so that we obtain

$$B_\nu(T) \approx B_\nu^{W}(T) = \frac{2h_P\nu^3}{c^2}\,e^{-h_P\nu/k_BT}\,, \tag{A.18}$$

called the *Wien approximation* of the Planck function.

The energy density of blackbody radiation depends only on the temperature, of course, and is calculated by integration over the Planck function,

$$u = \frac{4\pi}{c} \int_0^\infty d\nu\, B_\nu(T) = \frac{4\pi}{c}\, B(T) = a\, T^4 , \quad \text{(A.19)}$$

where we defined the frequency-integrated Planck function

$$B(T) = \int_0^\infty d\nu\, B_\nu(T) = \frac{a\,c}{4\pi}\, T^4 , \quad \text{(A.20)}$$

and where the constant a has the value

$$a = \frac{8\pi^5 k_B^4}{15 c^3 h_P^3} = 7.56 \times 10^{-15}\,\mathrm{erg\,cm^{-3}\,K^{-4}} . \quad \text{(A.21)}$$

The flux which is emitted by the surface of a blackbody per unit area is given by

$$F = \int_0^\infty d\nu\, F_\nu = \pi \int_0^\infty d\nu\, B_\nu(T) = \pi B(T) = \sigma_{SB} T^4 , \quad \text{(A.22)}$$

where the *Stefan–Boltzmann constant* σ_{SB} has a value of

$$\sigma_{SB} = \frac{ac}{4} = \frac{2\pi^5 k_B^4}{15 c^2 h_P^3} = 5.67 \times 10^{-5}\,\mathrm{erg\,cm^{-2}\,K^{-4}\,s^{-1}} . \quad \text{(A.23)}$$

A.4 The Magnitude Scale

Optical astronomy was being conducted well before methods of quantitative measurements became available. The brightness of stars had been cataloged more than 2000 years ago, and their observation goes back as far as the ancient world. Stars were classified into magnitudes, assigning a magnitude of 1 to the brightest stars and higher magnitudes to the fainter ones. Since the apparent magnitude as perceived by the human eye scales roughly logarithmically with the radiation flux (which is also the case for our hearing), the magnitude scale represents a logarithmic flux scale. To link these visually determined magnitudes in historical catalogs to a quantitative measure, the magnitude system has been retained in optical astronomy, although with a precise definition. Since no historical astronomical observations have been conducted in other wavelength ranges, because these are not accessible to the unaided eye, only optical astronomy has to bear the historical burden of the magnitude system.

A.4.1 Apparent Magnitude

We start with a relative system of flux measurements by considering two sources with fluxes S_1 and S_2. The *apparent magnitudes* of the two sources, m_1 and m_2, then behave according to

$$m_1 - m_2 = -2.5 \log\left(\frac{S_1}{S_2}\right) ; \quad \frac{S_1}{S_2} = 10^{-0.4(m_1 - m_2)} . \quad \text{(A.24)}$$

This means that the brighter source has a smaller apparent magnitude than the fainter one: the larger the apparent magnitude, the fainter the source.[1] The factor of 2.5 in this definition is chosen so as to yield the best agreement of the magnitude system with the visually determined magnitudes. A difference of $|\Delta m| = 1$ in this system corresponds to a flux ratio of ~ 2.51, and a flux ratio of a factor 10 or 100 corresponds to 2.5 or 5 magnitudes, respectively.

A.4.2 Filters and Colors

Since optical observations are performed using a combination of a filter and a detector system, and since the flux ratios depend, in general, on the choice of the filter (because the spectral energy distribution of the sources may be different), apparent magnitudes are defined for each of these filters. The most common filters are shown in Fig. A.2 and listed in Table A.1, together with their characteristic wavelengths and the widths of their transmission curves. The apparent magnitude for a filter X is defined as m_X, frequently written as X. Hence, for the B-band filter, $m_B \equiv B$.

Next, we need to specify how the magnitudes measured in different filters are related to each other, in order to define the color indices of sources. For this

[1] Of course, this convention is confusing, particularly to someone just becoming familiar with astronomy, and it frequently causes confusion and errors, as well as problems in the communication with non-astronomers – but we have to get along with that.

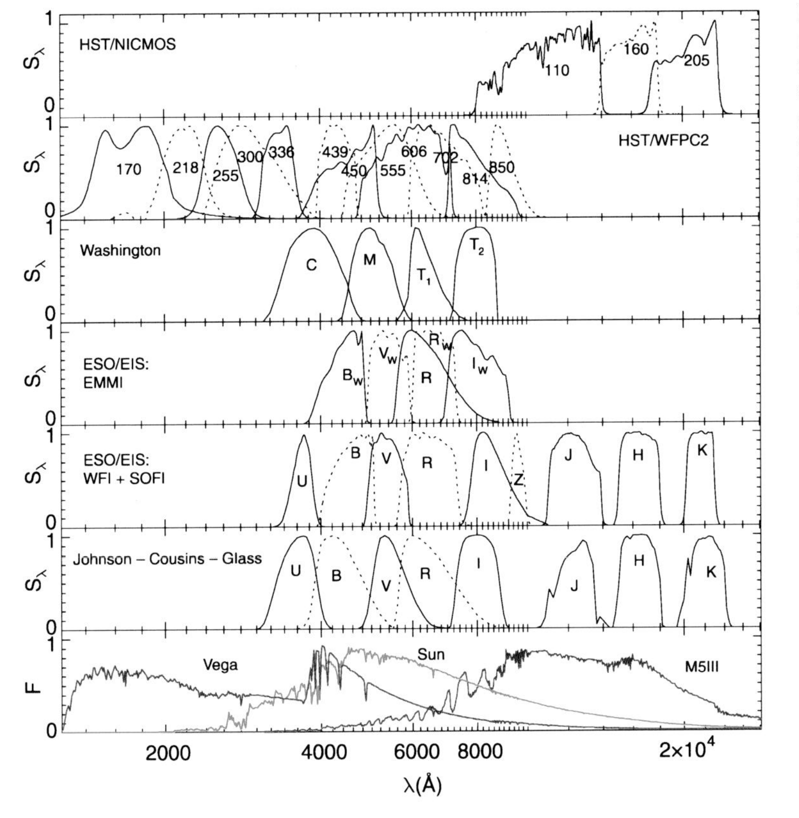

Fig. A.2. Transmission curves of various filter-detector systems. From top to bottom: the filters of the NICMOS camera and the WFPC2 on-board HST, the Washington filter system, the filters of the EMMI instrument at ESO's NTT, the filters of the WFI at the ESO/MPG 2.2-m telescope and those of the SOFI instrument at the NTT, and the Johnson–Cousins filters. In the bottom diagram, the spectra of three stars with different effective temperatures are displayed

Table A.1. For some of the best-established filter systems – Johnson, Strömgren, and the filters of the Sloan Digital Sky Surveys – the central (more precisely, the effective) wavelengths and the widths of the filters are listed

Johnson	U	B	V	R	I	J	H	K	L	M
$\lambda_{\rm eff}$ (nm)	367	436	545	638	797	1220	1630	2190	3450	4750
$\Delta\lambda$ (nm)	66	94	85	160	149	213	307	39	472	460

Strömgren	u	v	b	y	β_w	β_n
$\lambda_{\rm eff}$ (nm)	349	411	467	547	489	489
$\Delta\lambda$ (nm)	30	19	18	23	15	3

SDSS	u′	g′	r′	i′	z′
$\lambda_{\rm eff}$ (nm)	354	477	623	762	913
$\Delta\lambda$ (nm)	57	139	138	152	95

purpose, a particular class of stars is used, main-sequence stars of spectral type A0, of which the star Vega is an archetype. For such a star, by definition, $U = B = V = R = I = \ldots$, i.e., every color index for such a star is defined to be zero.

For a more precise definition, let $T_X(\nu)$ be the transmission curve of the filter-detector system. $T_X(\nu)$ specifies which fraction of the incoming photons with frequency ν are registered by the detector. The apparent magnitude of a source with spectral flux S_ν is then

$$m_X = -2.5 \log \left(\frac{\int d\nu \, T_X(\nu) \, S_\nu}{\int d\nu \, T_X(\nu)} \right) + \text{const.} \,, \quad \text{(A.25)}$$

where the constant needs to be determined from reference stars.

Another commonly used definition of magnitudes is the AB system. In contrast to the Vega magnitudes, no stellar spectral energy distribution is used as a reference here, but instead one with a constant flux at all frequencies, $S_\nu^{\rm ref} = S_\nu^{\rm AB} = 2.89 \times 10^{-21}\,{\rm erg\,s^{-1}\,cm^{-2}\,Hz^{-1}}$. This value has been chosen such that A0 stars like Vega have the same magnitude in the original Johnson V-band as they have in the AB system, $m_V^{\rm AB} = m_V$. With (A.25), one obtains for the conversion between the two systems

$$m_X^{\rm AB} - m_X^{\rm Vega} = -2.5 \log \left(\frac{\int d\nu \, T_X(\nu) \, S_\nu^{\rm AB}}{\int d\nu \, T_X(\nu) \, S_\nu^{\rm Vega}} \right) =: m_{{\rm AB}\to{\rm Vega}} \,. \quad \text{(A.26)}$$

For the filters at the ESO Wide-Field Imager, which are designed to resemble the Johnson set of filters, the following prescriptions are then to be applied: $U_{\rm AB} = U_{\rm Vega} + 0.80$; $B_{\rm AB} = B_{\rm Vega} - 0.11$; $V_{\rm AB} = V_{\rm Vega}$; $R_{\rm AB} = R_{\rm Vega} + 0.19$; $I_{\rm AB} = I_{\rm Vega} + 0.59$.

A.4.3 Absolute Magnitude

The apparent magnitude of a source does not in itself tell us anything about its luminosity, since for the determination of the latter we also need to know its distance D in addition to the radiative flux. Let L_ν be the specific luminosity of a source, i.e., the energy emitted per unit time and per unit frequency interval, then the flux is given by (note that from here on we switch back to the notation where S denotes the flux, which was denoted by F earlier in this appendix)

$$S_\nu = \frac{L_\nu}{4\pi D^2} \,, \quad \text{(A.27)}$$

where we implicitly assumed that the source emits isotropically. Having the apparent magnitude as a measure for S_ν (at the frequency ν defined by the filter which is applied), it is desirable to have a similar measure for L_ν, specifying the physical properties of the source itself. For this purpose, the *absolute magnitude* is introduced, denoted as M_X, where X refers to the filter under consideration. By definition, M_X is equal to the apparent magnitude of a source if it were to be located at a distance of 10 pc from us. The absolute magnitude of a source is thus independent of its distance, in contrast to the apparent magnitude. With (A.27) we find for the relation of apparent to absolute magnitude

$$m_X - M_X = 5 \log \left(\frac{D}{1\,{\rm pc}} \right) - 5 \equiv \mu \,, \quad \text{(A.28)}$$

where we have defined the *distance modulus* μ in the final step. Hence, the latter is a logarithmic measure of the distance of a source: $\mu = 0$ for $D = 10$ pc, $\mu = 10$ for $D = 1$ kpc, and $\mu = 25$ for $D = 1$ Mpc. The difference between apparent and absolute magnitude is independent of the filter choice, and it equals the distance modulus if no extinction is present. In general, this difference is modified by the filter-dependent extinction coefficient – see Sect. 2.2.4.

A.4.4 Bolometric Parameters

The total luminosity L of a source is the integral of the specific luminosity L_ν over all frequencies. Accordingly, the total flux S of a source is the frequency-integrated specific flux S_ν. The *apparent bolometric magnitude* $m_{\rm bol}$ is defined as a logarithmic measure of the total flux,

$$m_{\rm bol} = -2.5 \log S + \text{const.} \,, \quad \text{(A.29)}$$

where here the constant is also determined from reference stars. Accordingly, the *absolute bolometric magnitude* is defined by means of the distance modulus, as in (A.28). The absolute bolometric magnitude depends on the bolometric luminosity L of a source via

$$M_{\rm bol} = -2.5 \log L + {\rm const.} \quad \text{(A.30)}$$

The constant can be fixed, e.g., by using the parameters of the Sun: its apparent bolometric magnitude is $m_{\odot\rm bol} = -26.83$, and the distance of one Astronomical Unit corresponds to a distance modulus of $\mu = -31.47$. With these values, the absolute bolometric magnitude of the Sun becomes

$$M_{\odot\rm bol} = m_{\odot\rm bol} - \mu = 4.74 \,, \quad \text{(A.31)}$$

so that (A.30) can be written as

$$M_{\rm bol} = 4.74 - 2.5 \log \left(\frac{L}{L_\odot} \right) \,, \quad \text{(A.32)}$$

and the luminosity of the Sun is then

$$L_\odot = 3.85 \times 10^{33} \,{\rm erg\,s^{-1}} \,. \quad \text{(A.33)}$$

The direct relation between bolometric magnitude and luminosity of a source can hardly be exploited in practice, because the apparent bolometric magnitude (or the flux S) of a source cannot be observed in most cases. For observations of a source from the ground, only a limited window of frequencies is accessible. Nevertheless, in these cases one also likes to quantify the total luminosity of a source. For sources for which the spectrum is assumed to be known, like for many stars, the flux from observations at optical wavelengths can be extrapolated to larger and smaller wavelengths, and so $m_{\rm bol}$ can be estimated. For galaxies or AGNs, which have a much broader spectral distribution and which show much more variation between the different objects, this is not feasible. In these cases, the flux of a source in a particular frequency range is compared to the flux the Sun would have at the same distance and in the same spectral range. If M_X is the absolute magnitude of a source measured in the filter X, the X-band luminosity of this source is defined as

$$L_X = 10^{-0.4(M_X - M_{\odot X})} \, L_{\odot X} \,. \quad \text{(A.34)}$$

Thus, when speaking of, say, the "blue luminosity of a galaxy", this is to be understood as defined in (A.34).

B. Properties of Stars

In this appendix, we will summarize the most important properties of stars as they are required for understanding the contents of this book. Of course, this brief overview cannot replace the study of other textbooks in which the physics of stars is covered in much more detail.

B.1 The Parameters of Stars

To a good approximation, stars are gas spheres, in the cores of which light atomic nuclei are transformed into heavier ones (mainly hydrogen into helium) by thermonuclear processes, thereby producing energy. The external appearance of a star is predominantly characterized by its radius R and its characteristic temperature T. The properties of a star depend mainly on its mass M.

In a first approximation, the spectral energy distribution of the emission from a star can be described by a blackbody spectrum. This means that the specific intensity I_ν is given by a Planck spectrum (A.13) in this approximation. The luminosity L of a star is the energy radiated per unit time. If the spectrum of star was described by a Planck spectrum, the luminosity would depend on the temperature and on the radius according to

$$L = 4\pi R^2 \sigma_{\rm SB} T^4 , \quad (B.1)$$

where (A.22) was applied. However, the spectra of stars deviate from that of a blackbody (see Fig. 3.47). One defines the *effective temperature* $T_{\rm eff}$ of a star as the temperature a blackbody of the same radius would need to have to emit the same luminosity as the star, thus

$$\sigma_{\rm SB} T_{\rm eff}^4 \equiv \frac{L}{4\pi R^2} . \quad (B.2)$$

The luminosities of stars cover a huge range; the weakest are a factor $\sim 10^4$ times less luminous than the Sun, whereas the brightest emit $\sim 10^5$ times as much energy per unit time as the Sun. This big difference in luminosity is caused either by a variation in radius or by different temperatures. We know from the colors of stars that they have different temperatures: there are blue stars which are considerably hotter than the Sun, and red stars that are very much cooler. The temperature of a star can be estimated from its color. From the flux ratio at two different wavelengths or, equivalently, from the color index $X - Y \equiv m_X - m_Y$ in two filters X and Y, the temperature $T_{\rm c}$ is determined such that a blackbody at $T_{\rm c}$ would have the same color index. $T_{\rm c}$ is called the *color temperature* of a star. If the spectrum of a star was a Planck spectrum, then the equality $T_{\rm c} = T_{\rm eff}$ would hold, but in general these two temperatures differ.

B.2 Spectral Class, Luminosity Class, and the Hertzsprung–Russell Diagram

The spectra of stars can be classified according to the atomic (and, in cool stars, also molecular) spectral lines that are present. Based on the line strengths and their ratios, the Harvard sequence of stellar spectra was introduced. These spectral classes follow a sequence that is denoted by the letters O, B, A, F, G, K, M; besides these, some other spectral classes exist that will not be mentioned here. The sequence corresponds to a sequence of color temperature of stars: O stars are particularly hot, around 50 000 K, M stars very much cooler with $Tc \sim 3500$ K. For a finer classification, each spectral class is supplemented by a number between 0 and 9. An A1 star has a spectrum very similar to that of an A0 star, whereas an A5 star has as many features in common with an A0 star as with an F0 star.

Plotting the spectral type versus the absolute magnitude for those stars for which the distance and hence the absolute magnitude can be determined, a striking distribution of stars becomes apparent in such a *Hertzsprung–Russell diagram* (HRD). Instead of the spectral class, one may also plot the color index of the stars, typically $B - V$ or $V - I$. The resulting *color–magnitude diagram* (CMD) is essentially equivalent to an HRD, but is based solely on photometric data. A different but very similar diagram plots the luminosity versus the effective temperature.

In Fig. B.1, a color–magnitude diagram is plotted, compiled from data observed by the HIPPARCOS satellite. Instead of filling the two-dimensional parameter space rather uniformly, characteristic regions exist in

Peter Schneider, Properties of Stars.
In: Peter Schneider, Extragalactic Astronomy and Cosmology. pp. 425–429 (2006)
DOI: 10.1007/11614371_B

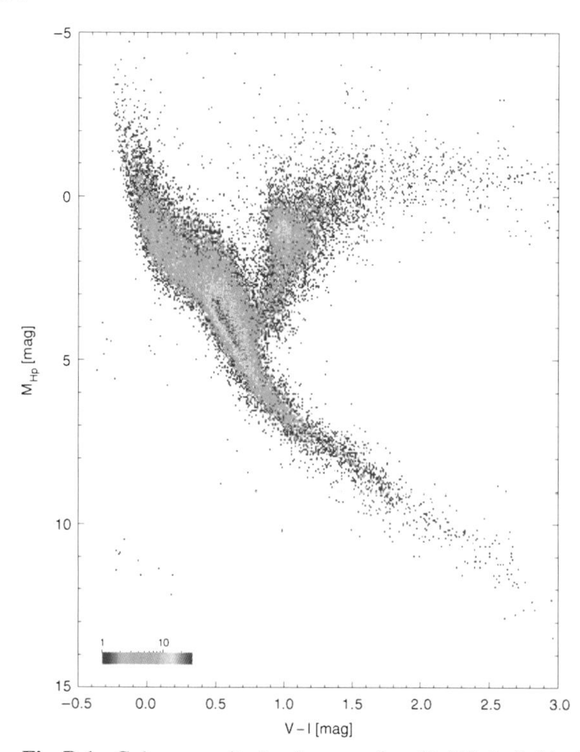

Fig. B.1. Color–magnitude diagram for 41 453 individual stars, whose parallaxes were determined by the HIPPARCOS satellite with an accuracy of better than 20%. Since the stars shown here are subject to unavoidable strong selection effects favoring nearby and luminous stars, the relative number density of stars is not representative of their true abundance. In particular, the lower main sequence is much more densely populated than is visible in this diagram

such color–magnitude diagrams in which nearly all stars are located. Most stars can be found in a thin band called the *main sequence*. It extends from early spectral types (O, B) with high luminosities ("top left") down to late spectral types (K, M) with low luminosities ("bottom right"). Branching off from this main sequence towards the "top right" is the domain of red giants, and below the main sequence, at early spectral types and very much lower luminosities than on the main sequence itself, we have the domain of white dwarfs. The fact that most stars are arranged along a one-dimensional sequence – the main sequence – is probably one of the most important discoveries in astronomy, because it tells us that the properties of stars are determined basically by a single parameter: their mass.

Since stars exist which have, for the same spectral type and hence the same color temperature (and roughly the same effective temperature), very different luminosities, we can deduce immediately that these stars have different radii, as can be read from (B.2). Therefore, stars on the red giant branch, with their much higher luminosities compared to main-sequence stars of the same spectral class, have a very much larger radius than the corresponding main-sequence stars. This size effect is also observed spectroscopically: the gravitational acceleration on the surface of a star (surface gravity) is

$$g = \frac{GM}{R^2} \,. \tag{B.3}$$

We know from models of stellar atmospheres that the width of spectral lines depends on the gravitational acceleration on the star's surface: the lower the surface gravity, the narrower the stellar absorption lines. Hence, a relation exists between the line width and the stellar radius. Since the radius of a star – for a fixed spectral type or effective temperature – specifies the luminosity, this luminosity can be derived from the width of the lines. In order to calibrate this relation, stars of known distance are required.

Based on the width of spectral lines, stars are classified into *luminosity classes*: stars of luminosity class I are called supergiants, those of luminosity class III are giants, main-sequence stars are denoted as dwarfs and belong to luminosity class V; in addition, the classification can be further broken down into bright giants (II), subgiants (IV), and subdwarfs (VI). Any star in the Hertzsprung–Russell diagram can be assigned a luminosity class and a spectral class (Fig. B.2). The Sun is a G2 star of luminosity class V.

If the distance of a star, and thus its luminosity, is known, and if in addition its surface gravity can be derived from the line width, we obtain the stellar mass from these parameters. By doing so, it turns out that for main-sequence stars the luminosity is a steep function of the stellar mass, approximately described by

$$\frac{L}{L_\odot} \approx \left(\frac{M}{M_\odot}\right)^{3.5} \,. \tag{B.4}$$

Therefore, a main-sequence star of $M = 10M_\odot$ is ~ 3000 times more luminous than our Sun.

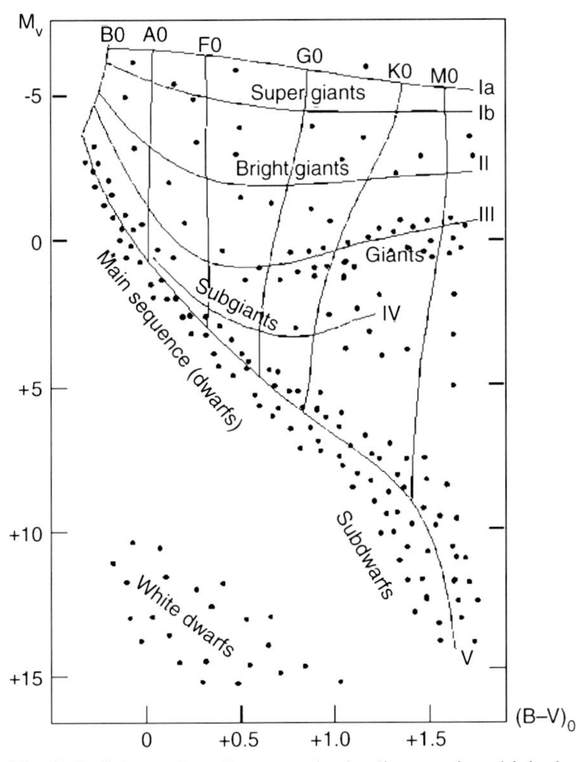

Fig. B.2. Schematic color–magnitude diagram in which the spectral types and luminosity classes are indicated

B.3 Structure and Evolution of Stars

To a very good approximation, stars are spherically symmetric. Therefore, the structure of a star is described by the radial profile of the parameters of its stellar plasma. These are density, pressure, temperature, and chemical composition of the matter. During almost the full lifetime of a star, the plasma is in hydrostatic equilibrium, so that pressure forces and gravitational forces are of equal magnitude and directed in opposite directions, so as to balance each other.

The density and temperature are sufficiently high in the center of a star that thermonuclear reactions are ignited. In main-sequence stars, hydrogen is fused into helium, thus four protons are combined into one ^{4}He nucleus. For every helium nucleus that is produced this way, 26.73 MeV of energy are released. Part of this energy is emitted in the form of neutrinos which can escape unobstructed from the star due to their very low cross-section.[1] The energy production rate is approximately proportional to T^4 for temperatures below about 15×10^6 K, at which the reaction follows the so-called pp-chain. At higher temperatures, another reaction chain starts to contribute, the so-called CNO cycle, with an energy production rate which is much more strongly dependent on temperature – roughly proportional to T^{20}.

The energy generated in the interior of a star is transported outwards, where it is then released in the form of electromagnetic radiation. This energy transport may take place in two different ways: first, by radiation transport, and second, it can be transported by macroscopic flows of the stellar plasma. This second mechanism of energy transport is called convection; here, hot elements of the gas rise upwards, driven by buoyancy, and at the same time cool ones sink downwards. The process is similar to that observed in heating water on a stove. Which of the two processes is responsible for the energy transport depends on the temperature profile inside the star. The intervals in a star's radius in which energy transport takes place via convection are called convection zones. Since in convection zones stellar material is subject to mixing, the chemical composition is homogeneous there. In particular, chemical elements produced by nuclear fusion are transported through the star by convection.

Stars begin their lives with a homogeneous chemical composition, resulting from the composition of the molecular cloud out of which they are formed. If their mass exceeds about $0.08 M_\odot$, the temperature and pressure in their core are sufficient to ignite the fusion of hydrogen into helium. Gas spheres with a mass below $\sim 0.08 M_\odot$ will not satisfy these conditions, hence these objects – they are called brown dwarfs – are not stars in

[1] The detection of neutrinos from the Sun in terrestrial detectors was the final proof for the energy production mechanism being nuclear fusion. However, the measured rate of electron neutrinos from the Sun was only half as large as expected from Solar models. This *Solar neutrino problem* kept physicists and astrophysicists busy for decades. It was a first indication of neutrinos having a finite rest mass – only in this case could electron neutrinos transform into another sort of neutrino along the way from the Sun to us. Recently, these neutrino oscillations were confirmed: neutrinos have a very small but finite rest mass. For their research in the field of Solar neutrinos, Raymond Davis and Masatoshi Koshiba were awarded with one half of the Nobel Prize in Physics in 2002. The other half was awarded to Ricardo Giacconi for his pioneering work in the field of X-ray astronomy.

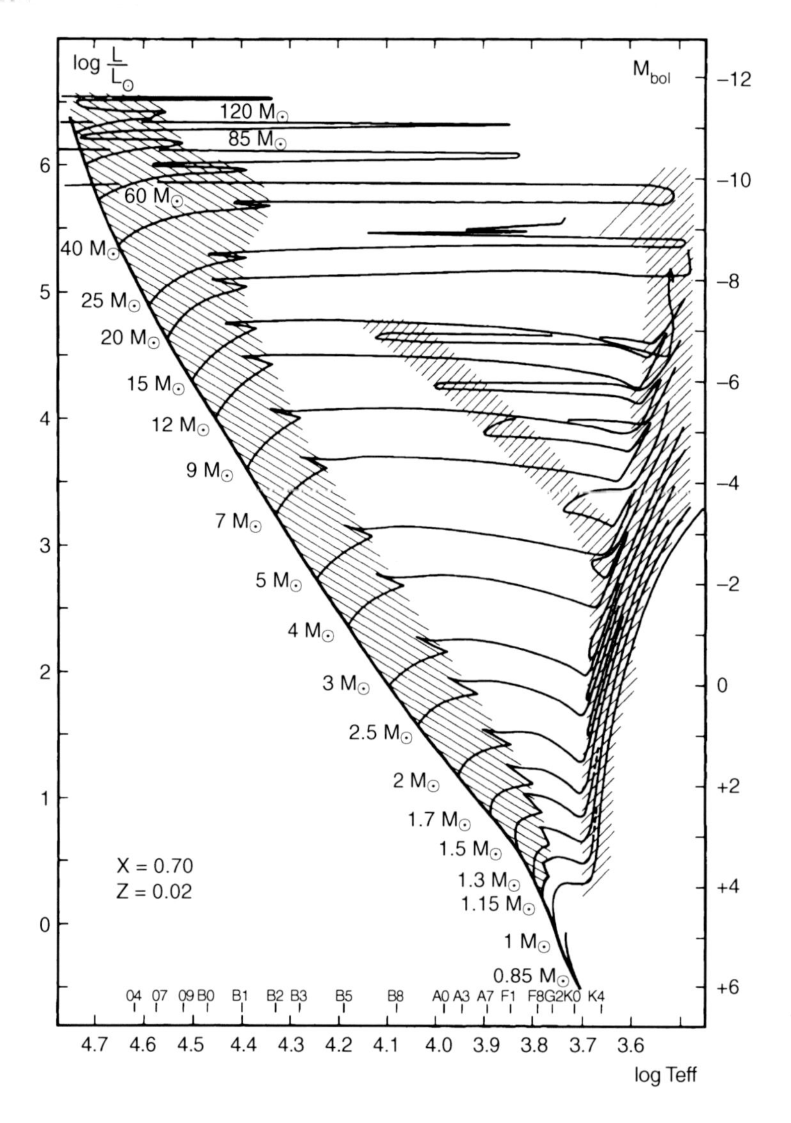

Fig. B.3. Theoretical temperature-luminosity diagram of stars. The solid curve is the zero age main sequence (ZAMS), on which stars ignite the burning of hydrogen in their cores. The evolutionary tracks of these stars are indicated by the various lines which are labeled with the stellar mass. The hatched areas mark phases in which the evolution proceeds only slowly, so that many stars are observed to be in these areas

a proper sense.[2] At the onset of nuclear fusion, the star is located on the zero-age main sequence (ZAMS) in the HRD (see Fig. B.3). The energy production by fusion of hydrogen into helium alters the chemical composition in the stellar interior; the abundance of hydrogen decreases by the same rate as the abundance of helium increases. As a consequence, the duration of this phase of central hydrogen burning is limited. As a rough estimate, the conditions in a star will change noticeably when about 10% of its hydrogen is used up. Based on this criterion, the lifetime of a star on the main sequence can now be estimated. The total energy produced in this phase can be written as

$$E_{\rm MS} = 0.1 \times Mc^2 \times 0.007 \,, \tag{B.5}$$

[2] If the mass of a brown dwarf exceeds $\sim 0.013 M_\odot$, the central density and temperature are high enough to enable the fusion of deuterium (heavy hydrogen) into helium. However, the abundance of deuterium is smaller by several orders of magnitude than that of normal hydrogen, rendering the fuel reservoir of a brown dwarf very small.

where Mc^2 is the rest-mass energy of the star, of which a fraction of 0.1 is fused into helium, which is supposed to occur with an efficiency of 0.007. Phrased differently, in the fusion of four protons into one helium nucleus, an energy of $\sim 0.007 \times 4m_{\rm p}c^2$ is generated, with $m_{\rm p}$ denoting the proton mass. In particular, (B.5) states that the total energy produced during this main-sequence phase is proportional to the mass of the star. In addition, we know from (B.4) that the luminosity is a steep function of the stellar mass. The lifetime of a star on the main sequence can then be estimated by equating the available energy $E_{\rm MS}$ with the product of luminosity and lifetime. This yields

$$t_{\rm MS} = \frac{E_{\rm MS}}{L} \approx 8 \times 10^9 \, \frac{M/M_\odot}{L/L_\odot} \, {\rm yr} \approx 8 \times 10^9 \left(\frac{M}{M_\odot} \right)^{-2.5} {\rm yr} \, . \tag{B.6}$$

Using this argument, we observe that stars of higher mass conclude their lives on the main sequence much faster than stars of lower mass. The Sun will remain on the main sequence for about eight to ten billion years, with about half of this time being over already. In comparison, very luminous stars, like O and B stars, will have a lifetime on the main sequence of only a few million years before they have exhausted their hydrogen fuel.

In the course of their evolution on the main sequence, stars move away only slightly from the ZAMS in the HRD, towards somewhat higher luminosities and lower effective temperatures. In addition, the massive stars in particular can lose part of their initial mass by stellar winds. The evolution after the main-sequence phase depends on the stellar mass. Stars of very low mass, $M \lesssim 0.7 M_\odot$, have a lifetime on the main sequence which is longer than the age of the Universe, therefore they cannot have moved away from the main sequence yet.

For massive stars, $M \gtrsim 2.5 M_\odot$, central hydrogen burning is first followed by a relatively brief phase in which the fusion of hydrogen into helium takes place in a shell outside the center of the star. During this phase, the star quickly moves to the "right" in the HRD, towards lower temperatures, and thereby expands strongly. After this phase, the density and temperature in the center rise so much as to ignite the fusion of helium into carbon. A central helium-burning zone will then establish itself, in addition to the source in the shell where hydrogen is burned. As soon as the helium in the core has been exhausted, a second shell source will form fusing helium. In this stage, the star will become a red giant or supergiant, ejecting part of its mass into the ISM in the form of stellar winds. Its subsequent evolutionary path depends on this mass loss. A star with an initial mass $M \lesssim 8 M_\odot$ will evolve into a white dwarf, which will be discussed further below.

For stars with initial mass $M \lesssim 2.5 M_\odot$, the helium burning in the core occurs explosively, in a so-called helium flash. A large fraction of the stellar mass is ejected in the course of this flash, after which a new stable equilibrium configuration is established, with a helium shell source burning beside the hydrogen-burning shell. Expanding its radius, the star will evolve into a red giant or supergiant and move along the asymptotic giant branch (AGB) in the HRD.

The configuration in the helium shell source is unstable, so that its burning will occur in the form of pulses. After some time, this will lead to the ejection of the outer envelope which then becomes visible as a *planetary nebula*. The remaining central star moves to the left in the HRD, i.e., its temperature rises considerably (to more than 10^5 K). Finally, its radius gets smaller by several orders of magnitude, so that the the stars move downwards in the HRD, thereby slightly reducing its temperature: a white dwarf is born, with a mass of about $0.6 M_\odot$ and a radius roughly corresponding to that of the Earth.

If the initial mass of the star is $\gtrsim 8\, M_\odot$, the temperature and density at its center become so large that carbon can also be fused. Subsequent stellar evolution towards a core-collapse supernova is described in Sect. 2.3.2.

The individual phases of stellar evolution have very different time-scales. As a consequence, stars pass through certain regions in the HRD very quickly, and for this reason stars at those evolutionary stages are never or only rarely found in the HRD. By contrast, long-lasting evolutionary stages like the main sequence or the red giant branch exist, with those regions in an observed HRD being populated by numerous stars.

C. Units and Constants

In this book, we consistently used, besides astronomical units, the Gaussian cgs system of units, with lengths measured in cm, masses in g, and energies in erg. This is the commonly used system of units in astronomy. In these units, the speed of light is $c = 2.998 \times 10^{10}\,\mathrm{cm\,s^{-1}}$, the masses of protons, neutrons, and electrons are $m_\mathrm{p} = 1.673 \times 10^{-24}$ g, $m_\mathrm{n} = 1.675 \times 10^{-24}$ g, and $m_\mathrm{e} = 9.109 \times 10^{-28}$ g, respectively.

Frequently used units of length in astronomy include the Astronomical Unit, thus the average separation between the Earth and the Sun, where $1\,\mathrm{AU} = 1.496 \times 10^{13}$ cm, and the parsec (see Sect. 2.2.1 for the definition), $1\,\mathrm{pc} = 3.086 \times 10^{18}$ cm. A year has $1\,\mathrm{yr} = 3.156 \times 10^7$ s. In addition, masses are typically specified in Solar masses, $1 M_\odot = 1.989 \times 10^{33}$ g, and the bolometric luminosity of the Sun is $L_\odot = 3.846 \times 10^{33}\,\mathrm{erg\,s^{-1}}$.

In cgs units, the value of the elementary charge is $e = 4.803 \times 10^{-10}\,\mathrm{cm^{3/2}\,g^{1/2}\,s^{-1}}$, and the unit of the magnetic field strength is one Gauss, where $1\,\mathrm{G} = 1\,\mathrm{g^{1/2}\,cm^{-1/2}\,s^{-1}} = 1\,\mathrm{erg^{1/2}\,cm^{-3/2}}$. One of the very convenient properties of cgs units is that the energy density of the magnetic field in these units is given by $\rho_B = B^2/(8\pi)$ – the reader may check that the units of this equation is consistent.

X-ray astronomers measure energies in electron Volts, where $1\,\mathrm{eV} = 1.602 \times 10^{12}$ erg. Temperatures can also be measured in units of energy, because $k_\mathrm{B} T$ has the dimension of energy. They are related according to $1\,\mathrm{eV} = 1.161 \times 10^4\,k_\mathrm{B}\,\mathrm{K}$. Since we always use the Boltzmann constant k_B in combination with a temperature, its actual value is never needed. The same holds for Newton's constant of gravity which is always used in combination with a mass. Here one has $G\,M_\odot c^{-2} = 1.495 \times 10^5$ cm.

The frequency of a photon is linked to its energy according to $h_\mathrm{P} \nu = E$, and we have the relation $1\,\mathrm{eV}\,h_\mathrm{P}^{-1} = 2.418 \times 10^{14}\,\mathrm{s^{-1}} = 2.418 \times 10^{14}$ Hz. Accordingly, we can write the wavelength $\lambda = c/\nu = h_\mathrm{P} c/E$ in the form

$$\frac{h_\mathrm{P} c}{1\,\mathrm{eV}} = 1.2400 \times 10^{-4}\,\mathrm{cm} = 12\,400\,\mathrm{\AA}\,.$$

Peter Schneider, Units and Constants.
In: Peter Schneider, Extragalactic Astronomy and Cosmology. pp. 431–431 (2006)
DOI: 10.1007/11614371_C

D. Recommended Literature

In the following, we will give some recommendations for further study of the literature on astrophysics. For readers who have been in touch with astronomy only occasionally until now, the general textbooks may be of particular interest. The choice of literature presented here is a very subjective one which represents the preferences of the author, and of course it represents only a small selection of the many astronomy texts available.

D.1 General Textbooks

There exist a large selection of general textbooks in astronomy which present an overview of the field at a non-technical level. A classic one and an excellent presentation of astronomy is

- F. Shu: *The Physical Universe: An Introduction to Astronomy*, University Science Books, Sausalito, 1982.

Turning to more technical books, at about the level of the present text, my favorite is

- B.W. Carroll & D.A. Ostlic: *An Introduction to Modern Astrophysics*, Addison Wesley, Reading, 1996;

its $\sim$ 1400 pages cover the whole range of astronomy. The text

- M.L. Kutner: *Astronomy: A Physical Perspective*, Cambridge University Press, Cambridge, 2003

also covers the whole field of astronomy. A text with a particular focus on stellar and Galactic astronomy is

- A. Unsöld & B. Baschek: *The New Cosmos*, Springer-Verlag, Berlin, 2002.

The recently published book

- M.H. Jones & R.J.A. Lambourne: *An Introduction to Galaxies and Cosmology*, Cambridge University Press, Cambridge, 2003

covers the topics described in this book and is also highly recommended; it is less technical than the present text.

D.2 More Specific Literature

More specific monographs and textbooks exist for the individual topics covered in this book, some of which shall be suggested below. Again, this is just a brief selection. The technical level varies substantially among these books and, in general, exceeds that of the present text.

Astrophysical Processes

- M. Harwit: *Astrophysical Concepts*, Springer, New-York, 1988,
- G.B. Rybicki & A.P. Lightman: *Radiative Processes in Astrophysics*, John Wiley & Sons, New York, 1979,
- F. Shu: *The Physics of Astrophysics I: Radiation*, University Science Books, Mill Valley, 1991,
- F. Shu: *The Physics of Astrophysics II: Gas Dynamics*, University Science Books, Mill Valley, 1991,
- S.N. Shore: *The Tapestry of Modern Astrophysics*, Wiley-VCH, Berlin, 2002,
- D.E. Osterbrock: *Astrophysics of Gaseous Nebulae and Active Galactic Nuclei*, University Science Books, Mill Valley, 1989.

Furthermore, there is a three-volume set of books,

- T. Padmanabhan: *Theoretical Astrophysics: I. Astrophysical Processes. II. Stars and Stellar Systems. III. Galaxies and Cosmology*, Cambridge University Press, Cambridge, 2000.

Galaxies and Gravitational Lenses

- L.S. Sparke & J.S. Gallagher: *Galaxies in the Universe: An Introduction*, Cambridge University Press, Cambridge, 2000,
- J. Binney & M. Merrifield: *Galactic Astronomy*, Princeton University Press, Princeton, 1998,
- J. Binney & S. Tremaine: *Galactic Dynamics*, Princeton University Press, Princeton, 1987,

Peter Schneider, Recommended Literature.
In: Peter Schneider, Extragalactic Astronomy and Cosmology. pp. 433–435 (2006)
DOI: 10.1007/11614371_D © Springer-Verlag Berlin Heidelberg 2006

- R.C. Kennicutt, Jr., F. Schweizer & J.E. Barnes: *Galaxies: Interactions and Induced Star Formation*, Saas-Fee Advanced Course 26, Springer-Verlag, Berlin, 1998,
- B.E.J. Pagel: *Nucleosynthesis and Chemical Evolution of Galaxies*, Cambridge University Press, Cambridge, 1997,
- F. Combes, P. Boissé, A. Mazure & A. Blanchard: *Galaxies and Cosmology*, Springer-Verlag, 2001,
- P. Schneider, J. Ehlers & E.E. Falco: *Gravitational Lenses*, Springer-Verlag, New York, 1992,
- P. Schneider, C.S. Kochanek & J. Wambsganss: *Gravitational Lensing: Strong, Weak & Micro*, Saas-Fee Advanced Course 33, G. Meylan, P. Jetzer & P. North (Eds.), Springer-Verlag, Berlin, 2006.

Active Galaxies

- B.M. Peterson: *An Introduction to Active Galactic Nuclei*, Cambridge University Press, Cambridge, 1997,
- R.D. Blandford, H. Netzer & L. Woltjer: *Active Galactic Nuclei*, Saas-Fee Advanced Course 20, Springer-Verlag, 1990,
- J. Krolik: *Active Galactic Nuclei*, Princeton University Press, Princeton, 1999,
- J. Frank, A. King & D. Raine: *Accretion Power in Astrophysics*, Cambridge University Press, Cambridge, 2002.

Cosmology

- M.S. Longair: *Galaxy Formation*, Springer-Verlag, Berlin, 1998,
- J.A. Peacock: *Cosmological Physics*, Cambridge University Press, Cambridge, 1999,
- T. Padmanabhan: *Structure Formation in the Universe*, Cambridge University Press, Cambridge, 1993,
- E.W. Kolb and M.S. Turner: *The Early Universe*, Addison Wesley, 1990,
- S. Dodelson: *Modern Cosmology*, Academic Press, San Diego, 2003,
- P.J.E. Peebles: *Principles of Physical Cosmology*, Princeton University Press, Princeton, 1993,
- G. Börner: *The Early Universe*, Springer-Verlag, Berlin, 2003,
- A.R. Liddle and D.H. Lyth: *Cosmological Inflation and Large-Scale Structure*, Cambridge University Press, Cambridge, 2000.

D.3 Review Articles, Current Literature, and Journals

Besides textbooks and monographs, review articles on specific topics are particularly useful for getting extended information about a special field. A number of journals and series exist in which excellent review articles are published. Among these are *Annual Reviews of Astronomy and Astrophysics* (ARA&A) and *Astronomy & Astrophysics Reviews* (A&AR), both publishing astronomical articles only. In *Physics Reports* (Phys. Rep.) and *Reviews of Modern Physics* (RMP), astronomical review articles are also frequently found. Such articles are also published in the lecture notes of international summer/winter schools and in the proceedings of conferences; of particular note are the Lecture Notes of the *Saas-Fee Advanced Courses*. A very useful archive containing review articles on the topics covered in this book is the Knowledgebase for Extragalactic Astronomy and Cosmology, which can be found at

http://nedwww.ipac.caltech.edu/level5.

Original astronomical research articles are published in the relevant scientific journals; most of the figures presented in this book are taken from these journals. The most important of them are Astronomy & Astrophysics (A&A), The Astronomical Journal (AJ), The Astrophysical Journal (ApJ), Monthly Notices of the Royal Astronomical Society (MNRAS), and Publications of the Astronomical Society of the Pacific (PASP). Besides these, a number of smaller, regional, or more specialized journals exist, such as Astronomische Nachrichten (AN), Acta Astronomica (AcA), or Publications of the Astronomical Society of Japan (PASJ). Some astronomical articles are also published in the journals Nature and Science. The Physical Review D and Physical Review Letters contain an increasing number of papers on astrophysical cosmology.

The Astrophysical Data System (ADS) of NASA which can be accessed via the Internet at, e.g.,

http://cdsads.u-strasbg.fr/abstract_service.html
http://adsabs.harvard.edu/abstract_service.html

provides the best access to these (and many more) journals. Besides tools to search for authors and keywords, ADS offers also direct access to older articles that have been scanned. The access to more recent articles is restricted to IP addresses that are associated with a subscription for the respective journals.

An electronic archive for preprints of articles is freely accessible at

http://arxiv.org/archive/astro-ph.

This archive has existed since 1992, with an increasing number of articles being stored at this location. In particular, in the fields of extragalactic astronomy and cosmology, more than 90% of the articles that are published in the major journals can be found in this archive; a large number of review articles are also available here. astro-ph has become the primary source of information for astronomers.

E. Acronyms Used

In this appendix, we compile some of the acronyms that are used, and references to the sections in which these acronyms have been introduces or explained.

2dF(GRS)	Two-Degree Field Galaxy Redshift Survey (Sect. 8.1.2)
ACBAR	Arcminute Cosmology Bolometer Array Receiver (Sect. 8.6.5)
ACO	Abell, Corwin & Olowin (catalog of clusters of galaxies, Sect. 6.2.1)
ACS	Advanced Camera for Surveys (HST instrument)
AGB	Asymptotic Giant Branch (Sect. 3.9.2)
AGN	Active Galactic Nucleus (Chap. 5)
ALMA	Atacama Large Millimeter Array (Chap. 10)
APEX	Atacama Pathfinder Experiment (Chap. 10)
AU	Astronomical Unit
BAL	Broad Absorption Line (-Quasar, Sect. 5.6.3)
BATSE	Burst And Transient Source Experiment (CGRO instrument, Sect. 9.7)
BBB	Big Blue Bump (Sect. 5.4.1)
BBN	Big Bang Nucleosynthesis (Sect. 4.4.4)
BCD	Blue Compact Dwarf (Sect. 3.2.1)
BH	Black Hole (Sect. 5.3.5)
BLR	Broad-Line Region (Sect. 5.4.2)
BLRG	Broad-Line Radio Galaxy (Sect. 5.2.3)
BOOMERANG	Balloon Observations Of Millimetric Extragalactic Radiation and Geophysics (Sect. 8.6.4)
CBI	Cosmic Background Imager (Sect. 8.6.5)
CCD	Charge Coupled Device
CDM	Cold Dark Matter (Sect. 7.4.1)
CERN	Conseil European pour la Recherché Nucleaire
CfA	Harvard–Smithsonian Center for Astrophysics
CFHT	Canada–France–Hawaii Telescope (Sect. 1.3.3)
CFRS	Canada–France Redshift Survey (Sect. 8.1.2)
CGRO	Compton Gamma Ray Observatory (Sect. 1.3.5)
CHVC	Compact High-Velocity Cloud (Sect. 6.1.3)
CIB	Cosmic Infrared Background (Sect. 9.3.1)
CLASS	Cosmic Lens All-Sky Survey (Sect. 3.8.3)
CMB	Cosmic Microwave Background (Sect. 8.6)
CMD	Color–Magnitude Diagram (Appendix B)
COBE	Cosmic Background Explorer (Sect. 8.6.4)
CTIO	Cerro Tololo Inter-American Observatory
DASI	Degree Angular Scale Interferometer (Sect. 8.6.4)
EdS	Einstein–de Sitter (Sect. 4.3.4)
EMSS	Extended Medium Sensitivity Survey (Sect. 6.3.5)
EPIC	European Photon Imaging Camera (XMM-Newton instrument)
EROS	Expérience pour la Recherche d'Objets Sombres (microlenses collaboration, Sect. 2.5)
ESA	European Space Agency
ESO	European Southern Observatory (Sect. 1.3.3)
FFT	Fast Fourier Transform (Sect. 7.5.3)
FIR	Far Infrared
FJ	Faber–Jackson (Sect. 3.4.2)
FOC	Faint Object Camera (HST instrument)
FORS	Focal Reducer / Low Dispersion Spectrograph (VLT instrument)
FOS	Faint Object Spectrograph (HST instrument)
FP	Fundamental Plane (Sect. 3.4.3)
FR(I/II)	Faranoff–Riley Type (Sect. 5.1.2)
FWHM	Full Width Half Maximum
GC	Galactic Center (Sect. 2.3, 2.6)

Peter Schneider, Acronyms Used.
In: Peter Schneider, Extragalactic Astronomy and Cosmology. pp. 437–439 (2006)
DOI: 10.1007/11614371_E

GRB Gamma-Ray Burst (Sects. 1.3.5, 9.7)
GUT Grand Unified Theory (Sect. 4.5.3)
Gyr Gigayear = 10^9 years
HB Horizontal Branch
HCG Hickson Compact Group (catalog of galaxy groups, Sect. 6.2.8)
HDF(N/S) Hubble Deep Field (North/South) (Sects. 1.3.3, 9.1.3)
HDM Hot Dark Matter (Sect. 7.4.1)
HEAO High-Energy Astrophysical Observatory (Sect. 1.3.5)
HRD Hertzsprung–Russell Diagram (Appendix B)
HRI High-Resolution Imager (ROSAT instrument)
HST Hubble Space Telescope (Sect. 1.3.3)
HVC High-Velocity Cloud (Sect. 2.3.6)
IAU International Astronomical Union
ICM Intracluster Medium (Chap. 6)
IGM Intergalactic Medium (Sect. 8.5.2)
IMF Initial-Mass Function (Sect. 3.9.1)
IoA Institute of Astronomy (Cambridge)
IR Infrared (Sect. 1.3.2)
IRAS Infrared Astronomical Observatory (Sect. 1.3.2)
ISM Interstellar Medium
ISO Infrared Space Observatory (Sect. 1.3.2)
IUE International Ultraviolet Explorer
JCMT James Clerk Maxwell Telescope (Sect. 1.3.1)
JVAS Jodrell Bank–VLA Astrometric Survey (Sect. 3.8.3)
JWST James Webb Space Telescope (Chap. 10)
KAO Kuiper Airborne Observatory (Sect. 1.3.2)
LBG Lyman-Break Galaxies (Sect. 9.1.1)
LCRS Las Campanas Redshift Survey (Sect. 8.1.2)
LHC Large Hadron Collider
LISA Laser Interferometer Space Antenna (Chap. 10)
LMC Large Magellanic Cloud
LOFAR Low-Frequency Array (Chap. 10)
LSB galaxy Low Surface Brightness galaxy (Sect. 7.5.4)
LSR Local Standard of Rest (Sect. 4.2.1)
LSS Large-Scale Structure (Chap. 8)
MACHO Massive Compact Halo Object (and collaboration of the same name, Sect. 2.5)
MAMBO Max-Planck Millimeter Bolometer (Sect. 9.3.2)
MAXIMA Millimeter Anisotropy Experiment Imaging Array (Sect. 8.6.4)
MDM Mixed Dark Matter (Sect. 7.4.2)
MIR Mid-Infrared
MLCS Multicolor Light Curve Shape (Sect. 8.3.1)
MMT Multi-Mirror Telescope
MS Main Sequence
MW Milky Way
NAOJ National Astronomical Observatory of Japan
NFW Navarro, Frenk & White (-profile, Sect. 7.5.4)
NGC New General Catalog (Chap. 3)
NGP North Galactic Pole (Sect. 2.1)
NICMOS Near Infrared Camera and Multi-Object Spectrometer (HST instrument)
NIR Near-Infrared
NLR Narrow-Line Region (Sect. 5.4.3)
NLRG Narrow-Line Radio Galaxy (Sect. 5.2.3)
NOAO National Optical Astronomy Observatory
NRAO National Radio Astronomy Observatory
NTT New Technology Telescope
OGLE Optical Gravitational Lensing Experiment (microlenses collaboration, Sect. 2.5)
OVV Optically Violently Variable (Sect. 5.2.4)
PL Period–Luminosity (Sect. 2.2.7)
PLANET Probing Lensing Anomalies Network (microlenses collaboration, Sect. 2.5)
PN Planetary Nebula
POSS Palomar Observatory Sky Survey
PSF Point-Spread Function
PSPC Position-Sensitive Proportional Counter (ROSAT instrument)
QSO Quasi-Stellar Object (Sect. 5.2.1)

RASS ROSAT All Sky Survey (Sect. 6.3.5)
RCS Red Cluster Sequence (Sect. 6.6)
REFLEX ROSAT-ESO Flux-Limited X-Ray survey
RGB Red Giant Branch (Sect. 3.9.2)
ROSAT Roentgen Satellite (Sect. 1.3.5)
SAO Smithsonian Astrophysical Observatory
SCUBA Submillimeter Common-User Bolometer Array (Sect. 1.3.1)
SDSS Sloan Digital Sky Survey (Sect. 8.1.2)
SFR Star-Formation Rate (Sect. 9.5.1)
SGP South Galactic Pole (Sect. 2.1)
SIS Singular Isothermal Sphere (Sect. 3.8.2)
SKA Square Kilometer Array (Chap. 10)
SN(e) Supernova(e) (Sect. 2.3.2)
SNR Supernova Remnant
SMC Small Magellanic Cloud
SMBH Supermassive Black Hole (Sect. 5.3)
STIS Space Telescope Imaging Spectrograph (HST instrument)
STScI Space Telescope Science Institute (Sect. 1.3.3)
SZ Sunyaev–Zeldovich (effect, Sect. 6.3.4)
TF Tully–Fisher (Sect. 3.4)
UDF Ultra-Deep Field (Sect. 9.1.3)
ULIRG Ultra-Luminous Infrared Galaxy (Sect. 9.2.1)
ULX Ultra-Luminous Compact X-ray Source (Sect. 9.2.1)
UV Ultraviolet
VLA Very Large Array (Sect. 1.3.1)
VLBI Very Long Baseline Interferometer (Sect. 1.3.1)
VLT Very Large Telescope (Sect. 1.3.3)
VST VLT Survey Telescope (Sect. 6.2.5)
WD White Dwarf (Sect. 2.3.2)
WIMP Weakly Interacting Massive Particle (Sect. 4.4.2)
WFI Wide Field Imager (camera at the ESO/MPG 2.2-m telescope, La Silla, Sect. 6.5.2)
WFPC2 Wide Field and Planetary Camera 2 (HST instrument)
WMAP Wilkinson Microwave Anisotropy Probe (Sect. 8.6.5)
XMM X-ray Multi-Mirror Mission (Sect. 1.3.5)
XRB X-Ray Background (Sect. 9.3.2)
ZAMS Zero Age Main Sequence (Sect. 3.9.2)

F. Figure Credits

Chapter 1

1.1 Credit: ESO

1.2 Credit: NASA, The NICMOS Group (STScI, ESA) and The NICMOS Science Team (Univ. of Arizona)

1.5 Source: http://adc.gsfc.nasa.gov/mw/mmw_product.html#viewgraph
Credit: NASA's Goddard Space Flight Center

Radio Continuum (408 MHz): Data from ground-based radio telescopes (Jodrell Bank Mark I and Mark IA, Bonn 100-meter, and Parkes 64-meter). Credit: Image courtesy of the NASA GSFC Astrophysics Data Facility (ADF). Reference: Haslam, C. G. T., Salter, C. J., Stoffel, H., & Wilson, W. E. 1982, Astron. Astrophys. Suppl. Ser., 47, 1. Online data access: http://www.mpifrbonn.mpg.de/survey.html

Atomic Hydrogen: Leiden-Dwingeloo Survey of Galactic Neutral Hydrogen using the Dwingeloo 25-m radio telescope. contact/credit: Dap Hartmann, dap@strw.strw.leidenuniv.nl References: Burton, W. B. 1985, Astron. Astrophys. Suppl. Ser., 62, 365 Hartmann, Dap, & Burton, W. B., "Atlas of Galactic Neutral Hydrogen," Cambridge Univ. Press, (1997, book and CD-ROM). Kerr, F. J., et al. 1986, Astron. Astrophys. Suppl. Ser. Online data access: http://adc.gsfc.nasa.gov/adc-cgi/cat.pl?/catalogs/8/8054

Radio Continuum (2.4–2.7 GHz): Data from the Bonn 100-meter, and Parkes 64-meter radio telescopes. contact/credit: Roy Duncan, ccroy@yowie.cc.uq.edu.au References: Duncan, A. R., Stewart, R. T., Haynes, R. F., & Jones, K. L. 1995, Mon. Not. Roy. Astr. Soc., 277, 36. Fuerst, E., Reich, W., Reich, P., & Reif, K. 1990, Astron. Astrophys. Suppl. Ser., 85, 691. Reich, W., Fuerst, E., Reich, P., & Reif, K. 1990, Astron. Astrophys. Suppl. Ser., 85, 633. Online data access: http://www.mpifr-bonn.mpg.de/ survey.html http://www.atnf.csiro.au/database/astro_data/ 2.4Gh_Southern

Molecular Hydrogen: Data from the Columbia/GISS 1.2 m telescope in New York City, and a twin telescope on Cerro Tololo in Chile contact/credit: Thomas Dame, tdame@cfa.harvard.edu References: Dame, T. M., Hartmann, Dap, & Thaddeus, P. 2001, Astrophysical Journal, 547, 792. Online data access: CO data (1987 Dame et al. composite survey) from ADC archives: ftp://adc.gsfc.nasa.gov/pub/adc/archives/catalogs/8/8039/

Infrared: Data from the Infrared Astronomical Satellite (IRAS) Credit: Image courtesy of the NASA GSFC Astrophysics Data Facility (ADF). Reference: Wheelock, S. L., et al. 1994, IRAS Sky Survey Atlas Explanatory Supplement, JPL Publication 94–11 (Pasadena: JPL). Online data access: IRAS pages at IPAC ADF/IRAS interface to all released IRAS data products: http://space.gsfc.nasa.gov/astro/iras/iras_home.html

Mid-infrared (6.8–10.8 microns): Data from the SPIRIT III instrument on the Midcourse Space Experiment (MSX) satellite. Contact/Credit: Stephan D. Price, Steve.Price@hanscom.af.mil Reference: Price, S. D., et al. 2001, Astron. J., 121, 2819. Online Information: http://sd-www.jhuapl.edu/MSX/

Near Infrared: Data from the Cosmic Background Explorer (COBE) Credit: Image courtesy of the NASA GSFC Astrophysics Data Facility (ADF). Reference: Hauser, M. G., Kelsall, T., Leisawitz, D., & Weiland, J. 1995, COBE Diffuse Infrared Background Experiment Explanatory Supplement, Version 2.0, COBE Ref. Pub. No. 95-A (Greenbelt, MD: NASA/GSFC). Online data access: COBE data from the COBE Home Page at the ADF http://space.gsfc.nasa.gov/astro/cobe/

Optical: Data from sites in the United States, South Africa, and Germany taken by A. Mellinger. Contact/Credit: Axel Mellinger, axm@rz.uni-potsdam.de Reference: Mellinger, A., Creating a Milky Way Panorama. http://canopus.physik.uni-potsdam.de/axm/astrophot.html

Peter Schneider, Figure Credits.
In: Peter Schneider, Extragalactic Astronomy and Cosmology. pp. 441–452 (2006)
DOI: 10.1007/11614371_F

X-Ray: Data from the X-Ray Satellite (ROSAT) Credit: Image courtesy of the NASA GSFC Astrophysics Data Facility (ADF). Reference: Snowden, S. L., et al. 1997 Astrophys. J., 485, 125. Online data access: ROSAT All-Sky Survey at MPE ROSAT data archives at the HEASARC: http://heasarc.gsfc.nasa.gov/docs/rosat/

Gamma Ray: Data from the Energetic Gamma-Ray Experiment Telescope (EGRET) instrument on the Compton Gamma-Ray Observatory (CGRO) Credit: Image courtesy of the NASA GSFC Astrophysics Data Facility (ADF). References: Hartman, R. C., et al. 1999, Astrophys. J. Suppl., 123, 79. Hunter, S. D., et al. 1997, Astrophys. J., 481, 205. Online data access: EGRET instrument team's Home Page EGRET data from the Compton Observatory SSC http://cossc.gsfc.nasa.gov/egret/index.html

1.6 Credit: S. Hughes & S. Maddox - Isaac Newton Group of Telescopes

1.7 Credit: M. Altmann, Observatory Bonn University

1.8 Source: http://www.astro.princeton.edu/ frei/ Gcat_htm/Sub_sel/gal_4486.htm
Credit: Z. Frei, J. E. Gunn, Princeton University

1.9 Source: Hale Observatories
Credit: J. Silk, The Big Bang, 2nd Ed.

1.10 Source: E. Hubble; Proc. Nat. Academy Sciences 15, No. 3, March 15, 1929
Credit: PNAS

1.11 Source: http://hubblesite.org/newscenter/ newsdesk/archive/releases/1996/35/image/b.
Credit: John Bahcall (Institute for Advanced Study, Princeton) and NASA

1.12 Source: http://hubblesite.org/newscenter/ newsdesk/archive/releases/1997/17/image/a.
Credit: Rodger Thompson, Marcia Rieke, Glenn Schneider (University of Arizona) and Nick Scoville (California Institute of Technology), and NASA

1.13 Source: http://heritage.stsci.edu/1999/41/ index.html.
Credit: STScI and The Hubble Heritage Project

1.14 Source: www.noao.edu/image_gallery/html/ im0118.html
Credit: NOAO/AURA/NSF

1.15 Source: http://chandra.harvard.edu/photo/0087/
Credit: Optical: La Palma/B. McNamara/X-Ray: NASA/CXC/SAO

1.16 Source: http://heritage.stsci.edu/1999/31/ index.html
Credit: STScI und das Hubble Heritage Project

1.17 Source: http://aether.lbl.gov/www/projects/cobe/ COBE_Home/DMR_Images.html
Credit: COBE/DRM Team, NASA

1.18 Source: http://www.nrao.edu/imagegallery/php/ level3.php?id=107
Credit: NRAO/AUI

1.19 Source: http://www.naic.edu/public/about/photos/ hires/ao004.jpg Courtesy of the NAIC - Arecibo Observatory, a facility of the NSF. Photo by David Parker / Science Photo Library

1.20 left: Source: http://www.mpifr-bonn.mpg.de/ public/images/100m.html
Credit: Max Planck Institute for Radio Astronomy

1.20 right: Source: http://www.nrao.edu/imagegallery/ php/level3.php?id=412
Credit: NRAO/AUO

1.21 Source: http://webdbnasm.si.edu/tempadmin/ whatsNew/whatsNewImages/s-
Credit: NRAO/AUO

1.22 Source: http://www.mpifr-bonn.mpg.de/staff/ bertoldi/mambo/intro.html
Credit: Max Planck Institute for Radio Astronomy

1.23 Source: http://outreach.jach.hawaii.edu/ pressroom/2003-scuba2cfi/jcmt.jpg
Credit: Joint Astronomy Centre

1.24 left: Source: http://www.spitzer.caltech.edu/ about/earlyhist.shtml
Credit: Courtesy
NASA/IPAC

1.24 right: Source: http://www.esa.int/esaSC/ 120396_index_1_m.html
Credit: ESA / www.esa.int

1.25 Source: R. Wainscoat
Credit: R. Wainscoat, Hawaii University

1.26 Credit: ESO

1.27 Source: http://hubblesite.org/newscenter/newsdesk/archive/releases/1996/01/image/d.
Credit: R. Williams (STScI), the Hubble Deep Field Team and NASA

1.28 Source: R. Wainscoat.
Credit: R. Wainscoat, Hawaii University

1.29 Credit: ESO

1.30 Credit: ESO

1.31 Credit: ESO

1.32 left: Source: http://www.mpe.mpg.de/PIFICONS/rosat-transparent.gif
Credit: www.xray.mpe.mpg.de / www.mpe.mpg.de

1.32 top right: Source: http://xrtpub.harvard.edu/resources/illustrations/craftRight.html
Credit: NASA/CXC/SAO

1.32 bottom right: sXMM: Source: http://sci.esa.int/science-e/www/area/index.cfm?fareaid=23.
Credit: ESA / www.esa.int

1.33 left: Source: http://cossc.gsfc.nasa.gov/images/epo/gallery/cgro/
Credit: NASA

1.33 right: Source: http://www.esa.int/esaSC/120374_index_1_m.html.
Credit: ESA / www.esa.int

Chapter 2

2.1 Source: http://belplasca.de/Astro/milchstrasse.html
Credit: Stephan Messner, Observatory Brennerpass

2.5 Source: http://www.ociw.edu/research/sandage.html
Credit: Allan Sandage, The Observatories of the Carnegie Institution of Washington

2.6 Source: Draine 2003, ARA&A 41, 241
Credit: Reprinted, with permission, from the Annual Review of Astronomy & Astrophysics

2.7 Credit: ESO

2.8 Credit: Unsöld Baschek, Der Neue Kosmos, Springer-Verlag, Berlin Heidelberg New York 2002

2.9 Source: Tammann et al., 2003, A&A 404, 423, p. 436.
Credit: G.A. Tammann, Astronomical Institute, Basel University, Switzerland

2.10 Credit: Unsöld Baschek, Der Neue Kosmos, Springer-Verlag, Berlin Heidelberg New York 2002

2.11 Credit: Schlegel, D.J., Finkbeiner, D.P. & Davis, M., ApJ 1999, 500, 525

2.12 Source: Reid 1993, ARA&A 31, 345, p. 355
Credit: Reprinted, with permission, from the Annual Review of Astronomy & Astrophysics

2.14 adopted from: Caroll & Ostlie, 1995

2.18 Introduction to Modern Astrophysics. Addison–Wesley

2.19 Credit: Englmaier & Gerhard 1999, MNRAS 304, 512, p. 514

2.20 Credit: Clemens 1985, ApJ 295, 422, p. 429

2.21 Credit: Wambsganss 1998, Living Review in Relativity 1, 12

2.23 Credit: Schneider, Ehlers & Falco 1992, Gravitational Lensing, Springer-Verlag, Berlin Heidelberg New York 2002

2.24 Source: Paczýnski 1996, ARA&A 34, 419, p. 424
Credit: Reprinted, with permission, from the Annual Review of Astronomy & Astrophysics

2.25 Credit: Wambsganss 1998, Living Review in Relativity 1, 12

2.26 Source: Paczynski 1996, ARA&A 34, 419, p. 425, 426, 427.
Credit: Reprinted, with permission, from the Annual Review of Astronomy & Astrophysics

2.27 Source: Alcock et al. 1993, Nature 365, 621
Credit: Charles R. Alcock, Harvard-Smithonian Center for Astrophysics

2.28 Source: Alcock et al. 2000, ApJ 542, 281, p. 284
Credit: Charles R. Alcock, Harvard-Smithonian Center for Astrophysics

2.29 Source: Alcock et al. 2000, ApJ 542, 281, p. 304
Credit: Charles R. Alcock, Harvard-Smithonian Center for Astrophysics

2.30 Credit: Afonso et al., 2003, A&A 400, 951, p. 955

2.31 Source: Paczynski 1996, ARA&A 34, 419, p. 435, 434
Credit: Reprinted, with permission, from the Annual Review of Astronomy & Astrophysics

2.32 Credit: Albrow et al. 1999, ApJ 512, 672, p. 674

2.33 Credit: W. Keel (U. Alabama, Tuscaloosa), Cerro Tololo, Chile

2.34 Sources: left: N.E. Kassim, from LaRosa et al. 2000, AJ 119, 207,
Credit: Produced at the U.S. Naval Research Laboratory by Dr. N.E. Kassim and collaborators from data obtained with the National Radio Astronomy's Very Large Array Telescope, a facility of the National Science Foundation operated under cooperative agreement with Associated Universities, Inc. Basic research in radio astronomy at the Naval Research Laboratory is supported by the U.S. Office of Naval Research

2.34 top right:
Credit: Plante et al. 1995, ApJ 445, L113

2.34 right center:
Credit: Image courtesy of NRAO/AUI; National Radio Astronomy Observatory

2.34 bottom right:
Credit: Image courtesy of Leo Blitz and Hat Creek Observatory

2.35 Credit: NASA/UMass/D.Wang et al.

2.36 Credit: Genzel 2000, astro-ph/0008119, p. 18
Credit: Reinhard Genzel, MPE

2.37 Credit: Schödel et al., 2003, ApJ 596, 1015, p. 1024

2.38 Credit: Schödel et al., 2003, ApJ 596, 1015, p. 1027

2.39 Credit: Reid & Brunthaler 2004, ApJ 616, 872, p. 875

Chapter 3

3.1 Credit: ESO

3.2 Credit: Kormendy & Bender 1996, ApJ 464, 119

3.4 Source: NASA, K. Borne, L. Colina, H. Bushouse & R. Lucas.
Credit: Kirk Borne, Goddard Space Flight Center, Greenbelt, MD 20771, USA

3.5 top left: Source: http://archive.eso.org/dss - dss 12.23.28.0.gif.
Credit: ESO

3.5 top right:
Credit: ESO

3.5 bottom left:
Credit: Leo I, Michael Breite, www.skyphoto.de

3.5 bottom right: Source: http://hubblesite.org/newscenter/newsdesk/archive/releases/2003/07/
Image Credit: NASA, ESA, and The Hubble Heritage Team (STScI/AURA) Acknowledgment: M. Tosi (INAF, Osservatorio Astronomico di Bologna)

3.6 Credit: Kim et al. 2000, MNRAS 314, 307

3.7 Credit: Bender et al. 1992, ApJ 399, 462

3.8 Credit: Schombert 1986, ApJS 60, 603

3.9 Credit: Davies et al. 1983, ApJ 266, 41

3.12 Credit: Kormendy & Djorgovski 1989, ARA&A 27, 235

3.13 Credit: Schweizer & Seitzer 1988, ApJ 328, 88

3.14 Source: http://www.ing.iac.es/PR/science/galaxies.html / Isaac Newton Group of Telescopes Image.
Credit: top left: NGC4826s.jpg / ING Archive and Nik Szymanek. Top center: m51_v3s.jpg / Javier Méndez (ING) and Nik Szymanek (SPA). Top right: m101s.jpg / Peter Bunclark (IoA) and

Nik Szymanek. bottom right: johan9ss.jpg / Johan Knapen and Nik Szymanek.

3.14 Bottom left, bottom center: Image
Credit: ESO

3.15 Credit: Rubin et al. 1978, ApJ 225, L107

3.16 Credit: van Albada et al. 1985, ApJ 295, 305

3.17 Credit: Aguerri et al. 2000, A&A 361, 841

3.18 Source: http://chandra.harvard.edu/press/01_releases/press_071901.html
Image Credit: X-ray: NASA/CXC/UMass/D.Wang et al. /
Optical: NASA/HST/D.Wang et al.

3.19 Source: Pierce & Tully 1992, ApJ 387, 47
Credit: Robin Phillips, www.robinphillips.net

3.20 Credit: Macri et al. 2000, ApJS 128, 461

3.21 Credit: McGaugh et al. 2000, ApJ 533, L99

3.22 Credit: Bender et al. 1992, ApJ 399, 462

3.23 Credit: Kormendy & Djorgovski 1989, ARA&A 27, 235

3.24 Source: http://hubblesite.org/newscenter/newsdesk/archive/releases/1997/12/
Credit: Gary Bower, Richard Green (NOAO), the STIS Instrument Definition Team, and NASA

3.25 Credit: Kormendy & Ho 2000, astro-ph/0003268

3.26 Source: http://www.astr.ua.edu/keel/agn/m87core.html
Credit: STScI, NASA, ESA, W. Keel and Macchetto et al., ApJ 489, 579 (1997) for providing the HST FOC data

3.27 Source: http://cfa-www.harvard.edu/cfa/hotimage/ngc4258.html
Credit: Harvard-Smithsonian Center for Astrophysics, the National Radio Astronomy Observatory, and the National Astronomical Observatory of Japan und C. De Pree, Agnes Scott College

3.28 Credit: Kormendy 2000, astro-ph/007401

3.29 From http://hubblesite.org/newscenter/newsdesk/archive/releases/1994/22/
Credit: Dr. Christopher Burrows, Ray Villard, ESA/ STScI and NASA

3.30 Credit: Jacoby et al. 1992, PASP 104, 599

3.31 Credit: Schechter 1976, ApJ 203, 297

3.32 Credit: Binggeli et al. 1988, ARA&A 26, 509

3.33 Reused with permission from I. K. Baldry, M. L. Balogh, R. Bower, K. Glazebrook, and R. C. Nichol, in Color bimodality: Implications for galaxy evolution, Rolan d E. Allen (ed), Conference Proceeding 743, 106 (2004). Copyright 2004, American Institute of Physics

3.34 Credit: Blandford & Narayan 1992, ARA&A 30, 311

3.35 Credit: Young et al., ApJ 241, 507

3.36 Credit: Narayan & Bartelmann 1996, astro-ph/9606001

3.37 top:
Credit: Young et al. 1980, ApJ 241, 507

3.37 bottom:
Credit: Harvanek et al. 1997, AJ 114, 2240

3.38 top:
Credit: Gorenstein et al. 1988, ApJ 334, 42

3.38 bottom:
Credit: Michalitsianos et al. 1997, ApJ 474, 598

3.39 Source: http://hubblesite.org/newscenter/newsdesk/archive/releases/1998/37/image/a.
Credit: Christopher D. Impey, University of Arizona

3.40 left:
Credit: Yee 1988, AJ 95, 1331: H. K. C. Yee Department of Astronomy, University of Toronto, Toronto, ON, M5S 3H8, Canada

3.40 right: Source: http://cfa-www.harvard.edu/castles/Postagestamps/Gifs/Fullsize/Q2237Hcc.gif
Credit: C. S. Kochanek

3.41 Credit: from Adam et al. 1989, A&A 208, L15

3.42 Source: http://www.jb.man.ac.uk/research/gravlens/lensarch/B1938+666/B1938+666.html/src_all.jpg King et al. 1998, MNRAS 295, L41

3.43 Credit: Langston et al. 1998, AJ 97, 1283

3.45 left:
Credit: Kundic et al. 1997, ApJ 482, 75

3.45 right:
Credit: Haarsma et al. 1997, ApJ 479, 102

3.46 Credit: Charlot, Lecture Notes in Physics Vol. 470, Springer-Verlag, Berlin Heidelberg New York, 1996

3.47 Credit: Charlot, Lecture Notes in Physics Vol. 470, Springer-Verlag, Berlin Heidelberg New York, 1996

3.48 Credit: Charlot, Lecture Notes in Physics Vol. 470, Springer-Verlag, Berlin Heidelberg New York, 1996

3.49 Credit: Charlot, Lecture Notes in Physics Vol. 470, Springer-Verlag, Berlin Heidelberg New York, 1996

3.50 Credit: Kennicutt 1992, ApJS 79, 255

Chapter 4

4.1 Source: http://www-astro.physics.ox.ac.uk/~wjs/apm_survey.html
Credit: Steve Maddox Nottingham Astronomy Group. The University of Nottingham. U.K.

Will Sutherland http://www-astro.physics.ox.ac.uk/index.html

George Efstathiou Director Institute of Astronomy, University of Cambridge, UK

Jon Loveday, University of Sussex, Brighton, UK with follow-up by Gavin Dalton, Astrophysics Department, Oxford University. U.K.

4.2 Source: http://www.obspm.fr/messier/xtra/leos/M005Leos.html Copyright: Leos Ondra

4.2 Credit: Hesser, J. E.; Harris, W. E.; Vandenberg, D. A.; Allwright, J. W. B.; Shott, P.; Stetson, P. B. 1987 PASP 99, 739.

4.3 Source: COBE,NASA http://lambda.gsfc.nasa.gov/product/cobe/firas_image.cfm.
Credit: We acknowledge the use of the Legacy Archive for Microwave Background Data Analysis (LAMBDA). Support for LAMBDA is provided by the NASA Office of Space Science

4.4 Credit: Windhorst, Rogier A.; Fomalont, Edward B.; Partridge, R. B.; Lowenthal, James, D., 1993 ApJ 405, 498

4.5 Credit: Cosmological Physics. J.A. Peacock. Cambridge University Press 1999.

4.6 Source: http://rst.gsfc.nasa.gov/Sect20/A1a.html
Credit: J. Silk, The Big Bang, 2nd Ed.

4.7 Credit: Cosmological Physics. J.A. Peacock. Cambridge University Press 1999.

4.11 Credit: Cosmological Physics. J.A. Peacock. Cambridge University Press 1999.

4.12 Credit: Aragon-Salamanca, Alfonso; Baugh, Carlton M.; Kauffmann, Guinevere, 1998, MNRAS 297, 427A

4.13 Source: Mass fraction of nuclei as a function of temperature. Deuterium and baryonic density of the universe. David Tytler, John M. O'Meara, Nao Suzuki, Dan Lubin. Physics Reports 333–334 (2000) 409–432.
Credit: Reprinted with permission from Elsevier

4.14 Source: Abundances for the light nucleu 4He, D, 3He and 7Li calculated in standard BBN. Deuterium and baryonic density of the universe. David Tytler, John M. O'Meara, Nao Suzuki, Dan Lubin. Physics Reports 333–334 (2000) 409–432.
Credit: Reprinted with permission from Elsevier

4.15 Credit: Penzias, A. A. & Wilson, R. W., 1965, Astrophysical Journal, vol. 142, p. 419–421: p. 419.

4.18 Source: "The Inflationary Universe", Ed. Perseus Books (1997), ISBN 0-201-32840-2
Credit: Alan Guth, MIT

Chapter 5

5.1 Credit: Netzer, Active Galactic Nuclei, Springer-Verlag, Berlin Heidelberg New York 1990

5.2 Credit: Francis et al. 1991, ApJ 373, 465

5.3 Credit: Francis et al. 1991, ApJ 373, 465

5.4 Credit: Morgan 1968, ApJ 153, 27

5.5 Credit: Laing & Bridle 1987, MNRAS 228, 557

5.5 Credit: Bridle et al. 1994, AJ, 108, 766

5.6 Credit: Bridle & Perley 1984, ARA&A 22, 319

5.8 Credit: Hewitt & Burbidge 1993, ApJS, 87, 451

5.9 Source: http://www.astr.ua.edu/keel/agn/vary.html
Credit: William C. Keel, University of Alabama, USA

5.10 Source: Hartman et al. 2001, ApJ 558, 583

5.11 Credit: Marscher et al. 2002, Nature, 417, 625

5.12 Source: http://www.stsci.edu/ftp/science/m87/bw4.gif
Credit: John Biretta, Space Telescope Science Institute. Press Release Text: Hubble Detects Faster-Than-Light Motion in Galaxy M87 January 6, 1999, J. Biretta (STScI)

5.14 Credit: Urry & Padovani 1995, PASP, 107, 803 (reprinted by permission of the author)

5.15 Credit: Fabian et al. 2000, PASP 112, 1145

5.16 Credit: Fabian et al. 2000, PASP 112, 1145

5.17 Credit: Streblyanska et al. 2005, A&A 432, 395

5.20 Credit: Malkan 1983, ApJ 268, 582

5.21 Credit: Clavel et al. 1991, ApJ 366, 64

5.22 Credit: Clavel et al. 1991, ApJ 366, 64

5.23 Credit: Clavel et al. 1991, ApJ 366, 64

5.24 Source: Allan Sandage, Observatories of the Carnegie Institution of Washington, 813 Santa Barbara Street, Pasadena, CA 91101, USA. and Andrew S. Wilson, Department of Astronomy, University of Maryland, College, USA

5.25 Credit: Sako et al. 2001, A&A 365, L168

5.26 Source: http://hubblesite.org/newscenter/newsdesk/archive/releases/2005/12/image/q.
Credit: John Bahcall (Institute for Advanced Study, Princeton), Mike Disney (University of Wales), and NASA

5.27 Credit: Miller et al. 1991, ApJ 378, 47

5.29 Credit: Pogge & de Robertis 1993, ApJ 404, 563

5.30 Source: http://hubblesite.org/newscenter/newsdesk/archive/releases/1992/27/image/b.
Credit: National Radio Astronomy Observatory, California Institute of Technology
Credit: Walter Jaffe/Leiden Observatory, Holland Ford/JHU/STScI, and NASA

5.31 Credit: Urry & Padovani 1995, PASP, 107, 803

5.32 Source: http://www.spacetelescope.org/images/html/opo9943e.html.
Credit: *NASA/ESA and Ann Feild (Space Telescope Science Institute)*

5.33 left: Source: http://hubblesite.org/newscenter/newsdesk/archive/releases/2003/03/image/c.
Credit for WFPC2 image: NASA and J. Bahcall (IAS)

5.33 right: Source: http://www11.msfc.nasa.gov/news/news/photos/2000/photos00-308.htm.
Credit: NASA/CXC/H. Marshall et al.

5.34 Source: http://hubblesite.org/newscenter/newsdesk/archive/releases/1999/43/
Credit: NASA, National Radio Astronomy Observatory/National Science Foundation, and John Biretta (STScI/JHU)

5.35 left: Source: http://heasarc.gsfc.nasa.gov/docs/objects/heapow/archive/active_galaxies/pks1127_chandra.html.
Credit: X-ray: NASA/CXC/A. Siemiginowska (CfA) & J. Bechtold (U. Arizona);
Radio: Siemiginowska et al. (VLA)

5.35 right: Source: http://chandra.harvard.edu/photo/2001/0157blue/ NASA/SAO/ R. Kraft et al.

5.36 Credit: Croom et al. 2004, MNRAS 349, 1397

5.38 Credit: Chaffee et al. 1988, ApJ 335, 584

5.39 Credit: Rauch 1998, ARA&A 36, 267

5.40 Credit: Sargent, Wallace L. W.; Steidel, Charles C.; Boksenberg, A. 1989, ApJS 69, 703

5.41 Credit: Turnshek 1988, in: QSO absorbtion lines: Probing the universe: Proceedings of the QSO Absorbtion Line Meeting, Baltimore, MD, Cambridge University Press, 1988

5.42 Source: http://www.mpia-hd.mpg.de/Public/Aktuelles/PR/2001/PR010809/pri0152.pdf
H.-W. Rix, Max-Planck-Institute for Astronomy, Heidelberg

Chapter 6

6.1 Source: http://pupgg.princeton.edu/ groth/
Credit: Seldner, Siebers, Groth and Peebles, 1977, A. J., 82, 249

6.2 Credit: Sharp, N.A. 1986, PASP 98, 740

6.3 left: Source: http://subarutelescope.org/Pressrelease/1999/01/index.html#hcg40
Credit: Subaru Telescope, National Astronomical Observatory of Japan (NAOJ)

6.3 left: Source: http://www.eso.org/outreach/press-rel/pr-2002/phot-18-02.html.
Credit: ESO

6.4 Credit: Eva Grebel, Astronomical Institute, University of Basel, Switzerland

6.5 Source: http://www.noao.edu/image_gallery/html/im0562.html
Credit: NOAO/AURA/NSF

6.6 Credit: Brüns et al. 2005, A&A 432, 45

6.7 Source: http://www.robgendlerastropics.com/M8182.html.
Credit: Robert Gendler

6.10 left: Source: http://www.astro.uni-bonn.de/~maltmann/actintgal.html.
Credit: M. Altmann

6.10 right: Source: http://hubblesite.org/newscenter/newsdesk/archive/releases/2002/22/image/Image.
Credit: NASA, J. English (U. Manitoba), S. Hunsberger, S. Zonak, J. Charlton, S. Gallagher (PSU), and L. Frattare (STScI) Science Credit: NASA, C. Palma, S. Zonak, S. Hunsberger, J. Charlton, S. Gallagher, P. Durrell (The Pennsylvania State University) and J. English (University of Manitoba)

6.11 Credit: Goto et al. 2003, astro-ph/0312043

6.12 Reused with permission from I. K. Baldry, M. L. Balogh, R. Bower, K. Glazebrook, and R. C. Nichol, in Color bimodality: Implications for galaxy evolution, Rolan d E. Allen (ed), Conference Proceeding 743, 106 (2004). Copyright 2004, American Institute of Physics

6.13 left: Source: http://heasarc.gsfc.nasa.gov/docs/rosat/gallery/clus_coma.html
Credit: S. L. Snowden USRA, NASA/GSFC

6.13 right:
Credit: Briel et al. 2001, A&A 365, L60

6.14 Source: http://wave.xray.mpe.mpg.de/rosat/calendar/1997/may.
Credit: Max-Planck-Institute for extraterrestrial Physics, Garching

6.15 Source: http://hubblesite.org/newscenter/newsdesk/archive/releases/1998/26/
Credit: Megan Donahue (STSCI) / Ground.
Credit: Isabella Gioia (Univ. of Hawaii), and NASA

6.16 Source: http://www.astro.uni-bonn.de/ reiprich/act/gcs/
Credit: Thomas Reiprich

6.17 Source: http://www.astro.uni-bonn.de/ reiprich/act/gcs/
Credit: Thomas Reiprich

6.18 Source: http://chandra.harvard.edu/photo/2002/0146/
Credit: NASA/IoA/ J. Sanders & A. Fabian

6.19 Credit: Peterson et al. 2003, astro-ph/0310008

6.20 Source: http://wave.xray.mpe.mpg.de/rosat/calendar/1994/sep Copyright: Max-Planck-Institute for extraterrestrial Physics, Garching

6.21 Source: http://chandra.harvard.edu/photo/2001/hcg62/
Credit: NASA/CfA/ J. Vrtilek et al.

6.22 Credit: Markevitch et al. 2002, ApJ 567, L2

6.23 Credit: Carlstrom et al. 2002, ARA&A 40, 643

6.24 Credit: Grego et al. 2001, ApJ 552, 2

6.25 Credit: Stanford et al. 2001, ApJ 552, 504

6.26 Source: http://www.xray.mpe.mpg.de/rosat/survey/sxrb/12/ass.html Image
Credit: M.J. Freyberg, R. Egger (1999), "ROSAT PSPC All-Sky Survey maps completed", in Proceedings of the Symposium "Highlights in X-ray Astronomy in honour of Joachim Trümper's 65th birthday", eds. B. Aschenbach & M.J. Freyberg, MPE Report 272, p. 278–281

6.27 Credit: Finoguenov et al. 2001, A&A 368, 749

6.28 Source: http://www.astro.uni-bonn.de/ reiprich/act/gcs/
Credit: Reiprich & Böhringer 2002, ApJ 567, 716

6.29 Credit: Lin et al. 2004, ApJ 610, 745

6.30 Source: http://serweb.oamp.fr/kneib/hstarcs/hst_a370.html.
Credit: Jean-Paul Kneib

6.31 Source: http://www.eso.org/outreach/press-rel/pr-1998/pr-19-98.html
Credit: ESO

6.32 Credit: Fort, B. & Mellier, Y. 1994, A&AR 5, 239

6.33 top: Source: http://hubblesite.org/newscenter/newsdesk/archive/releases/ 1995/14/image/a
Credit: W. Couch (University of New South Wales), R. Ellis (Cambridge University), and NASA

6.33 left: Source: http://hubblesite.org/newscenter/newsdesk/archive/releases/ 1996/10/image/a
Credit: W.N. Colley and E. Turner (Princeton University), J.A. Tyson (Bell Labs, Lucent Technologies) and NASA

6.34 Source: http://hubblesite.org/newscenter/newsdesk/archive/releases/2003/01 /image/b.
Credit: NASA, N. Benitez (JHU), T. Broadhurst (Racah Institute of Physics/The Hebrew University), H. Ford (JHU), M. Clampin (STScI), G. Hartig (STScI), G. Illingworth (UCO/Lick Observatory), the ACS Science Team and ESA

6.35 Credit: C. Seitz, LMU München

6.35 Credit: Optical Image: HST/NASA, Colley et al.

6.36 Shear Field and Mass Reconstruction, C. Seitz, LMU München

6.37 Credit: Squires et al. 1996, ApJ 461, 572

6.38 bottom left:
Credit: Luppino & Kaiser 1997, ApJ 475, 20

6.38 right:
Credit: Hoekstra et al. 2000, ApJ 532, 88

6.39 bottom:
Credit: Trager et al. 1997, ApJ 485

6.40 bottom: Source: http://www.noao.edu/outreach/press/pr01/pr0111.html
Credit: Lucent Technologies' Bell Labs/NOAO/AURA/NSF

6.41 Credit: Gioia et al. 2001, ApJ 553

6.42 Credit: Margoniner et al. 2001, ApJ 548, L143

6.43 Credit: Gladders & Yee 2000, AJ 120, 2148

6.44 Credit: Spinrad et al. 1997, ApJ 484, 581

6.45 Source: http://www.astro.cz/apod/ap990722.html
Credit: P. van Dokkum, M. Franx (U. Groningen / U. Leiden), ESA, NASA

6.46 Credit: 327; Mullis et al. ApJ 623, L85, 2005

6.47 Source: http://www.eso.org/outreach/press-rel/pr-2002/pr-07-02.html
Credit: George Miley, Observatory Leiden University, The Netherlands

Chapter 7

7.1 Source: http://www.mso.anu.edu.au/2dFGRS/
Credit: The 2dF Galaxy Redshift Survey team, http://magnum.anu.edu.au/ TDFgg/

7.2 Source: http://cfa-www.harvard.edu/ huchra/zcat/
Credit: John Huchra

7.4 Credit: Tucker et al. 1997, MNRAS 285, L5

7.7 Credit: Eke et al. 1996, MNRAS 282, 263

7.8 Credit: Bahcall & Fan 1998, ApJ 504, 1

7.9 Credit: Springel et al. 2005, astro-ph/0504097

7.10 Source: http://www.mpa-garching.mpg.de/galform/virgo/int_sims/index.shtml.
Credit: "The simulations in this paper were

carried out by the Virgo Supercomputing Consortium using computers based at Computing Centre of the Max-Planck Society in Garching and at the Edinburgh Parallel Computing Centre. The data are publicly available at www.mpa-garching.mpg.de/galform/virgo/int_sims"

7.11 Source: http://www.mpa-garching.mpg.de/galform/virgo/hubble/index.shtml.
Credit: The simulations in this paper were carried out by the Virgo Supercomputing Consortium using computers based at the Computing Centre of the Max-Planck Society in Garching and at the Edinburgh parallel Computing Centre. The data are publicly available at http://www.mpa-garching.mpg.de/galform/virgo/hubble

7.12 Credit: Springel et al. 2005, astro-ph/0504097

7.13 Credit: Navarro, Frenk & White 1997, ApJ 490, 493

7.14 Credit: Navarro, Frenk & White 1997, ApJ 490, 493

7.15 Credit: Navarro, Frenk & White 1997, ApJ 490, 493

7.16 Credit: Lin et al. 2004, ApJ 610, 745

7.17 Credit: Moore et al. 1999, ApJ 524, L19

7.18 Credit: Moore et al. 1999, ApJ 524, L19

7.19 Credit: Fassnacht et al. 1999, AJ 117, 658

7.20 left:
Credit: Koopmans et al. 2002, MNRAS 334

7.20 right: Source: http://cfa-www.harvard.edu/castles/Individual/MG2016.html.
Credit: http://cfa-www.harvard.edu/castles/ (C.S. Kochanek, E.E. Falco, C. Impey, J. Lehar, B. McLeod, H.-W. Rix)

Chapter 8

8.1 Source: http://cfa-www.harvard.edu/~huchra/zcat
Credit: John Huchra

8.2 Credit: Lin et al. 1996, ApJ 471, 617

8.3 Credit: Peacock 2003, astro-ph/0309240

8.4 Credit: Peacock & Dodds 1994, MNRAS 267, 1020

8.5 left:
Credit: Peacock 2003, astro-ph/0309240

8.5 right:
Credit: Peacock 2001, astro-ph/0105450

8.6 top:
Credit: Hamilton 1997, astro-ph/9708102

8.7 Credit: Hawkins et al. 2003, MNRAS 346, 78

8.8 Credit: Conolly et al. 2002, ApJ 579, 42

8.9 Credit: Dekel 1994, ARA&A 32, 371

8.10 Source: http://www.eso.org/outreach/press-rel/pr-1999/phot-46-99.html.
Credit: ESO

8.12 Credit: Schuecker et al. 2001, A&A 368, 86

8.13 Credit: Filippenko & Riess 2000, astro-ph/0008057

8.14 Source: http://www-supernova.lbl.gov/public/figures/stretch_hamuy.gif.
Credit: S. Perlmutter

8.15 Credit: Riess et al. 2004, ApJ 607, 665

8.16 Credit: Riess et al. 2004, ApJ 607, 665

8.17 Credit: Riess et al. 2004, ApJ 607, 665

8.18 Source: http://www.cfht.hawaii.edu/News/Lensing/ Image
Credit: Canada-France-Hawaii Telescope Corporation

8.19 Source: http://www2.iap.fr/LaboEtActivites/ThemesRecherche/Lentilles/arcs/cosmicshearstatus.html.
Credit: Yannick Mellier, Institut d'Astrophysique de Paris

8.20 Credit: van Waerbeke et al. 2001, A&A 374, 757

8.21 Credit: Miralda-Escudé et al. 1996, ApJ 471, 582

8.22 Credit: Davé 2001, astro-ph/0105085

8.23 Credit: Weinberg et al. 1998, astro-ph/9810142

8.24 Credit: Hu & Dodelson 2002, ARA&A 40, 171

8.25 Credit: Hu & Dodelson 2002, ARA&A 40, 171

8.26 Credit: Bennett et al. 2003, ApJS 148, 97

8.27 Credit: de Bernardis et al. 2000, astro-ph/ 0004404

8.28 Credit: Netterfield et al. 2002, ApJ 571, 604

8.29 Credit: Wang et al. 2002, Phys. Rev. D 68, 123001

8.30 Credit: Bennett et al. 2003, ApJS 148, 1

8.31 Credit: Bennett et al. 2003, ApJS 148, 1

8.32 Credit: Spergel et al. 2003, ApJS 148, 175

8.33 Source: http://space.mit.edu/home/tegmark/sdsspower.html.
Credit: M. Tegmark

8.34 Credit: Contaldi et al. 2003, Phys. Rev. Lett. 90, 1303

8.35 Source: http://supernova.lbl.gov/ adapted from Knop et al. 2003, ApJ 598, 102

Chapter 9

9.1 Credit: Fan et al., 2003, AJ 125, 1649

9.2 Credit: Steidel et al., 1995, AJ 110, 2519

9.3 Source: http://www.astro.caltech.edu/~ccs/ugr.html.
Credit: C. Steidel, Caltech, USA

9.4 Credit: Steidel et al., 1995, AJ 110, 2519

9.5 Credit: Steidel et al., 1996, AJ 462, L17

9.6 Credit: Hu et al., 1999, ApJ Letters, 522, L9

9.7 Credit: Adelberger 1999, astro-ph/9912153

9.8 Credit: Benitez 2000, ApJ 536, 571

9.9 Credit: R. Williams (STScI), the Hubble Deep Field Team and NASA

9.10 Source: Ferguson et al. 2000, ARA&A38, 667
Credit: Reprinted, with permission, from the Annual Review of Astronomy & Astrophysics, by Annual Reviews www.annualreviews.org

9.11 Source: Space Telescope Science Institute, NASA
Credit: S. Beckwith & the HUDF Working Group (STScI), HST, ESA, NASA

9.12 Source: http://serweb.oamp.fr/kneib/hstarcs/hst_a2390.html
Credit: Jean-Paul Kneib, Laboratoire d'Astrophysique de Marseille

9.13 left: Copyright Stella Seitz, LMU München

9.13 right:
Credit: ESO

9.14 Credit: Kneib et al., 2004, ApJ 607, 697

9.15 Credit: Whitmore et al. 1999, AJ 116, 1551

9.16 Source: http://www.casca.ca/lrp/ch5/en/chap520.html.
The image taken from: "Canadian Astronomy and Astrophysics in the 21st century", "The Origins of Structures in the Universe" Courtesy Christine Wilson, McMaster University

9.17 left: Source: http://chandra.harvard.edu/photo/2001/0120true/index.html.
Credit: NASA/SAO/ G. Fabbiano et al.

9.17 bottom left: Source: http://chandra.harvard.edu/photo/2001/0012/index.html
Credit: X-ray: NASA/SAO/CXC, Optical: ESO

9.17 top right: Source: http://chandra.harvard.edu/photo/2001/0094true/
Credit: NASA/SAO/ G. Fabbiano et al.

9.17 bottom right: Source: http://chandra.harvard.edu/photo/2001/0120true/lum_functions.jpg.
Credit: SAO/CXC/ A. Zezas

9.18 Credit: Cimatti et al., 2002, A&A 391, L1

9.19 Credit: Smith et al. 2002, astro-ph/0201236

9.20 Source: Blain et al. 1999, astro-ph/9908111

9.21 Credit: F. Bartoldi, MPIfR

9.22 Source: Blain et al. 1999, astro-ph/9908111

9.23 Source: Blain et al. 1999, astro-ph/9908111

9.24 Credit: Hauser & Dwek, ARA&A 2001 39, 249

9.25 Credit: Hauser & Dwek, astro-ph/0105539

9.26 Source: Tozzi et al. 2001, ApJ 562, 42
Credit:

9.27 Source: http://chandra.harvard.edu/photo/2001/cdfs/Cdfs_scale.jpg.
Credit: NASA/JHU/AUI/
R. Giacconi et al.

9.28 Credit: Fan et al., 2004, AJ 128, 515

9.29 Credit: Barkana & Loeb 2000, astro-ph/0010468

9.30 Credit: Barkana & Loeb 2000, astro-ph/0010468

9.31 adapted from: Barkana & Loeb 2000, astro-ph/0010468

9.32 Credit: Hopkins et al. 2001, astro-ph/0103253

9.33 Credit: Bell 2004, astro-ph/0408023

9.34 Credit: Bell 2004, astro-ph/0408023

9.35 Credit: Mirabel et al. 1998, astro-ph/9810419

9.36 Source: http://antwrp.gsfc.nasa.gov/apod/ap990510.html.
Credit: J. Gallagher (UW-M) et al. & the Hubble Heritage Team (AURA/ STScI/ NASA)

9.37 Credit: Lacey & Cole 1993, Mon. Not. R. Astron. Soc. 262, 627–649

9.38 Source: http://www.mpa-garching.mpg.de/galform/gif/index.shtml.
Credit: G. Kauffmann, VIRGO Kollaboration, MPA Garching

9.39 Credit: Springel et al., 2005, astro-ph/0504097

9.40 Credit: Springel et al., 2005, astro-ph/0504097

9.41 Source: http://heasarc.gsfc.nasa.gov/docs/objects/grbs/grb_profiles.html
Credit: J.T. Bonnell, GLAST Science Support Center, NASA Goddard Space Flight Center, Greenbelt, Maryland, USA

9.42 Source: http://www.batse.msfc.nasa.gov/batse/grb/skymap/images/fig2_2704.pdf
Credit: Michael S. Briggs, NASA

Chapter 10

10.1 Source: http://www.cfht.hawaii.edu/Instruments/Imaging/CFH12K/images/NGC34 86-CFH12K-CFHT-1999.jpg.
Credit: Dr. Jean-Charles Cuillandre, Canada-France-Hawaii Telescope Corporation, Hawaii, USA

10.2 Credit: ESO

10.3 Credit: ESO

10.4 Source: http://jwstsite.stsci.edu/gallery/telescope.shtml
Credit: Courtesy of Northrop Grumman Space Technology

10.5 Source: http://www.eso.org/outreach/press-rel/pr-2003/pr-04-03.html.
Credit: ESO

Appendix A

A.1 Source: T. Kaempf & M. Altmann, Observatory of Bonn University

A.2 from Girardi et al. 2002, A&A 195, 391

Appendix B

B.1 Source: ESA Web Page of Hipparcos-Projekts

B.2 Source: http://de.wikipedia.org

B.3 from: Maeder & Meynet 1989, A&A 155, 210

Subject Index

D